They Fought Like Veterans
The Military History of the Civil War in the Indian Territory

Michael J. Manning

Honey Springs Battlefield State Historic Site
Rentiesville, Oklahoma

Portions of this book were previously presented in <u>Blue & Gray Magazine</u> as *They Fought Like Veterans: The Civil War in the Indian Territory, Volumes 1 & 2.* © 2011 & 2015, Blue & Gray, Inc. Dave Roth, Editor.

For Karen, Becky, Colin, Caleb, and Claire

Special Thanks to my sister, Debra Manning Van Alstyne, for her dedication in providing editing services to my publication

Prairie Star Music
& Publishing

Woodlawn, Tennessee

Table of Contents

Introduction

The cannons and musketry could be heard clearly coming from the area north of the Creek village of Tohopeka, located within the Horseshoe Bend of the Tallapoosa River in present-day Alabama. The date is March 27, 1814. A Cherokee brave named Whale slipped into the cold water of the river and quickly swam across to the village where the Creek "Red Sticks" had lined the riverbank with a multitude of canoes. These had been placed in case escape was needed if the U.S. Army commanded by General Andrew Jackson broke through the wooden barricade that stretched across the neck of the bend. As Whale reached the village bank, he took a canoe and paddled back across the river. There he was joined by other Cherokees who were taken across the river to get more canoes. This continued until between 300 and 500 Cherokee and "Lower" Creek allies, and white militia soldiers had crossed the river and began to attack Tohopeka from behind. Many Creek Red Stick warriors left the barricade to counter this new threat to the rear, which weakened the Creek defense against the 39th United States Infantry Regiment and the supporting Tennessee Militia units. The soldiers soon scaled the solid barricade across the bend, and a bloody melee began that only ended with darkness. Most of the Creek Red Stick warriors and many women and children had been killed in the fight. The Battle of Horseshoe Bend broke the power of the Upper Creeks or Red Sticks and compelled them to sign the Treaty of Fort Jackson in April 1814. William Weatherford or "Red Eagle" of the Upper Creeks was forced to sign over 20 million acres of Creek land to the

United States.

It is said that the issues that flared up during 1861 in the Indian Territory had their roots in the Creek War of 1813-1814 between the Upper and Lower Creeks, as well as with the mixed blood Cherokees. (The names "Upper" and "Lower" Creeks indicated the level of contact and acceptance of Anglo-American civilization.) The Upper Creeks tended to be full-blood Creek that kept to traditional values and customs. Although the origin is lost to history, there are many opinions concerning the name "Red Sticks" for the Upper Creeks, the most accepted being that runners between the Upper Creek towns would carry bundles of red sticks to spread war-related information. The Lower Creeks believed that their world was changing, and they would have to adapt to the new reality. The Lower Creeks usually lived much closer to white settlements, intermarried, and began taking on the trappings of their Euro-American neighbors. This included building schools and Euro-American-style homes, establishing farms, and buying and selling African slaves. Although Christian missionaries were active with the Lower Creeks as well, but few of these members of the cloth were welcomed at the Upper Creek towns. Instead, the Upper Creeks were being ministered to by "prophets," a mystical order that can be traced to a visit by Tecumseh. These prophets enhanced the tension between the Upper and Lower Creeks considerably by preaching that the Lower Creeks had turned their backs on traditional Creek ways and wanted to be like the "white man." With the United States distracted by the War of 1812, and spurred on by British agents in Florida, a civil war broke out between the Lower and Upper Creeks that lasted from 1813-1814. And although the Upper Creeks were defeated by the combination of the Lower Creeks, mixed-blood Cherokees (there were differences between full and mixed-blood Cherokees but this didn't lead to civil strife until later), and U.S. forces, animosity still burned between the two groups up to and during the Civil War.

The Civil War within the Indian Territory represented a peripheral aspect of the broader conflict, situated in a less prominent region of the Trans-Mississippi West. The area did not witness significant battles or the emergence of notable military leaders; indeed, many senior officers from both Union and Confederate armies were assigned to this region following unsuccessful campaigns further east. However, the intensity of animosity and violence in the Indian Territory was unparalleled compared to other Civil War regions. Most of the Five Civilized Tribes—Cherokee, Creek, Choctaw, Chickasaw, and Seminole—had resided in the territory for less than a quarter-

century. While initial support among these nations leaned toward the Southern Confederacy, consensus was lacking, leading to internal divisions. These splits closely mirrored earlier fractures between pro-treaty members (predominantly of mixed ancestry) and anti-removal members (primarily full-blooded), divisions that originated during the 1820s and 1830s after implementation of the Indian Removal Acts. Relocation moved these nations into what became known as Indian Territory, now largely comprising the State of Oklahoma. Many tribal members, having assimilated aspects of Southern culture prior to removal, had become slaveholders. Individuals from each tribe served on both sides of the conflict, often confronting one another in battle. By the war's end, the Indian Territory had been left nearly deserted. The strife of the Civil War severely fractured the Five Civilized Tribes, splitting allegiances between the Union and the Confederacy. Two United States Army expeditions ultimately restored federal control north of the Arkansas River and precipitated the Confederate withdrawal from Fort Smith, Arkansas, effectively severing supply lines from the East. Hostilities concluded quietly in June 1865 when the remaining Confederate forces under Brigadier General Stand Watie surrendered to the United States Government.

Some of the locations mentioned in the text still exist, such as Fort Gibson, Honey Springs, Fort Towson, and Fort Washita. Others are now underwater, (for example North Fork Town) after the U.S. Army Corps of Engineers began damming the large and smaller rivers to create water reservoirs after the Dust Bowl devastated much of western Oklahoma. Many sites are on private property and visitors are not necessarily welcome. Most of the historic roads and trails have been lost to time and place after being replaced by section line roads and crisscrossed by state and Federal highways. Also, the War in the Indian Territory sometimes had fluid boundaries and many military actions related to the area spread into Kansas, Missouri, Arkansas, and to a lesser degree Texas, especially where the Union and Confederate Indian brigades were involved.

This book is divided into three parts. The first three chapters describe the origins of the Five Civilized Tribes, their "Trails of Tears" to their new homes in today's Oklahoma, and the effects of the Southern secession movement upon these Indian nations. Since this is intended as a military history, the second and larger part of the book describes the military actions that occurred in and around the Indian Territory, the troops involved, and the battles they fought. The final section will deal with the sad consequences of the war upon the Five Civilized Tribes who had broken their treaties with the United States to align with the Confederacy. The new treaties with the United States would be strict and harsh.

I have been a student of the Civil War since I was in the 5th grade when my teacher, Steven Kent, at Parkview (now Robert Hill) Elementary School in Romeoville, Illinois, had the class put on a play about the war for the whole school. My role was as a Union soldier who had been wounded at Cold Harbor. It was

only a small part, but it did plant the seed that eventually started me on the road toward a history career. That road began as a Historic Site Attendant at Fort Gibson State Historic Site, Oklahoma, with the Oklahoma Historic Society and culminated with my role as the Chief Ranger of Fort Donelson National Battlefield, Tennessee, with the National Park Service. My time at Fort Gibson also included drawing up some early development plans for a Honey Springs Battlefield Park, the ground of which I knew very well, that became a reality between the years 2000 and 2020. My time with the National Park Service also included a six-year tour of duty as a Park Ranger at Horseshoe Bend National Military Park in Alabama. Still, it was those early days of service at Fort Gibson that ignited my 30+ years interest in the Civil War events that occurred in the Indian Territory, present-day eastern Oklahoma. Although there is a massive amount of information available for inclusion in the many facets of this period, this work is designed to only examine the Indian Nations and their influence on the events of the Civil War in the Indian Territory.

Fort Gibson, Indian Territory, Cherokee Nation, Post-War (1875)

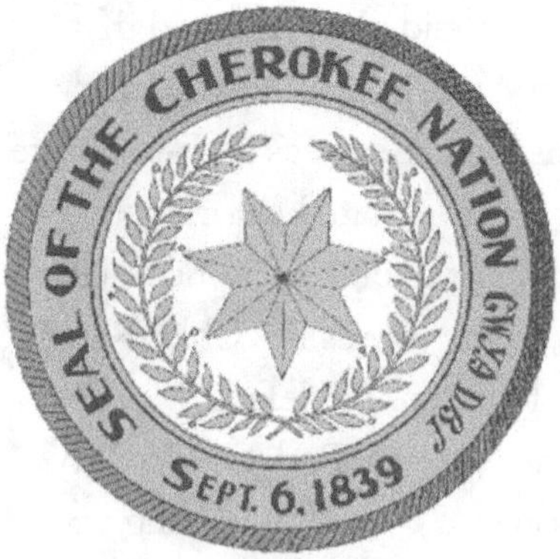
SEAL OF THE CHEROKEE NATION
SEPT. 6, 1839

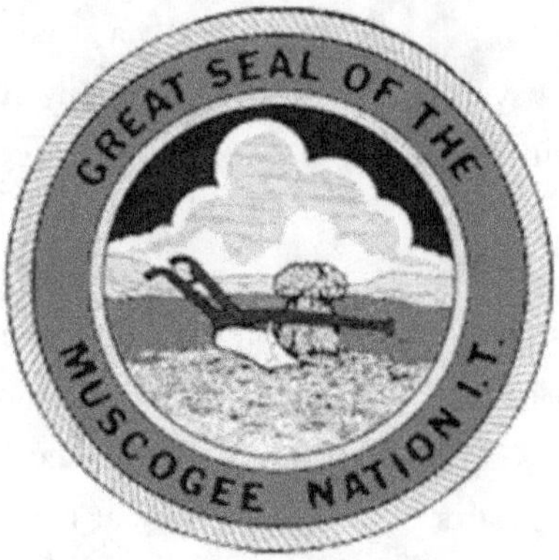
GREAT SEAL OF THE
MUSCOGEE NATION I.T.

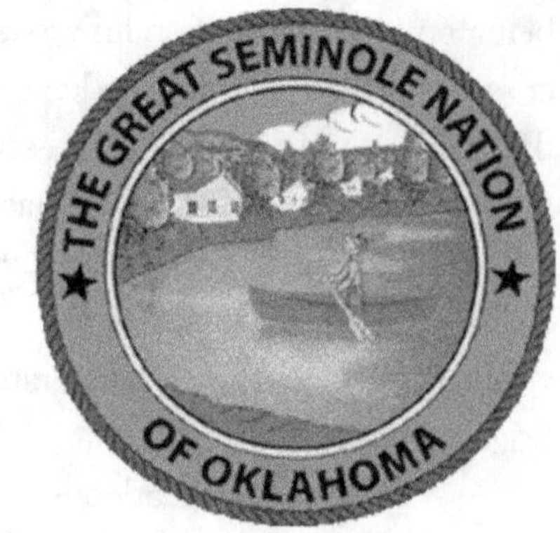
THE GREAT SEMINOLE NATION
OF OKLAHOMA

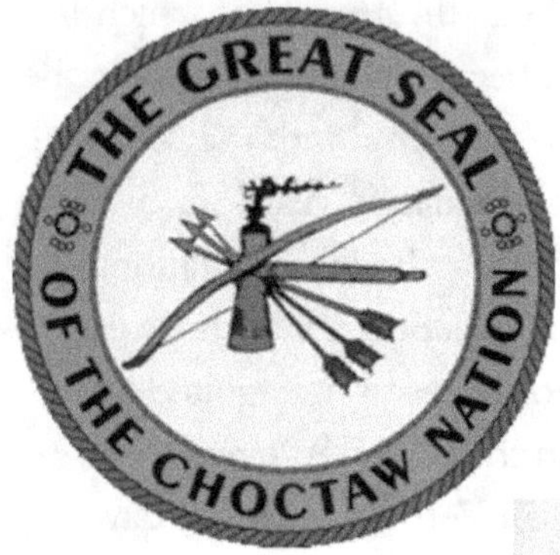
THE GREAT SEAL
OF THE CHOCTAW NATION

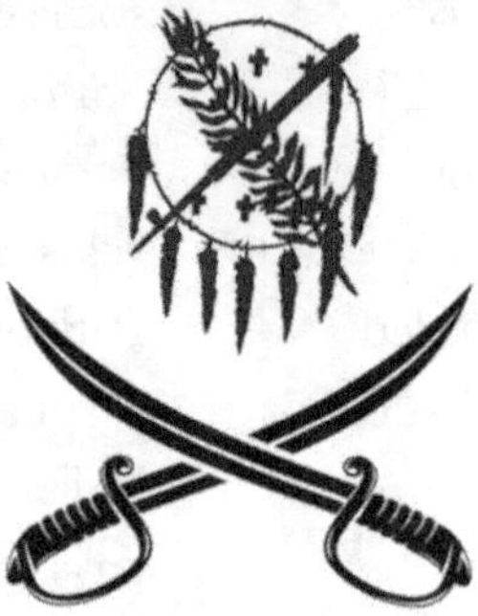

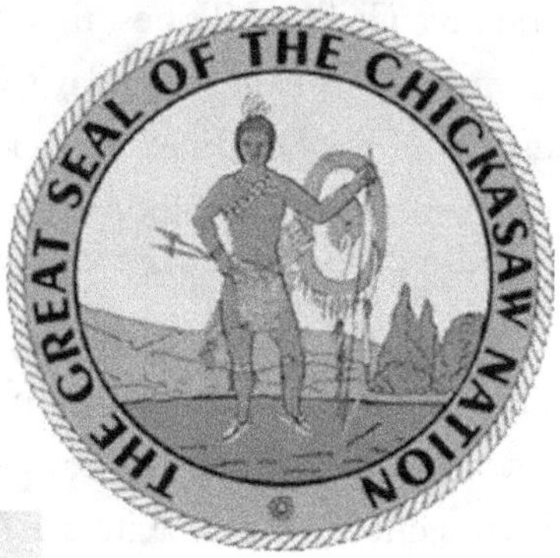
THE GREAT SEAL OF THE CHICKASAW NATION

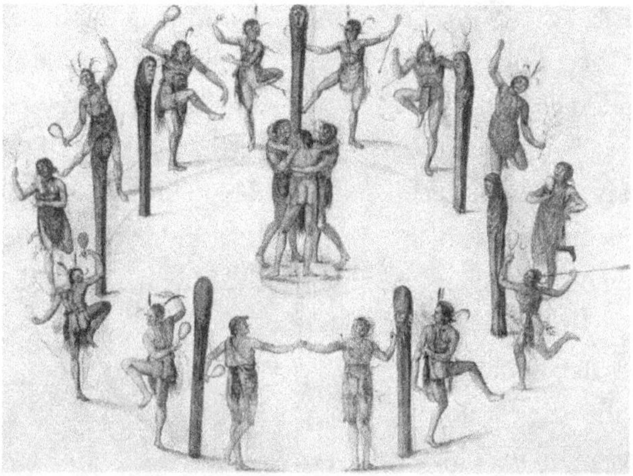

ALABAMA · TENNESSEE
Trail of Tears Corridor Commission
TRAIL OF TEARS
The United States government, unable to conclude an agreement with the duly authorized leaders of the Cherokee Nation, signed a treaty with a minority faction willing to cede the last remaining portion of the original Cherokee homeland on December 29, 1835. Despite the protests of the overwhelming majority of Cherokee people, the fraudulent "Treaty of New Echota" was ratified by the U. S. Senate by only a single vote on May 23, 1836. The Cherokees were given two years from that date to remove to the Indian Territory. When the time had expired only 2,000 of the nearly 17,000 Cherokee remaining in the east had departed their ancestral homeland.
In late May, 1838 General Winfield Scott and 7,000 federal and state troops arrived in the Cherokee Nation to enforce the removal. Cherokee families were forced from comfortable homes into 31 stockades and open military stations scattered throughout the Cherokee Nation in southeast Tennessee, western North Carolina, northwest Georgia and northeast Alabama. From the stockades the Cherokee were sent to the principal emigrating depots near Ross's Landing at Chattanooga, Tennessee, Fort Cass, near Calhoun, Tennessee, and a camp eight miles south of Fort Payne, Alabama.
(Continued on other side)

TRAIL OF TEARS
(Continued from other side)
In June 1838 the first three detachments of Cherokee captured by the Georgia Guard were forced to depart from Ross's Landing, Tennessee. Because of the high casualties of these first groups, permission was given to delay the removal of the other groups until fall when it would be cooler. Also, the Cherokee leaders petitioned General Scott that they be allowed to conduct their own removal. Permission was granted.
The remainder of the Cherokees began their trek west in the fall of 1838 in 13 detachments. After enduring an extremely severe winter, they arrived in the West in late winter and early spring of 1839. It has been estimated that from 2,000 to 4,000 of the 16,000 Cherokees died as a result of the forced removal.
The true story of the forced removal for the Cherokee people is one of survival. In spite of their hardships, they adapted and rebuilt their homes and government. Only 12 years after removal, the Cherokee Female Seminary opened on these grounds as the first public institution of higher learning for females west of the Mississippi River.

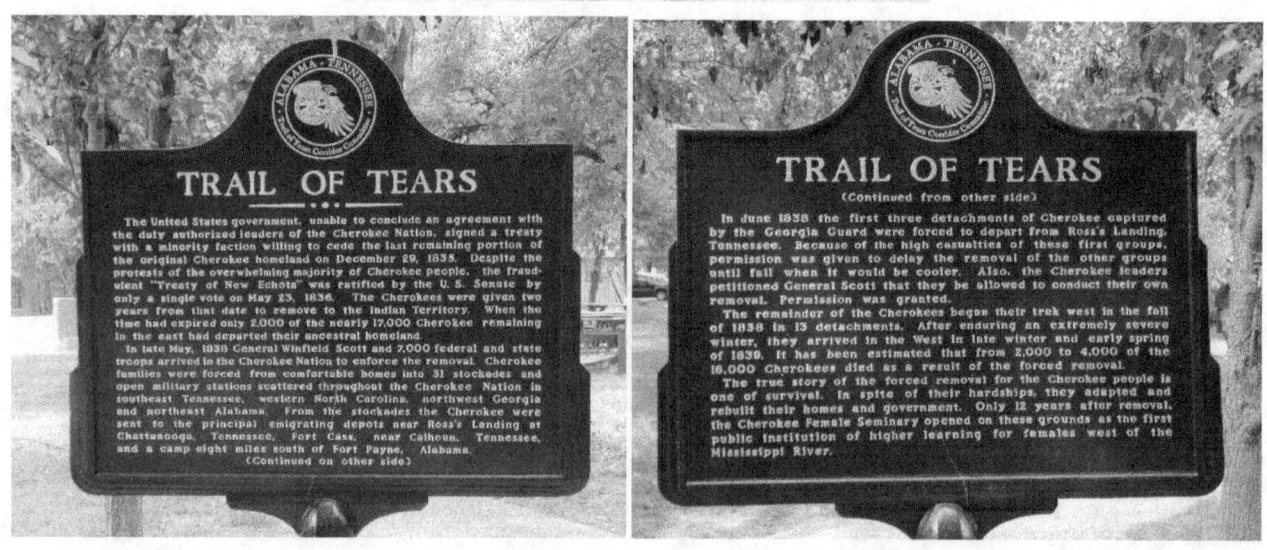

Chapter 1
The Five Civilized Tribes in the Southeast

The Five Civilized Tribes[1] Museum sits on a hilltop in Muskogee, Oklahoma, and through artwork and artifacts it tells the stories of these southeastern Indian tribes that had been relocated to the "Indian Territory" in the first half of the nineteenth century. The Five Tribes (Cherokee, Chickasaw, Choctaw, Creek, and Seminole) were considered "civilized" because they began to exhibit the traits and attitudes of their new Euro-American neighbors. By the 1830's the Indian Nations had established national boundaries and national constitutions which included public school systems and elected officials. The Nations also had representatives in the U.S. Congress. In many cases they eagerly embraced the white man's Christian religion and established farms and plantations, many of which were worked by enslaved Africans, modeled after the great southern plantations.[2] Although they did possess these advancements, they could not overcome the inherent racism or quench the insatiable desire of the Americans for land. This "clash of cultures" would severely affect the landscape of the American nation for many years to come. In the end it became a struggle between the forces of those supporting "state's rights" against those who believed in a strong central Federal Government, with little regard given to those most affected.

The Indian nations who occupied the North American continent believed they had always been a part of this land and felt a significant connection to it. Without a written history virtually all of their stories were communicated orally with different variations over time, and were many times filled with striking imagery and larger than life stories. Most of the Southeastern Indians were tied to the Mississippian mound-builder societies of Cahokia, Spiro, Ocmulgee, and Etowah,[3] which were in severe decline by the time of the first contact with Europeans of the Desoto Expedition during the 16th century. Their oral histories do not indicate much memory of these complex societies that built the great temple and funerary mounds. This may indicate that the modern Nations moved into the area after the Mississippian culture imploded, that they were forcibly displaced by the same tribes, or that perhaps they merged their declining numbers together. Most of the Nations were made up of many towns located near waterways or on flat and fertile lands. Even when these towns may have belonged to one tribe, did not mean that the towns necessarily got along. Their interactions and roles were quite complex, and they independently developed strengths, rituals, customs, and variations of their common language. These differences would lead to larger divisions by the time of the Civil War. As the contacts increased with first the British and then the Americans, there also developed a division between full-blood and mixed-blood tribal members. The mixed-blood members were the products of inter-marriage with their white neighbors and traders and were predominately present in the Cherokee and Creek nations due to their proximity to Euro-American settlements.

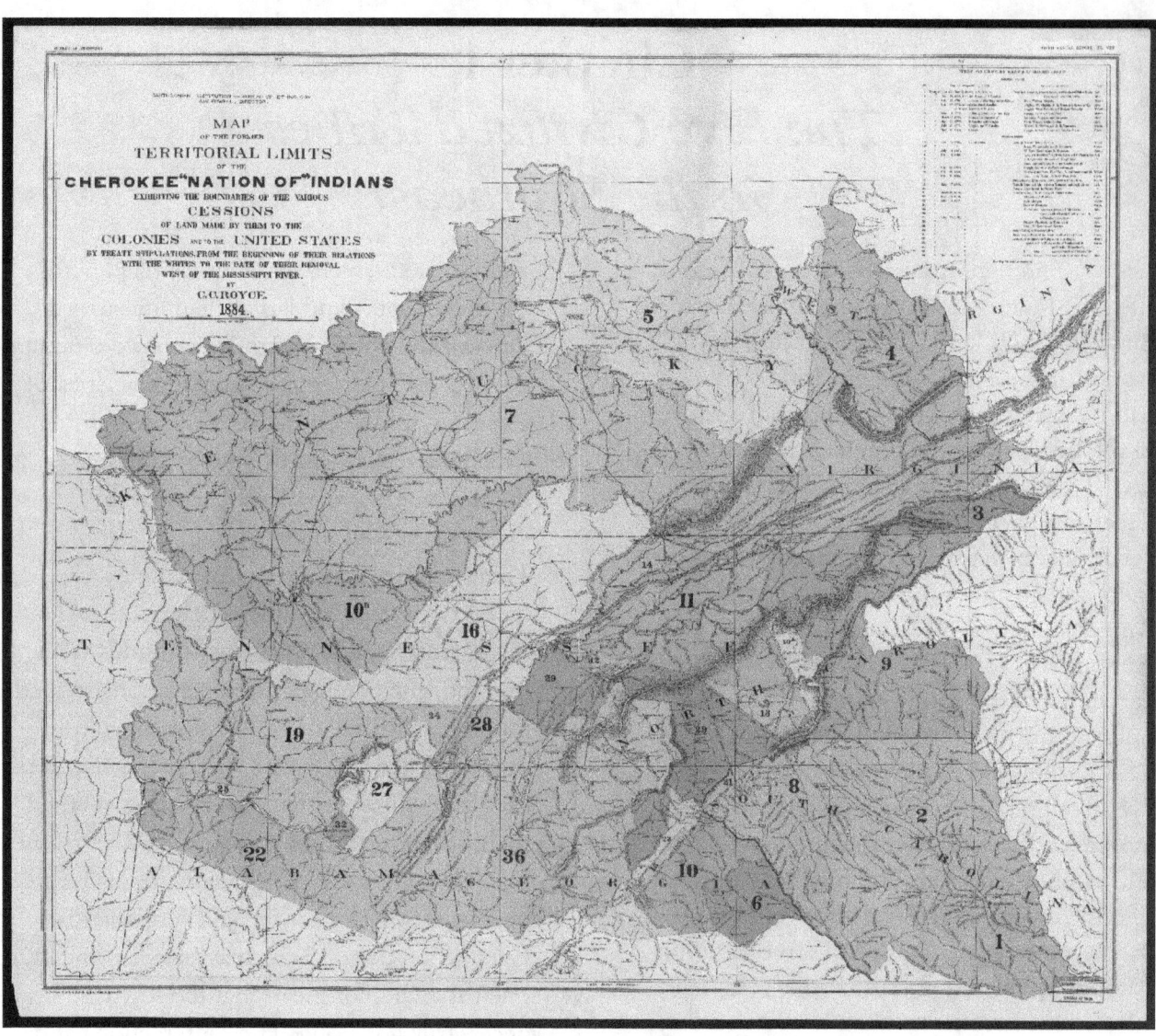

Cherokee Nation prior to Removal
Courtesy of the Library of Congress

The United States Government followed the English policy of recognizing each individual Indian tribe as an independent and sovereign nation. This meant that agreements between the Indian nations and the federal government were treaties between sovereign nations and required passage by the legislative arm of each respective side, for the United States, the Senate and for the tribes, the tribal council. However, for diplomatic treaties between sovereign nations, when one nation believes a change is required, the treaties are modified and, if considered necessary, broken by either side.[4] There were numerous Indian Intercourse Acts passed by Congress during the early years of the United States which outlined and regulated contact and activities between white settlers and the Indian nations. These Acts regulated trade, including intoxicants and weapons, prohibited white settlement on Indian lands, and gave the United States Army the authority to remove white settlers who encroached on those lands. The federal government also assigned an Indian agent to each tribe to ensure the precepts of the Acts were being followed.[5]

The Cherokee Nation was one of the largest and most "civilized" of these southeast Indian nations. Cherokee is a name that trader and historian James Adair[6] claimed meant "men possessed of divine power." Adair also believed that the Cherokee were one of the Lost Tribes of Israel, a popular belief at the time. He tried to show that the tribe used the organizational system of Ancient Israel: their division into tribes, their different languages and dialects, their festivals, feasts, and religious rites; their absolutions and anointings, their ideas concerning unclean things and persons, their practices of marriage, divorce, and punishment for adultery.[7] However, most modern ethnographers believe they drifted or were driven southward from the Iroquois areas to the north.[8] By the time of the first European contact, they occupied large portions of modern-day north Georgia, Alabama, Tennessee, and western North Carolina, a mixture of fertile flatlands and mountains including the Great Smoky Mountains.[9] At the time of the Cherokee's initial contact with Europeans in the 16th century they numbered approximately 22,000 members living in about sixty towns. The tribe also had developed three distinct dialects of the Cherokee language, the Lower or Elati, the Middle or Kituwha, and the Overhill or Otali dialects. These dialects were based on the three geographical divisions of the Cherokee Nation. The Cherokee were the first Indian nation to use a written language when Sequoyah, a mixed-blood descendant from the union of Nathanial Gist of Virginia and Cherokee-maiden Wurteh, developed a system of written symbols for the distinctive sounds of their language. This development led the Cherokee to begin printing a newspaper, the first Indian nation to do so.[10] The Cherokee were fairly quick to adopt Euro-American customs and practices, especially in agriculture, religion, inter-marriage, government, and African slavery. Their lands and farms were as profitable and looked as neat and well organized as the best of American farms. Their ancient religion aligned closely to Judea-Christianity, and Protestant missionaries were quick to point out these alignments to add converts.

The Creek or Muskogee Indian Nation was the second largest of the Civilized Tribes. The Creek/Muskogee oral histories relate that the Creeks migrated east, from the setting sun to the rising sun, although it is now believed they may be remnants of the Mississippian complexes.[11] The British and Americans called them Creeks, since they were living along the creeks and tributaries of the Ocmulgee River in modern-day Georgia and South Carolina. The tribe itself called themselves Muskogee and shared a unique language that could be loosely translated by other tribal members. They lived in towns called *talwa*, which consisted of private dwellings surrounding a public square which contained a *chunky yard*, a sunken area to play this game of hurled poles and stone discs. The

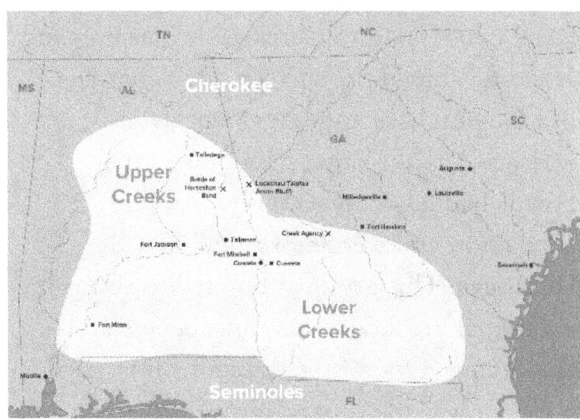

Creek Nation prior to Removal
Courtesy of georgiastudies.gpb.org

square also contained a *chokofa,* or hot house, a round clay and wood structure that was used for various worship practices or dancing. The square itself was the ceremonial and political center and contained the town house which was composed of four small structures forming a smaller square where the tribal leaders would meet. The number of *talwas* varied in number from fifty to eighty at any one time as Creek Nation spread across Georgia into Alabama. Each town, or *talwa,* considered itself an autonomous political entity that developed its own rules and customs and was led by a *micco,* or headman. It was only when the talwas began having issues with the British or Euro-Americans pressuring them for their lands in the late 18th and early 19th centuries that the Muskogee began to form themselves into a "Muskogee Confederacy." They formed into two distinctive groups; the Lower Creeks, made from talwas along the Chattahoochee, Ocmulgee, and Flint Rivers, and the Upper Creeks, made from the talwas along the Coosa, Tallapoosa, and Alabama Rivers.[12] The Lower Creek talwas had more contact with the British and Americans and were successful in developing a thriving trade with them. As time progressed, these Lower Creeks began to adopt the Euro-American civilized habits of dress, conversion to Christianity, farming and domestic animal production, inter-marriage, and the purchase and sale of enslaved Africans. On the other hand, the Upper Towns had less contact with the white population and were mostly full-blood Muskogee. This group attempted to retain the traditional Muskogee customs, beliefs and blood-lines. One of the traditions they tried to maintain was the four to eight-day *puskita,* or busk, and the Green Corn Festival that was

performed in the talwa square to worship the Great Spirit or Master of Breath, along with the earth and express thanks for the blessings it provided to the tribal members.

The first Christian missionaries to the Creek were the Monrovian Brethren in Georgia, but their presence was short-lived after they claimed to be pacifist and refused to assist the Georgia colony in its defense against the Spanish. During this early period the Creek had little respect or desire for the white man's culture or religion. As contact, trading, and intermarriage occurred the Creeks began to become more open to new ideas including Christianity, especially among the Lower Creeks, but beyond a Methodist mission near Fort Mitchell and a Baptist near present day Montgomery, Alabama, few missionaries ventured into the Creek Nation before their removal to the West.

During the American Revolution, the Lower Creeks had aligned with the Americans while the Upper Creeks made the mistake of aligning with the British during that conflict. These differences would come to a head during the War of 1812 when the 1813-1814 civil war broke out between the Upper and Lower Creeks. Led by Chiefs William Weatherford and Menewa, twenty-four Upper talwas took up the "Red Stick" against the Lower Creek towns. Although the British provided some material support to the Red Sticks, the United States threw its support firmly behind the Lower Creeks under mixed-blood William McIntosh. On March 27, 1814, the Red Stick rebellion was finally crushed at the Battle of Horseshoe Bend in central Alabama by an army consisting of U.S. Regulars, Lower Creeks, and Cherokee allies under the command of Major General Andrew Jackson. William Weatherford was forced to surrender the vast majority of the Upper Creek lands in Alabama to the Americans. This set the stage for their removal to the West.[13]

The Choctaw and Chickasaw Nation's creation mythologies were linked closely together when they emerged either from a cavern under the Gulf of Mexico and moved north through Mississippi or via a hill called "Nanih Waya" located in Winston County, Mississippi. The most accepted story is that they came from a "land beyond the setting sun" far to the Northwest and led by a principal chief carrying a sacred pole.[14] Two brothers, Chahtah, who carried the pole, and Chickasah, also a leader, eventually separated with their followers to become two related tribes.[15] The Choctaw Nation had become the largest of the Five Civilized Tribes by the time of removal in the early 1830s. The most striking element of this Nation was its governmental system. The Choctaw Nation was divided into three separate districts, the Northeastern, the Northwestern, and the Southern, of which was headed by a *mingo*, or principal chief, who was elected by the tribal members. There was no central leader of the entire Choctaw tribe, although at times one of the three principal chiefs, or *mingos*, such as Pushmataha, a beloved Chief of the Six Towns District, was able by his leadership skills to rise above the other two chiefs. The Choctaw Nation also held a national council each year that rotated among the three districts. This council developed and implemented policies for the entire tribal government. The Choctaws were especially adept in agricultural skills. They eventually produced so much corn and cattle that they were able to sell their surpluses to other Indian nations or to white settlers.[16] Non-

warlike and somewhat un-religious, this was the limit of their diplomatic and social interactions with others surrounding the Choctaw Nation. Eventually they developed into an isolationist state that simply wished to be left alone, although mixed-blood James Adair believed that they loved their home so much that they would die fighting for it.[17]

Although the Choctaws were not heavily influenced by their own religion, it was a staple of their lives in the towns and in the larger councils. Like most Indian nations they believed in the "Great Spirit" who guided the tribal members and provided the final "Happy Hunting Ground" after they left this world. After the success of the American Board of Commissioners for Foreign Missions in establishing schools in the Cherokee Nation, Protestant Christian missionaries began arriving in the early 19th century and opened missions and schools throughout the Choctaw Nation. Although the tribal members made good use of the schools the missionaries provided, the Choctaws were very lukewarm to the Christian path just as they were somewhat indifferent to their own religion. Missionaries had much better success converting tribal members of the Cherokee and Creek nations. An interesting note is that the Roman Catholic Church had also established two missions in the Choctaw Nation in the 1720s, including one in the large Six Towns District. Neither of these Catholic missions were able to instill a desire to convert since the Choctaws believed their own medicine was far superior to the white man's. The Catholic missions were abandoned soon after their establishment.[18]

The Chickasaws developed in a way similar to the Choctaws. Their origins are unknown, but most anthropologists believe

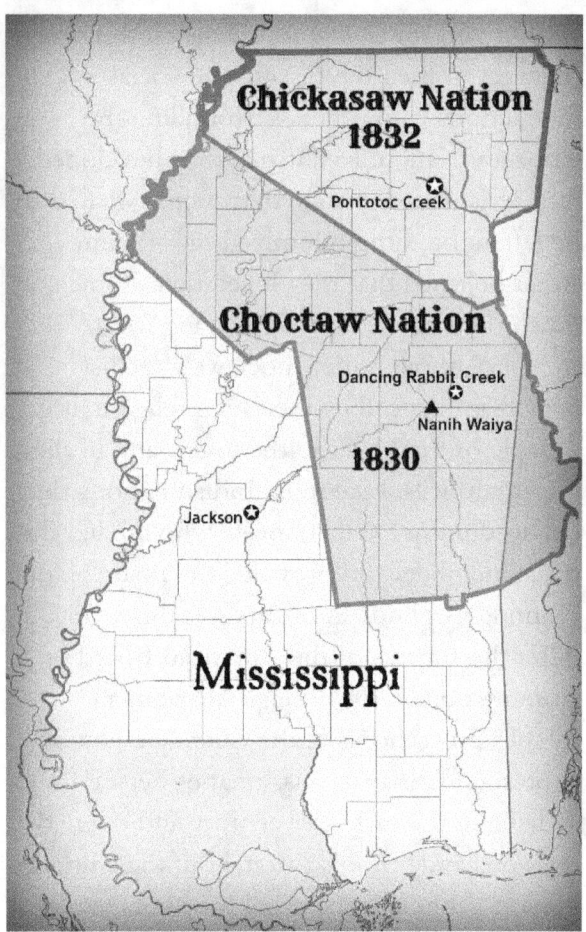

Choctaw & Chickasaw Nations prior to Removal
Map courtesy of Trinitarian Creek

that they are also descended from the Mississippian cultures that were encountered by Hernando DeSoto's 16th century expedition. The Chickasaw were divided by clans into three separate districts, each overseen by a minor chief. The principal chief of the tribe was known as the *micco*. Although they were also divided by the familiar situation of mixed and full-bloods, it was to a much lesser extent due to their relative distance from major white settlements where inter-marriage would normally occur. Even still, it was not uncommon for Chickasaw headmen to be named Colbert, Adair, or McLaughlin. The Chickasaw were also excellent hunters and agriculturists and provided well for their

communities as well as maintaining a steady trade with other tribes and white settlers. At the core of Chickasaw religious beliefs and traditions is the supreme deity Aba' Binni'li' (lit. 'the One sitting / dwelling above), the spirit of fire and giver of life, light, and warmth. Aba' Binni'li' is believed to live above the clouds along with several other lesser deities, such as the spirits of the sky and clouds, and evil spirits.[19] The Chickasaw differed from the Cherokee, Creek, and Choctaw in that they initially completely denied allowing Christian missionaries into their territories to establish missions and schools.[20] By 1800 the Chickasaw relented and allowed the missionaries into their country after pressure was exerted by the traders, merchants, Federal and state officials, and complicit mixed-bloods, to improve the economic opportunities within the Nation. The Presbyterians, Methodists, and Baptists eventually established missions, churches, and schools in the Mississippi territory before the removals in the 1830s. The Tribal Council continually complained to the mission's primary support leaders back East that their missionaries spent far too much time preaching and trying to suppress worship of the Chickasaw deities and not enough time teaching Chickasaw children. Although the intent of the missionaries was good, the execution of their practices gained few converts. One result was that no missionaries from the established Protestant missions accompanied them from their Mississippi homelands to their new homes in the Indian Territory.[21]

The United States Government fought three wars and spent over seven years and millions of dollars attempting to remove the Seminole Indians from Florida. The

Seminoles were technically an off-shoot of the Creek Nation that maintained many of the Creek cultural traditions such as celebrating the Green Corn Ceremony.[22] In reality, the Seminoles, a Creek word that meaning "runaway," were a European invention since they did not see themselves as one tribe but as a collection of many tribes that separated from various talwas in the southern Creek lands in Georgia and Alabama, including many refugees whose livelihoods were destroyed by the Horseshoe Bend battle. They united in an attempt to drive the Europeans out of their lands. Their move southward was the result of the near destruction of the aboriginal tribes of the Florida peninsula, including the Calusas, Tequestas, Timucuans, and Applachees, by the Spanish expeditions and colonies that brought diseases and forced labor on the tribes. By the early 18th century, most of these original tribes were almost completely wiped out. By this time, the Spanish had determined that their Florida colonies were unhealthy and unprofitable, so they began to abandon most areas of the peninsula, only maintaining military posts such as St. Augustine as bases to protect the Spain-bound treasure fleets from pirates and privateers.[23] The refugees from the various tribes from in and around Georgia and Alabama mixed with and intermarried into the remnants of the original tribes of Florida.

These "Seminoles" differed from their Creek forbearers in that, although they allowed for slavery, they allowed intermarriage with escaped African slaves from the colonies or States which lay to their north. The Seminoles were organized into at least six bands that took the name of the band's captain, such as Osceola, Jumper, and Alligator. Their overall government consisted of a head chief who was supported by a national council. As the various scattered groups banded together, they began to attack the remaining Spanish settlements and the newer settlements and farms established by the Americans. They came under the jurisdiction of the United States in 1819 when Spain ceded its Florida territory to the Americans. The white settlement of Florida began to grow as they found the land to be fertile and open in many areas for cattle grazing. This resulted in further attacks as conflicts over territory became greater, leading to a growing cry from the American settlers, especially those who were also losing escaped slaves to the Seminoles, for the forced removal of the tribe to the West.[24]

As the American settlements, especially in Georgia, pressed against the Indian's ancestral lands, pressure built to seize their lands and move them west of the Mississippi River. Unfortunately, the Federal government was never able to commit to an overall Indian policy. The government policy-makers were split into two primary groups: the gradualists and the separatists. These two groups fell into three policy factions: separation, concentration, and Americanization.

The Policy of Separation was based on a belief that it was unlikely the United States would ever need to settle the lands west of the Mississippi. As a result, the establishment of a permanent Indian frontier, first west of the Mississippi River, and later, west of the Arkansas and Missouri boundaries, would be all that was necessary. All native tribes east of that frontier line would be placed there on the western lands designated for them, which they would be free to develop as they saw fit.

President Andrew Jackson
Photo courtesy of the Library of Congress

Chief Justice John Marshall
Photo courtesy of the Library of Congress

John Ross
Principal Chief of the Cherokee Nation
Photo Courtesy of the Oklahoma Historical Society (OHS)

Stand Watie
Leader of the Cherokee Opposition Treaty Party
Photo Courtesy of the Oklahoma Historical Society

**Opothleyahola
Chief of the Upper Creeks**
Drawing courtesy of the Library of Congress

**Roley McIntosh
Principal Chief of the Creek Nation**
Photo Courtesy of the Oklahoma Historical Society

Major Ridge
Photo Courtesy of the Oklahoma Historical Society

The Trail of Tears by Elizabeth Janes
Courtesy of the Oklahoma Historical Society

The Policy of Concentration would keep moving the Five Civilized Tribes gradually westward away from those American settlements desiring separation to either give them a chance to integrate with that population later and without establishing a permanent Indian frontier. This policy would involve either moving the tribes from their homelands to new lands westward that were not adjacent to white settlers, or to maintaining their existing tribal boundaries and allowing the white settlements to surround them, the government would open schools and provide a means for the tribes to develop skills and industry.

The Policy of Americanization would leave the tribes as they were, grant them American citizenship, allot their lands and declare them as private property, and forcibly integrating the Indians into Euro-American culture.[25] This was the least popular of the three policies because the tribal nations and governments would be eliminated as unnecessary and because in these early years the Five Civilized Tribes were not considered to be civilized enough to immediately integrate into American society.

The Federal Government had initially attempted to stabilize and protect Indian lands. With the Cherokee lands, the government secured the Treaty of Hopewell in 1785 and the Treaty of Holsten in 1791, but the Euro-American population grew and demands for new lands increased. In an 1817 agreement with Washington, D.C., approximately one-third of the Cherokee tribal members agreed to relocate first to western Arkansas then finally into Northeastern Indian Territory, or Oklahoma. These emigrants became known as the "Old Settlers" and they occupied the area which

would become the Cherokee Nation.[26] All told, between 1721 and 1835 over three dozen land cessions significantly reduced the size of the Cherokee Nation.[27] In 1802 a compact was signed between the United States and the State of Georgia, with no consultation of the Cherokee Nation, that authorized the long process of removing the tribe out of the eastern United States. Thus, it was not necessarily the United States Government that insisted on removal, but the state and territorial governments that wanted the Indians removed and their lands opened up for white settlement and mineral development.

Although the federal government had treated the Indian Nations as autonomous entities, in 1823 the U.S. Supreme Court decided *Johnson v. McIntosh*, in which the Supreme Court under Chief Justice John Marshall ruled that the Indians could "occupy" the lands of the United States but could not hold title to these lands since these lands were ultimately owned by the Federal Government.[28] The Court further ruled that the Indian's "Right of Occupancy," which is: "A right of occupancy granted by the federal government to American Indian tribes based on their long-standing possession of the land. This means that the government recognizes the tribe's right to use and occupy the land, but not necessarily to own it." [29] Their occupation those lands was subordinate to the United States' "Right of Discovery," which is defined as: "The doctrine of discovery was used as the legal foundation for taking the land of Indigenous people by Europeans." This ruling illuminated the belief that Euro-American culture was supreme to that of the Native American culture. This ruling would have strong repercussions against the Indians

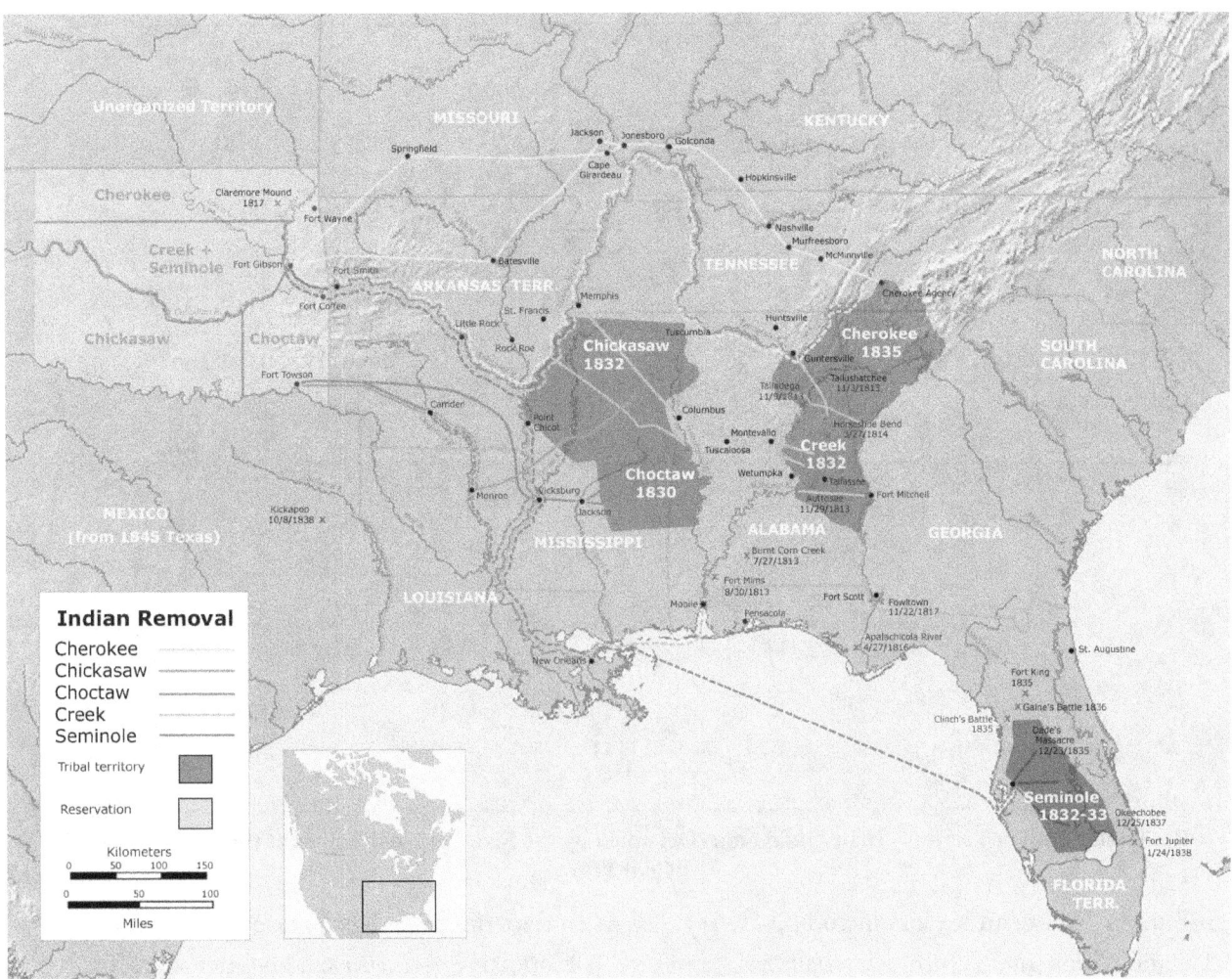

Five Civilized Tribes Removal Map to the Indian Territory
Map Courtesy of the National Park Service (NPS)

who wished only to be left alone. The immediate result of this ruling was that the States of Georgia, Alabama, Tennessee and Mississippi began to include Indian territories into their county divisions and to insist that the Indians abide by state and local laws.[30]

During the 1813-1814 Creek Civil War between the Lower and Upper Creeks, General Andrew Jackson was able to coerce many Cherokees and Lower Creeks into becoming his allies against the Upper Creeks. The resulting Battle of Horseshoe Bend in 1814 broke the strength of the Upper Creek's and brought about their surrender. In an unofficial promise to his Cherokees and Creek

allies, Jackson claimed he would protect their rights to remain on their lands in Georgia and North Carolina. Now, less than twenty years later, he is the instrument that breaks the promise that he himself had made.[31] Bowing to the pressure of Georgia's political machine, in 1830 now-President Andrew Jackson signed the Indian Removal Act which authorized the president to exchange unsettled lands west of the Mississippi with Indian lands within existing state borders.

By 1830 the Cherokees and Creeks had advanced enough to send legal teams and other prominent leaders to Washington to challenge the Indian Removal Act. Many

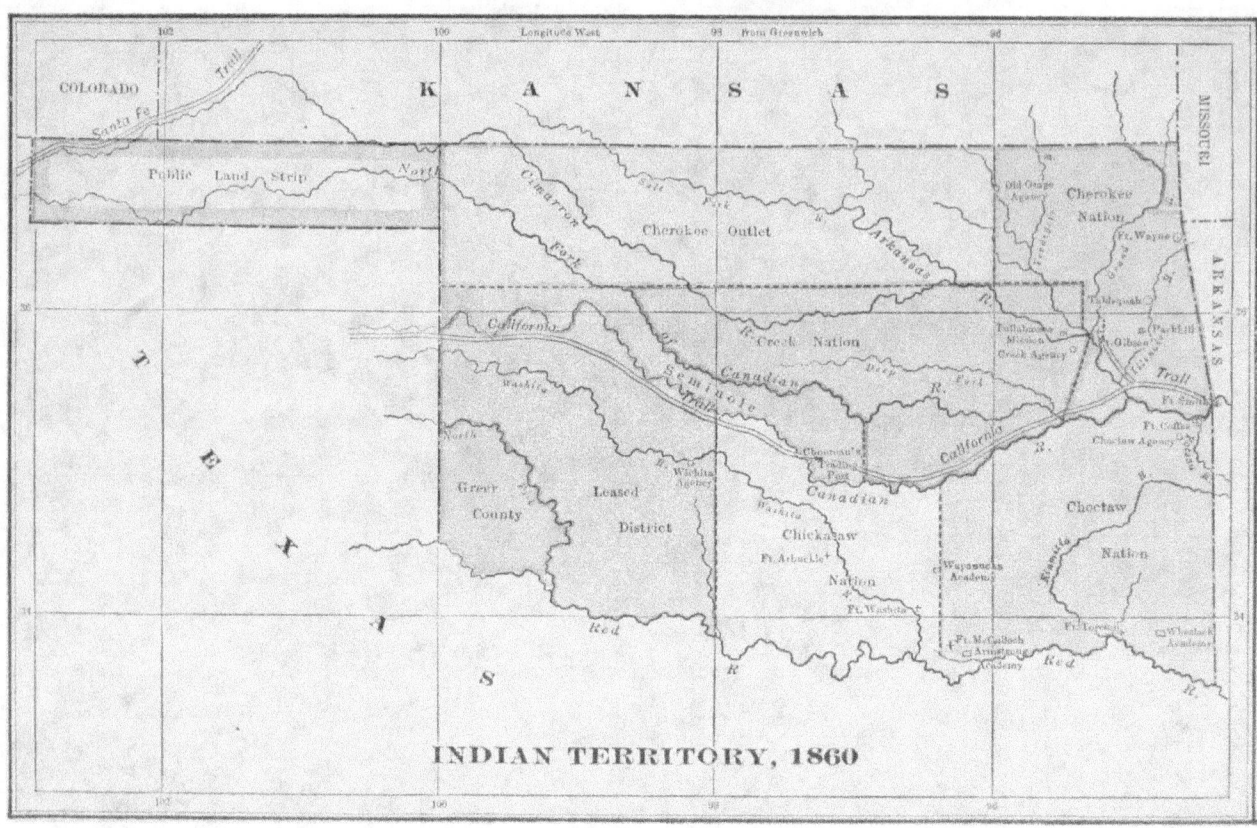

INDIAN TERRITORY, 1860

Designated Indian Territory (now Oklahoma) Occupied by the Five Civilized Tribes at the Beginning of the Civil War

prominent American leaders including Henry Clay and Chief Justice John Marshall supported the Indian's right to retain their ancestral lands. And since these disagreements were between the United States and the independent Indian Nations, the case was heard by the Supreme Court. The case of *Worcester v. Georgia* considered the fact that the State of Georgia had passed a law in 1828 that prohibited whites from living on Cherokee lands without a permit issued by the state. Reverand Samuel Austin Worcester operated a Methodist mission in the Cherokee Nation. When Worcester refused to submit to Georgia's permit regulation or abandon his mission, he was arrested by Georgia authorities, found guilty by an all-white jury, and imprisoned. The Cherokee Nation argued that Georgia had no authority to regulate

independent Indian nations within their respective territories. The Cherokees further argued that Worcester and his mission, along with all of the other missions, were allowed to remain as long as the Cherokee Nation permitted it. Under Chief Justice John Marshall's influence, the Court narrowly ruled that the Cherokees were a sovereign nation and as such, the State of Georgia had no authority in Indian affairs. Unfortunately, the *Worcester v. Georgia* ruling did not prevent the United States Government from initiating the removal of the Indian nations to the west.[32] President Andrew Jackson wrote to John Coffee: "*...the decision of the Supreme Court has fell still born, and they find that they cannot coerce Georgia to yield to its mandate.*" In other words, the Executive Branch of the government ignored the ruling of the Judicial Branch and insisted that the Court's opinion was moot because it had no power to enforce its edict. Soon Government agents were dispatched to the Indian Nations to negotiate treaties of removal in exchange for lands in the newly designated Indian Territory, consisting of modern-day Oklahoma and parts of Kansas and Arkansas.[33]

The arguments by which the president justifies his actions are presented in his message of December 1835:

"The plan for their removal and re-establishment is founded upon the knowledge we have gained of their character and habits and has been dictated by a spirit of enlarged liberality. A territory exceeding in extent that relinquished, has been granted each tribe. Of its climate, fertility and capacity to support an Indian population, the representations are highly favorable. To these districts the Indians are removed at the expense of the United States; and, with certain supplies of clothing, arms, ammunition, and other indispensable articles, they are also furnished gratuitously with provisions for the period of a year after their arrival at their new homes…

… . The pledge of the United States has been given by Congress, that the country destined for the residence of this people, shall be forever "secured and guaranteed to them.' A country, west of Missouri and Arkansas, has been assigned them, into which the white settlement are not to be pushed. No political communities can be formed in that extensive region, except those which are established by the Indians themselves, or by the United States for them, and with their concurrence. A barrier has thus been raised, for their protection against the encroachments of our citizens, and guarding the Indians, as far as possible, from those evils which have brought them to their present condition…"[34]

The Cherokee Nation leader, Principal Chief John Ross, backed by the vast majority of Cherokees, resisted any attempt to negotiate a removal treaty that was not the best possible for the Cherokee Nation. To prevent individual tribal members from selling Indian land, the Cherokee Council passed legislation making it illegal to sell or sign away any tribal lands without the Council's approval. The penalty for violation was death. In 1835 a small cabal of mixed-blood Cherokees under council member Stand Watie, along with Elias Boudinot and Major Ridge, representing approximately 10% of the tribal population, circumvented the official tribal government and negotiated the Treaty of New Echota with the United States. In an attempt to expedite the treaty process, Congress illegally regarded this treaty as the official treaty with the Cherokee Nation. Watie, Boudinot, and Ridge and their followers suffered the wrath of the majority of Cherokees and the cabal's leaders were sentenced to death by the Cherokee Council.

Down in Alabama, many of the same conditions existed within the Creek Nation. The repercussions of the Creek War of 1814 between the Lower Creek towns and the Upper Creek talwas still simmered just below the surface, although the Lower Creeks and

Upper Creeks had reformed into the Muskogee Nation. The leadership of the mixed-blood Lower Creeks had fallen to William McIntosh and Opothleyahola headed the full-blood Upper Creeks. Although the Cherokees attempted to sway the Creeks into jointly resisting removal, in the 1825 Indian Springs Treaty over five million acres of the tribal lands of the Creeks within Georgia were signed away by Chief William McIntosh, the leader of the mixed-blood Lower Creeks. It was also discovered that Chief McIntosh had received a payment of $25,000 for "improvements" that he had made to his property, the only Creek to receive such funds. The National Council condemned his actions and ordered McIntosh's execution. He was killed near Milledgeville, Georgia, by a large party of Creek Warriors in April 1825. The assassination convinced President John Quincy Adams that the Indian Springs Treaty had been coerced, and its implementation would create another civil war within the Creek Nation. William's half-brother Roley McIntosh became leader of the Upper Creeks. He and his supporters quickly relocated to the new Indian Territory. Opothleyahola was invited to Washington for negotiations where in 1826 they signed the Treaty of Washington with the United States dissolving the Creek Nation in Alabama, surrendered their remaining lands, and promised to relocate to the Indian Territory by mid-decade. By 1831, Opothleyahola and the Upper Creeks realized their ability to resist removal was rapidly declining and they began their movement west. Once in Indian Territory the two sides again merged with McIntosh and Opothleyahola as leaders of their respective factions.[35]

Although the Cherokee "Trail of Tears" suffering was the most publicized, the Creek removal was probably the worst of the five removals. There was a severe "Creek Rebellion" in 1836 when Creek chief Encah Emothla attempted to defend the tribal members who had accepted allotments in Alabama instead of moving west. White settlers ignored these designated allotments and attempted to push the Creeks from their deeded lands. After the Federal and State governments failed to take any action against the settlers, the Creeks attempted to drive the encroachers off their properties. In response, the white settlers demanded they be protected from the Creeks. The U.S. Army, with General Winfield S. Scott in command, forcefully gathered all Creek tribal members left in Alabama, placed them in concentration areas, and forcibly moved them to Indian Territory, completely ignoring the treaties and the allotted and deeded property of the tribal members. Almost twenty-five hundred Creeks were transported in shackles and leg irons all the way from Alabama to the Indian Territory in the middle of the bitter winter of 1836-1837.[36] When learning the details of the Indian removal process, Americans should not be less than shocked and ashamed by the treatment of the Indians, especially the Cherokees and Creeks, during this period. Most property was simply seized with no regard for personal or private property. Although it appears that the United States Government and the Army attempted to ensure proper treatment of the tribal members, state officials and government contractors repeatedly failed to fulfill their responsibilities and contracts. Chief John Ross of the Cherokee Nation was required on many occasions to pay for food and transportation with his personal funds.[37]

Similar to the Cherokees and Creeks, the Choctaw originally agreed to removal and signed the Treaty of Doak's Stand in 1820, under pressure from Andrew Jackson, special envoy for President James Monroe. Still, most members of the Choctaw Nation refused to leave their five-million-acre homeland in Mississippi and move to the 13-million-acre area set aside for them in the Indian Territory. In 1829 and 1830 the State of Mississippi passed a series of laws in violation of the Federal treaties and outlawed the Choctaw Nation government, which provided prison terms for any Choctaw chief who attempted to exercise their tribal authority and placed all Choctaws under the authority of State laws. When the Choctaws appealed to President Jackson for protection, he falsely stated that the Federal government had no authority to protect the Choctaw from the laws of Mississippi.[38] Faced with overwhelming odds, a small contingent of Choctaws in Mississippi signed the Treaty of Dancing Rabbit Creek with the United States in September 1830. Under the new treaty, the Choctaws surrendered all of their lands east of the Mississippi River. The Nation was given three year's grace to make the move, and most Choctaws, and their Christian missionaries reluctantly moved West during the winter of 1831-32. Those who remained behind were forced to accept allotments, become state citizens, and forfeit their annuity rights. forced to accept allotments, become state citizens, and forfeit their annuity rights. That the state government or white settlers did not eventually push the remaining tribal members out, would indicate that the desire for the Choctaw lands in Mississippi was not as strong as the desire for land-grabbing that the Cherokees and Creeks endured.[39]

The Chickasaw Nation in northern Mississippi faced the same issues as the Choctaws. These two tribes, as close in culture as in distance, were still determined to remain independent of each other. The Chickasaw leadership signed a treaty similar to that signed by the Choctaws and proceeded to move West. Their experience differed from the other nations in that, since they were the farthest from Washington and the closest to the Indian Territory, they stubbornly withheld the signing of their removal treaty, the Treaty of Pontotoc, until 1832. They then refused to relocate until they had found acceptable lands out west, and taking their time in doing so. In the meantime, in a blatant disregard to the Supreme Court's decision in the *Worcester* case, the State of Mississippi claimed them as state citizens and attempted to tax their properties and this spurred them into action. Finally, in 1837, they began their relocation west, which lasted until 1840. They settled within the Choctaw allotment until a Chickasaw territory was carved out of the Choctaw Nation a few years before the Civil War.[40] Due to better management of resources and a smaller distance to travel, the Chickasaw removal was the least traumatic of the removals.[41]

The United States was able to coerce the Seminoles to agree to relocation and did so in the Treaties of Payne's Landing and Moultrie Creek in 1833. Tribal members, led by dynamic leaders including Osceola, refused to relocate and hid deep in the Florida Everglades and the interior swamps. This forced the military to seek them out and arrange for relocation. This caused the 2nd Seminole War,[42] which caused quite a few extreme losses including the massacre of Major Francis L. Dade and his 110 soldiers while traveling on the Fort King Road in 1835. The Seminoles won the battle but ended up losing the war because by 1842 most Seminoles had been driven from their swamps and were sent to the Indian Territory. Small groups from Florida were periodically located and were voluntarily or involuntarily removed to the newly established Seminole Nation up until the Civil War.[43]

Although there were great divisions within the nations concerning their movement from the southeast westward to the Indian Territory by this series of "Trails of Tears", the Five Nations prospered quite well on their new lands since they were abundant with fertile soil, water, and rich in natural resources. Many of the old wounds regarding the removal treaties began to move from internal civil war to opposing political parties.

---

[1] The term "Five Civilized Tribes" came into use during the mid-nineteenth century to refer to the Cherokee, Choctaw, Chickasaw, Creek, and Seminole nations. Although these Indian tribes had various cultural, political, and economic connections before removal in the 1820s and 1830s, the phrase was most widely used in Indian Territory and Oklahoma. *Encyclopedia of Oklahoma History and Culture,* Oklahoma Historical Society, 2010,

[2] Able, Annie Heloise, *The American Indian as Slaveholder and Secessionist: An Omitted Chapter in the Diplomatic History of the Southern Confederacy.* Cleveland, Ohio: A.H. Clark Company, 1919. Section 1.

[3] Around 900–1450 CE, the Mississippian culture developed and spread through the Eastern United States, primarily along the river valleys. The largest regional center where the Mississippian culture was first definitely developed is located in Illinois along a tributary of the Mississippi and is referred to as Cahokia. It had several regional variants including the Middle Mississippian culture of Cahokia, the South Appalachian Mississippian variant at Moundville and Etowah, the Plaquemine Mississippian variant in south Louisiana and Mississippi, and the Caddoan Mississippian culture of northwestern Louisiana, eastern Texas, and southwestern Arkansas. Like the mound builders of the Ohio, these peoples built gigantic mounds as burial and ceremonial places. *Wikipedia*

[4] A modern example of this would be when France pulled out of the North Atlantic Treaty Organization (NATO) in the 1960s.

[5] Able, pp. 39.

[6] James Adair was a English explorer, trader, and amateur historian who lived and traded among the Southeastern Indian tribes for over forty years in the eighteenth century. His book "A History of American Indians" is still used as a primary source of information and knowledge of the tribes before they were heavily influenced by Euro-American culture.

[7] Ehle, John. Trail of Tears: The Rise and Fall of the Cherokee Nation, New York: Anchor Books. 1988. pp. 1.

[8] Cunningham, Hugh T. "The History of the Cherokee Indians," The Chronicles of Oklahoma, Oklahoma City: Oklahoma Historical Society, Volume 8, Number 3, September 1930. pp. 291-292.

[9] Adair, James. *The History of American Indians.* Pantianos Classics, Online Publisher, 2020. Original Printing, 1775.

[10] Woodward, Grace Steele. The Cherokees, Norman: The University of Oklahoma Press, 1963. pp. 86.

[11] Hudson. Charles, The Southeastern Indians, Knoxville: The University of Tennessee Press, 6th printing, 1992. pp. 112.

[12] This division would be significant in the War of 1812, the Creek Removal, and the Civil War.

[13] Owsley, Frank Lawrence. Struggle for the Gulf Borderlands: The Creek War and the Battle of New Orleans, Gainesville, Florida: University of Florida Presses, 1981. pp. 83-85.

[14] Gibson, Arrell Morgan. Oklahoma: A History of Five Centuries, 2nd ed. Norman and London: University of Oklahoma Press. 1981. pp. pp. 48.

[15] McReynolds, Edwin C. Oklahoma: A History of the Sooner State, Norman: University of Oklahoma Press, 1964. pp. 93-94.

[16] DeRosier, Arthur H. Jr. The Removal of the Choctaw Indians, Knoxville: The University of Tennessee Press, 1970. pp. 7-13.

[17] Adair, pp. 129-158

[18] Debo, Angie. The Rise and Fall of the Choctaw

Republic, Norman: The University of Oklahoma Press, 1934. pp.30.

[19] Gibson, Arrell M. "Chickasaw Ethnography: An Ethnohistorical Reconstruction." Ethnohistory, vol. 18, no. 2, 1971, pp. 99.

[20] Gibson, Arrell Morgan, The Chickasaws, Norman: University of Oklahoma Press, 1972. pp. 36, 41.

[21] Ibid, pp. 106-115.

[22] Hudson, pp. 367-368.

[23] Missall, John and Mary Lou Missall, The Seminole Wars: America's Longest Indian Conflict, Gainesville: University Press of Florida, 2004. pp. 3-7.

[24] Gibson, Arrell Morgan. Oklahoma: A History of Five Centuries, 2nd ed. Norman and London: University of Oklahoma Press. 1981. pp. 68-69.

[25] Satz, Ronald N. American Indian Policy in the Jacksonian Era, Lincoln: University of Nebraska Press, 1975. pp.1-6.

[26] Gibson, Arrell Morgan. Oklahoma: A History of Five Centuries, 2nd ed. Norman and London: University of Oklahoma Press. 1981. pp. 46-47.

[27] King, Duane H. "Cherokee," Hoxie, Frederick E. Editor. Encyclopedia of North American Indians: Native American History, Culture, and Life from Paleo-Indians to the Present. Boston and New York: Houghton Mifflin Company. 1996. pp. 105-108.

[28] Land transfers from Native Americans to private individuals are void. When a tract of land has been acquired through conquest, and the property of most people who live there arise from the conquest, the people who have been conquered have a right to live on the land but cannot transfer title to the land. Johnson & Graham's Lessee v. McIntosh, 21 U.S. 543 (1823).

[29] LSData - LSD.Law

[30] Weeks, Phillip. Farewell, My Nation: The American Indian and the United States, 1820-1890. Wheeling, Illinois: Harlan Davidson, Inc. 1990. pp. 11-24.

[31] McIntosh, Kenneth W. "Creek (Muskogee),"

Hoxie, Frederick E. Editor. Encyclopedia of North American Indians: Native American History, Culture, and Life from Paleo-Indians to the Present. Boston and New York: Houghton Mifflin Company. 1996. pp. 142-145.

[32] Ehle, pp. 240-258.

[33] Weeks, Phillip. pp. 30.

[34] Hill, Luther B. A History of the State of Oklahoma, 1908, Chicago: Lewis Publishing Company, 1908. Chapter VII

[35] McIntosh, Kenneth W. pp. 144.

[36] Gibson, Arrell Morgan. Oklahoma: A History of Five Centuries, 2nd ed. Norman and London: University of Oklahoma Press. 1981. pp. 60-61.

[37] Ibid, pp. 217.

[38] Hudson, pp. 454-456.

[39] Kidwell, Clara Sue. "Choctaw." Hoxie, Frederick E. Editor. Encyclopedia of North American Indians: Native American History, Culture, and Life from Paleo-Indians to the Present. Boston and New York: Houghton Mifflin Company. 1996. pp. 119-121.

[40] White Deer, Gary. "Chickasaw." Hoxie, Frederick E. Editor. Encyclopedia of North American Indians: Native American History, Culture, and Life from Paleo-Indians to the Present. Boston and New York: Houghton Mifflin Company. 1996. pp. 114-115.

[41] Gibson, Arrell Morgan. Oklahoma: A History of Five Centuries, 2nd ed. Norman and London: University of Oklahoma Press. 1981. pp. 62.

[42] The First Seminole War was in 1817, when General Andrew Jackson led an incursion into the Spanish Florida territory, over Spanish objections, to stop cross-border raids. Jackson's forces destroyed several Seminole and Black Seminole towns.

[43] Sattler, Richard A. "Seminole," Hoxie, Frederick E. Editor. Encyclopedia of North American Indians: Native American History, Culture, and Life from Paleo-Indians to the Present. Boston and New York: Houghton Mifflin Company. 1996. pp. 577.

Post-War Texas Road Crossing

Texas Road Interpretive Marker at Honey Springs Battlefield

Remnants of the Texas Road-Butterfield Mail Route-California Gold Road at Boggy Depot and Fort Washita

Chapter 2
Antebellum Indian Territory

Indian Territory was initially known as "Indian Country" after the Louisiana Purchase in 1803. In 1825 Secretary of War John Calhoun drew a line across a map of the Louisiana Purchase that set a boundary including parts of present-day Oklahoma, Kansas, Nebraska, and Arkansas. Congress officially designated the area as "Indian Territory" with the Trade and Intercourse Act of 1834, although both terms were used interchangeably. When the removals of the Five Civilized Tribes began in the 1820's, the territory was not vacant. There were also nomadic Plains Indian tribes; including the Cheyenne, Arapaho, and Comanche, that roamed the plains following the herds of American bison from south to north and back across the plains of central and western Indian Territory.[1] These lands were intended to be the new homes for the relocated Five Civilized Tribes from the Southeastern States as well for the "remnants" of Indian nations from the Northeast United States, including the Senecas and Shawnees, and the Quapaws from the Red River area. The Intercourse Act of 1834 also attempted to control the removal of the Five Civilized Tribes and designated the Indian lands, "that part of the United States west of the Mississippi, and not within the states of Missouri, Louisiana, or the Territory of Arkansas."[2] Due to the cultural acclimation of the Cherokee and other Southeastern tribes, the Federal government tried to provide them with land acceptable to them by allowing them to choose their new homelands within the Indian Territory. The Kansas-Nebraska Act of 1854 cut the top part of the Indian Territory off for the creation of those two new territories. The

Osage Indian Reservation was originally along the Kansas border with Indian Territory, but the Osage were moved south into the Territory after the Civil War.

The Indian Territory was quite beautiful and fertile but was cursed with blazing hot summers with temperatures commonly 100+ degrees F, and extremely fridged winters with temperatures dropping below zero quite often. These issues were coupled with an incessant dry wind condition common to the Great Plains.

The Indian Territory's geography is diverse, transitioning from plains in the west to hills and mountains in the east, with a mix of prairies, forests, and swamps. The state is bisected by two major river systems, the Arkansas and the Red, and includes various ecoregions, including the Western High Plains and Southwestern Tablelands. Oklahoma's climate ranges from humid subtropical in the east to semi-arid in the west. The eastern half of the Indian Territory was mostly wooded with a mixture of hickory, black oak, and pine, and well-watered by the rivers like the Arkansas flowing from the Rocky Mountains. The northeastern and central portion of Indian Territory was a spur of the Ozark Mountains with a mixture of black oak and hickory forests with large intermittent prairies and savannahs. These lands provided fertile soils and many open prairies for cattle grazing. This area was designated for the Cherokees, Creeks, Seminoles, and the "remnant" tribes.

The western end of the Ouachita Mountains was a rugged range that protruded from Arkansas and occupied a large portion of the Territory's southeast corner. These mountains were covered with loblolly and other southern pine forests. The region also contained wide, fertile, and temperate flatlands that would be conducive to raising money crops like cotton and tobacco. This area would be the new home for the Choctaw Nation.

In the South-Central region of the Indian Territory, the area immediately west of the Choctaw Nation, would become the new home of the Chickasaw Nation. The world's oldest known mountain range, now called the Arbuckle Mountains after General Matthew Arbuckle, an early Army commander in Indian Territory, appeared as a small range of east-west hills, north of the Red River and west of the region known as the "Cross-Timbers," a natural border that was described as a great thicket "composed of nettles and briars so thickly matted together—as almost to forbid passage."[3] There were numerous fresh-water springs in this region that provided relief to both native and white travelers. The terrain was mostly flat with periodic forests of hickory and oak. The area received a decent amount of rainfall in the fall and spring, which would provide good grazing

for cattle and fertile ground for various grain harvesting, especially wheat.

The western half of the Territory was mostly open prairie of big and little bluestem grasses that was periodically broken up by meandering streams bordered by cottonwood trees. Eroded mesas and buttes provided a spectacular visual relief to the traveler. The Wichita Mountains occupied a large portion of the southwestern area in the unassigned lands of the Indian Territory and were mostly occupied by the Plains Indian tribes, including the Comanches, Wichita, Cheyenne and Arapahos.

Travel across the Indian Territory was difficult. The 1834 Act also attempted to create what historians now call the "Permanent Indian Frontier," a line that traversed from Minnesota to the Red River at the border with the Texas Republic, basically along the 95th Meridian.[4] It was believed at the time, especially by those who held to the Separation Policy, that the United States would not grow past this line for many years, and that the line provided a safe boundary between white settlement and the relocated Indian nations. Congress directed the Army to build a military road along this line and establish a series of military posts along this road, linking them for mutual support, beginning with Fort Snelling in Minnesota down to Fort Towson, Indian Territory. Other than the Fort Leavenworth-Fort Towson Military Road and its branches built or improved by the U.S. Army, there were three primary travel land routes across the Indian Territory, one north-south, one east-west, and one northeast to southwest.

The Texas Road was the primary north-south route, originally called the Osage Trace because of its early use by the Osage and other Indian tribes. For the Americans, it was initially used as a transit route for early emigrants

Map of the Six Corners
Region of the Civil War in
the Indian Territory

moving from the midwestern states to Texas once the Spanish opened that territory to settlement. The Texas Road had two main branches. The main route began in the headwaters of the Neosho River, crossed the Arkansas River at Three Forks near current day Muskogee, Oklahoma, travelled southwest and crossed the forks of the Canadian River near North Fork Town. The road continued southwesterly, finally crossing the Red River just east of its confluence with the Washita River at what became known as Colbert's Ferry.[5]

The secondary route of the Texas Road began in Springfield, Missouri, moved southwest through Neosho, Missouri, and joined the Fort Leavenworth-Fort Towson Military Road south of Maysville, Arkansas. The two routes converged at Fort Gibson. After Texas was able to win its independence from Spanish Mexico in 1836 to later join the United States in 1845 as the twenty-eighth State, the Texas Road's use significantly increased. Texas-bound travelers and various traders serving the Indian Nations used the road as a convenient pathway that crossed the territory of four of the Five Civilized Tribes. Toll ferries and bridges across the major rivers and tributaries provided income to the Indians who owned and operated them. By the late 1850s more than one hundred thousand wagons per year were documented as using the Texas Road to cross the Indian Territory. The traffic helped build up the Creek trade center called North Fork Town, as well as Perryville and Boggy Depot in the Choctaw Nation. Driving cattle from Texas to the Kansas City stockyards became more prevalent in the years before the Civil War as the cattlemen tried to satisfy the desire for beef throughout the United States. The Texas Road later became significant during the Civil War as the main artery for both the Union and Confederate forces as they jockeyed for position within the Indian Nations.[6] In fact, the largest battle of the war in Indian Territory was fought astride the Texas Road at Honey Springs.

The California Road was the main east-west route across the Indian Territory. The road was developed as a second and more seasonal route to Sutter's Mill, California, and the Marshall gold fields. It was a better route for the gold-seekers due to its location; it was further south of the main California-Oregon Trail that was passable for only a limited time each year due to the severe Upper Plains weather. The new route followed the Canadian River through the Leased District of Indian Territory rather than along Platte River across the Nebraska Territory. In 1849 the Secretary of War assigned 1st Lieutenant J.H. Simpson of the U.S. Army Topographical Engineers to survey a route from Fort Smith, Arkansas, to the New Mexico Territory, following the Canadian River Valley. Soldiers of the 5th U.S. Infantry, under Capt. Randolph Marcy, blazed, cleared, and improved a wagon road on both sides of the Canadian River.[7] Near "Rock Mary" in western Oklahoma, the parallel roads converged into one, traversed the Texas Panhandle, and continued into the New Mexico Territory where the gold-seekers had a variety of trail choices to reach California.[8] In the late 1850s a decision was made by the federal government to improve the road surveyed by 1st Lieut. Simpson in 1849. This decision was spurred on by an 1853 Pacific-bound railroad survey led by 1st Lieut. Amiel Whipple that was laid out to follow the 35th Parallel westward to the Colorado River. Whipple reported that it would not be difficult to improve the route into a substantial wagon road.[9]

In 1858 Congress established the Pacific Wagon Road Program in which one of the preferred routes was the 35th Parallel route along the Canadian River. Secretary of War John B. Floyd selected Edward Fitzgerald Beale,[10] a former officer in the U.S. Navy and protégé of Commodore Robert F. Stockton,[11] as the superintendent of the road's evaluation and construction. This road would go from Fort Smith, Arkansas, westward to Albuquerque, New Mexico Territory. Beale and 130 soldiers and laborers headed west in late October 1858 to begin their work. They found that the well-used and heavily travelled road through the eastern half of Indian Territory needed little work. Beale, with his strong political connections, was able to secure six iron bridges, which were supplied by Pencoyd Iron Works in Philadelphia, to replace the temporary wooden bridges that his workmen had produced in the initial stages of the project. These iron bridges were necessary to protect the crossings since the wooden bridges risked being burned by roaming Plains Indian war parties. The six iron bridges were installed in late 1859 at those sites in the wetter (eastern) sections of Indian Territory where Beale thought the Army needed them: the Poteau River, Red Bank Creek, Little San Bois, Big San Bois, Longtown, or Frenchman's Creek, and Little River. These bridges were to provide reliable stream crossings from Fort Smith across 220 miles of Indian Territory west to the Leased District of Western Indian Territory. Two significant crossings of the Canadian River, one at North Fork Town, and another near Chouteau's Old Trading Post, were to remain ford or ferry crossings due to the width, currents, and terrain of those areas. Today, history has titled this pathway across the Indian Territory, surveyed by Lieutenants Simpson and Wipple, and blazed and improved by the U.S. Army, as the "Beale Wagon Road."[12]

The final major road providing access across the Indian Territory was the Butterfield Overland Mail Route that traversed the Choctaw and Chickasaw Nations. This overland route began in Tipton, Missouri, continued southwest into Fort Smith, Arkansas, down to El Paso, Texas, and west to California. The Butterfield Overland Mail route used the first six miles of the Beale Wagon Road and travelled another 200 miles within the Indian Territory from Fort Smith, Arkansas. From Geary's Station to Colbert's Ferry on the Red River the route followed the older, well-traveled Texas Road. At Colbert's Ferry, travelers would use flatboats to bring the coaches across the Red River. Along the route through the Choctaw and Chickasaw Nations, the Butterfield Overland Mail Company established 12 stage stations spaced about 15 miles apart for changes of horses, supplies, meals, and crew replacement.

The Butterfield Overland Mail used both four-horse Concord Celerity and elegant Concord Coach stagecoaches, providing various levels of comfort based on the traveler's ability to pay. Not surprisingly, the General Council of the Choctaw Nation passed legislation authorizing various individuals to build and operate bridges, turnpikes and gates, and unofficial stage stands at various places along the way. The owners charged tolls for every person, animal, and wheel that used these facilities.[13] Therefore, the Butterfield coach would be charged for four wheels, six to nine passengers, one or two drivers, and four horses or mules. This provided much needed hard currency for the cash-strapped Choctaws. The Butterfield Overland Mail ran twice a week through the Territory but ended its runs at the outbreak of the Civil War in 1861. Soon after the end of the war, the transcontinental telegraph and the railroad were completed so

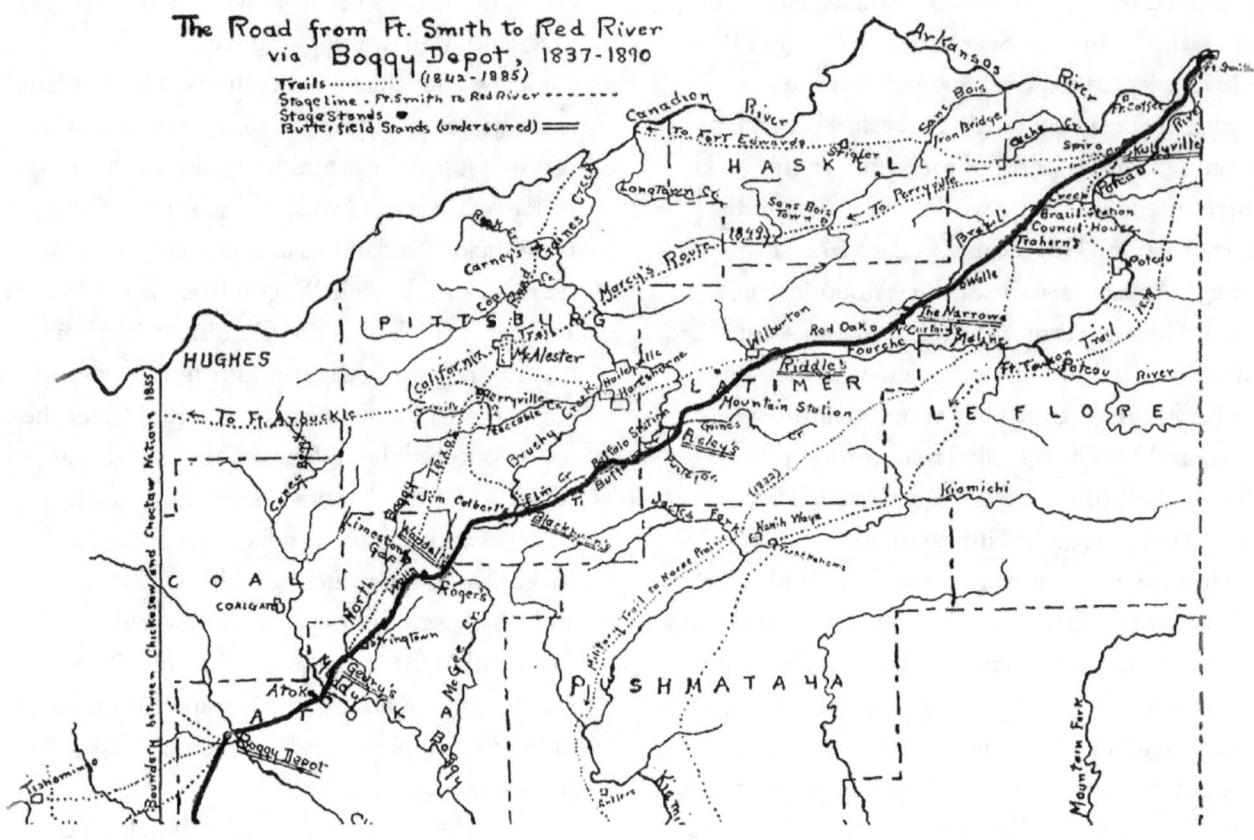

Butterfield Overland Mail Route through the Indian Territory
Map courtesy of the Oklahoma Historical Society

the Butterfield Overland Mail never resumed service.[14] Two interesting notes: first, during the Civil War the Butterfield Overland Mail route bisected two of the largest battlefields of the Trans-Mississippi West, Wilsons Creek, Missouri, and Pea Ridge, Arkansas. Secondly, the Butterfield Overland Mail did find a foothold in the regional mail service as part of the Wells Fargo Express Company.

In addition to the established roads, the rivers of the Indian Territory, fed by the melted snows of the Rocky Mountains, provided a significant, if somewhat limited, transportation waterway opportunity for the Plains Indians, the Army, traders, and emigrants. The largest river with navigation was the Arkansas River that flowed southeasterly from the eastern slopes of the Rockies, through the Kansas Territory, into Indian Territory, across Arkansas, and on into the Mississippi River. The Arkansas River was fed by the Canadian, Cimarron, Poteau, Neosho or Grand, and Verdigris Rivers as well as by many smaller tributaries. The site where the Arkansas, Grand, and Neosho Rivers converged became known as Three Forks. Three Forks was one of the first areas settled by the relocated Creeks, who established their new capitol at the Creek Agency located there. Steam riverboats were able to navigate most of these waterways for some distance during the wet season, which usually was January to June. Since these were Plains rivers, they were mostly wide, flat, and sandy, and most would dry to a trickle later in the year. Initially, self-propelled watercraft were employed by the Indians, traders, and trappers for navigation and transportation. They came up these streams in their canoes and keelboats, a shallow covered riverboat with a keel that is usually rowed, poled, or towed and is used to carry freight, as far as water would carry them, and then continued overland. The keelboats and flatboats, a cargo boat with a flat bottom for use in shallow water, were the primary vessels used to transit the rivers, but they were difficult to move upstream although the average keelboat and flatboat could carry a substantial amount of cargo and passengers upstream with a hard working crew.[15] (In fact, some of the first relocated Indians came to the Indian Territory by keelboat.) Eventually the steamboats on the Arkansas River began to service the Indian Nations and the Army posts as far inland as Fort Gibson on the Grand River, supplementing the cargo hauled by wagons on the newly constructed Fort Smith-Fort Gibson Military Road.

The other major waterway that connected Indian Territory to the Mississippi River was the Red River. It took many years for this river to develop any substantial commerce due to the "Great Raft," a huge accumulation of trees and other obstacles that lay in the river's path between Shreveport, Louisiana, and the Choctaw Nation. The cause of this natural obstacle course was the soft alluvial soil in the Red River flood plain that gave way during floods and brought trees crashing into the current. The blockage was, by some accounts, almost 150 miles in length. U.S. Army engineers were able to clear a substantial portion of the waterway by 1838. Their actions did open up the Red River for Indian Territory commerce, and steamboats were able to supply the southern military posts including Fort Towson, Fort Washita, and Fort Arbuckle. However, the remnants of the Great Raft obstructions were not completely removed from the Red River

until the 1870s.[16]

These pathways into the new Indian Territory were necessary for the Five Civilized Tribes to survive and later prosper. The territory lay in a very temperate zone with fertile soils and abundant water sources. All five of the tribes had developed agricultural skills while they lived in their homelands in the Southeast. Most had raised cattle, corn, cotton, tobacco, and other agricultural products prior to their removal to the West. After some environmental adjustment, due to the drier conditions and the increased winds, the Nations re-established their farms and ranches, built flour mills, developed salt works, and created a system of local and national trade that brought income to the Indians. The southern and central regions of Indian Territory were in the temperate zones for the growing of cotton and tobacco, products with significant market value. With the ports developing on the Red River, and to a lesser extent on the Arkansas River, the Indian Nations were able to find markets at major ports along the Mississippi River. Many Indians were able to accumulate a considerable amount of wealth, allowing them to build larger and larger plantations and buy more enslaved Africans to work on them.

Even before the Indian removals, there were several military posts already serving this frontier. The Post at Belle Point, at the confluence of the Arkansas and Poteau Rivers and within the Osage Reservation, was established in December 1817 by West Point-gray clad troopers of the U.S. Rifle Regiment under Major William Bradford. The post was established to maintain the peace between the war-like Osages and the newly arriving Five Civilized Tribes relocating from the East. The presence of the Osage Indians complicated the

Texas Road Marker near Okay, Oklahoma
Photo courtesy of Author

Beale Wagon Road / California Road
Photo by Author

Stand Watie and Major Ridge Marker at the Polsen Cemetery
Photo by Author

Enslaved African People made up approximately 14% of the total population of the Indian Territory. All Five Civilized Tribes allowed the buying and selling of slaves.

Reverend Samuel Worcester, Missionary to the Cherokee Nation
Photo courtesy of New Echota State Historic Site, Georgia

The Great Raft of the Red River

The Post at Belle Point Future Site of Fort Smith
Photo courtesy of the National Park Service

Fort Gibson State Historic Site
Photo by Author

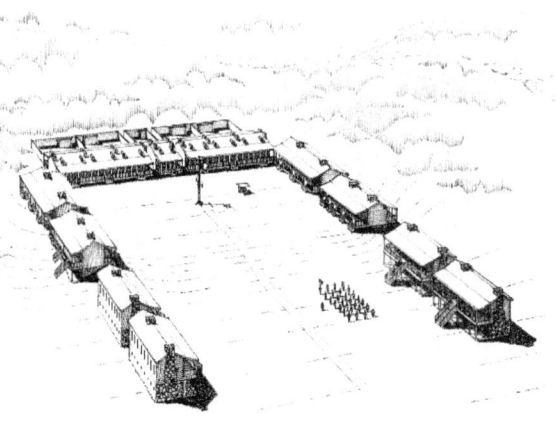

Fort Towson, Choctaw Nation
Photo courtesy of the Oklahoma Historical Society

Soldier drawing of Fort Coffee
Photo courtesy of the Oklahoma Historical Society

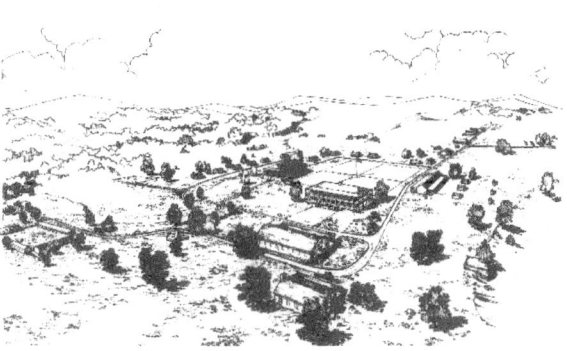

Fort Washita, Chickasaw Nation
Drawing courtesy of the Oklahoma Historical Society

Artist's Depiction of Old Fort Arbuckle
Photo courtesy of Author

Fort Arbuckle, Oklahoma

movement of the Five Civilized Tribes, especially the Cherokee and Creek Nations, into the designated Indian Territory. Considered to be "fringe-Plains Indians," the Osages believed they were the children of the "middle waters" and the universe of sky, earth, land, and water. They divided themselves into "Sky People" and "Earth People," each led by a chieftain. It is believed that the Osage came from the area around the Missouri River west of St. Louis and were from the same stock as the Quapaw, Ponca, and Kansa tribes. Linguistically, they were related to Siouan–Catawban tribes, a language family of North America that is located primarily in the Great Plains, Ohio, and Mississippi valleys, and southeastern North America with a few other languages in the east.. Although they were forest dwellers, they hunted American bison on the plains as the other Plains Indians did. Due to the pressure of white settlement in Missouri and Arkansas, the Osage signed a removal treaty in 1808 that established a reservation for them in southern Kansas. The Osage Indians lived on a reservation in southern Kansas until 1871, when they were forced to move to the Indian Territory. The Osage were not considered "civilized" by the same standard of the Southeastern Indians that had been moved into the territory. The agricultural and sedentary Cherokees, Creeks, and the smaller Indian tribal reservations, became easy fodder for the war-like Osages and their raids into the Indian Territory. This caused alarm among the Indians and the Army, which began to establish various posts to maintain separation and peace among the two sides.[17] Unfortunately, due to its distance from the Osage towns near and within Kansas, the new post on Belle Point was unable to control the Osages, so it was abandoned. In 1824, the Army moved up the Arkansas River to its confluence with the Verdigris and Grand Rivers and established Cantonment Gibson, and later, Fort Gibson. This fort was located closer to the centers of the Osage Nation and, as such,

could provide greater oversight of their activities. At the time of its establishment, it was the farthest West outpost of the young United States. It also became the rendezvous point for many of the Southeastern tribes moving into the Indian Territory.[18] Fort Gibson became the Indian agency distribution point for the Cherokees, Creeks, and, later, the Seminoles. The year 1824 also saw the establishment of Fort Towson on the Red River near its confluence with the Kiamichi River along the Texas border, which initially established the southern end of the Permanent Indian Frontier.[19] However, white settlers in Arkansas began to lobby the federal government for protection from their unfounded, yet perceived, threat of Indian attack from the Indian Territory. The presence of an Army post provided a financial benefit to the citizens living locally by providing employment, supply contracts, and improved infrastructure to the region. Arkansas politicians and powerful settlers prevailed and the Army re-established a more permanent post in 1833, adjacent to the former Post on Belle Point, which became Fort Smith, Arkansas.[20] Both of these military posts were garrisoned on and off until the end of the Civil War based on the needs of the Army.

A military road between Fort Smith and Fort Towson was surveyed and laid out by Captain Benjamin Bonneville, the western explorer of whom Washington Irving wrote, and his troops from the 7th U.S. Infantry. The road would be built for use in transporting troops and stores. This military-constructed road passed almost through the heart of what became the Choctaw Nation and was a great help to the Indians in reaching markets at Skullyville, the first capitol of the Choctaw Nation, and Fort Smith, where the Indians could exchange their products for clothing and some of the white man's luxuries.[21] Fort Towson also became the Indian agency distribution point for both the Choctaws and Chickasaws just as Fort Gibson was for the Cherokees, Creeks, and Seminoles.

Later, in 1838, Fort Wayne was established north of Fort Gibson to provide security along the northern end of the frontier line within the Indian Territory. This location was abandoned in 1839 and was re-established closer to the Arkansas line in response to the unrelenting concerns of the Arkansas citizens, who still claimed to fear Indian attacks from the Territory although none had ever been reported. This new location was abandoned in May 1842, and its garrison of Companies A & C, 1st U.S. Dragoon Regiment, moved up the Military Road and established Fort Scott, Kansas, on the Marmaton River.[22]

The Plains Indian tribes became a major concern for the mostly sedentary and agrarian Southeastern Indians. These nomadic "wild" Indians soon began to harass and attack the new arrivals, killing members and stealing livestock. As the Indian population of Indian Territory grew and conflicts between the Five Civilized Tribes and the Plains Indians increased, the Army was forced to establish many permanent and temporary military posts throughout the region to protect the relocated Southeastern tribes and keep peace on the frontier. The first of these forts built to protect the new arrivals was Fort Coffee, which was established in 1834 along the Arkansas River on Swallow Rock, near the Mississippian-built Spiro Mounds, and approximately 12 miles west of Fort Smith. Although the post's primary mission was to be an observation post to control whiskey smuggling into the Indian Territory, for most of its three-year existence it became a defensive post for the frontier.[23] At nearly the same time, the first Fort Arbuckle was established at the confluence of the Arkansas and the Cimmaron

Rivers in what is now western Tulsa County, Oklahoma. This "first" Fort Arbuckle in Indian Territory was built by troops of the 7th U.S. Infantry with Major George Birch in command. The stockade fort was placed between the Osage and the Creek settlements to maintain the peace between these two tribes. The post was garrisoned only a few months and was abandoned when it was deemed no longer necessary after the Osage and Creek reached an informal peace agreement.[24]

The next major post constructed by the U.S. Army was Fort Washita, built along the Washita River about 18 miles above its mouth with the Red River which was begun in 1841 and completed in 1843 by Companies A and I, 2nd U.S. Dragoon Regiment under Captain George A.H, Blake. Fort Washita was located in that portion of the Choctaw-Chickasaw country occupied mostly by the Chickasaws after the Doaksville Treaty of 1837. This post was built with a mission to protect the Chickasaw Nation members who were arriving from their homeland in Mississippi. Initially, the Chickasaws refused to leave Mississippi to move into their new territory out West due to the danger of the five thousand or so wild and aggressive Plains Indians, refugees from the almost continual Texas Indian wars, who were occupying these lands. After the fort was constructed and garrisoned, the Plains tribes were pushed back to what became known as the Leased District, west of the Chickasaw Nation and the designated area for the nomadic tribes. The post became the guardian of the nearby Chickasaw Agency,[25] the official office for the Indian agent, and continued to protect the tribe from the Plains Indians coming from Texas and lands further west.

The second Fort Arbuckle was constructed in 1850-1851 due to the anticipated dissolution of the joint Choctaw and Chickasaw Nations, which was accomplished by the second Treaty of Doaksville in 1855. Captain Randolph Marcy and Company D, 5th U.S. Infantry were the first troops to begin construction. The fort was built just north of the Arbuckle Mountains near the dividing line between the two nations and in 1857 it became the headquarters of the 7th U.S. Infantry Regiment. Its mission was to keep order among the Plains Indians, protect the Chickasaw members from those same Indians, and help protect white emigrants heading to California.[26] This was the final Army post established within the territory of the Five Civilized Tribes in the years before the Civil War. Although Fort Cobb was built in 1858 on the Canadian River, its mission was to control the Comanches in the Leased District far to the west and only played a small role in the Civil War.

A significant issue that needed to be addressed by the U.S. Government was the designation of boundaries between the various incoming Five Civilized Tribes and the Indian nations that were already present in the Indian Territory, including the large Osage tribe. In an effort to negotiate the initial boundary disputes and attempt to pacify the nomadic Plains Indians, in 1832 President Andrew Jackson sent Governor Montfort Stokes of North Carolina, Henry L. Ellsworth of Connecticut, and Reverand John F. Schermerhorn from New York, to the Indian Territory for this purpose. Governor Stokes' long interest in Indian affairs fit him well as the Chairman of what was to be known as the Stokes Commission. Commissioner Ellsworth was the first to arrive on October 8, 1832, and he established the Commission's base of operations at Fort Gibson. As they traveled, the Commissioner was accompanied by writer Washington Irving

and English scientist Charles Joseph Latrobe. Both went on to write and draw some of the finest descriptions of the flora, terrain, wildlife, and the native inhabitants of the early Indian Territory.

This early expedition to the western portion of Indian Territory was to be significant to the Army since it was the first show of force made to some of the western tribes in many years. On this initial fact-finding expedition, Commissioner Elsworth was also accompanied by three companies of Mounted Rangers under the command of Captains Jesse Bean, Lemuel Ford, and Nathan Boone, son of frontiersman Daniel Boone, This mounted force was responsible for the protection and security of the group. Unfortunately, these rangers bore little resemblance to regular soldiers. In fact, Washington Irving described the rangers as acting like they were on an unorganized hunting trip.[27]

By February 1833 Commissioner Stokes arrived and the commission successfully settled a boundary dispute between the Cherokee and Creek along the Verdigris and Arkansas rivers. In March 1833 the commissioners assigned lands in northeastern Oklahoma, in present-day Ottawa County, to remnants of the Seneca and Shawnee Indians, which had recently been relocated from Ohio. In May 1833 the commission authorized land adjacent to the Seneca and Shawnee tribes to two hundred homeless Quapaw Indians who were living with the Caddo Indians along the Red River. Turning to their main assignment, pacifying the western tribes (most notably the Comanche, Kiowa, Wichita, and Osage), the commissioners faced a daunting task. Suspicions of the Commission's intentions, hostility among the tribes, especially the Osage, and dissension among the Commission members themselves, created a somewhat hostile atmosphere, making it difficult for any treaty negotiations to be successful.[28]

In 1834 another expedition was formed at Fort Gibson to proceed west and contact the Comanches, Wichita, and other western tribes in an effort to convince them to agree to stop attacking and stealing from the relocated Eastern tribes. This expedition was led by General Henry Leavenworth, Commander of the Western District, and was accompanied by Colonel Henry Dodge and his newly formed 1st U.S. Dragoons Regiment. Artist George Catlin rode along with the intention of drawing the likenesses of the Western Plains tribal members they were expected to encounter during the mission. In history texts this mission is identified as either the "Dodge-Leavenworth Expedition" or the Dragoon Expedition of 1834."[29] Either way, the expedition departed Fort Gibson in June, bringing along Cherokee, Seneca, Osage, and Delaware scouts along with two teenaged Indian girls, one Kiowa and one Pawnee, who had been captured by the Osages. The girls were to be returned to their families in an act of goodwill. After an arduous two-month journey in which sickness had ravaged the Dragoons, 70% of who were sick at any one time, the expedition came into the region in which the Plains Indians were most active, the Southwest portion of the Indian Territory. Just prior to contacting the Wichita Indians, General Leavenworth fell from his horse while chasing a bison calf and received serious internal injuries which led to his illness and he remained at the camp. Col. Dodge was given command of the expedition and ordered to continue toward the Wichita village. Despite his attempt to catch up with Col. Dodge, Leavenworth never recovered and died on July 21, 1834, and was buried somewhere along the Canadian River. Dodge

was able to contact the Wichita, Kiowas, and the Comanches and convinced their chieftains to come to Fort Gibson for a meeting with the remnants of the Stokes Commission, even though its authority had expired. Although the Plains tribes arrived for a conference in September 1834, the U.S. Government representatives did not have the authority to conclude a treaty with the participants. Nevertheless, the Plains tribes did verbally agree to stop the harassment of whites and the eastern Indians. However, it would take another year for the necessary Federal authority for the treaties to be granted. The treaties were finally signed in 1835 and 1837.[30]

During the 1830s-1850s, the Five Nations re-established their govern-ments, schools, and began to civilize their new homelands. They built homes, churches, farms, ranches, mills, and retail establishments. To improve transportation, they built or improved roads, bridges, and ferries. The removal-inspired tribal divisions within the Nations remained, but the vast distances of the Indian Territory allowed these factions to remain distant from each other except during the periodic Council meetings. The newcomers were somewhat welcomed by other local tribes including the Osage, Kaw, and Kansas Indians. It is interesting that each of the Five Civilized Tribes, many of which had adopted white-Southern culture, permitted African slavery. At the beginning of the Civil War, approximately 14% of the Indian Territory's population consisted of enslaved persons.[31]

Although most members of the Five Nations were now in the Indian Territory, they were not united in any way. They were still treated by the United States Government as separate and equal sovereign entities. And even within the Five Nations there were great divisions, especially in the Cherokee and Creek Nations dating back to the signing of the removal treaties. In these two tribes, the anger ran very deep, simmering under the surface most times but periodically flaming into violence. As mentioned in the previous chapter, Major Ridge, Elias Boudinot, and Stand Watie, who together only represented approximately 10% of the Cherokee Nation members, signed the Treaty of New Echota in 1835, which surrendered the tribal lands in Georgia and North Carolina. Although the Cherokee National Council and Principal Chief John Ross vehemently appealed this decision to President Andrew Jackson, the appeal fell on deaf ears. This betrayal hardened the attitudes of all tribal members, the Treaty Party, the Non-Treaty or National Party, and the Old Settlers (Cherokee who voluntarily moved west before the Removal treaty), all of which were represented by Sequoyah.

The dissension made it difficult when they attempted to create a new Cherokee Constitution to replace the one developed in 1828 when they were still in their homelands. During a break in these meetings, Major Ridge, his son John Ridge, and Elias Boudinot, were killed near their respective new homes in the Indian Territory, probably under the orders, or at least the knowledge of, Chief Ross and the National Council. Stand Watie was also targeted but was able to avoid execution. (Watie eventually took revenge on John Ross and his home during the Civil War.)[32] General Matthew Arbuckle, who commanded the U.S. Army forces in the Indian Territory at Fort Gibson, insisted that the assassins be delivered to the Army for prosecution. Chief Ross refused to comply and told General Arbuckle that it was an internal tribal issue over which the U.S. Government had no jurisdiction or right of involvement.[33] But the assassinations also tainted Chief Ross' influence with the new U.S. President, Martin Van Buren. Van Buren refused to receive Ross when he came to Washington to re-negotiate the treaties with him. Instead, Van Buren met with Stand Waite and members of the Treaty Party. Van Buren's decision may have been due to a letter from former president Andrew Jackson to Stand Watie and John A. Bell in which he reported he had written to President Van Buren on their behalf. Jackson reiterates his support and friendship of the Treaty Party and condemns "the outrageous & tyrannical conduct of John Ross & his self-created council." Jackson further states "I trust the president will not hesitate to employ all of his rightful power to protect you and your party from the tyrannical & murderous scemes [sic] of John Ross."[34] It should be kept in mind that Jackson's letter was likely an attempt to cover his own tracks since he had negotiated the Cherokee removal treaty with an unauthorized, minority delegation from the Cherokee Nation. Because Van Buren was fairly new in office, he may not have been completely aware of the internal divisions and sensitive politics of the Indian Nations. By late 1839 the United States, the National Party and the Treaty Party agreed to an Act of Union under which the Cherokee Nation would operate until the Reconstruction period after the Civil War.[35]

The Creek Nation's attempt to establish a constitutional government was also difficult with treaty and non-treaty factions vying for power. When the remainder of the Upper Creeks arrived in the new Creek Nation in the 1830's, the previously relocated Lower Creeks insisted that the newcomers submit to the constitutional government and laws that they had established when they first arrived in Indian

Territory. The Upper Creeks agreed, but both factions kept a safe distance from each other.[36] In the new Creek Nation, the Lower Creek towns, under Chief Roley McIntosh, developed mainly along the Arkansas River, centering on the confluence of the Verdigris River around the new Creek Agency at Three Forks. The Upper Creeks followed Opothleyahola, moving south and developing their towns along the Canadian and Deep Fork Rivers in the area around present-day Eufaula, Oklahoma.[37] Even with these distances the difficulties between the two groups continued. An 1837 Report of the Commissioners of Indian Affairs stated,

> "The nation was divided into two parties, rivaling each other in animosity and bitter hatred, ex-cited with jealousy and discord, and requiring great exertions on the part of government officers to prevent bloodshed and bring about an amiable understanding."[38]

In 1839 there was an agreement of unity made between the Upper and Lower Creeks that recognized the leaders of each of the two factions but provided for a unified National Council. The Lower Creek faction leader Roley McIntosh would serve as the nation's Principal Chief. Still, the Lower Creeks lamented that after all of their negotiations with the U.S. Government, they mistakenly believed that because they had supported and fought with President Jackson in the earlier Creek War, they would be able to either secure their lands in the east or, at least, negotiate a better removal treaty. They were wrong. These relations did not improve before the Civil War erupted.

Due to the significantly smaller number of mixed-bloods among the Choctaws and Chickasaws owing to their comparative isolation from American settlers, these tribes did not experience the same level of dissension as their new neighbors to the north and east. The largest issue confronting these tribes was created by the 1837 Treaty of Doaksville, which required that the Choctaws absorb the smaller and weaker Chickasaw Nation into its designated territory. The Choctaws were required to adopt a new constitution to replace the one developed in 1834. The previous three districts within the Choctaw Nation had to be expanded to four to accommodate the Chickasaw Nation into its boundaries. Unfortunately for the Chickasaws, the area they purchased from the Choctaws was remote and was regularly overrun by Kickapoo, Wichita, and Comanche war parties. In response, the Army then established Fort Washita in the southern portion of the Choctaw Nation to help protect the Chickasaw settlers and maintain peace in the region.[39] In 1850 the Chickasaws seceded from the Choctaw Nation and established their own homeland in the western Choctaw region between the Canadian and the Red Rivers. Both nations created a series of national constitutions over the years prior to the Civil War that established their respective governments, capitols, schools, and other agencies as needed.[40]

Between 1836 and 1859 the members of the Seminole Nation were hunted down, captured, and taken from the Florida Everglades and swamps and transported to the Indian Territory. After many years of searching, the final organized group to be relocated was that of Billy Bowlegs. Approximately 3,400 Seminoles were actually transported, but many remained hidden in those swamps. When they first arrived at either Fort Gibson or Webber's Falls, most of the Seminole tribal members remained in the immediate area. Under the terms of the 1833 Treaty of Fort Gibson, the Creek were obligated to absorb the Seminoles into their territory and subject them to the laws of the Creek Nation. The issue of slavery became a significant problem for relations between the Creeks and the arriving Seminoles. The Creeks claimed that the African males and females who accompanied the Seminoles were actually runaway slaves from the Creek Nation and demanded that they be returned to their Creek owners. The Seminoles refused, and the Creek Nation dropped the issue. But this was the first of the issues confronting the two separate, but related tribes. The Seminoles arrived as virtual prisoners of war by the U.S. Army since they had to be forcefully brought to Indian Territory. As a result of this, they were the least contented of the relocated Indians. The Seminoles refused to submit themselves to the Creeks and looked to the Army to protect them from the Creeks. Finally in 1845 a treaty between the Creek and Seminole Nations provided for a homeland for the Seminoles in the region around the Little River in the western Creek Nation.[41] The Seminoles did not permanently receive their own territory until a treaty was signed with the Creeks and the United States in 1855 that finally established a separate Seminole Nation and tribal government.[42] It took many years of small groups moving around the Cherokee and Creek Nations before the Seminoles moved to the western part of the Creek Nation that had been designated for the Seminole Nation. The windy, dry, and extreme temperature environment of the Indian Territory was vastly different from their hot and humid former homelands. It took time for them to figure out the agricultural and hunting practices that were appropriate for the treeless plains.

The Indian Nations were in many ways comparable to their white neighbors in the Southern states, especially among the mixed-blood population, in that much of the agricultural work done on the large farms and plantations was done by enslaved labor. They were also comparable with their southern neighbors in that only a small number of tribal members actually owned the vast majority of these enslaved laborers. One Choctaw plantation owner had over 500 enslaved persons working his lands. Most of the enslaved in Indian Territory lived and worked among the mixed-blood members who had settled the fertile river valleys and continued to develop Euro-American lifestyles and mannerisms. The Five Civilized Tribes altered their social organization, political structure, and economic system to accommodate the institution of slavery, which was a direct a result of the effects of European colonization of the Americas.[43]

On the other hand, the full-bloods of the Nations owned few enslaved persons. They lived mostly in the forests away from the main settlement areas and attempted to maintain a traditional livelihood based on hunting, trapping, subsistence farming, and the exchange of pelts with the settlements. Although it appears that the Indians treated their enslaved better than their white counterparts, each of the Indian nations developed slave codes with severe consequences, especially for the teaching of abolitionism. The Seminoles were unique in that they allowed their enslaved persons to marry into the tribe. They also allowed most of the enslaved to live together in separate towns, with their only allegiance to their masters being the payment of tribute each year. Many smaller Indian farmers would hire enslaved workers from the larger plantations. For the small and mid-sized landowner, this was a cheaper option than the actual purchase of a person. It also provided the slave-owners with another steady flow of cash or barter for other needed products from these other properties.[44]

All Five Civilized Tribes had some level of Christian missionary activity. Due to the rabid anti-Catholic rhetoric that persisted in the American South, the activity was primary limited to Protestant denominations who were able to compete for the annual $10,000 in congressional appropriation for the Indian Civilization Fund. This money was earmarked for educating the tribal members in classical subjects, such as math, spelling, biology, and history, as well as practical arts including agriculture, carpentry, sewing, and cooking skills. Most of these missionary efforts had begun prior to the Removals, and many of the committed Protestant missionaries moved west with their flocks. The Baptists, Presbyterian, Congregationalists, and Methodist churches spent the most time and effort in building churches and schools, providing secular and religious education, and providing access to printing presses for both secular and religious uses. Religious societies also provided funding for teachers, physicians, ministers, and practical arts instructors, with funds donated by various congregations around the country. The Baptists were organized and overseen by the Baptist Mission Board in Boston, Massachusetts, and operated primarily within the Cherokee Nation. The Presbyterians and Congregationalists were overseen by the American Board and later by the Presbyterian Board of Foreign Missions. Both of these missionary groups and the funds from these societies built static missions and schools. Because these groups were rooted in New England, these missionaries tended to be abolitionist when it came to the slavery question. The Methodists on the other hand were funded and managed by regional Methodist conferences that existed near the Indian Territory, such as Arkansas and Missouri.

These conferences took over for the Tennessee and Mississippi conferences after the Five Tribes moved West. These conferences eventually merged into the Indian Methodist Conference. Since these conferences were in Southern states, their missionaries tended to be pro-secessionist and pro-slavery. The Methodists shied away from static missions and instead relied upon circuit-riding ministers, much as they did in the Eastern United States.[45]

The impact of the missionaries on the Five Civilized Tribes was mixed. The Cherokees embraced the religious teachings of the Methodists and the Baptists, and they became a stable and competent part of the Nation. Reverend Samuel Austin Worcester, who had been imprisoned in Georgia, established a significant mission at Park Hill near the Cherokee capitol at Tahlequah. He became a leading figure within the Cherokee Nation and helped to provide assistance and guidance to the Cherokee government in its development. This is countered by the Choctaws who, as mentioned in the previous chapter, were indifferent to religion, both their own and Christianity. The remaining three nations, Creek, Chickasaw, and Seminole, accepted Christian missionaries and belief in varying degrees between the extremes presented by the Cherokee and Choctaw Nations. In fact, the Creeks had a period of strong anti-Christian sentiments, causing the missionaries serving the Nation to be driven from their missions. (They were invited back in the years prior to the start of the Civil War.)[46]

In the period of 1820 to 1860 the Five Civilized Tribes had been forcefully removed from their homelands, transported to a strange new land by the military, and had been expected to build new nations and lives without the support that had been promised them by the

United States Government. Although the tribes were mostly successful in adapting, there were many years in which the weather did not cooperate, bringing floods or droughts, or in which the government's promised annuities and food support did not arrive in a timely manner or, sometimes, not arrive at all. There were many times when things were lean, resulting in the Indians becoming destitute, with hunger becoming widespread. Still, they held on the best that they could. A positive result was that U.S. Army had done a great job of surveying and building roads across and through the Indian Territory and clearing the rivers so the commercial shipping interests could develop water-borne traffic to bring supplies in and

products out of the Indian Nations. The Army had also built posts to provide safety to the relocated Indians as well as to allow settlements to develop around them that would bring needed products and services to customers. By 1861 the Five Civilized Tribes were able to re-establish themselves, create constitutions and government, arrange for the educational and spiritual needs of tribal members, and create a growing economy that they hoped would someday reduce their dependence on the Federal government. While all of these activities were happening, a dark menace began to shadow the entire United States. Even the Five Civilized Tribes would not be able to escape the grips of the oncoming Civil War.

[1] Heidler, David S. and Heidler, Jeanne T. Indian Removal, New York & London: W.W. Norton Co. 2007. pp.46.

[2] U.S. Congress Act of June 30, 1834, ch. 161, 4 Stat. 729.

[3] Pelzer, Louis, ed. "A Journal of Marches by the First United States Dragoons, 1834-1835" Iowa Journal of History and Politics, Vol. VII, July 1909. pp. 346

[4] McPherson, James M. The Battle Cry of Freedom: The Civil War Era, New York: Oxford University Press, 1988. pp. 45.

[5] Gibson, Arrell Morgan. Oklahoma: A History of Five Centuries, 2nd ed. Norman and London: University of Oklahoma Press. 1981. pp. 130-131

[6] Weaver, Bobby D. "Texas Road," The Encyclopedia of Oklahoma History and Culture, 2010

[7] Lieut. James H. Simpson, "Report and Maps of the Route from Fort Smith to Santa Fe," in Report from the Secretary of War (1850), 31st Cong., 1st Sess., Sen. Ex. Doc. No. 12.

[8] Gibson. pp. 107

[9] Jackson, W. Turrentine, Wagon Roads West: A Study of Federal Road Surveys and Construction in the Trans-Mississippi West, 1846-1869. New Haven & London: Yale University Press, 1964. pp. 244. Also 1st LT A. W. Whipple, "Itinerary," Report of Explorations for a Railway Route, Near the 35th Parallel of North Latitude, front the Mississippi River to the Pacific Ocean, Vol. 111, Part I, 33d.

Conp., 2d Sess., House Ex. Doc. 91, S. Ex. Doc. 78 (Washington, 1856)

[10] Edward Beale had an interesting life. Starting as a U.S. Naval Officer, he became established in early California and owned significant properties near Fort Tejon. It was his prompting of then Secretary of War Jefferson Davis to experiment with camels as the primary beast of burden for the Army in the Southwest. In fact, the western portion of the road was an attempt to improve the conditions for the camels to travel. He acquired 70+ camels from the Middle East and led the newly established Camel Corps for the Army. Unfortunately, the idea did not take root and was phased out by the beginning of the Civil War.

[11] Robert Field Stockton was a United States Navy commodore, notable in the capture of California during the Mexican–American War. He was a naval innovator and an early advocate for a propeller-driven, steam-powered navy. Stockton was from a notable political family and also served as a U.S. senator from New Jersey. From Wikipedia

[12] Messer, Carroll, "Beale's Wagon Road to the Pacific Coast: Western Camel Road and Eastern Iron Bridge Road." Self-published Paper via Texas A&M University, 2021

[13] Acts and Resolutions of the General Council of the Choctaw Nation, 1858, published by authority of the General Council, by Josephus Dotson, printer for the Nation (Fort Smith, Ark., 1859).

[14]Everett, Dianna, "Butterfield Overland Mail," The Encyclopedia of Oklahoma History and Culture, 2010. Also Grant Foreman, "The California Mail Route through Oklahoma" Chronicles of Oklahoma, Vol. IX, No. 3, September, 1931)

[15] Foreman, Grant. "Early Trails Through Oklahoma," Chronicles of Oklahoma, Volume 3, No. 2. June, 1925. pp. 100.

[16] Gibson, 101-104.

[17] Wilson, Terry P. "Osage," Encyclopedia of North American Indians: Native American History, Culture, and Life From Paleo-Indians to the Present, Hoxie, Frederick E. Editor. New York: Houghlin Mifflin Company, 1996. pp. 449-450.

[18] Agnew, Brad. Fort Gibson: Terminal on the Trail of Tears, Norman & London, University of Oklahoma Press, 1980. pp. 3.

[19] Fort Jesup, Louisiana was added to this frontier line when the situation between the Texans and Mexican Government heated up during the 1830's.

[20] Bearss, Edwin C. and Gibson, Arrell M. Fort Smith: Little Gibraltar on the Arkansas. Norman and London: University of Oklahoma Press. 1969. pp. 3-7.

[21] Culberson, James. "The Fort Towson Road: A Historic Trail," Chronicles of Oklahoma, Volume 5, No. 4, December, 1927. pp. 414.

[22] Oliva, Leo E. Fort Scott, Topeka: Kansas State Historical Society, 1984. Pp. 12-13.

[23] Litlefeld,Jr. Daniel F, and Lonnie E. Underhill, "Fort Coffee and Frontier Affairs, 1834-1838." Chronicles of Oklahoma, article, Autumn 1976; Oklahoma City, Oklahoma. pp. 314-315.

[24] Faulk, Odie B., Franks, Kenny A., Lambert, Paul F. Editors. Early Military Forts and Posts in Oklahoma. Oklahoma City: Oklahoma Historical Society. 1978.
--Hughes,J. Patrick, "Forts and Camps in Oklahoma Before the Civil War," pp. 39

[25] Faulk, Odie B., Franks, Kenny A., Lambert, Paul F. Editors. Early Military Forts and Posts in Oklahoma. Oklahoma City: Oklahoma Historical Society. 1978.
--Howard II, James A. "Fort Washita," pp. 54-55.

[26] Hughes, pp. 49-52

[27] Irving, Washington. A Tour on the Prairies, New York: Skyhorse Publishing, 2013.

[28] Steffen, Jerome O. "Stokes Commission," The Encyclopedia of Oklahoma History and Culture. 2010.

[29] Two soldiers of later notoriety who served in this expedition were Lieutenant Colonel Steven W. Kearny and 1st Lieutenant Jefferson Davis. Both Kearney and Davis would later serve as U.S. Army officers in the Mexican War. Kearney later became a General in the U.S. Army and Jefferson Davis became the President of the Confederate States of America.

[30] Agnew, pp. 121-141.

[31] Krauthamer, Barbara. "Slavery" The Encyclopedia of Oklahoma History and Culture, Oklahoma Historical Society Online

[32] Foreman, Grant & Ross, Allen. "The Murder of Elias Boudinot" Chronicles of Oklahoma, Volume 12, No. 1, March 1934. pp. 19-22.

[33] Foreman, Grant. The Five Civilized Tribes, Norman: University of Oklahoma Press, 1934. pp. 301.

[34] Letter from Andrew Jackson to John A. Bell and Stand Watie, October 5, 1839. Dale, Edward Everett, & Gaston Litton, eds. Cherokee Cavaliers: Forty Years of Cherokee History as told in the Correspondence of the Ridge-Watie-Boudinot Family, Norman: University of Oklahoma Press, 1939. pp. 17.

[35] Foreman, pp. 301-303.

[36] Gibson. pp. 50-51.

[37] Foreman, pp. 152-155.

[38] Ibid, pp. 163.

[39] Faulk, Odie B., Franks, Kenny A., Lambert, Paul F. Editors. Early Military Forts and Posts in Oklahoma. --Howard II, James A. "Fort Washita," pp. 54-55.

[40] Gibson. pp. 76-77.

[41] Ibid, pp. 79.

[42] McReynolds, Edwin C. Oklahoma: A History of the Sooner State, Norman: University of Oklahoma Press, 1964. pp. 186-191.

[43] Perdue, Theda. Slavery and the Evolution of Cherokee Society, 1540-1866, Knoxville: The University of Tennessee Press, 1979. pp. *xiii-xiv*.

[44] Gibson, pp. 98-101.

[45] Ibid, pp. 84-90.

[46] Ibid, pp. 91-97.

Tullockchishko-Choctaw Stickball
Photo courtesy of the Smithsonian American Art Museum

**British Governor of the Georgia Colony
James Oglethorpe meeting with Creek Indians**
Photo courtesy of the Georgia Historical Society

Pushmataha, regional chief of the Choctaw Nation
Photo courtesy of Library of Congress

Creek Chieftain William McIntosh
Photo courtesy of the Georgia Historical Society

Three 18th Century Cherokees in traditional attire
Drawing courtesy of the National Park Service

Chapter 3
The Secession Movement in the Indian Territory

When the United States began to drift toward division and civil war during the 1850s, the Five Civilized Tribes were forced to examine their collective consciences and begin choosing sides for the upcoming showdown between the North and South. Events were accelerating politically during this period and the federal government was seemingly paralyzed in any attempts to slow down the division of the Union of States based on the slavery issue. Although many in the 20th and 21st Centuries falsely claim that it was solely a "states-rights" issue, and admittedly there were some issues that fell under this heading, such as nullification and tariffs, clearly slavery was the overwhelming factor that caused the Civil War in America.

In an effort to spur construction of a northern-connected trans-continental railroad route from Chicago across Nebraska Territory, Senator Steven A. Douglas of Illinois introduced the Kansas-Nebraska Act of 1854. The Act organizing this large territory north of the Indian Territory, is regarded one of the worst pieces of legislation ever put into federal law. The Act nullified the Missouri Compromise of 1820 which had prohibited slavery north of the 36-30° line of latitude.[1] This legislation would ultimately usher in an early start of the war by allowing "popular sovereignty," meaning that the residents of those territories could vote to allow or prohibit slavery in those areas. The immediate result was that citizens in other states with abolitionist beliefs, known as Free-Soil or Jayhawkers, or citizens who were pro-slavery,

Border Ruffians Invading Kansas
Photo courtesy of Kansas Historical Society

called "Border Ruffians," flooded into the Kansas territory and began to spar with each other. This began a period known as "Bleeding Kansas" in which both sides of the slavery question began committing violence and atrocities on an ever-increasing level well into the Civil War years. One bloody incident gained national recognition for the Kansas issue on May 19, 1858. A party of about 30 Pro-Slavery Missourians seized 11 Kansas Free-State men from Trading Post, a small community in Linn County, Kansas. They marched these men to a deep ravine just west of the Missouri-Kansas state line near the Marais des Cygnes River. Lining up their captives, they callously shot them down, killing five and wounding five. One feigned death and survived. Several of the Missourians refused to shoot the unarmed prisoners which may have spared the lives of the six survivors. The "Marais des Cygnes Massacre," as it became known, horrified Northerners and Southerners alike. A poem written by John Greenleaf Whittier, the "Le Marais du Cygne," immortalized the event and led others to take up arms against the pro-slavery forces.[2]

The nightmare of Bleeding Kansas spread to Virginia when in 1859, John Brown, a Kansas Free-Soiler, and a group of his followers attempted to capture the U.S. Arsenal at Harpers Ferry, Virginia, confiscate the weapons, and arm the enslaved persons throughout the South. Brown was captured by a force of U.S. Marines under the command of Army Colonel Robert E. Lee, and hanged for treason by the State of Virginia, but he was hailed as a hero by many abolitionists and westerners. Many of these Kansas issues also bled southward across the border into the Indian Territory. In fact, many of the prominent abolitionists from Kansas would soon lead Union troops and armies into the Indian Territory once Kansas was accepted as a free state in 1861.[3]

The election of 1856 brought President James Buchanan of Pennsylvania into office. Although Buchanan was a Northerner, he held deep seated pro-southern and pro-slavery loyalties and filled his cabinet with powerful Southerners. Never before in the history of the United States was there a man in the presidency who was as ill-suited and ill-equipped to lead the nation through the worst crisis since its founding. Buchanan reacted with a "deer in the headlights" mentality as he watched the nation break apart. In his Inaugural Address, the President referred to the territorial question as "happily, a matter of but little practical importance" since the Supreme Court was about to settle it "speedily and finally." Two days later Chief Justice Roger B. Taney delivered the Dred Scott decision, asserting that Congress had no Constitutional power to deprive persons of their property rights over slaves in the territories. Southerners were delighted, but the decision created a furor in the North. Buchanan decided to end the troubles in Kansas by accepting the pro-slavery Lecompton Constitution that had been developed by the Border Ruffian Kansans and by urging the admission of the territory as a slave state. Although he exerted his Presidential authority to try and achieve this goal, the admission of Kansas as a slave state failed to pass either the Kansas voters or the House of Representatives in 1858. In pushing this legislation, he further angered the Republicans and alienated members of his own Democratic Party, who were almost as equally divided on the slavery issue. Kansas remained a territory until statehood as a free state was granted in 1861.

Presiding over a rapidly dividing Nation, Buchanan did not comprehend the political realities of the time. The new President did not realize how sectionalism had realigned political parties due to the Compromise of 1850, to the point that the Whigs didn't stand for anything of consequence. This brought the Republican Party, an anti-slavery party, into existence in 1856 from the remnants of the Whigs and the abolitionists. Relying on Constitutional doctrines to close the widening rift over slavery, Buchanan also failed to understand that the North would not accept Constitutional arguments that favored the South or the expansion of slavery. The President also didn't know or didn't care that his own administration was adding to the sectional confusion by beginning to prepare the Southern states for secession. In fact, under President Buchanan's nose, Secretary of War John B. Floyd, (who as a Confederate general would in 1862 gain notoriety by deserting Fort Donelson, Tennessee, on the eve of its surrender) made it a point to transfer arms and ammunition from the

U.S. arsenals in northern states to those in the southern states. This was a clandestine effort to equip the South when secession finally took place.

Buchanan's predecessor, Franklin Pierce, had appointed Jefferson Davis to the role of Secretary of War, and Davis, soon to be President of the Confederate States, had also attempted to ship as many arms, ammunition, and equipment to the southern arsenals as quietly as possible. Ordinance development was a strength of the antebellum U.S. Army, along with engineering and logistics.[4] In 1860, the U.S. Regular Army consisted of just over 16,000 soldiers, but by the time the states began to secede, Southern arsenals had approximately 175,000 modern firearms on hand. Of these, about 140,000 were smooth-bored muskets and the remainder were rifles. Approximately 104,000 of these shoulder arms had been transferred South during 1859-1860. Of the eight U.S. Arsenals in the Southern states, each was commanded by a Southerner appointed by Secretary Floyd.[5]

Although somewhat isolated and mostly indifferent, these sectional problems and secession ideas were already present and active in the Indian Territory by 1860. When the U.S. Department of the Interior was established in 1849, one of the duties of the new department was to take over the Office of Indian Affairs that had been supervised by the War Department until that time. The Secretary of Interior was responsible for appointing and supervising the U.S. Indian Agents assigned to the various Indian tribes to ensure treaty agreements were adhered to as well as acting at the Federal contact for the tribal leadership. The Indian Office stated that the:

"chief duty of an agent is to induce his Indian to labor in civilized pursuits. To attain this end every possible influence should be brought to bear, and in proportion as it is attained ... an agent is successful or unsuccessful."[6]

The agent was also responsible for the distribution of the annual annuities due to the tribal members. Many white Americans saw the role of Indian agent as largely inefficient and dishonest in monetary and severalty, which is property owned by individual right, not shared with any other, dealings with various Indian tribes.[7] President Buchanan's Secretary of the Interior was Jacob Thompson of Mississippi, a dedicated states-rights and pro-slavery advocate. (In fact, he was so deeply invested in the Southern cause that he became Inspector General of the Confederate States Army immediately after resigning as Secretary of Interior in 1861.) Even during his service with Interior, he was active in preparing the Southern states for secession. One action directed at the Indian Territory was that each of the U.S. Indian agents he appointed for the Five Civilized Tribes were Southerners who shared the same pro-slavery and pro-Southern sentiments as he did. His preferred choices always seemed to be committed adherents of the Southern cause and, if possible, from Arkansas and Texas.

Interior Secretary Thompson selected A.B. Greenwood of Arkansas as U.S. Commissioner of Indian Affairs, and Elias Rector, formerly the U.S. Marshal for Arkansas, became the Southern Superintendent of U.S. Indian Affairs for the Five Civilized Tribes in 1857, serving until he resigned in 1861 at the beginning of the Civil War. Among those who worked under Rector's supervision was the illiterate and unprincipled Robert Cowart of

☐ Union states

▨ Border states (part of the Union)

☐ Confederate states

☐ Territories

▨ West Virginia
(seperated from Virginia in 1861
and joined the Union in 1865)

The United States and the Confederacy in 1861

Georgia, agent for the Cherokee Nation, States-Rights advocate Douglas Cooper of Mississippi, agent for the Choctaw and Chickasaw Nations, and ardent secessionist William H. Garrett of Alabama, agent for the Creek Nation, and Massachusetts-born, but Arkansas resident and Southern-supporting Albert Pike, was selected as the Commissioner of Indian Affairs for the State of Arkansas. All of these agents were deeply involved in pressuring the Indian nations in the Indian Territory to plan to align with the Southern Confederacy if the Southern states seceded from the Union. (Both Cooper and Pike would later become generals in the Confederate States Army.)[8]

In November 1860 voters in the United States went to the polls to select a new president as well as the other slate of politicians who were up for election. President Buchanan kept a

campaign promise to only serve one term, which provided him an opportunity to stay in the background and let the chips fall as they may. Republican Abraham Lincoln won the Presidency against a split Democratic ticket that had both a Northern and a Southern candidate. This is especially noteworthy because Lincoln was not on the ballot in most southern states. Unfortunately, Lincoln would not be sworn in until March 1861, nearly four months after the election.

President Buchanan took this four-month opportunity to retreat first into the White House and then to his home at Wheatland in Pennsylvania, to sit back and watch as the Southern states began to secede one by one, and the Union fell apart. Buchanan seemed happy to just leave the whole mess for Lincoln to deal with. As each state seceded, the "state troops" would seize the Federal forts, shipyards, and arsenals. Instructions from the War Department in Washington, led by traitorous John B. Floyd, were often vague, contradicting, or non-existent. Most instructions that were issued were to not challenge the States from occupying the Federal properties but to simply get a signed receipt for the transfer of property. Soon the Federal arms, ammunition, and equipment in the arsenals and forts, including those moved from Northern sites, were in the hands of the Southern-state authorities. At this point, these States would begin to recruit and arm their citizens as part of the new Confederate States of America.

The few loyal clerks in the War Department attempted to manage the constant communications from the field and tried to find answers or instructions to pass on to those trying to manage from the field. The Army itself was contending with the resignations of many officers from the Southern states who believed loyalty to their home state was greater than their loyalty to the United States.[9] An example of the issues facing field commanders is illustrated in a February 6, 1861, report from Captain James Totten, 2nd U.S. Artillery, commanding the Little Rock, Arkansas, Arsenal, who wrote to the War Department:

> "Sir, I have the honor to enclose herewith a copy of a communication just received from H.M. Rector, Governor of the State of Arkansas, demanding the surrender of this arsenal to the State authorities.
>
> As I have already written and telegraphed you for the information of the President, I am perfectly in the dark as to the wishes of the administration, from want [of] instructions how to meet

such a crisis as at present. If I had positive orders to cover the case in point I should obey them implicitly, but I have nothing whatsoever;..."[10]

Due to the increasing pressure from the State authorities, Captain Totten surrendered the U.S. Arsenal on February 8, 1861. Unfortunately, all of Totten's reports had gone directly to Adjutant General Samuel Cooper, a Virginian who would resign and become a general in the Confederate States Army within the month. It is a credible suspicion to believe that Cooper was purposely avoiding a Virginia-born, yet pro-Union, General-in-Chief Winfield S. Scott by corresponding with and reporting directly to fellow conspirator Secretary of War John Floyd. (In the succeeding months U.S. Army soldiers took such a dim view of the former Adjutant General's treasonous activities that when they were building the network of defenses around Washington, D.C., they destroyed Cooper's house and used its bricks and wood in building a new fort, dubbed "Traitor's Hill.)"[11]

Elsewhere in the Trans-Mississippi area, and especially in Indian Territory, things were as dark and confusing for the Federal officials as they were seemingly bright and hopeful for the Confederates. For the Five Civilized Tribes, it was turning into a tumult that they had really wished to avoid. In less than 30 years they had been driven from their homes in the Southeast U.S., taken by force on long and arduous Trails of Tears, and were settled on lands to which they had no connection. Finally, having become somewhat settled, the tribes were again thrown into confusion by the coming Civil War. This would bring all of it to an end when the Five Civilized Tribes, some by choice and some by force, broke their treaties with the United States Government and, in many cases reluctantly, aligned with the new Confederacy.

The leadership of the Five Civilized Tribes watched as the United States began to fall apart over the slavery and the States-Rights issue, especially through the 1850s. Since the Five Nations' economies and antebellum agrarian lifestyles were tied firmly with the South, including the use of African slavery, it was not surprising that many of the tribes were quick to entertain alignment with the secessionist movement flowing through the South. Many tribal members were supportive of the movement and were vocal about their sympathies. In the winter of 1860, before the actual outbreak of hostilities, adherents of the Southern cause, among the most effectual and influential of whom were the official Indian agents of the United States assigned to the Indian tribes, were active in promoting the doctrines of secession among the Cherokees and other tribes of Indians in the Territory. Secret societies were organized, especially among the Cherokees. Stand Watie, the recognized Cherokee leader of the old Ridge or Treaty party, was at the head of the Southern party which was heavily influenced by a chapter of the Knights of the Golden Circle. The Knights of the Golden Circle, a profoundly pro-Southern and pro-secession society, was founded by George W.L. Bickley in Lexington, Kentucky in the mid-1850s. The primary mission of the national organization was to influence the Southern states to secede, to form their own confederation, and then to invade and annex the other areas of the Golden Circle as a haven for slave holders. The proposed new country's northern border would roughly coincide with the Mason–Dixon line and would include expansion into Mexico and Cuba. This organization in the Cherokee Nation promoted progressive ideas and was dedicated to promoting the interests of slavery and punishing abolitionists within the Cherokee Nation.

A counter organization was formed among the loyal or conservative-inclined portion of the Cherokee Nation, who looked to Principal Chief John Ross as their leader. John Ross had been born in 1790 near Lookout Mountain, Tennessee. His father, David Ross, was Scottish, and his mother, Mary (McDonald) Ross, was three-quarters Scottish and one-quarter Cherokee. John Ross was elected principal chief of the eastern Cherokees in 1828 and became leader of the Cherokee party opposed to westward removal; but after his efforts failed, he led his people in 1838-1839 to their new home in the Indian Territory. He was chosen principal chief of the united western and eastern Cherokees in 1839 and had served continuously since that time.[12] This new society termed themselves the "Keetoowah" or "Night Hawks," a name by which the Cherokees were said to have been known in their ancient confederation with other tribes. The society was formed under the influence of Baptist missionaries and was abolitionist driven. The distinguishing badge of membership in the Keetoowah or Night Hawks was a set of crossed pins worn in a certain position on the coat, vest or hunting shirt. This gave rise to the common designation of "Pin Indians" or "Pins" in referring to the conservative loyalist or National Party.[13]

A unique situation existed between the Keetoowah and the Knights of the Golden Circle in that each was influenced by the actions of the Freemasonry lodges within the Cherokee Nation to which most Cherokee leaders belonged. In fact. Commissioner of Arkansas Albert Pike was an active Free Mason who would rise through the ranks to the 32nd Order, the highest in the Masons.[14] The other four tribes were also divided among various levels but no others as much as the Cherokees and the

Creeks, the others being deeply affected by their secessionist Indian agents who gently maneuvered the Indian nations under their care to the Southern cause. On the other hand, there were large groups of other tribal member in all of the Indian nations that saw the issue as "the white man's problem." These Indians simply wished to just stay out of the fray and let the Americans from the North and South fight it out among themselves. This group consisted primarily of the non-slave-holding full bloods who felt they needed to stand by their commitments and abide by the treaties they had signed with the United States.[15] They were also concerned about the annuities and supplies provided under their treaties and wondered whether, during a war, they would still be able to receive these necessary items. They further believed that if they supported one side over the other it would only hurt the tribes in the end. They were right.

Gaining control of the Indian Territory was crucial to the new Confederacy, especially the Southwestern states. The Indian Territory provided large numbers of horses and cattle, large supplies of grain and lead deposits, it provided a land bridge between the eastern and western Confederacy, and it provided a buffer between Union Kansas and Confederate Texas. The Confederate leaders knew that mixed-blood tribal members owned many enslaved Africans and feared that the new Lincoln Administration would seize their "human property" without payment. (In fact, there were some in the incoming Lincoln administration, William H. Seward for one, who openly advocated abolishing the Indian Territory and opening it up for white settlement.) Because most products coming in and out of the Indian Territory traveled by either the Arkansas or Red Rivers to the Mississippi River, virtually all of the tribes'

trade was with the South. Also, the annuity funds for the Five Civilized Tribes were mostly invested in Southern banks and business enterprises.[16]

On January 5, 1861, under the influence of Southern-supporting Governor Cyrus Harris, and U.S. Indian Agent Douglas H. Cooper, the Chickasaw Legislature proposed a convention of the Five Civilized Tribes, at a time and place to be designated by the chief of the Creek Nation. The Chickasaw Council stated that the convention was "for the purpose of entering into some compact . . . for the future security and protection of the rights and Citizens of said nations, in the event of a change in the United States. . ."[17] The Choctaw Council, also under the influence of Agent Cooper, followed a month later and resolved:

> "That in the event a permanent
> dissolution of the American Union
> takes place . . . we shall be left to follow
> the natural affections . . . which
> indissolubly bind us . . . to the destiny of
> . . . the Southern States, upon whom we
> are confident we can rely for the
> preservation of our rights of life, liberty,
> and property, and the continuation of
> many acts of friendship, general
> counsel, and material support."[18]

The Creeks for their part hoped to re-establish the Inter-Tribal Council to further strengthen the 1859 Asbury Mission Compact. In this agreement, all Five Civilized Tribes had agreed that they would stand as one against the Federal government in matters that affected any member-Nation. The Five Civilized Tribes convention was to be held on February 17, 1861, at North Fork Town, Creek Nation, which was chosen due to its central location.

Unfortunately, due to high waters on the Canadian River, neither the Chickasaws nor Choctaws were able to attend this council. Regardless, the Cherokee delegation was instructed by Chief John Ross to:

> "guard against any premature movement . . . Should any action of the Council be thought desirable, a resolution might be adopted, to the effect, that we will in all contingencies rest our interests on the pledged faith of the United States, for the fulfillment of their obligations."[19]

The gathered Cherokee, Creek, and Seminole delegations opted to do nothing at this time and to simply wait to see what would happen. Nevertheless, one unusual ordinance passed by the gathered Creek Council was that all free black persons within the Creek Nation were to be returned temporarily to enslavement and were required to attach themselves to a Creek owner before March 10. They could not be traded or sold to non-Creek citizens. It is not known or recorded why the Creek Council took this action as there did not seem to be an obvious need to return these individuals to enslaved status.[20]

On December 20, 1860, the State of South Carolina declared itself independent of the United States of America. By February 1, 1861, six other Southern states had seceded from the Union beginning with Mississippi, and followed by Florida, Alabama, Georgia, Louisiana, and Texas, and had formed the Confederate States of America. President James Buchanan and his lame-duck administration again did nothing. On February 8 and 22, 1861, the Confederate Congress, meeting at Montgomery, Alabama, directed the Committee on Indian Affairs to consider the advisability of sending agents to the Southern Indians. On March 4, 1861, the Confederate Congress authorized President Jefferson Davis to send an agent to negotiate treaties with the Indian tribes west of Arkansas.[21] President Davis ended up sending two special envoys to the Five Civilized Tribes. Originally, he had appointed David Hubbard of Alabama as the Commissioner of Indian Affairs, a new division within the Confederate War Department, with instructions to go to the Indian Territory and pledge to each of the Five Civilized Tribes that the goal of the Confederate States of America was:

> "to protect them and defend them against the rapacious and avaricious designs of their common enemy whose real intention was to emancipate their slaves and rob them of their lands."[22]

Due to sickness, however, Commissioner Hubbard was unable to complete his assignment.

Earlier in March, Albert Pike had been appointed Special Commissioner to the Indian tribes west of Arkansas and south of Kansas by President Davis under the authority of the Confederate State Department. Pike was to report directly to Secretary of State Robert Toomes.[23] A native New Englander, Pike was well known and respected in the Indian Territory as a writer, lawyer, teacher, poet, and editor. There was no official reason given for the two separate and independent assignments, but one can surmise that it occurred because the Confederate Government was new and was trying to organize itself; so more than likely this was one of many unintentional instances of duplication of services. Another possibility is that President Davis was not yet sure who he could trust within his administration.

Drawing and Site of the Marais des Cygnes Massacre, Kansas
Photos by Author

George Hudson
Principal Chief of the Choctaw Nation
Photo courtesy of the Smithsonian National Postal Museum

Governor Cyrus Harris
Principal Chief of Chickasaw Nation
Photo courtesy of the Chickasaw Nation

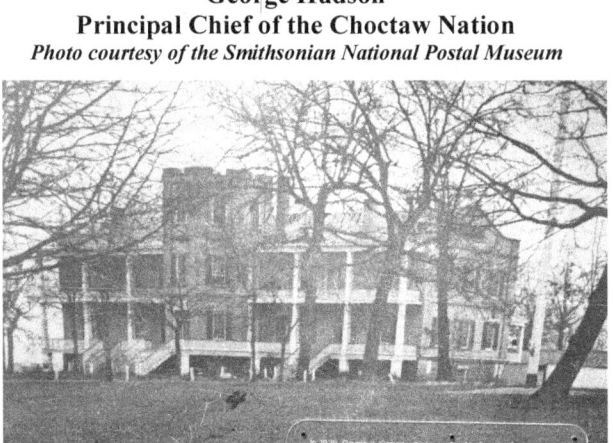

Little Rock Arsenal, Arkansas in 1863
Photo by Author / MacArthur Military Museum

Little Rock Arsenal, Arkansas in 2025
Photo by Author

Major John Jumper
Principal Chief of the Seminole Nation
Photo courtesy of the Seminole Nation

1st Sergeant John Chupco
Union Principal Chief of the Seminole Nation
Photo courtesy of the Seminole Nation

James E. Harrison
Texas Commissioner to the Five Civilized Tribes
Photo courtesy of Library of Congress

Major General Winfield S. Scott
General-in-Chief, United States Army
Photo courtesy of the Gilder Lehrman Collection

Lt. Col. William Emory, USA
1st U.S. Cavalry
Photo courtesy of Harvey County Historical Museum

2Lt William Averell
U.S. Regiment of Mounted Rifles
Photo courtesy of U.S. Army History Center

Col. Douglas Cooper, CSA
Choctaw & Chickasaw Nations
Photo courtesy Oklahoma Historical Society

Brig. Gen. Albert Pike, CSA
Photo courtesy of Scottish-Rite Temple

Brig. Gen. Benjamin McCulloch, CSA
Commander of Indian Territory
Photo courtesy Texas Historical Commission

Early in February 1861, Texas, now a Confederate state, passed an ordinance "to secure the friendship and co-operation of the Choctaw, Cherokee, Chickasaw, Creek, and Seminole Nations of Indians." Texas appointed James E. Harrison, James Bourland, and Charles A. Hamilton as special commissioners and charged them to persuade the Indian Nations to join Texas in separating from the Union.[24] The special commissioners crossed into Indian Territory via the Red River approximately 20 miles southwest of Fort Washita, the U.S. Army post on the Washita River, approximately 18 miles above the Red River. They met with Chickasaw Governor Cyrus Harris, who accompanied the Texas commissioners to a meeting with a convention of the Choctaw and Chickasaw Nations in Boggy Depot, Choctaw Nation. Commissioner Harrison spoke to the assembly, outlining the reasons the State of Texas left the Union and proposed that their Nations do the same and join the new Confederacy. Upon hearing of the February Inter-Tribal Council meeting at North Fork Town, the commissioners hastened to that location but missed the meeting by two days.

Upon their arrival, Baptist missionary H.F. Buckner of the Asbury Mission arranged a meeting with Creek leaders Chilly McIntosh, Daniel H. McIntosh, and Judge Benjamin Marshall. However, the two principal chiefs of the Creek Nation, Motey Kennard and Echo Harjo, were still in Washington, D.C., along with their U.S. Agent William Garrett, attempting to get assurances that their treaty expectations would be protected in the event of separation of the North and South. In their absence, the temporary leaders were unable to commit the Creeks to an agreement with the Texans. After this meeting, the Texans then entered the Cherokee Nation, "calling on their principal men and citizens" and "conversing with them freely." At Park Hill, near Tahlequah, they met with Chief John Ross and:

> "We called on him officially. We were not unexpected, and were received with courtesy, but not with cordiality . . He was very diplomatic and cautious. His position is the same as that held by Mr. Lincoln in his inaugural; declares the Union not dissolved; ignores the Southern Government." [25]

Chief Ross apparently had an inclination as to how the entire situation would work out in the end, having many years of Cherokee-U.S. Government relations experience. He was also being pressured by the pro-Southern Treaty Party and the Knights of the Golden Circle to align with the Confederacy.[26] Yet, as principal chief, he had the final decision.

Traveling back to the Creek Nation, the Texas commissioners again visited with the other Indian leaders. At the next North Fork Town convention on April 8, the Choctaws and Chickasaws were again absent, prevented once again by high waters from attending, the rivers and creeks being full and impassable. Commissioner Harrison addressed the assembled Creeks, Cherokees, and Seminoles, as well as a few Sac & Fox Indians from Kansas, another relocated tribe, and Quapaws from near the Cherokee Nation. His views were "cordially received" by the assembly. The commissioners returned to Texas convinced that the "Creeks are Southern and sound to a man, and when desired will show their devotion to our cause by acts." The commissioners also believed that the Seminole, Choctaw, and Chickasaw Nations were ready to join the Southern cause; however, no Indian nation had yet joined the Confederacy.[27]

At the time Fort Sumter was attacked in April 1861, only three federal military posts were still active in the Indian Territory: Forts Washita, Arbuckle, and Cobb. Fort Smith on the Arkansas side was also active, but Forts Gibson and Towson had been abandoned in the preceding years as the Indian problems had moved west. The mission of the remaining posts was to guard the "Washita River Line" located in the Chickasaw and Choctaw Nations. Fort Washita was garrisoned by troops C and I of the 1st U.S. Cavalry and Company E of the 1st U.S. Infantry. Fort Arbuckle was garrisoned by Companies A and B of the 1st U.S. Cavalry. Fort Cobb was garrisoned by Companies B, C, D, and F of the 1st U.S. Infantry. The objective of these forces in April 1861 was to respond to a threatened uprising of the Comanches on the southern Great Plains. These dedicated regulars continued to perform their assigned mission while the nation they were serving dissolved around them.

The U.S. Regular Army forces in Indian Territory were under the command of Lieutenant Colonel William H. Emory, an officer of the Topographical Engineers serving with the 1st U.S. Cavalry, stationed at Fort Smith, Arkansas. Emory had been called back to Washington D.C. in March 1861 to brief the new Lincoln-led War Department about the secession conditions in Arkansas and Indian Territory. Realizing the precarious situation that the Regular Army were in with the secession of Texas, and soon-to-follow Arkansas, the War Department issued Special Orders, No. 40, on March 13, that ordered Emory to move his headquarters west from Fort Smith to Fort Cobb and take charge of the Indian Territory from there. After further discussion, Lt. Col. Emory was issued a new set of orders on March 18, 1861, to abandon Fort Cobb and Fort Arbuckle and concentrate his entire command at Fort Washita. Although Emory had been given the option to abandon Fort Cobb, he needed to ensure that the Comanche and Wichita tribes under its protection understood the reasons why the Army was leaving. The two tribes needed to make self-defense arrangements against the nomadic Indians: the Southern-supporting Choctaw and Chickasaw leaders refused to let them move into their territories for protection since to do so was a violation of their treaties.[28] On April 6 Emory issued an order to the Fort Cobb commander to immediately send two companies of the 1st. U.S. Infantry to Fort Arbuckle:

"The permission given the Commanche and Wichita Indians to move with the troops was vigorously objected to by the Indian agents at Fort Smith, headquarters of the Southern Superintendency. Emory was informed by Superintendent Rector that it would 'give great dissatisfaction' to the Choctaws and Chickasaws to bring the Fort Cobb Indians within their territory. Furthermore, they were 'hutted and planting and without means' to move even if permitted. Matthew Leeper, agent of the Leased District, journeyed to Fort Smith to protest against the removal of his wards. Emory bowed to the storm and rescinded the invitation."[29]

After his return from Washington, having been stranded on the Steamer *Arkansas* due to low water on the Arkansas River, Emory found the situation in Arkansas becoming so precarious that on April 13 he decided to withdraw all troops and supplies from Fort

Smith, other than one company of cavalry, and move the entire garrison to Fort Washita. Emory was able to perform this activity although his command was plagued by the widespread resignations of his officers to join the newly forming Confederate States Army. However, Fort Smith was almost immediately taken over by Arkansas state troops, and the remaining Regular Army cavalry company rushed to rejoin Emory's main column. Before he arrived at Fort Washita, Emory learned that Texas troops were moving toward that post, resulting in him ordering the evacuation of Fort Washita and ordering that garrison to make haste to join his column at Fort Arbuckle.[30]

On April 17 Texas troops crossed the Red River and occupied Fort Washita. On that same date the War Department in Washington, realizing the danger looming from Texas, sent new orders to Lt. Col. Emory at Fort Smith. At this time, although Arkansas had not yet seceded from the Union, the War Department did not trust the telegraph or mails to properly deliver the orders. Instead, they decided to send a courier, Second Lieutenant William W. Averell of the Regiment of Mounted Rifles, who was in the Capitol on convalescent leave after being wounded in an Indian battle near Fort Craig, New Mexico Territory in 1859. Lt. Averell left Washington D.C. carrying orders to Lt. Col. Emory directing him to consolidate all his forces and abandon the Indian Territory by quickly moving them northward to Fort Leavenworth. The orders stated:

HEADQUARTERS OF THE ARMY, Washington, April 17, 1861.
Lieut. Col. WM. H. EMORY, First Cavalry, Commanding Fort Arbuckle:

SIR : On receipt of this communi-cation, you will, by- order of the General-in-Chief, with all the troops in the Indian country west of Arkansas, march to Fort Leavenworth, Kans., taking such useful public property as your means of transportation will permit. The troops may or may not be replaced by Arkansas volunteers. The action of that State will not affect your movement. Capt. A. Montgomery, A. Q. M., will be left at Fort Smith, to take charge of public property, and as staff officer of volunteers who may be mustered into the service of the United States.

I am, sir, very respectfully, your obedient servant,

E. D. TOWNSEND,
Assistant Adjutant- General.[31]

This set of orders was to give Lt. Col. Emory wide discretion in moving the United States forces around the Indian Territory. Dressed in civilian clothes because of the heightened tensions in the Southwest, Lt. Averell traveled by train as far as Rolla, Missouri, arriving on April 21. The next morning, he boarded a stagecoach, probably a Butterfield Overland Mail coach, which was full of secession-excited Southern gentlemen. For part of the stagecoach's journey, Lt. Averell was obliged to drive the stagecoach due to the drunkenness of the driver. At Evansville, Arkansas, he learned that Fort Smith had been taken over by Arkansas state forces. He was somehow able to keep his identity and mission hidden even though he was questioned numerous times as to the reason for his travel. The stagecoach arrived at Van Buren, Arkansas, was ferried across the Arkansas River, and arrived at Fort Smith on the morning of April

27. At Fort Smith he was originally expected to have transportation provided by the post and assistance in delivering the orders to Lt. Col. Emory. When he arrived at the Fort Smith military post, he found that, as reported, Arkansas troops were conducting training on the parade ground. When he inquired about the location of the U.S. garrison, the only information he was able to collect was that they had "gone West." He also discovered that the post quartermaster, Captain A. Montgomery, who had been given responsibility for providing Averell's transportation, was in the post guardhouse under arrest. Realizing his perilous situation, Lt. Averell used the last of his money and his gold watch to procure a horse, saddle, and bridle. Unfortunately, the horse was unbroken to the saddle. Upon mounting the horse, the horse took him on a wild ride through the drilling soldiers. However, he was able to control the horse enough to get across the Poteau River and into the Indian Territory. Unfortunately, his inquiries at Fort Smith had aroused suspicions, and his true identity and mission became known to the Arkansas state authorities, and a large force of Arkansas secessionists headed out in pursuit.[32] Now in Indian Territory, Lt. Averell followed the Butterfield Overland Mail Road towards Fort Washita. At Holloway's Station, he was overtaken by four of his pursuers, but he was able to convince them he was not the person for whom they were searching for. They soon discovered the ruse, however, and continued their pursuit. Averell discovered their renewed pursuit and decided to abandon the Fort Washita Road and cross over the San Bois Mountains, off-trail, and get onto the Fort Arbuckle Road (California Road) near Perryville. The secessionists apparently were no match for this frontier-experienced professional soldier. With the help of a Choctaw man who provided him with food and well as acting as a guide, Averell was able to reach the new road after a couple of days. He was then able to recruit a new Indian guide who led him to the Fort Washita-Fort Arbuckle Road and, after a wild storm in which "the Indian lost his way and I lost the Indian," he had to swim the swollen Big Blue River. The Indian guide found Averell the next morning, and on May 2 they found Lt. Col. Emory and his command camped along the Military Road, and there he hand-delivered the orders from Washington.[33] Emory informed him that if he had gone to Fort Washita he probably would've been captured since Texas troops had already occupied the post. Lt Col. Emory appreciated Lt. Averell's commitment to his mission and allowed him to ride in an Army ambulance to help heal his damaged body. Lt. Averell stated:

> "I accompanied Colonel Emory's command to Fort Arbuckle, where we arrived May 3, and found Major Sacket, Captains Crittenden, Williams, and others who had been left with a small force in charge of the post when the main body went to Washita. The trains were loaded to their utmost capacity, and on the 4th of May the flag was lowered with military honors, Fort Arbuckle was abandoned, and we marched northward,"[34]

Lt Col. Emory and his command left Fort Arbuckle and moved about five miles to the east side of the Washita River. Earlier, on May 3rd they had been joined by the two companies of the 1st U.S. Infantry from Fort Cobb, guided by Black Beaver, a well-known Delaware scout and guide who lived near the

Map of Lt. Averell's Ride and the U.S. Army Evacuation of the Indian Territory

Fort Cobb and the Wichita Agency with his band of Delawares. He had guided many U.S. Army expeditions and exploring parties throughout the West. Black Beaver agreed to guide the U.S. troops to Fort Leavenworth, Kansas. On May 4, after the solemn flag lowering ceremony at Fort Arbuckle, the command was joined by the wagon train carrying as much military and commissary supplies as they could hold. Even still, Sergeant Charles E. Campbell of Company E, 1st U.S. Infantry, and a small detachment were left behind to guard the remaining stores. The entire command moved north along the east side of Washita River. They were to rendezvous with the remaining two companies of the 1st U.S. Infantry from Fort Cobb on the California Road, approximately 35 miles northwest of that post. On the morning of May 5, Texas troops occupied Fort Arbuckle, and Sgt. Campbell signed over the public property to the Texas commander, Captain S.T. Benning of the Fannin County Company. Benning reported that he in turn signed over all United States property over to the Chickasaw Nation.[35] Also on May 5, Lt. Col. Emory was notified that the command was being followed by Texans out of Fort Washita and that the advance guard was closing on their position. Emory ordered Captain Samuel Sturgis, accompanied by Lt. Averell, to take his cavalry company and confront the pursuers. The Texans wisely surrendered to the battle-line deployed company of frontier-hardened Regulars of the 1st U.S. Cavalry and were quickly brought to Emory. He stated in his report that they were "gentlemen acting under erroneous impressions." Emory released them the next morning to return to their comrades.[36]

On May 9, 1861, Emory's column finally rendezvoused with the two remaining companies of the 1st U.S. Infantry. The command now contained eleven companies, 750 soldiers, plus 150 wives, children, teamsters, and other non-combatants. Led by Black Beaver they changed direction north from the California Road towards Fort Leavenworth, Kansas, crossing the Canadian River at what would become the Silver City Crossing of the Chisholm Trail in the post-Civil War years. Upon arrival at Fort Leavenworth on May 31,

Emory sent this brief message to the War Department:

> HEADQUARTERS TROOPS FROM TEXAS FRONTIER,
> Fort Leavenworth, May 31, 1861.
> I arrived here this morning and turned over the command to Major Sacket in good condition; not a man, an animal, an arm, or wagon has been lost except two deserters.

Lt Col. Emory became the only federal officer to bring his troops out of Confederate territory without the loss of a single man. The Indian Territory was left to the Confederacy.[37] The Confederate agents took full advantage of the vacuum left by the U.S. Army departing from the Indian Territory. Because the Indian agents were all Southerners, they did not bother to fully explain why the U.S. Army had abandoned them. The seceded Southern states were already enrolling thousands of State troops, who were then being reorganized into the new Confederate States Army. In contrast, the U.S. Army only consisted of 16,402 soldiers on January 1, 1861, and these troops were stationed throughout the United States and its territories. The vast majority of these troops were located west of the Mississippi and spread among the dozens of small military posts along the Oregon,

California, and the Santa Fe Trails, in Utah, California, and the Pacific Northwest. Texas had the largest number of line troops assigned. The 48 batteries of the four artillery regiments were split between the western frontier and the coastal forts and along the Canadian border. None of the 99 infantry companies (Company G of the 8th U.S. Infantry had been broken up and distributed among the other 9 companies) or 50 mounted companies were east of the Mississippi. On February 19, 1861, Georgia-born, pro-secession, Major General David Twiggs surrendered the 36 Army companies (five artillery, ten mounted, and 21 infantry) under his command to the Texas authorities upon their demand. This deception had clearly been planned in advance as Twiggs was known to and was often seen in the presence of powerful secessionists in Texas. Texas had not yet officially seceded from the Union, so this was not something he was forced to do: whatever forces the Texas authorities could manage to muster would not have been able to defeat 2,200 frontier regulars. While Twiggs had asked the War Department for instructions during the preceding months and had not received an answer from the Buchanan Administration, he could not, or would not, abandon his Southern sympathies. Instead, he agreed to a humiliating agreement of surrender. Twiggs had been officially relieved of command by General Scott on January 28, 1861, by Special Orders No. 20, and command of the Department of Texas was transferred to Colonel Carlos A. Waite, commanding officer of the 1st U.S. Infantry.

Unfortunately, the order relieving Twiggs of command, probably sent by regular mail on purpose, did not arrive in San Antonio, Texas, until after Twiggs had already signed the agreement of surrender. Although the surrender agreement stated that the U.S. Army regulars were permitted to take their weapons and as much of their ammunition and supplies as they could transport, the Texans refused to let the U.S. Army units move north into the U.S. held areas. Instead, the Texans insisted that the regulars march southward toward the Gulf of Mexico and be transported North by ship via the ports of Brazos Santiago and Indianola, Texas. This was intentionally done to increase the time those troops would be unavailable to defend Washington or other Northern locations: they had to march from distant posts, some from more than a thousand miles away, then wait for hired ships to transport them back to Northern ports. Of the more than 2,200 troops in Texas, only about 1,200 had been transported out of Texas before the Confederates fired on Fort Sumter in South Carolina on April 12, 1861. The remaining U.S. soldiers were made prisoners of war by the Confederacy and paroled after signing agreements to not fight against the Southern Confederacy.[38] These parole agreements were technically mute since the parole was not referred to in the Texas surrender.

In Washington, General-in-Chief Winfield Scott was pulling units from the entire frontier to establish some sort of defense of the Capital and other vulnerable government facilities. It was not just the Indian Territory that was affected by troop withdrawals, a fact the secessionist Indian agents again failed to pass on to the Five Civilized Tribes under their care. Those same Indian agents, even while still collecting a salary from the U.S. Treasury, used the withdrawal as proof that the Federal government was abandoning the tribes. This was coupled with the sight of the U.S. Army quickly marching out of the military posts and the Texas forces moving in. This was a concrete sign to the Indians that, although the United States Government had abandoned them, the Confederate States would provide and protect them from the marauding Comanche, Wichita, and Kiowa war-parties.[39]

On May 21, 1861, the Provisional Congress of the Confederate States of America, meeting in its second session in Montgomery, Alabama, passed "An Act for the Protection of Certain Indian Tribes" which was signed into law by President Davis. This bill, although most of its contents have been lost to history, apparently gave the Confederate Office of Indian Affairs the authority to negotiate a series of treaties with the Indians located "West of Arkansas and South of Kansas."[40] Three later financial amendments to this Act are preserved in the Journal of the Congress, 1861-1865, provided that the Confederate States would assume the financial responsibilities of the Indian nations under its care.[41] President Jefferson Davis described this now-lost document as "a declaration by Congress of our future policy in relation to those Indians" that was promptly "transmitted to the Commissioner and he was directed to consider it as his instructions in the contemplated negotiations." This statement is included in all of the treaties entered into by the Confederacy with the Indian Tribes. Earlier, on May 13, the Indian Territory was designated as a military district, and command was given to Brigadier General Benjamin McCulloch, who in the 1830s had come to Texas from Tennessee with David Crockett. He was a veteran of the Battle of San Jacinto with the Texas Republican Army and had served under U.S. General and later President Zachary Taylor in the Mexican War. It was he who had personally demanded the surrender of the Department of Texas from General Twiggs in February after he had been commissioned as a colonel in the Texas Provisional Army. McCulloch was authorized three regiments for service in the Indian Territory, one mounted from Texas, one mounted from Arkansas, and one infantry regiment from Louisiana. He was also authorized to recruit and equip any regiments raised by the Five Civilized Tribes, or by any of the other tribes, for Confederate service. His mission was to protect the Indian Territory from invasion from Kansas or elsewhere. The Texas regiment was to be organized in Dallas, Texas, and the Arkansas and Louisiana regiments organized at Fort Smith, Arkansas.[42]

On that same date the Confederate War Department, under Secretary of War LeRoy Pope Walker, an Alabama-born lawyer turned politician, sent orders to Douglas H. Cooper, the former U.S. and now Confederate agent to the Choctaw and Chickasaw Nations, authorizing him to raise a mounted regiment from the Indians under his care. The regiment was to consist of ten companies of 64-100 soldiers each. Once the regiment was raised to those amounts, the Confederate War Department would ensure that all equipment and supplies would be forwarded to his new command. His regiment would be part of the Indian Territory Military District and would report to Brig. Gen. McCulloch. This was the first of the six Confederate regiments and various battalion-sized units raised from within the Five Civilized Tribes.[43] This was an easy task for Cooper since the Choctaw and Chickasaw Nations were firmly in support of the Confederacy.[44] Cooper was a native of Mississippi and a Mexican War veteran. In 1853 he had been appointed U.S. Indian Agent to the Choctaw Nation where he was afforded the opportunity to exercise great influence over his charges. He was also a crook, one of those dishonest Indian agents who sought wealth via the Federal government and the Indian annuities he managed. He had been appointed by his friend, then-Secretary of War Jefferson Davis, who had been his commanding officer in the Mexican War, and was now protected by the same man who was President of the Confederate States.[45] In a May 1861 letter from Cooper to Indian-hating, yet still Indian Commissioner, Major Elias Rector he stated:

> *Private & Confidential*
> Fort Smith, May 1, 1861
> *Copy*
>
> Major Elias Rector
> Dr. Sir: I have concluded to act upon the suggestion yours of the 28th Ultimo contains. If we work this thing shrewdly we can make a fortune each, satisfy the Indians, stand fair before the North, and revel in the unwavering confidence of our Southern Confederacy.
>
> My share of the eighty thousand in gold you can leave on deposit with Meyer Bros, subject to my order. Write me soon.
>
> Cooper[46]

It is believed that much of this gold came from the illegal sale of Indian horses, cattle, and livestock to U.S. Army agents in New Mexico. It is not known if Cooper ever received his cut.

Unlike the Choctaws and Chickasaws, unity of purpose and commitment to the Southern Confederacy was not the case for the Creek, Seminole, or especially, the Cherokee Nations. These three tribes had significant divisions that would be hard to overcome if the Confederate States wanted treaties of alliance for the Southern cause. Most of the divisions in the tribes were still aligned between the treaty and the non-treaty factions or the mixed-blood verses full-bloods of the three nations. This was the tight rope that Confederate Indian Commissioner Albert Pike had to cross as he began his trek into the Indian Territory in late May 1861 to meet with and negotiate alliance treaties with the Five Civilized Tribes. He was also directed to treat all of the smaller tribes residing south of Kansas and west of Arkansas and secure their support. On May 20, Pike sent a message to Confederate Secretary of State Robert Toomes describing his apprehensions about negotiating the treaties with the Indian nations, especially the Cherokee, without clear instructions regarding exactly what the Confederacy would and would not guarantee. He did not want the Indians to agree to a treaty to have it later rejected by the Confederate Government. Pike stated:

> "I very much regret that I have not received distinct authority to give the Indians guarantees of all their legal and just rights under treaties. It cannot be expected they will join us without them, and it would be very ungenerous, as well as unwise and useless, in me to ask them to do it. Why should they, if we will not bind ourselves to give them what they hazard in giving us their rights under treaties!"[47]

And further:

> "As you have told me to act at my discretion, and as I am not directed not to give the guarantees, I shall give them, formal, full, and ample, by treaty, if the Indians will accept them and make treaties. General McCulloch will join me

in this, and so, I hope and suppose, will Mr. Hubbard, and when we shall have done so we shall, I am sure, not look in vain to you, at least, to affirm these guarantees and insist they shall be carried out in good faith."[48]

Commissioner Pike decided to try and crack the toughest nut first by immediately approaching the Cherokee Nation and Chief John Ross. Prior to Pike's expected visit, Chief Ross had called a meeting of the Executive Council of the Cherokee Nation to discuss their situation. The Council majority stated clearly that they wanted to remain neutral and not become embroiled in the "white man's war." To further expound on his expectations of the Cherokee Nation, on May 17, 1861, Chief John Ross issued a "Proclamation of Neutrality." In this document he stated:

> "I, John Ross, Principal Chief, hereby issue this my proclamation to the people of the Cherokee Nation, reminding them of the obligations arising under their treaties with the United States, and urging them to the faithful observance of said treaties by the maintenance of peace and friendship toward the people of all the States....
> …The peculiar circumstances of their condition admonish the Cherokees to the exercise of prudence in regard to a state of affairs to the existence of which they have in no way contributed; and they should avoid the performance of any act or the adoption of any policy calculated to destroy or endanger their territorial and civil rights. By honest

adherence to this course they can give no just cause for aggression or invasion nor any pretext for making their country the scene of military operations, and will be in a situation to claim and retain all their rights in the final adjustment that will take place between the several States. For these reasons I earnestly impress upon the Cherokee people the importance of non-interference in the affairs of the people of the States and the observance of unswerving neutrality between them…"[49]

The proclamation caused anger and resentment from the mixed-blood Cherokees, especially Stand Watie, leader of the Treaty Party, who wished to raise troops and fight for the Confederacy. The mixed-bloods were a powerful yet minority group that held much of the wealth of the Cherokee Nation. Although Chief Ross was a very wealthy, slave-holding, mixed-blood himself, he had almost complete support of the majority full-bloods, including the Keetoowah Society, or "Pin Indians." He also had the full support of the avowed-abolitionist and Baptist missionary Evan Jones who had a strong influence among the full-blood Cherokee community.[50] This assured Chief Ross that he would remain Principal Chief for the foreseeable future.

When Pike arrived on June 1, 1861, he was given the same reception as the Texas commissioners had been given: the Union was not dissolved and the Cherokee Nation would abide by the treaties it signed with Washington. Even after Pike was joined by Brig. Gen. McCulloch for military-related support, Chief Ross insisted that the nation was to remain neutral and not support either side. Pike had with him six or seven pages of arguments as to why the Cherokee Nation should align with the Confederacy. After days of not obtaining the results he wanted, Pike left the Cherokees behind and travelled south to the much more welcoming Choctaw and Chickasaw Nations. Still, he continued his correspondence with Stand Watie and other members of the minority Treaty Party.[51]

The Choctaw Nation initially looked like an easy win for the Confederacy after their declaration of support back in February 1861. In the meantime, Choctaw Principal Chief George Hudson became convinced that the best approach was to follow the Cherokee Nation and Chief John Ross in maintaining a neutral stance regarding the war. This stance was strengthened after meeting the "Net Proceeds," the Choctaw Nation representatives stationed in Washington, who stated that the United States wished for them to remain neutral, and that the Federal government would strive continue to maintain its treaty obligations. Even from under a threat from a Texas vigilante committee, Chief Hudson called for a special session of the Council on June 1 at the capital in Doaksville, where he planned to present his statement of neutrality. Indian Agent Douglas Cooper was forewarned about this action and reached out to Robert M. Jones, one of the most ardent secessionists in the Choctaw Nation, to attend. In a very heated session in which Jones repeatedly attacked all those opposed to secession, Chief Hudson was either threatened or convinced to change his stance regarding neutrality and to support the secession movement. He proposed sending a set of commissioners to treat with the Confederacy and agreed to raising a regiment for the new Confederate States Army. Later, Robert Jones would be selected as the Choctaw delegate to

the Confederate Congress in Richmond. On June 10 the Choctaw Nation issued its Declaration of Independence from the United States of America.[52] The declaration in part reads:

> "Whereas the general council of the Choctaw Nation, on the 10th day of June, 1861, by resolution declared that in consequence of the dissolution of the United States, by the withdrawal of eleven States formerly comprising a part of said Government, and their formation into a separate government, and the existing war consequent thereon between the States, and the refusal on the part of that portion of the States claiming to be, and exercising the functions of the Government of, the United States to comply with solemn treaty stipulations between the Government of the United States and the Choctaw Nation... Now, therefore, I, George Hudson, principal chief of the Choctaw Nation, do hereby publish and proclaim that the Choctaw Nation is, and of right ought to be, free and independent;..."[53]

The Choctaw Nation was ready to sign a treaty with the Confederate States of America.

The Chickasaw Legislature, much like the Choctaw Legislature had done in February, on May 25 announced their independence from the United States and pledged their support to the Confederate States of America. Their House and Senate proclamation stated:

> "Whereas the Government of the United States has been broken up by the secession of a large number of

States composing the Federal Union —
that the dissolution has been followed
by war between the parties; and whereas
the destruction of the Union as it
existed by the Federal Constitution is
irreparable, and consequently the
Government of the United States as it
was when the Chickasaw and other
Indian nations formed alliances and
treaties with it no longer exists;"

The document contains nine resolutions that
range from a pledge of allegiance to other
Indian nations that will align with them to pleas
to the Chickasaw warriors to support and
defend the nation from the "Lincoln hordes and
Kansas robbers."[54] Chickasaw Governor Cyrus
Harris was instrumental in pushing the
secession movement through the Chickasaw
Nation. He had deep Southern allegiance and, as
shown earlier, went out of his way to
accommodate the Texas commissioners.

The Creek Nation had a problem
similar to the Cherokees that dated back to the
War of 1812 and the Creek War of 1814. The
Lower Creeks, or mixed-bloods, were a minority
within the tribe but held most of the wealth.
The Upper Creeks, or full-bloods, were in the
majority but still held deep-seated anger with the
Lower Creeks for helping General Andrew
Jackson slaughter the Upper Creeks at the Battle
of Horseshoe Bend in 1814. This same group of
Lower Creeks under Chief William McIntosh
signed away most of the Creek Nation's lands in
the following years against the will of the
National Council. By the time of the Civil War,
the Lower Creeks were still led by the powerful
McIntosh family. The Upper Creeks were led by
Chief Opothleyahola, or "Gouge" as he was
known to the whites, a mixed-blood leader who
had differed with the McIntosh's since the

Removal. He had been born in the late 1700's in
the tribal town of Tuckabatchee in the Creek
homeland of Alabama. Although Opothleyahola
was not currently the chief representing the
Upper Creeks, at an estimated 80-90 years old,
he still held a powerful sway over that group
and over much of the Creek Nation. The
McIntosh faction was ready and willing to sign a
treaty with Albert Pike, but Opothleyahola
refused to meet with the Confederate
representative. He believed that the Creek
Nation should remain neutral and stay out of
the white man's fight and called a council of all
the tribes, eastern and western, to meet in the
Antelope Hills in the far west of the Indian
Territory. Opothleyahola wanted all tribes to
shun the Confederacy, abide by their treaties,
and simply remain neutral for the duration. The
meeting was set for late June 1861. This played
right into the hands of the Lower Creeks and
the Confederate delegation.

In the first week of July a council was
called at North Fork Town in the Creek Nation
with delegates from the Choctaw, Chickasaw,
and Seminole Nations to meet and discuss their
relations with the Confederacy. In the absence
of Opothleyahola and his full-blood delegation,
the Lower Creek leaders, including Principal
Chief Motey Kinnard and Chilly McIntosh,
attended to represent the Creek Nation. Stand
Watie and other members of the Cherokee
Treaty Party had planned on attending but did
not, fearing retribution from Chief Ross and his
supporters. Firebrand Robert M. Jones of the
Choctaw Nation was elected leader of the
meeting. The first order of business for the
gathering was to sign a treaty among themselves
to establish a permanent inter-tribal council
called the "United Nations of Indian Territory,"
which established a "Grand Council" consisting
of six delegates and the principal chiefs of each

member Nations. They were to meet at least once a year at North Fork Town to report on the conditions of their tribes and to suggest legislation and laws to consider that would be binding on all member Nations.

Commissioner Pike was present and again outlined what the Confederate States were willing to include in any negotiated treaties. These terms were much better than they had ever been offered by the United States, considering the Confederates were willing to grant almost anything to the Five Civilized Tribes at this point simply to get their support. Since most of the representatives present came from the wealthy, mixed-blood factions of the Nations, they were open to the offers being presented by the Confederate government. Under Pike's authority granted by the May 1861, Confederate Provisional Congress' "An Act for the Protection of Certain Indian Tribes," the first treaty was signed on July 10, 1861, with the Creek Nation. The treaty was signed without Opothleyahola and the full-blood's consent, although at least three names from that group were forged onto the document, they were Ok-ta-ha-hassee Harjo (better known as Sands), Tallise Figico, and Mikko Hutke. Upon Opothleyahola's return, the full-bloods refused to sign the document and left North Fork Town for their own towns. Technically, Pike needed the consent of the whole tribe, but he chose instead to ignore this condition and sent the treaty on to Richmond with only the mixed-blood's and the forged full-blood's signatures.

On July 12 the Choctaw and Chickasaw Nations signed their treaties with Pike before the meeting came to an end. At this point Pike traveled over to the Seminole Nation Council House, near present-day Wewoka, Oklahoma, to meet with their National Council. The council approved and they signed their treaty

with the Confederacy on August 1. Pike now had four of the Five Civilized Tribes under treaty with the Confederate States.[55] Back in the Creek Nation, Opothleyahola and the Loyal Creeks met in council on August 5 and declared the Confederate treaty null and void and, declaring the office of principal chief vacant, voted Opothleyahola into that roll. This action was ignored by the powerful, yet minority, mixed-blood faction.[56]

Each of the treaties, even the one that was later signed by the Cherokee Nation, were basically the same with exceptions being discussions of boundaries, census, leadership, etc. that were tribe specific. The Confederacy became the protector of each nation and assumed the annuity obligations of the United States. All lands of the Indian Territory were annexed to the Confederate States, but the individual Nations were granted a guaranteed title to these lands in perpetuity. The Confederacy reserved its rights-of-way for roads, telegraph lines, and railroads. They also were able to construct and garrison military posts as needed and establish a postal system across the boundaries. The treaties required each of the tribes to raise regiments for the Confederate Army but promised to get tribal permission before sending these troops outside the Indian Territory.[57] The Confederate States also insisted that slavery laws be established or kept in place since the "institution of slavery in the said nations is legal and has existed from time immemorial; that slaves are taken and deemed to be personal property."[58] The Five Civilized Tribes were also authorized to send delegates to the Confederate States Congress in Richmond, Virginia.

Stand Watie, who became the most prominent and picturesque Indian officer in the Confederate Army, signer of the Removal treaty and undisputed leader of the Southern Rights Party, was born in 1806 in the old Cherokee Nation in Georgia near the site of the present city of Rome. His father was a full-blood Cherokee, but his mother was half-white. Contemporaries state that he was short and squat but still had a commanding presence and tended to do what he wanted rather than adhere to the wishes of the Cherokee National Council. On July 2, 1861, Watie began recruiting Cherokees into a battalion for Confederate Army service with him as its colonel, utilizing the remaining facilities of Fort Wayne, the abandoned U.S. Army post. This began while the Cherokee Nation was still officially neutral. The old post's location was very convenient to both Watie's home on Honey Creek as well as being just a few miles from Camp Jackson, near Maysville, Arkansas, where Brig. Gen. McCulloch had his headquarters. Watie also knew that if he raised enough support he could end up the Principal Chief under a Confederate-Cherokee tribal government. Brig. Gen. McCulloch was at first hesitant to allow these Cherokees into the Confederate Army since the Cherokee Nation was still officially neutral. Only Albert Pike was authorized to negotiate any treaties, but McCulloch believed military needs came before diplomatic courtesy. Early in September General McCulloch reported to the Secretary of War:

> "I have, previous to this time, employed some of the Cherokees, under Col. Stand Watie, to assist me in protecting the northern borders of the Cherokees from the inroads of the jayhawkers of Kansas. This they have effectually done, and at this time are on the Cherokee neutral lands in Kansas....I hope our Government will continue this gallant

man and true friend of our country in service, and attach him and his men (some 300) to my command. It might be well to give him a battalion separate from the Cherokee regiment under Colonel Drew. Colonel Drew's regiment will be mostly composed of full-bloods, whilst those with Col. Stand Watie will be half-breeds, who are educated men, and good soldiers anywhere, in or out of the Nation."[59]

Watie again tried to increase his public standing when he and his supporters took a bold step and attempted to raise a Confederate flag above the Cherokee National Capitol in Tahlequah in defiance of Chief John Ross' Neutrality Proclamation. On July 12 Stand Watie received a commission as a Colonel in the Provisional Confederate Army. His battalion-sized unit of approximately 300 or so troops were unofficially brought into the Confederate army and were to be attached to Brig. Gen. McCulloch's command. Its mission was to patrol the northern border with Kansas to protect the Cherokee Nation from invasion or depredations from the Jayhawkers in Kansas. In response, Chief Ross and the Executive Council authorized and began raising a 1,200-man Home Guard Regiment with Councilman, and Ross' nephew-in-law, John Drew as its colonel. This regiment's officers would be selected and appointed by the principal chief. This ensured most of the regimental officers were either supporters or actual Pin Indians.[60]

Events outside of the Indian Territory were having far-reaching effects on the Five Civilized Tribes. Back in Virginia, the first major battle at Bull Run in July 1861 was, tactically, a Federal loss, but it did stop the Confederate army from crossing the Potomac because the

South's losses were heavy as well. Out in Missouri, the Battle of Wilsons Creek in August 1861 was also tactically a Federal loss, but it ensured that Missouri would not leave the Union: the Missouri State Guard forces eventually abandoned the state and took up positions in Confederate Arkansas. Stand Watie was hailed as a hero in the Wilsons Creek battle by some Confederate newspapers for leading his troops against Brig. Gen. Franz Sigel's Union division and capturing all his artillery. Watie's fame and glory reached high levels in the Cherokee Nation.[61] Chief Ross feared with Watie's new-found fame that the hero would sweep into the National Council and be named Principal Chief, while at the same time jailing Ross and his supporters. Unfortunately, none of it was true. Stand Watie was not at Wilsons Creek, nor were any of his troops, although there may have been a few individual Cherokees with the Confederates. Someone invented the entire story, and Watie obviously did not correct the reports. Neither did Brig. Gen. McCulloch or Maj. Gen. Sterling Price.[62] This was not information that John Ross had available, and he soon began to realize that the United States had indeed surrendered control of the region to the Confederacy. This surrender of control when coupled with the fact that the Cherokee Nation was being destroyed by its internal divisions, caused Chief John Ross to open negotiations with the Confederacy.

The initial Confederate victories were heralded by the South and caused celebration in the Confederate-supporting Indian tribes as well as bringing concerns among those who were supporting the Union. Efforts by the United States Government to counteract the propaganda of the Southern states among the Indian nations proved futile. In fact, after the withdrawal of the Federal troops in May 1861,

communication between the Indian Territory and the North "almost entirely ceased." Indian agents newly appointed from the Northern states were unable to reach their posts and those previously appointed from the Southern states soon went over to the Confederacy.[63] Although all five tribes eventually signed treaties of alliance with the Confederacy, that did not mean that all the tribal members agreed with or supported their tribe's action. All the Nations had substantial numbers of tribal members who wished to remain with the Union or at least remain neutral in the conflict. These tribal members found it more and more difficult to stay within their national boundaries with the prevailing anti-Union attitude. The danger was especially high in the Creek, Seminole, and Cherokee Nations where the factions were still committing acts of violence, even after the signing of the treaties with the Confederacy.

Meanwhile, further north in the Cherokee Nation at Tahlequah and Park Hill, Chief John Ross was being continually bombarded with letters and notes asking him to re-consider his neutral stance and sign a treaty with the Confederate States. He shocked Brig. Gen. McCulloch, who had respected the Cherokee Neutrality Proclamation by not sending any Confederate forces into the Cherokee Nation, when he stated in one such reply that the Nation's neutrality was actually one-sided. Ross explained that the Cherokee Nation was required by the treaty they had signed years before with the United States to allow the Army access to build and garrison military posts within the tribe's boundaries. The result was that the Cherokee Nation had no authority to prohibit access by the U.S. Army. Still, events within the Nation were getting violent between the Pin Indians and the Treaty Party, which renamed itself the "Southern

Rights Party." Ross feared a complete breakdown and a civil war developing among the Cherokees. Under pressure by the Southern-supporting forces within the tribe, in late August 1861 the Cherokee Executive Council met in Tahlequah with about four thousand tribal members present. This council was called for the purpose of giving the Cherokee people an opportunity of expressing their opinions and taking whatever actions they deemed necessary.[64] Spurred on by the crowd, the Council voted on August 25 to invite Albert Pike and Ben McCulloch to Tahlequah to negotiate a treaty with the Confederacy. It is believed that Chief Ross had no choice since he was under a death threat from Stand Watie and the Southern Rights Party. And with none of the promised Federal government help on the horizon, he probably believed this was his only option.[65] Evan Jones, the former Baptist missionary to the Cherokees, who had now left the nation, was "perfectly astounded at the announcement of the defection of John Ross and the Cherokees." From Lawrence, Kansas, he wrote to U.S. Indian Commissioner Dole: "I have no doubt the unfortunate affair was brought about under stress of threatened force, which the Cherokees were by no means able to resist."[66]

In August 1861 John Ross invited Albert Pike to Park Hill for a meeting, also inviting the smaller tribes including the Seneca, Shawnee, Osage, and Quapaw. They met beginning on October 1, 1861, and all negotiated treaties of alliance with the Confederate States of America. The Osage were the first to sign on October 2 and the Seneca and Shawnee, and the Quapaw signed on October 4. Finally, Chief John Ross signed a treaty for the Cherokee Nation on October 7, 1861. Now, all the tribes in the eastern half of Indian Territory had broken their treaties with the United States and were now aligned with and had placed all of their hopes and trust in, the new Confederate States of America.[67]

After the National Council's vote, Ross actually offered his new regiment to Brig. Gen. McCulloch as one of two for Confederate service. Ross' offer was refused because the Cherokee had not yet signed their treaty with the Confederacy,[68] even though the Confederate Army had already accepted Stand Watie's force into service and had also commissioned Douglas Cooper and his Choctaw-Chickasaw regiment long before those tribes had signed a treaty of alliance. McCulloch must have had his doubts about the loyalty of the men in Ross' Home Guard regiment. After the treaty was negotiated and signed by the Cherokee Nation and the Confederate States in October 1861, the regiment was accepted, but an issue developed over the numbering of the two regiments. Under the Confederate treaty, the Cherokee were authorized to raise two regiments of mounted troops for Confederate service. Since Colonel Drew had a full regiment, his became the 1st Cherokee Mounted Rifles. Stand Watie's battalion-sized unit was to be brought up to regimental strength and designated the 2nd Cherokee Mounted Rifles. Watie and his supporters objected that since their unit was raised first, they should be the 1st regiment. Officially, the regimental numbers stayed as authorized, but in practice the units were simply referred to as Drew's or Watie's Regiment.[69] Thomas Fox Taylor, a Cherokee lawyer, was selected as lieutenant colonel of Stand Watie's regiment, and his nephew, E.C. Boudinot, became the regiment's major.

The Indian Territory had been designated as a military district of the Confederate Army, and command of the district was given to Albert Pike, with a commission as Brigadier General in the Provisional Confederate Army, dated August 15, 1861. During the autumn Pike received permission to travel to Richmond to convince the Confederate Congress to ratify the treaties he had negotiated with the Five Civilized Tribes, the Seneca and Shawnee, Quapaw, Osage, and the Plains Tribes. He also went to plead for arms and supplies for his growing Indian Brigade since they had few arms, accoutrements, or camp equipment, and little in the way of subsistence stores. During his absence, Colonel Douglas Hancock Cooper was in command of the Indian Territory District.

As previously mentioned, the Confederate treaty with the Choctaw and Chickasaw Nation had allowed for the raising of a mounted regiment consisting of members of the two tribes. At least two months before the signing of the treaty, the Confederate States were already taking official actions with the two Nations. A letter was addressed to "Major Douglas H. Cooper, Choctaw Nation" by Secretary of War L. P. Walker, Confederate States, dated Montgomery, Alabama, May 13, 1861, empowering Cooper to raise a mounted regiment of Choctaws and Chickasaws to be commanded by him in co-operation with Brigadier General Ben McCulloch, who on the same day was assigned "to the command of the district embracing the Indian Territory lying west of Arkansas and south of Kansas."[70] Walker further stressed that the Confederate Government had deemed it "expedient to take measures to secure the protection of these tribes in their present country from the agrarian rapacity of the North." The Chickasaw Nation legislature, meeting in Tishomingo, passed a resolution on May 25 declaring its independence from the United States and proclaiming its

support for the Southern Confederacy. On the same date, the legislature passed another resolution signed by Governor Cyrus Harris, adopting Douglas H. Cooper as a member of the Chickasaw Nation, thus sealing his role as the colonel of the new regiment.

In the Choctaw Nation a similar independence resolution was passed by its legislature and signed into law by Principal Chief George Hudson on June 12. Their resolution went further in that it required military service for all male Choctaw residents aged 18 to 45 and required an immediate call up of at least 700 men for service in the new Regiment of Choctaw and Chickasaw Mounted Rifles. This unit was forming under the direction of Colonel Cooper at their new training grounds at Buck Creek, approximately ten miles southwest of the Choctaw capital at Scullyville.[71] Responding promptly, the Choctaws formed the bulk of the regiment, the Chickasaws being able at that time to furnish only about twenty men. A surplus of three companies—two Choctaw and one Chickasaw—was shortly afterward incorporated in a separate battalion.[72] On July 25 Cooper claimed in a letter to Albert Pike that if needed the Choctaw and Chickasaw Nations could furnish 10,000 warriors.[73] Unfortunately, all new trainees then encountered a problem that would plague the Confederate Indians for the entire war: a lack of arms, ammunition, uniforms, accoutrements, and other critical supplies.

Among the Confederate Creeks a dispute erupted over the command of the 1st Creek Mounted Rifles Regiment. After recruitment began in July and August 1861, Brig. Gen. Albert Pike, commander of the Indian Territory District, wanted Indian Agent William Garrett to be given command of the regiment. Brig. Gen. Ben McCulloch objected stating "from what I know of his habits, a worse appointment could not be made."[74] The Creek leadership stepped in and insisted that the colonel of the new regiment be elected from among the troops. At this point Pike and McCulloch backed down. The Creek soldiers elected Daniel N. McIntosh, son of the late Chief William McIntosh, as the colonel of the regiment. After recruitment for the regiment was completed, they began to form the 1st Creek & Seminole Battalion with Chilly McIntosh, the older son of the late chief, in command as a lieutenant colonel and Seminole Chief John Jumper as the battalion's major. But again, there were supply problems. General Albert Pike wrote to the Secretary of War:

> "The Creek and Choctaw regiments were raised in August and the Cherokee regiment in October; but it was a long time before Colonel Cooper's regiment was even partially armed. No arms were furnished the others; no pay was provided for any of them, and with the exception of a partial supply for the

Choctaw regiment, no tents, clothing, or camp and garrison equipage were furnished to any of them."[75]

The Seminole also split along old standing treaty lines that dated back to their treaty of removal from Florida to the Indian Territory. The pro-treaty faction, led by would-be Southern gentleman John Jumper, generally supported alignment with the Confederacy while the non-treaty party, led by Billy Bowlegs and Alligator, supported remaining with the Union and maintaining the tribe's neutrality. Jumper was able to induce a large number of Seminoles into agreeing with the Confederate treaty, so Bowlegs, Alligator, and War-Chief Halleck Tustenuggee led their supporters out of the Seminole Nation and into the relative safety of Opothleyahola's camp near Brush Hill in the Creek Nation.[76]

[1] United States Military Academy. The West Point History of the Civil War. Rogers, Clifford J., Ty Seidule, and Samual J. Watson, Editors. New York: Simon & Schuster, 2014. pp. 19. (USMA)

[2] Todd Mildfelt and David D. Schafer. Abolitionist of the Most Dangerous Kind: James Montgomery and His War on Slavery. Norman: University of Oklahoma Press, 2023. pp. 62.

[3] Monaghan, Jay. Civil War on the Western Border, 1854-1865. Lincoln and London: University of Nebraska Press. 1955. pp. 97-16

[4] USMA, pp. 43.

[5] Ness, George T. Jr. The Regular Army on the Eve of the Civil War. Baltimore: Toomy Press, 1990. pp. 175-186.

[6] U.S. Department of the Interior, Bureau of Indian Affairs, web page. www.doi.gov/bia.

[7] Unrau, William E. "The Civilian as Indian Agent: Villain or Victim?" Western Historical Quarterly. 3 (4) pp. 405-420.

[8] Abel, Annie Heloise. The American Indian as Slaveholder and Secessionist: An Omitted Chapter in the Diplomatic History of the Southern Confederacy. Cleveland, Ohio: A.H. Clark Company, 1919. pp. 62-90.

[9] Ness, pp. 239-251.

[10] United States War Department. War of the Rebellion: A Compilation of the Official Records of the Union and Confederate Armies, Washington D.C.: U.S. Government Printing Office. 1888-1901. Totten to Cooper, pp. 640.

[11] Christine Jirikowic; Gwen J. Hurst; Tammy Bryant. "Archeological Investigation at 206 North Quaker Lane (44AX193)" (PDF). p. 2. Archived from the original (PDF) on October 25, 2021. Retrieved August 13, 2021 – via City of Alexandria, VA.

[12] Meserve, John Bartlett. "Chief John Ross," Chronicles of Oklahoma (Oklahoma City), XIII 1935, pp. 421-37.

[13] Hill, Luther B. A History of the State of Oklahoma, with assistance of Local Authorities,

Volumes 1 & 2, New York and Chicago: The Lewis Publishing Company, 1908. pp. 19. Also: Minges, Patrick Neal. The Keetoowah Society and the avocation of religious nationalism in the Cherokee Nation, 1855-1867, Unpublished Thesis, Union Theological Seminary, 1999. pp. 1.

[14] Of note, Albert Pike was the only Confederate General to have a statue in Washington, D.C., although it had little to do with his Civil War service. He was one of the most influential Free Masons in the organization in the mid-19th Century. It was heavily damaged during the illegal Black Lives Matter riots in 2020. The remains of the statue stand near 3rd and D Streets NW.

[15] Debo, Angie. The Road to Disappearance: A History of the Creek Indians, Norman: The University of Oklahoma Press, 1941. pp. 142.

[16] Gibson, Arrell Morgan. Oklahoma: A History of Five Centuries, 2nd ed. Norman and London: University of Oklahoma Press. 1981. pp. 117.

[17] Act of Chickasaw Legislature, Jan. 5, 1861.

[18] United States War Department. War of the Rebellion: A Compilation of the Official Records of the Union and Confederate Armies, Washington D.C.: U.S. Government Printing Office. 1888-1901. Series I, Volume I, 682. (Further designated as OR)

[19] Chief Ross to Cherokee Delegation, Feb. 12, 1861. Indian Office General Files, Cherokee, 1859-1865.

[20] Debo, pp. 143.

[21] Journal of the Provisional Congress of the Confederate States, Vol. I, pp. 70 and 81.

[22] Walker to Hubbard, May 14, 1861, OR, Series I, Vol. I, pp. 577-578.

[23] Pike to Toombs, May 29, 1861, ibid., p. 785.

[24] Abel, pp.88-95.

[25] OR., Series IV, I, pp. 323. Abel, pp. 91-95.

[26] Abel, pp. 68.

[27] Trickett, Dean. "The Civil War in the Indian Territory, 1861," Chronicles of Oklahoma, Volume 17, No. 3. September, 1939. pp. 317-318.

[28] Ibid, pp. 665.

[29] Trickett, pp. 319. OR, Series I, LIII, 487; ibid., Series I, I, 656-57.

[30] Ibid.

[31] OR, Series I, Vol. I, pp. 667.

[32] Wright, Murial H. "Lieutenant Averell's Ride at the Outbreak of the Civil War." Cantrell, M.L. and Harris, Mac. Editors. Kepis & Turkey Calls: An Anthology of the War Between the States in Indian Territory. Oklahoma City: Western Heritage Books. 1982. pp. 22-26.

[33] Ibid

[34] OR, Series I, Vol. LIII, pp. 496.

[35] OR, Series I, Vol. I, pp. 653.

[36] Ibid

[37] Emory, William. Report, May 19, 1861, OR Series 1, Volume 1: 648-649. It should be noted on this message Emory refers to himself as "Late Lieutenant-Colonel." With all of the responsibility and leadership he exhibited, the truth was he was from Maryland and considered himself a Southerner. He had been asking to be removed and Major Sackett placed in command of the Indian Territory troops since he expected Maryland to secede along with the other southern states. When he arrived at Fort Leavenworth, he submitted his resignation from the Army. In a twist of fate Maryland did not secede and it took Emory almost a year to regain his U.S. Army commission. He went on to have a successful career in the Union Army and was brevetted as a Major General of Volunteers by the end of the Civil War. William H. Emory, biography, Wikipedia

[38] Ness, George T. Jr. The Regular Army on the Eve of the Civil War, Baltimore: Toomy Press, 1990. pp. 249; Newell, Clayton R. & Charles R. Shrader. Of Duty Well and Faithfully Done: A History of the Regular Army in the Civil War, Lincoln: University of Nebraska Press, 2011. pp. 3-13.

[39] Abel, pp. 186.

[40] Confederate States of America, Journal of the Congress, 1861-1865, Washington: Government Printing Office, 1904. pp. 263.

[41] Ibid, pp. 244.

[42] OR. Series I, Vol. III, pp. 575.

[43] Ibid. pp. 574-575.

[44] Gibson, pp. 119.

[45] White, pp. 60.

[46] Douglas H. Cooper to Major Elias Rector, Letter. United States Department of the Interior, Indian Office, General Files, Southern Superintendency, 1861-1865. I 435.

[47] OR. Series I, Vol. III, pp. 580-581.

[48] Ibid, pp. 581.

[49] OR, Series I, Vol. VIII, pp. 489-490.

[50] Dale, Edward Everett, & Gaston Litton, eds. Cherokee Cavaliers: Forty Years of Cherokee History as told in the Correspondence of the Ridge-Watie-Boudinot Family, Norman: University of Oklahoma Press, 1939. pp. 99.

[51] Abel, pp. 156.

[52] Debo, Angie, The Rise and Fall of the Choctaw Republic, Norman: The University of Oklahoma Press, 1934. pp. 81-83.

[53] OR. Series I, Vol. III, pp. 593.

54 Ibid, pp. 585-587.

55 Debo, Angie. The Road to Disappearance: A History of the Creek Indians, Norman: The University of Oklahoma Press, 1941. pp. 142-147.

56 Abel, pp. 194.

57 Gibson, pp. 120.

58 Abel, pp. 166-167. Article XXXII, Choctaw and Chickasaw Treaty. OR, Series IV, Vol. I, pp. 256.

59 OR, Ser. I, Vol. III, pp. 692.

60 Franks, Kenny A. Stand Watie and the Agony of the Cherokee Nation. Memphis: Memphis State University Press. 1979. pp. 117-119.

61 Woodward, Grace Steele. The Cherokees, Norman: The University of Oklahoma Press, 1963. pp. 265.

62 Franks, Kenny A. Stand Watie and the Agony of the Cherokee Nation. Memphis: Memphis State University Press. 1979. pp. 118.

63 Trickett, Dean. "The Civil War in the Indian Territory, 1861 #1," Chronicles of Oklahoma, Volume 17, No. 3. September, 1939. pp. 401.

64 Britton, Wiley. The Union Indian Brigade in the Civil War. Kansas City: Franklin Hudson Publishing, 1922. pp. 28.

65 White, Christine Schultz, and White, Benton R. Now the Wolf Has Come: The Creek Nation in the Civil War, College Station: Texas A&M University Press, 1996. pp. 25-26.

66 Interior, Indian Affairs. Report of the Commissioner of Indian Affairs, 1861, op. cit., 655. Coffin to Dole, Oct. 2, 1861.

67 Abel, Annie Heloise. The American Indian as Slaveholder and Secessionist: An Omitted Chapter in the Diplomatic History of the Southern Confederacy. Cleveland, Ohio: A.H. Clark Company, 1919. Pp. 226-227.

68 Ibid, pp.156-157.

69 Confer, Clarissa W. The Cherokee Nation in the Civil War, Norman: University of Oklahoma Press, 2007. Pp. 55-58.

70 Wright, Murial H. "General Douglas Hancock Cooper, CSA." Chronicles of Oklahoma, article, Summer 1954; Oklahoma City, Oklahoma. pp. 162.

71 Wright, pp. 163.

72 Trickett #1, pp. 143.

73 Cooper, Douglas H. Letter to President Davis, OR, Ser. I, Vol. III pp. 614.

74 OR, Ser. I, Vol. III, pp. 623-24.

75 OR, Ser. I, Vol. III, pp. 720.

76 Ibid, pp. 34.

Chapter 4
Opothleyahola's Flight

During this early period of the Civil War, the United States Government paid little attention to the events happening in the Indian Territory. The United States was at war with itself. The pleas for help from a paralyzed Federal government were coming from every corner of the North. Confronted with the very real possibility of the new Confederate States Army, currently organizing near Richmond, Virginia, marching across the Potomac and capturing the U.S. Capitol, President Abraham Lincoln focused his energy on ensuring that the government could protect and hold the line. This was the primary reason for abandoning or reducing the size of the frontier military post garrisons and bringing as many Regular Army troops as possible to Washington to confront any threat from Virginia. Lincoln's call for 75,000 volunteers from the individual states was quickly filled by the Northern states but fell on deaf ears to the Southern governors, even those of states that had not yet seceded. The Army also had to garrison its coastal and border fortifications with experienced artillery regiments to prevent any foreign power from taking advantage of the war to attempt an invasion. Further west and north, Fort Leavenworth, Kansas, had been under threat from the Missouri State Guard units that were training at St. Joseph, Missouri, in May 1861. Colonel Dixon S. Miles, commanding officer of the 2nd U.S. Infantry at Fort Leavenworth, stated he feared the Missouri State Guard or Confederate troops would attempt to bypass his location and, by way of the Oregon Trail, would attack and capture Fort Kearney,

Old Fort Kearney
by William Henry Jackson
Painting courtesy of National Park Service, Scotts Bluff NM

Nebraska Territory. This post was a huge target with its substantial stores of quartermaster, commissary, and ordinance stores, including its numerous 12-pounder howitzers.[1] For this reason he wanted to keep as many Regular Army forces in the area as possible.

In the Creek Nation, full-blood chief Opothleyahola, responding to threats on his life and fearing for his safety, fled to his plantation on the Deep Fork near its confluence with the North Fork of the Canadian River. He was joined by Oktarharsars Harjo, or Sands, an Upper Creek leader an second-level chief, and many other Upper Creek followers who responded to his call for support and joined him at his encampment. This group could have numbered as many as four thousand including the possibility of 1,000-1,700 warriors.[2] In December Chief Ross of the Cherokee Nation stated that he wrote to Opothleyahola hoping he would cooperate with the Confederate States. He reported:

"I...dispatched a messenger to Opothleyoholo (*sic*)...and advised him to submit to the treaty made with the Creeks, and to be advised by Colonel Cooper, who was his friend, and had used his utmost exertions to bring about peaceful relations with the parties in the Creek Nation. Opothleyoholo(*sic*) replied that he was at peace with the South, with Colonel Cooper and the Cherokees and desired to remain so. He was willing also to submit to all proper treaties, but that a party in his own nation was against him and his people, who would not allow him to be at peace."[3]

Back at the Deep Fork encamp-ment, Creek leaders Opothleyahola and Oktarharsars Harjo wrote to President Abraham Lincoln on August 15 as the leaders of the Creek Nation. This standing was based on the Upper Creek council meeting on August 5. They stated their desire to abide by the Treaty of 1856 and to either remain neutral or provide limited support to the Union. The letter stated:

> Creek Nat.
> Aug.15, 1861
> "Now I write to the President our Great Father who removed us to our present homes, & made a treaty, and you said that in our new homes we should be defended from all interference from any people and that no white people in the whole world should ever molest us unless they come from the sky but the land should be ours as long as grass grew

or waters run, and should we be injured by anybody you would come with your soldiers & punish them, but now the wolf has come,..."

> "We his people want it to be so again, and we want you to send us word what to do. We do not hear from you & we send a letter..."

> "I was at Washington when you treated with us, and now White People are trying take our people away to fight against us and you. I am alive. I well remember the treaty. My ears are open & my memory is good. This is the letter of Your Children by
> OPOTHLEHOYOLA
> OUKTAHNASERHARJO"[4]

A large number of disaffected loyal Creeks, Seminoles, Cherokees, and Chickasaws soon gathered in a council near the Little River in the western Creek Nation. They were joined by many Plains Indians who also wished to remain either neutral or assist the United States since many had deep hatred for Texas due to their experiences against the Texas Rangers. In early September 1861, this Council sent Micco Hutke, Bob Deer, and Joe Ellis, their interpreter, to deliver Opothleyahola and Oktarharsars Harjo's letter, as well as an oral message from the Council, to the U.S. representatives in Kansas. The oral message reinforced the statements and claims made by the Confederate commissioners regarding the United States abandoning them and how they will turn around, take their slaves and property, and drive the Indians from their homes. The messengers initially met with Kansas Indian Agent E. H. Carruth, an educator with service in both the Cherokee and Creek nations, at

Iola, Kansas. His instructions were to arrange an interview "at Fort Lincoln on the Osage or some point convenient thereto" between Kansas Senator James Lane and representatives of the Indian tribe. Carruth escorted them to Shawnee Mission, Kansas, and U.S. Indian Agent James B. Abbott forwarded the letter and message to Washington. It was suggested that there be a series of councils held in Humbolt, LeRoy, or Fort Scott for those Indian nations that wished to be protected and supported by the United States. The Indian delegation was unable to contact any persons of authority and found the situation in Kansas to be disheartening. The Indian agents suggested returning in October for meetings, so the delegation turned around and headed back to the Indian Territory to report back to their respective tribes. Agent Carruth sent a message to the Indian leaders in the Territory:

"Your letter by Mikko Hutke is received. You will send a delegation of your best men to meet the commissioner of the United States Government in Kansas. I am authorized to inform you that the President will not forget you. Our Army will soon go south, and those of your people who are true and loyal to the Government will be treated as friends. Your rights to property will be respected. The commissioners from the Confederate States have deceived you. They have two tongues. They wanted to get the Indians to fight, and they would rob and plunder you if they can get you into trouble. But the President is still alive. His soldiers will soon drive these men who have violated your homes from the land they have treacherously entered...."[5]

The delegates traveled as far north as Lawrence, Kansas, where Evan Jones, the Baptist missionary to the Cherokees, had a long talk with Mikko Hutke. Jones offered the Indian $25 to deliver a letter to Chief John Ross and bring back an answer, but he declined to undertake it. "I suppose he was afraid of being intercepted with documents in his possession," commented Jones in a letter to Commissioner Dole.[6]

The primary reason the situation seemed hopeless to the Indians was that Kansans had been fighting a form of the Civil War for nearly eight years between the Jayhawkers, or Free-Soilers, and the pro-slavery Border Ruffians from Missouri and various other Southern states. The leader of the Free-State Movement was James Henry Lane, a former congressman and lieutenant governor of Indiana who had come to Kansas in 1855 ostensibly to farm and raise hemp. During his earlier political career, he had been a staunch pro-slavery Democrat who had voted for the Kansas-Nebraska bill as a congressman. His stance on slavery ruined his political career in Indiana and he sought new prospects out west. Lane was a shrewd opportunist and soon he became known as the "Grim Chieftain" of Kansas as he maneuvered his way to leadership of not the pro-slavery Democrats, but as a free-soiler Republican. Operating under the Free-State Wyandotte Constitution, which coun-tered the Buchanan-supported Lecompton Constitution, the state legislature elected Lane as Kansas's first U.S. senator in 1859, and he finally took his seat in 1861 when the former territory became a state.[7]

Highlighting his military service experience in the Mexican War, in June 1861, Senator Lane was able to convince President Lincoln to appoint him a brigadier general of volunteers, with authority to raise two regiments. The mission of this "Kansas Brigade" was to defend the state from invasion coming from Missouri, Arkansas, or Texas, via the Indian Territory. With the majority of the Five Civilized Tribes and the smaller ones pledging their support to the Confederacy as well, he viewed this as the most probable route any Texas forces would use to invade Kansas and Nebraska. The fear was that a Confederate force could block not only the Oregon-California and the Sante Fe Trails, but also the transcontinental telegraph lines that would cut off the Western United States from the rest of the Union. Lane was also the first to suggest the recruitment of Indians into his Kansas forces from the tribes already present in Kansas, the Kansa, Kaw, and Osage. His initial idea was to place them along the Kansas-Missouri border in the Cherokee Neutral Lands to discourage foraging or military raids from Missouri.

In May 1861 President Lincoln appointed William Coffin of Indiana as the new Southern Superintendent of Indian Affairs, replacing Elias Rector who had defected to the Confederacy. The same Indian delegation, with the addition of Oktarharsars Harjo and some Seminole and Chickasaw tribal members, traveled to LeRoy, Kansas, in early November for an arranged meeting with Senator Lane and the Indian agents, including Lane's Indian agent, E.H. Carruth. Senator Lane had departed for Washington before the council meeting and, as a result, there were only Indian agents available. They met with Superintendent Coffin, Agent Carruth, and then met with the new U.S. Indian Agent to the Creek Nation, Major George A. Cutler, who had been unable to take his post due to the troubles in the Indian Territory. Oktarharsars Harjo made a speech to the assembly reflecting on his story of loyalty to the Union and his confusion as to why there had not been any help sent to the Loyal Creeks by the United States. He further said:

> "Never knew that Creek had an agent here until he come and see him and that is why I have come among this Union people. Have come in and saw my agent and want to go by the old Treaty. Wants to get with U.S. Army so that I can get back to my people as Secessionists will not let me go. Wants the Great Father to send the Union Red people and Troops down the Black Beaver road and he will guide them to his country and then all of his people would be for the Union."[8]

Agent Cutler decided to take them to Fort Scott and, although Senator Lane had left for Washington, Cutler hoped to still catch him there. Colonel James Montgomery at Fort Scott, commander of the 3rd Kansas Infantry and Lane's successor and fellow Jayhawker,[9] advised that the delegation be taken to the Federal capital where they might meet with the Senator. The delegation traveled to Fort Leavenworth, where the new depart-ment commander, General David Hunter, concurred with the views of Montgomery. Agent Cutler and the delegation immediately left for Washington. The Indian agents hoped that by sending them to Washington, the three Creek messengers could see for themselves that the whole country was on a war-footing. They would also see the defenses being constructed and then have the opportunity to meet with President Abraham Lincoln, the "Great Father." One of the messengers later wrote:

> "On arriving there, face to face, we informed our Great Father of the situation that our country was in and were informed by our Great Father that our treaties were and should be respected; and we were further assured that he would send us help as soon as he could."[10]

"The result of that journey," Cutler afterward wrote, "has strengthened their confidence and belief in the power and stability of the Government."[11]

The Lincoln administration responded officially in November 1861, with instructions for all Loyal Indians to evacuate the Indian Territory to Fort Row, Kansas, a post being built along Walnut Creek by the local militia company. There they could meet with their Indian agents and arrange for supplies and housing. This is not the answer the delegation was hoping for. There would not be a Union army marching down the Black Beaver Road to drive off the secessionists, at least, not yet.

As the Creek and other refugees began their movement to meet at Opothleyahola's homestead near Brush Hill, they received the news that John Ross and the Cherokee Nation had signed a treaty with the Confederate States. Opothleyahola was devastated by this report since Ross had pleaded with him and his Creek followers to maintain a neutral stance. Further, in a September 19, 1861 letter, Ross then promised that those same loyal followers were

welcome in the Cherokee Nation to wait out the war. Feeling betrayed by Chief Ross and the Cherokee Nation, who had pleaded with the Upper Creeks to follow the Cherokee route of neutrality, Opothleyahola believed that Ross had broken his trust and refused to negotiate, not realizing the danger Ross was facing.

In October 1861, Albert Pike wrote to Opothleyahola from Park Hill in the Cherokee Nation offering him a colonel's commission and a pardon to all of his Creek followers if they would submit to Confederate authority. He also offered to organize a battalion of Opothleyahola's followers for service in the Confederate Army, with service limited to duty in the Indian Territory. The old chief refused the offer and continued gathering his followers at his homestead near North Fork Town, Creek Nation. Scores of refugees from all Five Civilized Tribes, as well as from the other smaller tribes in the Indian Territory, began arriving and set up a large camp near Brush Hill. The number soon grew into the thousands, and this pro-Union, or neutral-leaning, gathering caught the attention of the Confederate authorities. The area became so crowded that new camps were established, the largest being along Hilliby Creek west of the main encampment. Other groups gathered at other western Creek towns including Greenleaf Town and Thlophlocco. The pro-Confederate Indian factions sent spies into the camps and reported on the conditions and moods of the Indians to the tribal leaders and the Confederate chain of command. Apparently, these undercover agents went so far as to spread misinformation and rumors detrimental to Opothleyahola's followers. They claimed that Opothleyahola and his supporters had burned down North Fork Town and Fort Gibson and planned a large raid upon the Creek Nation Council House, an attack on the Creek Agency at Three Rivers, and hunt down and kill the McIntosh's.[12] Once again, none of this was true, but it did create a sense of apprehension and fear among the pro-Southern Creek. The Confederate command, convinced this large pro-Union gathering was dangerous, opted to attack and destroy Opothleyahola's gathering.

The Loyal Creeks soon learned that the Confederates and their Lower Creek allies planned on breaking up their encampments on the Deep Fork and elsewhere. Faced with forced submission or massacre by the Confederate Army and the pro-Southern Indians, on November 5, 1861, Opothleyahola started his Creek and Seminole followers on a long slow march northward towards Kansas. The caravan included warriors, women, children, the aged, a large number of African slaves and freedmen, and large droves of cattle and other livestock. Some rode horses or on wagons, but most walked while carrying their small measure of belongings on their backs.[13] They moved first from their gathering place at Opothleyahola's homestead near North Fork Town northwest along the Deep Fork. As they followed this path, the Loyal Creeks were joined by the disbursed groups in Thlophlocco and Greenleaf Town groups until they reached the Hilliby Creek encampment area north of present-day Boley, Oklahoma, when they all turned north and headed toward the confluence of the Cimmaron (Red Fork) and Arkansas Rivers, following the Dawson Road, an 1834 marked trail that followed an ancient Osage trail that led from the Little River region northward to the Osage Crossing of

the Arkansas.[14] They reached the Big Pond camping site a few miles southeast of Depew, Oklahoma, on or around November 14, 1861.

The Loyal Seminoles had joined with the caravan at the Brush Hill encampment. They were led by a celebrated chief, Halleck Tustenuggee, who with a band of seventy warriors had fought with United States troops in 1843, the last battle of the Seminole Wars in Florida. Pascofa and other Seminole chiefs had also joined the Loyal Indians. In command of all the warriors—Creek and Seminole—was a Creek called the Little Captain (Keptene Uchee). Although he led the warriors in the three battles fought by Opothleyahola's forces, he remains today a virtually invisible figure known only by name.[15]

Col. Cooper soon learned of the flight of Opothleyahola's followers, and with Brig. Gen. Pike being in Richmond, he took it upon himself as the acting department commander to gather a force and prevent their going to Kansas. The pro-Southern, McIntosh-led, Creeks and Seminoles were also determined to stop the exodus of Opothleyahola and his followers. They called upon Cooper, and he was able to gather together a force made up of six companies of his 1st Choctaw and Chickasaw Regiment, the 1st Creek Regiment, the Creek and Seminole Battalion, approximately 900 men. Cooper stated that his intention with Opothleyahola and the Loyal Creeks was the "submission to the authorities of the [Creek] nation or drive him and his party from the country." After an armed clash between Southern-supporting and Union-supporting Creeks occurred in mid-October along present-day Battle Creek, north of the North Canadian River near Thlophlocco, Cooper claimed that he had attempted to set up negotiations with Opothleyahola to discuss a peaceful settlement between his Loyalist Creeks and the McIntosh faction. Cooper arrived at Thlophlocco on October 29, 1861, and attempted to meet with the Loyal Creek leaders but none agreed to do so. Cooper wrote to Col. Drew of the 1st Cherokee regiment stating,

> "It is exceedingly vexatious to be detained here by party feuds amongst the Creeks, but it is unavoidable, inasmuch as the Creeks would probably refuse to march northward and leave the matters unsettled at home."[16]

Again, knowing Cooper's desire for ill-gotten riches, it is possible he was trying to gain information to discover the stash of Creek gold and currency that was rumored to be with Opothleyahola. Another rumor of the treasure was that Opothleyahola had buried the fortune on his homestead, since the treasure was not found with the Loyal Creeks after they arrived in Kansas. For many years fortune hunters dug holes all over Opothleyahola's former homestead looking for the lost Creek treasure. If there ever was a buried treasure on the plantation, it now lies beneath the peaceful waters of Lake Eufaula.

Other than the Confederate Creeks, few other tribal nations felt any real desire to stop the Loyal Creeks since they did not see them as a threat because they were leaving the Indian Territory. Chief Ross called it "the Creek feud," and wanted to keep out of it although he had earlier tried to settle the differences.[17] Cooper would not accept this and was determined to stop or destroy the Loyal Creek band.

With estimates of the Loyal Creek band approaching six thousand persons with upwards of two thousand warriors, Col. Cooper determined that his 900 Indian soldiers would not be strong enough to deter or stop the group from leaving the Indian Territory. He also received incorrect reports that Opothleyahola planned on attacking his camps. Although he had been joined by Welch's Texas Cavalry Company, under the command of Captain Otis G. Welch, Cooper still requested reinforcements from Brig. Gen. McCulloch. He had sent Col. Drew and the 1st Cherokee Mounted Rifles north to Coody's Bluff. Cooper obviously was not familiar with the Cherokee Nation and believed that Coody's Bluff was on the Arkansas River near Tulsey Town, modern day Tulsa. Instead, the location was far to the north on the Verdigris River near the Kansas border. He stated in a directive to Colonel Drew:

> Head Quarters Indian Brigade
> Deep Fork Near Mshers
> Nov. 5th. 1861.
> Colonel,
> Your Regiment having been mustered into the Service, you will march, with the least possible delay, up the Neosho, to support Col. Stand Watie-pentrate [sic] Kansas (if possible), and carry into effect the instructions heretofore given you. I learn, verbally, from Majr Clark who brought despatches from Genl. McCulloch arrived day before yesterday that the Genl. supposed you had already marched for Kansas -Genl. McCulloch having placed at my disposal such of the Texas Regiments

now on the march for North Fork Town a. may be needed for the defence*[sic]* of the Indian Country I have directed Lt. Wells to dispense with the services of such additional Indian forces as may have offered themselves under my call unless specially required by Genl. McCulloch. I shall be in the Cherokee Country as soon as possible with the forces under my command, and will Communicate with you - Hopoithlayahola's *[sic]* people are said to be moving towards Walnut Creek.

> I am Col'n your Obt. Servt
> Coln John Drew Comdg.
> Cherokee Regiment
> Camp at [?][18]

Col. Cooper also informed Lieut. Col. W. Ross:

> HeadQuarters, Indian Dept.
> Camp Pike, Creek Nation,
> Nov. 10, 1861.
>
> Sir, I have received your communication, dated at Fort Gibson, Nov. 8th. and fully concur in the opinions therein expressed, and have done all in my power, to effect a friendly settlement of the Creek difficulties. You are mistaken in regard to Hopoithlayahola's *[sic]* pacific intentions, as from reliable information, I am perfectly satisfied that he is now meditating an attack upon my camp, in conjunction with Doct. Jamison, and 1000 Jayhawkers, at this time near the Arkansas River. If you can make a rapid march, in the direction of "Goody's," (which I

suppose to be on the California road up the Arkansas) and then get in rear of the Kansas force, it would be of material aid to me, and an advisable movement.

> Very respectfully, Yr. obt. Servt.
> Douglas H. Cooper
> Col. Comdg. Indn. Dept.
> Lt. Col. W. P. Ross,
> Cherokee Regt.[19]

Stand Watie and his 2nd Cherokee Mounted Rifles were also in that area to the east of Drew's area of operations. As a result of this mistake, two of Cooper's largest units were out of place for his operations against Opothleyahola and the Loyal Creeks. Moreover, he only had six companies of his own 1st Choctaw & Chickasaw regiment and, although he had sent for the remaining companies at his training camp at Buck Creek, due to a shortage of forage stocks, they had not yet arrived by the time his forces moved out.

Albert Pike, sensing the danger of another internal civil war, wrote to Secretary of War Benjamin:

> "When I was informed of Opothleyoholo's *[sic]* intentions to fight, I could do no more than request Colonel Drew and Colonel Cooper to march to the assistance of Colonel McIntosh . . . "The Cherokees and Creeks are neighbors and the former are very desirous of maintaining their present friendly relations. They have long had a treaty between themselves by which they can settle in each other's country, and many of each nation are domiciled and married in

the country of the other. The Cherokees naturally fear that if they fight any part of the Creeks the feud will last between them for many years after our difficulties are settled."[20]

Meanwhile, the 9th Texas Cavalry had been enroute to rendezvous with Col. McCulloch's command in Arkansas when they were re-routed to support Cooper's Confederate Creeks near North Fork Town. The 9th Texas was a new unit that had been recruited from around Sherman, Texas. On October 14 they were mustered into the Confederate States Army and elected Clarksville, Texas, merchant William Sims as its colonel. He was described as having "a voice equal to the modern fog horn," and was "a born commander," conscientious and dutiful toward his men. Not everyone agreed, however, as James C. Bates noted in his diary on December 13, 1861: "If ever a Reg was cursed with a bigoted self conceited for Col ours is the one." Sea captain and soldier of fortune, William Quayle, was elected as the regiment's lieutenant colonel. One of his trooper's described Quayle as "the finest appearing horseman I had .. ever beheld; he was the military man of the regiment."[21] The regiment had departed Sherman with close to a thousand troops and followed the Texas Road northward, camping at Boggy Creek, Choctaw Nation, enroute. By the time they reached North Fork Town on November 11, nearly half of the command had succumbed to sickness. They set up camp at the Asbury Mission north of town. As the weary men were feeding their horses, a courier from Cooper came galloping into camp urgently requesting their help. Col. Sims complied by selecting 500 of his men who had not yet

succumbed to illness and sent this detachment out under Lieutenant Colonel Quayle the following morning. Col. Sims and the sick would remain behind with the regiment's wagon train and would move forward to one of Cooper's base camps at Concharty when enough troops recovered.[22] The detachment arrived at Camp McCulloch near Tahlequah after dark on November 12 after a nearly 60-mile trek. A.W. Sparks of the 9th Texas Cavalry recalled: "Gen. Cooper's Indians received us with a great joy as we marched through their camp and fired us a grand salute, which we returned with about the same unmilitary regularity. After passing through the camps, we were encamped for the remainder of the night." Sparks later learned that Cooper had them march in formation through the Indian camps to inspire confidence in the Confederate Creeks.[23] Sparks continued:

> "During the [next] day and early night the rations were all cooked, after a manner, and the remainder of the night was spent in watching our Indian friends prepare for battle in a war dance. Many people seem to think the Indian war dance is a frivolous affair; but from my observation it is really a very serious ceremony, that is just as necessary for the Indian before his battle for life or death as is prayer for the Christian on like occasion."[24]

The Texan from Titus County described how the Creeks staged a war dance for the edification of the wide-eyed Texas boys, a few of whom were so seized by the spirit of the occasion that they joined in too. All were electrified by the spectacle. He stated:

"A ring was filled with painted Indians,"... "all marching in a side-like manner, stepping high and fast, while they chanted a strange song.... The glimmer of the low campfires made the painted bodies of the Indians look like demons rising from infernal regions."[25]

That same night each soldier, Texan or Indian, was issued a red and a blue string to be tied around the left arm to identify each other during the battle since few of the soldiers had been issued anything resembling a standard uniform.[26] Sparks recalled the Indian uniforms:

"The Indian warriors, as I noticed were well supplied with rations, and rode small ponies and were dressed in a garb ranging from a common gent's suit to a breech clout and blanket, most of the full bloods wore only the latter, their faces were painted in such a manner that many of them were frightful to even look upon,... Their arms were as varied as their apparel and were old rifle, guns and bows and arrows mostly, and in this motleyed (sic) assembly we marched upon the enemy,..."[27]

On November 15, 1861, Colonel Cooper and his forces left Camp McCulloch, the same day Opothleyahola's Loyal Creeks and Seminoles moved north from Big Pond. He wrote to Col. Drew,

Headquarters, Indian Department

Camp McCulloch, C.N.
Nov. 14. 1861

Col . I shall march from this post, tomorrow morning, with all my available force except such as it is necessary to leave as a guard for my train. It will become necessary to move the train as soon as Col. Sims comes up with the balance of his Regt. in consequence of the failure of forage. It will cross the Arkansas when it moves, above Pole Cat, or at Rider Fields', Concharty settlement. I have 500 of the Texas Regt. with me.

Very. Respectfully
Yr. Obt.
Douglas H. Cooper,
C.S.A.
Com. of Indian Dept.
Col. John Drew, C.S.A.
Comdg. Cherokee Regt.
Fort Gibson Ch. Nation
By Command
R. W. Lee, A. A. Adjt.
Genl.[28]

From this point Cooper's movements are not clear. The primary question for this writer is which of the Loyal Creek encampments did Cooper find on November 16? Some accounts state that he approached the encampment at Opothleyahola's plantation near Brush Hill, an air distance of 50 miles from Camp McCulloch, a long stretch for a military cavalry unit. It is reported that it took four days to catch up with the Loyal Indians once they left Camp McCulloch on November 15. They discovered the encampment deserted on November 16, were at Sell's Store on November 17, and attacked the Red Fork encampment late in the day on the 19th. If the Confederates followed the Brush Hill group as they moved up the Deep Fork and then north on the Dawson Road through the Hilliby Creek and Big Pond encampments and, if they caught up with the Loyal Indians at the Red Fork of the Arkansas River on November 19, the dates would line up fairly well.

Unfortunately, another version of the events has the Confederates passing Fort Gibson, crossing the Grand and Verdigris Rivers near that point, and moving upstream along the north bank of the Arkansas River. In this version they crossed the Arkansas River at Choskey and headed west across the prairies, past the present-day town of Beggs, and into Sell's Store, located on Brown's Creek, a distance between 40-50 miles,[29] about a day and a half travel. But traveling this route would not place them anywhere close to the known encampments of the Loyal Indians unless the first encampment encountered was the Big Pond, which would be unlikely since Cooper reported finding the camp on November 16 and he was not reported to be at Sell's Store until November 17. In addition, when Cooper discovered the first encampment he was informed that the Loyal Indians had been gone for a week. Opothleyahola and his followers left the Big Pond encampment on November 15, the same day Cooper left Camp McCulloch.[30]

A third version has the Confederate expedition leaving Camp McCulloch on November 15, arriving near Opothleyahola's Brush Hill plantation encampment that evening. On November 16, they find the camp vacant and begin following the Loyal Indians' trail along the north side of the Deep Fork. Unfortunately for Cooper, the Indians

he was following were mostly full-bloods who still knew, and had taught their young warriors, how to disguise a trail. To disguise a trail of upwards of four to five thousand people with wagons and livestock was quite an accomplishment. This was a skill lost on the mostly white mixed-bloods. Several times Cooper had to stop his pursuit and send out scouts to reacquire the Loyal Indians' trail.[31] They then proceeded northward to Camp Porter, adjacent to the Creek Council Hill, north of Hitchita, Oklahoma, where the McIntosh's had their base camp for the Creek regiments. This was about a half-day's ride from Brush Hill, but they were probably again delayed by the Loyal Indians having erased and disguised their trail. This would place them at Camp Porter on the evening of November 16. The 1st Creek regiment joined the column at this point and all continued moving northwest along the Loyal Indians' trail until veering off towards Sell's Store. This would be a full-day's ride and would have brought the Confederates to Sell's Store on the evening of November 17, which lines up with the record. A half day's ride west of Sell's Store would bring them at around mid-day of November 18 to the Big Pond encampment that Opothleyahola's Loyal Indians had abandoned on November 15. From this point it is between 30-35 miles to the Red Fork Crossing. Depending on the speed and terrain this would bring them near to the Round Mountain and Red Fork River crossing late in the day on November 19. The first shots were exchanged when Cooper's Confederates caught up with the Loyal Indians after capturing some stragglers from Opothleyahola's band.[32]

Upon close examination of the distances that Opothleyahola and his followers were travelling, it seems that they were not in a great hurry to get to Kansas. From November 5 until Cooper's force caught up with them on November 19, Opothleyahola and his group had only travelled approximately 75 miles over a fourteen-day period. Even with his elderly, children, wagons, and animal herds slowing him down, he still was only moving at a pace of less than six miles per day. The reason for this slow movement is unknown. The J.T Cox map, drawn in 1864, shows six encampments from Brush Hill, near North Fork Town, to the Red Fork camp, but these may have only been major stops, not every camping spot. All of these camps are shown to be north of the Deep Fork River. It is possible that all of the groups were advised by Opothleyahola either to join up with the larger group as it moved through or simply to have groups meet at the Hilliby Creek or the Big Pond encampments. The reasons cited include that Opothleyahola did not believe the Confederate threat to be as great as it was, or that he was reluctant to bring his followers to Kansas, perhaps opting to either fight to stay in the Creek Nation or moving his group further west to the un-ceded lands of Indian Territory. Maybe he was still looking and hoping for that Union army coming down the Black Beaver Road.

Cooper's forces discovered Opothleyahola's camp after spotting campfire smoke at dusk on November 19 at the crossing of the Red Fork of the Arkansas River, near the present-day city of Mannford, Oklahoma. The Texas troops were feeling the pangs of hunger by this time. Although they had tried to prepare ten days' worth of rations while still at Camp McCulloch, they had extreme difficulties since all of their cookware had been left with the regiment's wagon train which was now almost 50 miles behind them. The Indian troops seemed to fare well: they probably had a better supply chain since they were local, but the Texans were reduced to eating burr oak acorns found near the waterways. The Texans' horses were beginning to suffer as well since the Loyal Indians had burned the prairies, with their nutritious big and small bluestem grasses, as they moved north and west. The result was that the soldiers were forced to share their meager rations with their horses. As the Confederates got closer to the Loyal Indian's camps, they were able to slaughter some stray cattle that had drifted from the herd moving north. This kept the hunger at bay for a day or so, but it would take more than a few stray cows to feed a 500-man cavalry unit for more than just a short time.

Opothleyahola placed his Red Fork encampment approximately four miles north of the Red Fork crossing among the hills in that region. Realizing the proximity of their pursuers, Opothleyahola had divided up his followers in a standard Indian military tactic. He sent the non-combatants out of the camps and across the Arkansas River by various crossings into the Cherokee Nation. The old chieftain kept Little Captain's warriors near the Red Fork crossings as a diversion to halt the advancing Confederates. In their haste the Loyal Indians were forced to abandon some mostly empty wagons, a buggy, about eight yoke of oxen, some livestock, and some subsistence provisions.

For most of two days the Texans had observed an Indian scout on a white horse ahead of them and they attempted to capture him on numerous occasions, but he was able to elude their pursuits. He also made his presence obviously visible to the Texans, not

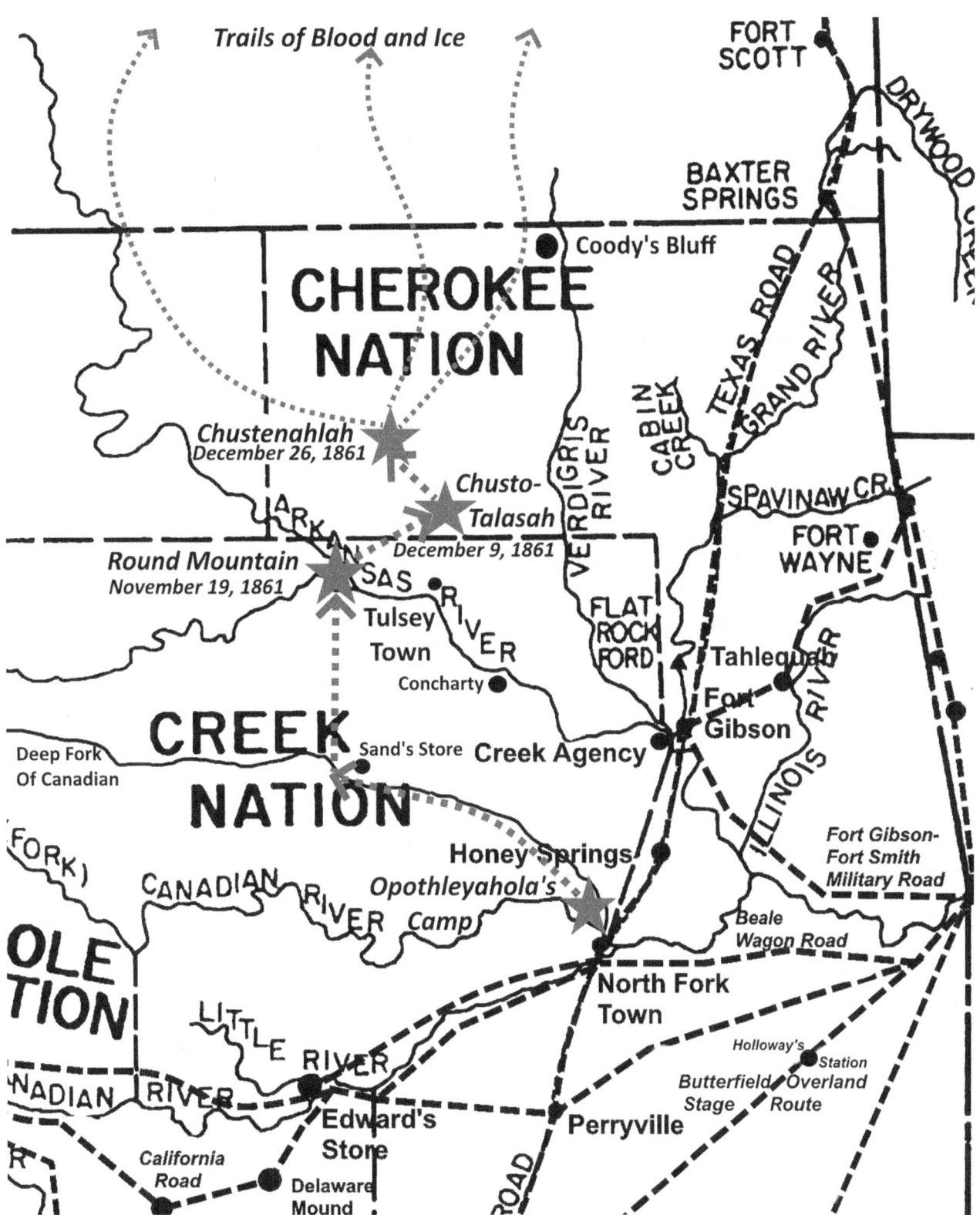

Map of Opothleyahola's attempt to escape from the Indian Territory along with the Battle Sites of Round Mountain, Chusto-Talasah, and Chustenahlah

making any real attempt to remain hidden. As the Confederates approached the Red Fork encampment the scout was seen more often and closer than ever before. A.W. Sparks remarked:

> "I have neglected to mention that during the march for the past few days that a Vidette, most always to be seen on our left flank had, several times been the occasion of a lively chase and many had been the shots fired at him and his horse. I had more than once chased him only to be easily left, which was mortifying to my pride for I rode a horse 'Old Napoleon' that to my mind was equal to the best... [His] horse was white, very white, and the rider appeared to be small and without beard and of a light complexion, and carried arms that were long range. Both rider and horse appeared to be proof against any arms that we carried, and our superstition had led us to believe he was an enchantment and it was shared alike with our Indian allies..."[33]

One of the locations that the "Vidette" probably used as a lookout point was a flat topped, round hill that is located in the far southwest corner of Tulsa County. It is believed that this rounded hill was the landmark that gave its name to the Battle of Round Mountain. Near the Round Mountain, the trail splits: the primary Dawson Road heading straight north to the Old Fort Arbuckle crossing of the Arkansas River, and the other branch swinging west to the Red Fork crossing and into what the J.T. Cox map identifies as the main Loyal Indian camp.

Col. Cooper reported that by 4 PM he had crossed the Red Fork and encountered an abandoned encampment. The author believes that it is possible the Confederates had encountered a previous encampment on Salt Creek that may have been mistaken for the Red Fork by commanders unfamiliar with the region, especially considering the amount of light available at that time of day in late November. The Salt Creek crossing was a little more than four miles from the Red Fork crossing. With five or six thousand Loyal Indians camping, the individual camps must have been spread over large areas without definitive boundaries. As the command moved northwest from Salt Creek crossing, they could see the smoke from the main encampment on the hills to their front.

The 9th Texas Cavalry had ridden out ahead of Cooper's main force and, although it did take the precaution of placing guards on their flanks, Lt. Col. Quayle failed to curb his over-anxious troopers and conduct a thorough leader's reconnaissance. They didn't realize they were being led into an ambush. To be fair, life as a sea captain did not prepare him for fighting Indians. Trooper Sparks remarks that he was with the forward squadron of 70 men of consisting of Capt. Brinson's Company D and Capt. Stewart's Company I, with Stewart in command of the small squadron, when they began to deploy down a slope into an area where it looked like the creeks were merging into a V formation. Regardless, it was getting dark and they were in unfamiliar territory, and they wanted to catch the Vidette who was staying just beyond reach. As the scout approached the crossing they noticed him leaning to one side and then another on his white horse as only Indian riders seemed to be able to do. The pursuers then saw fire coming from both sides of him,

Author's Interpretation of the Battle of Round Mountain
November 19, 1861
Map courtesy of the U.S. Geological Survey / Additions by Author

Osage Trail Crossing

Opothleyahola's Camp

Ambush

Old U.S. Crossing

Skirmishers

Abandoned Camp

Dawson Trail

Skirmishers

Cooper's Advance

Round Mountain

Skirmishers

Senator James Lane of Kansas
Photo courtesy of KSHS

Colonel James McIntosh
2nd Arkansas Mounted Rifles
Photo courtesy of the NPS / Pea Ridge NMP

BATTLE OF ROUND MOUNTAIN

Round Mountain in Southwest Tulsa County, Oklahoma
Photo by Author

which quickly set the prairie grasses on fire. In their bloodlust they rashly decided to attack the distant camp in the deepening darkness. They finally crossed the Red Fork and headed towards the camp, still being led by the Vidette. They entered an area with tree-lined creeks on either side shaped like a V with a small space at the point. The entire area is crossed by numerous creeks and ravines, many of which would have qualified for that formation. They charged quickly down the slope towards the space when suddenly they were met by a severe volley of rifle fire and arrows from Opothleyahola's warriors who had been deployed by Little Captain in the tree lines somewhere beyond the Red Fork crossing. The Indian scout had led them into a crossfire ambush. In their confusion the Texans quickly lost their momentum and, after a failed attempt to reform on the nearby prairie, began to retreat towards Cooper's main force. Lt. Col. William Andrew Quayle of the 9th Texas stated:

> "We found the enemy in considerable numbers, supposed to be 1500 strongly posted. We were fired on by them, and formed as quick as the circumstances of the attack would admit, and returned their fire until it was evident that they were attempting to surround us."[34]

Which their numbers and position enabled them to do. I thought it best to retreat, and were followed by the enemy nearly 2 miles, returning fire as we went, until we were reinforced by [Col.] Cooper."[35]

Capt. M.J. Brinson, also of the 9th Texas Cavalry reported to Col. Cooper:

"The attack was brought on by the second squadron about sunset, composed of about 70 men. I was promptly aided on my right by Captain Berry and on my left by Captain McCool, who formed in my own, or second squadron. After firing from three to five rounds I perceived the enemy in strong position and force, numbering some 1,500 Indians, and flanking my small force upon the right and left, I had necessarily to fall back to the main command, some 2½ miles, under a heavy retreating fire. The whole command-in which I fought my own squadron, Captain Berry's company, a part of McCool's, and a part of Captain Williams' company - I am confident did not amount to exceeding 150 men."[36]

Lt. James Bates of the 9th Texas Cavalry, Company H, observed:

> "We had scarcely formed when we heard firing ahead but it was now so dark we could not distinguish an Indian from a white man at three paces... The Choctaws had been formed on our left and rather in advance of us... Our men ran into an ambush where they were fired on in front and on both flanks. The Creeks fled and left the Texans to take care of themselves..."[37]

Sgt. George Griscom of Company D, 9th Texas Cavalry wrote in his diary:

> "The advance finds & drives in their pickets, the command divides, charge about five miles... keep up a furious

rate to Round Mountain Creek where they were posted in a horseshoe of timber… the enemy tried to stampede us by sending dogs through our ranks… the enemy tried to turn our left flank in heavy force but Col. Cooper with his Choctaws met them & a bloody fight of 15 minutes turned them back. We fell back about 2 mile…"[38]

The 9th Texas Cavalry retreated until they were met by the 1st Choctaw and Chickasaw regiment who attempted to advance against the Loyal Indians. Capt. Robert A. Young, commanding Company K, 1st Choctaw and Cherokee regiment, wrote:

"We advanced within 150 yards of the enemy and dismounted. While dismounting we were fired upon. My men dismounted in good order, and I ordered them to advance and fire – the enemy keeping constant fire on us. The prairie was on fire on my right and we advanced to the attack I could see very distinctly the enemy passing the fire, and I supposed a large body of men (200 or 300) but they were 300 yards away, and the prairie was burning very rapidly…"[39]

A portion of the other Confederate units were able to cross the Red Fork crossing and charged into the empty camp. They were quickly driven back over the crossing by the other portion of Little Captain's warriors. The soldiers encamped on the prairie south of the crossings. Opothleyahola's warriors silently deserted their camps and crossed the Arkansas River to join up with the remainder of their tribal members.

Judge James R. Gregory, who served in the 9th Kansas Cavalry, lived just a short distance from the battlefield and knew many of the participants. He stated that he believed he understood Opothleyahola's strategy during the battle. He wrote in a newspaper article, "Creeks in the Civil War,"

"One body of the Union Creeks was camped on the Arkansas River near the old Skiatook place (then in the Cherokee Nation but now in the Osage Nation) and the others on the North Fork River, above mentioned. General Cooper with his forces proceeded to attack the Creek Camp on the North Fork River. The Union Creeks, under the command of Chief Opothleyahola marched in one-fourth circle around the right flank of Cooper's army to the Northeast, attempting to form a junction with the Union Creeks on the Arkansas River. Before the junction was effected General Cooper's army overtook this faction of the Union Creeks crossing the Cimarron just at dusk. A battle ensued, which was fought after darkness set in. After stopping the advance of the Confederates, the Union Creeks proceeded on the same night to form the junction which they had in contemplation on the outset, and which they accomplished the following day. General Cooper did not follow the Union Creeks the next day, but retired toward Choslta (sic) to wait reinforcements…"[40]

This ended the Battle of Round

Mountain, the first of many Civil War battles in the Indian Territory. The 9th Texas lost twenty men with little to show for their effort. In the night Opothleyahola and his followers retreated to a friendly Cherokee village near the Big Bend of the Arkansas River to which they had been invited prior to the battle. One prize Cooper tried to claim was Opothleyahola's personal buggy, but Opothleyahola was known as a marvelous horseman and it is unlikely he had been riding in a buggy.[41] Cooper claimed to have killed or wounded 110 of the Loyal Indians, which is a wild exaggeration considering his forces fired so few shots after being ambushed. The Confederates reported seven killed and four wounded.

It is very difficult to pinpoint the exact location of the Round Mountain battlefield. Historians have been unable to agree on where the battle actually took place. It was a minor skirmish occurring late in the afternoon in late November when darkness comes early. Few of the participants who reported the action were from the region, especially the Texans, and probably did not know one creek or river from another. The Twin Mounds location west of Yale, Oklahoma, favored by many early historians, has been discarded since it was too far to the west to be within Opothleyahola's area of operation. The Round Mound location south and west of the Keystone Dam in far southwestern Tulsa, northeast Creek, eastern Pawnee, and southeast Osage Counties has been thought by many historians to be the correct location of the battle. What makes this area questionable is that it is east of the Cimarron River, or Red Fork, of the Arkansas River, and most accounts and historic maps, except for the 1864 J.T. Cox map, place the battle

west of the river. The best evidence is that the battle was fought from the Salt Creek crossing to the Red Fork crossing, then north to Section 13, Township 20 North, Range 8 East. Although this author disagrees with her troop movement interpretations, especially the Texans crossing the Arkansas River twice, this is the location that celebrated Oklahoma historian Murial Wright determined the battle to be, someplace west of State Route 48, in Pawnee County, Oklahoma.

An interesting story that appeared in a July 1, 1923, Dallas Morning News article was by Captain June Peak, who as a seventeen-year-old served on Col. Cooper's wagon train that accompanied the expedition during the Battle of Round Mountain. He claimed:

> "We met one morning in October [November], at Round Mountain. The day was spent in skirmishing, without any losses or advantage to speak of on either side. We went Into [sic] camp for the night on a level prairie, covered with sedge grass waist high, beginning to dry considerably. Making a corral of our wagons, we placed our stock within It [sic]. We retired with the understanding that the battle would begin early in the morning. It was a serene night. At 1 o'clock we all of one accord leaped to our feet. The prairie was on fire in hundreds of places around us, and a fierce wind which had sprung up was carrying wisps of blazing grass hundreds of yards and starting new fires. The weird beauty of the landscape revealed by the widespreading [sic] conflagration was perhaps not wholly lost on even the most fearful of our panic-stricken

train. Our poor mules gave vent to their distress in sounds that seemed to be compounded of bray, bellow and squeal. In our efforts to save our wagons and teams we had no leisure to return the fire of the enemy who were raining bullets and arrows into our confused rout. We abandoned the whole of our provisions, and left in our wake a dozen or so wagons, scores of mules, and fifteen or twenty dead and wounded men. Fortunately for us, Opothyola [sic] did not follow up his advantage. We were more than two hours getting out of the fire...."[42]

This does make an interesting account if true since Cooper related that he had left his wagon train behind at Concharty. This incident is not shown in any of the official documents or is it mentioned by any other participant. But it may have been one of those incidents that Col. Cooper was never notified of or just chose to leave out of his report.

Because of the dismal performance of his command during the battle of Round Mountain, Cooper failed to follow up on Opothleyahola's retreat and instead quickly retreated back towards his supply train at Concharty. He explained this action by stating that he had received orders from General Ben McCulloch, Confederate commander of the Indian Territory, to move back east to support operations against the threat of a Union invasion of Arkansas. Cooper reported this in his January 20, 1862, official report of his operations in the Indian Territory:

"In consequence of notice received from General McCulloch that

Fremont was at Springfield with a very large force ; that his advance guard had marched, and that probably his main body would move South the next day; that he (General McCulloch) would obstruct the roads and fight from the line down, but might be obliged to fall back to Boston Mountains, and he having directed me to take position near the Arkansas line, so as to co-operate with him, in connection with the fact that the forage of the country had been destroyed by the enemy and the horses of my command worn down by rapid marches, it was considered improper to pursue the enemy farther, and I returned with the troops to my train at Concharta (sic), which was reached on the 24th of November, 1861."[43]

The orders to move towards the Arkansas and Missouri state lines are questionable. On November 19, 1861, the same date as the Round Mountain battle, Brig, Gen. McCulloch wrote to the Confederate Secretary of War, J.P. Benjamin that the Union army in Southwest Missouri had gone back to their respective bases, and he wanted his units to go into winter quarters.[44] It seems improbable that McCulloch would have ordered Col. Cooper to abandon his operation if there was no threat from General John Fremont's Union army in southwest Missouri. Since the report went directly to the Confederate Secretary of War and was not endorsed or signed off by McCulloch, Cooper probably used this reasoning to justify needing time to get himself and his units re-organized.

93

Opothleyahola and the Loyal Indians gathered themselves together in the following days and slowly moved north and east, finally settling along Bird Creek in the Cherokee Nation, north of Tulsey Town (Tulsa) by the time December came. It is believed that Opothleyahola was still waiting for the Federal assistance from Kansas that had been promised during the meetings during November, even though President Lincoln asked them to move north into Kansas. Also, in a strange twist of fate, the area that Opothleyahola's camp was known locally as "Horseshoe Bend." It seemed like a poor choice of a defensive area coupled with the fact that many of the older Upper Creeks must have remembered what had happened the last time they were in a horseshoe-shaped riverbend during the Creek War in 1814. Fortunately for the Loyal Indians, Colonel Cooper was not Andrew Jackson and did not possess the requisite skills needed to conduct a possible multi-front battle. But this Horseshoe Bend had a few advantages over the one back in Alabama. The banks on both sides of Bird Creek were steep and unstable, which gave rise to one of the battle's names of "Caving Banks," and had few fordable locations. The rear areas of the camp were protected by Hominy Creek to the north and Delaware Creek to the south that also had steep banks with few fords. The Loyal Indians centered their defenses on a house near Bird Creek and fairly close to the middle of their line. The house had a small corn crib and was surrounded by a 400 to 500-yard-long rail fence that dipped into one of the recesses of the prairie. In preparation for the attack they expected, Little Captain and his warriors were to deploy in the heavy tree lines and ravines on the other side of Bird Creek facing the prairies to the east. The Loyal Indians also planned to make some offensive moves if the opportunity presented itself.

Soon Col. Cooper and his Confederate Indian forces were rested and re-supplied but the size of his command had dwindled to less than half of the period before the Round Mountain battle. The 1st Choctaw and Chickasaw Regiment, under the temporary command of Major Mitchell Laflore had 430 soldiers ready for duty; the Choctaw Battalion had 50, the 1st Creek Regiment under Col. D.N McIntosh had 285 present; and 15 Creeks under Captain James M.C. Smith. This gave Cooper a total of 780 men available at Concharty. He also had 500 men of Col. Drew's 1st Cherokee Regiment still stationed at Coody's Bluff and the remnants of the 9th Texas Cavalry were camped along the west bank of Verdigris River at Tullahassee Mission, a sturdy brick building built by the Baptists that had been used as a seminary in the years preceding the war. The Texans were ordered to move north towards Drew's unit at Coody's Bluff. Col. Cooper set his command towards Tulsey Town on November 29. Upon reaching the town, an escaped prisoner from the Loyal Indian camp reported to the Confederates the location of Opothleyahola's followers and their intention to make an immediate attack on Cooper's forces. The commander then sent a series of orders to his subordinate commanders:

> "Colonel Drew was ordered to march from Coody's and form a junction with my command somewhere on the road to James McDaniels. Colonel Sims, then at Mrs. McNair's, on Verdigris, was ordered to join me at David Van's. From some

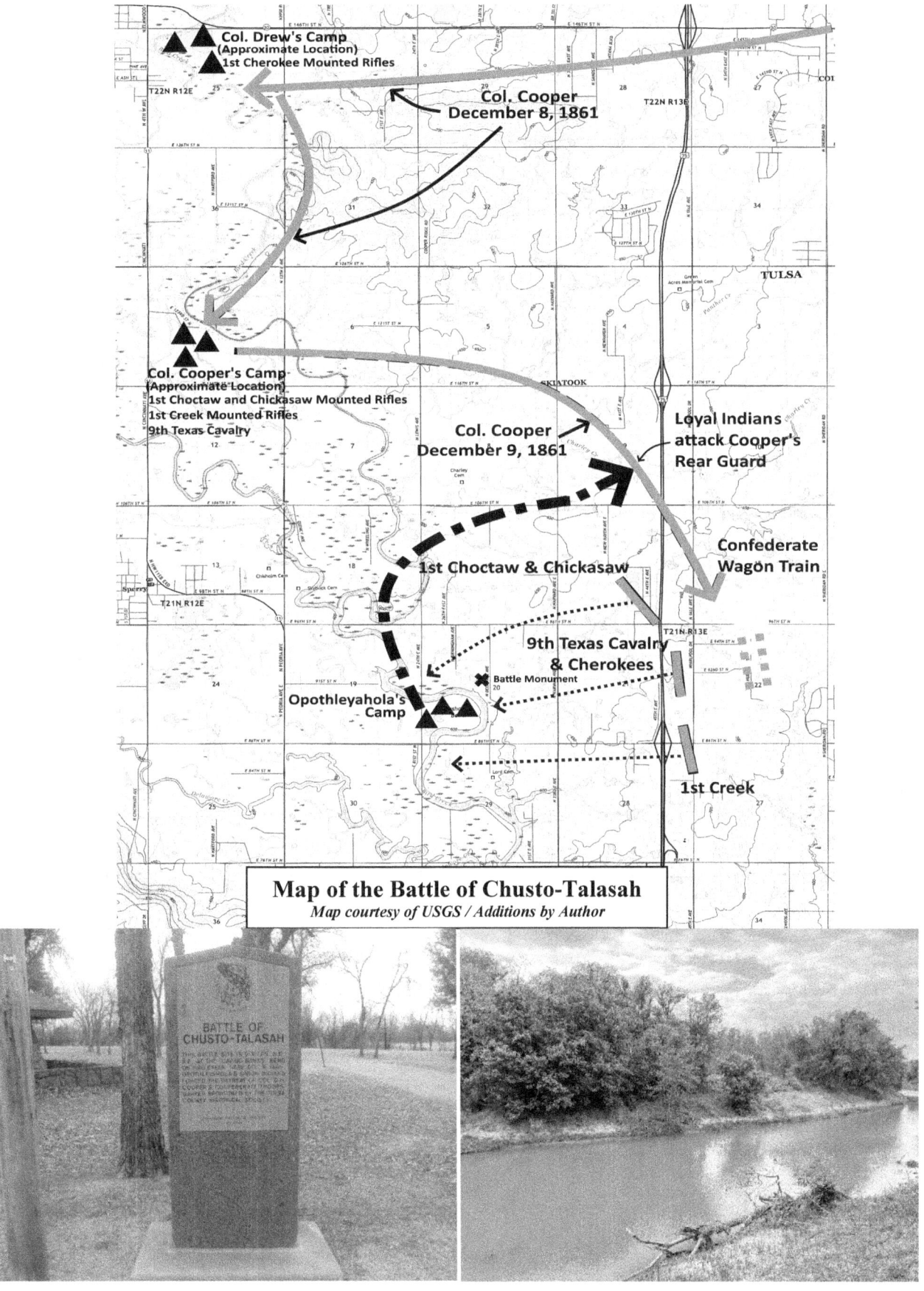

Map of the Battle of Chusto-Talasah
Map courtesy of USGS / Additions by Author

misunderstanding Colonel Drew instead marched direct to Melton's, 6 miles northeast from Hopoeithleyohola.[sic] While following the direction contained in his reply I marched north from Van's to Musgrove's on Caney. Thus he arrived in the immediate vicinity of the enemy twenty-four hours or more in advance of the main body. On the 8th of December, about 12 o'clock, I found him encamped on Bird's Creek."[45]

The exact location of places given in Cooper's report and identified only by owner's names (James McDaniels', Mrs. McNair's, David Van's, Melton's, Musgrove's, etc.) cannot now be determined. It can be difficult to attempt to track these movements on modern maps since most of the locations for particular homeowners have been lost to history. On the other hand, it may be those locations were used as codes since most of Opothleyahola's followers were Creek, and they would not know the locations of particular Cherokee residences in case the written orders were obtained by the Loyal Indians. Regardless, although Cooper was able to connect with Col. Drew's Cherokees, all of the other commands missed their rendezvous with each other. It is at this point that events begin to break down for the Confederates.

Col. Cooper met with Col. Drew on the afternoon of December 8 at the Cherokee encampment. Drew informed Cooper that he had been contacted by Opothleyahola who wished to avoid Indian on Indian blood. The Loyal Creek leader wanted to negotiate a peace agreement at a conference with the Confederates the next day, but wanted to

meet only with Cherokee representatives. Col. Drew was in a tough situation. His regiment consisted of mostly full-bloods and "Pin" Indians, and they had no stomach for fighting the Creek full-bloods. In their view the Creeks were neighbors and posed no danger to the regiment or to the Cherokee Nation. Cooper disregarded this argument and authorized Major Thomas Pegg of the 1st Cherokee Regiment, (some sources state it was Captain James S. Vann),[46] to go to Opothleyahola's camp and arrange the meeting. Maj. Pegg was accompanied by Captains J.P. Davis and George W. Scraper, and Reverand Lewis Downing. In the meantime Cooper moved his force to the west side of Bird Creek, about two miles south of Drew's encampment. When the Cherokee delegation arrived at the Loyal Indian's camp, they were not allowed to see Opothleyahola. Instead, they were met by a large gathering of warriors painted for battle. Major Pegg stated he and his delegation were held until they were promised that the women and children would be allowed to evacuate before any more fighting took place.[47] When the delegation returned to the 1st Cherokee's encampment, of the 480 troops that had been present, they only found about 60 soldiers left. After the delegation had departed the Cherokee camp, a rumor circled through the camp that Opothleyahola and his followers were planning on attacking their camp. At this news, approximately 400 Cherokee soldiers deserted to the Loyal Indian camp while the others either remained or evacuated back to Fort Gibson. They stated that they would not fight against full-blood Creeks and Seminoles whose only offense was loyalty to the United States.[48] Some left with such haste that they abandoned equipment, weapons, and horses in some cases. Word filtered down to Col.

Cooper about the desertion of the vast majority of the 1st Cherokee regiment, and he dispatched a squadron from the 9th Texas Cavalry under Lt. Col. Quayle to proceed to the Cherokee camp and confirm the situation. Enroute the Texans encountered Col. Drew and 28 of his soldiers making their way south to Cooper's camp. Teamsters also arrived in Cooper's camp with a portion of the 1st Cherokee's train.[49] With the threat of an overnight attack by the Loyal Indians, troops were ordered to sleep under arms and ready to deploy. Private Edward Folsom of the 1st Choctaw and Chickasaw Regiment afterwards wrote:

> "We lay in line of battle all night expecting them to attack us. The next morning the Creeks met them and the fight commenced… The northern Creeks ran to a large creek called Bird Creek and commenced fighting. We were in the prairie while they fought from the timber. Col. Sims dismounted his men and away they went I never did see such charge until they reach the brush and the firing was heavy. They could not see anybody. It was not long before they all came running out Col. Sims trying to check them, but could not."[50]

Private Sparks later recalled an incident that long night:

> "…but about midnight we were called into line of battle and told that the regiment of Cherokees under Col. Drue had deserted and all gone to the camp of Hopothleholu (sic) and as they knew our position and force, we

were momentarily expecting an attack. I was too sleepy to stay awake and our officers walked the line to see that every man was in readiness to meet the expected, when Lieut. Haynes of Company I passed along and found me asleep. He was a large man and of fine physique and had endurance equal to any Indian, and to impress me with importance of my vigilance, he in short order, jerked a limb from a sapling and proceeded with a courtmartial [sic], not exactly in military style, but much after the style of a parent with his little boy, he gave me a good whipping, an act that I have always regarded as one of great kindness, it drove sleep from my eyes."[51]

By the morning of December 9, it was apparent that an attack by Opothleyahola was not imminent. At daylight Adjutant R.W. Lee and a small party went north to reconnoiter the Cherokee camp and found it complete and untouched. Col. Drew with some Cherokees and Choctaws and some from the Texas regiment returned to the camp and packed up the camp equipment and brought it to the 1st Cherokee's wagon train near Cooper's camp. At noon it was determined that without the 1st Cherokee regiment the Confederates did not have the strength in numbers to attack Opothleyahola's encampment. Cooper was also worried that his line of communication and supply to the Koweta Mission was threatened with being compromised by the Loyal Indians. The command left their encampment and began moving south with their wagon train in the lead. Cooper led his troops in a wide arc that was at least two miles east from the Loyal Indians' position on Bird Creek in an attempt to avoid contact that would bring about a general engagement. Cooper sent Captain Abram Foster of Company K, 1st Creek Regiment, with his and another company on a reconnaissance towards Park's Store on Shoal Creek to determine if Opothleyahola's force had moved from the mountains down to the creek bed. Captain Foster reported that the Loyal Indians in fact were in the Horseshoe Bend and ready to fight. Cooper and his command team went forward to observe the situation.

The strategy of Opothleyahola, Little Captain, Alligator, and the other war captains was clearly observed and was outlined by Col. D. N. McIntosh in his official report:

> "Without any doubt our enemies had the following advantages over us:
>
> 1st. From all appearances it was a premeditated affair by them. They had placed their forces in a large creek, knowing by marching across the prairie that we would be likely to pass in reach of the place.
>
> 2d. The grounds they had selected were extremely difficult to pass, and in fact most of the banks on the creek were bluff and deep waters, so that no forces could pass across only at some particular points, which were only known to them.
>
> 3d. This place was fortified also with large timber on the side they occupied, and on our side [the] prairie extended to the creek, where the enemies were

bedded, lying in wait for our approach."[52]

Once Colonel Cooper observed the defenses of Opothleyahola's camp he was discouraged. He stated in his report:

> "The position then taken up by the enemy at Chusto-Talasah, or the Caving Banks (the Creeks call the place Fonta-hulwache, Little High Shoals), presented almost insurmountable obstacles to our troops.
>
> The creek made up to the prairie on the side of our approach in an abrupt, precipitous bank, some 30 feet in height, at places cut into steps, reaching near the top and forming a complete parapet...The opposite side, which was occupied by the hostile forces, was densely covered with heavy timber, matted undergrowth, and thickets, and fortified additionally by prostrate logs."[53]

At a point where the Confederate rear guard crossed a fairly large creek (probably Charlie's Creek) close to the edge of the prairie, a large party of nearly 200 Loyal Indians sprung out from the creek's tree line and fiercely attacked the mounted column. Captain Robert Young of Company K, 1st Choctaw and Chickasaw regiment, wheeled around to support the rear guard units. Captain Young reported:

> "On the morning of December [November] 19 I was ordered to bring up the rear with my squadron, and about 6 miles from camp the rear

guard sent me a message that they were attacked by the enemy. I immediately wheeled the squadron and went back to their assistance and got about half a mile, [when] I discovered the enemy retreating towards the creek… and Colonel Cooper rode up and ordered me to charge. After pursuing about 2 miles we came to the creek and I dismounted my men and advanced into the swamp, but not finding the enemy, I ordered the men to return to their horses and mount."[54]

Cooper had no battle plan yet decided to attack the Horseshoe Bend defenses directly. Had he conducted at least a minor leader's reconnaissance he would have better understood the terrain, especially the numerous swamps that bogged down many of his troops in the upcoming fight. Instead, he formed his units in three columns: The 1st Creek Mounted Rifles on the left, the 9th Texas Cavalry and the remnants of the 1st Cherokee Mounted Rifles in the center, and the 1st Choctaw and Chickasaw Mounted Rifles on the right. The three columns advanced at a gallop across the prairie until they reached the thin timberline of Bird Creek when, at about 100 paces, the Loyal Indians released a volley from their prepared defenses that stopped the charging cavalrymen. Most Confederates eventually dismounted and attempted to engage the Loyal Indians but, in standard Indian fashion, they would fire from a secure location then quickly move to another covered location. The fighting was hot and even hand-to-hand in some locations. Since there were no uniforms, it was difficult to know friend from foe, even with the red

and blue strings on the Confederates and the corn shucks on the Loyal Indians. Clouds of black powder smoke filled the ravines and creek bottom adding to the confusion. Eventually, the Confederates were able to cross Bird Creek and the fighting became especially heavy around the abandoned log farmhouse and corncrib. Both sides advanced and retreated across this portion of the Horseshoe Bend numerous times. Captain Young further stated:

"…I received orders from Colonel Cooper to form on the left of the Texas regiment, and in order to get to the left of the Texas regiment I had to pass down the creek, and discovered the regiment coming up to my right, and about the same time discovered the enemy to my right in a bend of the creek, formed around a house. I formed and charged. We routed them from this position and followed them into the swamp 200 yards. They flanked us, and we fell back to the house in order to prevent them from surrounding us. We advanced on them a second time, and were compelled to fall back to the house in consequence of their flanking around. We had only 80 men in the squadron, while the enemy had 400 or 500, fighting us with all the advantages of the creek on us and a complete natural ambuscade to protect them."[55]

Colonel Sims of the 9th Texas Cavalry reported:

"At the commencement of the engagement on the 9th instant with

Hopoeithleyohola's [sic] forces on Bird Creek, Cherokee Nation, in obedience to your commands I proceeded to divide the detachment of my regiment, amounting to about 260 men, into two divisions, [Lieutenant-Colonel Quayle] advanced with his command on to the creek, to the left of the Choctaw regiment. Not finding the enemy there, he returned and charged a ravine on the right of the Choctaws, which he succeeded in taking, under a heavy fire from the enemy. Driving them from their position, he marched on and charged another ravine still farther on the right, but when he got into the ravine the Indians, who had possession of its mouth, opened a raking fire upon his men. He ordered them to charge down the ravine, which they did, and put the enemy to rout. A party of Indians still kept up a heavy fire upon them from the right, who were at first supposed to be Choctaws, as they were wearing our badges, but they were deserted Cherokees and Creeks. In the last charge with Colonel Quayle there were about 20 Choctaws, who acted with the greatest bravery…"[56]

Colonel Sims continued:

"…With the men under my command, to wit, parts of four companies,… after having dismounted I charged to the right of Colonel McIntosh's command and put the Indians to flight without firing a gun. I then ordered my men to mount their horses and moved down, with the Creeks still remaining on their right, about half a mile, where we dismounted, charged into the creek bottom, and put the Indians to flight… We then mounted our horses; it was then reported that the enemy was again advancing. We again dismounted and charged down the creek, putting the Indians completely to rout. We then mounted our horses and advanced up the creek about 1 mile, dismounted, and joined the remainder of my command on the right, who were then fighting on foot in a ravine. We there withstood a heavy fire from the enemy for some time, which finally abated." [57]

It seems that not all was as well for the 9th Texas as Col. Sims described. Private Edward Folsum of Company K, 1st Choctaw and Chickasaw's later recalled what he witnessed:

"The northern Creeks ran to a large creek called Bird Creek and commenced fighting… Col. Sims dismounted his men and away they went. I never did see such [a] charge until they reached the brush the firing was heavy. They could not see anybody. It was not long until they all came running out. Col. Sims trying to check them, but could not."[58]

This pell-mell retreat is not mentioned in any of the official reports.

After four hours of battle, neither side had achieved anything close to a decisive victory, exchanging ground and attempting flanking maneuvers, but Opothleyahola's forces still held their ground when the firing ceased. The fighting ceased as darkness fell on the battlefield. Opothleyahola's followers silently slipped away and moved northwestward from Chusto-Talasah along Hominy Creek about twenty miles to a prominent hill a few miles west-northwest of present-day Skaitook, Oklahoma. At this location they dug in to await supplies of food and ammunition from Kansas. This area was known as Chustenalah, or "natural barrier of rocks." That night a group of 100 Cherokees passed the Confederates and joined Opothleyahola's force, making up for whatever losses they may have experienced. The Confederates retreated back to their wagon train and camped. That night three inches of snow fell on them.

Of course, Col. Cooper claimed a victory since he held the field at the end of the battle. Unfortunately, Cooper did not understand that the Loyal Indians had no need or desire to continue occupying the Horseshoe Bend encampment since they were already moving to a new location. The battle is listed as a draw although Opothleyahola's forces did damage the Confederates enough that they had to retreat to Fort Gibson for reinforcements, ammunition, and other supplies. Col. Cooper estimated the total losses for Chusto-Talasah as 15 killed and 37 wounded. The Loyal Indian's losses are unknown. Cooper's Texas troops were proud of the many Loyal Indian scalps they had taken. Neither the Loyal or Confederate Indians claimed any scalps although some were surely taken by both sides. The dead of both sides were buried on the battlefield although their locations are long forgotten. Cooper also lost most of the 1st Cherokees to the other side. Using the excuses of running low on ammunition and wanting to prevent

more of the pro-Union Cherokees from assisting Opothleyahola, Cooper ordered most of his force back to the protection of Fort Gibson, although he left eight companies of the Choctaw and Chickasaw regiment at Choska.[59]

During the Chusto-Talasah battle about 40 Loyal Indian prisoners were taken. Unfortunately for the prisoners, the Texans turned them over to the Confederate Indians upon their return to Tulsey Town. Private Sparks observed a cruel ceremony:

"It was here that I saw the scene that I have heard described in savage life of running the "gauntlet." I did not see the commencement for I only knew what was going on, when I heard the report of arms and saw the great commotion among the Indians and rode to the scene as soon as I could get to them. From what I saw the Indians had formed themselves in a manner to command the way for the runners and if fleetness save them, they were out of the hands of their tormentors. I saw but one take the run, he was a long slim Indian prisoner, I do not know his tribe and I do not know whether it was his first run or not, but am of the opinion that he had made the race with his companions and had run through unhurt and had been recaptured and given a second run, as many had been killed before I got there, but let that be as it may, I saw him run and 'Oh my God!' it was a run not only for life, but from cruel captors. As he started, clubs and tomahawks were hurled at him, knives and stones, then arrows from bows, and after that, guns were fired at him... the runner made it out with a knock-down and ran like a ghost and a great howling multitude after him… he hid himself amid the roots of the oak under the embankment. He was followed, dragged out and dispatched among a howling din of his captors. My heart was touched with such cruelty."[60]

On December 11, 1861, Col. Cooper sent an express message to Col. James McIntosh (no relation to the Creeks), commander of the regular Confederate forces, who was in winter quarters with his brigade at Van Buren, Arkansas. Cooper requested ammunition and other supplies from Fort Smith. He also confided that he needed reinforcements consisting of white soldiers to help him stiffen up his Indian Brigade. Col. J. McIntosh responded quickly that he would send significant reinforcements to Cooper's brigade along with ensuring that ammunition and other supplies were issued from Fort Smith. In the meantime, Cooper made arrangements to meet with Principal Chief John Ross at Park Hill to discuss the situation in the Cherokee Nation and how to reorganize and re-equip Col. Drew's 1st Cherokee Mounted Rifles. Following a different route to the post, Colonel Sims's column arrived at Fort Gibson on December 13, where his 250 measles-stricken men were quartered in the abandoned garrison buildings. For ten days, the men were allowed to rest and try to keep warm, "awaiting upon the sick and burying the dead." Private Sparks recalled that "our flag was nearly always at half-mast and a funeral procession was of daily occurrence."[61]

Col. James McIntosh was an interesting character. He came from a military family; he was born at Fort Brooke, now Tampa, Florida, in 1828. His father had been killed during the Mexican War and his brother was serving as a Union officer. He graduated in last place from West Point in 1849 and was serving as a Captain with the 1st U.S. Cavalry of the Regular U.S. Army when he resigned on May 7, 1861. He was soon commissioned a Colonel in the Regular Confederate Army and was in command of the 2nd Arkansas Mounted Rifles Regiment.[62] He was regarded as ambitious and aggressive. McIntosh arrived in Indian Territory on December 20, 1861, with thirteen hundred and eighty Texas and Arkansas veterans of the Battle of Wilsons Creek in Missouri consisting of parts of the 6th and 11th Texas Cavalry, the South Kansas-Texas or 3rd Texas Cavalry regiment, and his own 2nd Arkansas Mounted Rifles regiment. The plan was to use Cooper's and McIntosh's troops in a pincer movement with Opothleyahola's forces in between. McIntosh assigned Major John Whitfield's battalion of the 27th Texas Cavalry to Cooper. With these reinforcements Cooper was to move his command up the Arkansas River to block Opothleyahola's forces from retreating while McIntosh moved up the Verdigris River to attack the Loyal Indians from the front. Col. McIntosh's units began their march at noon on December 22 and spent most of that day fording the Grand and Verdigris Rivers. The water and banks of the latter river were so steep that his troops used windlasses to winch them down, across and back up the opposite side. McIntosh then stated in his report to Brig. Gen. McCulloch that he proceeded towards Opothleyahola's new encampment at Chustenalah without waiting for Cooper, claiming the latter had been delayed when his wagon train teamsters deserted. Cooper's report does not mention this event although he was delayed at Choska waiting for the promised ammunition to arrive. The delayed ammunition arrived on December 23, and Cooper's force advanced towards Tulsey Town the next day. Cooper also sent a message to Col. Stand Watie to advance his 2nd Cherokee Mounted Rifles to join up with Col. McIntosh's command.

A private Texas soldier recalled an interesting Christmas story. On the night of Christmas Eve, Cooper and his command were camped in a grove of trees along a stream that emptied into the Arkansas River. In the 9th Texas' camp, some of the men wanted to celebrate the Christmas holiday. In a festive spirit they began to make loud noise as only partying soldiers can make. Private Sparks continues:

"Col. Sims at once sent us orders to retire without further noise. This order was received with great protest. Not even free on Christmas. So we thought best to send a delegation to Col. Sims and ask him to allow us a little recreation, as it was Christmas. This writer was with the delegation who went to Col. Sims' quarters with the request. I well remember him reclining before his campfire, half dressed and wrapped with his blanket, with saddle, sword and pistols all within easy reach,.. On making our errand known to him he only arose to a sitting posture on his couch, and his answer was: "No! Sons, No! remember you are soldiers, and I, as your commander, have promised to

keep you at all times in a manner that you shall be able to render to your country the most effective service, and while we rejoice with the season, we must make no demonstration, for we are in the front of a savage enemy and know not when he may strike at us. No! go to your beds and sleep, and husband all your energies for the hard service that is yet before us."[63]

On the evening of Christmas Day, Col. McIntosh's column made camp in the area of Mingo Creek, approximately 12 miles from the Loyal Indian's stronghold, when a party of Opothleyahola's force was observed close to the Confederate camp. McIntosh ordered one regiment to follow but warned them to avoid an ambush, which was a standard Indian military tactic. The force moved westward in the direction they believed the Loyal Indian force was located but stopped short that evening when darkness came and then returned to camp.

By looking at the map and the distances Opothleyahola's followers were moving, it is obvious that they did not intend to go to Kansas unless they absolutely had to. Although the Lincoln administration wanted them to move to Fort Row on Walnut Creek in Kansas, it is apparent that Opothleyahola was killing time, believing that help was on the way from the Union forces up north. Unfortunately, the Union army did not have any help to send to them. Most troops on both sides had moved into winter quarters and were not actively campaigning. Regardless, the site that Opothleyahola and his war captains had picked out was a very good location to defend their new encampment. It was on a steep and rocky hill

(300+ feet) and heavily forested with scrub oak and thickets. The massive hill overlooked a narrow valley at the junction of Shoal (today's Battle Creek) and Quapaw Creeks a few miles west of the present-day city of Skaitook, Oklahoma. They created a fortified position on the massive outcropping where any attacker would be forced to cross Shoal Creek then advance 200-300 yards of open prairie under fire from above. Unfortunately, for the Loyal Indians, their leaders and the war captains had failed to properly make use of many of the advantageous terrain features that the area provided. These failures and shortcomings may be traced to fatigue as well as a lack of arms, ammunition, food and other supplies that had been lost or used up during the journey and the battles they were forced to fight. These failures would lead to a bloody defeat for Opothleyahola and the Loyal Indians on the day after Christmas, 1861.

On Christmas Day the Confederates under McIntosh approached the Osage Hills located northwest of present-day Tulsa. As the column progressed they encountered increasing signs and evidence of the Loyal Indians roaming and scouting in the area. That night McIntosh ordered his men into a closely guarded bivouac in a grove of blackjack, where a blizzard, freezing rain and sleet storm made life miserable and sleep impossible for the entire night. Bugler Albert Blocker of the 3rd Texas Cavalry later recalled:

"…the sky became overcast, … and the wind began to rise out of the northwest, … and turned loose on us a regular northern blizzard, with snow and icicles thrown in. About midnight, all of us … were 'frozen out' and had to get up and face the cutting wind,

which at times would almost lift us off our feet. Our fires were no benefit to us, as the wind had blown them out. We folded our blankets around us and got on the south side of trees, logs or anything that would break the force of the bitterly-cold wind, and sat it out until daylight."[64]

The night had been the coldest that many could remember. The morning of December 26 dawned very cold with periods of sleet hammering all combatants. McIntosh decided to keep his wagon train at the Mingo Creek camp with a 100-man security force commanded by Capt. Elstner, the acting brigade quartermaster. The Confederate commander also ordered his troops to cook and pack four days rations for the upcoming advance. The Confederates left the camp at Mingo Creek and advanced towards the Loyal Indians' position and arrived near that place at 12 noon. Col. McIntosh failed to wait for the arrival of Col. Stand Watie and his 2nd Cherokee Mounted Rifles regiment who had been ordered by Col. Cooper to link up with the column before the attack. Watie learned that McIntosh's force was about 6 miles in advance, but the Confederate Indians failed to catch up until the battle was over. In the meantime, Col. McIntosh sent the 3rd Texas Cavalry (South Kansas-Texas) regiment forward with a Capt. Short's Company E sent out as an advanced guard with flankers to protect the advancing Confederates from an ambush. As Company E crossed Shoal Creek they came under fire from the Loyal Indians up on the mountain. The advance company held its ground while Col. McIntosh sent the 6th Texas Cavalry under Lieut. Col. Griffith

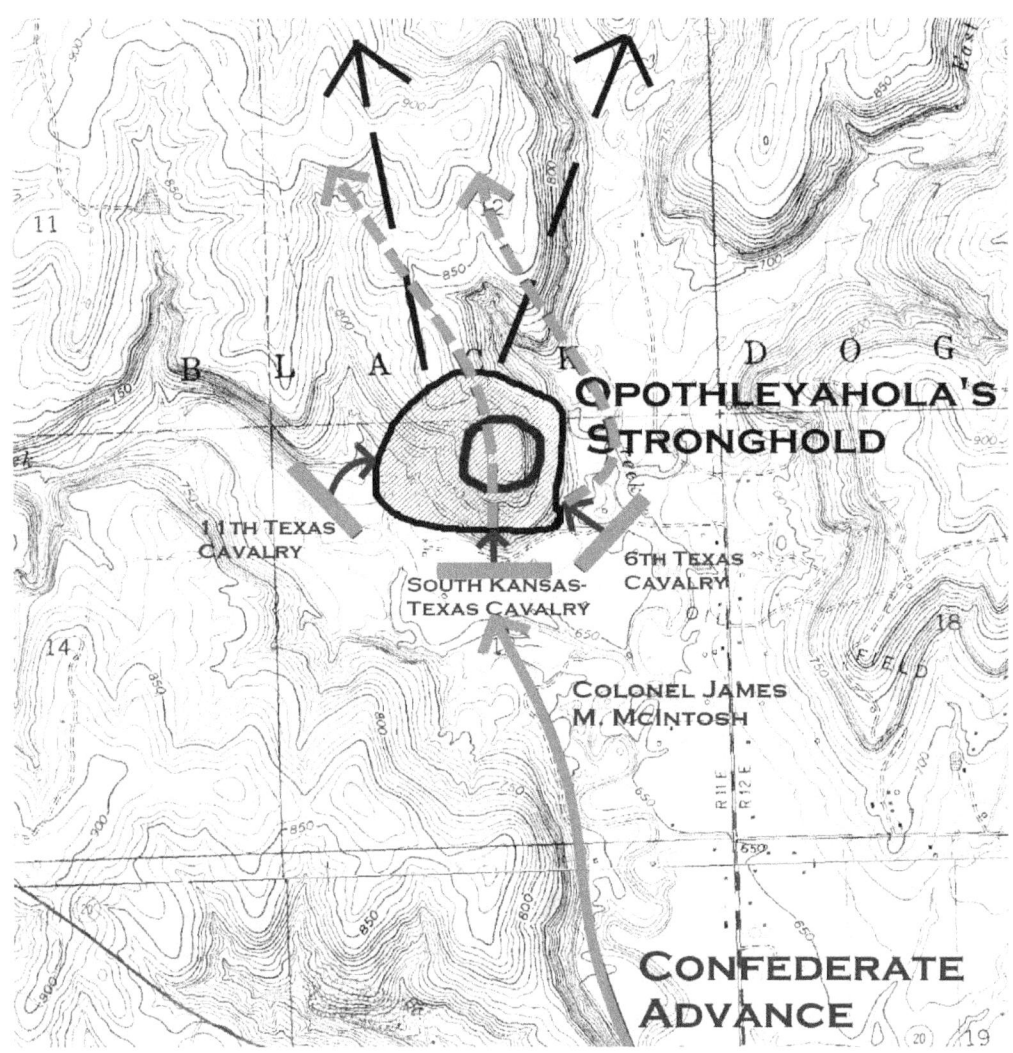

Map of the Battle of Chustenalah
December 26, 1861
Map courtesy of USGS / Additions by Author

Opothleyahola's Stronghold on Chustenalah Mountain in Osage County, Oklahoma
Photo by Author

around to the right, up today's Quapaw Creek. He was to cross the creek, dismount, and attack the left flank of the stronghold. Colonel Young was ordered to take his 11th Texas Cavalry to the left, up Shoal Creek, and attack the stronghold's right flank. Col. McIntosh maintained command of the center of the battleline consisting of the 3rd Texas Cavalry, under Lieut. Col. Walter Lane, four companies of the 2nd Arkansas Mounted Rifles, and Capt. H.S. Bennett's Independent Texas Company. Once in position the Texans and Arkansans halted and started fires to warm themselves and to melt the ice that covered all of their equipment, especially the horses. After a short period, all of these Confederate forces moved forward across the open prairie towards the Loyal Indian's positions on the hill. The Confederates could see the mounted Loyal Creeks on the hillside and others on foot along the slopes. They recognized the Seminoles under Halleck Tustenuggee on foot near the center of the Loyal Indian's line fighting from among the trees and rocks. After about an hour of skirmishing, the Confederate commander at about 2 PM ordered a charge of all units to the summit of the stronghold. A yell from over a thousand Confederate voices was heard as the charge was called, and all of McIntosh's troops swept forward towards Opothleyahola's position on the steep hill. The Confederates swept up the hillside and over the top, the 3rd Texas being the first to reach the summit. Col. Lane claimed that his regiment and the units attached to his center division struck the strongest of Opothleyahola's positions. Col. Lane recalled:

"…Many of the enemy made for their stronghold on the top of the hill, where there was a natural breastwork of rocks, and fired over the rocks at us. Many of my men, without making any halt-, gained the heights by the few narrow entrances on the side where it was alone accessible, while others dismounted and scaled the rock, and here for a short time a desperate struggle ensued. Many shots were fired when the contending parties were only in a few steps of each other, and in some instances they were engaged in a hand-to-hand struggle. Soon the point was cleared by us, and the enemy retreated in great confusion, some of them making a stand for a short time in the deep gorges and rocky defiles of the mountains…"[65]

It was not as easy for the soldiers of Companies A and B, who found themselves stalled by the steepness of the slope in their sector. Major Chilton quickly ordered the men to dismount and led the two companies up the bluff on foot, leaving their mounts in the rear under the care of horse-holders. Captain H.L. Taylor of the 3rd Texas Cavalry remembered,

"Col. Walter Lane, one of the oldest and best Indian fighters on earth, without waiting orders to advance, had his gallant little bugler, Charlie Watts, to sound charge; and as [o]ne man Col. Lane led our boys to the assault. The Indians were scattered in full retreat. We followed them in a running fight and would have killed most of them had it not been for the splendid tactics practiced by them."[66]

On the left Col. Young moved up

Shoal Creek and advanced up the hill attacking the stronghold's right flank. Col. Young of the 11th Texas Cavalry reported:

> "...I took up my position on the left, according to your instructions, at the commencement of the action. I remained there until the woods were on fire, and being satisfied that the enemy did not intend an attack on our left, I moved my regiment in the direction of the mountains, on the right... I moved my regiment in an oblique direction through the mountains, where, after going some 2 miles, we came up with the enemy, strongly posted among the rocks and timber. We immediately charged them, carrying everything before us. After this the enemy, being completely routed..."[67]

In response to the sweeping Confederate charge from three sides, the Loyal Indians fell back from their fighting positions towards their encampments on the far side of the hill. One portion made a brave attempt to protect their principal encampment and supply area from Lieut. Col. Griffith's 6th Texas Cavalry's advance but were overrun by the hard charging Confederates. Griffith later recalled:

> "...I was ordered by you to move my command up on the right of and parallel with Colonel Lane's command. This executed brought me to Hominy Creek, when 1 was further ordered to dismount my men and form a line... I ordered my men to their horses, formed, and rapidly advanced in a flanking movement you

intended for me to make up the valley for half a mile, crossed over to the west, or battle side of the creek, proceeded a short distance up, and discovered the enemy upon the opposite bank. I charged across the creek, put the enemy to rout, continued up the valley something like a half mile farther, cutting off all the straggling and then flying Indians in that direction.

> I then turned to the left in a northwestward direction over the rocky hills and gorges that made into the larger gorge that was then in between Colonel Lane's command and mine. Continuing this course, I crossed over five or six rocky hills, on three of which, behind the rocks, the enemy were in position in considerable numbers. My men gallantly charged in succession,.."[68]

The 6th Texas continued to pursue to the northwest passing onto the heights far beyond the stronghold. Small pockets of the Loyal Indians fought for the next two to three hours but in the end but were eventually neutralized, either by death or capture. In a final effort the Loyal Indians ignited the scrub brush and prairie on fire to slow down the advancing Confederates to no avail. Private Hillary Taylor of the Third Cavalry remembered it, "the flames would sweep rapidly over the gorges, and we were in the midst of it. We would then have to retreat as rapidly as possible ... and ... wait for the fire to cool down before we could advance." At one of the final defenses, Griffith had his troops dismount and engage the Loyal Indians at about 125 yards. The remnants of

Opothleyahola's followers streamed to the north and west to avoid the Confederates. Many were caught and many were killed in the first hours after the battle. Private Doug Cater was impressed by the Loyal Indian's last stand. He remembered:

> "some of those Indians were very brave and daring and would not leave, but continued to shoot. . . . One big feathered-cap fellow stood out from the trees and continued shooting until he fell. I had shot both barrels of my gun and one of my holster pistols at him before he fell."[69]

McIntosh's force captured the remains of the Loyal Indian's women and children, supplies, oxen, horses, cattle, and other livestock. Those that escaped had done so without any supplies or food, having fully believed that their stronghold was impenetrable. The morning's sleet soon turned to snow and high winds, further exacerbating an already difficult situation.

Col. McIntosh's buglers sounded recall at approximately 4 PM. Col. Watie and his Confederate Cherokees arrived just as the battle was ending. In the days that followed Stand Watie's Mixed-Blood 2nd Cherokee Regiment attempted to run down the remnants of Opothleyahola's force. He was successful in killing or capturing the vast majority of them. In all, over seven hundred Loyalist Indians, including Seminole chief Alligator, perished in the last battle and in the days following. John Chupco, a Seminole ally of the Creek leader, lamented a few years later that at Chustenahlah:

> "…we lost a great many of our law men, and capable men to do business, and a great many of our young men, and women and children. We left them in cold blood by the wayside. At that battle we lost everything we possessed, everything to take care of our women and children with, and all that we had."[70]

By late January 1862, the last of the Loyal Indians able to escape had crossed into the relative safety of Kansas on what became known as the "Trail of Blood and Ice." The Confederate forces in Indian Territory had won their first campaign of the war, and the Territory was firmly in their control.[71] One treasure that fell into the Confederate's hand was a collection of three letters, dated September 1861 from E.H. Carruth, the Indian agent in Kansas, that were found in Opothleyahola's encampment. These letters were addressed to Opothleyahola, the chief of the Choctaws and Chickasaws, and the chief of the Wichita. These letters promised assistance from the United States Government to the Loyal Indians.[72]

One of the primary failures of Opothleyahola's war captains was failure to use the heights surrounding the battle area. Any attacking units would be forced into the narrow valley of the Quapaw Creek from the south to approach the front of Opothleyahola's stronghold. Little Captain and the other war captains failed to take advantage of a height to the west of the probable line of advance, today's Patriot Hill, to establish a crossfire ambush on any force approaching the main stronghold. An even greater cross-fire opportunity existed along Shoal, or Battle Creek where the 11th Texas Cavalry advanced to get to the stronghold's right flank. There were strong and steep hillsides along both sides of the creek that would've placed that unit in extreme danger. Had the 11ᵗʰ Texas been placed into disarray, the battle might have gone a different way. But the Loyal Indians were not soldiers. Opothleyahola and his war captains were not trained military officers. They did the best they could with the resources they had available. It may never be known why Opothleyahola and his followers failed to retreat to the relative safety of Union Kansas, especially after the first two battles. Interviews were conducted after the war but even with the information those provided, an actual reason has never been discovered.

The complete massacre of Opothleyahola's followers, who were in fact civilians and had not violated any law, was almost complete. Once Stand Watie and his mixed-blood Cherokee regiment arrived they systematically killed and captured many of the refugees. The reports submitted by all of the pursuing Confederates, white and Indian, were very formal and proper, unemotionally stating only the facts. An interesting facet to Douglas Cooper's unfounded personal drive to destroy Opothleyahola's followers is that they did not pose any threat to the Confederate-aligned Indians. Cooper insisted that the dissident group either surrender or he would compel submission or drive them out of the Indian Territory. To leave was exactly what Opothleyahola and his followers wanted: to find a neutral and safe, haven. Although there were warrior-aged men in the encampments, the vast majority of the refugees were old men, African slaves, women, and children, who were hardly a threat to the Confederate States. It is believed that he took offense to any Indian leader who did not share his vision for an Indian

Territory under the Confederacy. He used the unconfirmed rumors, most which were planted by pro-Southern Indians, to justify his attacks on "civilians" to the Confederate War Department. In reality, Opothleyahola never made any offensive move against the Confederate Indians but simply offered an alternative route for the affected Indians and did not pose any military risk to the Confederacy.[73] It was the bad blood between the Creek McIntosh's and the mixed-bloods and Opothleyahola and the full-bloods that dated back to the Creek War of 1813-1814, and was compounded by the removal treaties and the killing of Chief William McIntosh that kept the hatred burning among both sides.

From reading the correspondence from Co. Cooper and Col. James McIntosh, they speak of each other in generous terms. But reading between the lines and observing the actions of both parties, a person can see that this was not a joint operation. Col. McIntosh probably believed that after Cooper's two defeats and the loss of control of his Indian troops, especially the 1ˢᵗ Cherokee Mounted Rifles, that the Confederate Indian Brigade was not a competent fighting force. Cooper kept asking for white soldiers to firm up his Indian soldiers, so Col. McIntosh probably already held a low opinion of the Indian troops. As a Regular U.S. Army officer serving on the frontier, he probably had no great love of any Indian, civilized or not. McIntosh went so far as to order Cooper and the Indian Brigade in a wide circle advance while the Texas and Arkansas units took a direct route and attacked without any coordination with Cooper. McIntosh also failed to wait a couple of hours for Col. Watie and the 2ⁿᵈ Cherokee Mounted Rifles. History indicates that Col.

McIntosh simply did not want any Indian troops under his command.

One cannot help but be astonished by the level of brutality that marked the first year of the Civil War in Indian Territory. The Confederate forces, both Indian and white, were making war on the Loyal Indians whose only crime was a desire to remain neutral and leave the Territory. The Loyal Indians were not soldiers. They were family men, women, children, and elderly who were trying to get out of the way of the war but the lines had already been drawn between the groups before the move west under the new treaties. Life remained hard on those staying behind in the Five Nations during the war since mini-civil wars broke out within the tribes, usually pitting Full-Bloods against Mixed-Bloods, the aggressor being whoever, Union or Confederate, happened to be in command of the area.

In a final report, Col. James McIntosh stated they had captured and turned over to the commissary at Fort Gibson, 190 head of sheep, between 800-900 head of cattle, and 250 Indian ponies. Col. McIntosh finally ordered his Texans, including the 9th Texas, and his Arkansas regiment to return to winter quarters in Arkansas. As he moved towards his destination, Col. McIntosh met with Col. Cooper on the trail and after exchanging greetings and information, each proceeded towards their designated winter quarters.

[1] Miles to Townsend, May 10, 1861. United States War Department. War of the Rebellion: A Compilation of the Official Records of the Union and Confederate Armies, Washington D.C.: U.S. Government Printing Office. 1888-1901. Series I, Vol. III, pp. 369.

[2] Trickett #1, pp. 151.

[3] Joseph B. Thoburn, ed., "The Cherokee Question," Chronicles of Oklahoma (Oklahoma City), II (1924), pp. 186-187. Ross' Address to Drew's Regiment, December 19, 1861.

[4] Debo, Angie. The Road to Disappearance: A History of the Creek Indians, Norman: The University of Oklahoma Press, 1941. Pp. 147-148.

[5] The original of this letter was found in Opothleyahola's camp after the battle of Chustenahlah, Dec. 26, 1861. OR, Ser. I, Vol. VIII, pp. 25.

[6] "Report of the Commissioner of Indian Affairs," in Report of the Secretary of the Interior, 1861 (Washington: Government Printing Office, 1862), 655. Jones to Dole, Oct. 31, 1861. Trickett, #3, pp. 152-153.

[7] Monaghan, Jay. Civil War on the Western Border, 1854-1865. Lincoln and London: University of Nebraska Press. 1955. pp. 25-26.

[8] Debo, pp. 148-149.

[9] Colonel James Montgomery was a Jayhawking zealot. During the Bleeding Kansas period he led a number of raids into the Missouri border areas to burn plantations and free any enslaved Africans.

[10] Debo, pp. 149.

[11] Report of the Commissioner of Indian Affairs, Report of the Secretary of the Interior, 1862, Washington: Government Printing office, 1863, pp. 138. Cutler to Coffin, Sept. 30, 1862. Trickett #3, pp. 153.

[12] Bahos, Charles, "On Opothleyahola's Trail: Locating the Battle of Round Mountains," Chronicles of Oklahoma, Spring 1985. Pp. 60-63.

[13] Confer, pp. 34-36.

[14] Warde, pp. 69-71.

[15] Foreman, Grant. The Five Civilized Tribes, Norman: University of Oklahoma Press, 1934, pp. 236. Trickett, 1861-4, 269.

[16] Cooper to Drew, letter, October 29, 1861

[17] John Ross to John Drew, letter, October 20, 1861, Grant Forman Papers, Gilcrease Museum

[18] Wright, Murial H. "Colonel Cooper's Report on the Battle of Round Mountain," Chronicles of Oklahoma, article, Winter 1961; Oklahoma City, Oklahoma. pp. 377.

[19] Wright, pp. 378.

[20] OR, Ser. I, Vol. VIII, pp. 719-720.

[21] Hale, Douglas, "Rehearsal for Civil War: The Texas Cavalry in the Indian Territory, 1861." Chronicles of Oklahoma, article, Autumn 1990; Oklahoma City, Oklahoma. pp. 231.

[22] Hale, pp. 235.

[23] Sparks, A.W. Recollections of the Great War:

The War Between the States as I Saw It, Tyler: Lee & Burnett, Printers. 1901. pp. 14.

[24] Sparks, pp. 16.

[25] Hale, pp. 236.

[26] Sparks, pp. 16.

[27] Sparks, pp. 17.

[28] Wright, pp. 380.

[29] Wright, pp 381.

[30] Wright, pp. 381.

[31] Confer, pp. 36.

[32] This scenario is the opinion of the author based on reports, maps, terrain and times, as well as from personal physical knowledge of the battlefield area.

[33] Sparks, pp. 18.

[34] William A. Quayle. Report, January 1862, Special Microfilm Collections, Roll IAD-5, Oklahoma Historical Society.

[35] Ibid.

[36] Brinson, M.J. report to Douglas Cooper, November 25, 1861. OR, Ser. I, Vol. 8, pp. 14.

[37] Bates, James C. A Texas Cavalry Officer's Civil War: The Diary and Letters of James C. Bates. Richard Lowe, editor, Baton Rouge: Louisiana State University, 1999, pp. 22. Edwards, Whit. The Prairie was on Fire: Eyewitness Accounts of the Civil War in Indian Territory. Oklahoma City: Oklahoma Historical Society, 2001. pp. 4-5.

[38] Grisom, George L. Fighting with Ross' Texas Cavalry Brigade, CSA: The Diary of George L. Grisom, Adjutant, Ninth Texas Cavalry Regiment. Homer L. Kerr, editor, Hillsboro, Texas: Hill Junior College Press, 1976. pp. 5-6. Edwards, pp. 5.

[39] Young, R.A. to Douglas Cooper, report, OR, Ser. I, Vol. VIII, pp. 14-15.

[40] Gregory, Judge James R. Galveston News, November 27, 1901.

[41] Cooper, Douglas H. Report, January 20, 1862, OR, Ser. I Vol. VIII, pp. 5-7.

[42] Adair, W.S. "Civil War Repeated in the Indian Territory," Dallas Morning News, July 1, 1923.

[43] Cooper, Douglas H. Report, January 20, 1861, OR, Ser. I Vol. VIII, pp. 7-11.

[44] McCulloch, Benjamin, to J.P. Walker, CSA Secretary of War, Ser. I. Vol. VIII, pp. 686.

[45] Cooper, Douglas H. Report, January 20, 1861, OR, Ser. I Vol. VIII, pp. 7-11.

[46] Bates, pp. 23.

[47] It should be noted that some historians, Murial Wright among them, believe that Opothleyahola was not with the Loyal Indians but had already departed either to the next defensive area or was already on his way to Kansas.

[48] Drew, John. Report, December 18, 1861, OR, Ser. I Vol. VIII, pp. 16-18.

[49] Drew, pp 17.

[50] Folsom, Edward A. "Reminiscences of E.A. Folsom," n.d. E.E. Dale Collection, Box 218, F17, University of Oklahoma/Western History Collections. Also Edwards, Whit, pp. 8.

[51] Sparks, pp. 21.

[52] McIntosh, D.N. Report, December 16, 1861, OR, Ser. I Vol. VIII, pp. 16.

[53] Cooper, Douglas H. Report, January 20, 1861, OR, Ser. I Vol. VIII, pp. 9.

[54] Young, R.A. to Douglas Cooper, report, OR, Ser. I, Vol. VIII, pp. 15.

[55] Ibid.

[56] Sims to Cooper, OR, Ser. I Vol. VIII, pp. 18-19.

[57] Ibid

[58] Folsom, n.d. Edwards, Whit. pp. 8.

[59] Cooper, Douglas H. Report, January 20, 1861, OR, Ser. I Vol. VIII, pp. 9.

[60] Sparks, pp. 26-27.

[61] Hale, pp. 245. Sparks, pp. 27.

[62] Warner, Ezra J. Generals in Gray: Lives of Confederate Commanders, Baton Rouge and London: Louisiana State University Press. 1987. Pp. 202-203.

[63] Sparks, pp. 25.

[64] Lale, Max S. "The Boy-Bugler of the Third Texas Cavalry: The A.B. Blocker Narrative." Military History of Texas and the Southwest, Vol. XIV, No. 2. 1978.

[65] Lane, Walter P. Report, December 26, 1861. OR, Ser. I Vol. VIII, pp. 28-29.

[66] Lane, Walter P. Report, December 28, 1861. OR, Ser. I Vol. VIII, pp. 649-651.

[67] Young, W.C. to Cooper, report, OR, Ser. I, Vol. VIII, pp. 26.

[68] Griffith, John S. to J. McIntosh, OR, Ser. I, Vol. VIII, pp. 27-28.

[69] Cater, D.J. "The Battle of Chustenalah," Confederate Veteran, June 1930. pp. 233.

[70] Foreman, Carolyn Thomas, "John Jumper" Chronicles of Oklahoma, Summer 1951, pp. 143

[71] Griffith, John S. to J. McIntosh, OR, Ser. I, Vol. VIII, pp. 12-13.

[72] Carrurth, E.H. to Opothleyahola, Chief of the Choctaw and Chickasaws, Chief of the Wichitas. Letters, OR, Ser. I, Vol. VIII, pp. 26-27.

[73] Confer, pp. 59-60.

The Telegraph/Wire Road through Pea Ridge
National Military Park *Photo by Author*

Historic location of Leetown on the Pea Ridge
battlefield *Photo by Author*

Historic views of Fort Davis, located near Muskogee, Oklahoma
Photos courtesy of the Oklahoma Historical Society

Chapter 5

Albert Pike's Confederate Indian Brigade and the Battle of Pea Ridge

After Brig. Gen. Albert Pike had been selected as the commander of the Confederate Department of the Indian Territory in November 1861, he decided to move from Fort Gibson and chose a new location to make the military and civil headquarters of the Confederacy for the Indian Territory. Fort Gibson had been abandoned by the U.S. Army in 1857, so its buildings were not in good shape and many Cherokee families had taken up residence on the old military reservation. Instead, Pike chose a location approximately seven miles west of Fort Gibson on the south side of the Arkansas River, one or two miles upstream from the Grand River. It lay between the current city of Muskogee, Oklahoma, and the Arkansas River. He designated the site as Cantonment Davis, and later, Fort Davis, in honor of Confederate President Jefferson Davis. The headquarters buildings were built on the south side of the Arkansas River because Pike believed it would stop, or at least delay, any attack from the other side of the river. It was constructed by white soldiers and civilians contracted from Arkansas and Texas. They built fortifications and erected a large number of log and plank buildings for barracks, kitchens, a post hospital, quartermaster and commissary stores, stables, officer's quarters, and other purposes. Thirteen structures surrounded the parade ground and the flagpole stood atop a prehistoric burial

mound. Two wells were also dug to provide water to the garrison. The fort was shaped like a U, with an open end facing to the east. It overlooked the Texas Road crossings of the Arkansas River and the east slope fell away gently toward the Arkansas River and Fort Gibson, leaving an unobstructed view across the river to the old fort and to the surrounding prairies. The height of the compound kept much of the interior of the fort screened from Fort Gibson or other observers. The west side was also a gentle slope where large numbers of troops could be mustered out of sight of Fort Gibson. Pike arrived on February 24, 1862, and took command of Fort Davis. From here he planned to oversee his new command and conduct the operations of his Indian Brigade. Although Pike intended all five of his regiments to be stationed at Fort Davis, most times only a small number of Indian soldiers were actually garrisoned there due to military commitments throughout the Indian Territory. Confederate Secretary of War Judah Benjamin designated Maj. N.B. Pierce as the Department of Indian Territory's chief of commissary and subsistence, and Maj. G.W. Clarke as the post quartermaster.[1] One contemporary familiar with the situation stated that the Confederate government invested nearly one million dollars in Fort Davis.

It is surprising that the 1st and 2nd Cherokee Mounted Rifles would work together at this critical time. There were still many lingering hard feelings between the two regiments, resulting from not only the traditional division between mixed-blood and Pin Indians, but from the four hundred Indian soldiers of the 1st Cherokees that had deserted en-masse prior to the Battle of Chusto-Talasah. In February 1862 the 2nd Cherokees were stationed in the Delaware District of the Cherokee Nation, attempting to prevent any more Cherokees from fleeing to Kansas. At some point during the early part of the month, Col. Stand Watie's nephew, Charles Webber, killed and scalped Chunestootie, one of Col. John Drew's regiment who had gone over to Opothleyahola's side. Col. Watie was not pleased that the murder had occurred and stated "it does not tend to reconcile the factions already too bitter for the good of the country." Watie also declared that Chunestootie had been a vocal supporter of the pro-Union faction and had "been for years hostile to southern people and their institutions." It was further claimed that Chunestootie belonged to a radical group that swore to murder "any and all who should attempt to raise a southern flag." The murder angered everyone in the 1st Cherokees, and Col. Drew stated that it was a "barbarous crime." Col. Watie refused to take any corrective action since it was "futile at this time to think of settling this affair according to the usual course of the law." Although no actions were taken by either side, a burning anger simmered beneath the surface of the men of both regiments. That energy would need to be put on hold as they approached their next challenge.[2]

There had also been discussions between Chief Ross of the Cherokee Nation, the National Council, and Col. Douglas Cooper regarding the deserters from the 1st Cherokee Mounted Rifles who had left just before the Battle of Chusto-Talasah. Of the 400 or so Indian soldiers who deserted, only about 150 actually joined Opothleyahola. The others, including Chunestootie, simply returned home. Col. Stand Watie wanted the deserters rounded up and executed. Any decision was postponed on Chief Ross's insistence since a measles epidemic was sweeping through the Confederate camps. Chief Ross needed to re-constitute the 1st Cherokee regiment and pardoned any of the deserters who rejoined the unit. Ross needed the 1st Cherokees to counter Watie's 2nd Cherokees.[3]

The Confederate Army in the Trans-Mississippi region had not been idle over the Winter of 1861-1862. After their victory at the Battle of Wilson's Creek in August 1861, the Confederates and the Missouri State Guard under Maj. Gen. Sterling Price remained in the southwest portion of Missouri with headquarters at Springfield. Price remained hopeful he could re-conquer all of Missouri and officially bring it into the Confederacy. On January 10, 1862, the Confederate War Department created the Trans-Mississippi District, which included most of Arkansas and Missouri, parts of Louisiana, and the Indian Territory. The new District fell under Department #2, which was commanded by Maj. Gen. Albert S. Johnson. Maj. Gen. Earl Van Dorn was selected as the new District commander. He established his official headquarters at Little Rock, Arkansas, but spent most of his reign in Pocohontas, Arkansas.[4] The overall mission of the Confederate forces in this district was to push the Union army out of Missouri with the ultimate goal of capturing St. Louis, thereby cutting off any support from the eastern to the western states.[5] The Confederates hoped that by capturing the city they could use it as a base of operations to further infiltrate into the upper Missouri and Ohio Rivers.

In an effort to counter the Confederate action, the U.S. Army Department of the Missouri, under Maj. Gen. Henry Hallack, decided to build up its forces in southwest Missouri. Hallack charged Maj. Gen. Samuel R. Curtis, an 1831 West Point graduate and Mexican War veteran, with pushing the remaining Confederates out of Missouri. Curtis proceeded to Rolla, Missouri, and began moving his troops down the Telegraph/Wire Road towards Springfield in early February 1862. Price's Missouri State Guard units fell back each time when confronted with Curtis's Union forces. On February 13 the 4th Iowa Infantry entered Springfield, proceeded to the county

courthouse, and promptly raised the United States flag, all without a shot being fired. All Confederate and State Guard units had evacuated the city upon the Union Army's approach. The next day, the Union forces continued their march to the southwest towards the Arkansas-Missouri border. The Southern units did not attempt to oppose their movement, and they eventually fell back into northwestern Arkansas, with Union cavalry under Colonel Eugene A. Carr right behind them. A small skirmish occurred on February 16 at Cross Timber Hollow, or Potts Hill, on the Arkansas-Missouri line where the Telegraph Road entered the ten-mile long Cross Timber Hollow. Brig. Gen. Curtis's Union advance finally overtook the rear guard of Maj. Gen. Price as he retreated southward. This marked the official beginning of the Pea Ridge Campaign. This was an interesting occurrence as it was the 1st Missouri Cavalry under Price's Missouri State Guard facing the 1st Missouri Cavalry of Curtis's United States Army. Although the units faced off against each other and a few shots were fired, there were no reported casualties and the Missouri State Guard unit retreated.

Another small skirmish occurred at Dunagan's Farm, Arkansas, approximately one-half mile south of Little Sugar Creek, on February 17, 1862. Confederate Brig. Gen. Rains with the 4th Arkansas, the 3rd Louisiana, and Captain Churchill Clark's Missouri Battery, attempted to delay the advance of the Union Army under Brig. Gen. Curtis to give Price enough time to reach the Confederate Cross Hollow cantonment safely. Following a cavalry charge by the 1st and 6th Missouri (US) regiments, the battle-line deployed Confederate forces pulled back. Curtis reported 13 killed, and 15-20 wounded. The

Confederates reported a loss of only three or four. This was the "bloodiest" skirmish between the troops of Curtis and Price thus far. Eventually, the Missouri State Guard units fell back all the way to Cove Creek, deep in the Boston Mountains southwest of Fayetteville. Here, Maj. Gen. Price was met by Brig. Gen. Ben McCulloch, who now commanded all Confederate forces in Arkansas. There had been an issue of rank between Price, a major general of the Missouri State Guard, and McCulloch, a brigadier general of the Confederate States Army. McCulloch believed his rank as a Confederate officer was greater than Price's rank as a Missouri state officer. McCulloch had already travelled back to Richmond in an attempt to settle the question. Moreover, Indian Territory historian Annie Able has stated that McCulloch had little or no tolerance for the Missouri leaders, Governor Claiborne Jackson and Sterling Price. McCulloch saw their plans as impractical and much too focused on guerilla warfare, much like the Border War between Missouri and Kansas.[6] To put the issue to rest, Trans-Mississippi commander, Maj. Gen. Earl Van Dorn, arrived at the Confederate camp on March 3 and took command of the entire force. Eventually the entire Confederate force was moved northward to Camp Stephens adjacent to the confluence of Brush and Little Sugar Creeks along the Bentonville Detour Road.

Union General Curtis gathered his scattered forces and took up a position astride the Telegraph/Wire Road, east and west along Little Sugar Creek and dug in, expecting a Confederate attack. Van Dorn, realizing the futility of a frontal assault, on the night of March 6, 1862, opted to do a flank march north along the Bentonville Detour, around Curtis's right flank, up to Cross Timber Hollow and get behind the Union army. Unfortunately for Van Dorn, Curtis's scouts discovered the movement, which gave the Union army a chance to turn around and face the new threat. For the next two days, March 7-8, 1862, these two forces would clash near Elkhorn Tavern and Leetown in the shadow of Pea Ridge. The victor would determine the fate of Southwest Missouri and Northwest Arkansas.

In the weeks before the start of the Battle of Pea Ridge, Maj. Gen. Van Dorn tried to gather as many Confederate forces as possible for his attempt to drive Curtis's Union force out of Arkansas and Missouri. Van Dorn organized his command into two divisions. The first division consisted of the Missouri State Guard under Maj. Gen. Sterling Price. The second division was commanded by Brig. Gen. McCulloch. This second division was divided into two brigades: a cavalry brigade under Brig. Gen. James McIntosh, and an infantry brigade under Col. Louis Hebert. In addition to sending for Generals Price and McCulloch, on February 14, Van Dorn sent an order to Brig. Gen. Albert Pike, directing him to bring his Indian Brigade to join his forces in Arkansas. The location to meet was Elm Springs, Arkansas, although this location was not identified in Van Dorn's letter to General Price, only that the location had been sent to Pike.[7] This order created two different problems. First, Pike initially believed that his position as commander of the Department of the Indian Territory was independent of the Trans-Mississippi District and did not believe he was necessarily subject to Van Dorn's orders. Brig. Gen. Pike had good reason to question this since Special Order #234, issued on

November 22, 1861, created the Department of the Indian Territory with Albert Pike in command.

Special Orders, No. 234.)
Adjt. and Insp. General's Office,
Richmond, Va., November 22, 1861.

7. The Indian country west of Arkansas and north of Texas is constituted the Department of Indian Territory, and Brig. Gen. Albert Pike, Provisional Army, is assigned to the command of the same. The troops of this department will consist of the several Indian regiments raised or yet to be raised within the limits of the department.

By command of the Secretary of War:

JNO. WITHERS,
Assistant Adjutant- General.
Hdqrs. First Division,
Western Department,[8]

Since under the Confederate War Department it seemed that a "department" was a greater command than a "district," this discrepancy was addressed with Special Order #8, issued on January 10, 1861, that clearly stated that the Indian Territory was under the jurisdiction of the Trans-Mississippi District.

Special Orders (No. 8.)
Adjt. and Insp. Gen.'s Office,
Richmond, January 10, 1862.

XIX. That part of the State of Louisiana north of Red River, the Indian Territory west of Arkansas, and the States of Arkansas and Missouri,

excepting therefrom the tract of country east of the Saint Francis, bordering on the Mississippi River, from the mouth of the Saint Francis to Scott County, Missouri (which tract will remain in the district of Major-General Polk), is constituted the Trans-Mississippi District of Department No. 2, and Maj. Gen. Earl Van Dorn is assigned to the command of the same. He will immediately repair to Bowling Green, Ky., and report for duty to General A. S. Johnston, commanding Department No. 2.

By command of the Secretary of War:

JNO. WITHERS,

Assistant Adjutant-General,[9]

The second issue was more problematic. The treaties that Pike had negotiated with the Five Civilized Tribes all had stipulations that the Indian regiments were not to be taken out of the Indian Territory without the express consent of the tribal governments and the individual Indian soldiers. In addition, Pike did not necessarily regard the Indian troops as soldiers that would be available, or suitable, to participate in offensive operations outside of the Indian Territory. He saw them as a defensive "home guard" whose mission was to deter and delay any Union invasion into the Indian Nations until regular Confederate forces could respond.[10] At this late date, Pike would also have difficulties in getting the authorizations he would need from the tribal governments to bring his Indian Brigade out of the Indian

Territory and into Arkansas. Even with the tribal permissions, the soldiers of the 1st Chickasaw and Choctaw and 1st Creek regiments refused to leave the Indian Territory without first being paid. It was not until February 25 that Pike was able to get permission from the Tribal governments and monies to pay the disaffected regiments. The process of paying the 1st Chickasaw and Choctaw regiment at Fort Gibson took most of three days to complete. Pike then moved to Park Hill, where he paid the 1st Creek regiment. On March 3 Pike received the following messages from Maj. Gen. Van Dorn:

> Headquarters Trans-Mississippi District, Boston Mountains, March 3, 1862.
> Brig. Gen. Albert Pike:
> Commanding Indian Brigade
>
> General: I am instructed by Major-General Van Dorn to inform you that he will move from here to-morrow morning with the combined forces of Generals Price and McCulloch in the direction of Fayetteville. He wishes you, therefore, to press on with your whole force along the Cane Hill road, so as to fall in rear of our army.
>
> You will please, during your march, keep out your scouts, especially toward your left. Your troops will march light and be ready for immediate action. Your baggage train will follow your column slowly, making marches of not more than 5 or 6 miles per day. Should you have passed Evansville before this dispatch reaches you, please change direction at once and get into the Cane Hill road. It is expected that you will make such efforts as will insure your being in position, and send two couriers per day to keep the general commanding informed of your position and progress.
> I am, general, very respectfully, yours,
> > D. H. MAURY,
> > > Assistant Adjutant-General.
> > > Headquarters,
> > > Trans-Mississippi District
> > > Boston Mountains, Ark.,
> > > March 3, 1862.

> Brig. Gen. Albert Pike,
> Commanding Indian Brigade:
>
> General: This morning I sent you instructions concerning the movement of your brigade.
>
> The general commanding desires that you will hasten up with all possible dispatch and in person direct the march of your command, including Stand Watie's, McIntosh's, and Drew's regiments.
>
> The route indicated this morning in the order to you and to those colonels is such that they may not reach their position by the time desired. I am therefore directed to modify those orders, so that your command will be near Elm Springs (marching by the shortest route) day after to-morrow afternoon.
> By order of Maj. Gen. Earl Van Dorn
> > D.H. MAURY,
> > > Assistant Adjutant-General.
> > > Headquarters Trans-Mississippi District

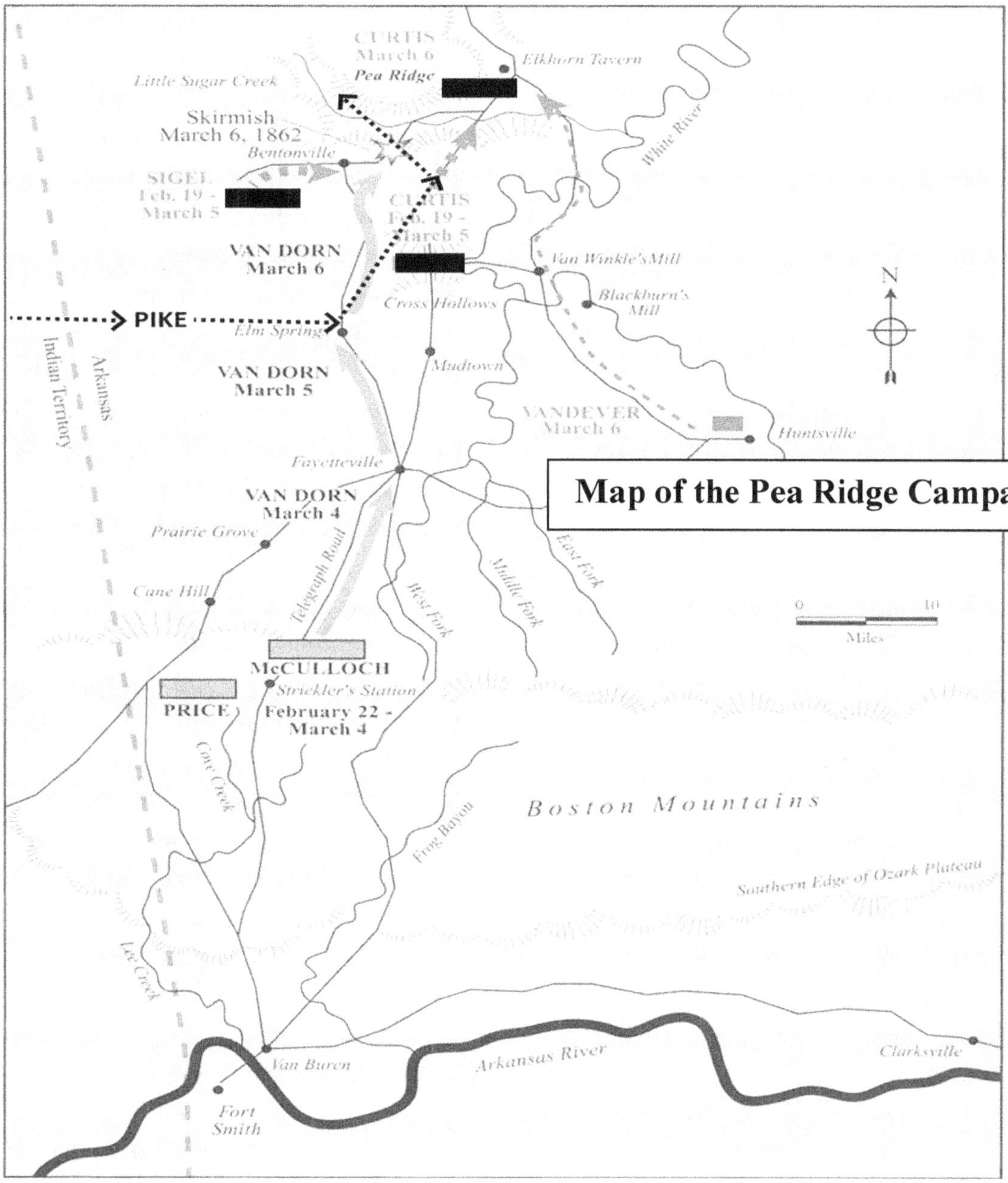

Map of the Pea Ridge Campaign

Maj. Gen. Samuel Curtis, USA
Army of Southwest Missouri
Photo courtesy of the Library of Congress

Maj. General Earl Van Dorn, CSA
Trans-Mississippi District
Photo courtesy of NPS / Pea Ridge NMP

Maj. Gen. Sterling Price,
Commander,
Missouri State Guard
Photo Courtesy NPS / Pea Ridge NMP

In case these orders were lost or were late, Van Dorn had these orders sent directly to the Indian Brigade's colonels:

> Headquarters Trans-Mississippi District,
> Boston Mountains, March 3, 1862.
>
> Colonels Drew, McIntosh, and Stand Watie:
> Colonels: Major-General Van Dorn, commanding the Confederate forces in this vicinity, directs me to inform you that you will move along the road from Evansville to Fayetteville, so as to be within 5 or 6 miles of Fayetteville to-morrow evening and in rear of our army, which will move from here in the morning.
>
> You will during your march keep out scouts toward your left especially, and you will report at once to Brigadier-General McIntosh, C. S. Army, your progress and your position. You will march light, ready for immediate action, and you will leave your heavy baggage to follow you slowly.
>
> You will, if possible, procure corn on the road or have it hauled to your halting place (after you) to-morrow night. Send a special courier at once to report to General Yan Dorn at General Price's headquarters your receipt of this order and your present position.
> I am, sir, very respectfully, your obedient servant,
> D. H. MAURY,
> Assistant Adjutant-General.[11]

Regardless, the Indian Brigade was to be assigned to Brig. Gen. McCulloch's division. Pike took what troops he had, O.G. Welch's Texas Squadron and a battalion of the 1st Chickasaw and Choctaw regiment. They marched first to Evansville, on the Arkansas-Indian Territory line, and the next day to Cincinnati, Arkansas. At this point Pike met up with Col. Stand Watie's 2nd Cherokee Mounted Rifles. By Thursday, March 6, Pike had overtaken Col. John Drew's 1st Cherokee Mounted Rifles near Smith's/Osage Mills. That night the Indian Brigade, approximately one thousand strong, camped within two miles of Confederate Camp Stephens on Little Sugar Creek. Col. D.H. McIntosh's 1st Creek and Col. Douglas Cooper and the remainder of the 1st Chickasaw and Choctaw regiments failed to arrive in time to support the Indian Brigade's actions during the upcoming battle.[12] It is not known why these regiments were delayed since they had been paid and had plenty of time to catch up with the remainder of the brigade.

Before the Indian Brigade's departure for Arkansas, Chief Ross of the Cherokee Nation sent a letter to Brig. Gen. Pike addressing some of his concerns. Ross stated:

> Executive Department, Park Hill, Feb 25th, 1862.
> To Brig. Genl. A. Pike, Comdg Indian Department.
>
> Sir: I have deemed it my duty to address you on the present occasion- You have doubtless received my communication enclosing the action of the National Council with regard to the final ratification of our Treaty- Col. Drew's Regiment promptly took up the line of march on

the receipt of your order from Fort Smith towards Fayetteville… There are so many conflicting reports as to your whereabouts and consequently much interest is felt by the People to know where the Head Qrs. of your military operations will be established during the present emergencies –… But I am sorry to say I have been dissuaded from going at present… I have at all times in the most unequivocal manner assured the People that you will not only promptly discountenance, but will take steps to put a stop to such proceedings for the protection of their persons and property.

I shall be happy to hear from you - I have the honor to be your obt Serv*

John Ross, Print Chief, Cherokee Nation.[13]

What Chief Ross was worried about were rumors planted by discontented elements within Col. Watie's 2nd Cherokee Mounted Rifles regarding Drew's soldiers who might perpetrate atrocities within the Cherokee Nation while Watie's regiment was away. Ross was equally concerned about the condition of both Cherokee regiments, believing neither was really set and equipped for a campaign. The bottom line of this letter was Ross showing Pike that he was committed to the treaty he had signed with the Confederate States.

Pea Ridge, or Elkhorn Mountain, terminating on the east in a rocky hill fronting Elkhorn Tavern and the Telegraph Road, lies in an east-west direction, parallel to Sugar Creek, which is about three miles to the

south. Between the two is a rough and, at that time, wooded country, broken here and there by small prairies and fenced clearings. The Union Army was entrenched, facing south, on Sugar Creek. A mile to their north and rear was the village of Leetown.[14]

This action was to be a baptism of fire for Brig. Gen. Albert Pike. During the winter campaigns against Opothleyahola, he had been in Richmond negotiating the treaties of the Five Civilized Tribes to the Confederate Congress. He had absolutely no knowledge or experience in military matters or tactics. The Confederate commanders held very little faith in the Indian Brigade. In fact, Pike himself regarded his troops poorly. He described them as "entirely undisciplined, mounted chiefly on ponies, and armed very indifferently with common guns and shotguns."[15] If nothing else, Pike understood the weaknesses of his brigade, knowing that they would not fight like white troops in line formation across a field. They were accustomed to fighting like skirmishers and would flee at the first shell or two dropped by a "shooting wagon," i.e. artillery. He never expected them to be used as offensive troops.[16]

On the morning of March 7 Albert Pike received information from Maj. Gen. Van Dorn that the Army would begin its move at 8:00 am and that he was to follow Brig. Gen. McCulloch's division up the Bentonville Detour behind the Union lines. Unfortunately, Pike did not receive the order until 9:30 am. By that time, McCulloch's supply train had taken up the Indian Brigade's place in line which created difficulties for Pike's troops to get around them on the narrow roads. This difficulty was coupled with the Union troops having cut down trees to block the roads to the north. Pike encountered Col. Sims of the 9th Texas and was directed to follow his unit from the Bentonville Detour approximately four and a half miles to the south towards a small town called Leetown. The Indian Brigade fell in behind Brig. Gen. McCulloch's and newly promoted Brigadier General James McIntosh's units. The Indian Brigade, marching across the fields in traditional Indian clothing including feathered headdress, breechcloths, and war paint, riding on small Indian ponies, must have been an interesting sight to both Confederate and Union soldiers who had never seen an Indian before. Pike's units were to dismount and advance with the infantry. Brig. Gen. James McIntosh's brigade led the way, advancing in five columns southward along Twelve Corners Road, with Pike's Indian Brigade right behind. McCulloch ordered Pike to conform with Brig. Gen. James McIntosh's brigade and to line up on the right of that brigade. Pike dismounted the 2nd Cherokee Mounted Rifles, and the Indian Brigade pushed to the right towards the northern end of Foster's Lane, approximately 300 yards north of the farm's field. A squadron consisting of Companies A and B, 3rd Iowa Cavalry, under Lt. Col. Henry Trimble, had been on a reconnaissance up Foster's Lane when they encountered the advancing Indian Brigade. Trimble attempted to swing his hugely outnumbered cavalrymen to the right to counter the threat but was quickly overwhelmed and, after a short fight, fled south along Foster's Lane. As the Indian Brigade finally left the woods into the clearing of Foster's Farm, they observed the three thousand Texas and Arkansas cavalry formed into line and performed the largest cavalry charge that had been seen so far in the war across Foster's empty wheat field. Dozens of Confederate bugles sounded "charge!" and

there arose a screaming a combination of the rebel yell from the Arkansans and the Comanche war whoop from the Texans. The Confederate force encountered the three-gun battery of Capt. Gustavas Elbert of the 1st Missouri Flying Artillery supported by the cavalry brigade of Col. Cyrus Bussey, 3rd Iowa Cavalry, approximately 300 yards to their front, centered on the Foster Farm. Bussey's small Union brigade was outnumbered at least six to one, but they fought valiantly until they were finally overwhelmed by the Confederate onslaught, abandoning Elbert's battery. Col. Bussey's cavalry brigade broke for the rear, abandoning any intention of re-taking the three-gun battery.[17] Brig. Gen. Pike wrote in his report of the battle actions of his Indian Brigade:

> "...we discovered in front of us, at the distance of about 300 yards, a battery of three guns, protected by five companies of regular cavalry. A fence ran from east to west through the woods, and behind this we formed in line, with Colonel Sims' regiment on the right, the squadron of Captain Welch next to him, and the regiments of Colonels Watie and Drew in continuation of the line on the left. The enemy were in a small prairie, about 250 yards across, on the right of which was the fenced field, and on our left it extended to a large prairie field, bounded on the east by a ridge. In rear of the battery was a thicket of underbrush, and on its right, a little to the rear, a body of timber... The enemy opened fire into the woods where we were, the fence in front of us was thrown down, and the Indians (Watie's regiment on foot and Drew's

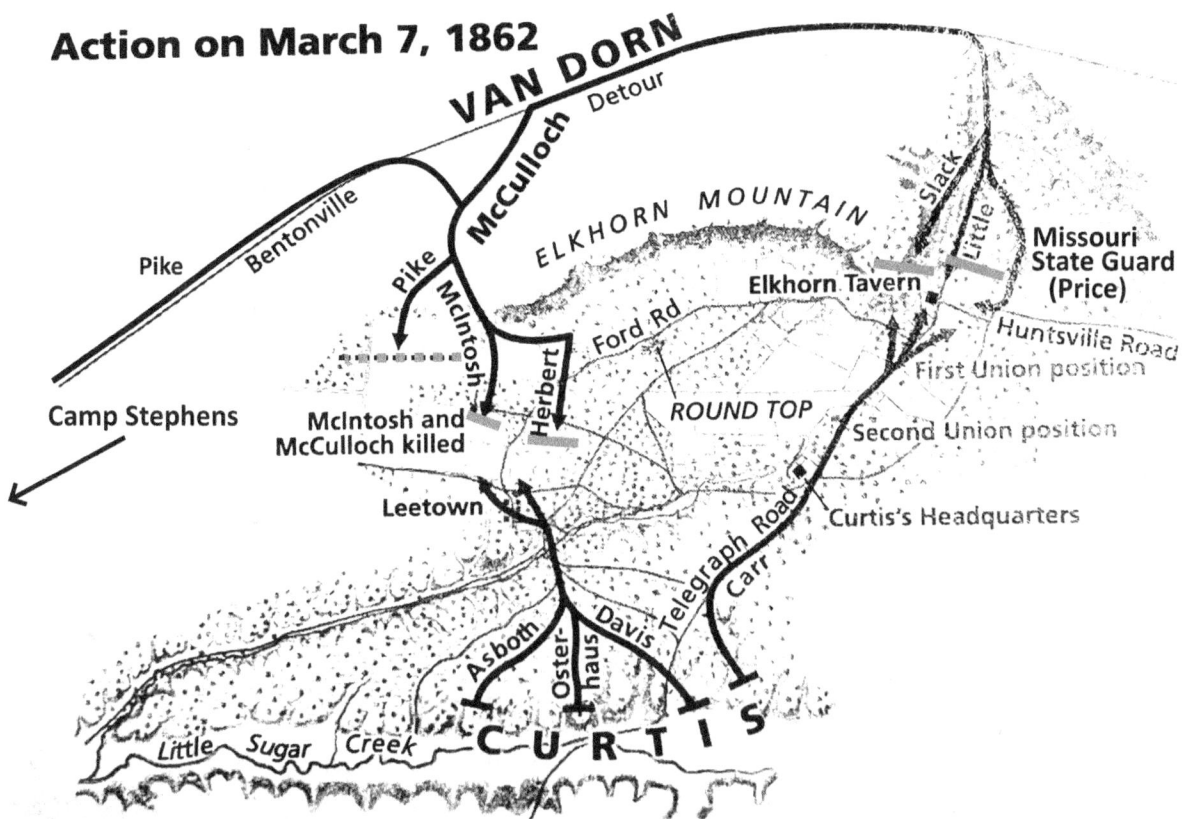

on horseback), with part of Sims' regiment, gallantly led by Lieutenant-Colonel Quayle, charged full in front through the woods and into the open ground with loud yells, routed the cavalry, took the battery, fired upon and pursued the enemy, retreating through the fenced field on our right, and held the battery, which I afterwards had drawn by the Cherokees into the woods. Four of the horses of the battery alone remained on the ground, the others running off with the caissons, and for want of horses and harness we were unable to send the guns to the rear…"[18]

Later historians, including William Shea and Earl Hess, have determined that the Indian Brigade did not participate in the cavalry charge and did not capture the battery but did help to celebrate the event.

It was at this point that things turned for the worse for the Confederates and especially the Indian Brigade. An unknown number of Confederate Indian soldiers had gone on a rampage of murder and mutilation on the prisoners and the wounded of the 3rd Iowa Cavalry in the woods north of the Round Prairie. At least eight Iowan's were scalped. One Confederate officer stated, "the Indians swarmed around the guns like bees, in great confusion, jabbering and yelling at a furious rate." Even Brig. Gen. Pike observed that his Indian troops were wandering around "in the most confusion, all talking, riding this way and that, and listening to no orders from anyone." The 3rd Iowa Cavalry suffered 24 killed, 17 wounded, and 9 missing.[19] Private Sparks of the 9th Texas Cavalry wrote:

"On the taking of the battery the Indians under Col. Pike were highly elated and many of them straddled the guns and rode them in joy over the victory. The gunners were all killed and nearly all the horses, only team enough being left to move one gun."[20]

Meanwhile, the retreating Union cavalrymen fled down Foster's Lane towards the Oberman's Field where Col. Nicholas Greusel's Union brigade, the second brigade under Brig. Gen. Peter Osterhaus, was posted. As the fleeing Union cavalrymen passed their line, this brigade stood firm. In an effort to keep the Confederate forces milling about the Foster Farm unsettled, Col. Greusel ordered the Missouri Light Artillery, under Captain Welfley, and the 4th Ohio Battery, commanded by Captain Hoffman, both batteries of 12-pounder howitzers, to open a harassing fire on the farm on the other side of the tree line. After the first few salvos of exploding shells fell among the Confederates, the Indian troops panicked and fled back into the woods where they had defeated the Iowans. Brig. Gen. Pike further reported:

"At this moment the enemy sent two shells into the field, and the Indians retreated hurriedly into the woods out of which they had made the charge. Well aware that they would not face shells in the open ground, I directed them to dismount, take their horses to the rear, and each take to a tree, and this was done by both regiments, the men thus awaiting patiently and coolly the expected advance of the enemy, who now and for two hours and a half afterwards, until perhaps twenty minutes before the action ended, continued to fire shot and shell into the woods where the Indians were from their battery in front, but never advanced."[21]

Although Pike was able after many hours to prompt the Cherokees to come out and move the three guns of Elbert's battery into the woods, the Indian Brigade was technically worthless to Van Dorn's Confederate Army from this point until the end of the Battle of Pea Ridge.

In addition, during this break, General McCulloch, the former Texas Ranger, decided to conduct a leader's reconnaissance of the Union position after Major Ross of the 6th Texas Cavalry had observed Union infantry on the south side of Oberson's Field. He sent skirmishers from the 16th Arkansas Infantry forward through the belt of trees between Foster's and Oberson's fields. He was unaware that Companies B & G of the 36th Illinois Infantry were posted on the south edge of the forest belt. When McCulloch on horseback stepped into an open space between the trees, the men of Company B opened fire and dropped the General from his horse with a shot through the heart. His death was hidden from the remainder of his brigade to prevent a drop in morale. This also left Brig. Gen. Pike as the senior officer on the Leetown battlefield.[22] Unfortunately, Van Dorn had a very low opinion of Pike dating back to an attack that Van Dorn had perpetrated in 1858 while commanding the 2nd U.S. Cavalry on a Comanche-Wichita Indian camp that resulted in the deaths of seventy men and women. Pike, as an Indian agent, was very vocal about Van Dorn's impetuosity for not investigating to discover that the friendly Wichita were negotiating a peace treaty with the Comanche. The feud continued into the Civil War years. Regardless, Van Dorn, unaware that General McCulloch had been killed, had failed to share critical information or include Pike in any of the planning of the battle.

Brigade commander Brig. Gen. James McIntosh, who had decisively defeated

**View and Direction of the Confederate Cavalry
Charge on March, 7 1862**

**Elkhorn Tavern
Pea Ridge National Military Park**
Photo by Author

Capt. Elbert's battery from Confederate Position

Capt. Elbert's Battery facing Confederate Line

All photos by Author

**View towards Foster's Lane where the Iowa
troopers were scalped**
Photo by Author

**Rally Point for Albert Pike's Confederate Indian
Brigade on March 7, 1862**
Photo by Author

Opothleyahola and the Loyal Indians at Chustenahlah, when informed of Gen. McCulloch's death, told his cavalry commanders to wait for orders. He advanced with his former regiment, the 2nd Arkansas Mounted Rifles, who were dismounted to act as infantry, to the belt of trees that separated Foster's and Oberson's fields. After the regiment had mostly made its way through the trees and brush, McIntosh, still mounted, rode around the right flank of and in advance of the regiment. As he and the regiment approached the same rail fence, although further east, that Gen. McCulloch had approached, they were met by a volley from the companies of the 36th Illinois Infantry stationed on the right. Brig. Gen. James McIntosh fell dead with a bullet in the heart.[23]

At approximately 3:00 pm Pike encountered the 16th Arkansas and 17th Arkansas Infantry regiments, commanded by Col. John Hill and Col. Frank Rector, the 1st Arkansas Mounted Rifles (dismounted), commanded by Col. Thomas Churchill, and the 4th Texas Cavalry Battalion (dismounted),, commanded by Major John Whitfield, all units belonging to Col. Louis Hebert's brigade. Major Whitfield informed Gen. Pike that Generals McCulloch and McIntosh had been killed.

Brig Gen. Albert Pike was now in command of all three brigades. He attempted to rally General McIntosh's cavalry regiments into a fighting unit with little success. Pike was able to coax the remnants of the Indian Brigade out of the woods and into formation behind the white regiments. He placed the infantry and the dismounted cavalry regiments of Col.'s Hill, Rector, Churchill, and Maj, Whitfield's battalions in a line of battle behind the fence on the south side of and centered

on Foster's farm with the mounted cavalry units and the remnants of the Indian Brigade on the north end of the field. Pike soon realized his mistake and brought the infantry back to the north end of the field.

Meanwhile, the remainder of Col. Louis Hebert's infantry brigade, consisting of the 4th, 14th, and 15th Arkansas Infantry, and the 3rd Louisiana Infantry regiments, were on the division's left flank. Although spending much of their time floundering in Morgan's Woods below Little Mountain, they were successfully pushing back Union Maj. Gen. Jefferson C. Davis's (no relation to the Confederate president) right flank, consisting of only the 37th and 59th Illinois Infantry, with his four infantry regiments. But without any support from the other brigades, his losses became too great to continue the push. Eventually Union brigade commander Col. Thomas Pattison was able to bring two Union regiments, the 18th and 22nd Indiana Infantry, on a road around Hebert's left flank in a surprise move that resulted in the rout of the Confederate brigade and the capture of its commander.

Due to the vacuum of lost commanders and a lack of information sharing by the other commanders, Pike was indecisive as to how to proceed. He did not know the overall plan, nor did he know where or what the other Confederate or Missouri State Guard units were doing. He admitted that he was completely ignorant of the roads and terrain in which the battle was being fought. Runners were dispatched to find Maj. Gen. Van Dorn with the information that General's McCulloch and McIntosh had been killed and asking for orders for what was now unofficially Pike's division. Without waiting for a reply or discovering what had become of

Action on March 8, 1862

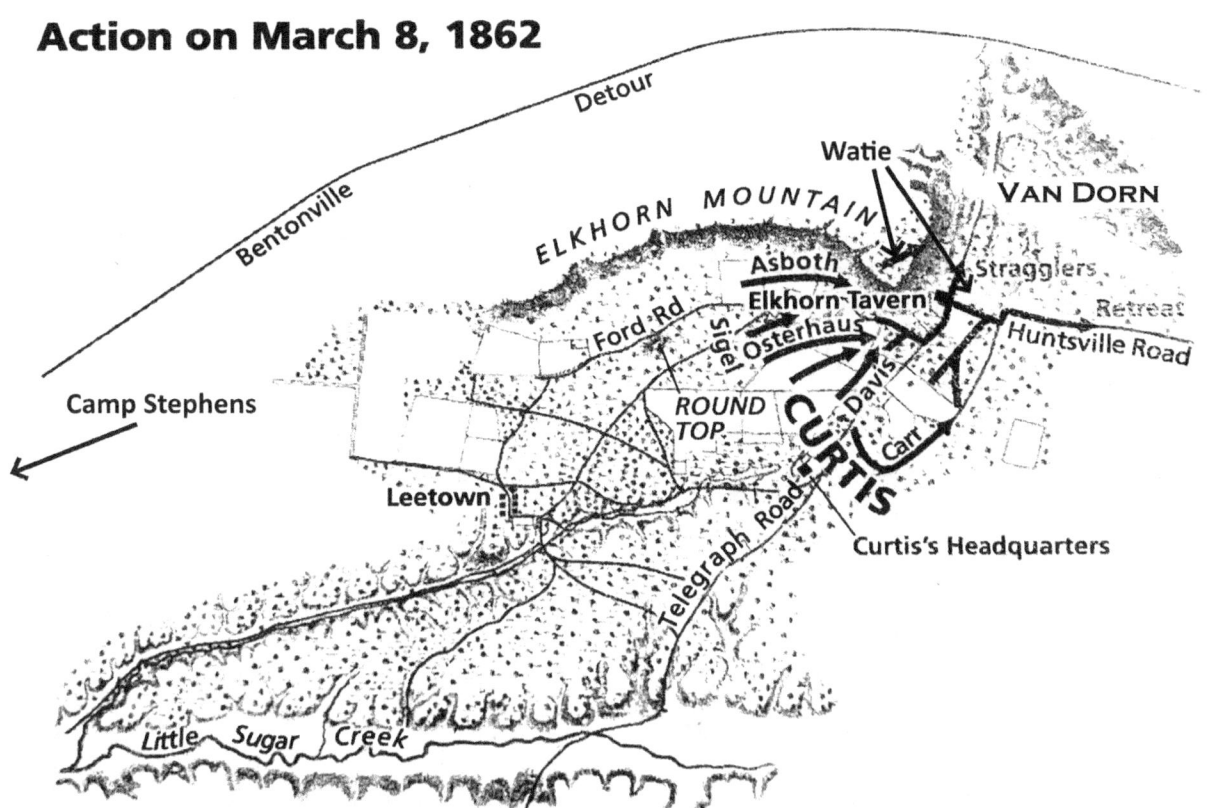

Col. Hebert's brigade, Pike called together all of the regimental commanders and informed them that they would proceed towards Twelve Corner Church on the Bentonville Detour. They would then turn right and move towards the Telegraph-Wire Road and join the remainder of the Confederate Army at Elkhorn Tavern. Pike stated in his report:

> "…At this time the firing on the field had ceased, and I saw coming into the road at the farm house a large body of cavalry and Good's battery. It was evident enough that the field was left to the enemy, and as we were not in sufficient numbers to resist them and the ground afforded no defensive position I determined to withdraw the troops and lead them to General Van Dorn. Indeed, the officers assured me that the men were in such condition

that it would be worse than useless to bring them into action again that day. I accordingly sent orders to the artillery and cavalry to join me. What had become of the other troops engaged no one could inform me. I concluded they had retreated towards Camp Stephens, gaining the road by which we had come in the morning. Colonel Stone and Captain Good came to me, and I informed them of my purpose. Placing the squadron of Captain Welch in front, the infantry marching next, followed by Good's battery, with the Cherokees on the flanks, and, as I supposed, Colonel Stone's regiment in the rear, we gained the Bentonville road, and marched on it in perfect order to the Telegraph road…"[24]

Unfortunately for Pike, only about two

thousand of the division's troops followed him, the 16th and 17th Arkansas Infantry, the 1st Arkansas Mounted Rifles, the 4th Texas Cavalry Battalion, the 2nd Cherokee Mounted Rifles, and Good's Texas Battery. About twelve hundred turned left at the Twelve Corners Church and proceeded back to Camp Stephens and the supply trains. These troops included the 6th Texas Cavalry, 1st Texas Cavalry Battalion, the 1st Cherokee Mounted Rifles, and Provence's Arkansas Battery. Further, the commanders of the 9th and 11th Texas Cavalry, the 2nd Arkansas Mounted Rifles, the 1st Arkansas Cavalry Battalion, and Gaines's and Hart's Arkansas Batteries refused to obey Gen. Pike's orders and remained on the field at the Ford Road. Pike's order to withdraw failed to reach Colonel Drew and his 1st Cherokee Mounted Rifles regiment was the last to leave the field, marching to Camp Stephens, where they caught up with the retreating train. There they met Colonel Douglas Cooper, with the rest of the 1st Choctaw and Chickasaw regiment and battalion, and Col. D. N. McIntosh, with 200 men of the 1st Creek regiment, who had finally arrived but too late for the battle. They all remained with the train until it reached Elm Springs, then marched with their own train to Cincinnati.

Col. Elkanah Greer of the 3rd Texas Cavalry regiment, whose unit had been sent to the top of Little Mountain, was forgotten during the fighting. He brought his force down to the Foster farm where he was informed of McCulloch and McIntosh's deaths and discovered that, because Pike had left the field with seven of the division's regiments, he was now in command of the right wing of the Confederate Army. He gathered the force that had remained behind as well as the remnants of Col. Hebert's brigade. Making the difficult decision that he could not hold the Leetown battleline against a growing Union force, he decided to burn the carriages of the three captured Union artillery pieces. By 6:00 pm Greer's force too had begun to move towards the Telegraph-Wire Road and Elkhorn Tavern via the Twelve Corners Church Road. While enroute, Major Ross appeared with orders from General Van Dorn to hold the Leetown line at all costs. Greer informed Van Dorn that the Leetown battleline no longer existed.[25]

Brig. Gen. Pike arrived at Van Dorn's headquarters at the Elkhorn Tavern after dark. There, he was informed that Maj. Gen. Sterling Price had driven Union Col. Eugene Carr's brigade on the Confederate far left flank back during the day. However, later that night, Van Dorn learned that the ammunition was almost exhausted and that the ordnance officer could not find his wagons. None of the Confederate commanders knew that the ammunition train had returned to Camp Stephens. Van Dorn made the decision, however, to continue the battle the next day.

The only remaining action involving the Indian Brigade was the posting of the 2nd Cherokee Mounted Rifles under Col. Stand Watie, to the hilltops on Big Mountain and above the Elkhorn Tavern as scouts and observers of the final day's battle. On the morning of March 8, Pike gave Welch's Texas squadron permission to join one of the Texas regiments. The battle on March 8 was short. Although Van Dorn had posted much of his artillery on the Elkhorn Mountain above the Tavern, the severe lack of artillery ammunition limited their effectiveness. The Union artillery pummeled the Confederate positions for two hours before a five-pronged Union infantry assault swept against the faltering Southern lines. Gen. Van Dorn began to withdraw his army at 10 o'clock over the Huntsville Road, leading east. When the 36th Illinois Infantry, "with its dark-blue line of men and its gleaming bayonets," swarmed over the hill, Col. Watie and his 2nd Cherokee Mounted Rifles retreated along the ridge and eventually made their way back to Camp Stephens. Two hundred of Watie's Indians were detailed by Brig. Gen. Martin Green of the Missouri State Guard, commander of the train, to escort the ammunition wagons at Camp Stephens to General Van Dorn, but the army had left before they arrived. Watie's Indians rejoined the train, by a circuitous route, at Walnut Grove, southwest of Fayetteville. General Pike narrowly escaped capture by riding north on the Telegraph-Wire Road, along which the artillery was retreating. After several unsuccessful attempts to induce a battery to make a stand, he turned into the Bentonville Detour Road, where he was joined by two other officers. Pursued by Union cavalry, they took to the woods, skirted Pea Ridge, and rode westward between the Pineville and Bentonville roads. Eventually, over a period of days, the remnants of the Indian Brigade drifted into Cincinnati, Arkansas, and began to regroup.[26] The losses of the Indian Brigade are unknown as they were not reported by Brig. Gen. Pike, but that is not unusual. The Indian troops tended to come and go as they pleased, so an accurate count would not be possible in most circumstances.

The retreat of Van Dorn's Confederate Army began as disorganized but evolved into an almost complete rout because there were no advanced plans for rendezvous points, many commanders got separated from their units, and, leaderless, the soldiers simply fled south as fast as they could. One Cherokee schoolgirl in Van Buren, Arkansas, stated that the retreating Confederate Indians were disturbing the town's routine because all public buildings, including schools and stores, were being set up as makeshift hospitals. She said that they were a "a terrible looking crowd, bringing their wounded with them."[27] Rumors swarmed that both Generals Price and Van Dorn had been captured and that the Union cavalry was sweeping down upon the remnants and stragglers of the Confederate Army. None of these rumors were true. Van Dorn and Price eluded capture and were able to retreat safely, and the Union cavalry, like the rest of the Union Army, was almost as broken and disorganized as the Confederates. However, it had better command and control over its units.

General Pike drew heavy criticism for not only the poor performance of the Indian troops, but also for his incompetent performance while in command near Leetown. In Pike's defense, he had been left out of the communication loop with Generals McCulloch, Price, and Van Dorn, and did not know the overall plan. In Van Dorn's decisions for troop levels and commands that should accompany him east, the Indians were not included and were sent back to Indian Territory.

The defeat substantially reduced the Confederacy's ability to support operations in Missouri, Northwest Arkansas, and the Indian Territory, as they began to shift regular units east of the Mississippi.[28] Van Dorn lost the battle although he had the larger army, was closer to his supply sources, and was in territory familiar to his subordinate commanders.

On Sunday, March 9, following the

battle, Maj. Gen. Van Dorn sent a Confederate burial party back to the field under a flag of truce. The Confederate commander requested that they be permitted to collect and bury the Confederate dead on the battlefield. Permission was granted, and the work was completed. Before the detail returned to the Confederate lines, the officer in charge of the party was given a letter by General Curtis to deliver to Van Dorn. Curtis's letter said:

> "The general regrets that we find on the battlefield, contrary to civilized warfare, many of the Federal dead who were tomahawked, scalped, and their bodies shamefully mangled, and expresses a hope that this important struggle may not degenerate to a savage warfare."

Colonel Maury, Van Dorn's adjutant, replying March 14, assured Curtis that the Confederate commander will:

> "...most cordially unite with you in repressing the horrors of this unnatural war...He hopes you have been misinformed with regard to this matter,... the Indians who formed part of his forces having for many years been regarded as civilized people....he desires me to inform you that many of our men who surrendered themselves prisoners of war were reported to him as having been murdered in cold blood by their captors, who were alleged to be Germans."[29]

Curtis had not named the Indians, but Van

Dorn naturally made that assumption.

To respond to the claims regarding the Germans, Maj. Gen. Franz Sigel wrote to Maj. Gen. Curtis, and his response thus forwarded to Van Dorn:

> "While Capt. Elbert's three pieces were taken by the enemy, and our men serving the guns were surrounded, they were shot dead by the rebels, although seeking refuge behind the horses."

Col. Cyrus Bussey, 3rd Iowa Cavalry, commander of the Union cavalry force on the north end of the Foster Field, stated in his official report:

> "Hearing it reported by my men that several of the killed had been found scalped, I had the dead exhumed, and on personal examination I found that it was a fact beyond dispute that 8 of the killed of my command had been scalped. The bodies of many of them showed unmistakable evidence that the men had been murdered after they were wounded that first having fallen in the charge from bullet wounds, they were afterwards pierced through the heart and neck with knives by a savage and relentless foe. I then had the bodies returned, each in a separate grave, properly marked."[30]

Maj. Gen. Curtis later forwarded to the Federal Joint Committee on the Conduct of the War four affidavits signed by members of the 3rd Iowa Cavalry, which had been driven back in the charge made by Pike and McIntosh at noon March 7. A detail of the 3rd

Iowa returned to the site of the battle and dug up the recently buried soldiers of the 3rd Iowa Cavalry. In one of the affidavits, Adjutant John W. Noble stated, "from personal inspection of the bodies of the men of the Third Iowa Cavalry, who fell upon that part of the field, I discovered that eight of the men of that regiment had been scalped."

Brig. Gen. Pike himself knew that atrocities had been committed by his Indian Brigade, primarily the 1st and 2nd Cherokee Mounted Rifles. He had given permission to his Indian troops to fight "in their own fashion." Writing from Dwight Mission, an established Congregational mission located in the Cherokee Nation, Pike expressed his horror at seeing in the action of March 7 a person unknown to him, "and who immediately passed beyond his sight," shoot a wounded enemy begging for mercy." On March 15 Pike issued an order to the Indian troops, in which he said:

> "…The commanding general has also learned with the utmost pain and regret that one, at least, of the enemy's dead was found scalped upon the field. That practice excites horror, leads to cruel retaliation, and would expose the Confederate States to the just reprehension of all civilized nations… Against forces that do not practice it, it is peremptorily forbidden during the present war…"

Maj. Gen. Van Dorn's orders for the Indians were that they should make their way back as best they could to their own country and there operate "to cut off trains, annoy the enemy in his marches, and to prevent him as far as possible from supplying his troops from Missouri and Kansas." The Trans-Mississippi commander made it clear that the Indian Territory troops were not to call upon his Army of the West. On March 24 Van Dorn had Col. Maury write to Brig. Gen. Pike:

> "…It is not expected that you will give battle to a large force, but by felling trees, burning bridges, removing supplies of forage and subsistence, attacking his trains, stampeding his animals, cutting off his detachments, and other similar means, you will be able materially to harass his army and protect this region of country. You must endeavor by every means to maintain yourself in the Territory independent of this army. In case only of absolute necessity you may move southward. If the enemy threatens to march through the Indian Territory or descend the Arkansas River you may call on troops from Southwestern Arkansas and Texas to rally to your aid… You may reward your Indian troops by giving them such stores as you may think proper when they make captures from the enemy, but you will please endeavor to restrain them from committing any barbarities upon the wounded, prisoners, or dead who may fall into their hands…"[31]

Maj. Gen. Van Dorn was done with his Indian Brigade.

In response to the severe backlash against the Indian Brigade, on April 30, 1862, the Cherokee National Council, adopted a resolution expressing their opinion that the war should be conducted on the "most humane principles which govern the usages of war among civilized nations" and "recommended to the troops of this nation...to avoid any acts toward captured or fallen foes that would be incompatible with such usages." In the post-war years, the members of the two Confederate Cherokee regiments blamed each other for the atrocities in an effort to create better chances of receiving increased benefits in the new treaties.

The atrocities committed by the Cherokee troops reverberated across the nation. The New York Tribune published an article on March 27, 1862, harshly and sarcastically criticizing Pike:

"The Albert Pike who led the Aboriginal Corps of Tomahawkers and Scalpers at the battle of Pea Ridge, formerly kept school in Fairhaven, Mass., where he was indicted for playing the part of Squeers, and cruelly beating and starving a boy in his family. He escaped by some hocus-pocus law, and emigrated to the West, where the violence of his nature has been admirably enhanced. As his name indicates, he is a ferocious fish, and has fought duels enough to qualify himself to be a leader of savages. We suppose that upon the recent occasion, he got himself up in good style, war-paint, nose-ring, and all. This new Pontiac is also a poet, and wrote 'Hymns to the Gods' in Blackwood; but he has left Jupiter, Juno, and the rest, and betaken himself to the culture of the Great Spirit, or rather of two great spirits,

whisky being the second."[32]

The Confederate loss at the Battle of Pea Ridge ended any hopes for the Southerners to bring Missouri into the Confederacy. The Confederate Army in Northwest Arkansas was so depleted that it would be many months before they could re-build to the point of attempting to drive the Union army out of that area of the state. Maj. Gen. Earl Van Dorn was sent to Tennessee with most of the Arkansas troops to counter the advances by the Union army under Brig. Ulysses Grant, after his capture of Fort Henry and Fort Donelson. Van Dorn's Army of the West arrived too late to help at Shiloh, and he was soundly defeated in the Second Battle of Corinth in October 1862. Maj. Gen Earl Van Dorn would meet his end at Spring Hill, Tennessee, on May 7, 1863, not by battle with Union forces, but by a pistol shot fired by a Dr. Peters, a very jealous husband who claimed that Van Dorn had "violated the sanctity of his home."[33]

After the Confederate Army's defeat at Pea Ridge and the subsequent withdrawal of the Confederate and Missouri State Guard troops from western Arkansas, General Pike also began to have misgivings about the security of his Arkansas River line based on Fort Davis and Fort Smith. The Fort Davis post was difficult to support with the commissary and quartermasters only fulfilling one-third of the supplies needed by the Indian Brigade. The Indian Territory had a very limited industrial capacity so almost everything needed to be purchased and shipped in from other places in the Confederacy. Brig. Gen. Pike came to believe that Fort Davis was untenable and decided to move his headquarters further south.

Although he still maintained a small garrison at Fort Davis under the command of Col. Douglas Cooper, he moved his headquarters to a new post he named Fort McCulloch, after the Confederate general killed at Pea Ridge. The new fort was a large earthen fortification built in the southern side of the Choctaw Nation on a bluff overlooking Nail's Crossing of Blue Creek, on the Texas Road near Boggy Depot. Pike believed that he had a better chance of defending the Indian Territory along the Canadian River versus the Arkansas River. It was also closer to the Indian Brigade's supply sources in Texas and the Red River. Unfortunately, Pike's supply problems were never solved as Confederate shipments of arms, ammunition, and other equipment shipped from the eastern part of the Confederacy were routinely appropriated by Confederate units while in transit. Col. Watie took his 2nd Cherokee Mounted Rifles up to Cowskin Prairie in the northeastern part of the Cherokee Nation where he intended to conduct raids into Union-held areas of Missouri and Kansas. Col. Drew brought his 1st Cherokee Mounted Rifles back towards the Tahlequah and Park Hill area. The remaining regiments went back to their respective home nations. The Union realized these apparent weaknesses and began making plans to exploit them.[34]

[1] United States War Department. War of the Rebellion: A Compilation of the Official Records of the Union and Confederate Armies, Washington D.C.: U.S. Government Printing Office. 1888-1901. Vol. LIII, Supplement, pp. 764, 770.

[2] Franks, Kenny A. Stand Watie and the Agony of the Cherokee Nation. Memphis: Memphis State University Press. 1979. pp. 123.

[3] Warde, Mary Jane. When the Wolf Came: The Civil War and the Indian Territory, Fayetteville: The University of Arkansas Press, 2013. pp. 79.

[4] Special Orders #8, OR, Series I, Vol. VIII, pp. 734.

[5] Bond, John W. "The Pea Ridge Campaign," The Battle of Pea Ridge, 1862, Eastern National Park Booklet, Date Unk. pp. 4.

[6] Abel, Annie Heloise. The American Indian as Participant in the Civil War, Cleveland, Ohio: A.H. Clark Company, 1919. Pp. 16.

[7] Van Dorn to Price, OR, Series I, Vol. VIII, pp. 751.

[8] Special Orders #234, OR, Series I, Vol. VIII, pp. 690.

[9] Special Orders #8, OR, Series I, Vol. VIII, pp. 734.

[10] Able, pp. 23-24.

[11] Maury to Pike, Drew, McIntosh, and Watie, OR, Series I, Vol. VIII, pp. 763-764.

[12] Clifford, Ray A. "The Indian Regiments in the Battle of Pea Ridge." Kepis & Turkey Calls: An Anthology of the War Between the States in Indian Territory. Oklahoma City: Western Heritage Books. 1982. 66-67.

[13] John Ross to Pike, John Ross Papers #9 in the U.S. Indian Office

[14] Tricket, Dean. "The Civil War in the Indian Territory, 1862-2" The Chronicles of Oklahoma (December 1941), Vol. XIX, No. 4, p. 391.

[15] Pike, Albert, Letter. May 4, 1862. OR, Series I, Vol. XIII, pp. 819.

[16] Shea, William L. and Earl J. Hess. Pea Ridge: Civil War Campaign in the West, Chapel Hill and London: The University of North Carolina Press, 1992. pp. 25.

[17] Shea and Hess. pp. 97-100.

[18] Pike, Albert. Report, March 14, 1862. OR, Series I, Vol. VIII, pp. 287.

[19] Shea and Hess. pp. 102

[20] Sparks, A.W. Recollections of the Great War: The War Between the States as I Saw It, Tyler: Lee & Burnett, Printers. 1901, pp. 29.

[21] Pike, Albert. Report, March 14, 1862. OR, Series I, Vol. VIII, pp. 288.

[22] Shea and Hess. pp. 109-111.

[23] Shea and Hess. pp. 113-114.

[24] Pike, Albert. Report, March 14, 1862. OR, Series I, Vol. VIII, pp. 289.

[25] Shea and Hess. pp. 143-146.

[26] Tricket, Dean. "The Civil War in the Indian Territory, 1862-2" The Chronicles of Oklahoma (December 1941), Vol. XIX, No. 4, p. 394.

[27] Confer, Clarissa W. The Cherokee Nation in the Civil War, Norman: University of Oklahoma Press,

2007. Robinson, Elle Coody, Indian Pioneer Papers, 107:466, University of Oklahoma.

[28] Clifford, Roy A. "The Battle of Pea Ridge," *Kepis & Turkey Calls: An Anthology of the War Between the States in Indian Territory*: pp. 65-73.

[29] Maury to Curtis, *OR*, Series I, Vol. VIII, pp. 195.

[30] Noble to Curtis, *OR*, Series I, Vol. VIII, pp. 236.

[31] Maury to Pike, *OR*, Series I, Vol. VIII, pp. 796.

[32] Abel, pp. 31. *Footnote*

[33] Warner, Ezra J. <u>Generals in Gray: Lives of Confederate Commanders</u>, Baton Rouge and London: Louisiana State University Press. 1987. pp. 315.

[34] Corbett, William P. "Confederate Strongholds in Indian Territory: Forts Davis and McCulloch." *Early Military Forts and Posts in Oklahoma*. Faulk, Odie B., Franks, Kenny A., Lambert, Paul F. Editors. Oklahoma City: Oklahoma Historical Society. 1978: 65-77.

Chapter 6
Forming the Union Indian Brigade

The winters in Kansas are long and cold, and the wind feels like it has blown directly from the Arctic Circle. Although the southeastern part of the state has more hills and trees than western Kansas, the wind can blow just as hard. Couple this with the snow and ice and it transforms the landscape into a yellow and gray and frozen world. It is a time for sheltering out of the wind and gathering by the warmth of the fire in the hearth and awaiting the warmer winds of spring. This is what greeted the Loyal Indian refugees after Confederate forces drove them out of their homes in the Indian Territory.

During the remainder of winter and spring of 1862, as many as ten thousand Union-leaning Indians from the Five Civilized Tribes made their way northward to the safety of United States forts in Kansas.[1] Although the Loyal Indians traveled over the "Trails of Blood and Ice" by various routes, most aimed for Fort Row, a Kansas militia-built stockade fort on the banks of Walnut Creek near its confluence with the Verdigris River. This log fort was originally built to protect Free-Soilers, or anti-slavery citizens, from Border Ruffians, or pro-slavery citizens, attacks and was now to be the gathering point for the Loyal Indians coming out of the Indian Territory. The U.S. Indian agents in Kansas had been informed by the Sac & Fox Indians, whose reservation was in the region, that the Loyal Indians were enroute, but the agents failed to properly prepare for their arrival in early January 1862. U.S. Indian Southern Superintendent William Coffin had ordered all of the respective Indian agents to Fort Row to

Fort Row

assist the refugees upon their arrival. General David Hunter at Fort Leavenworth sent his Chief of Commissary and Subsistence, Capt. J.W. Turner, down to the gathering point to determine what was needed. But soon the Indian agents were embroiled in a lengthy discussion with the Army over which department was responsible for providing rations and supplies to the new arrivals from Indian Territory. Both the Interior Department and War Department claimed that they had neither the money nor resources to feed and house the refugees. The Loyal Indians began arriving at Fort Row tired, hungry, and cold. They would remain this way for the time being. Agent Cutler stated in his report:

> "And over the snow-covered roads, they travelled all night and the next day, without halting to rest. Many of them were on foot, without shoes, and very thinly clad. . . In this condition they had accomplished a journey of about three hundred miles; but quite a number froze to death on the route,

and their bodies with a shroud of snow, were left where they fell to feed the hungry wolves. . . Families who in their country had been wealthy, and who could count their cattle by the thousands and horses by hundreds, and owned large numbers of slaves, and who at home had lived at ease and comfort, were without the necessaries of life."[2]

The refugee Indians camped on the prairies between the Arkansas and Verdigris Rivers. These lands belonged to the previously relocated New York Indians and were a safe haven for the Loyal Indians for the time being since white settlers were prohibited from remaining there. General Hunter had also sent his Chief Surgeon A.B. Campbell to the refugee camp to survey the medical situation. He reported:

"In compliance with instructions from Major-General Hunter, contained in your order of the 22d. ultimo, I left this place on the 22d. and proceeded to Burlington, where I learned that the principal part of the friendly Indians were congregated, and encamped on the Verdigris river, near a place called Roe's Fork, from twelve to fifteen miles south of the town of Belmont. I proceeded there without delay. By a census of the tribes taken a few days before my arrival, there was found to be of the Creeks, 3,168; slaves of the Creeks, 53; free negroes, members of the tribe, 38; Seminoles, 777; Quapaws, 136; Cherokees, 50; Chickasaws, 31; some few Kickapoos and other tribes, about 4,500 in all. But the number was being constantly augmented by the daily arrival of other camps and families. . ."

- A. B. CAMPBELL, surgeon, U.S.A., to James K. Barnes, surgeon, U.S.A., medical director, Department of Kansas, dated Fort Leavenworth, February 5, 1862.[3]

Superintendent Coffin received emergency relief of $10,000 from the Interior Department and was able to purchase some cheap blankets for distribution, and the Army contributed some worn out and condemned tents for the refugees. Coffin and his son, who was acting as his assistant, contributed additional funds from their salaries.[4] Coffin wrote in his report to the Secretary of the Interior on February 13, 1862:

"...destitution, misery and suffering amongst them is beyond the power of any pen to portray, it must be seen to be realised [*sic*]— there are now here over two thousand men, women, and children entirely barefooted and more than that number that have not rags enough to hide their nakedness, many have died and they are constantly dying. I should think at a rough guess that from 12 to 15 hundred dead..."[5]

The hunger and sickness were compounded by the stench of the 1,500-2,000 dead Indian ponies that were spread across the prairie. Another 2,000 were starving and would be dead in a matter of weeks.

Even with medical care, many refugees suffered from frostbite-related amputations and sickness. Starvation was a constant problem. The Army and the Indian agents were unable to provide enough

Regional Map showing Opothleyahola's Flight and the Trails of Blood and Ice and Colonel Doubleday's advance on Cowskin Prairie

subsistence to accommodate the constant arrivals from the Indian Territory, which at times brought the refugee number up close to ten thousand. The Loyal Indian warriors attempted to hunt game to supplement their meager government rations but soon the game disappeared due to over-hunting.[6] The leaders of the Sac & Fox Nation, whose reservation was north of the refugee camp, offered asylum to the Loyal Indians, but they were insistent about going back home to the Indian Territory.

According to Sergeant Wiley Britton of the 6th Kansas Cavalry, an author with first-hand knowledge, wrote that the newer white settlers in Kansas learned of their needs and, with deep sympathy, stepped forward to help. The Kansans had experienced a very good crop of corn and wheat and much of this production, as well as extra clothing and shelter, was diverted to the starving Indians at very reasonable prices. Even with this one bright spot, it is estimated that over one thousand Loyal Indians perished from starvation or a variety of sicknesses.[7] The ground was frozen, so the dead were placed in logs, buried in hacked-out shallow graves, or just left to the elements. Eventually, most of the Loyal Indians were moved up to Fort Belmont, another militia-built defensive position further north about 15 miles. Opothleyahola's daughter died during the winter and was buried at Fort Belmont. Later they were moved on to LeRoy, Kansas, another 20 miles, where Senator James Lane had established a post that would be occupied by Union forces if Fort Scott was captured by the Confederates. Many of the Loyal Indians, especially the women and children, remained at LeRoy until the spring of 1864.[8]

U.S. Indian Commissioner William P. Dole did not learn of the evacuation of Indian Territory or the plight of Opothleyahola's followers until he arrived at Fort Leavenworth in late January 1862. On Saturday, February 1, Commissioner Dole held a conference with Creek leader Opothleyahola and Seminole leader Tustenuggee at the Planters' House in Leavenworth, Kansas. Coffin, Cutler, and the newly appointed agent of the Seminoles, George C. Snow, were present. Dole and Opothleyahola engaged in a spirited conversation:

> "*Mr. Dole*—[The] Government did not expect the Indians to enter this contest at all. Now that the rebel portion of them have entered the field, the Great Father will march his troops into your country. Col. Coffin and the agents will go with you on Monday, and will assist you in enlisting your loyal men. Your enlistment is not done for our advantage only; it will inure to your own benefit. The country appreciates your services. We honor you. You are in our hearts.

> "One party tells us that John Ross is for the Union, and one that he is not.

> "*Opothleyahola*—Both are probably right. Ross made a sham treaty with Albert Pike, to save trouble. Ross is like a man lying on his belly, watching the opportunity to turn over. When the northern troops come within the ring, he will turn over.

> "Dole—You did not, and our people remember you. But we hope you will manifest no revenge.

"*Opothleyahola*—The rebel Indians are like a cross, bad slut. The best way to end the breed is to kill the slut.

"*Dole*—The leaders and plotters of treason only should suffer.

"*Opothleyahola* —That's just what I think. Burn over a bad field of grass and it will spring up again. It must be torn up by the roots, even if some good blades suffer. The educated part of our tribes is the worst…"[9]

These conversations gave a final push towards organizing and arming the Loyal Indians for their participation in the re-taking of the Indian Territory from the Confederates. The push by Senator James Lane (he also held a blank brigadier general of volunteers commission from Lincoln that he had not yet accepted)[10] to form an Indian brigade made up of local Kansas tribes to protect the Kansas-Missouri border counties had fallen on deaf ears. Now the influx of thousands of Loyal Indians, many of whom were warriors or of warrior-age, gave re-birth to the idea. After enduring a harsh and vicious winter on the Kansas prairie, the Loyal Indians began to strongly and vocally express to their leaders that they wished to return home to the Indian Territory. They believed that they had a better chance of survival on the lands that they knew than on the bitter-cold windswept bluestem prairies where they were now forced to live. Few had come to Kansas with the intention of staying there. In fact, it was obviously never part of Opothleyahola's plan to evacuate the Indian Territory into Kansas. His plan had been to find a secure place within the Cherokee

Nation to stay and sit out the war. The Loyal Indian leaders stated to each of the changing commanders of the Union forces in Kansas that all they wanted was a military force to accompany them and protect them as had been promised all along.

The first seven months of the Civil War brought many changes to the command structure of the U.S. Army. In Washington D.C., the new General-in-Chief of the United States Army, George B. McClellan, was taking the reins from a now-retired Winfield Scott. McClellan was making elaborate plans for his advance up the peninsula of Virginia towards Richmond with his newly formed Army of the Potomac. Out on the western border, Maj. Gen. David Hunter, the former Paymaster of the Army and newly appointed Regular Army Colonel of the new 6th U.S. Cavalry, relieved Maj. Gen. John C. Fremont of command of the Western Department on November 2, 1861.[11] Fremont had been a California senator and was the failed and controversial Republican presidential candidate who had lost the 1856 election to James Buchanan. He was also partly to blame for the Union loss at Wilson's Creek since he failed to provide the needed support to Brig. Gen. Nathanial Lyon prior to the battle. Fremont was relieved by McClellan due to his attempt to emancipate the slaves of rebellious slave holders in Missouri. (Gen. Hunter would be accused of the same thing in South Carolina later in 1862) Within a week the Western Department of the Army was disestablished and split into the Department of Kansas, which included the Indian Territory and was headquartered at Fort Leavenworth, Kansas, and the Department of the Missouri, headquartered at St. Louis, Missouri. Maj. Gen. Hunter was placed in command of the newly designated

Department of Kansas, and Maj, Gen. Henry Hallack, known throughout the Army as "Old Brains" due to his earlier extensive academic research and writing on military doctrine and tactics, was placed in command of the new Department of the Missouri.[12] With pressure from the White House and the War Department for advancement into the Confederacy, Maj. Gen. McClellan wanted the Union force in Kansas to plan an invasion Texas via the Indian Territory. Adjutant General Lorenzo Thomas sent this message:

> Adjutant-General's Office,
> Washington D.C.
> November 26, 1861.
> Maj. Gen. D. Hunter, U. S. Army,
> *Commanding Department of Kansas, Fort Leavenworth, Kans.:*
> Sir: The General-in-Chief thinks an expedition might be made to advantage from your department west of Arkansas against Northeastern Texas. He accordingly desires you to report at an early day what troops and means at your disposal you could bring to bear on that point.
> I am, sir, &e.,
> L. THOMAS,
> Adjutant-General[13]

It was Senator Lane who had primarily developed the initial plan along with Brig. Gen. Benjamin Butler, commander of the Department of New England. The plan was for a column, under the command of Lane, to move south out of Kansas, through the Indian Territory, and into northeastern Texas. A second column, under Butler, would land on the Texas Gulf Coast, move northward into northeastern Texas, and join forces with the Kansas column. It was hoped that this movement would cut off Texas and its vast resources from the remainder of the Confederacy. Maj. Gen. Hunter replied, much to the anger of Maj. Gen. McClellan:

> Headquarters
> Department of Kansas
> Fort Leavenworth, Kans
> December 11, 1861
> *[Brig. Gen. Lorenzo Thomas:]*
> General : "…In reply I have the honor to report that I think the expedition proposed by the General-in-Chief altogether impracticable. We have a hostile Indian force, estimated at 10,000, on the south, and Price's command, some 20,000, on our east and north. To cope with this force we have only about 3,000 effective men, scattered over an extended frontier. So far from being able to make successful expeditions into the enemy's country with our present force, I think we shall be very fortunate if we prevent his having possession of the whole of Kansas. …The possession of Leavenworth would be a great feather in the enemy's cap, and really there is nothing to prevent his having it any day he may see fit. We will give him a hard fight, but he will have ten to our one…"
> I have the honor to be, very
>   respectfully,
> your most obedient servant,
> D. HUNTER
> Major General, Commanding[14]

General-In-Chief George McClellan quickly responded to Maj. Gen. Hunter's message:

[Unofficial.]

Headquarters of the Army,

Washington,

December 11, 1861.

Maj. Gen. D. Hunter,

Commanding Department of Kansas :

General : Your telegram to General Thomas surprised me exceedingly. Realizing as I do the very trying nature of the circumstances in which you are placed, I have attributed it to momentary irritation, which your cooler judgment will at least lead you to regard as unnecessary. Immediately after you were assigned to your present department I requested the Adjutant-General to inform you that it was deemed expedient to organize an expedition under your command to secure the Indian territory west of Arkansas, as well as to make a descent upon Northern Texas, in connection with one to strike at Western Texas from the Gulf. The general was to invite your prompt attention to this subject, and to ask you to indicate the necessary force and means for the undertaking... I would again call your attention to this very important subject, stating the necessary force shall be placed at your disposal.[15]

Maj. Gen. Hunter had obviously misunderstood the message, and Adjutant General Thomas had not fully explained the General-In-Chief's intentions. McClellan simply wanted a status of the forces Maj. General Hunter had available in his command and what levels of troops he would need to execute the mission. Still offended, Hunter felt he was going to simply be a figurehead while Senator Lane, after he accepted his commission, would command the troops and Dole would manage the Indians. Hunter wrote to President Lincoln and said, "I am very deeply mortified humiliated, insulted, and disgraced..." by the letter and sharp response from Maj. Gen. McClellan. Lincoln responded angrily:

"I am, as you intimate, losing much of the great confidence I placed in you... from the flood of grumbling dispatches and letters I have seen from you... You constantly speak of being placed in command of only 3,000. Now tell me, is this not mere impatience? Have you not known all the while that you are to command four or five times that many? I have been, and am, sincerely your friend; and if, as such, I dare to make a suggestion I would say you are adopting the best possible way to ruin yourself."[16]

It should be noted that these requests for offensive movements were made before the U.S. Army in the Department of Kansas was further burdened with the Loyal Indian refugees flowing into Kansas in January and February 1862.

Maj. General Hunter had reason to suspect devious moves by Senator Lane. The Senator had been busy undercutting Hunter in Washington by claiming he had Hunter's full support for the plan and was promoting himself the position as the leader of the Indian Expedition. Once President Lincoln caught wind of Lane's false representations of Maj. Gen. Hunter's wishes and his promotion of his own plan for a 30,000-man force to invade northeastern Texas via the Indian Territory. Lincoln sent a message to the secretary of war:

> "It is my wish that the expedition commonly called the "Lane Expedition" shall be as much as has been promised at the Adjutant-General's Office under the supervision of General McClellan and not any more. I have not intended and do not intend now that it shall be a great exhausting affair, but a snug, sober column of 10,000 of 15,000. General Lane has been told by me many times that he is under the command of General Hunter and assented to it as often as told. It was a distinct agreement between him and me when I appointed him that he was to be under Hunter."[17]

Although Lane had secured a congressional resolution calling for his installation as commander of the expedition, Lincoln stood his ground and refused to submit to the pressure from Congress. In response, a very angry Hunter decided to try and outsmart Lane by announcing to the Kansas newspapers that there was not to be a "Lane Expedition." Hunter announced that he himself would lead the Indian Expedition into the Indian Territory.

> General Orders, > No. 11.)
> Headquarters Department of Kansas,
> Fort Leavenworth, Kans.,
> January 27, 1862.
> 1. In the expedition about to go south from this department, called in the newspapers General Lane's Expedition, it is the intention of the major-general commanding the department to command in person, unless otherwise expressly ordered by the Government.
> 2. Transportation not having been supplied, we must go without it. All tents, trunks, chests, chairs, camp-tables, camp-stools, &c., must be at once stored or abandoned. The general-commanding takes in his valise one shirt, one pair drawers, one pair socks, and one handkerchief, and no officer or soldier will carry more. The surplus room in the knapsack must be reserved for ammunition and provisions. Every officer and soldier will carry his own clothing and bedding.
> 3. The general commanding has applied to the Government for six brigadier-generals, that his command may be properly organized. Until their arrival it is necessary that he appoint acting brigadier-generals from the senior colonels. To enable him to do this, in accordance with the order on the subject, each colonel will immediately report the day on which he was mustered into the service of the United States.
> D. HUNTER,
> Major-General,
> Commanding.[18]

When Lane arrived at Fort Leavenworth from Washington, he was stunned by news that he was not to be the commander of the expedition. This act infuriated Lane, and he began to search for an opportunity for revenge. He did so by sabotaging and using his congressional friends to find ways of diverting the funding, supplies, and staffing away from the expedition.[19] Lane also refused his brigadier general commission and finally returned to Washington to re-take his seat in the Senate. When the western commands were shifted again in March, Maj. Gen. Hunter was ordered back east and was assigned to command the Department of the South.[20]

On February 15, 1862, the Army stopped supplying the Indian refugees and placed the entire burden on the Interior Department. Secretary of the Interior Caleb Smith again authorized relief funds to be used in supplying the refugees. Unfortunately, this created another problem, one that seemingly always haunted any government funds issued without proper oversight. Contractors continually overcharged and undersupplied the Indian refugees and the Interior Department. They provided shoddy material and tainted and moldy food products. To counter this, Indian Affairs Commissioner Dole hired Dr. William Kile of Illinois from the staff of Senator James Lane as a special agent who would be responsible for the purchase and distribution of supplies for the Loyal Indian refugees. Dr. Kile actually visited the encampments to determine needs and requisitioned the needed supplies. His investigations showed that a great portion of the government funds already expended had been misused and pocketed by dishonest contractors, agents, and members of the military. From this point the flow of supplies and materials to the refugee encampments improved dramatically.[21]

Due to the conflict between Maj. Gen. Hunter and Senator Lane, President Lincoln called off the Indian Expedition in late February 1862, believing it to be inconsequential. Still, there were forces in Washington and in Kansas who believed that by sending a military expedition into the Indian Territory and holding it, the refugee Loyal Indians could accompany the Army and re-establish themselves on their lands. This action could give the Loyal Indians a chance to plant crops and husband livestock that would help them in becoming self-sufficient and thus relieve the government of continually providing provisions. This would far out-shadow Senator Lane's idea for using Indian troops to patrol the Kansas-Missouri border. Moreover, with the Confederate defeat at the Battle of Pea Ridge on March 7, and their subsequent abandonment of northwestern Arkansas, the only challenge to a Union invasion of Indian Territory was Brig. Gen. Albert Pike's Confederate Indian Brigade. On March 13, 1862, Commissioner Dole recommended to the Secretary of the Interior that Union regiments be raised from the Loyal Indians to invade and restore the Indian Territory to United States control. The Indian regiments would remain in the Indian Territory as a "home guard." Dole requested from Secretary Smith:

> "Procure an order from the War Department detailing two Regiment of Volunteers from Kansas to go with the Indians to their homes and to remain there for their protection as long (as) may be necessary, also to furnish two thousand stand of arms and ammunition to be placed in the hands of the loyal Indians."[22]

Washington's response was unusually prompt. In less than a week, Dole had been promised two white regiments and two thousand armed Loyal Indians. General Henry Halleck, commander of the Department of the Mississippi, was more concerned with keeping the "jayhawkers," free-soilers who terrorized, robbed, and freed slaves from slave-owners, from making anti-slavery raids into western Missouri which was creating discontent among the citizens there, especially since Missouri had not left the Union and was still an official State. To avoid any misunderstandings, Judge James Steele, a confidential special agent from the U.S. Indian Office, was sent to General Halleck with the order from Washington to begin recruitment and form an expedition to retake the Indian Territory.

> Adjutant-General's Office,
> Washington, April 4, 1862.
> Maj. Gen. H. W. Halleck, U. S. A.,
> Comdg. Department of the
> Mississippi, Saint Louis, Mo.:
> General : The Secretary of War, with the concurrence of the Secretary of the Interior, has granted authority to Robert W. Furnas and John Richay, esqrs., to raise two regiments from such loyal Indians as have been driven from their own country into Kansas by other Indians in rebellion against the United States Government. These regiments are to be raised for the purpose of restoring their lands to the loyal Indians and offering them protection while planting their crops… it is the desire of the Secretary of War that you

> furnish two regiments of volunteers to aid these Indian troops in effecting the purpose for which they are to be raised.
>
> I am, general, very respectfully, your obedient servant,
> L. THOMAS,
> Adjutant- General.[23]

Although he was reluctant to act, by April 1862 Halleck had begun to collect the required arms and started recruiting the Indian regiments. Recruitment officers swept into the refugee Indian encampments with the promise of a new blanket, musket, and ammunition. Many Indian men were quick to sign up simply to get away from the awful conditions of the refugee camps.[24] The Indian troops were to receive pay equal to white soldiers, with the tribal chiefs paid a modest amount more. Their families would receive the same death benefits as the other troops.[25] An unusual arrangement that the United States Army took in regard to the Indian regiments is that, although the commanding officers were white, some of the field-grade officers and most of the company officers were Indians, which is in stark contrast to the later-recruited black regiments. It was also a relief to the situation for the Indian Office. Since the men in the recruited regiments would be in the Army, the War Department would be responsible for providing food and shelter for those Indians.

The Union Indian Expedition had four primary goals. The first was to return the Loyalist Indians to their homes, especially those in the Cherokee Nation, and set them up to be self-sufficient. The second mission was the re-capture of Fort Gibson for use as a headquarters and supply depot for U.S. Army operations in the Indian Territory. The third assignment was to capture Fort Smith, Arkansas, which was considered the mother post of Indian Territory. Its capture would give Federal forces uninterrupted use of the Arkansas River for supplies and troop movements. A final goal, which was probably the most important, was to be a show of force to the Five Civilized Tribes that the United States Government was still present and would re-conquer the Indian Territory from the Confederacy. They would give the tribes the opportunity to peacefully withdraw their support for the Confederate States and return to the jurisdiction of the United States.[26]

General Halleck had previously created the District of Kansas within the Department of the Mississippi and placed Brigadier General James W. Denver[27] in command on April 2. Denver was also tapped to command the Indian Territory expedition. Denver was President Lincoln's personal choice to lead the upcoming expedition. He was a Virginian and Mexican War veteran who had relocated to Kansas from California during the late 1850s. Denver had served under President Buchanan as the Commissioner of Indian Affairs and secretary for the Kansas Territory and had an intimate knowledge of Kansas politics. The selection did not please Senator Lane and his disciples in Kansas, probably because he came from a Democratic administration, and they made their positions known to President Lincoln. Denver had only been in command for five days and had just started his preparations for the expedition when he was notified that he was to be replaced. His replacement was to be Brigadier General Samuel D. Sturgis, a Pennsylvania native and officer in the Regular Army, and a Mexican War veteran who, as mentioned in an earlier chapter, was the

Maj. Gen. David Hunter, USA
Department of Kansas
Photo courtesy of the Library of Congress

officer who evacuated Fort Smith just before its occupation by Arkansas state forces.[28] It seems that Maj. Gen. Halleck still did not understand the delicate political situation in Kansas regarding command slots. Halleck stated in a letter to Secretary of War Edward Stanton:

> Hon. E.H. Stanton
> *Secretary of War, Washington*
> "…In detailing General Denver for the command in Kansas I followed the advice of the officers of General Hunter's staff. They gave it as their opinion that he was best suited for the place, and as I had very little personal acquaintance with him I felt bound to follow the best advice I could obtain some political influences connected with this matter. Not being a politician, this did not occur to me. I am a little surprised, however, that

politicians in Congress should be permitted to dictate the selection of officers for particular duties in this department. Under such circumstances I cannot be responsible for the results. Nevertheless I shall permitted to dictate the selection of officers for particular duties in this department. Under such comply with the President's wishes, and place some other officer in command in Kansas as soon as I can spare one for that purpose…"[29]

Brig. Gen. Sturgis assumed command of the District of Kansas on April 10. After the scalpings and atrocities that had occurred at Pea Ridge, Sturgis was firmly opposed to the idea of arming the Indians and ordered the recruitment for the Indian regiments to be halted and those already enlisted to be released. Anyone who violated this rule would be arrested. The denial of enlistment and firearms for the Loyal Indian recruits went firmly against the orders given by the War and Interior Departments. U.S. Indian Affairs Commissioner Dole complained loudly enough to the Secretary of the Interior and President Lincoln that the Department of Kansas was re-established on May 2 as an independent command after being absorbed into the Department of the Mississippi. Brig. Gen. Sturgis was quickly recalled to the east and was absorbed into the Washington D.C. defense forces. James G. Blunt, who was currently serving as the Lt. Col. of the 3rd Kansas Infantry, was named a Brigadier General of Volunteers and the commander of the new department. Blunt was an unusual choice for such an important command assignment. This native of Maine was a

physician by trade and an ardent abolitionist, being a close supporter of John Brown. He had little practical military experience but was a trusted confidante of the powerful Senator Lane. (Lane had again lobbied President Abraham Lincoln for himself to command of the expedition but had been refused) Blunt had many admirers and detractors, did not care for military decorum or customs, serving most days in civilian clothing, not a uniform. He was basically an angel and a devil at the same time. He was an aggressive leader who loved to fight and get close in combat. On the other hand, he had numerous ethical issues. One chaplain of the 10th Illinois Cavalry, Reverend Francis Springer, was much offended by Blunt's activities. The good reverend stated:

> "Though his head is large and well-proportioned to a stout chunky body, the manifestations of it in the way of brains are plainly surpassed by the man's love of good eating, intoxicating liquors, and free women… He is not a man of reflection and reason, but only of impulse."[30]

He was basically just the right person for the job in Kansas. General Blunt immediately reversed Sturgis's orders and began recruitment and arming the Union Indians again.:

General Orders,
Hdqrs. Department of
Kansas, No. 2.
Fort Leavenworth, Kans.,
May 5, 1862.

William P. Dole
U.S. Commissioner of Indian
Affairs
Photo courtesy of the Library of Congress

Brig. Gen. James G. Blunt,
Department of Kansas
Photo courtesy of the Kansas State
Historical Society

Colonel Robert W. Furnas, USA
1st Indian Home Guard
Photo courtesy of the National Archives

Colonel John Ritchie, USA
2nd Indian Home Guard
Photo courtesy of the Kansas State
Historical Society

Colonel William Phillips
3rd Indian Home Guard
Photo courtesy KSHS

Lt. Col. Albert C. Ellithorpe
1st Indian Home Guard
Photo courtesy of the Kansas State
Historical Society

Fort Scott National Historic Site
Photos by Author

Baxter Springs Depot, Kansas
Photo by Author

I. General Orders, No. 8, dated Headquarters District of Kansas, April 25, 1862, is hereby rescinded.

II. The instructions issued by the Department at Washington to the colonels of the two Indian regiments ordered to be raised will be fully carried out, and the regiments will be raised with all possible speed.

By order of
Brig. Gen. James G. Blunt
Thos. Moonlight,
Captain and Assistant
Adjutant-general.[31]

Since Blunt would need to remain at Fort Leavenworth to oversee the department and ensure supplies made it to the Indian Expedition, Col. Charles Doubleday of the 2nd Ohio Cavalry was designated as the Indian Expedition commander. Blunt was unsure of his ability to command such a large department, and he felt he had been placed in an unpleasant and embarrassing situation. He later wrote:

"…Of the troops in my command, the greater portion of them were Kansas regiments, all of which had become more or less disaffected in consequence of the unauthorized interference of the governor with their organizations, while the fact that military matters in Kansas had been conducted very much in the manner of a political canvass, rendered the administration of the affairs of the department anything but pleasant to an inexperienced commander. My assignment to this command was the signal for a combined attack of all my personal and political opponents, as also the opponents of all with whom I had held intimate personal or political relations, and to make my position still more difficult, this crusade against me was headed by the governor of the state, from whom, in his official capacity, I had a right to expect cooperation, but whose acts seemed to indicate more of a desire to embarrass and complicate military operations than to contribute to their success…"[32]

The selection of Brig. Gen. Blunt and Col. Doubleday did not sit well with the Loyal Indians. They wanted Senator Lane to lead the Union Indian Brigade and the expedition. They believed that he was a "great war chief" and had a "big heart" for the Indians. Opothleyahola himself had appealed to President Lincoln in late January 1862 by stating, "Our object…is to beg that General Lane be placed in command of the expedition." The Loyal Creek leader further stated, "General Lane is our friend…," as he appealed on behalf of the six thousand Loyal Indians in Kansas.[33] Unfortunately, the conflict and issues between General Hunter and Senator Lane had damaged each of their reputations, and neither would command the expedition. But since Brig. Gen. Blunt had been hand-picked by Senator Lane from the remnants of "Lane's Kansas Brigade" to command the Department of Kansas, it was an accepted fact that Lane would still have influence on the Indian Expedition.

After the flurry of changes in command and orders and counter-orders at the department and district levels, the recruitment of the 1st Indian Home Guard began in earnest in mid-May 1862 at LeRoy and Humbolt, Kansas. When completed by May 22, the regiment would consist of ten companies, eight of Loyal Creeks, and two of Loyal Seminoles. One of the companies of Creeks included a group of "Yuchies," an Indian group affiliated with the Creeks but that had always been regarded as a separate entity. The regiment was to be commanded by Col. Robert W. Furnas, an Ohio native who had re-located to Brownsville, Nebraska Territory, and published a couple of newspapers, the "*Nebraska Farmer*" and the "*Nebraska Advertiser.*" Furnas was mustered in on April 18, 1862, at Fort Leavenworth after leaving behind a wife and seven children. His second-in-command was Lt. Col. Stephen H. Wattles, another newspaperman also from the Nebraska Territory. Many of the prominent men from the Loyal Indians were selected as officers in the regiment including Tuckabatchee Harjo, Opothleyahola's nephew, who was selected as a company commander; Tulsy-Fixico, another Upper Creek leader, who received a commission; and Halleck Tustenugee, a veteran of the Seminole Wars, was commissioned as a captain. John Chupco, the assistant chief of the Seminoles, was commissioned into the regiment as was Sonaki Mikko, or Billy Bowlegs, and became a company commander. Also scattered throughout the regiment were African Creeks and Seminoles, former enslaved or freedman who had accompanied the refugees from Indian Territory. Since many of these individuals could speak both Creek and English, many would serve as interpreters between the white commanders and their Indian troops.[34] The 1st IHG now consisted of 1,009 enlistees assigned to the following units:

Commanding Col. Robert Furnas
Lt. Col. Stephen Wattles
Maj. William Phillips
Company A.   Seminole
    Capt. Billy Bowlegs
Company B.   Seminole
    Capt. A-ha-luk-tus-ta-na-ke
Company C.   Creek
    Capt. Tus-te-ne-ke-ema-ela
Company D.   Creek
    Capt. Tus-te-nuk-ke
Company E.   Creek
    Capt. Jon-neh (John)
Company F.   Creek
    Capt. Mic-co-hut-ka (White Chief)
Company G.   Creek
    Capt. Ah-pi-noh-to-me
Company H.   Creek
    Capt. Lo-ga-po-koh
Company I.   Creek
    Capt. Jan-neh (John)
Company J.   Creek
    Capt. Lo-ka-la-chi-ha-go[35]

Among the earliest white officers selected for service in the 1st Indian Home Guards was Albert C. Ellithorpe, an "Old Settler" from Chicago, Illinois who had been a businessman, inventor, and a sometimes-adventurer who longed to find a place within the Union Army as an officer. Through a personal appeal to President Lincoln, a pre-war friend, he was finally selected to be a first lieutenant in the new 1st Indian Home Guard. A dedicated officer, he was responsible for almost 60% of the enlistments into the new regiment. Ellithorpe would find a journal that had been lost by a Confederate soldier during the Battle of Locust Grove, and because most of the journal was unused, he began using it himself recording his experiences as an officer in the Union Indian Brigade. Much of what we know of their actions and daily life of the Indian Brigade comes from this personal journal.[36]

The 2nd Regiment of Indian Home Guards (IHG) took longer to recruit and was not as evenly distributed as the First Regiment as to company tribal makeup. It was organized at Big Creek and at Five-Mile Creek, Kansas in May 1862. It included two companies of Osage, two companies of Cherokee, and one company each from the Delaware, Seneca, Kickapoo, Quapaw, and Shawnee. At the time of the expedition's departure the 2nd IHG had not yet been recruited to full strength. This regiment was commanded by Colonel John Ritchie, who was born in Uniontown, Ohio, moving to Kansas Territory in 1854. In addition to farming and real estate development in and near Topeka. Ritchie was described by the *Daily Times* as "an ultra-Abolitionist, woman's rights man, teetotaler and general advocate for reform." He had commanded the 5th Kansas Cavalry in the Kansas Brigade before being given the colonelcy of the 2nd Indian Home Guards.[37] The Indian Expedition's Wagon Boss R.M. Peck, formerly a sergeant in the Regular Army, made this observation of the new commanding officer of the 2nd IHG:

"This man, John Ritchie, of Topeka, Kans., was a fanatic, a monomaniac on the subject of slavery, who through political influence was appointed Colonel of the 2d Indian

Regiment, a position he was totally unqualified for. His chief recommendation seemed to be a violent hatred of every one who favored slavery… it was said that he tolerated and encouraged the killing and scalping of rebel Indians, and atrocious treatment of their families, in his march through the country. Our Government never allowed any such cruelties…"[38]

Col. Ritchie had been delayed by going south into the Osage country to gather recruits for his command. He did not arrive to take command of his regiment until after most of the 1st IHG had been recruited. The regiment's adjutant, Sergeant George Dole, was burdened with getting the regiment recruited and organized without the assistance of the commanding officer. Very little information is available on the 2nd Indian Home Guard's personnel. The Civil War Soldiers & Sailors Database, operated by the National Park Service, shows the following arrangement of commands within the 2 IHG:

Commanding Col. John Ritchie
Lt. Col. Fred W. Schaurte,
　　　David B. Corwin
Maj. Moses BC Wright
Company A.   Unknown Tribe
　　　Capt. James McDaniel
Company B.   Unknown Tribe
　　　Capt. John Cochran,
　　　Robert Lombard (dates unk)
Company C.   Unknown Tribe
　　　Capt. Ned Jim,
　　　Joe Besaillion (dates unk)
Company D.   Unknown Tribe
　　　Capt. Leaf Fall,

Archibald Scraper (date unk)

Company E.   Unknown Tribe

Capt. Chetopa

Company F.   Unknown Tribe

Capt. William Pryor

Company G.   Unknown Tribe

Capt. John Belne

Budd Gritts (dates unk)

Company H.   Unknown Tribe

Capt. George Scraper

Company I.   Unknown Tribe

Capt. Dirt Throw Tiger

Company J.   Unknown Tribe

Capt. Frog Spring

Colonels Furnas and Ritchie soon began to have trouble recruiting enough Indian men into the new Indian Home Guard regiments. Superintendent Coffin suggested that they use artillery demonstrations to awe potential recruits with how much power in "them waggons [sic] that shoots" provided to the Army. Three demonstrations were held at Iola, St. Paul, and LeRoy, Kansas. Superintendent Coffin reported to Commissioner Dole:

"…nearly five thousand Indians a very fine show men of artillery which was received by the Indians with entire satisfaction and applause, it has made a grand impression upon them and has also strengthened their confidence in the success of the have unbounded confidence in their "waggons [sic] that shoot's [sic] I doubt very much whether those Indians could have been induced to go at all had it not been for that Battery going along with them."[39]

One problem confronting both the Union and Confederate Indian troops was procuring adequate firearms, supplies and equipment. Neither side regarded its Indian regiments as a high priority and never made a determined effort to see that they were supplied or equipped as well as the white troops. This was a consistent attitude toward non-white troops that served throughout either army. Letters and messages flew between the western departments and their respective governments in Washington and Richmond, but the situation never improved significantly, especially for the Confederates. One problem the Confederate Indians endured was that whatever supplies were sent their way were many times confiscated by other Confederate units further east, especially in Arkansas. This appropriation by Confederate commanders also included promised treaty annuity goods, which resulted not only in hardship among the Indian people, but also in turning many away from supporting the Confederate cause.[40] The Union Army tended to fare much better in this regard. The Army did issue standard blue uniforms to the Indian Home Guard (IHG) regiments although they tended to be issued the odd sizes that would not fit the white soldiers. It was common to see an IHG soldier with a uniform that was far too large or much too small. Sgt. Wiley Britton of the 6th Kansas Cavalry remarked about the outfitted Indian regiments:

"Shortly after these regiments were organized the men were furnished clothing by the government the same as other soldiers. It was quite amusing to the white soldiers to see the Indians dressed in the Federal uniform and equipped for the service. Every thing seemed out of just proportion. Nearly every warrior got a suit that, to critical tastes, lacked a good deal in fitting him. It was in a marked degree either too large or too small. In some cases the sleeves of a coat or jacket were too short, coming down about two thirds the distance from the elbows to the wrists. In other cases the sleeves were too long, coming down over the hands."

"…At the time these Indian troops were organized the Government was furnishing its soldiers a high-crowned stiff wool hat *[Hardee Hat]* for the service. When, therefore, fully equipped as a warrior, one might have seen an Indian soldier dressed as described, wearing a high-crowned stiff wool hat, with long black hair falling over his shoulders, and riding an Indian pony so small that his feet appeared to almost touch the ground, with a long squirrel rifle thrown across the pommel of his saddle. When starting out on the march every morning any one with this command might have seen this warrior in full war-paint,.."[41]

Wagon Boss Peck also noted how the Loyal Indians took to wearing their uniforms:

"It was interesting to notice the use made by the full bloods of the (to them) munificent gifts of Uncle Sam, when clothing and equipments [sic] were issued to them. Probably few of them had ever before possessed more clothing than they carried on their

persons at one time, and as soon as they received their outfits they would put all on, one article after another, even to the soldiers overcoats; then buckling on their belts with cartridge box and bayonet, wrapping their heavy double U.S. blankets around over all, with a new musket in hand, would strut about camp apparently very proud of being a United States soldier. It seemed to be a matter of duty with them."[42]

To arm the upwards of two thousand soldiers of the Indian Home Guard regiments, they were issued the long-barreled .54 caliber M1817 "Common or Indian Rifle." It was the only firearm that was available in sufficient numbers on hand at Fort Leavenworth to arm a large part of the regiments. This rifle used a round bullet that was quite effective at close range and had been converted to a percussion cap design instead of the older style flint and powder pan. The primary issue with these rifles was that each soldier had to mold his own bullets. Sgt. Britton states that the Indian troops preferred this rifle to the standard Army rifle being issued at the time.[43] Indian Superintendent Coffin was concerned about the distribution of the firearms to the Indian soldiers. He believed that many of the firearms would end up in the hands of most of the Indian men, soldier or not.

Senator James Lane, serving as a Lincoln-appointed brigadier general of volunteers (although he had not yet signed his commission), had earlier been given authority to raise two Kansas regiments. Lane began organizing troops for defense along the Kansas-Missouri border. He quickly began recruiting and, within a short time, the Third, Fourth, Fifth, Sixth, and Seventh regiments were ready for service. Lane took command of the some 1,500 untrained and undisciplined militiamen at Fort Scott, Kansas. This "Kansas Brigade" participated in actions in western Missouri and in Kansas, raided and confiscated the property of Missouri slaveholders (including slaves), and assisted fugitive slaves to escape.[44] As mentioned in an earlier chapter, after the Battle of Wilson's Creek, Missouri on August 10, 1861, the Union army retreated back towards Rolla, Missouri. With the Kansas border exposed, Missouri State Guard General Sterling Price threatened Fort Scott and the "free-soilers" of Kansas. Lane led his Kansas Brigade against General Price in the Battle of Dry Wood Creek, located 12 miles east of Fort Scott, Kansas, on September 2, 1861. Heavily outnumbered in the two-hour battle, the Kansas force was able to stop Price's forces from attacking Fort Scott. Lane was forced to abandon his mules and withdraw to Fort Scott and then to Kansas City. Rather than try and capture Fort Scott and being surprised by the stiff resistance of the Kansas Brigade, the Missouri State Guard switched course to attack Union forces at Lexington, Missouri. Though his troops lost the battle, Lane continued, fighting through the towns of Paninsville, Butler, Harrisonville, and Clinton, Missouri, before he ended his campaign by the burning of Osceola on September 23, 1861. This unauthorized "jayhawker" attack resulted in the destruction and plundering of much of the town, including the summary execution of nine of its residents. (Guerilla leader William Quantrell stated that his sacking of Lawrence, Kansas, was a retaliation for this attack.) The troops continued to pursue Price's men for a time, but Lane was severely criticized for his actions in Osceola

and was soon sent back to Kansas. Lane was most severely condemned by General Henry Halleck, who was at the time commander of the Department of the Mississippi.[45] The remnants of the Kansas Brigade spent most of the winter of 1861-1862 milling around Fort Scott, with many in a constant state of intoxication. When Agent Cutler had brought the representatives of the Creek Nation to Fort Scott prior to their departure for Fort Leavenworth in the fall of 1861, the Creek leaders were shocked by the appearance of the men of the Kansas Brigade. They observed the Kansas militiamen as disrespectful, undisciplined, loud, violent, dirty, and drunk. They were the exact opposite of what the Creek leaders would encounter at Fort Leavenworth and its garrison of soldiers of the Regular U.S. Army.[46] The Kansas Brigade was also crippled with desertions and the effects of far too many men showing up with officer commissions from Kansas Governor Charles Robinson. Maj. Gen. Hunter, prior to his transfer to the Department of the South, determined that the best course of action before the Indian Expedition would be to break up the Kansas Brigade and reform it into re-designed or new regiments. These regiments would be the auxiliary white troops that were required to accompany the Indian Home Guard regiments into the Indian Territory. In response to the War Department's request for information, General Hunter issued this order:

> General Orders, 26)
> Headquarters Dept, of Kansas,
> Fort Leavenworth, Kans.,
> February 28, 1862.
> I. Pursuant to instructions from
> Headquarters of the Army, dated

Adjutant-General's Office, Washington, D. C., February 1, 1862, and by and with the consent of His Excellency the Governor of Kansas, the following reconstruction of the volunteer forces of the State of Kansas is hereby made, in accordance with the acts of Congress heretofore promulgated in General Orders, series of 1861, from the War Department.

> By order of Major-General Hunter:
> CHAS. G. HALPESTE,
> Assistant Adjutant-General.

* The remainder of this order, in detail, recognizes the existing First and Fourth Infantry, the Seventh Cavalry, and Clark's battalion; and reorganizes the Third and Eighth Infantry, and the Fifth, Sixth, and Ninth Cavalry.[47]

The main problem with most of the Kansas Brigade regiments is that they had been raised as part cavalry and part infantry. There were also many men who believed they were only to be a Home Guard and would not see service outside of Kansas. Among the discontented men, there were many unfit for military service on account of age or other disabilities. In some companies there were enough discontented men to become almost mutinous, and their representations were taken up by higher authority to the War Department. In response, the companies with the discontented men were ordered mustered out and broken up, with those wishing to remain transferred to other regiments. In the 6th Kansas Volunteers, three complete companies of infantry were mustered out, and the remaining companies of cavalry were organized as a new cavalry regiment and filled up by transferring to it the cavalry companies from the 4th and 5th regiments. These regiments had mostly infantry companies and they were transferred to the newly organized 10th Kansas Infantry. In addition, the 3rd and 4th regiments, Kansas Volunteers, consisting of mostly infantry companies, were broken up, and the men were either mustered out or transferred to other regiments. The new organizations received new arms and equipment appropriate to the arm of the service into which they were mustered, to take part in the campaign that was being prepared to advance into the Indian country. After the reorganization, the following units were gathered at Fort Scott for possible inclusion into the Indian Expedition:

Infantry
1st Kansas Infantry
> Colonel George Deitzler
10th Kansas Infantry
> Colonel William Weer
9th Wisconsin Infantry
> Colonel Frederick Salomon
12th Wisconsin Infantry
> Colonel George Bryant
13th Wisconsin Infantry
> Colonel Maurice Malony
Cavalry
6th Kansas Cavalry
> Colonel William Judson
9th Kansas Cavalry
> Colonel Edward Lynde
3rd Wisconsin Cavalry
> Colonel William Barstow
2nd Ohio Cavalry
> Colonel Charles Doubleday
Artillery
2nd Indiana Battery
> Captain John Rabb

At this time all units were located either at Fort Scott or in other military posts under the temporary command of acting brigadier general George W. Deitzler, who had been severely wounded in the Battle of Wilsons Creek while leading his regiment, the 1st Kansas Infantry. In late May, before the expedition was started, the 1st Kansas Infantry had been originally marked for the 1862 New Mexico Campaign. Instead, the regiment was ordered to West Tennessee along with the 13th Wisconsin Infantry. Colonel Deitzler went with his regiment.[48]

Colonel Doubleday of the 2nd Ohio Cavalry was originally given command of the expedition. Born in Leicestershire, England, January 28, 1829, Doubleday came to the United States early in life and received a common school education in Ohio. He went to California in the early days of the "gold fever" and led a life of adventure as a military mercenary and fought in Nicaragua's civil war in the 1850s. He returned to the United States prior to the start of the Civil War and settled in northern Ohio. He was probably the most experienced soldier in the Indian Expedition. Doubleday began gathering supplies and organizing the troops available for the expedition. He insisted to General Blunt that he firmly believed they could reach Fort Gibson without any difficulty.[49] On June 1, Col. Doubleday reported to Brig. Gen. Blunt that he had ordered his 2nd Ohio Cavalry (eight companies), the 9th Wisconsin Infantry (four companies), the 10th Kansas Infantry (three companies), the 6th Kansas Cavalry (one company), and the 2nd Indiana Battery from the staging area at Humbolt to Baxter Springs, on Spring River near the confluence of Shoal Creek, the recently designated supply point for the expedition located in the Cherokee Neutral Lands of far southeastern Kansas. A second column consisting of three companies of the 2nd Ohio Cavalry and four companies of the 10th Kansas Infantry would depart Fort Scott escorting a supply train containing 100,000 rations for the upcoming expedition. However, they would need far more rations than they were carrying. Wagon Boss Peck noted:

> "When rations were issued to these "noble red men" for 10 days at a time, as is the rule among white soldiers, they would turn loose to cooking and eating, like gluttons, and keep it up day and night, till the whole amount was consumed- 10 days' rations generally lasting them about three days- and then go without, or steal or starve the rest of the time."[50]

This column would rendezvous with the first column at the Baxter Springs depot. It had been reported that Col. Stand Watie and his 2nd Cherokee Mounted Rifles were located in that area.[51]

There had already been one small skirmish between Federal forces and Confederate Indians. On April 26, 1862, about 150 men of the First Missouri Cavalry (US) skirmished at Neosho, Missouri, with 200 to 300 Cherokees, Chickasaws, and Choctaws under Colonel Stand Watie. Watie had received information that a Union force was near Elk Mills in Southwest Missouri, along the Elk River just over the line from the Cherokee Nation. Since he only had forty troops with him, Watie awaited reinforcements from his regiment, the 2nd Cherokee Mounted Rifles, and from a detachment of Missouri State Guard (MSG) under Col. John Coffee. The additional personnel arrived the next morning and found that the Federals had fallen back to Neosho, Missouri. With 125 Cherokees and Missourians, Watie ordered a two-prong attack on the Federals, his troops on the left and the MSG on the left. Watie stated in his report that he dismounted his command and advanced on the Union 1st Missouri Cavalry (US) from a distance of two miles. Unfortunately, the MSG did not fulfill their role in the attack and Watie called a retreat after having killed a few Union pickets.[52]

Major J. M. Hubbard of the First Missouri reported a victory to General Samuel Curtis, claiming 32 of the enemy killed or wounded, 62 prisoners and 76 horses captured as well as a large quantity of arms.

> Report of Lieut Col. Colly B. Holland,
> Phelps Missouri Infantry.
> Cassville, May 1, 1862.
> Major Hubbard, commanding First Missouri Cavalry, with 146 of his men, fought and routed Colonels Coffee and Stand Watie and 200 Indians at Neosho on the 26th, killed and wounded 30, and took 62 prisoners and 70 horses and a large quantity of arms.
> C.B. HOLLAND
> Capt. J. C. Kelton. Assistant Adjutant-General[53]

At the same time, Confederate Colonel Douglas Cooper reported to General Albert Pike:

> GENERAL: I have to inclose [sic] Col. Stand Watie's official report of an engagement between a small party of his regiment and about 300 Federal troops near Neosho, Mo., which resulted in our favor and the retreat of the Federals.
> Too much praise cannot be awarded Col. Stand Watie and his brave men for their ceaseless vigilance on the northern line of the Cherokee Nation and their gallantry in attacking and routing a superior force of regular, well-drilled Federal troops.[54]

Watie reported killing 31 of the enemy, capturing three, and wounding several others while he lost only two men killed and several slightly wounded. This incident highlights the claims and counterclaims in the multitude of skirmishes that characterized the war in Missouri. Often there was no clear winner in these smaller engagements and the opposing force's losses were routinely over-stated.

At some point near the end of May and the beginning of June, Brig. Gen. Blunt replaced Col. Doubleday as commander of the Indian Expedition. He chose Col. William Weer, a Philadelphia-born attorney, prosecuting attorney, county judge, Kansas territorial attorney general, and former brigadier general of Kansas Militia, who resided in Wyandotte, Kansas. No official reason for the replacement is noted although it is probable that since Weer was a fellow active "jayhawker" in Kansas, he was closer in line with the politics of Senator Lane and Gen. Blunt than the English-born Ohioan Doubleday. Weer was also a drunk.[55] Weer also insisted that the Indian Home Guard regiments be a part of the expedition since the recovery of the Indian lands was the primary reason for the movement into the Indian Territory. He also impressed upon Col.

Doubleday that whatever activities that were occurring on the Missouri side of the line was to be the responsibility of the new Missouri State Militia (US) or other Union troops from the Army of the Southwest in Springfield.[56]

On May 4, 1862, Brig. Gen. Pike wrote a long report to Maj. Gen. Van Dorn regarding the situation he was encountering in the Indian Territory. Much of Pike's report deals with his continual explanation and defense of his actions and the actions of the Indian Brigade at Pea Ridge. He continues with commissary, quartermaster, fiscal, and pay issues, of which there were many. Pike would have been a successful adjutant or inspector general instead of a field general. On the operational side he has little to report beyond the locations of his units and what he believed they were doing. He wrote from his new headquarters at Fort McCulloch:

"…The Cherokee and Creek troops are in their respective countries. The Choctaw troops are in front of me, in their country, part on this side of Boggy and part at Little Boggy, 34 miles from here. These observe the roads to Fort Smith and by Perryville toward Fort Gibson. Part of the Chickasaw battalion is sent to Camp McIntosh, 11 miles this side of the Wichita Agency, and part to Fort Arbuckle, and the Texan company is at Fort Cobb.

I have ordered Lieutenant-Colonel Jumper with his Seminoles to march to and take Fort Larned, on the Pawnee Fork of the Arkansas, where are considerable stores and a little garrison. He will go as soon as their annuity is paid.

The Creeks under Colonel McIntosh are about to make an extended scout westward. Stand Watie, with his Cherokees, scouts along the whole northern line of the Cherokee country from Grand Saline to Marysville, and sends me information continually of every movement of the enemy in Kansas and Southwestern Missouri…"[57]

Col. Drew of the 1st Cherokee Mounted Rifles had become nearly convinced of the secessionist way of thinking, although many of his men had not and were hoping to cross sides. His regiment remained around the Cherokee capital at Tahlequah, protecting the Cherokee Nation's valuable properties and treasury. Meanwhile, Colonel Stand Waite of the 2nd Cherokee Mounted Rifles had been actively raiding and interrupting the Union supply lines in southwest Missouri and northeastern Indian Territory for most of the spring of 1862. In fact, there was a significant skirmish at Neosho, Missouri, on May 31 when an encampment consisting of six companies of the 14th Missouri Militia and one company of the 10th Illinois Cavalry under Col. John Richardson, was raided by the 2nd Cherokee Mounted Rifles, under Capt. R.C. Parks, Watie not being present, and Col. John Coffee's unit of the Missouri State Guard. The camp was located on the flatland just north of the town. Although pickets had been posted on most approaches to the camp, they neglected to post one on a small hill southwest of the encampment. Capt. Parks and his command came through the brush from that small hill and surprised the Union troops as they performed their routine morning tasks. The Confederates deployed their battlelines with their right anchored on the town's west side and advanced to the northeast, with the Cherokees dismounted and Coffee's Missourians mounted. Col. Richardson was successful in attempting to get his companies in battlelines, but he was soon wounded in his right arm and his horse was shot from under him and fell upon his leg. No other officers stepped up to command the demoralized troops who began to run from the battlefield. Col. Stand Watie stated in his regimental report:

> Headquarters
> First Cherokee Regiment,
> Camp near Elk Mills, Mo.
> June 1, 1862.
> On the morning of the 31st our troops, who had remained all the preceding [night] in the immediate vicinity of Neosho, attacked the enemy, who were not dreaming of their presence. The troops of my regiment and the greater portion of those with Colonel Coffee dismounted. The enemy were taken completely by surprise. At the first fire of our troops they attempted to form, returned a volley at random, then broke and fled in the utmost confusion, our troops advancing rapidly upon them all the time. Colonel Coffee's cavalry, which had charged simultaneously with our infantry, kept up the pursuit for miles... Fourteen tents, 5 wagons and teams, arms, horses, some commissary stores and ammunition, and, in fact, all the enemy's baggage, fell into the hands of the Confederates. There was 1 man killed on our side, who

belonged to Colonel Coffee's regiment.

> I am, colonel, your obedient servant,
> STAND WATIE,
> Colonel, Commanding
> First Regiment Cherokee Cavalry.[58]

Despite the activities and operations of the Confederate Indian Brigade, and even with advance notice, they were not prepared for any invasion from Kansas by the Union Army. No plans or responsibilities had been laid out to the various commands as to what actions to take in case of an invasion. In fact, the only assignment the Indian Brigade had been given was to hold off the Union Army until regular Confederate Army units could respond. Unfortunately, in the spring of 1862, there were not any Confederate units to respond. The Indian Brigade would be on its own.

Colonel Weer spent much of the first half of the month of June 1862 traveling back and forth among Fort Scott, LeRoy, and Humbolt, Kansas. He was desperately trying to gather enough supplies and get the Quartermaster and Commissary Departments to commit to an effective logistics supply chain for the expedition. He also spent much time attempting to get the Indian Home Guard regiments motivated and moving. Weer claimed that the Indian troops had "a thousand and one excuses" for not moving from their camps. The Colonel decided the best thing to do was to move the regimental camps away from the camps of the soldiers' families and their tribal chiefs. He discovered that whenever a company commander gave an order to an Indian soldier, the soldier would go to and ask his tribal chief if he should obey it or not. This was a severe break in military discipline and protocol. Weer gathered all of the captains and other field officers together to instruct them on how to give an order and how to ensure it is carried out. He further explained their duties and responsibilities and pointed out that they, as officers, were responsible for the actions of their soldiers. Wagon Boss R.M. Peck observed that the Osage Indians, who he considered to be nothing more than "blanket Indians," could be troublesome. He stated:

> "One company of these was enlisted in the Indian Brigade at Humboldt, but it was found so difficult to reduce them to anything like military discipline that they were shortly afterwards disbanded and allowed to return to their country, of which the Osage Mission (Catholic), on the Neosho River, about 40 miles south of Humboldt, was considered the headquarters…"
>
> "The Osages could not be induced to wear the soldier trousers. Some of them cut off the legs and used them as leggings, but the body and seat of the trousers they threw away…"[59]

Weer was also frantically writing to Brig. Gen. Blunt requesting that the white officers who were absent from the Indian regiments be relieved of whatever duty or furlough that they were engaged. He rightly believed that he was exhausting himself trying to tie all of the strings together and needed his officers back to help carry the burden.[60]

On June 14, 1862, the 1st and 2nd Indian Home Guard regiments marched away from their camp at LeRoy, Kansas, with Capt. Norman Allen's 1st Kansas Volunteer Battery leading the way. Wagon Boss Peck observed:

> "One at the head of the column would utter a prolonged, shrill note, and as soon as he ceased the whole body of warriors would give forth in chorus a short, sharp bark like a dog. This was repeated several times. Then the marching column would take to gobbling like turkeys. And this war whooping and gobbling they kept up till they passed several miles on the road."[61]

The 1st IHG had around a thousand soldiers, about 360 who were mounted, and the 2nd IHG had between 500-600 soldiers. It took two full days to make the 35-mile march, and they arrived in poor condition. The summer heat, coupled with marching with heavy packs and weapons, caused these recently emaciated Indian refugee soldiers to have difficulties on the trail.

During the time that Col. Weer was scrambling around trying to set up his supply of provisions, Col. Doubleday was remaining active and taking aggressive actions. Upon receiving word that Watie's regiment was camped on the Cowskin Prairie in the far northeast corner of Indian Territory and making raids into southwest Missouri, he decided to strike. Colonel Doubleday believed that he could get between Watie's camp on the Cowskin Prairie and the rest of Indian Territory, cut him off and then destroy him. Leaving his base camp at Baxter Springs on June 6, Doubleday led a 1,000-man force made up of 2nd Ohio Cavalry (eight companies), the 9th Wisconsin Infantry (four companies), the 10th Kansas Infantry (three companies), the 6th Kansas Cavalry (one company), and the 2nd Indiana Battery, forward toward Watie's camp. The column arrived at the point of Carey's Ford (Sec. 10, T24N-R23E) along the Fort Scott-Fort Gibson Military Road, adjacent to the reported location of the Confederate camp. Col. Doubleday reported to Col. Weer:

> Report of Col. Charles Doubleday,
> Second Ohio Cavalry.
> Headquarters U. S. Troops,
> On Spring River,
> June 8,1862—6 p. m.
> "… I ordered the First Battalion, Second Ohio Cavalry, across the river, to take position south of the rebel encampment, advancing at the same time with my artillery, supported by infantry, in skirmishing order, through the woods to the crossing, which was effected by the entire force by 9 p. m… Not having daylight I could not accurately ascertain their precise positions, except in the camp of Stand Watie, which was in a grove. I ordered the artillery to the front, and from the distance of about 500 yards threw a few shot and shell into their camp..."[62]

Doubleday's attack completely surprised Watie's troops and, when the Indiana battery began dropping shells into the camp, Watie's force hastily retreated under the cover of darkness. In doing so, they abandoned five or six hundred head of cattle and horses. Capt. Rabb of the 2nd Indiana Battery reported:

> "Four pieces of the battery formed a part of the expedition which marched to Round Grove, below Cowskin Prairie. Came upon the enemy in

force, under Col. J.T. Coffee and Stand Watie about dark. The enemy was routed after our firing 6 rounds of shot and shell. Captured a large amount of stock, equipage, and munitions of war."[63]

Regarding the artillery Pvt. Luman Tenney of Company H, 2nd Ohio Cavalry wrote:

> "…The shelling was splendid. The shells would bound from tree to tree and burst with a thundering noise… The Battery took a position on the hill favorable for shelling the enemy, and was supported by the Kansas Infantry…"[64]

Pvt. Isaac Gause of Company E, 2nd Ohio Cavalry remembered:

> "…After crossing the river we met and engaged Stanwaity [sic] at Round Prairie. After a short skirmish he retreated, and we captured their beef herd, with ponies and pack-mules, twelve hundred in number. Lt. [Henry] Rush of our company was detailed to deliver the cattle to the beef contractor at Ft. Scott. We moved at noon…"[65]

Confederate Col. Stand Watie reported:

> "…The Federal forces supposed to be 600 strong dashed into Cowskin Prairie yesterday and drove in our pickets. I have moved my train onto Spavinaw Creek…"[66]

Neither side reported any casualties, and it was not known if the skirmish accomplished anything useful. Although Watie's forces were scattered, they remained in the general area around Cowskin Prairie.[67] The newly-appointed expedition commander immediately sent couriers after Col. Doubleday's column to have him return all advancing forces to Fort Scott so Col. Weer would be able to organize the expedition the way he wanted it done. Col. Doubleday turned around and proceeded back to the Baxter Springs Depot. Newspapers in Kansas laid the blame on Col. Weer for the failure of the raid since he had not captured Stand Watie due to his orders preventing Doubleday from pursuing the Confederates any further south.[68] The accusations and recriminations followed in the *Daily Conservative* and *Leavenworth Republican* newspapers, which was a common occurrence during the Civil War. But there was a positive effect with the Col. Doubleday's raid. The routing and flight of Stand Watie's command at the Cowskin Prairie apparently gave hope to the pro-Union Cherokees, mostly members of the Keetoowah Society under Old Salmon. Salmon sent word to Col. Weer that there were two thousand Loyal Cherokees waiting to join any Union invasion of the Cherokee

Nation. Weer also reported to Capt. Moonlight on June 13 that upwards of twenty Osage lodges had fled north for Union protection. He was able to enlist many of these Osage warriors into the 2nd Indian Home Guard.[69]

By June 25 Col. Weer was able to gather the Union forces comprising the Indian Expedition at the supply depot at Baxter Springs, Kansas to begin the southward movement.

The 6th Kansas Cavalry Regiment joined the Indian Expedition just prior to its departure. It had been organized at Fort Scott, Kansas, in July 1861. The regiment began as three companies of home guard infantry, followed quickly by five additional companies, one of which was cavalry. Under the reorganization of the Kansas Brigade all infantry companies were replaced by cavalry. The regiment was placed under the command of Colonel William R. Judson. With this final piece of the puzzle inserted, the Union Indian Expedition was ready to begin its march southward to retake the Indian Territory, followed by a multitude of Loyal Indian refugees who only wished to return to their homes.

[1] Opothleyahola, Letter, January 28, 1862, *OR*, Series I Vol. VIII, pp. 534.

[2] Abel, Annie Heloise. The American Indian as Slaveholder and Secessionist: An Omitted Chapter in the Diplomatic History of the Southern Confederacy. Cleveland, Ohio: A.H. Clark Company, 1919. pp. 261.

[3] Ibid, pp. 260.

[4] Tricket, Dean. "The Civil War in the Indian Territory," The Chronicles of Oklahoma (March 1941), Vol. XIX, No. 1, pp. 66

[5] Indian Office General Files, Southern Superintendency, 1859-1862, C 1526

[6] William P. Dole, report, *OR*, Series II, Vol. IV, p. 9; U.S. Army Surgeon A. B. Campbell, report, *OR*, Series II, Vol. IV, p. 7.

[7] Britton, Wiley. The Union Indian Brigade in the Civil War. Kansas City: Franklin Hudson Publishing, 1922. pp. 45-46. *(Brigade)*

[8] Abel, *Slaveholder*, pp. 266.

[9] Trickett, pp. 64.; Frank Moore, ed., The Rebellion Record (New York: G. P, Putnam, D. Van Nostrand, 1861- 68), IV, 59-60 (Doc.).

[10] Nichols, David A. Lincoln and the Indians: Civil War Policy and Politics. St. Paul: Minnesota Historical Society Press, 1978. pp. 36.

[11] Fremont, General Orders #28, November 2, 1861. *OR*, Series I, Vol. III, pp. 559.

[12] Garesche, General Orders #97, November 9, 1861, *OR*, Series I, Vol. III, pp. 567.

[13] Thomas to Hunter, *OR*, Series I, Vol. VIII, pp. 379.

[14] Hunter to Thomas, *OR*, Series I, Vol. VIII, pp. 428.

[15] McClellan to Hunter, *OR*, Series I, Vol. VIII, pp. 428.

[16] Lincoln to Hunter, *OR*, Series I, Vol. LIII, pp. 511.

[17] Lincoln to Secretary of War, *OR*, Series I, Vol. XIII: pp. 533. Nichols, David A. pp. 44.

[18] Lane to Covode, *OR*, Series I, Vol. VIII, pp. 529-530.

[19] Nichols, David A. pp. 43.

[20] Warner, Ezra J. Generals in Blue: Lives of Union Commanders, Baton Rouge and London: Louisiana State University Press. 1987. pp. 243-244.

[21] Able, Annie Heloise. The American Indian as Participant in the Civil War, Cleveland, Ohio: A.H. Clark Company, 1919. pp. 83-84.

[22] Able, *Participant*, pp. 99.

[23] Thomas to Halleck, *OR*, Series I, Vol. XIII: pp. 659-660.

[24] Britton, pp. 61.

[25] Nichols, David A. pp. 41.

[26] Watie, Report, *OR*, Series I, Vol. XIII, pp. 94-95.

[27] James Denver established Arapahoe County, Colorado, and the city of Denver was named in his honor.

[28] General Orders No. 4, *OR*, Series I, Vol. XIII: pp. 683.

[29] Halleck to Stanton, *OR*, Series I, Vol. XIII: pp. 647-648.

[30] Furry, William, ed. The Preacher's Tale: The Civil War Journal of Rev. Francis Springer, Chaplain, U.S. Army of the Frontier. Fayetteville: University of Arkansas Press, 2001. Pp. 101-102.

[31] Dole, *OR*, Series I, Vol. XIII: pp. 624-625.

[32] Blunt, James G. "Civil War Experiences," Kansas State Historical Quarterly, May 1932: 219.

[33] Nichols, David A. pp. 46-47.

[34] Johansson, M. Jane, Editor Albert C. Ellithorpe, The First Indian Home Guards, and the Civil War on the Trans-Mississippi Frontier. Baton Rouge: Louisiana State University Press, 2016. pp. 16-19.

[35] Able, *Participant*, pp.108-109.

[36] Johansson, M. Jane, *Ellithorpe*, pp. 1. The journal currently resides at the Wilsons Creek National Battlefield, Missouri

[37] Kansaspedia, Online Encyclopedia of the Kansas State Historical Society, "John Ritchie"

[38] Peck, R.M. "Wagon Boss and Mule Mechanic," National Tribune, 1904.

[39] Coffin to Dole. June 13, 1862. Letters Received, U.S. Office of Indian Affairs: MC 234, Roll 834.

[40] Pike, *OR*, Series I, Vol. XIII, pp. 819-823.

[41] Britton, Wiley. The Civil War on the Border, Volumes I & II, New York and London: G.P. Putnam's Sons, 1899. pp. 299. (*Border*)

[42] Peck, R.M.

[43] Spencer, John D. The American Civil War in the Indian Territory, New York and Oxford: Osprey Publishing, Ltd. 2006. pp. 20.

[44] Phillips, Christopher. "Lane, James Henry" *Civil War on the Western Border: The Missouri-Kansas Conflict, 1854-1865*. The Kansas City Public Library

[45] Phillips, Ibid.

[46] White, Christine Schultz, and White, Benton R. Now the Wolf Has Come: The Creek Nation in the Civil War, College Station: Texas A&M University Press, 1996. pp. 73.

[47] General Order #26, *OR*, Series I, Vol. XIII, pp. 617.

[48] Britton, *Brigade*, pp. 50.

[49] Able, *Participant*. Pp. 119.

[50] Peck, R.M.

[51] Doubleday to Blunt, *OR*, Series I, Vol. XIII, pp. 408.

[52] Watie to Cooper, *OR*, Series I, Vol. XIII, pp. 63.

[53] Holland to Curtis, *OR*, Series I, Vol. XIII, pp. 61.

[54] Cooper to Pike, *OR*, Series I, Vol. XIII, pp. 62.

[55] Able, *Participant*, pp. 119-122.

[56] Weer to Doubleday, *OR*, Series I, Vol. XIII, pp.418-419.

[57] Pike, *OR*, Series I, Vol. XIII, pp. 821-822.

[58] Watie, Report, *OR*, Series I, Vol. XIII, pp. 94-95.

[59] Peck, R.M.

[60] Weer to Moonlight, *OR*, Series I, Vol. XIII, pp. 434.

[61] Peck, R.M.

[62] Doubleday to Weer, *OR*, Series I, Vol. XIII, pp. 102.

[63] Rabb, Report, *Supplement OR*, Vol. XV, Part 2, pp. 546.

[64] Tenney, Lumen Harris. War Diary of Lumen Harris Tenney, 1861-1865, Cleveland: Evangelical Publishing House, 1914. pp. 17; Edwards, Whit. The Prairie was on Fire: Eyewitness Accounts of the Civil War in Indian Territory. Oklahoma City: Oklahoma Historical Society, 2001, pp. 17.

[65] Gause, Issac. Four Years with Five Armies: Army of the Potomac, Army of the Missouri, Army of the Ohio, and the Army of the Shenandoah. New York: Neale Publishing Company, 1908. pp. 85; Edwards, pp. 17.

[66] Watie, June 9, 1862. Grant Forman Collection, Box 11, Gilcrease Institute, Tulsa, OK

[67] Heath, 82.

[68] Weer to Moonlight, *OR*, Series I, Vol. XIII, pp. 446.

[69] Weer to Moonlight, *OR*, Series I, Vol. XIII, pp. 431.

Chapter 7
The March of the
Union Indian Expedition

A person born and raised in a cooler, northern area will have difficulty imagining the heat and humidity of the southeastern Great Plains, like modern-day eastern Oklahoma, in the midst of summer. The temperatures hover well over one hundred degrees, and the humidity is usually ninety-plus percent. There is an almost constant hot and dry wind that blows across the region, coming from the deserts of western Texas and New Mexico. The months of spring tend to be very wet with numerous storm fronts and super cells dropping heavy rains across wide swaths of the open and savannah prairies. This is tornado season when these violent storms rock what we now call 'tornado alley.' But by mid-June, the weather begins to dry and settle into a very hot summer with that ever-present wind. This weather is normal for the area. The rivers and creeks dry up into small trickles or pools and drift southward towards the Arkansas or Red Rivers. After the Dust Bowl of the 1930's, the U.S. Army Corps of Engineers began damming up most of the primary rivers in Oklahoma to store the winter and spring runoff so there would be water available during the hot dry months of summer. But this solution was not available in the Indian Territory of the 1860s. The Five Civilized Tribes had adapted their lifestyles to accommodate the lack of water in the summer during their time after the relocation. Unfortunately, this was the situation that the Union Indian Expedition was marching into with a substantial number of soldiers who had never experienced such conditions in their lifetimes.

Col. Weer received information that Stand Watie and a Missouri unit were camped in the vicinity of Round Grove and Cowskin Prairie in the Cherokee Nation. On June 28 the Indian Expedition began its long march southward towards the Indian Territory. Colonel Fredrick Salomon, a Prussian native who had distinguished himself in Missouri early in the war, had been given command of the 9th Wisconsin Infantry. His regiment consisted of German-Americans, many of whom did not speak English. Weer now placed him in command of the First Brigade and issued this order to him:

Headquarters Indian
Expedition,
Camp, Baxter Springs, Kans.,
June 27, 1862.
Colonel Salomon,
Commanding, First Brigade:
Colonel: I am instructed by the
colonel commanding to say that on
to-morrow you will march the main
body of your brigade by the way of
Hudson's Ferry down the west side
of Grand Biver to a suitable point on
Cowskin Prairie and there await
further orders…

You will please send the
Second Indian Home Guard
Regiment of your brigade, across
Spring River, thence to move
southward to the point indicated by
you as the place of rendezvous.
Instruct them to scour thoroughly the
country between Grand Biver and
the Missouri State line, arresting or
driving before them all rebels in that
portion of the country…

Great care must be observed
that no unusual degree of
vindictiveness be tolerated between
Indian and Indian…

JAMES A. PHILLIPS,
First Lieutenant and
A. A. A. G.[1]

The Second Brigade was placed under
the command of Col. William Judson, a Scots
native who became a correspondent for the
New York Tribune and found himself sent to
Kansas in 1855 on a special assignment. He
became a friend of Senator Lane, became
active in the Free-Soil movement, and ended
up both as a member of the Senator's staff
and as a correspondent for the *Daily*

Conservative, the Free-Soil newspaper in
Leavenworth.[2] The Second Brigade left a day
later than the First and Col. Judson was also
issued orders similar to the one given to Col.
Salomon:

Headquarters Indian Expedition,
Camp, Baxter Springs, Kans.,
June 28, 1862.
Col. W. R. Judson,
Commanding Second Brigade:
Colonel: I am directed by the colonel
commanding to say that on to-
morrow morning at daybreak you
will march the main body of your
brigade by the way of Hudson's
Ferry down the west side of the
Grand River until you join Colonel
Salomon's First Brigade. You will
probably find him encamped on
Cowskin Prairie. You will please
send that portion of the First Indian
Home Guards that are mounted, of
your brigade, across Spring River,
thence to move southward to the
place of rendezvous of the First and
Second Brigades, Indian Expedition.
Instruct them to scour the country
between Grand River and the
Missouri State line, particularly that
portion that the Second Indian
Regiment have failed to visit… I
would invite your careful attention to
the delicate position your command
will occupy in its relation to the
Indians. The evident desire of the
Government is to restore friendly
intercourse with the tribes and return
the loyal Indians that are with us to
their homes…

JAMES A. PHILLIPS,
First Lieutenant, A. A. A. G.[3]

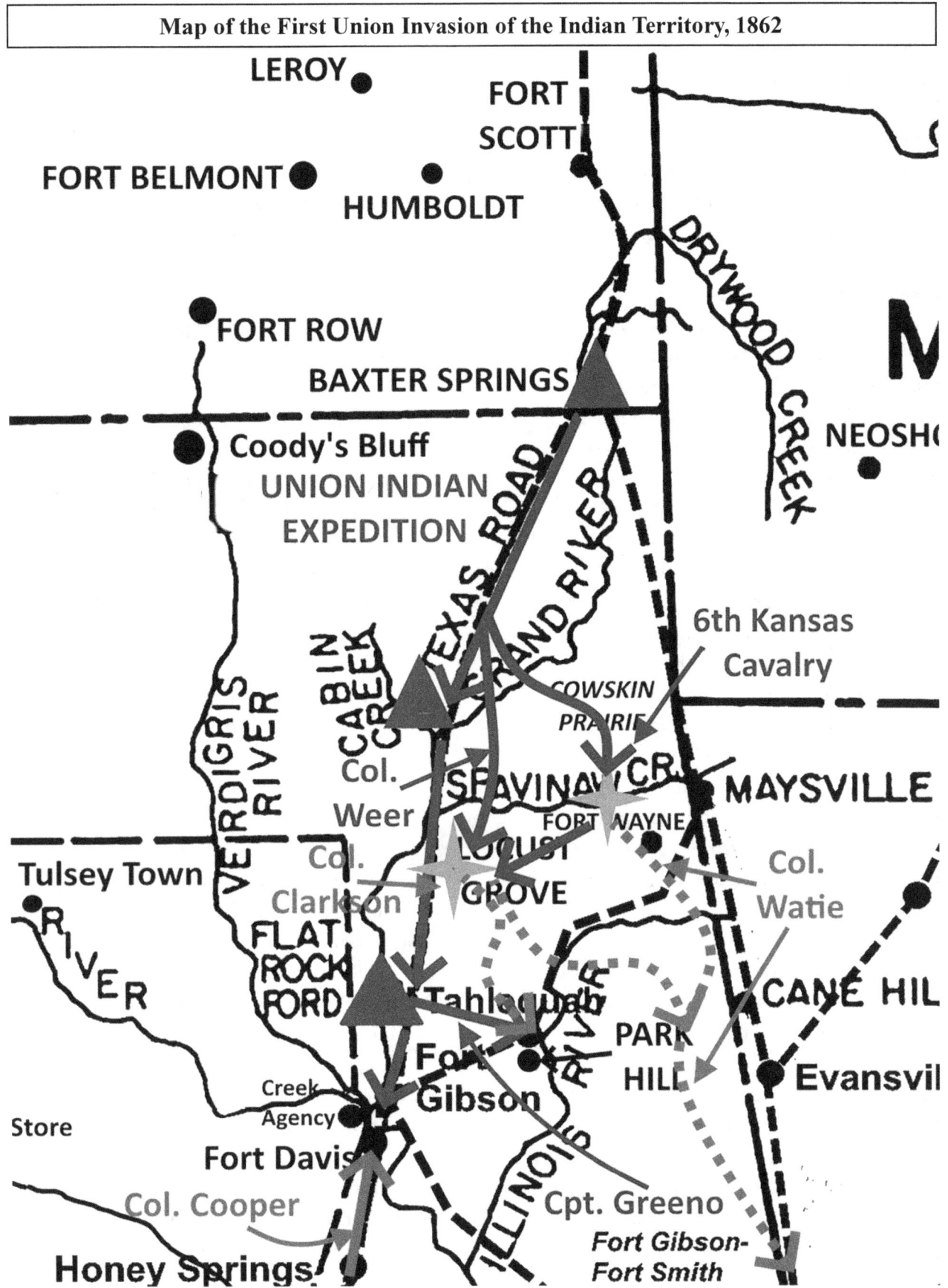

Map of the First Union Invasion of the Indian Territory, 1862

As shown by the two different sets of orders, Col. Weer was concerned about instigating any issues with the Indians who had remained in the Indian Territory, especially between the Cherokees, Creeks, and Seminoles. Many of the Indian soldiers had bragged about taking revenge against those who had driven them out of their homes. Instead, Weer wanted to have a transition and cooling-off period. To assist in this effort, the Indian Office sent two special Indian agents, E.H. Carruth and H.W. Martin, whose mission was to look to the interests and welfare of the Indians. They were to closely examine the political and economic conditions within the Indian Territory and assess the quantity and quality of livestock and crops that could be useful to the Loyal Indians. Reverend Evan Jones also accompanied the Indian Expedition, carrying a special message to Cherokee Principal Chief John Ross from the U.S. Indian Affairs Southern Superintendency that pledged the continued interest of the United States Government in providing support to the Cherokee Nation. [Note: Many of the dates indicated by the report writers during the Union Indian Expedition do not match or agree with other events or known timelines. The author has recorded the dates as the reporting officers recorded them.]

The First Brigade moved south on the Fort Gibson-Fort Scott Military Road on the west side of the Spring River. The 2nd Indian Home Guard regiment also moved southward but crossed to the east side of the Spring River and then along the east side of the Grand River to meet with the remainder of the First Brigade at or near Cowskin Prairie. They were to sweep the country between the rivers and the Missouri state line and clear the region of any combatants. The remainder of the First Brigade continued down the Military Road then crossed to the south side of the Neosho River at Hudson's Crossing, a distance of about fourteen miles.[4] Here a large number of other Indian refugees, men, women, and children, from the Cherokee, Creek, Seminole, and Osage tribes, most of whom had been along the southern line of Kansas for some time, met and began to follow the troops of the expedition.[5] Sgt. Britton wrote:

"The refugee Indian families that had been in Southern Kansas west of Baxter Springs since mid-winter came over and followed in the rear of the army after it entered the Indian Territory, and when it was encamped at Hudson's Crossing of the Neosho River, the white soldiers saw hundreds of families, women and children, bathing nude in the warm, shallow water of the stream, apparently unconscious of what we call shame. They were mostly Creeks and Seminoles."[6]

The next day the First Brigade continued down the Military Road on the west side of the Grand River and advanced into the Round Grove (modern day Grove, Oklahoma). They had a difficult crossing of the Grand River at Carey's Ford[7] because the steep eastern embankment was difficult for the wagon teams to climb. As a result, it took most of the day to cross. Because the high temperatures were hard on the Union soldiers from far northern states, the brigade would begin their daily march between three and four in the morning, hoping that by stopping around noon they could rest during the hottest part of the day.

The Second Brigade under Col. William Judson departed the Baxter Springs Depot early on the morning of June 29. As ordered, Judson sent the mounted portion of the 1st Indian Home Guard, commanded by Maj. William Phillips, across to the east side of the Spring River with instructions to scout the areas between the rivers and the Missouri state line that might have been missed by the 2nd Indian Home Guards the previous day. They were to meet up with the rest of the Expedition at Cowskin Prairie. The remainder of the brigade followed the track of the First Brigade to the Round Grove.

Col. Weer did receive a slight rebuke from Brig. Gen. Blunt for operating outside of the Department of Kansas. When Col. Doubleday was still in command of the Expedition, he had sent a detachment of his 2nd Ohio Cavalry under Lt. Col. Ratcliff to Neosho, Missouri, apparently as an advanced picket. Weer then heard that Missouri State Guard Brig. Gen. James Rains was enroute to attack the detachment with a large force. Weer dispatched four companies of the 9th Wisconsin Infantry and a section of the 2nd Indiana Battery to assist. When the report proved false, Col. Weer ordered all of the troops at Neosho south to the Cowskin Prairie. The reason for Blunt's concern was that Missouri was outside of the Department of Kansas and was outside of his command jurisdiction, although Brig. Gen Brown in Springfield had requested reinforcements from Col. Weer. Blunt also wanted to emphasize that the Indian Expedition was not to enter Arkansas or Texas, both Confederate states. Weer's purpose was to designate a base of operations and a depot of subsistence to assist the reestablishment of the Loyal Indians in the Indian Territory and provide for their needs in the short term.[8]

Col. Weer had received information that a Confederate force of Missourians with a large wagon train containing arms, ammunition, and provisions was located near Locust Grove in the Cherokee Nation. They were camped on a ridge approximately five miles south of the ford near the Locust Grove post office.[9] Weer also learned that Col. Watie and his 2nd Cherokee Mounted Rifles had fallen back to Watie's mills along Spavinaw Creek about 25 miles south of Round Grove. He had information that Stand Watie's and other Confederate troops operating in the Cherokee Nation were to join with this Confederate unit and were to be concentrated at some point for the purpose of checking the advance of the Union Indian Expedition. Now the two Confederate forces that the Indian Expedition faced were separated by many miles with rough terrain between the two. To make these separate strikes, Colonel Weer saw that it might be necessary to make one- or two-night marches. In an effort to make his strike forces more mobile, he sent all his baggage and supply trains, part of his artillery, the 2nd Ohio Cavalry, and the 9th and 12th Wisconsin Infantry regiments, from Round Grove back over Carey's Ford to the west side of Grand River. They were instructed to march down on the west side of that river on the Military Road to Cabin Creek[10] and await further orders. Col. Weer divided the remaining troops into two separate units. He, himself, would lead a battalion of the 9th Kansas Cavalry, a detachment of the 10th Kansas Infantry, under Captain Quigg, in wagons to hasten the movement of the units, a detachment of 200 mounted Creek troops from the 1st Indian Home Guard, and a section of the 1st Kansas Battery. His intent was for his unit to march down on the east side of Grand River and, if possible, strike Stand Watie's command on Spavinaw Creek, and surprise the Confederate Missourians at Locust Grove. Weer's second prong was to send Lt. Col. Jewell with a large portion of the 6th Kansas Cavalry east towards Maysville, Arkansas, then south towards Spavinaw Creek, finally joining with the other unit at Locust Grove. All units departed on their assignments late in the evening on July 2 for a long night march.[11]

During this same period, the Confederate command in Indian Territory was in turmoil. Two different generals believed they were the commander of Confederate troops in the Indian Territory. Albert Pike had been given command of the Department of the Indian Territory in November 1861 and, as far as he knew, he was still in command. In late May 1862 there was a change of command in the Trans-Mississippi Department issued by the Confederate War Department:

General Orders (No. 39).
Adjt. and Inspector General's Office,
Richmond, May 26, 1862.

IV. The boundary of the Trans-Mississippi Department will embrace the States of Missouri and Arkansas, including the Indian Territory, the State of Louisiana west of the Mississippi, and the State of Texas.
By command of the Secretary of War:
S. COOPER,
Adjutant and Inspector
General.[12]

This order clearly placed the Indian Territory under the Trans-Mississippi Department. Pike did not understand that the War Department in Richmond did not consider the Indian Territory to be an independent "department," but only a "district." This order was issued by General Van Dorn:

Special Order No. 100.
Headquarters Army of the West,
Camp Clark, May 27, 1862.

X. Brig. Gen. Albert Rust is relieved from duty with this army and will report at Little Rock for orders to Major-General Hindman, who has been placed in command of the Trans-Mississippi District.
By order of Maj. Gen. Earl Van Dorn:
M. M. KIMMEL,
Major and Assistant
Adjutant-General[13]

As shown, there is ambiguity in the two sets of orders. Nowhere does Richmond designate Gen. Hindman to command the Trans-Mississippi District. Perhaps it was a backdoor or a lost communication. Maj. Gen. Thomas Hindman, the new Confederate Trans-Mississippi District Commander, stated upon his appointment on May 31, 1862, "I have come here to drive out the invader or to perish in the attempt..."[14] But Pike refused to acknowledge that his command of the Indian Territory had been eliminated or that he fell under the command of General Hindman. On June 23 he issued this order:

Chapter 7: The March of the Union Indian Expedition

General Orders, No. —.
Hdqrs. Dept, of Indian Territory
Fort McCulloch,
June 25, 1862.
I. The Indian country was created a department by order of the Secretary of War on the 22d day of November, 1861, and Brig. Gen. Albert Pike was assigned to the command of it and of all the Indian troops that then were or thereafter might be raised in the department. That order has never been rescinded, or if rescinded no notice of it has ever been received.[15]

Pike later stated that he was willing to serve under General Hindman but believed that the new commander had more than enough issues to deal with without adding the Indian Territory into the mix.[16]

Even before the placement of Hindman into the district command, there was already bad blood between the two generals. The conflict between Pike and Hindman dated back to the mid-1850s when Pike was a leader in the Whig party that had devolved into the American, or "Know Nothing" Party, after the passing of the Kansas-Nebraska Act in 1854. Both Pike and Hindman had settled in Arkansas and became involved in the state's politics. A state's rights champion and secessionist since the 1850s, Hindman was totally committed to Southern independence. He was a staunch Democrat who believed the Know Nothings were enemies of "civil and religious liberty." The heavy influence of abolitionists and free-soilers in the ranks of Northern Know Nothings convinced him that the American Party was trying to harm or impede the interests of the South and that they were just a step away from being anti-slavery Republicans. Although members of the American Party in Arkansas reaffirmed their commitment to slavery and the right to expand it into the territories, their attacks in the face of Hindman's relentless crusade to crush the Know Nothings, assured eventual Democratic control in the state. This dedication is illustrated by an 1856 incident that took place in Helena, Arkansas. Hindman and his good friend Patrick Cleburne (who would rise to fame as a Confederate general as well) engaged in a gunfight with three members of the Know Nothing party in the middle of the street in broad daylight. Both men received serious chest wounds, but Hindman recovered fairly quickly while Cleburne hovered near death for many days before recovering.[17] Hindman and his fellow party brethren were able to rally most voters to the Democratic standard by using a combination of good organizational operations and many successful field campaigns. The Democratic Party swept to landslide victories in the 1856 state and county elections. This situation definitely placed Pike and Hindman at odds in the years preceding the War.

At the same time that the Union Indian Expedition was embarking from Baxter Springs, the eyes of the Confederate Army in the Trans-Mississippi District were on General Curtis's Union Army moving on Little Rock, the capital of Arkansas. The state government was complaining to Richmond that they believed the Confederacy was slowly abandoning them to the Union since most Confederate Army troops had been sent east of the Mississippi River after their defeat at Pea Ridge. Now a Union Army was approaching the capital, and there were few troops to oppose them. Hindman began pulling troops from all areas, including the Indian Territory, to help defend Little Rock. On May 31 Hindman ordered Pike to move his entire "infantry force of whites," together with wagons and ammunition and one six-gun battery to Little Rock "without the least delay." After dragging his feet for a week, on June 8 Pike reluctantly sent the requested troops to Little Rock.[18]

The news of the Union Indian Expedition reached the Confederate commanders in June as well. The new district commander wanted Pike to send all of his forces from areas south of the Arkansas River, north to meet the Union Indian Expedition. In compliance, Pike sent Col. Douglas Cooper north to take command of "all the troops there or may be sent there," The Indian Territory commander insisted that Cooper was to be in command and that all commanders operating west of Arkansas, south of Kansas, and north of the Canadian River, would report to him. Pike also insisted that at no time could an officer of the Missouri State Guard exercise authority over duly-enlisted soldiers or commissioned officers of the Confederate Army within the Indian Territory.[19] The Confederates already had Col. Watie's 2nd Cherokee Mounted Rifles and Drew's 1st Cherokee Mounted Rifles in the Cherokee Nation, although Drew's regiment was still in camp around Tahlequah and Park Hill and had not been very active. This was in stark contrast with Watie's regiment, which had been conducting raids throughout the Missouri-Kansas region. Pike had received a substantial amount of criticism for moving his headquarters south of the Canadian River and building Fort McCulloch in the Choctaw Nation. In a letter to Stand

Watie on April 1, 1862, Brig. Gen. Pike outlines his reasons for basically abandoning Fort Davis and the Arkansas River Line:

> Colonel:
> "...I should acquaint you with my intentions in regard to the defense of the Indian Country... If the forces of Generals Van Dorn and Price had held the western part of Arkansas and controlled the roads running westward from Fort Smith, I would have placed myself on the south side of the Canadian River and invited an attack there... Fort Gibson and Fort Davis became equally intenable,[sic] when our forces abandoned the position north of the Boston Mountains;.. When those positions were abandoned, the positions at Gibson and Davis became worthless. The Canadian became the next line, behind which to withdraw our supplies."[20]

Although Pike, the assumed commander of Indian Territory, had ordered Cooper to take charge of all Confederate Indian forces north of the Canadian River and to oppose Weer, Hindman ordered Colonel James Clarkson, the commander of Clarkson's Missouri Cavalry Battalion, Independent Rangers, to assume command of all Confederate forces in the area and to oppose the advancing Union Army. Hindman's General Orders No. 26 charged Clarkson to defend Confederate Indian allies "against federal enemies, as well as marauders and vagrants among our own white population."[38] Apparently, neither Pike nor Hindman knew of the other's orders. This failure to communicate also led to Colonel Watie's command being left in the Cowskin Prairie without orders to move until they were forced to move by the approach of the Indian Expedition.[21]

Colonel Clarkson, a former postmaster of Lawrence, Kansas, had been a rabid pro-slavery "Border Ruffian" before returning to Missouri and receiving a commission in Missouri State Guard. A veteran of both the battles of Wilsons Creek and Lexington, he had the unfortunate pleasure of being one of the first to come in direct contact with the invading Union Army. Earlier, on March 20, 1862, Maj. Gen. Earl Van Dorn authorized James Clarkson "to muster into service and organize a battalion of cavalry of six companies--for six months if they furnish their arms and equipment, otherwise for the war" and ordered him to report back for further orders as soon as he raised his battalion. He quickly organized Clarkson's Missouri Cavalry Battalion, Independent Rangers, composed primarily of men he had soldiered with since the beginning of the war.[22] Clarkson's original mission had been to gather and enlist "disaffected" Missourians and traverse the Indian Territory into southwest Kansas. This mission was to block the Santa Fe Trail and the Cimmaron Cut-off at a point west of Fort Larned to prevent communications between Fort Leavenworth and Fort Union in the New Mexico Territory. He was approved to raise his single battalion strength to a full regiment of mounted men--using conscripts, if necessary. His command of approximately 500 men, along with 60 wagons with supplies for their Santa Fe Trail mission, were the only regular Confederate troops that Hindman could spare. His command was brought up to

nine companies when he consolidated Major Thomas Livingston's small battalion. Clarkson's Battalion had moved north and west in late June from Fort Smith into the Indian Territory, with a wagon train of freshly acquired supplies intended for the Confederate Indian Brigade from Pike's storehouses. The Confederates under Clarkson had been moving back and forth within the Cherokee Nation with no known purpose. They were as far north as the Cowskin Prairie, near Stand Watie's mills, and on Finches' Prairie. By July 1 the entire force was camped on a high ridgeline south of Locust Grove in the Cherokee Nation.[23] Clarkson had received further orders from Maj. Gen. Hindman to postpone his Kansas mission and stay in the Indian Territory to take command of the defense against the Union Indian Expedition. Pike still believed he was in complete command of Confederate units operating within the Indian Territory and another dispute broke out between him and Hindman.[24] Pike was having difficulty believing his independent command was being dissolved. This war of words continued while the Union Indian Expedition was about to strike.

Colonel Watie had finally received orders, and he moved south away from the Cowskin Prairie area. Col. Clarkson later stated that he had ordered Watie to the west side of the Grand River to prevent the Expedition from advancing on his left flank. He reported that Watie was "visiting" almost fifty miles further north and on the east side of the river. Watie was, in fact, encamped in the Spavinaw Creek area, near where he and his family operated various mills. He had been warned of the Union advance and intended to strike Jewell's advance guard at the Spavinaw

Creek Crossing. He sent Major Elias Boudinot's force of 100-men a half mile west of the mill to contest the crossing of Spavinaw Creek. Jewell had moved eastward to Maysville, Arkansas and discovered traces of Watie's movements and followed these tracks. This small affair almost resulted in the capture of Col. Watie. Wiley Britton wrote:

"...in the afternoon of the first day's march he came [Col. Jewell] upon the fresh trail of the enemy. He was informed by an Indian family that Stand Watie had passed, marching south, only an hour or so before, with three or four hundred mounted men. This news caused a ripple of excitement, and Colonel Jewell's cavalry at once struck up a fast trot in the pursuit, and after about two hours came in sight of Stand Watie, where he had stopped at an Indian house for supper. He had heard that the Federal column was moving south and had taken the precaution to leave a guard in the road, who warned him of the nearness of the Federal advance in time to enable him to mount his horse and gallop off in sight of his pursuers. He soon overtook his command, but the Federal cavalry were right at his heels, the foremost troopers firing at him every time they came in sight of him and his attendants, killing one of his men. He made no effort to form his men in line, but every man seemed bent on saving himself..."[25]

A report from Maj. Michael Woods, of Clarkson's Missouri Cavalry Battalion gives a somewhat different perspective to this encounter:

"We received from Col. Watie intelligence that a body of the enemy had driven in his pickets and were advancing on his mill... 100 men were detailed... placed in charge of Major [Elias C.] Boudinot, who was directed to dispute the crossing of Salina Creek and if necessary to fall back upon our main body which was drawing up in line of battle some three miles from the creek. The enemy in attempting to cross the creek were effectively repulsed and driven back by the force under Major Boudinot... becoming satisfied that the enemy would not attack us and feeling it all-important to form a junction with the train, I took responsibility to move in that direction. We had scarcely marched a mile before I received reliable information of the capture of the train... I immediately changed my course and led by way of Tahlequah to this point [Clarksville, Texas] where I shall wait the arrival of Col. Cooper and report to him for duty."[26]

Maj. Woods probably had difficulty in justifying his retreat of over 200 miles from Saline Creek, Cherokee Nation, to Clarksville, Texas, coupled with a long wait since Col. Cooper was still at Fort Davis. As the pursuit continued, Watie's troops began to break up into smaller groups in Indian-fashion. Watie stated that he deployed his troops in a line of battle four miles south of the crossing, but the records do not back up this claim. Jewell correctly believed that the Confederate Indians were so shaken up that they would fail to warn Col. Clarkson's command south of Locust Grove. Jewell halted his command for a short rest to give the units time to close up before they continued south to join up with Weer at Locust Grove. If it had actually been Watie's intention to challenge Jewell's advance, he did a poor job with its execution since the Union force was neither stopped nor delayed.

A few miles west, after an overnight march on the road past the Grand Saline Creek and leading directly towards Locust Grove, on the morning of July 3, Weer divided his command again. This time he sent three hundred Union soldiers from the 1st Indian Home Guard, under Lt. Col. Wattles, and the 9th Wisconsin Infantry, under Major Bancroft, and the 10th Kansas Infantry, under Capt. Quigg, up against Clarkson's ridge-line encampment at Locust Grove. He had taken the western road closest to the Grand River, when possible, and overland when necessary. If they passed an occupied residence, they would take the male resident into custody to prevent him from alerting the Confederates. The detachment depended on Indian guides who knew exactly where Col. Clarkson's command was encamped. The Confederate camp was on a steep, rocky hill above a local spring with only a narrow east-west road traversing the small valley on the north side of the hill. The camp was covered by dense woods on each side of the hill and, as the Union force approached, the pickets were easily captured. Weer deployed the section of the 1st Kansas Battery which was protected by the detachment of the 10th Kansas. The mounted 1st Indian Home Guard was the first to advance with Wattles commanding the left

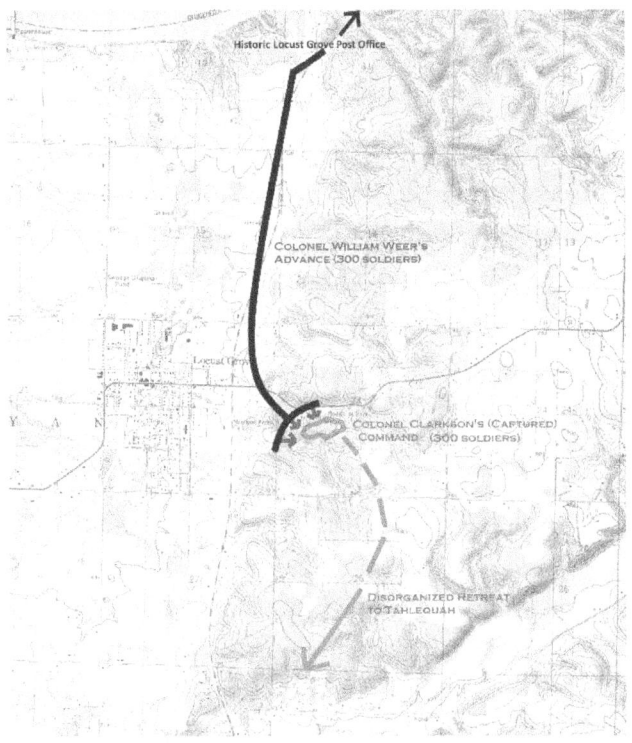

Battle of Locust Grove
July 3, 1862
Map courtesy of USGS / Additions by Author

and Adjutant Ellithorpe on the right. Ellithorpe recorded in his new diary:

> "…The charge was made up a steep hill some 50 rods, the enemy commenced fire as soon as we rose over the brow of the hill, the fire was rapid but their range was to[sic] high, the balls passing over our heads some two feet. Upon our rapid approach at a cavalry charge the enemy broke ranks & soon fled in terrible confusion, leaving everything pertaining to their command…"[27]

The Confederate forces were completely surprised and routed. Due to the heat, most Confederates were sleeping undressed and had to scramble to get clothing and weapons together. As the Union Indians swept up the hillside and over the top, the Confederates were quickly surrounded and forced to surrender after a fight lasting about two hours. Private Theodore Gardner of the 1st Kansas Battery recalled:

"The doughboys and cannoneers were piled promiscuously into wagons. There was no road. A number of Indian cabins were passed on the trip, in which the men were immediately mad [sic] prisoner to prevent alarm of the enemy. With the first blush of dawn our cavalry took the enemy pickets and rushed pell-mell upon the unsuspecting Rebs who were just starting campfires for their morning meal."[28]

Reverend Hervey F. Buckner, a civilian Confederate States representative to the Creek Agency wrote:

"...They slipped around Stand Watie's regiment and captured his train without a fight; also they treated Clarkson's regiment the same way. ...they rather caught us with our breeches down. You ought to have seen the stampede and how our women and children skedaddled towards Dixie."[29]

The Union force killed approximately 30 Confederate soldiers and captured most of the Missouri troops, estimates vary but probably around 110 including Colonel Clarkson. The Union troops also captured the sixty wagons of ammunition and salt, sixty-four mule teams, and a large quantity of military provisions that had been intended to sustain the Confederates during their Santa Fe

Trail mission. The Confederates who escaped capture fled back home towards Missouri or Arkansas, or followed Watie's retreating troops to Tahlequah, spreading fear through most of the Cherokee Nation.[30]

A victorious Col. Weer reported to Brig. Gen. Blunt:

Headquarters Indian Expedition
Camp near Grand Saline,
 July 6, 1862
Captain: As promised, I send you a more detailed account of the affair of the 3d instant. Its locality I find to be known as Locust Grove, that being the name of a post-office there... The artillery was, however, planted in battery, defended by a detachment of the Tenth Kansas, and was only prevented from paying its respects to the enemy from fear of destroying our own men, who were engaged with the enemy in the woods in scattered parties. The suddenness of the attack and the brushy nature of the ground caused the fight to be one in which each participant thrown more or less on his individual resources...[31]

The U.S. Indian Agents assigned to the Expedition had a different view of what happened on the hillside of Locust Grove. They reported to the Commissioner of Indian Affairs:

"The Creek Indians were first in the fight, led by Lieutenant Colonel Wattles and Major Ellithorpe. We do not hear that any white man fired a gun unless it was to kill the surgeon of the 1st Indian regiment. We were since informed that one white man was killed by the name of McClintock, of the 9th Kansas regiment. In reality, it was a victory gained by the 1st Indian regiment; and while the other forces would, no doubt, have acted well, it is the height of injustice to claim this victory for the whites. . ."[32]

When Col. Clarkson was able to submit a report in February 1864, after his long imprisonment as a prisoner-of-war, on his conduct during the battle, he placed the blame on Col. Watie for not providing adequate surveillance of the river crossings. He stated:

"I ordered Colonel Watie to cross to the west side of Grand River and watch the enemy. I being on the east side of that river. This order was not obeyed and the consequence was the enemy came down the west side of the river, crossed about one mile above me, and captured myself and forty-eight men together with all the baggage and trains."[33]

While it is true that Watie did not move his troops to the west side of the Grand River, doing so would have had little effect on the outcome since both advancing Union columns were already on the east side of the river. Clarkson may have been under the impression that the main body of the Union Indian Expedition moving down the Military Road towards Cabin Creek was the force he believed intending to attack his camp. The author believes that Stand Watie was far more concerned about protecting his family properties and his business enterprises in the Grand Saline, Honey Creek, and Spavinaw Creek areas than he was about any other issues regarding his fellow Confederates. Although he had been ordered to the west side of the Grand River, he chose to remain in the areas around his home and businesses which included a sawmill and lumber operation, a salt works, and a general merchandise store at Millwood (unknown location). Eventually, Watie's venture properties and his homes would be destroyed by Union Indian troops.

Unfortunately for the Union Indian Expedition, the glory was not as great for Col. Weer as he reports. Adjutant Ellithorpe was thoroughly exasperated by Col. Weer and he expressed his disillusionment with the commander in his diary:

"...The great mishap of this expedition so far is the inefficient & unreliable character of the officer in command (Col Wm Weere)[sic] No dependence can be placed upon him, as was fully demonstrated in this battle, for while he well knew during the whole nights march that we were approaching the enemy, he made frequent visits to his bottle & the nearer we approached the enemy the greater his excess, & by the time the attack was made he was in no condition to give an order, in fact he did not approach the battle ground untill [sic] some time after the battle was over & then in a state of intoxication. While we were yet

Battlefield and Spring of Locust Grove
Photos by Author

Cabin Creek Crossing of Texas Road
Photo by Author

Flat Rock Crossing of the Texas Road
Photo by Author

Colonel William F. Cloud
2nd Kansas Cavalry
Photo courtesy of the Cloud County
Historical Museum

Col. Frederick C. Salomon
9th Wisconsin Infantry
Photo courtesy of Wisconsin
Historical Society

Lt. Col. William Penn Adair
1st Cherokee Mounted Rifles
Photo Courtesy of the Cherokee
Nation

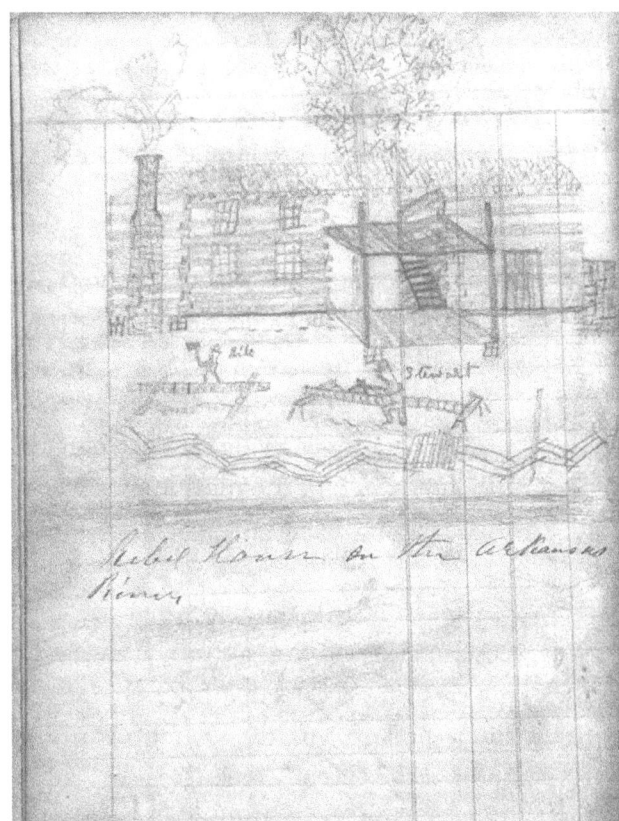

Adj. Ellithorpe's Drawing of a typical Native American Dog-Trot Cabin
Photo courtesy of NPS / Wilsons Creek NB & M. Jane Johanson

Modern Road over Vann's Ford at Sandtown
Photo by Author

Direction of Ellithorpe's Advance from Vann's Ford
Photo by Author

Members of the Five Civilized Tribes enlisting into the United States Army
Photo courtesy of the Oklahoma Historical Society

Post-War Photograph of Fort Gibson showing Commissary Building, Barracks, and Post Bakery
Photo courtesy of the Oklahoma Historical Society

encompassed by enemys[sic] on three sides, he threw himself into his ambulance and became entirely unconscious & past being aroused…

Efforts were mad[e] to arouse the Colonel to no purpose…"[34]

This issue was not unknown to others associated with the Indian Expedition. Similar complaints had been made against Col. Weer before the beginning of the night march. Sgt. Britton wrote:

"The regimental commissary sergeants made out requisitions on the division commissary for rations to issue to the regiments to which they belonged, and the requisitions had to be approved by the division commander, and it so happened that one time when the command was to make a night march so as to strike the enemy at daylight the next morning, that one commissary sergeant was directed to have the rations issued to the companies of his regiment by dark. He took the requisition to Colonel Weer's headquarters to have him approve it; but was unable to see him, that officer being drunk in his tent; the errand was repeated three or four times, and finally just before midnight his approval was secured; his drunken condition caused a delay of several hours in the movement of the troops whose requisition for commissary supplies was to be filled."[35]

Isaac Gause of the 2nd Ohio Cavalry observed Col. Weer first-hand:

"…The column moved out after dark one night, with Colonel Ware [sic], who was then in command of the expedition. I do not know what he had to eat, but I know he had a ten-gallon keg strapped on a mule, and of course that means he did not lack for drink. And there was plenty of evidence of it before morning, in the bungling moves made on the prairie in the dark, and in the morning we were hardly out of sight of the camp…"[36]

After the battle the two wings of Weer's command and the captured wagons were moved to the west side of the Grand River and joined up with the main body of the Expedition at the Cabin Creek Crossing on the Fort Scott-Fort Gibson Military Road. After relieving Col. Clarkson of his regimental books, personal arms, and papers, Weer marched him and his captured Missourians twelve miles to temporary camp on the Grand Saline. At Grand Saline, the Union troops destroyed the Cherokee salt works that supplied the Indian Territory and much of western Arkansas with salt. The destruction of the Grand Saline salt works was a great blow to the Cherokee Nation. The areas of eastern Indian Territory held some of the richest deposits of salt in the country. It was always in great demand and was a large constant source of income for the salt works operators and the Nation. Col. Auguste P. Chouteau (whose house was destroyed by the Union Indians), an early settler of the area, had owned a salt spring on the east bank of the Grand River, and he sold the property to Sam Houston in 1830. When Houston discovered that he could not own land in the Indian

Territory, the property transferred to John Rogers, a Cherokee. Rogers opened a salt works and named the area Grand Saline. In 1843 the Cherokee Nation removed the operators from these salines, declaring these to be tribal property. Salt garnered a large price in Indian Territory before, during, and after the Civil War. During the war, salt became scarce, and the price was exorbitant. Union and Confederate commanders confiscated the salt works to supply their troops. Lewis Ross, the brother of Principal Chief John Ross, was granted one of the exclusive leases to operate the salt works around the Grand Saline. Moses Lonian was a young, enslaved African who had been purchased along with his father by Lewis Ross from a white man in Bentonville, Arkansas. The father was made the foreman and overseer of the salt work's enslaved workers. He told of his and Lewis Ross's first encounter with the Union Army:

"…Sometime in 1862 the Northern soldiers came down from Kansas and made a drive up and down Grand River, and meeting no opposition from the Southern soldiers they set every slave they could find free Louis Ross had heard. They were in the country and kept watching for them to come, but they rode in on him one day when he least expected them. He broke, ran down toward the branch on foot, with the soldiers on horseback running behind him/shooting as they went. Boss was so scared he did not stop for a deep hole of water in the branch but plunged right into it and went in over his head. He was a very large man and the soldiers told my father that he hit that water so hard his own weight carried him to the far bank where he grabbed some brush and pulled himself out. My father said Ross could not swim a lick and had he not have reached the bank as he did, would, have drowned. The soldiers also told my father that they were not trying to hit Ross. They had already received orders to shoot over the heads of the Indians if any of them ran and let them get away. They were then to help themselves to what they could find, and they did. They ran-sacked the place from cellar to garret, and made the slaves load the loot into wagons and haul it off…"[37]

Another enslaved child recalled a similar experience. In an interview with Chaney McNair for the Slave Narratives, she said:

"…I'se nine or ten years old then. The Northern soldiers came and took Marster Williams prisoner and all us slaves up to Fort Scott, Kansas. I remember it. They come to the house one day and say, "You all get ready to go north." It was June in 1862. They take us in wagons and on horseback. They went to different plantations and take as many slaves as they could get. They did a lot of robbin' too; took an awful lot of stock. I can't remember going hungry on the trip, but we had an awful time gettin' water. Sometimes we drink muddy water out of the creeks. Don't know how long it took us, see I'se just a little girl, but I do remember how tired I got and sleepin' under a wagon at night. I didn't know what it's all about…"[38]

Chaney McNair's owner, the "Marster William," was actually William Penn Adair, Confederate Lt. Col. and quartermaster of the 1st Cherokee Mounted Rifles. He was captured at his Greenbriar home near the town of Adair. Lt. Col. R.C. Parks of the 2nd Cherokee Mounted Rifles, an enemy of Adair, claimed he was drunk at the time. Regardless, Lt. Col. Adair ended up at Camp Douglas in Chicago, Illinois, as a prisoner of war. He was later exchanged in 1863 and became the colonel of the newly-revised 2nd Cherokee Mounted Rifles, the position to which R.C. Parks had believed he would be appointed to, which was a source of friction between the two men. (An interesting note is that the early 20th century, Cherokee Nation-born, humorist Will Rogers's full name was "William Penn Adair Rogers.")

The next day was the Fourth of July, and the Union troops celebrated Independence Day by dividing up the goods and supplies captured in the wagon train. Military items such as powder and weapons were kept by the Expedition, and the clothing and other items were divided up among the Indian troops. The cannons were fired, and the leadership conducted a full-dress review of the military forces out on the prairie as a show of force for the Indians in the area. The Confederate prisoners of war, including Col. Clarkson, were marched to Fort Scott and then to Fort Leavenworth with a guard commanded by Lt. Col. Corwin, the ardent abolitionist. Adjutant Ellithorpe notes that "Col. Clarkson seemed much down cast & from his appearance deeply regrets ever conspiring to destroy the Govt-.."[39] He was probably also unhappily expecting long-winded anti-slavery speeches from Corwin. The escort was also taking three companies of newly enlisted Cherokees who needed to be mustered in and equipped. Col. Ritchie informed General Blunt that his 2nd Indian Home Guard regiment was now fully manned and nearly all were mounted.[40] During the day and evening of July 4 the alcoholic beverages were heartily consumed by most of the troops. Pvt. Lumen wrote in a letter home:

> "…The Fourth of July was duly celebrated at Cabin Creek Camp. We did no marching, and perfect license was given to all to drink and carouse as much as they chose. One officer even told his men that the one who wasn't drunk that night should be ducked in Grand river… When the 1st Brigade had their fight near here, Col. Weir, our commander, was so intoxicated that he could neither receive the report of the battle or give any orders. One reason everybody liked Col. Doubleday so well was, that he never drank…"[41]

Beginning on July 5, the Union Indian Expedition packed up and moved south along the Military Road. By July 10 they were camped at the Flat Rock Creek crossing, approximately 20 miles from Fort Gibson. The Expedition was to camp at this site for about two weeks. The weather had become hot and dry, the green grass had turned brown, and the water sources began to slow to a trickle. This was not unusual for the summer months in the Indian Territory, but the Northern troops were mostly unaccustomed to this weather. The large amount of stock used by the mounted troops and the wagon teams began to eat the dry grass to its roots and had to be constantly moved to fresher pastures. Food stocks for the troops had also begun to run low, and the water sources were becoming drier and more contaminated by soldiers and livestock. Col. Weer and other officers became somewhat obsessed with the arrival of the resupply wagon train and their line of communications with the supply depots at Fort Scott and Baxter Springs.

By the summer of 1862 Cherokee Chief John Ross had made it known to the Confederate government that he was unhappy with the way his people had been treated by their inactions. Ross had outlined some grievances including the destitute situation of many of the families of Confederate Indian troops due to lack of consistent pay, the lack of protection of the borders of Missouri and Kansas, and control of roving bands of white men who were committing violent crimes and destroying property.[42] On July 14, 1862, Colonel Weer sent out patrols on various missions. The first of these patrols was led by Capt. Harris Greeno with Company C, 6th Kansas Cavalry, accompanied by two mounted companies from the 1st Indian Home Guard. Dr. Rufus Gillpatrick, an agent of the Federal government who had earlier developed many contacts in the Cherokee Nation, and the two U.S. Indian Office agents, Carruth and Martin, also joined the column. Capt. Greeno was a physician and practitioner who was gifted with a great amount of tact and diplomacy, which was essential when dealing with the Indian nations. He was also a well-practiced orator. Earlier, on July 7, Weer sent a message via Dr.

Gillpatrick, under a flag of truce, to Cherokee officials requesting a meeting. Weer's message stated:

> Headquarters Indian Expedition,
> Camp on Wolf Creek,
> July 7, 1862.
> His Excellency John Ross,
> Chief of the Cherokee People:
>
> Sir : The bearer of this communication is an accredited agent of the United States Government, and as such bears to you this official note. I am here with an armed force of regularly enlisted soldiers, instructed and prepared to enforce the observance of treaty obligations by the Cherokee people…
>
> I am here to injure no one who is disposed to do what treaties made by his nation bind him to do; but am here to protect all faithful members of the tribe…
>
> I desire an official interview with yourself, as the Executive of the Cherokee people. I accordingly request this interview between us at my camp, promising you a safe return to your home.
> I am, respectfully, your
> obedient servant,
> WM. WEER,
> Colonel, Commanding[43]

Chief Ross sent an official message back to Weer refusing to meet and reiterating the commitment of the Cherokee Nation to the Southern Confederacy:

Executive Department,
Park Hill, Cherokee Nation,
July 8, 1862.

Col. William Weer,
United States Army, Commanding:
"…in reply I have to state that a treaty of alliance, under the sanction and authority of the whole Cherokee people, was entered into on the 7th day of October, 1861, between the Confederate States and the Cherokee Nation, and published before the world, and you cannot but be too well informed on the subject to make it necessary for me to recapitulate the reasons and circumstances under which it was done. Thus the destiny of this people became identified with that of the Southern Confederacy…

…I cannot, under existing circumstances, entertain the proposition for an official interview between us at your camp. I have therefore respectfully to decline to comply with your request.

I have the honor to be, colonel, your obedient servant,
JNO. ROSS,
Principal Chief Cherokee Nation.[44]

Regardless of the Chief's official message, later a secret message was received from the Union-leaning Ross, stating he would be forced to surrender the Cherokee Nation if the capital at Tahlequah and Park Hill was occupied by United States troops. Under these circumstances, Capt. Greeno's detachment was to go to the Cherokee capital at Tahlequah and continue to Chief Ross's residence at Park Hill. He would deliver the official letter of Union support from the U.S.

Indian Southern Superintendent Coffin. Greeno's Union force did not encounter any resistance in its approach to Tahlequah late on July 14, and the cavalrymen surrounded the town without incident. Greeno immediately sent Dr. Gillpatrick into the town to assess the mood of the residents. Gillpatrick found that most of the residents who remained were Union-leaning, and he received information that a large number, 200-300, of Col. John Drew's 1st Cherokee Mounted Rifles regiment was camped around Park Hill, adjacent to Rose Cottage, Chief Ross's home. Officially these Confederate troops were there to protect Chief Ross, the Cherokee treasury, and the nation's archives. Wagon Boss Peck described the Cherokee capital when he saw it:

> "Tahlequah, the capital of the Cherokee Nation, 18 miles east of Fort Gibson, was a small village, containing, for an Indian town, some very respectable brick houses; and although now deserted, except for a few families of women and children, it looked as though [in] its prosperous times of peace it might have contained a population of 200 people. Two miles west of the town, on the road to Fort Gibson, we passed a large brick building which, before the war, was their Male Seminary. Three miles south of the Male Seminary is another little hamlet called Park Hill, near which is the Female Seminary, a building similar to the other."[45]

Greeno's force camped at Chapel Springs between Tahlequah and Park Hill that night. On that same night, Chief Ross received a message from Colonel Douglas Cooper at

Fort Davis demanding a proclamation in the name of the President of the Confederate States of America calling on all Cherokee Indians over 18 and under 35, to come forward and defend the Cherokee Nation from the invasion, as had been stated in the October 1861 treaty. With the Union Army so close, Ross opted to hold off on the proclamation until he discovered their intentions. Ross also received a letter from Brig. Gen. Pike asking for $50,000 in gold from the Cherokee treasury to add to the $50,000 he had received from the Confederate government. These funds were to help him pay those Cherokees "true to the alliance of the Cherokee government and true to the alliance of the Cherokee people with the Confederate States."[46] Cooper also sent an order to Chief Ross's son, Lt. Col. William Ross of the 1st Cherokee Mounted Rifles, to gather his forces and proceed directly to Fort Davis.

On July 15, Capt. Greeno arrived at Park Hill and met with Chief Ross and those supporters who held positions in the Cherokee government. He also discovered the nearly 300 Confederate Indian soldiers from Col. Drew's regiment. The Union detachment quickly rounded up all of the Confederate Cherokee soldiers, as well as many of their officers, including Lt. Col. Ross and Maj. Thomas Pegg, without any resistance. Greeno promptly made all of the gathered Confederates prisoners of war. Several hundred Cherokees, many of them leading men of the Nation, had gathered at Tahlequah and Park Hill, to discuss the situation and they were unsure what actions they would take. Greeno then put his oratory skills to good use by addressing the gathered leaders of the Cherokee Nation. He stated:

"Leaders and people of the Cherokee Nation, I am here under orders of the commanding officer of the United States forces encamped at Cabin Creek. It is an expedition of several thousand men, well supplied with cavalry, artillery and infantry, and sent by the Government into the Indian country to restore peace and tranquility among the Cherokee people, and to protect those who have lived up to treaty relations with the Government. You have heard of the swiftness of our movements and of the first blow we have struck the enemy in the capture of Colonel Clarkson's command at Locust Grove a few days ago...

...Last year the Federal Government did not have its forces organized until after the enemy, white and Indian soldiers, had overrun your country, taken your property, and by threats and intimidation forced many of your people to take sides with the South against their will and judgment. But I am able to tell you now that the Federal Government is rapidly getting its forces organized and equipped for an aggressive campaign on all fronts..."[47]

Greeno continued his address by describing how the fortunes for the United States had been sharply turning and there were numerous successes happening in areas around the western theater, including the capture of Fort Henry and Fort Donelson on the Tennessee River, of the victory at Shiloh, and the destruction of the Confederate ram fleet at Memphis. He also included the naval successes on the Mississippi River. The captain reiterated that the United States was a forgiving and honorable country and would not seek revenge upon the Indian nations who had been misled by the Confederate leadership.[48]

Greeno accepted the surrender of the Cherokee Nation by Chief Ross and promptly placed him under house arrest. He also paroled Ross's supporters, including a majority of the officers and men of the Colonel John Drew's 1st Cherokee Mounted Rifles regiment.[49] Three of John Ross's sons joined the Union Indian Home Guards. Confederate Colonel Drew later lamented:

> "There is a general move and rallying of the Cherokee northward. They are being rapidly armed and received into Federal service. Our unhappy country is now to become the general battleground. God only knows what is to be our fate as a Nation."[50]

Those who wished protection were escorted to the Union Army camp at Flat Rock Creek. A mustering officer was quickly requested from Fort Scott, Kansas, and the paroled officers and men who wished were quickly formed into the 3rd Indian Home Guard regiment. The new regiment was to be commanded by a newly promoted Colonel William A. Phillips, a native of Scotland, who as a journalist, became involved in the activities of the jayhawkers and abolitionists in Kansas and became a friend of Senator Lane.[51] Capt. Greeno left a strong guard over Chief Ross and his Rose Cottage home at Park Hill for his confinement and protection. His detachment returned to the Flat Rock encampment on July 17.

Chief Ross immediately came under suspicion by his pro- Stand Watie enemies of prearranging the entire scenario when he had earlier told Col. Weer that he would be forced to surrender the Cherokee Nation if they were occupied by Union troops, knowing there were no effective Confederate troops anywhere near Tahlequah and Park Hill.[52] At this point Chief Ross's main concern became the safety of his family, and he feared the Union-leaning Cherokees would take revenge upon those who had supported the Confederate treaty. An all-out civil war between the factions was very possible, and reports had begun to come in of violent incidents taking place between the two opposite sides. Both the Keetoowah, or Pin Indians, and the Southern Rights party had armed vigilantes roaming the Cherokee Nation killing, burning and looting the people and property of their rivals. Union troops were also guilty of this, not caring which side of the fence a particular Cherokee family supported. The Union soldiers were instructed to not shoot or kill any Indians unless they were forced to for self-defense. Property was a different matter since foraging was an established military process. Looting by soldiers in a hostile area is very difficult to control since officers cannot be everywhere. The Union Army did have a provost marshal guard that would scour the camps and inspect wagon trains for contraband or stolen items. The most valuable items confiscated would usually end up in the personal baggage of a high-ranking officer.[53] Moses Lonian described how the enslaved persons were afraid to load up their owner's property into the wagons because they believed that they would suffer severe consequences. Mr. Lonian was terrified of Lewis Ross who he claimed whipped his father and the other slaves regularly. Lonian said that the soldiers kept telling them that they were free and deserved everything that their owners had in payment for their many years of unpaid service. So, the Lonian family loaded up their owner's wagons with the owner's valuable property, crossed the Grand River at Salina, and followed the Military Road all the way to Kansas, but always with an eye to their back trail out of fear.[54]

The second patrol was sent out to reconnoiter Fort Gibson and Fort Davis. The patrol consisted of two companies of the 6th Kansas Cavalry and was commanded by Major William T. Campbell. In his after-action-report, Campbell reported that they had occupied Fort Gibson, the old army post near the junction of the Grand and Arkansas Rivers that had been in use from 1824 until it was de-activated in 1857. There were a few Confederate troops at Fort Gibson, but they immediately fled across the Arkansas River to Fort Davis on the Union force's approach. Campbell's men raised the United States flag over the old fort and fired a few shots in salute.

After their small show of force, Maj. Campbell and his detachment spent the night about four miles from Fort Gibson where he found good grass and water. After receiving a situation report from Campbell, Col. Weer advanced with 600 mounted men to the detachment's position, arriving at 2:00a.m. At first light, they reoccupied Fort Gibson and Weer then ordered Campbell and his detachment to advance to the crossing of the Arkansas River. As was common during the Civil War, Campbell reported a wildly over-estimated number of Confederate troops at Fort Davis.[55] The Union and Confederate

troops soon exchanged gunfire across the Arkansas River. One Union soldier and a horse were wounded.

The Confederates in this fight were some of the troops that responded to Col. Cooper's order to gather at the fort before they attempted to stop the Federal column. But Cooper had difficulty collecting enough troops for the assignment since he had significant problems with military discipline among his Indian units. Many of the units failed to arrive in time or simply ignored his orders. (Even Col. Drew himself did not respond to the call to respond to Fort Davis. He was with relatives at Webber's Falls who had contracted measles, and he was one of the fifteen patients.) This was a common problem for both Union and Confederate commanders of Indian troops. The soldiers of the Indian regiments tended to come and go as they pleased and they shunned the trappings of military discipline. The situation was a little better in the Union Army because they were more tightly organized and regulated than the Confederate units. The Confederate regiments served under Native American officers for the most part. The Union Indian regiments were commanded by white officers, but a majority of the field and company level officers were Native Americans. This was in stark contrast to the later recruited Black regiments which were staffed solely by white officers. This clearly reflects the 19th century Euro-American view that Indians, as primitive as they were in many aspects, were still on a higher social and competency level than African-Americans.

The entire Union force returned to the Expedition's encampment at Flat Rock Creek by a circuitous route and arrived late in the day. The ease in which the Union troops moved around gave the impression that the Confederates had already given up the idea of keeping the Union Army completely out of Indian Territory and had abandoned the area north of the Arkansas River. After receiving the report from Campbell's reconnaissance and additional information, Weer wrote to General Blunt that he intended to return with the entire Expedition, force a crossing of the Arkansas, and take Fort Davis. This never occurred due to developing circumstances. If Weer had known how disorganized and unprepared the Confederates were at Fort Davis, the Union Indian Expedition forces would have overwhelmed and driven off whatever forces tried to oppose them. The Civil War in the Indian Territory might well have ended that day, but aggressive action was not one of Col. Weer's strengths. There were no obvious Robert E. Lee's or Ulysses S. Grant's west of the Missouri-Arkansas line.

A third patrol was conducted by Adjutant Albert Ellithorpe of the 1st Indian Home Guard with 250 mounted soldiers of that regiment. He was to scout the country between the Grand River and the Verdigris River. While approaching Vann's Ford of the Verdigris[56] at about 4:00p.m., near the hamlet of Sandtown, he discovered signs of Confederate movements on the west side of the crossing. Ellithorpe wrote in his diary:

> "Upon close examination I found signs of the enemy in a thickly timbered bottom on the west bank of the river. the [sic] brush was so thick that I dare not undertake to penetrate without first learning the strength of the enemy, & to do this was a difficult task, not having any canon [sic] to shell the woods… I detailed

30 men of the most daring to approach as follows—Placing themselves one hundred yards apart & on a line with the river they advanced to the river they advanced to the bank cautiously,.. –after dark I advanced my whole force cautiously, leaving a rear picket & guard… then pushed the pickets across the river to the opposite bank…"[57]

Ellithorpe insisted that his command was to stay silent, keeping the enemy from knowing their presence. Pickets informed him that the Confederates were posted in force about a mile from the crossing. While waiting for daylight to attack, heavy rains began to fall on the area, and the Union troops feared that their weapons and powder would get wet. Sacrificing necessity over comfort, Ellithorpe ordered that the troops should wrap their muskets in their blankets to keep them dry. He continued in his diary:

> "At daybreak I commenced a cautious advance through the woods in battle line, with pickets extended two hundred yards. half [sic] an hour brought us to within sight of the enemys [sic] camp.
>
> The pickets fell back to the main line & then I ordered a charge with instructions not to fire untill [sic] the enemy were fully aroused (their camp was asleep) the men moved at a double-quick. The woods were dense & the alarm soon ran through the camp… I ordered to fire, upon the first voly [sic] the[y] fled in great consternation leaving everything behind…"[58]

Ellithorpe reported that the Confederate force consisted of approximately 300 Indians and Texas Rangers. He estimated the enemy's losses at 10 killed and 30 wounded. There was no official report of this incident, and it is unknown to which force the Confederates belonged. Ellithorpe believed that it was the advance party of a larger attack intended to strike the Expedition's encampment on Flat Rock Creek.[59]

The Union Indian Expedition had been a resounding success so far. They had delved deeply into the Indian Territory. They had defeated each Confederate force that attempted to challenge them and captured two Confederate supply trains in the process. Capt. Greeno had received the unconditional surrender of the Cherokee Nation and had arrested the principal chief. The Union troops had also re-taken Fort Gibson and verbally and physically challenged the Confederates occupying Fort Davis. The Union Army also gained approximately thirteen hundred Cherokee recruits to fill out the under-staffed 2nd Indian Home Guard and a large portion of a new regiment, most of whom had been part of Col. Drew's Confederate regiment. Finally, Adjutant Ellithorpe's force had thwarted a possible major attack on their Flat Rock Creek encampment by a Confederate force via the Sandtown crossings of the Verdigris River. During the month that the Union Army had occupied the Indian country north of the Arkansas River, hundreds of Indian families had returned to their homes in the confidence that they could live in peace and have the protection promised them.[60] However, in spite of the huge success of the Union forces, at this point the Expedition began to fall apart.

Colonel Weer was aware that he had his supply line stretched rather far and was convinced that he was being cut off by Watie and that he was mistakenly facing a large Confederate force at Fort Davis. Although General Blunt had told Weer to keep the Union Indian Expedition together with a singular purpose, he began sending frantic messages to Weer advising him to watch his supply and communication line with Fort Scott, fearing the Indian Expedition would be surrounded and cut off. As mentioned earlier, the weather was having a very negative effect on the northern soldiers, especially those from Wisconsin who were having difficulties with the heat. There were rumors that the command was down to only three days rations, and Col. Weer did in fact reduce the Expedition to half-rations. The resupply train had not arrived by July 18; it was actually at Hudson's Crossing waiting for an escort, and there was fear among the command staff that they would run out of food. However, this seems unlikely since they had captured both Col. Clarkson's and Col. Watie's supply trains filled with provisions. The local area also had a large number of beef herds that were owned by the local Cherokees or by the Nation itself. The main problem was the heat, lack of water and the complications of dysentery caused by consumption of unsalted beef. With the thousands of men and an unknown number of livestock drawing from the same shrinking pools of available water, it is not surprising that the water sources were becoming contaminated. It is not known or recorded what actions the commanders took to find reliable and clean water sources. This is a situation where true leadership was needed but was severely lacking within the command staff.

These problems came to a head when it was reported that sometime around July 15 Col. Weer had taken to his bed with his bottle, inebriated to the point where he was unable to exercise any reasonable level of leadership. Nevertheless, he was clear-headed enough to call a council of war of the brigade and regimental commanders on the evening of July 17. After a very heated and contentious meeting, the council voted to conduct a "retrograde," or withdrawal movement, to either find the supply train or return to Kansas if needed. Weer disagreed with the council's recommendation but did not give a clear answer to what the next steps should be. He may not have known what steps to take because his orders were to establish the refugee Indians, set up a defense system for them using the Indian Home Guards, establish a commissary for supplies, and ensure the refugee Indians were able to plant a corn crop. Since this was a heavy burden, he needed to be sober and begin making plans and assignments. Weer may have been in the process of this when he was arrested by Colonel Frederick Salomon of the 9th Wisconsin Infantry who assumed command in a clear case of mutiny. Salomon stated that Weer was intemperate, disloyal, and insane but could not clearly establish any instances in which this was apparent or when it affected the Expedition. Col. Salomon issued a *pronunciamento* which tried to explain the reasons for his drastic and illegal action:

Headquarters Indian Expedition, Camp on Grand River, July 18, 1862.

To commanders of the different corps constituting Indian Expedition:

"…Suffice to say that we are 160 miles from the base of operations, almost entirely through an enemy's country, and without communication being left open behind us. We have been pushed forward thus far by forced and fatiguing marches under the violent southern sun without any adequate object. By Colonel Weer's orders we were forced to encamp where our famishing men were unable to obtain anything but putrid, stinking water… Yesterday a council of war, convened by the order of Colonel Weer, decided that our only safety lay in falling back to some point from which we could reopen communication with our commissary depot. Colonel Weer overrides and annuls the decision of that council and announces his determination not to move from this point.[61]

After the arrest, Salomon issued General Orders No. 1 in which he has assumed command of the Union Indian Expedition, reassigned various staff officers, placed Lt. Col. Ratcliff of the 2nd Ohio Cavalry in command of the First Brigade. He finally ordered all of the white regiments back north the next morning. He did not include the 1st and 2nd Indian Home Guards, who he left with vague orders to be a "corps of observation" on or near the Verdigris and Grand Rivers.[62]

The U.S. Indian Agents, Carruth and Martin, were infuriated. They immediately wrote to Brig. Gen. Blunt:

Camp on Grand River, July 19, 1862.

188

General James G. Blunt,
Commanding:

Sir : This morning our whole camp was thrown into confusion by the arrest of Col. William Weer, commanding, and the retreat of all the white forces, leaving the three Indian regiments behind to fight the enemy's forces, amounting to from 3,000 to 10,000 men. It will leave all that portion of the nation through which our army has passed defenseless. The families of the men who have flocked to the Union standard will be ruthlessly murdered, it is feared, and justly, by the gangs of cut-throats which will infest the country. Our Government should stop this.

We beg of you in behalf of the Cherokee Nation, especially that portion of it, whites and Indians, who have for months slept in thickets and canes, to do something speedily to arrest the desolation that will follow the shameful retreat of our army while in sight, already demoralized by fear and Union feeling…

The arrest of the colonel commanding is here considered a mutiny… Besides, there are many families of white missionaries already threatened with punishment, who, because they expressed joy at our arrival, may be murdered… We have the honor to be, very respectfully, your obedient servants,

E. H. CARRUTH,
United States Indian Agent.

H. W. MARTIN,
Special Indian Agent.[63]

General Blunt immediately had recognized that Col. Salomon's command change was actually mutiny. Blunt later wrote:

"…This expedition penetrated as far south as Tahlequah (the capital of the Cherokee nation), defeating and capturing several small rebel forces, and was in every respect as successful as could have been anticipated, until disagreements and difficulties arose among officers, that finally culminated in mutiny and the forcible arrest of the commanding officer (Col. Weer) by his subordinate (Col. Saloman [sic], of the Ninth Wisconsin) and the assuming of the command by the latter, and the abandonment of the Indian country…"[64]

On the morning of July 19, the Union Indian Expedition began its retrograde movement back to the north, leaving the Indian regiments to fend for themselves with little in the way of instructions.[65] Salomon did not even notify the Indian regiments until July 22 that they had left. Col. Furnas of the 1st Indian Home Guard tried to remind Col. Salomon that they still had 200 mounted troops under Adjutant Ellithorpe at Vann's Crossing with reinforcements sent by Col. Weer enroute to counter the threat from the Verdigris River crossings. When the reinforcements from the 2nd Indian Home Guard arrived, a large number of that unit deserted. Furnas believed that, without the stability of the white troops, the Indian soldiers lost confidence and any discipline they had acquired. In addition, Col. Weer had sent another 200 troops from the new 3rd Indian Home Guard under Major Foreman to Fort Gibson to watch for any movement from Fort Davis. Once again, Col. Salomon gave vague orders to watch the Arkansas River fords and promptly began his withdrawal up the Military Road, leaving the three Indian Home Guard regiments without any supplies, munitions, medical supplies, or rations. Col. Furnas assumed command of the Indian troops and formed the First Indian Brigade. Adjutant Ellithorpe issued this order for Col. Furnas:

> General Orders No. 1.
> Hdqrs. First Indian Brigade,
> Camp Corwin, on the
> Verdigris, July 19, 1862.
>
> I. In consequence of the retreat of all white troops from the command of the Indian Expedition I hereby assume command of the Indian regiments in the field.
> II. The commanders of the First, Second, and Third Indian Regiments will be present at these headquarters at precisely 8 o'clock p. m. for council of war.
> By order of Col. R. W. Furnas commanding First Indian Brigade.
> > A. C. ELLITHORPE,
> > Lieutenant and Acting
> > Assistant Adjutant- General.[66]

Left on their own and still in need of fresh water and good forage, Furnas split up the brigade to find sufficient sources to satisfy their needs, sending the 3rd Indian Home Guard regiment as far north as the Pryor Creek crossing, a distance of about 20 miles. All of the Indian regiments were suffering from high levels of desertion. Many of the Indian troops could not understand why they were deprived with limited rations when there was an ample and bountiful supply of beef and other foodstuffs all around them. It seemed to the Indian troops that if the food did not come on a wagon then it did not count as rations to the white officers. The Union Indian Brigade remained in their respective encampments until they received orders from Col. Salomon on July 23 to withdraw up to the Horse Creek Crossing, over 50 miles north of the encampment on Flat Rock Creek. The Indian Brigade arrived there on July 25. The encampment, called Camp Wattles, stretched from Horse Creek to Wolf Creek, ensuring the availability of water from the rivers and the creeks. Furnas met with Salomon at this camp and asked for at least half of Allen's Kansas battery to be kept with him for duty in the Indian Territory. The three Union Indian Regiments under Col. Furnas remained in various areas along the Verdigris and Grand Rivers, but desertions by the Indian troops became so widespread that they backtracked as far as Baxter Springs. This action highlighted the difference between the expectations of the Indian troops and the white troops. Many members of the Indian regiments only wanted to fight to reclaim their homes and once there, they felt their obligation ended.

On the ground in the Cherokee Nation, the Union Indian Brigade under Colonel Robert Furnas tried to accomplish the vague mission of being a corps of observation between the Grand and Verdigris Rivers. On July 27 further instructions were issued to Col. Furnas from Col. Salomon via Adjutant Ellithorpe regarding his operations in the Cherokee Nation. In the letter of instruction Ellithorpe continually stressed for Furnas to strictly go by the orders given. He wrote:

> "…In my interview with the Genl [Salomon] I can but think that his plan of operations is in the main good. The cherokee [sic] country will be held firmly & Gen Salomon will give you timely information of his movements. You will find it the best of policy to adhere strictly to his orders… The Genl approves of your plans of possessing Fort Gibson, Park Hill, Taliqua [sic], the ford across the Verdigris &c…" Col. Jewel will send you white men to assist you, communicate further with him on that part. Every thing is tinder footed & you must tread cautiously but above all things obey orders strictly…"[67]

The tone of the letter seems to be that Ellithorpe was warning Furnas that he needed to tread lightly and watch his step since Col. Salomon was on such tenuous ground and was replacing anyone who supported Weer during the mutiny. The instructions did not reach Col. Furnas until he had already accomplished the goals set out in the letter.

On July 25 Furnas and the Union Indian Brigade were located at the Flat Rock Creek crossing on the Grand River from where he sent out a strong patrol under Major William Phillips of the 3rd Indian Home Guard with between three and four hundred soldiers. By forced and night marches they were to sweep into Tahlequah, then on to Park Hill to check on Chief Ross. Finally, they were going to conduct a reconnaissance of Fort Davis by way of Fort Gibson. At Park Hill Phillips received information that a Confederate force was going to move on that location from Fort Davis. Hoping to contact this force, Phillips divided up his force to take advantage of three different routes that passed through the Bayou Menard between Tahlequah and Park Hill going towards Fort Gibson. The three routes converged in Bayou Menard about seven miles from the fort. Lt. John Haneway's detachment took the advance on the northeastern road, Maj. Phillips and his command took the southeastern road, and the center road was covered by a large detachment under the command of Lt's. Robb, Howard, Blunt, and Phillips. Lt. Haneway's detachment was the first to approach the crossroad meeting place in the early evening of July 27.[68]

Colonel Douglas Cooper was monitoring the situations created by the Union Indian Expedition from Fort Davis, as he had been ordered to do by General Albert Pike. There is no clear information as to the number of Confederate troops, white or Indian, that Cooper had available at Fort Davis to counter any other Union advances. There is little information to indicate if Cooper had any of the 1st Choctaw and Chickasaw regiment with him at the Confederate fort. The location of the 1st Creek regiment is not known but was most likely scattered around the Creek Nation with larger detachments around Council Hill. Adjutant Ellithorpe estimated he had faced three to four hundred Confederate Indians and Texas Rangers at Vann's Ford on the Verdigris River, but he was unable to identify whose troops they were. They were located in the Creek Nation so these troops may have been parts of the Creek regiment. What is known is that around four hundred troops of the 2nd Cherokee regiment, under the command of Lt. Col. Thomas F. Taylor, along with small detachments of other Confederate Indian regiments, were enroute to Park Hill from Fort Davis via Fort Gibson, intending to take revenge upon the Union-supporting Cherokees and Chief John Ross. They were approximately seven miles east of the Fort approaching the crossroads when they ran into Lt. Haneway's small detachment.[69] Haneway's far outnumbered detachment fell back onto the Park Hill road and the Confederate force, not knowing there were two other wings, thrust forward on Haneway's small unit. As the Confederates moved forward Haneway's command was joined by the center and left detachments who had heard the sound of battle and galloped quickly down the roads toward the crossroad to support the lieutenant and his small group. The two wings struck the flanks and surprised and almost surrounded Lt. Col Taylor's command. As the Union command attacked the flanks and center of the Confederate unit they joined together and pushed back at the Confederate force in a sharp fight. A final charge by the Union troops broke the Confederate line and they fell back in a rout, leaving their dead and wounded behind.[70] The Union initially counted thirty-two dead on the field, including Lt. Col. Taylor, and captured twenty-five Confederate Indians. Later information from Col. Cooper put the total lost as one hundred twenty-five, killed, wounded, or missing. Maj. Phillips reported only one man wounded, a private in Company F. This must have been a great blow to

Colonel Stand Watie and the rest of the 2nd Cherokee regiment since they were severely defeated by the Union 3rd Indian Home Guard. This regiment consisted almost entirely of former Confederate Cherokees from Colonel John Drew's 1st Cherokee regiment that had deserted to join the Union Army. Regardless, Col. Cooper had received information on the Expedition's encampment on Horse Creek crossing that wildly overestimated the number of Union troops. Upon receiving this information Cooper withdrew all Confederate forces south of the Arkansas River. Even with this victory under his belt, Maj. Phillips and his unit were far in advance with no rations except beef. He fell back and rejoined the remainder of his regiment at Flat Rock Creek crossing.[71]

When the retreating white regiments of the Expedition reached Cabin Creek, they encountered the 2nd Kansas Cavalry regiment under Colonel William F. Cloud who was intending to reinforce the Union Army. Col. Salomon apparently had already given up any intention of returning to the Indian Territory because the supply train waiting at Hudson's Crossing did not advance with the 2nd Kansas Cavalry. Instead, Col. Salomon sent the regiment to Tahlequah and Park Hill to escort Chief Ross out of the Indian Territory since his life was in danger, especially because the majority of the Union forces had withdrawn from the Cherokee Nation. At Tahlequah, Chief John Ross packed up the official seal, the Cherokee treasury of approximately $70,000 in gold, and the Nation's archives and evacuated the Cherokee Nation along with the other parolees, escorted by the 2nd Kansas Cavalry. The Ross party filled upwards of a dozen wagons and buggies. With Ross's departure Col. Cloud issued a proclamation to the Cherokee people explaining the situation:

> Park Hill Ind Territory
> Aug'3d/1862
> To the Members of the Cherokee Nation.
> --The personal safety of your Principal Chief and the safety of the Archives of your Nation require that I move him, and them within our lines, where they will be safe from the strife of war.
> --The United States Government is able to protect its friends, and punish its Enemies, all Cherokees and members of the other Nations and Tribes are invited and expected to assist in maintaining their original nationalities as guaranted them by their Treaties with the U States Government. All marauding and guerilla parties are hereby ordered to immediately desist from their depredations. All personal and real property left behind by the Chief of the Nation or other parties must remain unmolested under the penalties of the severest punishment.
> --All those who are now or have been in arms against the Government of the United States will be freely forgiven if they at once forsake their disloyal practices and purposes, and immediately use their utmost endeavors to assist in restoring peace & harmony.
>
> W. F. Cloud
> Colo. 2d Kan. Vol. U S. A
> Commanding 1st Brigade

Ind. Expedition[72]

The Expedition encountered the re-supply train still at Hudson's Crossing, and instead of returning south to continue their assigned mission, they continued the withdrawal to the north towards Baxter Springs. At Baxter Springs, Col. Salomon received instructions from Brig. Gen. Blunt via Dr. Gillpatrick that indicated that he was not pleased with the way the Union Indian Expedition had withdrawn from the Indian Territory:

> "…I have entertained fears, based upon rumors, that you were withdrawing all the force from the advanced position the command had previously occupied. This must not be done. The country as far south at least as Arkansas River must be held by our forces. A retrograde movement at this time would stampede all the families of loyal Indians who look to our army for protection. You will therefore send the Sixth and Tenth Kansas Regiments, with the other two sections of Allen's battery, to support the three Indian regiments now occupying the advance…"[73]

Col. Salomon simply ignored Blunt's order and continued his withdrawal to Fort Scott by claiming he was under threat of a Confederate force of 8,000 to his east and rearward. The entire command arrived at Fort Scott on or around August 3. Blunt then realized that the command staff of the remnants of the Indian Expedition was out of control and, on August 8, he left his headquarters at Fort Leavenworth and traveled to Fort Scott to take command of the troops in the field. Blunt later wrote:

> "…As soon as I received intelligence of this affair, and that Col. Salomon, with the command, was falling back to Fort Scott, upon the false plea that a large rebel force was flanking him on the east, I despatched [sic] a messenger directing him to halt the command wherever the order reached him, to send certain troops to reinforce or support the Indian regiments that had not yet abandoned the Indian country, and with the remainder of the command await further orders, assuring him at the same time, that there was no enemy threatening him on his flank, or elsewhere,..
> …On my arrival at Fort Scott, to my great surprise, I found the entire command at that place, notwithstanding Col. Salomon had received my order at Baxter's Springs, sixty-five miles south of Fort Scott…"[74]

At this point Gen. Blunt needed to address Col. Salomon's mutiny and failure to follow orders. The other officers of the Expedition were unhappy with the way Salomon had taken control of the Expedition from Col. Weer. Part of this attitude may have been that Weer was a Kansas officer as were most of the other officers. Salomon was a German immigrant who commanded a Wisconsin regiment. Col. Salomon was blessed in that he was serving in the American Army. If he had taken this action back home in Germany or

Prussia, he would've been immediately executed without the bother of a trial. Still, Blunt attempted to take action as he later reported:

> "…Upon my assuming command of the troops in the field, I found them in a disorganized and demoralized condition, resulting from the mutinous proceedings before referred to. A general wrangling among officers and charges and countercharges had followed this occurrence. For the purpose of investigating the conduct of officers accused of being implicated in this insubordination and mutiny, I convened at Fort Scott a general court martial, but on learning that a large proportion of the officers were in one way or another involved in the affair... I therefore dissolved the court, restored such officers as had been placed under arrest..."[75]

An unusual twist to this entire mutiny action is that the U.S. Senate had confirmed a brigadier general of volunteers commission to Frederick Salomon to date from July 16, 1862. This would have definitely complicated any attempt to punish Salomon since he was technically a brigadier general at the time of his mutinous conduct and, thus, was a superior officer to Weer. On the other hand, because Salomon was ignorant of his promotion, the actions that he took were in fact illegal. This complication may have also been a reason for Blunt to simply sweep the whole incident under the rug. A court martial record would have helped in establishing facts, fault, and responsibility for the failure of the Expedition, perhaps by denying Salomon his promotion and cashiering both Salomon and Weer from the service. Without this record of fact-finding, and without any record of Col. Weer writing an official statement, we are left without a clear picture of what actually transpired at the Flat Rock Creek encampment on the evening of July 18, 1862.

As to why it was Col. Salomon who took the drastic action of arresting Weer and taking command of the Expedition, this author believes that the heat and humidity of the Indian Territory, coupled with a lack of clean water, took more of a toll on the 9th Wisconsin Infantry than on the other troops. Since they were an infantry regiment, they were not utilized as much as the cavalry units. These infantry soldiers were probably serving time standing guard over the camp and supplies and sweltering in the unfamiliar and hot summer southern plains sun that is usually coupled with a steady, hot, southwest wind. They wore heavy wool uniforms and, unlike cavalry, had to carry all of their equipment on their backs the entire distance on foot. There was a significant lack of trees or shade on the open prairie. Modern-day eastern Oklahoma is far more forested than it was in the mid-19th century, trees limited mostly to the banks of the waterways. Any trees that may have been present were quickly cut down and used as fuel for cooking fires. Moreover, inactivity and being left out of most of the action is one of the worst problems for a military unit. The Regular Army had already learned these hard lessons during their time pulling duty at lonely remote posts across the West. Since few, if any, of the officers of the Union Indian Expedition had served in the Regulars, they did not recognize the signs. Salomon was their commander and was probably hearing the

complaints all day long from his subordinates all while enduring the heat and lack of water as well. The collapse in command also showcases the problems with a political army led by physicians, lawyers, and journalists who failed to operate and abide by military standards of conduct, instead depending more upon their connections with powerful politicians. The Department of Kansas was in dire need of experienced officers of the Regular Army who could at least keep the appointed commanders informed as to how to operate within the confines of the military establishment. However, experienced Regular Army officers were in short supply with the vast majority serving east of the Mississippi River in the Virginia and Tennessee campaigns.

General Blunt also had to contend with the presence of Chief Ross and his entourage at Fort Scott. On August 13 he wrote a letter of introduction to President Abraham Lincoln on behalf of the Cherokee Chief. He had earlier forwarded all of the documents and proclamations that the Cherokee Nation had submitted or had received from the Confederate States. Blunt reiterated that he believed Ross to be an honorable man who was forced into the Confederacy due to the division of the Cherokees and the lack of action on the part of the United States. Chief Ross departed Fort Scott to meet with the President in Washington, and would only return to his home in the Cherokee Nation one last time after the war.[76]

Over on the Confederate side things were still in disarray. The feud between Brig, Gen. Pike and Maj. Gen. Hindman continued and soon came to a heated breaking point. On July 12, Pike submitted his resignation to

Richmond in protest. A few days later Maj. Gen. Theophilus Holmes, an elderly former U.S. Army officer who Robert E. Lee reportedly stated had performed poorly during the Peninsula Campaign in Virginia, was named as Gen. Van Dorn's replacement as the commander of the Trans-Mississippi Department of the Confederate States Army. Perhaps not knowing all of the facts involved in the conflict between the two general officers, Holmes dove right into the middle of the conflict and sided with Hindman. Upon learning of this, Hindman immediately dismissed Pike from his position as commander of the Department of the Indian Territory and later absorbed it into his own newly created District of Arkansas. Although Pike's resignation had not yet been accepted by the War Department, his command in the Indian Territory was over. It was probably best this way since he had shown that he held very little control over his Indian Brigade.

The Confederate high command was proven right when Pike's actions, or lack thereof, during the Union Indian Expedition, did not in any significant way hinder the Union advance into the Indian Territory. Pike did not deploy any of the remaining Confederate Indian units to assist Col. Watie and Col. Drew in halting the Union Army's advance into the Cherokee Nation. Although he did send Col. Cooper to Fort Davis, he did not send any of the Choctaw and Chickasaw regiments nor any of the Creek units to support the Cherokees. Instead, Pike's conflict with Hindman blinded him to the actual danger confronting the Confederate Indians. If Col. Salomon had not interfered and abandoned the Indian Territory, the Union Army could have marched straight through the Confederate-held areas, north and south of the Arkansas River. In the end it could have been that Pike had simply lost his nerve after his narrow escape from the battle of Pea Ridge. It was after that event that Pike abandoned Fort Davis, built Fort McCulloch near Boggy Depot, and never again strayed north of the Canadian River. When the Expedition entered the Cherokee Nation, it was his job to gather his forces and at least delay the Union Army's advance until regular Confederate forces could be deployed from Arkansas or Texas. Instead, he sent Col. Cooper to Fort Davis to try to put together a defense with few responding to his call for assistance. Although Hindman did order Col. Clarkson to support the Cherokees during the invasion with his oversized battalion, his defeat and capture brought an end to Confederate resistance to the Union invasion.

The Union Indian Expedition that had started with such high hopes quickly ended in an embarrassing failure. The Confederate Army soon reoccupied the Cherokee capitol. With Ross' departure the Cherokee National Council was called into session at Tahlequah. On August 21 this Confederate-leaning Council elected Stand Watie as the new principal chief of the Cherokees, claiming that John Ross had abandoned the position.[77] And at Fort Scott the recriminations continued: although Salomon had committed a clear case of mutiny, not only was he not censured for his rash action, but he also kept his promotion to brigadier general. Col. Weer was later reinstated with the command of one of the brigades of the new District of Kansas before he was later cashiered out for drunkenness and monetary irregularities.[78] The bottom line was that Union officers had no great love for Indians, wild or not. They believed it was beneath them to assist in moving the Indians back into Indian Territory and protect them from Confederate forces. To be fair, most who joined the Union Army during the war wanted to go fight the Confederates, not Indians. But the Army still had the responsibility of keeping the overland trails and telegraph services safe from attack, be it Confederates or Indians. Most white officers in the Union Indian Brigade took the positions to receive a commission and spent much of their time trying to get into a better unit later on. The same situation occurred when the U.S. Colored Troops units began to be organized in 1863. Although most actions in the Cherokee Nation were favorable for the Union Army, the presence of the Confederate forces were too much for Colonel Furnas and the Union Indian Brigade and they soon abandoned Camp Wattles and retreated back to Baxter Springs, Kansas. Soon afterward Colonel Furnas asked to be relieved of command of the brigade due to his dislike and disrespect for the Indian troops under his command. He was reassigned to service in the Northwest.[79] Sgt. Britton stated that the Expedition had one very positive result: the creation, arming, and equipping of the Union Indian Brigade itself, with three full regiments that could be called upon to protect and serve in the Indian Territory.[80]

[1] Weer to Salomon, *OR*, Series I, Vol. XIII, pp. 452.

[2] Abel, *Participant*, footnote. pp. 126.

[3] Weer to Judson, *OR*, Series I, Vol. XIII, pp. 456.

[4] USGS Topo Map. Wyandotte Quad: Sec. 22, T27N, R23E

[5] Britton, *Border Vol. 1*. pp. 298.

[6] Britton, *Brigade*. pp. 62.

[7] USGS Topo Map. Wyandotte Quad: Sec. 10, T24N, R23E

[8] Blunt to Weer, *OR*, Series I, Vol. XIII, pp. 461.

[9] USGS Topo Map. Pryor Quad: Sec 14, T20N, R20E

[10] USGS Topo Map. Pryor Quad: NE ¼, Sec. 12, T23N, R20E

[11] Britton, *Border Vol. 1*. pp. 300-301.

[12] Cooper, General Order, *OR*, Series I, Vol. XIII, pp. 829.

[13] Kimmel, Special Order, *OR*, Series I, Vol. XIII, pp. 829.

[14] Thomas W. Kremm and Diane Neal, "Crisis of Command: The Hindman/Pike Controversy over the Defense of the Trans-Mississippi District," The Chronicles of Oklahoma, article, Spring 1992; Oklahoma City, Oklahoma. pp. 26.

[15] Pike, General Oder, *OR*, Series I, Vol. XIII, pp. 844.

[16] Kremm & Neal, pp. 27-29.

[17] Shea, William L. Fields of Blood: The Prairie Grove Campaign, Chapel Hill: The University of North Carolina Press, 2009. pp. 3.

[18] Pike to Hindman, *OR*, Series I, Vol. XIII, pp. 936-943.

[19] Hewitt, General Order, *OR*, Series I, Vol. XIII, pp. 844-845.

[20] Dale, Edward Everett, & Gaston Litton, eds. Cherokee Cavaliers: Forty Years of Cherokee History as told in the Correspondence of the Ridge-Watie-Boudinot Family, Norman: University of Oklahoma Press, 1939. pp. 115-117.

[21] Pike, *OR*, Series I, Vol. XIII, pp. 839-840.

[22] Bowen, Nancy Bunker. An Uncivil Warrior: Missouri's Col. James J. Clark. Unpublished Manuscript.

[23] Abel, *Participant*, pp. 129-130.

[24] Britton, *Border Vol. 1*. pp. 300.

[25] Britton, *Border Vol. 1*. pp. 301.

[26] Michael W. Buster, *Supplemental OR*, Volume II, Part III, pp. 467. Edwards, Whit. The Prairie was on Fire: Eyewitness Accounts of the Civil War in Indian Territory. Oklahoma City: Oklahoma Historical Society, 2001. pp. 21.

[27] Johansson, M. Jane, *Ellithorpe*, pp. 25.

[28] Theodore Gardner, "The First Kansas Battery: An Historical Sketch, with Personal Remembrances of Army Life, 1861-65," *Collections of the Kansas State Historical Society* 14 (1915-1918): 235-240.

[29] Hervey Buckner, Letter, n.d. Hervey Buckner Collection, Box 96:47, Oklahoma Historical Society/Archives Division.

[30] Weer to Moonlight. Report, *OR*, Series I, Vol. XIII, pp. 137-138.

[31] Ibid, pp. 137.

[32] Commissioner of Indian Affairs, Report, 1862. pp. 162.

[33] Clarkson, James J. Report, February 29, 1864, James J. Clarkson Collection, Missouri Historical Society. Edwards, Whit. The Prairie was on Fire: Eyewitness Accounts of the Civil War in Indian Territory. Oklahoma City: Oklahoma Historical Society, 2001. pp. 20.

[34] Johansson, M. Jane, *Ellithorpe*, pp. 26.

[35] Britton, *Brigade*. pp. 63.

[36] Issac Gause, Four Years with Five Armies: Army of the Potomac, Army of the Missouri, Army of the Ohio, Army of the Shenandoah. New York: Neale Publishing Company, 1908. pp. 89-90

[37] Lonian, Moses. Interview, Indian-Pioneer Collection, Western History Collection, Works Progress Administration, University of Oklahoma, 1937. #6652.

[38] McNair, Chaney, Interview, Indian-Pioneer Collection, Western History Collection, Works Progress Administration, University of Oklahoma, 1937. #5680.

[39] Johansson, M. Jane, *Ellithorpe*, pp. 27.

[40] Ritchie to Blunt, *OR*, Series I, Vol. XIII, pp. 463-464.

[41] Lumen Harris Tenney, War Diary of Lumen Harris Tenney, 1861-1865. Cleveland: Evangelical Printing House, 1914. pp. 20

[42] Confer, Clarissa W. The Cherokee Nation in the Civil War, Norman: University of Oklahoma Press, 2007. pp. 78-79.

[43] Weer to Ross, *OR*, Series I, Vol. XIII, pp. 464.

[44] Ross to Weer, *OR*, Series I, Vol. XIII, pp. 486-487.

[45] Peck, R. M.

[46] Pike to Drew, July 14, 1862, Grant Forman Papers, Gilcrease Institute, Gaines, W. Craig. The Confederate Cherokees: John Drew's Regiment of Mounted Rifles. Baton Rouge: Louisiana State University Press, 1989. pp. 109.

[47] Britton, *Brigade*. pp. 69-70.

[48] Britton, *Brigade*. pp. 71-73.

[49] Greeno, H.S. Reports x2, July 15, 1862, *OR*, Series I, Vol. XIII, pp. 473.

[50] John Drew, Letter, August 6, 1862, Manuscript No. 6740, Box 2, University of Virginia, Alderman Library, Charlottesville, Virginia.

[51] Monaghan, Jay. *The Civil War on the Western Border, 1854-1865*. Lincoln and London: University of Nebraska Press, 1955: 253.

[52] Able, *Participant*. pp. 137-138.

[53] Peck, R.M.

[54] Lonian, pp. 7-9.

[55] Campbell to Weer, *OR*, Series I, Vol. XIII, pp. 161.

[56] USGS Topo Map. Muskogee Quad: NE ¼, Sec. 22, T17N, R17E

[57] Johansson, M. Jane, *Ellithorpe*, pp. 29-30.

[58] Ibid

[59] Ibid

[60] Britton, *Brigade*. pp. 73.

[61] Saloman, *OR*, Series I, Vol. XIII, pp. 475-476.

[62] Saloman, *OR*, Series I, Vol. XIII, pp. 476.

[63] Carruth, Martin to Blunt, *OR*, Series I, Vol. XIII, pp. 478

[64] Blunt, *Experiences*, pp. 223.

[65] Salomon, *OR*, Series I, Vol. XIII, pp. 475-477.

[66] Ellithorpe, GO#1, *OR*, Series I, Vol. XIII, pp. 481.

[67] Johansson, M. Jane, *Ellithorpe*, pp. 36-37.

[68] Phillips to Furnas, *OR*, Ser. I, Vol. XIII, pp. 181-182.

[69] USGS Topo Map. Muskogee Quad: Sec. 1, T15N-R20E; Sec. 6, T15N-21E

[70] Gaines, W. Craig. The Confederate Cherokees: John Drew's Regiment of Mounted Rifles. Baton Rouge: Louisiana State University Press, 1989. Pp. 116-117.

[71] Britton, *Brigade*. pp. 82.

[72] Abraham Lincoln papers: Series 1. General Correspondence. 1833-1916: William F. Cloud to Cherokee Nation, Sunday, August 03, 1862 (Moving their chief and archives)

[73] Blunt to Saloman, *OR*, Series I, Vol. XIII, pp. 531-532.

[74] Blunt, *Experiences*, pp. 223.

[75] Blunt, *Experiences*, pp. 224.

[76] Blunt to Lincoln, *OR*, Series I, Vol. XIII, pp. 565.

[77] Franks, Kenny A. Stand Watie and the Agony of the Cherokee Nation. Memphis: Memphis State University Press. 1979. pp. 129-131.

[78] Moonlight, *OR*, Series I, Vol. XIII, pp. 595.

[79] Furnas, R. W. Report, July 25, 1862, Ibid: 511-512.

[80] Britton, *Brigade*. pp. 79.

Chapter 8
The Missouri – Arkansas Border War

It was a typical hot and dusty mid-September day in Washington, D.C. when John Ross, Principal Chief of the Cherokee Nation, accompanied by William P. Dole, U.S. Commissioner of Indian Affairs, sat in the President's White House outer office, which was filled with clerks, lobbyists, and Army officers dashing about fulfilling their duties. Amid this chaos they were waiting to be admitted to a meeting with President Abraham Lincoln. It was Friday, September 12, 1862, and Chief Ross was at the end of a long journey from his home in the Indian Territory after being brought out by troops of the 2nd Kansas Cavalry and escorted to Fort Leavenworth, Kansas, in early August. He and his family traveled first to Philadelphia, Pennsylvania, then to Wilmington, Delaware, and finally to Lawrenceville, New Jersey. Ross and his party were fortunate in being able to occupy homes owned by the Staplers, Ross's wife Mary's family homes that she had inherited as members of her family passed on. General Blunt and Indian agents at Fort Leavenworth had encouraged Ross to go on to Washington to gain President Lincoln's sympathies for the Cherokees. The official visit to President Lincoln had been arranged by the Secretary of the Interior. and Commissioner Dole was to accompany him. Once the meeting began, the Chief explained to the President that the dislocated Cherokees in Kansas were concerned about their families that were still trapped in the Cherokee Nation. Since the withdrawal of the Union Indian Expedition in August, the Nation had been rocked with violence and turmoil. He further detailed Cherokee needs and stressed that the Cherokees had placed themselves under the

**Drawing of the Battle of Newtonia
by Wagon Boss R.M. Peck**

protection of the United States under previous treaties. Ross emphasized that when protection was withdrawn by the evacuation of the Regular Army forts in Indian Territory, the Cherokee had no choice except to align with the Confederacy or fall into an internal civil war. He further explained that although the Cherokee were divided, the majority of tribal members wanted to remain faithful to their treaties with the United States. He further explained that the Cherokee Nation was also surrounded by Confederate Texas and the Confederate-allied Choctaw and Chickasaw Nations to the south, Confederate Arkansas to the east, and the Confederate-allied Creeks to the west. The Cherokees were further burdened and bordered by the people of questionable loyalty in Southwest Missouri. Their only support would come from Union Kansas, and that had been a disaster. Due to these difficulties, they were unable to assume their true position until the summer of 1862 when a "great mass" of Cherokee people rallied spontaneously around the authorities of the United States.

There is no clear record of what President Lincoln's response to the Chief of the Cherokees was except that he noted that he had very little knowledge of what protections had been promised in the previous treaties between the United States and the Cherokee Nation. In this situation, both the Secretary of the Interior and the Commissioner of Indian Affairs failed the Cherokees by not previously providing the President with a copy of the Cherokee treaty and a synopsis of the situation in the Indian Territory, which forced them to explain the entire situation cold. So, it is also understandable that the President may have been less than accommodating to Chief Ross, since, at the time of the meeting, the Union Army of the Potomac was trying to re-organize after being severely beaten at the Second Battle of Manassas on August 30. In addition, Robert E. Lee's Army of Northern Virginia was currently moving up the Shenandoah Valley and crossing the Potomac River. This action would result in the surrender of the Union garrison at Harper's Ferry, the Battle of South Mountain, and would culminate in the Battle of Antietam on September 17. Compared to the issues the President was dealing with at the time, the Cherokee refugee problem probably did not rank high on his list of priorities. Lincoln suggested that Chief Ross write a letter outlining the situation and fully explaining the reason the Cherokee Nation violated its treaty with the United States and signed one with the Confederacy. Back in Lawrenceville, Chief Ross wrote to the President:

Lawrenceville, New Jersey
September 16th 1862

President Lincoln. Sir / During the interview which I had the honor to have with your Excellency the 12th. Inst. you requested that the objects of my visit should be communicated in writing. I therefore beg leave, very respectfully, to represent,

1st. That the relations which the Cherokee Nation sustains towards the United States have been defined by Treaties entered into between the Parties from time to time, and extending through a long series of years.

2nd. Those Treaties were Treaties of Friendship and Alliance. The Cherokee Nation as the weaker party placing itself under the Protection of the United States and no other Sovereign whatever, and the United States solemnly promising that Protection.

3rd. That the Cherokee Nation maintained in good faith her relations towards the United States up to a late period and subsequent to the occurrence of the war between the Government and the Southern States of the Union and the withdrawal of all protection whatever by the Government.

4th. That in consequences of...the overwhelming pressure brought to bear upon them the Cherokees were forced for the preservation of their

Country and their existence to negotiate a Treaty with the "Confederate States"

5th. That no other alternative was left them surrounded by the Power & influences, that they were, and that they had no opportunity freely to express their views and assume their true position until the advance into their Country of the Indian Expedition during the last summer.

6th. That as soon as the Indian Expedition marched into the Country the great Mass of the Cherokee People rallied spontaneously around the authorities of the United States and a large majority of their warriors are now engaged in fighting under their flag.

…The advance of the Indian Expedition gave the Cherokee People an opportunity to manifest their views by taking [as] far as possible a prompt and decided stand in favor of their relations with the U.S. Govt. The withdrawal of that Expedition and the reabandonment [sic] of that People & Country to the forces of the Confederate States leaves them in a ruin! What the Cherokee People now desire is ample Military Protection for life and property; a recognition by the Govt. of the obligations of existing Treaties and a willingness and determination to carry out the policy indicated by your Excellency of enforcing the Laws and extending

to those who are loyal all the protection in your power… For the satisfaction and encouragement of my own People and of the Indian Nations who live near them, I beg leave very respectfully to suggest that you will issue a Proclamation to them… which will enable me to make assurances in behalf of the Govt. in which they can confide… I have the honor to be, Sir, With Sentiments of high regard

> Yr. Obt. Serv't.
> John Ross,
> Princl. Chief
> Cherokee Nation"[1]

Amidst the pressures and tumult of Washington on the days following the bloodiest day of the Civil War at Antietam, Maryland, on September 17, all within a day's ride from the White House. This was coupled with General McClellen's subsequent slow pursuit of Lee's Confederate Army, but President Lincoln still found time to respond to Chief Ross's letter. Lincoln admitted that due to the fall-out of the military actions around Washington, he had not been able to research the claims made in Ross's letter. As a result, Lincoln did not commit the United States to anything beyond protecting those Cherokees who were loyal to the Union. The President stated:

> Executive Mansion,
> Washington,
> Sept. 25, 1862

Sir:
Your letter of the 16th. Inst. was received two days ago. In the multitude of duties cares claiming

my constant attention I have been unable to examine and determine the exact treaty relations between the United States and the Cherokee Nation. Neither have I been able to investigate and determine the exact state of facts claimed by you as constituting a failure of treaty obligation on our part, and justifying or excusing, the Cherokee Nation to make for making a treaty with a portion of the people of the United States in open rebellion against the government thereof— This letter, therefore, must not be understood to decide anything upon these questions. I shall, however, cause a careful investigation of them to be made— Meanwhile the Cherokee people remaining practically loyal to the federal Union will receive all the protection which can be given them consistently with the duty of the government to all parts of the whole country. No more than this can safely be promised, even to the loyal white people of Missouri, or other border states— I sincerely hope the Cherokee country may not again be over-run by the enemy; and I shall do all I consistently can to prevent it.

Your Obt. Servt.

A. Lincoln.[2]

The war had been tough on Chief Ross and his family. For the next three years, John Ross and his family were refugees. The exiled Chief spent most of his time in Washington, D.C., pleading the case before elected officials. The older sons, James, Allen, Silas, and George, served in the Third Regiment of Union Indian Home Guards organized from former Confederate forces under Colonel John Drew. Allen was captured by Confederate troops in 1862 while returning to Park Hill to take badly needed supplies to his family. He was sent to various prison camps and died a prisoner in 1864. The home of John Ross's daughter Jane was attacked by Confederate marauders in 1863, and her husband, Andrew Nave, was shot and killed while trying to flee. Still there was hope on the horizon. The actions of Chief Ross and of the Indian Affairs officials did make an impact on President Lincoln. In fact, the commander of the new Department of the Missouri, Major General Samuel Curtis, received a letter from President Abraham Lincoln requesting him to reoccupy the Cherokee Nation using the Indian regiments. Presidential interest in the affair brought the Union forces in Kansas into action.[3]

EXECUTIVE MANSION,

October 10, 1862.

Major-General CURTIS, *Saint Louis, Mo.:*

I believe some Cherokee Indian regiments, with some white forces operating with them, now at or near Fort Scott, are within your department and under your command. John Ross, Principal Chief of the Cherokees, is now here an exile, and he wishes to know, and so do I, whether the force above mentioned could not occupy the Cherokee country consistently with the public service.

Please consider and answer.

A. LINCOLN.[4]

General Curtis replied with a less than satisfying answer to Lincoln's inquiry:

> SAINT LOUIS, *Mo.,*
> *October* 10, 1862.
> His Excellency ABRAHAM
> LINCOLN
> President of the United States:
> My forces have driven the enemy to Pineville, near the Indian line. I yesterday ordered an advance, driving them into the Territory and beyond. I doubt the expediency of occupying ground so remote from supplies, but I expect to make rebels very scarce in that quarter pretty soon.
> SAML. R. CURTIS,[5]

General Curtis's response must be taken in context. He had previously disapproved the use of Indian troops when he had been ordered to raise the initial Indian Home Guard regiments after the scalpings by the Confederate Indian Brigade at Pea Ridge. Regardless, the wheels began to move on the formation and execution of a second Union invasion of the Indian Territory.

Meanwhile, the saga of the General Hindman and General Pike feud continued. After the Union retreat from Indian Territory in July and August, the Confederate forces in Indian Territory once again fell into disarray. The commander of Indian Territory, General Albert Pike, continued his personal battle with his superior, Major General T.C. Hindman, commander of the Trans-Mississippi District. Pike insisted that the Confederate government was breaking its treaty promises with the Indians that had sworn allegiance to the Confederacy. He pointed to the lack of proper supplies and equipment for the Indian

regiments. Pike's forces were in such despair for supplies and equipment that Pike paid $20,000 of his own money to provide for the troops. The feud with Hindman and with the Confederate leadership in Richmond, resulted in Pike submitting his resignation and its acceptance by General Hindman on July 28.[6] Believing his resignation had been accepted in Richmond, on July 31, Pike issued a farewell address circular proclamation to the "Chiefs and People of the Cherokees, Creeks, Seminoles, Chickasaws, and Choctaws," notifying them that he was resigning his position as department commander. He outlined his personal and official reasons for resigning but did not place the blame for all the Confederacy's failures in the Indian Territory on himself or on the Confederate States. He instead placed the blame on the fate of the war itself.

> "Remain true, I earnestly advise you, to the Confederate States and yourselves. Do not listen to any men who tell you that the Southern States will abandon you. They will not do it. If the enemy has been able to come into the Cherokee country it has not been the fault of the President; and it is but the fortune of war... Be not discouraged, and remember, above all things, that you can have nothing to expect from the enemy. They will have no mercy on you, for they are more merciless than wolves and more rapacious… And whatever may be told you about me, you will soon learn that if I have not defended the whole country it was because I had not the troops with which to do it; that I have cared for

your interest alone; that I have never made you a promise that I did not expect, and had not a right to expect, to be able to keep, and that I have never broken one intentionally nor except by the fault of others."[7]

Behind the scenes, Colonel Douglas Cooper had been actively subverting his commander in an effort to replace the non-aggressive, defense-minded General Pike and win a general's star for himself. Cooper received the circular on August 6 and immediately took offense to its contents. He went so far as to try to retrieve as many copies of it as possible. Apparently, Pike's issuance of the circular violated an obscure Confederate Army regulation against an officer addressing the President via a printed material or one that mentioned troop movements. Cooper reasoned that his commander was deranged or treasonous and ordered his arrest on August 7 so he could be brought to General Hindman's headquarters in Little Rock. Hindman approved of this action and added a violation of Section 29 of the Indian Intercourse Act, which prohibited any individual from influencing any Indian or Indian tribe to be critical of the Confederate States or their treatment of the Indians.[8] Hindman reported:

"…Under these circumstances it seemed that the interest of the service would be promoted and his own desires gratified by complying with General Pike's request. I therefore forwarded his resignation to Rich-mend [sic], with my approval, and at the same time relieved him from duty.

On the receipt of my order to that effect he issued and distributed a printed circular, addressed to the Indians and equally likely to reach the enemy, in which, under pretense of defending the Confederate Government, he evidently sought to excite prejudice against it, and endeavored thoroughly to disgust and dishearten our Indian allies by suppressing or perverting facts where their publication would be beneficial to our cause and openly proclaiming them when they should have been concealed…

Col.(now Brig. Gen.) D. H. Cooper, who was next in rank and had succeeded to the command, deemed it his duty to place General Pike in arrest, and so informed me, inclosing a copy of the circular, and expressing the opinion that the author was insane or a traitor. I approved his action, and ordered General Pike sent to Little Rock in custody. I also forwarded Colonel Cooper's letter to Richmond, with an indorsement, asking to withdraw my approval of General Pike's resignation, that I might bring him before a court-martial on charges of falsehood, cowardice, and treason…"[9]

General Hindman was positive that his superior, General Holmes, would support his actions against Pike since he had successfully manipulated the Robert E. Lee-dejected general to fully siding with him. Pike was delivered to Little Rock and was allowed to stay with family until the issue of his resignation was resolved. In Richmond, President Davis positively refused Pike's resignation, which left his status in limbo. Pike applied for a leave of absence from General Holmes, which was granted. The leave would last until the matter was resolved, so Pike then proceeded to Sherman, Texas. At Sherman, Pike received information regarding orders from Secretary of War Randolph that, stated unequivocally that military supplies destined for the Indian Territory were not, under any circumstances, to be diverted to any other command and that commanders were not to interfere with Pike's unique command of the Indian Territory. This letter proved that Pike's allegations concerning the Confederate supply chain violations were valid and that his superiors in the Trans-Mississippi were complicit in diverting materials, as well as treaty annuities, to other white commands outside the Indian Territory.[10] The letter also strengthened Pike's claim that the Department of the Indian Territory was an independent command. It was a small victory for Pike in his mission to protect and serve the Indian nations.[11] If he had received this information when it was issued in mid-July, he likely would not have resigned and never would have issued the circular that was the cause of so much dissension within the Trans-Mississippi. Holmes, Hindman and Cooper, fearing that Pike would be completely reinstated, took and ran with a rumor that Pike had aligned himself with pro-Union Texans in order to sabotage the Trans-Mississippi war effort by diverting ammunition trains. Without any evidence for this action, once again Pike was arrested on Holmes's order on November 14, this time in Tishomingo, capital of the Chickasaw Nation. A detachment of General Jo Shelby's Texas brigade was to escort Pike to Little Rock. Unfortunately for the Confederate command, General Pike's resignation had finally been accepted and approved by the President and Secretary of War on November 5, so he was no longer under the jurisdiction of the Confederate Army, so Pike was quickly released. Fearing a retaliation by the Confederate Indians due to his popularity among them, the Confederates made sure they released Pike in Arkansas, away from the Indian Territory. With Pike's resignation and subsequent arrest, any semblance of respect for the Indian nations or the treaties that had been signed with Richmond was gone. Confederate Trans-Mississippi commanders now simply ignored them.[12]

For all of General Hindman's faults, he was the type to get the job done. Since most of his troops in Arkansas were cut off from supplies and equipment from the eastern part of the Confederacy, he found solutions to the shortages. First, by the use of rumor and planted information, he convinced the Union command that he had forces numbering over twenty thousand in and around Little Rock, when in fact he had less than half that amount. This kept the Union Army on land and the Union Navy on the White River a safe distance from the state capital. He also found and updated various muskets and other arms. He was able to set up lead mines for supplying bullets, especially in Granby, Missouri, small factories for making percussion caps, and set men to discovering and digging saltpeter for the making of powder. Hindman was somewhat of a ladies man and using his charm he was able to corral many women into sewing circles to make and mend uniforms and socks.[13]

Historic and Modern View of the Ritchey Mansion in Newtonia
Photos courtesy of Author and Newtonia Battlefield Association

Historic photograph of Matthew Ritchey's stone barn, which sheltered Confederate forces during the First Battle of Newtonia.

Historic Photo of stone wall along Mill Street
Photo courtesy of Missouri Department of Natural Resources

Colonel Sampson Folsom, CSA
1st Choctaw Mounted Rifles
Photo courtesy of the Choctaw Nation

General James S. Rains
Missouri State Guard (MSG)
Photo courtesy of Wikipedia

Colonel Joseph "Jo" Shelby,
MSG/CSA
Photo courtesy of NPS / Wilsons Creek

Newtonia (Water Tower) from Union line on the "Hill"
Photo by Author

View of the Union line and Artillery position from the direction of the City Cemetery
Photo by Author

Newtonia Branch where the 9th Wisconsin (-) crossed to attack the Ritchey Barn
Photo by Author

Newtonia Cornfield where Melee between the Union and Confederate Indian Brigades
Photo by Author

Modern views of the City Cemetery and the view of Capt. Howell's Confederate Battery towards the Union artillery on the "Hill"
Photos by Author

A new issue that both sides had to contend with was the passage of the Partisan Ranger Act by the Confederate Congress in April 1862. This new law allowed the recruitment of partisans, which accelerated the already growing guerilla warfare problems in the Trans-Mississippi West. Soon this type of warfare became the primary style of warfare in western Arkansas and the Indian Territory due to its sparse population and heavily wooded and rugged mountains. This made hit-and-run operations the preferred method of engaging the enemy or just killing and looting regardless of who was at the receiving end. This act stated:

CHAP. LXIII.--*An Act to organize bands of Partisan Rangers.*
April 21, 1862.
Bands of Partisan Rangers.
The Congress of the Confederate States of America do enact, That the President be and he is hereby authorized to commission such officers as he may deem proper with authority to form bands of Partisan Rangers, in companies, battalions or regiments, either as infantry or cavalry, the companies, battalions or regiments to be composed each of such numbers as the President may approve.

Pay, rations and quarters.
    SEC. 2. *Be it further enacted*, That such Partisan Rangers, after being regularly received into service, shall be entitled to the same pay, rations and quarters during their term of service, and be subject to the same regulations as other soldiers.

The Rangers entitled to full value of the arms, etc., captured.
    SEC. 3. *Be it further enacted*, That for any arms and munitions of war captured from the enemy by any body of Partisan Rangers and delivered to any Quartermaster at such place or places as may be designated by a Commanding General, the Rangers shall be paid their full value in such manner as the Secretary of War may prescribe.

APPROVED April 21, 1862.[14]

The Partisan Ranger Act led to the recruitment of unconventional soldiers into the Confederate Army. Partisan Rangers were designed to have the same rules, supplies, and pay as the regular soldiers of the army, but they would be acting independently, detached from the rest of the army. The Partisan Rangers were supposed to gather intelligence and take supplies from the Union Army. Anything they brought back was to be turned into the quartermaster, a military officer in charge of providing food, clothing, and other necessities; in return, they would get paid.[15] Back within the Eastern theater, some Partisan Ranger groups like Mosbey's Rangers operated most of the time in close cooperation with the Confederate commanders. Back in the Trans-Mississippi West, instead of assisting the Army, the partisan groups quickly developed into roving bands of bandits, under leaders with questionable or non-existent commissions, that took what they wanted from whom they wanted, regardless of under which flag they served. They also became vessels of revenge upon anyone who had crossed them individually before or during the war. Unfortunately, this Act brought forth some of the worst atrocities committed during the war and highlighted some of the worst offenders including William Quantrill, William "Bloody Bill" Anderson, and future outlaw Cole Younger. With the absence of any large bodies of Confederate Army units within western Arkansas or southwestern Missouri, these roving bands operated freely without any oversight or control and became the de facto Confederate Army in the region.

The Union command saw these new Confederate militiamen and Partisan Rangers as nothing more than pre-war bushwhackers in uniform. On the other hand, the Confederates in Missouri saw the Kansas troops as "jayhawkers' in blue. Of course, neither side had anything resembling a complete regulation uniform, either blue or gray. The Union Army did a much better job outfitting its soldiers because the supply and logistics chain was much more efficient for the North. At times there were shortages of cotton and wool for the northern factories because, before the war, industry depended on southern cotton. With that supply cut off, the Union was forced to approach foreign markets for cotton at high prices since most of the available supply came from India and Egypt. In the Confederacy, they had a substantial supply of cotton but few factories with which to produce uniforms and related items. With the Union Navy blockading the southern ports, the Confederates were unable to ship their cotton to foreign markets for the much needed currency, weapons, and medicines. The Union Army also simply used the blue uniforms that had been in use for decades and had many storehouses stocked with uniform items, especially at some of the

western military posts like Fort Leavenworth and Fort Larned. But with the Union's western forces there were some wide variations and additions to the standard blue, four-button frock coat and sky-blue kersey trousers with suspenders. The uniform of the Indian Home Guard regiments was discussed in an earlier chapter, and they eventually took those issued items and modified them for their own uses (for example, cutting off the brass buttons and sewing them onto their broad hats or decorating the uniform with various feathers and beads). The Union command was usually accepting of these variations because it was important to an Indian soldier to be seen as an individual along with being a soldier. Plus, these modified uniforms were a substantially better means of identifying a Union Indian than the cross-section of an ear of corn pinned to their civilian frock coat that had been used during the Opothleyahola campaigns of 1861. Most Union soldiers had been issued standard round kepi hats, but they were mostly used only in a garrison environment. In the field they almost universally wore broad felt or straw hats to reduce the burning southern plains sun. Although most Union soldiers had been issued the standard brogan, or Jefferson, shoe for field use, mounted troops and officers tended to have knee-high boots that were most likely purchased from the local economy. Of course, the Indian soldiers preferred moccasins or other homemade footwear. From photographs of Union soldiers of the Trans-Mississippi regiments, we see they added many individualized items such as long knives and pistols and homemade cartridge boxes and canteens. The western Union soldier looked much different than his fellow soldiers belonging to the Army of the Potomac.[16]

As mentioned earlier, the Confederate States had difficulty outfitting its entire army in anything like a standard uniform. Although the vast majority of officers, especially those in the eastern commands, had well-designed and identifiable uniforms, most of the enlisted troops had mixtures of gray or butternut uniforms. Many simply wore civilian clothing from home or captured blue Union uniform items. As the war went on, Confederate uniform standards became less and less regulated. If it was cold, they did not care if the overcoat they were wearing was gray or blue. The Confederate Trans-Mississippi troops from Arkansas and Texas began getting partial uniforms by mid-to- late 1862. Many of the Texas troops were being supplied with uniforms produced at the Texas State Penitentiary in Huntsville, the largest working clothing mill in Texas. Up to that time and continuing until the end of the war, the "battle shirt" was the most common "uniform" among western Confederate soldiers and the Partisan Rangers. For the Confederate Army, it was simply a regular hunting shirt that had some type of military designation placed upon it such as colored stripes and further enhanced by the addition of neckerchiefs of a certain color or collar devices with unit identification. Partisans, including Quantrill's forces, probably did not wear any identifying markings. The Confederate Indian regiment's uniform stocks were almost always pilfered before they reached the Indian Territory supply depot at Fort Smith, Arkansas. As a result, most the Indians belonging to the Indian Brigade went through the war in civilian or tribal war clothing. Part of Brig. Gen. Pike's issue with

the Confederate War Department concerned the uniform issue. He became enraged when he discovered that Texas troops had commandeered 6,500 sets of uniforms, 3,000 pairs of drawers, and 1,000 tents at Fort Smith intended for the Indian Brigade. In the early campaigns, the Confederate Indians used red, white, and blue strings as identification on their sleeves. When or if they ever stopped using them is not known. Regardless, the Trans-Mississippi Confederate enlisted soldier was more than likely issued a butternut gray, homespun uniform accompanied by a wide variety of accoutrements, both issued and homemade, with broad slouch hats that may or may not have had an insignia of some type. The remainder wore civilian clothing from home or whatever items they could find or capture.[17]

The effects of the Partisan Ranger Act quickly began to materialize in western Missouri. For most of the summer of 1862, many Confederate and Missouri State Guard recruiters were dispatched northward from Arkansas into Missouri to replenish the depleted ranks of Trans-Mississippi forces. Some of these officers included Captain Jo Shelby, Colonel John T. Coffee, Colonel Upton Hays, Colonel John T. Hughes, Colonel Gideon W. Thompson and Colonel DeWitt C. Hunter. Along with these Confederate and Missouri State Guard (CS) officers, various guerrillas and bushwhackers, most notably those under William Quantrill and George Todd, had gathered in Missouri and assisted these recruiters as they worked in the region. In an effort to legitimize Quantrill, Confederate authorities were able to secure a commission of Captain for him.[18] Most of these commands were working independently

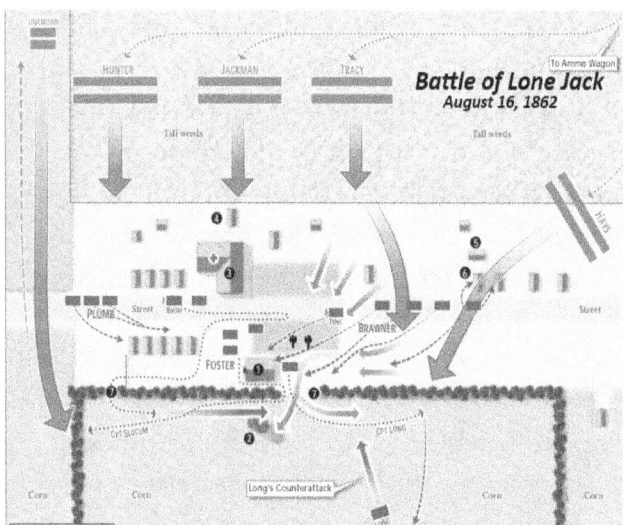

and there was no clear sense of seniority yet established. A large group under Missouri State Guard Colonel John T. Hughes and William Quantrill planned to move to areas in northwest Missouri to recruit for the Confederate forces. All of the partisan groups were in desperate need of ammunition so, when they met with other officers or partisan leaders, they decided to attack the small Union garrison at Independence, Missouri. On August 10, 18-year-old 1st Lt. Cole Younger and another partisan conducted a reconnaissance of the town and the garrison. Lt. Col. James T. Buel of the 7th Missouri Cavalry (US) was in command of the garrison, which included 7th Missouri Cavalry (cos. B & D) - Capt. James Breckenridge, Lt. James M. Vance, 2nd Battalion Missouri State Militia Cavalry (Cos. B, D, & E) - Capt. Jacob Axline, Capt. Franklin Cochran, Capt. Aaron Thomas, and the 6th Regiment Missouri Enrolled Militia (one company) - W.H. Rodewald. Buel ignored warnings of imminent attack and left his command widely dispersed. On August 11, Col. Hughes's Confederate force, including William Quantrill and his partisans, attacked Independence before dawn, in two columns using different roads. They drove

through the town to the Union Army camp, delivering a deadly volley into the sleeping men. Unfortunately, the Confederate charge at daybreak left the Confederate commander, Col. Hughes, dead with a shot through the forehead. Union Captain Breckenridge suggested surrender, but Captain Jacob Axline formed the Federal troops behind a rock wall. Lt. Col. Buel attempted to hold out with part of his force in the bank building he used as his headquarters. Capt. Axline was preparing to move his force towards the square to try to reach Buel. Before Axline could set his force in motion, he received orders from Buel to surrender. Quantrill threatened that he was going to set fire to a small building next to the headquarters building and roast Buel and his men alive in the spreading conflagration. In the sure knowledge that Quantrill could and would carry out this threat, Buel hauled up the white flag and surrendered his whole command.[19] Most of the Union command in Independence was captured, with only a few groups of men making good their escape. The Confederate victory was costly, however, resulting in the death of ten experienced officers. Confederate guerilla leader George

Todd freed the prisoners at the jail, including City Marshal James Knowles and Captain Aaron Thomas of the 2nd Battalion Missouri State Militia Cavalry who had both been interred at the jail. Apparently, Knowles had guided Thomas's force in a successful ambush of Todd's command in an earlier engagement, killing several of them. Todd and his men summarily executed Knowles and Thomas. This First Battle of Independence resulted in approximately 344 known Union casualties; total losses for the Confederate side are unknown. The raiders quickly paroled most of the captured Union troops and evacuated the town when they received information that two large Union Army striking forces were enroute.[20]

A similar encounter happened on August 15, 1862, at the town of Lone Jack, Missouri. Union Maj. Emory S. Foster led a 740-man combined force from Lexington to Lone Jack, Other forces were dispatched from Kansas under General James G. Blunt (2,500 men) and from Missouri under General Fitz Henry Warren (600 men), but neither of these commands would arrive in time to prevent the disaster in Lone Jack. Maj. Foster received intelligence that 1,600 Confederate troops, probably a mixture of Missouri State Guard (MSG) militiamen and newly-enrolled partisans, under Col. John Coffee and Lt. Col. John C. Tracy, were camped near town and prepared to attack them. At about 11:00 p.m., Foster and his men attacked the Confederate camp and dispersed the enemy. Unfortunately, when Foster deployed his two bronze artillery pieces, the sound alerted the other MSG units in the area of the Union force's position as well as displayed their intent to fight.[21] Foster's men returned to town to rest along the main street, having spent several days on the road. The Missouri State Guard (MSG) commanders, Col. Upton Hayes, Lt. Col. Sydney Drake Jackman, Lt. Col. John Charles Tracy, and Col. DeWitt Hunter, serving under Colonel Jeremiah "Vard" Cockrell, gathered west of the town and intended to attack at dawn. Surprise was lost when Foster's pickets sighted the gathering of the MSG forces in the early morning light and gave the alarm. This discovery gave the Union troops time to deploy a defense along a hedgerow of Osage-orange trees along the main street. Confederate Private C.B. Lotspeich of Lt. Col. Jackman's command remembered lying prone in the field and listening to the Union forces as they began stirring. "We lay there and could hear them give every command… hear them putting on accouterments, loading their guns [and] forming lines of battle."[22] The fighting began at dawn and continued with attacks and counterattacks with especially hard fighting around the two pieces of artillery, which switched hands numerous times. Just as the Confederate forces were beginning to retire, Col. Coffee, whose command had been attacked the night before and had regained their composure, attacked the left flank of the Union force. After five hours of fighting and the wounding of Foster while leading a charge to re-take the battery, next-in-command Capt. Milton H. Brawner, ordered a retreat. Brawner reported that the men left the field in good order and returned to Lexington. With the death of all of the artillery horses, the cannon were hastily spiked or disabled and hidden before the Union troops departed. The approach of Union forces including General Blunt and General Fitz Henry Warren forced the combined Southern forces to withdraw on

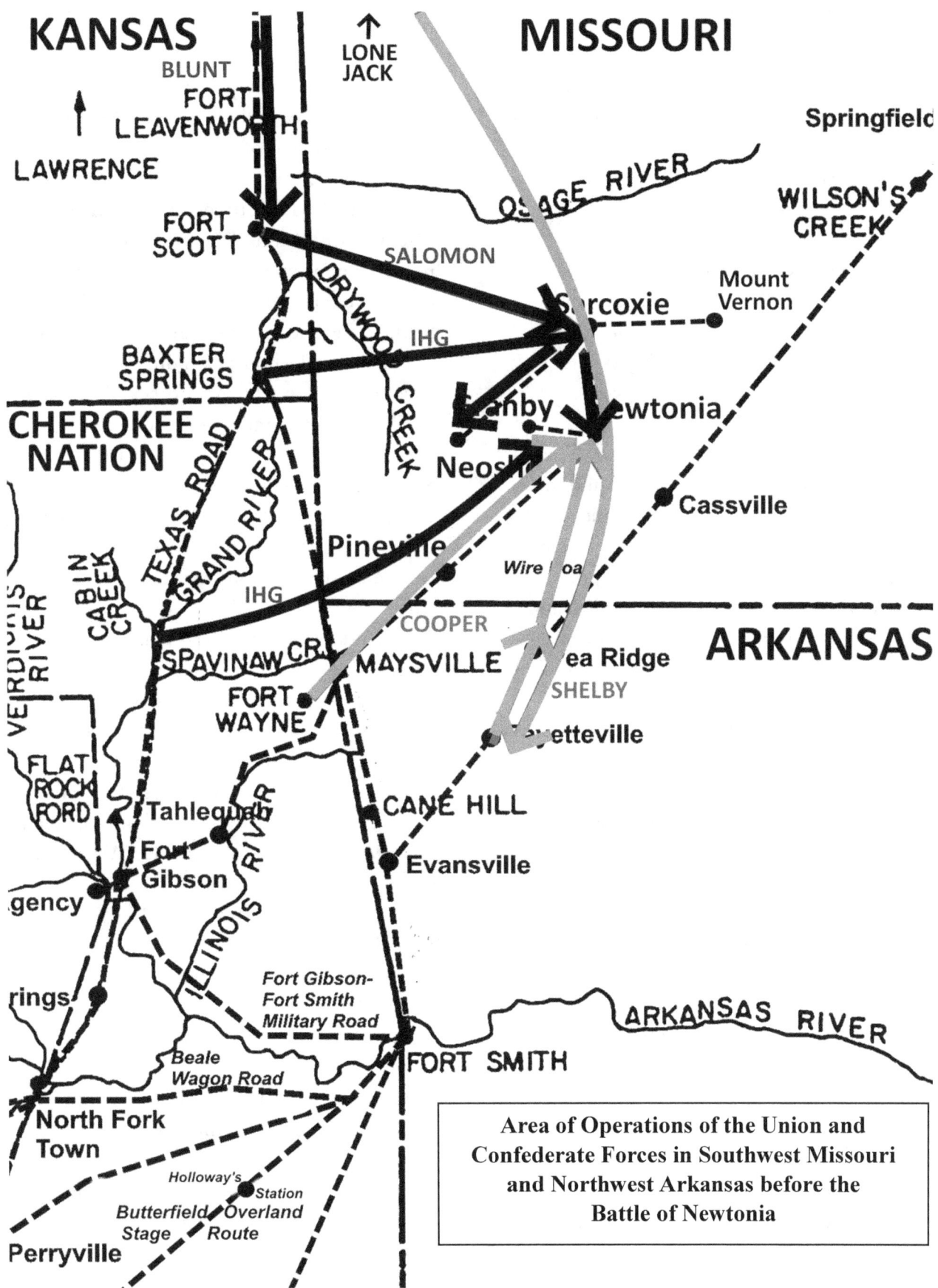

Area of Operations of the Union and Confederate Forces in Southwest Missouri and Northwest Arkansas before the Battle of Newtonia

August 17.[23] General Fitz Warren occupied the town that day but discovered that the Confederates had found the two bronze cannons. Future U.S. Senator and Secretary of War Steven Elkins wrote this about the Battle of Lone Jack:

> "…Foster thought the Confederates were the guerrilla hands who raised the black flag, and never gave any quarter. So, he refused to surrender, and every one of his officers was picked off. The guerrillas were victorious."[24]

Later reports indicate that Quantrill and most of his partisans were still in Independence, looting and burning the community. Union losses at Lone Jack were variously reported as between 43 and 65 killed, 154 wounded, and 75 missing / captured. Confederate losses were about 55 killed and an unknown number of wounded.[25] (in the story "True Grit," Rooster Cogburn lost his eye in "a little scrape at Lone Jack.") Another interesting story from the Lone Jack battle is that the Confederates had captured the severely wounded Maj, Foster and were holding him in a cabin in town. A ranking Confederate officer, Dr. Josiah Hatcher Caldwell, had been a bitter enemy of Foster before the war and intended to kill him as he lay wounded. 1st Lt. Cole Younger stood between Caldwell and the wounded Foster to prevent him from harming the wounded soldier, eventually physically pushing the officer out of the cabin. Years later, Cole Younger was captured after he and the rest of the James-Younger gang tried to rob the banks in Northfield, Minnesota. Major Foster by that time was editor of a St. Louis newspaper and used his position and printing press to push for leniency for Cole Younger at trial.[26]

After the disappointing Union losses at Independence and Lone Jack, General Blunt was advised of further aggressive movements of the Confederates and partisans in Missouri and determined to commence offensive operations. The week after their return, the regiments involved in the previous battles were rested and reequipped at Fort Scott and in the local area. At the same time, Major General Thomas Hindman encountered the retreating forces of Colonels Shelby, Cockrell, and Coffee in Benton County, Arkansas. Hindman decided to retake Southwest Missouri from the Union Army. The three or four thousand recruits brought out of Missouri were organized into regiments, and a cavalry brigade formed, of which Colonel Jo Shelby was given the command. By adding Brig. Gen. Cooper's 1st Choctaw regiment, the 1st Choctaw and Chickasaw regiment, and the 1st Cherokee Battalion, Hindman had upwards of seven thousand men with which to invade Missouri.[27] The new 1st Cherokee Battalion was commanded by Major Joel M. Bryan, who had been authorized by Brig Gen Albert Pike to raise one hundred men or more as Partisan Rangers during the summer of 1862 in accordance with the Partisan Ranger Act. Company A was mustered in at Fort Davis on July 20, 1862. Bryan proceeded to recruit another four companies and was ordered by Hindman to organize the force into a battalion, which was done on September 13, 1862. Records indicate that the enlistment term was three years.

Meanwhile, alerted to Confederate activity in Southwest Missouri, Union Major General Samuel R Curtis, who was newly appointed to the Department of the Missouri, contacted Brigadier General John M. Schofield in Springfield on September 27. Curtis, hoping to avoid the collapse of the Union presence in Southwest Missouri, instructed Schofield to consolidate his troops and be prepared to march. He also sent a letter to General Blunt in Kansas to bring "all available troops" and combine them with those of General Schofield and with General Egbert Brown, of the Missouri Militia (US). Wiley Britton explained why Southwest Missouri was so important to both sides:

"... the state was considered the granary for the Southern Army west of the Mississippi and could furnish it lead from the Granby Mines to make small arms ammunition to an unlimited extent. As the Southern Army has been driven out of the state the latter part of the Winter, the people of Southwest Missouri had raised good crops of corn, oats, and apples to their orchards, all of which made a tempting prize for the Confederate leaders to get possession of..."[28]

During this time, General Schofield had convinced himself that he had been demoted from command of the Department of the Missouri to the District of Missouri. He had graduated from West Point in 1853 along with Phillip Sheridan and John Bell Hood but was dismayed by the slow process of promotion within the peacetime Regular Army and took a leave of absence to teach physics and astronomy at Washington University in St. Louis. When the war started, he went back onto active duty, serving under

Confederate Choctaw Soldiers in standard civilian clothing
Photo courtesy of the Oklahoma Historical Society

Brig. Gen. Nathanial Lyon at Wilsons Creek. He was a good administrator but was too much given to the political intrigues of the Army. He spent most of his time wrangling for promotion and better assignments while at the same time sabotaging the careers of others that he felt were competitors or in his way to higher commands. Regardless, the War Department in Washington had determined that they needed someone with more experience running such a large area as the expanded Department of the Missouri, so he was replaced by Maj. Gen. Samuel Curtis, the victor of Pea Ridge. Schofield was given field command of the new District of Missouri.[29]

After the debacle of the Indian Expedition, on August 24 Brig. Gen. Blunt issued new general orders reorganizing his command into three brigades. Due to Col. Furnas's departure to the Department of the Northwest, the Union Indian Brigade was broken up. The new First Brigade would be commanded by Brig. Gen. Frederick Salomon and would include the 9th Wisconsin Infantry, the 2nd Ohio Cavalry, the 2nd Indian Home Guard, Major C.W. Blair's 2nd Kansas Battery, and Captain Stockton's Battery. The Second Brigade, under the command of the restored Col. William Weer, included the 10th Kansas Infantry, the 6th Kansas Cavalry, 3rd Indian Home Guard, and Capt. Allen's 1st Kansas Battery. The Third Brigade would be commanded by Col W.F. Cloud and included the 11th Kansas Infantry, 2nd Kansas Cavalry, 1st Indian Home Guard, and Captain Rabb's 2nd Indiana Battery. He further ordered the brigade commanders away from Fort Scott to the south and east to obtain a good supply of forage and water while they awaited their new mounts and equipment.[30] The deployed brigades were also expected to scout various areas of Southwest Missouri in anticipation of the Confederate advance from Northwest Arkansas. This movement was confirmed by Brig. Gen. James Totten, who was in command of the Army of the Southwest at Springfield, Missouri. He stated that the Confederates were concentrating around Mount Vernon or Neosho.[31] The two Union brigades from Kansas, the First Brigade under General Frederick Salomon, and the Second Brigade under Colonel William Weer were to join with the Missouri State Militia (MSM) of General Egbert Brown at Sarcoxie under Schofield's orders. In early September Blunt's command took up the line of march for an active campaign in Southwest Missouri and Northwest Arkansas. Col. Weer was frustrated by the lack of actual coordination among the groups, feeling that Gen. Schofield was trying to run the operation by messenger and telegraph. He wrote to Gen. Blunt on September 4:

"…Whether I shall run, or advance, or do anything desperate, I also do not know, as I have no less than four brigadier-generals giving me orders at the same time… One thing I do know, if you or somebody else do not come out here and take command of all these scattered forces we will be cut up in detail."[32]

Things were just as confusing for most of the Confederate commanders. Since General Hindman had been recalled to Little Rock by General Holmes, General James Rains was in command of the Confederates in Northwest Arkansas and Southwestern Missouri. He spent most of his time in command at the Elkhorn Tavern on the Pea Ridge battlefield playing cards and drinking whiskey. Hindman had left with him orders "to make no aggressive movement, but if assailed to hold the line occupied as long as practicable." For some reason he decided to send all of his cavalry to Newtonia. At about the same time General Rains had ordered the 31st Texas Cavalry, commanded by Colonel Trezevant C. Hawpe, and the 1st Cherokee Battalion to Newtonia, Missouri. They were to occupy the town as a base for Confederate operations as well as open up the local mill for the production of flour for the Army.[33] Col. Hawpe had recruited the 31st Texas Cavalry regiment himself from the streets of Dallas, Texas. He attracted his recruits by putting an advertisement in the Dallas Herald inferring that Texas was going to be invaded by proclaiming "Who will defend Texas!" The scheme worked, and he soon had a one-thousand-man regiment formed with himself as the Colonel.[34] By the end of August the Texans and Cherokees had the mill in

Newtonia producing flour from corn, wheat, and barley impressed from local farmers. The other Confederate units were spread out among many of the small towns in Southwest Missouri including Neosho and Pineville. The Confederates were dispersed in such small parties that it was difficult for the Union command to determine where and when they planned to attack. Cooper with his Indian Brigade was enroute from the Cherokee Nation and held up at the abandoned Fort Wayne just across the border of Indian Territory from Maysville, Arkansas. During the period in between the resignation and the arrest of General Pike, Cooper was given command of the Confederate Indian troops in Indian Territory. Looking out for his post-war career, he had also begun to push for being named Southern Superintendent of Indian Affairs for the Confederate government. But like Colonel Weer on the Union side, his intemperate habits were well-known among his contemporaries and superiors, which would come back to haunt him.

In Sarcoxie, Missouri, Colonel Weer caught wind of intelligence that Col. Stand Watie and his 1st Cherokee Mounted Rifles regiment had begun a long trek northward towards Kansas and was planning on meeting with Confederate partisan Captain T.R. Livingston somewhere west of Baxter Springs. Their destination was supposed to be the Catholic Osage Mission where they hoped to draw the Osage away from the United States and into the arms of the Confederacy. The intelligence stated that Watie was moving up the Grand River Valley into Kansas where it becomes the Neosho River Valley.

Apparently, on September 26 Watie captured, then released two Osage scouts who immediately notified Gen. Blunt at Fort Scott

of Watie's plan. Blunt ordered Col. Cloud and the 2nd Kansas Cavalry to intercept Watie before he got to the mission. Unfortunately, there is no record of Watie's unit arriving at Osage Catholic Mission, nor did Col. Cloud report any contact between the two forces. It is possible that the "Osage scouts" were paid by pro-Confederate sources to create alarm and divert Union forces from Southwest Missouri.[35]

The Second Brigade of Blunt's Kansas troops had departed Fort Scott on September 5 and arrived at Carthage, Missouri on September 9. They went into camp on the south shore of the Spring River about a mile east and south of the center of town. Between patrols and military maintenance duties, the soldiers were able to enjoy cooling themselves in the calm and cool waters of the Spring River. Col. Phillips and his 3rd Indian Home Guards (IHG) had earlier been sent southward out of Baxter Springs to patrol the Indian Territory border and the southwest corner of Missouri to include Elk Mills. While enroute to rendezvous with the remainder of the Second Brigade at Carthage, the 3rd IHG entered Neosho on September 4 and drove off the small Confederate force that was occupying the town. The regiment remained in Neosho for a number of days, patrolling and seeking intelligence on Confederate operations in the area, but much of the population were Southern sympathizers and provided false or misleading information. After days of skirmishing with small parties of Confederates and feeling his regiment was exposed and too far from sufficient support for the rest of the brigade, Col. Phillips evacuated Neosho on September 13 and began a retrograde movement to join with Col. Weer. Blunt ordered the First Brigade,

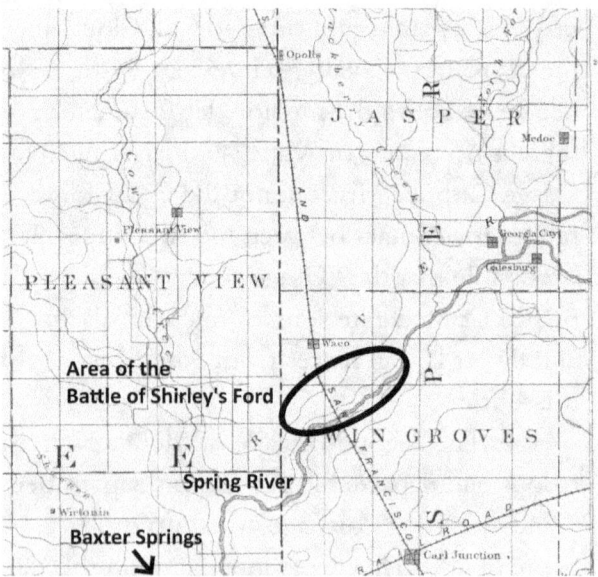

**Kansas / Missouri Border Area around the Site
of the Battle of Shirley's Ford**
Map courtesy of USGS / 1889

under Gen. Salomon, minus Blair's battery, to Lamar, Missouri, on the same date. They arrived on September 15. The Third Brigade, under Col. Cloud, was ordered to proceed into the Indian Territory as far as the Arkansas River unless his brigade was needed in Missouri to support the First and Second Brigades. Blunt's intention may have been to use the Third Brigade as a means of keeping Col. Cooper's Confederate Indian Brigade in the Cherokee Nation and away from the other two brigades that were facing Gen. Hindman's force in Southwest Missouri.[36]

General Totten in Springfield received continued intelligence that the Confederates under Hindman intended to attack the city and began to place calls for assistance to both Gen. Schofield and Gen. Blunt. Totten withdrew his Missouri State Militia units from Mount Vernon back to Springfield. Schofield promised five regiments, four of which he would send that day as far as the end of the railroad tracks at Rolla. From that point they would need to force march the remaining one hundred and ten miles.[37]

The 2nd Indian Home Guard, under the command of the very unpopular, insubordinate, and staunch abolitionist Colonel John Ritchie left their camp at the Baxter Springs Depot in early September and moved approximately 15+ miles up the Spring River to Shirley's Ford,[38] across the line in Missouri. As many as twelve to fifteen hundred non-combatants including soldiers' families, and other Indian refugees, made up of old men, women, and children, followed the column since they feared being left alone so close to the Cherokee Nation boundary. This was actually a good tactical location since it covered a significant crossing of the Spring River and was in position to prevent any Confederate forces from getting around the left flanks of the brigades of Brig. Gen. Salomon and Col. Weer who were encamped just southwest of Carthage, Missouri. But the camp was in shambles. Weer reported to Gen. Blunt: "His camp is, from what I can learn, a motley assemblage." Ritchie should have forced the refugees back but failed to do so. He also failed in using his mounted troops to scout the areas around his encampment for signs of the enemy. Then, in the early morning of September 20, a strong Confederate force fired upon Ritchie's pickets and attacked his position. The refugees fled to the camp for protection, making a "Bull Run retreat." He ordered a strong guard for his supply wagons and ordered the refugees and the wagon train back to the Cow Creek crossing, about five miles downstream. Ritchie then formed his approximately two hundred unmounted troops as infantry in the center, placed his mounted troops on the flanks, and advanced in the direction of the growing battle. He sent Major M.B.C. Wright with a strong detachment on a wide flank

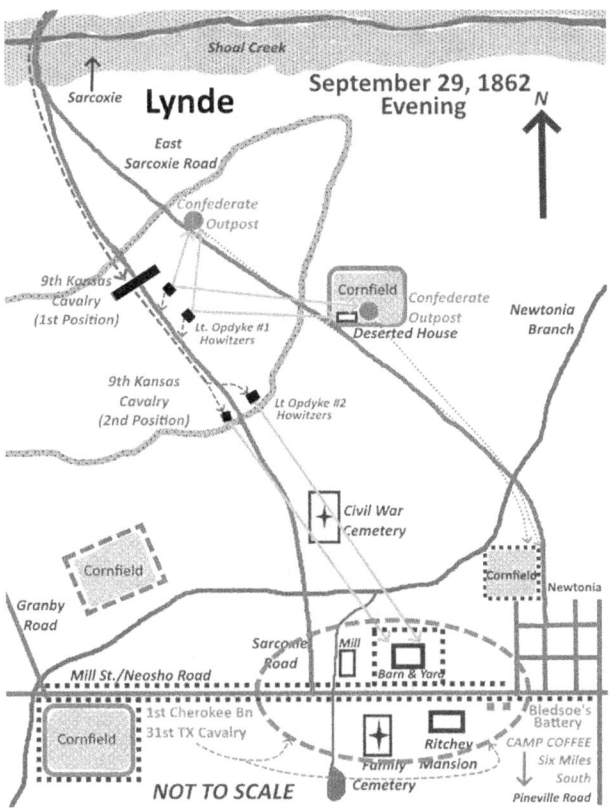

around the attackers. The Confederate units were deployed in a long battleline and were later identified as the 31st Texas Cavalry, and Major Livingston's and Col. Jackman's Partisan Rangers, who had been operating in Jasper and Newton Counties. They also reported Stand Watie's Cherokees, but they were probably misidentifying the new 1st Cherokee Battalion that had been in Newtonia. Ritchie placed his two hundred unmounted infantry into a concealed ravine facing the Confederates, took 100 mounted men and went forward to draw the enemy line forward onto a bit of high ground. As the Confederates did so, Ritchie ordered his entire force forward and they charged with the "war-whoop," after which the enemy hastily retreated, leaving approximately twenty-two dead.[39] The Union reported between twelve and twenty killed, and nine wounded. Col. Cloud and the Third Brigade

arrived at Cow Creek the next day to take command. A disturbed Wiley Britton wrote that Col. Ritchie was a poor commander and human being:

> "Colonel Ritchie's conduct in this section was not such as to commend it to the favorable consideration of rightminded men. He ordered or permitted the burning of houses and the destruction of private property along Spring River without the slightest pretext of a military necessity. On the day of the engagement at Shirley's Ford, he ordered or permitted to be shot five citizen prisoners, two at least of whom were unquestionably Union men. And he permitted the plundering and robbing of the families of Union soldiers of Colonel Weer's command. For certain of his actions that could hardly be considered the actions of a sane man, he was placed in arrest and recommended by Colonel Weer for dismissal from the service."[40]

The First and Second Brigades of the Army of Kansas (the new name) were ordered from Carthage to Sarcoxie, a distance of about twelve miles, by Gen. Salomon who was nominally in charge until Gen. Blunt arrived. Based on some information Salomon had received that the Confederates were enroute to attack Springfield, he marched the command about eight miles towards Sarcoxie but then moved them almost forty miles north and east to Turnback Creek. At this point Salomon determined that the information was incorrect, reversed direction, and finally arrived at Sarcoxie. The First Brigade encamped in the town of Sarcoxie and the Second Brigade camped on Jenkin's Creek, approximately five or six miles west of town. Col. Weer was concerned about how scattered the Union force was in Southwest Missouri and wrote to Gen. Blunt about his concerns on September 24:

+

> "…Whether I shall run, or advance, or do anything desperate, I also do not know, as I have no less than four brigadier-generals giving me orders at the same time. I am at this place by order of General Salomon. He is at Sarcoxie. One thing I do know, if you or somebody else do not come out here and take command of all these scattered forces we will be cut up in detail…" and from the best information I can get, if we had together the troops now lying idle between here, Fort Scott, and Springfield, we can easily whip the rebel force in the Southwest and end the war in this region…"[41]

On September 29 Salomon decided to send out a reconnaissance-in-force from the First Brigade toward Newtonia where he believed the Confederates were maintaining a small garrison. He sent Companies D, E, F, and H, 9th Kansas Infantry, under the command of Colonel Edward Lynde, along with two 12-pounder mountain howitzers toward Newtonia to determine where the Confederates were located and the size of the garrison. Col. Weer was sent towards Neosho on the same type of mission. A third column of unknown size or organization was sent towards Granby but returned without any

contact with the Confederates. Newtonia was a little college town of four or five hundred people and was located on a small stream called Newtonia Branch, which flowed north across the prairie in the direction of Shoal Creek. Wiley Britton, who was from Newton County, Missouri, described the road approaches to Newtonia:

> "…on the west and northwest of it [Newtonia], one half to three quarters of a mile off, there was a high-rolling ridge of prairie, perhaps seventy-five feet above the town, that extended to Shoal Creek timber. The Sarcoxie road came over the ridge from the northwest, and the Granby and Neosho roads came over the ridge from the west, and these roads entered town from the west side."[42]

As the Sarcoxie Road entered the town on the northwest side, it passed the city cemetery on the east side. This location would be somewhat significant during the upcoming battle. In addition, there was a second road that branched off of the Sarcoxie Road on the south side of Shoal Creek that passed by a large cornfield and entered the town from the north. Mill Street, or historic Neosho Road, runs from east to west through the length of and, at the time, was the southern boundary of the town. Along Mill Street the central property was a mansion owned by Colonel Mathew H. Ritchey, a prominent Union supporter, who had it constructed of handmade bricks. The mansion was completed in 1852 and was surrounded by a stone fence on the south side of Mill Street.

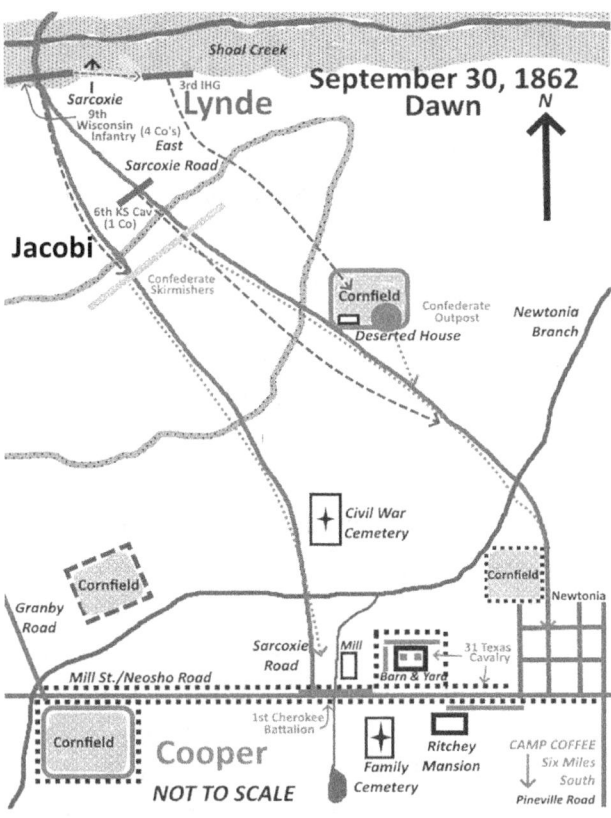

The steam-powered grist mill was located on the spring-fed Newtonia Branch just west of the Ritchey Mansion, across the road and south of a plum orchard. Since the mill's steam boilers were fed by a spring on the property, the mill exchanged hands many times during the war since it could be operated year-round. The Ritchey barn and surrounding barnyard was located directly across Mill Street and just west of the mansion. The barnyard was also surrounded by a stone fence that could be used for cover by the Confederate infantry and artillery as a strong defensive position. The Ritchie Farm complex ended up playing an integral part in the outcome of the battle by providing an excellent defensive position for the Confederate forces.[43]

The Confederates had gathered over four thousand soldiers around Newtonia. General Hindman needed to hold Newtonia

for as long as possible so he could gather his scattered troops together for an attack on Springfield. With Hindman's absence and Rains's disinterest, Colonel Douglas Cooper, as ranking officer, took command of the outpost's defense. Cooper had brought up the 1st Choctaw and Chickasaw and the 1st Choctaw regiments up from Old Fort Wayne in the Cherokee Nation via Scott's Mill in far southwest Missouri. On September 26 he joined with Col. Shelby's Missouri Cavalry Brigade, consisting of Lt. Col. B. Frank Gordon's 5th Missouri Cavalry, Col. John T. Coffee's 6th Missouri, and Col. Beale Jean's Jackson County Cavalry regiments, at Big Spring at the head of Indian Creek about six miles south of Newtonia adjacent to the Pineville Road.[44] Lt. Col. Beal Jeans had replaced Col. Upton Hays as commander of the Jackson County Cavalry after Hays was killed by a Union picket in Newtonia on September 13 or 15 (the records do not agree on the date). The Confederates had set up a training area at that location named Camp Coffee. Also at Camp Coffee was Col. Almerine Alexander's 34th Texas Cavalry and Col. J.G. Stevens's 22nd Texas Cavalry. Most of the commands arrived at Newtonia on September 29 and joined with the 31st Texas Cavalry and the 1st Cherokee Battalion who were already present, operating the local mill and conducting local patrols. In addition, Captain Joseph Bledsoe's Missouri Light Artillery with its two guns were sent and deployed in town. Early on that day Cooper received information that "a body of Pin Indians and Federals" were headed for Granby, Missouri. Granby was a location containing large lead mines that were desperately needed for the Confederate Army.[45] He ordered Col. Stevens and his 22nd

Texas Cavalry regiment to Granby for a reconnaissance, but they failed in making any contact with Union forces. Cooper also received reports of Union scouts probing his defensive positions north of Newtonia.[46]

The Union scouting party reported by Cooper was the 9th Kansas Cavalry, led by Col. Lynde, that Gen. Salomon had sent out the morning of September 29. In the afternoon Col. Lynde and his command first struck the Confederate pickets approximately eight miles south of their camp at Sarcoxie. They pushed the pickets across Shoal Creek and onto the open prairie beyond, using the main Sarcoxie Road as a guide, until they were about one and a half miles from the town of Newtonia. At this point they encountered a strong Confederate outpost centered on a deserted house and cornfield to Lynde's left flank along with another enemy strong point to his left rear. He ordered his two mountain howitzers, commanded by 1st Lt. Henry Opdyke, to fire upon the house and cornfield, which he did with exploding shells that put the Confederates into flight back to town. Major Bancroft was sent to the right with two companies, and Major Pomeroy with one company was swung to the right to protect the small battery. Col. Lynde continued moving his small force forward slowly until they were within three-quarters of a mile from the Confederate line in town. The Union gunners fired numerous solid howitzer rounds at the stone barn and compound without any obvious effect for ninety minutes. Scouts captured two Confederates who stated that there were at least twenty-six hundred troops and two cannons in Newtonia. Col. Lynde determined he was facing unfavorable numerical odds and soon began to retire his force back up the Sarcoxie Road across the open prairie and over Shoal Creek. The

command rested on the prairie beyond the creek, then proceeded back to the main camp south of Sarcoxie.[47]

Gen. Salomon heard the cannon fire from the direction of Newtonia and ordered Lt. Col. Arthur Jacobi of the 9th Wisconsin Infantry to take Companies D and G, 9th Wisconsin Infantry, one company from the 6th Kansas Cavalry, one company from the 3rd Indian Home Guard, and three cannons (one 6-pounder brass, two 3-in rifled guns) of the 25st Ohio Light Artillery commanded by 1st Lt Julius Hadley, forward to support Col. Lynde's forward units. This reinforcement group departed at 3 p.m. and encountered Col. Lynde's command approximately three miles south of Sarcoxie enroute to report to Gen. Salomon. The two commanders exchanged information, and Lt. Col. Jacobi advanced his force, with flankers deployed to the right and left, into the timber north of Shoal Creek. They advanced until they reached Mathew Ritchey's farm (present-day Ritchey, MO, not to be confused with the Ritchey property in the town of Newtonia), a location three and a half miles north of Newtonia and went into camp. At 11 p.m. he was reinforced by Companies E and H, 9th Wisconsin Infantry, who passed on instructions from Salomon to observe and collect information but to avoid a general engagement.[48] This was due to the instructions Gen. Blunt sent to him stating:

"Unless you are confident of your ability to make a successful fight you will not risk a battle, but fall back slowly, endeavoring to draw the enemy on until you form a junction with the re-enforcements coming up."[49]

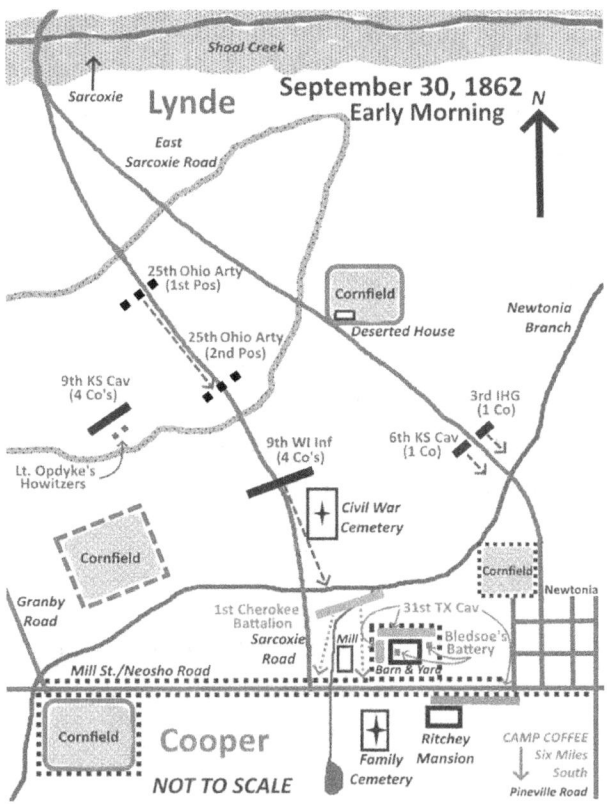

Texas Cavalry that had been sent back to Granby the previous evening to protect the lead mines from Union occupation. At about 7 a.m. Confederate pickets northwest of Newtonia reported Union infantry advancing in their direction.[52]

Col. Lynde had his 9th Kansas Cavalry up and enroute to Newtonia by 3 a.m. He reached the Ritchey prairie at 6 a.m. where the 9th Wisconsin Infantry had bivouacked the previous night but found they had already departed for Newtonia. Lt. Col. Jacobi had departed the campsite at daybreak and had deployed his four infantry companies in the tree line on the south side of the Shoal Creek timber. Again, he used his company from the 3rd Indian Home Guard as flankers and scouts. He conducted a leader's recon-naissance out onto the prairie and observed a strong Confederate outpost on the left in the cornfield that had been dispersed by Lt. Opdyke's 12-pounder howitzers the previous evening. He also noted that the Confederates had brought up troops and had deployed them within the town, making use of the numerous stone buildings and walls.

It was at this point in the battle that the inexperience of the Union commanders became self-evident. They should have kept to their simple mission of observation and intelligence gathering. A seasoned commander would have understood that attacking a force of unknown size, across an open prairie, who are deployed within a town lined with stone walls and occupied stone structures, is not an offensive tactic that will be successful. The Wisconsin infantry's open ranks on this open prairie left them vulnerable to both artillery fire and cavalry attack. It is doubtful that any of the Union or Confederate infantry regiments had mastered the complex infantry

Regardless of what information the two Confederate prisoners told Col. Lynde, Maj. J.M. Bryan of the 1st Cherokee Battalion reported the Confederate defense numbered no more than five hundred soldiers that morning.[50] Although most of Cooper's troops had stayed in town overnight, most then returned to Camp Coffee early on the morning of September 30, except the 31st Texas Cavalry, the 1st Cherokee Battalion, the 5th Missouri Cavalry, and the Jackson County Cavalry. For defensive purposes Cooper placed the 31st Texas Cavalry on the right around the mansion and stone barn and deployed the 1st Cherokee Battalion on the left along the Mill Street stone wall.[51] Supporting these units was Captain Bledsoe's battery of two cannons which were deployed within the Ritchey property. Cooper himself was with Colonel Alexander's 34th Texas Cavalry enroute to Granby to relieve the 22nd square to defend against cavalry attacks.[53] They now had enough information to inform General Salomon of the situation and simply wait for reinforcements. Instead, Colonels Lynde and Jacobi decided to attack an enemy of unknown size behind substantial physical defenses after crossing a two-mile wide prairie in full view of the Confederates.

Colonel Jacobi deployed Capt. Mefford's company of the 6th Kansas Cavalry through the tree line to the left to engage the cornfield outpost. This action alerted Cooper's Confederates that the Union Army was advancing. Jacobi sent the 3rd Indian Home Guard company from the flanks to assist Capt. Mefford in cutting off the soldiers of the cornfield outpost as they were fleeing towards the town. He then ordered Capt. Gumal Hesse and the four companies of infantry (approximately two hundred ranks) forward to the high ground about a mile northwest of the town. Lt. Hadley and his three-gun battery was deployed at a high point along the Sarcoxie Road that overlooked the town from about one and a half miles. At this point Col. Lynde arrived with his four companies of the 9th Kansas Cavalry, and he assumed command from Lt. Col. Jacobi. After about nine shots, Lynde conferred with Hadley and determined that the artillery needed to be advanced closer to the Confederate positions. The guns were moved forward another five hundred yards along the road. The four companies of the 9th Wisconsin Infantry advanced to the wooded ravine of the Newtonia Branch that went through town. Lynde sent his four 9th Kansas Cavalry companies with the two howitzers under Majors Bancroft and Pomeroy to the far right to a small rise near the Granby Road, almost directly west of the town. Capt.

Mefford's company and the Indian Home Guards covered the left flank of the Union deploy-ment. The Confederate guns seemed to get the range of the new location of the Ohio artillery, so Lt. Hadley requested to move the guns off to the right about a thousand yards. This was accomplished and, it seemed after this point across the north and northwest side, the Confederate gunners were much less effective. Col. Lynde re-deployed Capt. Flesher and Company E, 9th Kansas Cavalry, over to the left to support Capt. Mefford's company of the 6th Kansas Cavalry on the left.[54]

When Col. Hawpe saw the deployed Union force coming towards Newtonia, he immediately sent a messenger to Camp Coffee to inform Col. Cooper about the attack. It must have been a difficult decision because only thirty minutes before he had sent a messenger to report that he had not sent out any scouting patrols overnight to watch the approaches to the Confederate defenses. Again, he deployed the 1st Cherokee Battalion to the west observed any enemy movements that morning. He also failed to mention that he to cover his left flank along the stone wall and the Granby, Neosho, and Sarcoxie Road approaches. He placed his own 31st Texas Cavalry dismounted around the Ritchey barn and barnyard as well as across the front and to the right of the mansion along the stone wall. He placed Capt. Bledsoe's two-gun battery in the barnyard. At least one gun was physically placed inside the barn. Cooper was actually not that far away as he was about halfway between Camp Coffee and Newtonia with Col. Alexander and his 34th Texas Cavalry enroute to relieve Col. Steven's 22nd Texas Cavalry at Granby. Cooper immediately increased speed to assist Col. Hawpe. He sent

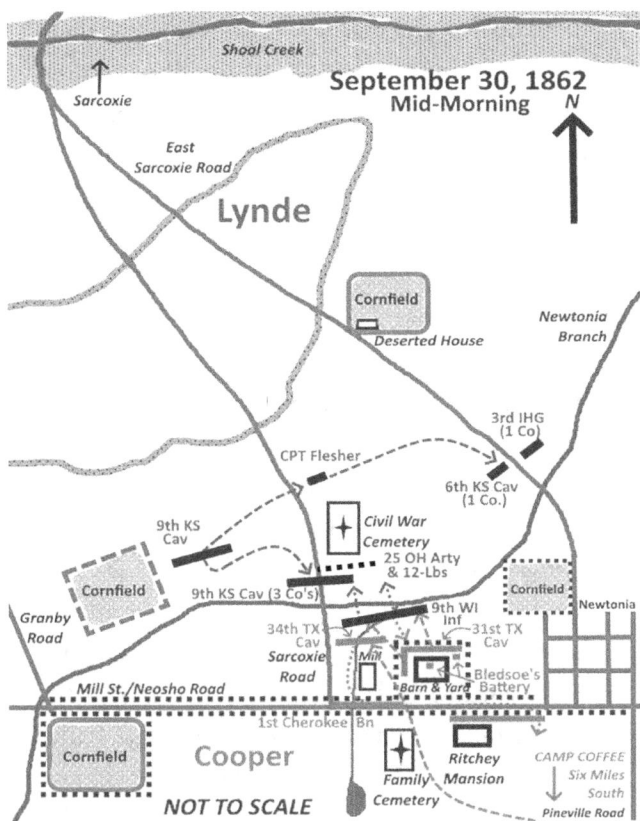

Col. Alexander's troops down to the Newtonia Branch below the gristmill, and they dismounted and deployed as infantry. Lt. Hadley observed them moving into position, so they were met with a hail of canister and grapeshot from the Ohio artillery. Realizing the difficult position he was in, Col. Alexander remounted his regiment, and they fell back towards the Ritchey house. Cooper redeployed them to support Bledsoe's battery on the right and along the stone wall to the right along Mill Street. The Wisconsin infantry had crossed the ravine and began to occupy some of the town's buildings on the north side. The Wisconsin troops possessed long-range Springfield rifles, and they began to pick off many of Capt. Bledsoe's artillerymen. The Ohio and Missouri batteries exchanged shots until Capt. Bledsoe's cannons had expended all of their ammunition. At this point Cooper directed them to a spot of high ground

approximately one hundred-fifty yards southeast of the Ritchey house. Without direct orders from Col. Hawpe, his command suddenly leaped over the stone wall and charged the Wisconsin infantry coming up from the ravine. Col. Hawpe reported

> "…After several shots were exchanged between the batteries the Federal infantry came up the ravine to within a few hundred yards of the wall, when a young captain belonging to Colonel Coffee's command, wholly unknown to my regiment and representing himself as aid to Colonel Cooper, came up, cursed my men, called them cowards, and ordered them to come out from behind the wall and charge. That portion of my men who were next to t his would-be aide to Colonel Cooper, hearing the order and believing him to be what he represented himself, instantly obeyed the order; and I, seeing a portion of my men charging the enemy, and believing they were acting under orders from Colonel Cooper, ordered those who still remained where I had first placed them to charge also. After a severe conflict with the infantry under heavy firing from the Federal batteries, which were only a few hundred yards distant, they fell back to the place first assigned them, and were soon followed up by the infantry to within gun-shot, when they were fired upon by my regiment…"[55]

It is believed that the officer who ordered Hawpe's command to charge was Capt. John

T. Crisp, of Col. Coffee's regiment, who had been serving as an aide to Col. Cooper. Nevertheless, it is unknown if the order actually came from Cooper or was something he did on his own. Regardless, the charge failed, and Col. Hawpe's command retreated towards the stone barn. There was a young woman named Mary Grabill living on a small farm across which the 9th Wisconsin troops advanced toward the stone wall. She recalled after the war:

> "...After a sharp encounter, a company of German troops pushed in through the field back of our garden, forced their way through the yard, and making a shield of the hedge in front, fought from there..."[56]

Captain Hesse's infantry companies continued moving forward towards the Ritchey barn after Col. Hawpe's troops fell back behind the stone wall of the barnyard. When the Union infantry got within a hundred yards or so, the Confederates behind the wall opened fire, halting Capt. Hesse's advance. Col. Lynde reported:

> "...I soon after saw the infantry close to the stone wall already described, from which soon leaped a perfect stream of fire right into the ranks of the Infantry, they returning the fire nobly and slowly retired..."[57]

Colonel Lynde at this point realized he would not be able to take the town with the five hundred or so troops he brought into the fight, especially with the losses he had already experienced. He decided to bring his troops back and begin a slow withdrawal over the open prairie to the cover of the Shoal Creek timber, approximately three miles behind them.

The messenger sent back to Camp Coffee by Col. Hawpe reported to Col. Shelby that the town was under attack. Col. Shelby quickly ordered the 1st Choctaw and Chickasaw (Col. Cooper's regiment) towards Newtonia. The regiment was currently under the command of Lt. Col. Tandy Walker, who was a mixed-blood Choctaw, born in Mississippi in 1814. He was a former governor of the Choctaw Nation and a strong supporter of southern rights. He was made a lieutenant colonel in 1861 and was second in command to Col. Douglas H. Cooper of the 1st Choctaw and Chickasaw Mounted Rifles. He served as commander when Cooper was serving as a brigade commander. Col. Shelby also sent his own regiment, the 5th Missouri Cavalry, now commanded by Lt. Col. Frank Gordon, towards Newtonia as well. Moving northward at a gallop on the Pineville Road, they reached Newtonia at approximately 8:30 a.m. Walker and his Choctaws and Chickasaws did not even slow down as they entered the town from the south but charged directly into the 9th Wisconsin Infantry, who were just starting to withdraw from their attack on the stone barn and barnyard. Col. Cooper joined his regiment in its charge as they passed his headquarters at the Ritchey mansion. Lt. Col. Gordon and his regiment came up soon, but owing to a delay to determine the identity of a body of troops approaching from the west that turned out to be the 22nd Texas Cavalry from Granby, they swept to the right side of Walker's regiment. One of Gordon's soldiers saw the Confederate Indians charge and later wrote:

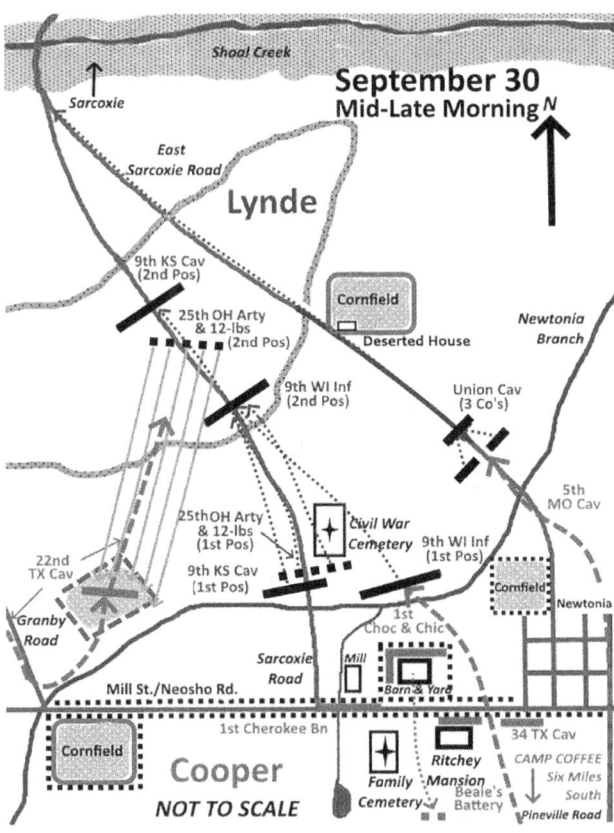

"...Oh, horrors! That frightful war whoop. The most blood curdling, ear-splitting yell went up that I had ever heard; similar to our modern church solos. Then like an avalanche those furious warriors went at them with demon-like savagery, keeping up that unearthly howl comparable only, in my imagination, to the unhappy shriek of lost souls coming up from the dismal depths of endless torture. Well, that was too much for our friends in blue. The rebel yell would have been like sacred music compared to it..."[58]

Gordon's force drove directly at the Union left flank being held by Captains Mefford,

Flesher, and the 3rd Indian Home Guard detachment. The two-prong attack by two full regiments of Confederate cavalry was too much for the small Union forces in their way. Col. Cooper reported that the attack by Walker and Gordon sent the Union troops flying to the rear. Col. Lynde reported that Captains Mefford's and Flesher's cavalry companies, and Capt. Hesse's infantry companies, although vastly outnumbered, fell back in an orderly fashion by forming and reforming as needed. The Wisconsin infantry would deploy in line with the 9th Kansas Cavalry behind them. The infantry would fire a volley and fall back behind the cavalrymen and reload while the cavalrymen fired their weapons, and over again. Eventually they were completely overwhelmed after being outflanked and nearly surrounded by the Confederates. Things were not any better on the Union right. Col. Stevens and his 22nd Texas Cavalry had heard the sound of battle from their station in Granby. They mounted up and quickly dashed towards Newtonia, arriving just as Col. Lynde's Union force began its retreat. Stevens was met on the Granby Road by Capt. Crisp, the same officer who presumably had ordered Col. Hawpe's regiment to charge the Wisconsin troops, who again relayed an order from Col. Cooper to attack the Union right flank. Stevens formed his command in a column of platoons and advanced across a cornfield toward the Union right flank. Lieutenants Hadley and Opdyke had earlier moved their artillery pieces toward Newtonia on the Sarcoxie Road and deployed them adjacent to the city cemetery. After Col. Lynde had given the order to retreat, they limbered up their guns and began moving back up the road towards the Shoal Creek crossing. As they reached the higher ground on the open prairie near where the battle started, Lt. Hadley observed the advancing 22nd Texas Cavalry approximately 350 yards to the southwest, breaking through a fence and moving through the cornfield in their direction. The artillerymen unlimbered their guns again, swung them to the right, and brought them into action against the Texas troops, firing eleven rounds of canister and two solid shots at 250 yards. This broke up the Texan's advance and the Union artillery resumed its retreat towards Sarcoxie.[59] The 9th Wisconsin Infantry and the 9th Kansas Cavalry tried to slow the three-prong attack by the 22nd Texas on the right, the 1st Choctaw and Chickasaw in the center, and the 5th Missouri on the right. The 9th Kansas had been issued long-range Sharp's carbines, and they were able to keep the attackers at a distance. Capt. Mefford's 6th Kansas company did its best to keep the Confederates at bay, but they were only armed with sabers and revolvers. As the Union troops reached the edge of the timber, the road narrowed, which caused a bottleneck as everyone attempted to get through the small opening in the timber. The Union troops were suddenly surrounded when the Kansas cavalry troops drew their sabers and cut their way through the

Confederates who were cutting them off from the road. The artillery and most of the cavalrymen escaped by way of the hole created by the saber charge and carbine volley. Unfortunately, the infantry, who had numbered about two hundred at the beginning of the battle, were almost annihilated. Approximately 166 Wisconsin infantrymen were killed or captured by the Confederate Indian regiment. A correspondent with the *Leavenworth Daily Times* was present and described the final actions of the Union defenders:

> "…Limbering up their guns, the battery moved off. The cavalry and infantry, forming on the left and right, fell back, fighting as they went, the entire distance to the timber, pursued and almost surrounded by apparently countless numbers, who kept up an incessant firing, often at a distance of only a few yards. Fortunately little damage was done until the edge of the timber had been gained; but here some confusion took place owing to the narrowness of the road; and the enemy making a determined rush, the whole party was surrounded. The cavalry and artillery dashed through and escaped, but the infantry were almost entirely cut to pieces or taken…"[60]

As the Union troops began to retreat through the timber, the Confederate cavalry kept up its pressure on them, eventually turning the retreat into almost a rout with soldiers throwing away their weapons and failing to follow orders from their officers as they fled towards Sarcoxie. It was reported that nine Union soldiers who had been killed by the Choctaw and Chickasaw regiment had been stripped of their clothing but were not mutilated as at Pea Ridge.[61]

General Salomon in Sarcoxie had heard the sound of battle coming from Newtonia and realized that Col. Lynde and Lt. Col. Jacobi had encountered the butternut-clad Confederates. He began to make plans to send a reinforcement column and ordered the First and Second Brigades to readiness and, in addition, sent word to Col. Brown of the Fourth Brigade, Missouri State Militia, to advance on Newtonia from his encampment

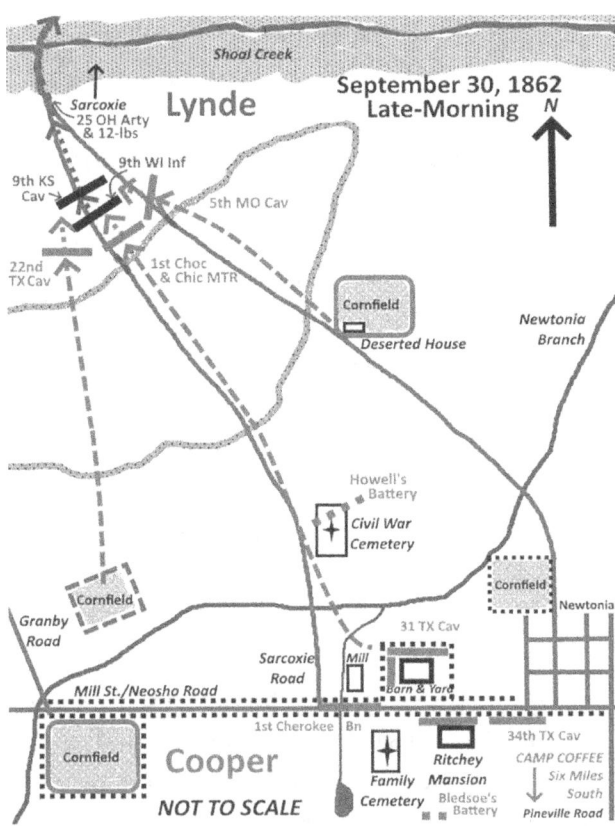

at Mount Vernon. Salomon ordered the remainders of the 6th Kansas Cavalry, with two mountain howitzers, the 3rd Indian Home Guard, and the second section of the 25th Ohio Artillery, under the overall command of Col. William Judson, to immediately proceed towards Newtonia as he organized the relief column. Salomon and Col. Weer took the remaining companies of the 9th Kansas Cavalry that were in camp, two companies of the 3rd Wisconsin Cavalry, two companies of the 2nd Ohio Cavalry under Captain Smith, the 10th Kansas Infantry, two companies of the 9th Wisconsin Infantry, Captain Allen's 1st Kansas Battery, and Rabb's 2nd Indiana Battery, and started to the relief of Colonel Lynde's force. Major Blair's 2nd Kansas Battery that had joined the command at Sarcoxie, part of the 9th Wisconsin Infantry, and three or four companies from other regiments, including

about four hundred Union Indians, were left to guard the camp and trains.

After approximately ten miles south on the Sarcoxie Road, Colonel Judson and the first relief column encountered the remnants of Col. Lynde's and Lt. Col. Jacobi's defeated commands along with three guns of Lt. Hadley's battery and Lt. Opdyke's two mountain howitzers belonging to the 9th Kansas Cavalry. They advised that the Confederates had strong forces within the timber south of Shoal Creek. Judson entered a clearing where he reported finding ten killed and wounded in the Shoal Creek timber, all of whom had been stripped of their clothing. To prevent the wild hogs from damaging the bodies, the Kansans built three or four rail pens and placed the bodies in them for protection. Col. Judson further reported:

> "…There we caught the first glimpse of the enemy and followed him to the prairie, where he formed his line of battle 3 miles out from Newtonia on the Sarcoxie road. I at once ordered my men into line and directed Lieutenant Benedict to bring his mountain howitzers into position on the gallop; then threw a few shells, and the enemy fell back. My men followed them with a shout to the town, where the lieutenant again commenced shelling them, when the enemy opened his battery upon us within short range with three guns, using shell and round shot pretty freely…"[62]

The three Confederate cavalry regiments that had pushed Lynde's and Jacobi's force into the Shoal Creek timber fell back onto the prairie when the 6th Kansas Cavalry approached them from the creek's crossing. Col. Stevens' 22nd Texas Cavalry was covering the Confederate withdrawal and formed into a line of battle on the prairie approximately three miles from Newtonia. When Judson's troops cleared the woods onto the prairie, they too formed into a battle line. Judson ordered Lt. Brainard Benedict, who commanded the battery of two 12-pounder mountain howitzers, to form in the center of the line and open fire. After just a few rounds were fired, the Texans turned and retreated towards town. At about 10 a.m. Judson advanced his command and small battery to the high ground that had been held by Lynde and Jacobi earlier in the day. Col. Cooper had earlier brought up Capt. Howell's 4-gun battery and placed them in the cemetery northwest of Newtonia. Benedict turned his small battery towards the cemetery and fired a few rounds at the Confederate artillery posted there. As the three Confederate cavalry regiments passed into town, Howell's battery opened fire on Judson's advancing force and Capt. Benedict's small battery. The Confederate gunners were fairly accurate, and Judson realized he was in a bad position, so he and his force backed away out of effective range of the Confederate guns. He maintained a small outpost on the high ground to observe the enemy movements. Once they had re-deployed, Judson sent a message back to Gen. Salomon giving a status report and asking for reinforcements. The 6th Kansas Cavalry then spent the next four hours skirmishing with Col. Cooper's Confederates until they were finally joined by the slow-moving Col. Phillips and the 3rd Indian Home Guard at 2 p.m. At the same time a reinforced 25th Ohio Artillery under Captain Job Stockton arrived on the

field along with Lt. Hadley's 3-gun section. They took position in the center of the battle line with Col. Judson's 6th Kansas Cavalry on the right and Col. Phillip's 3rd Indian Home Guard on the left. At approximately 3:30 p.m., Brig. Gen. Salomon arrived with the First and Second Battalions of the 10th Kansas Infantry and the remaining six companies of the 9th Wisconsin Infantry. These infantry units were accompanied by Capt. Allen's Kansas Artillery. Salomon immediately took command and began redeploying his forces. He placed the First Battalion of the 10th Kansas Infantry on the center-right and the remnants of the 9th Wisconsin Infantry he placed on the center-left. The artillery was posted in the center and the Second Battalion of the 10th Kansas Infantry and Lt. Hadley's section of the 25th Ohio Artillery were placed behind the center in reserve. He moved the 6th Kansas Cavalry and its two 12-pounder howitzers to the far right to a piece of high ground near the Granby Road. Salomon then sent the 3rd Indian Home Guard to the far left and ordered them to approach the town by way of the wooded ravine that contained the Newtonia Branch. As soon as the Union had its forces deployed, they began a severe artillery fire on the heavy stone buildings and walls of the town. One Kansas soldier recalled:

> "…The artillery – Allen's, Stockton's, and a section (I believe) of Blair's batteries, with six howitzers – took positions and open fire on the town, which was immediately answered by the rebel battery of six guns only. It was a beautiful sight, with just enough excitement to give it a "delicious

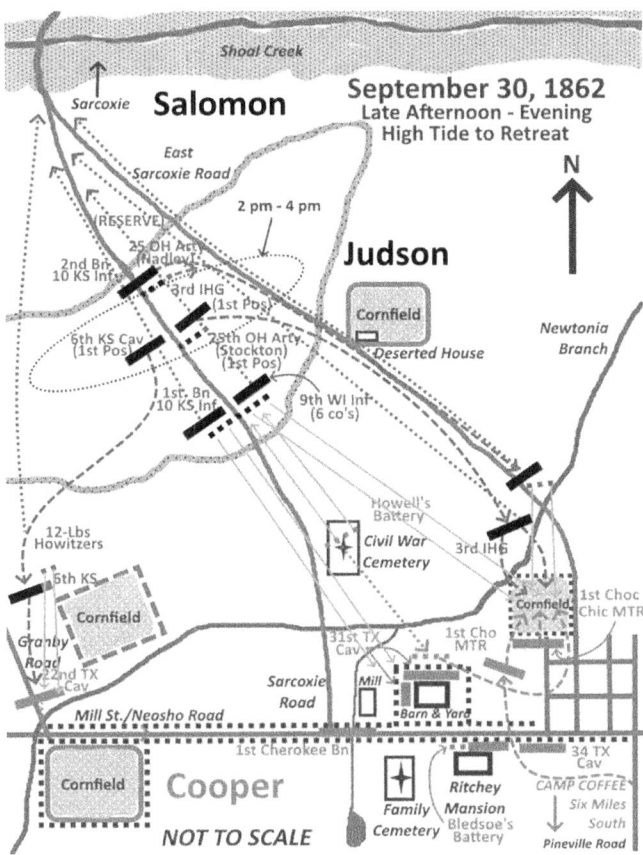

flavor." It is a beautiful sight, worth risking to see, to witness a fight between artillery, when the whole thing is spread before you in all its terrible realities. The thundering of our own guns, the spiteful reply of the enemy, the peculiarly whizzing music of the shells and shot, as they fly through the air, and the crash of the destructive missiles as they plow up the ground, or, perchance, crash through some animal, gives an excitement better felt than expressed. It is true that when you hear the shells whizzing through the air you cannot tell where they will fall – whether on your own head or some other luckless spectator – but still the sight is worth seeing…"[63]

Colonel Cooper had realized that Howell's battery was far too exposed in the city cemetery and redeployed them to the Ritchey barn. Capt. Bledsoe had sent his caissons back to Camp Coffee where they were refilled and brought back to Newtonia. He deployed them to the west of the mill behind the stone wall along Mill Street. The 31st and 34th Texas Cavalry, the 5th Missouri Cavalry, and the 1st Cherokee Battalion were deployed around the Ritchey Mansion and supporting Howell's artillery around the barn and barnyard. The Jackson County regiment and the 22nd Texas Cavalry were posted far to the west along Mill Street near the Granby Road. Lt. Col. Walker and the 1st Choctaw and Chickasaw Mounted Rifles regiment had been placed on the north side of town adjacent to a cornfield to protect the right flank.[64]

Captain Stockton's battery on the ridge was under continuous fire from Confederate artillery positioned in the town. The Union gunners were concentrating on the Confederate guns around the stone barn. Stockton observed a Confederate gun pointing out of a barn window, firing, and being wheeled back to reload. He ordered Lt. Edward Hubbard's section of 3-inch rifled guns to concentrate on taking out that Confederate gun. Hubbard's gunners placed two exploding shells through the window, hitting and silencing the cannon, killing some and dispersing the other cannoneers.[65]

The artillery fire from both sides was having a devastating effect on the towns-people as well. Mary Grabell looked out of her house and saw some of the same German Wisconsin soldiers she had seen attacking earlier but were now being escorted under guard as prisoners of war. She also saw twenty-year-old Martha Ritchey wandering around the battle zone looking for her father, Mathew Ritchey. She and other women and children were searching in vain for a safe place to hide. The groups would drift into the Grabill house in search of refuge. They tried to hide in the stone fireplace, the cellar, the floor in the middle of the room, finally giving up trying to find a safe spot and simply hoping for the best. Mary remarked she ended up "standing in the doorway in the familiarity which bred a contempt of danger, in a desire to see what I could see."[66]

Colonel Cooper, realizing he far outnumbered the Union force on the high ground, sent Lt. Col. Jean's Jackson County Cavalry and Lt. Col. Gordon's 5th Missouri, supported by Capt. Bledsoe's battery, around to the left to attack the right flank of the Union cavalry near the Granby Road. Just as the Confederates were beginning their attack, Stevens's was diverted to identify a large cavalry unit moving into their left rear. Stevens identified the unit as the 1st Choctaw Mounted Rifles commanded by Col. Sampson Folsom, a mixed-blood Choctaw from near Fort Towson. He was a slave-owner, had served as a delegate for the Choctaws in Washington, D.C. in the 1850s, and was a signer of the 1855 Chickasaw Constitution. He was also a violent alcoholic who had committed numerous murders while he was intoxicated. Regardless, he had enough clout to wrangle a Confederate commission and soon raised a regiment of Choctaw Indians. The Choctaws had force-marched the thirty-five miles from Scott's Mill starting that morning. Cooper sent them to join with his own 1st Choctaw and Chickasaw regiment at the cornfield north of town. Stevens and his command then joined with Jeans to attack the Union right flank. They formed into a line of

battle with Jeans's Missourians in the front line and Steven's Texans in the next rank and advanced across the prairie near the Neosho Road and Granby Road junction. Seeing this attempted flanking movement, Lt. Col. Lewis Jewell, commanding Judson's 6th Kansas Cavalry, moved off the ridge southeastward along the Granby Road. One of the sections of the 25th Ohio Artillery joined Jewell's advance, took a position two hundred yards in front of the cavalrymen, and opened an enfilading fire on Jeans's and Stevens's Confederates. The Union guns caught the Confederates off guard, and they broke and scattered back towards their defenses along Mill Street. After seeing the Missourians and Texans break for the rear, Jewell's men drew sabers and charged after the fleeing Confederates. Wiley Britton described the scene:

> "…After a round from his Sharp's carbines. Colonel Jewell ordered his men to draw sabres and charge, and away they flew over the prairie, raising a cloud of dust in their rear. When the Confederates saw the Federal sabres flashing in the sunlight, and the dust arising from the horses' feet of the advancing foe, they instantly wheeled about and galloped back to town. Colonel Jewell pursued them to within about two hundred yards of the stone fence,.."[67]

Stopping short of the Confederate defenses, the Kansans reverted to their long-range Sharps carbines and exchanged small arms fire with the defenders behind the walls. At this point Bledsoe's battery opened fire on the 6th

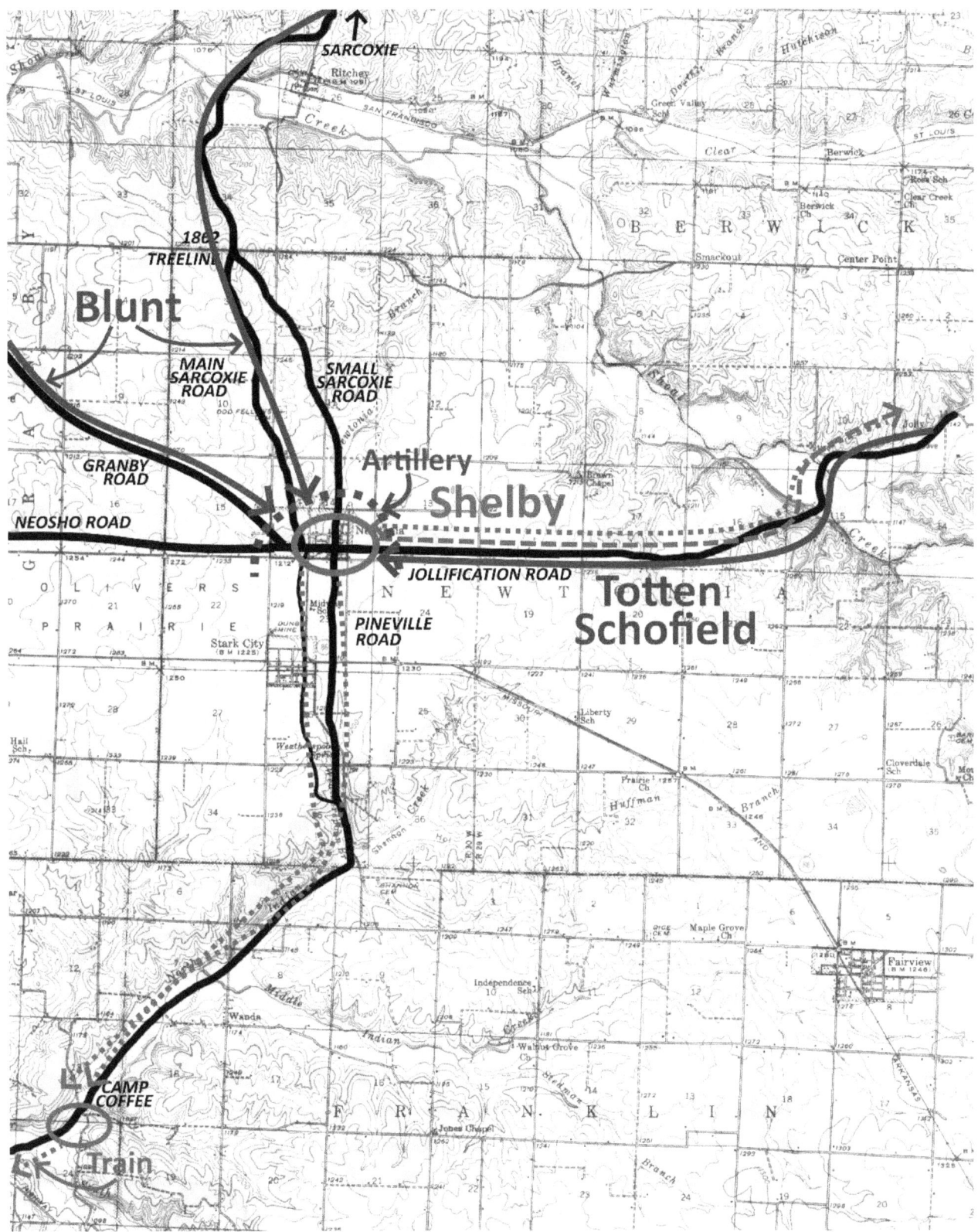

Area of Operations indicating Colonel Hall's expected and actual Routes

Map courtesy of USGS / Additions by Author

Kansas Cavalry, and they pulled back to their original position.[68]

Over on the Confederate right flank, the 3rd Indian Home Guard regiment was able to advance up the Newtonia Branch towards the town mostly unobserved by using cover and camouflage provided by the trees, plum thickets, and split-rail fences. They managed to get to the northern edge of a cornfield where the 1st Choctaw and Chickasaw regiment was posted and opened fire on the Confederate Indians. The Union Indians were beginning to push the Confederate Indians across the cornfield when Lt. Col. Folsom and his 1st Choctaw regiment arrived to support the other Indian regiment. The cornfield became a melee of Indian versus Indian, with both sides at times reverting to traditional battle styles, including war whoops and hand-to-hand fighting. General Salomon directed his reserve section of the 25th Ohio Artillery under Lt. Hadley to move forward onto Col. Phillips's left flank to support the Union Indians. Stockton reported:

> "…I immediately ordered Lieutenant Hadley to proceed with the left half of the battery to a point commanding the left of Colonel Phillips' line, and directed the fire of the right half of the battery to my left, checking the advance of the large rebel force moving against the Indian regiment. As soon as the rebels commenced to waver the Indians commenced to advance and drove them until their ammunition gave out, when they retired under cover of my guns. I now discovered large masses of troops posted in the corn field directly in front of the position lately occupied

by Colonel Phillips' Indians and directed my whole fire upon their condensed masses. The slaughter was terrible, and the officers could be seen by the aid of the glass endeavoring to keep their men in that position, but two percussion shells from my rifled pieces bursting in the midst of what appeared to be a regiment by division closed in mass they scattered, and rushing upon a fence, crushed it flat to the ground. Other masses, posted near the stream and about the houses, suffered greatly from both our solid shot and spherical-case…"[69]

Salomon also committed the Second Battalion of the 10th Kansas Infantry to support the 3rd Indian Home Guard. In turn, Cooper sent a two-gun section of Howell's battery to the cornfield as a close support. When Cooper observed the 10th Kansas Infantry advancing towards the cornfield, he ordered Howell's other section to direct their fire towards the Union reinforcements. Cooper reported:

> "…The battle was now raging in all parts of the field. Their masses of infantry could be plainly seen advancing in perfect order, with guns and bayonets glittering in the sun. The booming of cannon, the bursting of shells, the air filled with missiles of every description, the rattling crash of small-arms, the cheering of our men, and the war-whoop of our Indian allies, all combined to render the scene both grand and terrific…"

At this point of the fight, Lt. Col. Buster and his Missouri Indian Battalion arrived on at the battlefield. They been delayed due to an internal issue in which one of his soldiers had killed a fellow soldier. In any case, he arrived just in time to turn the tide in the cornfield fight. Along with Gordon's 5th Missouri Cavalry, Cooper's forces were able to push the Union troops back onto the prairie. General Salomon had been counting on the arrival of Colonel George Hall and his brigade of Missouri State Militia that he had sent for early that morning. Unfortunately, Hall did not understand what Salomon wanted him to do beyond head for Newtonia. Apparently, Salomon expected Hall to arrive at Newtonia from Mount Vernon via the Jollification Road. The actual message from Salomon was brief without any specific details regarding this:

> Advance immediately toward
> Newtonia. Heavy fighting in
> advance.
> A. Blocki
> Assistant Adjutant-General
> Colonel Hall
> Commanding Fourth Brigade,
> Missouri State Militia
> P.S.—Send answer by bearer where
> you will strike Newtonia road.

Salomon had left Hall without any stated purpose or plan of action. Hall reported to General Brown at Springfield:

> "…This was the only communication ever received by me from General Salomon concerning the movement of the troops under my command at that time. It left me in utter ignorance of General Salomon's intentions. I did not know whether he intended to advance with his forces from Sarcoxie toward Newtonia. I was entirely without information of the intention of General Salomon. I ordered the different regiments, battalions, and companies of the brigade to march. We marched toward Newtonia by the way of Jollification. We marched about 8 miles. I had not heard any firing since we started. I had not received any communication from General Salomon. I was ignorant of his movements or intentions. I had no means of knowing the result of the morning's engagement, but had good reason to believe that the enemy had maintained his position. By marching to Newtonia by way of Jollification the whole force of the enemy would be directly between my command and the troops of General Salomon, whether they were at Sarcoxie or advancing from Sarcoxie to Newtonia. I therefore, upon consultation with the principal officers of your brigade, determined to march west until I struck the road leading from Sarcoxie to Newtonia. Here we came upon some stragglers belonging to Salomon's command. From them I learned that General Salomon and Colonel Weer, with their brigades, had marched toward Newtonia…"[70]

The Jollification Road was a direct route to the east side of Newtonia, joining Mill Street just east of town. With only one understrength brigade, Colonel Hall did not want to approach Newtonia directly without some knowledge of where General Salomon's forces were. The road that Col. Hall opted for from the village of Jollification (modern-day Jolly, Missouri) followed the path of Shoal Creek to the northwest. This road joined the

Sarcoxie Road at the crossing near the Ritchey farm, about halfway between the two towns. Hall and his Missouri State Militia brigade arrived at the junction at about 4 p.m. The brigade then moved southward along the Sarcoxie Road toward Newtonia until they met General Salomon at the edge of the prairie around sundown. At this point, Hall was directed by Salomon to deploy his brigade to cover the on-going retreat of Salomon's command. In his report to General Schofield, Salomon implied that if Hall had followed his plan, he would have hit the right flank of the Confederate defenses just as the 3rd Indian Home Guard and the 10th Kansas Infantry were hitting Cooper's Confederate Indians in the cornfield. Instead, he was required to order a general retreat back across the open prairie. Unfortunately, Salomon failed to share his plan with Hall, or perhaps messengers carrying the plan failed to find Hall's brigade. Either way, Hall was in the dark.

The Union retreat began as an organized withdrawal but fell apart the further back they traversed. Salomon ordered Capt. Stockton and his 25th Ohio Artillery to deploy near the location the Sarcoxie Road entered the Shoal Creek timber. As he readied his guns, Stockton reported that the Confederates arrived at a location approximately one mile from his location, deployed into a line of battle, and opened fire on the retreating Union soldiers. When Col. Hall's brigade arrived on the field, they formed Capt. Murphy's Company F, 1st Missouri Artillery, battery along with the Ohio guns and opened a heavy fire on the advancing Confederates, stopping their advance. Wiley Britton observed this action:

"…Captain Murphy's battery was carefully masked behind thickly leaved clumps of small black-jack and post oak near the prairie, and was supported by the fresh troops and by part of the troops that had been withdrawn from the field. The enemy were marching up rapidly in line as if they felt certain that they were following a demoralized foe. As it was now getting dark the advancing Confederates could not see the Federal troops in the timber, but the Federal troops could easily see the Confederates in the open prairie. The Confederate cavalry therefore steadily advanced to within a hundred or so yards of the Federal batteries, when suddenly they belched forth a stream of grape and canister shot and shell that fairly made the earth tremble. Round after round of canister was fired by Captain Murphy's six guns into the ranks of the enemy, knocking horses and riders into confused heaps upon the prairie, and throwing their horses and riders into wild confusion. General Cooper's artillery followed his cavalry closely, and from the ridge in the prairie shelled the Federal position in the woods for a short time. The terrific shock of the Federal batteries was more than the Confederate cavalry could stand, and in a few moments they broke and were sent flying back over the prairie in the direction of Newtonia, leaving their killed and wounded on the field…"[71]

Hall continued to report his observations to General Brown:

> "…General Salomon, on the battlefield, requested me to cover his retreat with my brigade. Gen-Salomon's troops were retreating in great confusion. The enemy in force were advancing to attack the rear of General Salomon's column. His rear must have been captured and destroyed but for the timely arrival of your brigade. I drew up your brigade in line of battle in front of the enemy and between the enemy and the retreating force of Brigadier-General Salomon… About dark the enemy withdrew their forces from the field. Having secured the retreat of the forces of General Salomon I drew off the men under my command… General Salomon's troops scattered from near the battlefield to Sarcoxie. I saw them all safely back to Sarcoxie…"[72]

A great many of the Union soldiers had exhausted the water in their canteens and were happy to refill them at the Shoal Creek crossing. The individual soldiers and smaller organized units drifted back into Sarcoxie late that night or early the next morning after having caught a few hours of sleep along the road. A heavy rainstorm occurred in the night, which drenched the retreating Union troops and muddied the roads, making it difficult to move the wagons, artillery, and ambulances. Wiley Britton remarked that the troops were "not demoralized, but confident of success when the attack should be renewed in a few days." Indian Agent Carruth stated in his report to Washington:

> "…In this Contest our own regiments now freely acknowledge them [Indians] to be valuable Allies and in no case have they as yet faltered, until ordered to retire, the prejudice once existing against them is fast disappearing from our Army and it is now generally conceded that they will do good service…"

The *Detroit Free Press* of January 28, 1863, printed a report by a Missouri correspondent praising the Union Indian Brigade in the Army of the Frontier operating in Missouri and Arkansas.

> "…These men are mostly Cherokees and Creeks. They are truly a study, with their variety and grotesqueness of costume. Their peculiar features and the traditions and associations that linger around their history … they have clung with a pertinacity and heroism with a self-sacrificing fidelity and patriotism that might well put to shame the loud-mouthed professions of the radicals of the North, who claim so vast a superiority over them in civilization. They are quite tractable, perfectly obedient, and take readily to military discipline. Yet they do not see the propriety of remaining in ranks during a fight. They say it looks very well on parades, but is too dangerous in fighting. They appeared well on review, and gave the General three war whoops as a salute…"

The conduct of the Union Indian Brigade during the battle and in covering the Union retreat made a positive impression on the Union troops.

On the Confederate side, Col. Cooper reported that he did not want to follow the retreating Union forces into the dark forest, knowing that their artillery could clearly see them as they advanced. The Confederate commander also stated that he ordered Capt. Howell to drop a few shells randomly into the forest and claimed that these shells caused a wide scene of confusion on the Kansas troops. This seems unlikely since Maj. Buster of the Missouri Battalion, who was acting as rear guard and covering their retreat, reported to Cooper that the 1st Choctaw and Chickasaw regiment was still in the woods, failing to receive word of the Confederate withdrawal. Regardless, both sides were through fighting for the day and fell back to their respective camps for rejuvenation and resupply.[73]

Both the Union and Confederate commanders reported low casualties for their own sides and greatly exaggerated the losses of their foe. The Union side fared worse with a loss of at least 250 killed, wounded, or missing out of a total between 3,500 and 4,000. This includes the loss of four companies of the 9th Wisconsin Infantry when only ten were able to avoid capture and make it to the woods after the morning retreat. The Confederates reported a total of 78 casualties out of approximately 5,000 involved. That number is probably closer to 100 since the Confederate records are incomplete, especially numbers from the Confederate Indian Brigade.

Over the next couple of days, the Union force did little beyond sending out a few small patrols and posting pickets in vulnerable areas. Taking Col. Weer's advice, on October 1 General Blunt departed Fort Scott and travelled quickly to Sarcoxie. Blunt recalled:

> "…About midnight, I met a messenger from Gen'l Soloman[sic] with despatches [sic] stating that he had an engagement the day previous with rebel forces under Generals Cooper and Shelby, at Newtonia, in which he (Solomon)[sic] had been defeated and driven back to Sarcoxie. With a small escort I pushed rapidly forward, leaving the Third brigade to follow with as little delay as possible…"

Blunt arrived in Sarcoxie on the evening of October 2 and reported to General Schofield early the next morning. Schofield was the ranking officer who had arrived the previous day with his command from Springfield. Blunt gathered the official reports from his subordinate commanders and attempted to determine what had happened over the previous days. In this meeting it was determined that they would strike Cooper's force at Newtonia again but from two different sides. Blunt explained:

> "…As it was to be presumed that the enemy would be expecting an attack in front, and would have the approaches by the direct route guarded, we agreed that, with my command I should move to the right while Schofield was to move to the left, come in on the east of Newtonia, and throwing his cavalry -- of which he had a large force -- in

their rear, cut off their retreat, after I had broken their lines and routed them…"

Colonel Cooper had moved his headquarters up between Shoal Creek crossing and Granby. Col. Shelby was sent up the Sarcoxie Road, north of the crossing, and instructed to find a defensible place where the remainder of the division could deploy. Cooper learned that the Union reinforcements from Springfield, under the field command of General Totten, were currently located in the western parts of Lawrence and Barry Counties. When Cooper was denied reinforcements from General Rains, he decided to bring Shelby back to Newtonia. Rains also ordered Cooper to begin withdrawing from Newtonia as soon as possible, and he began to make plans to do so.

On the evening of October 3, General Schofield issued Special Field Orders, No. 12, that instructed Generals Blunt and Totten on the plan of action and the routes that would be used to approach Newtonia. The two Union divisions left their camps enroute to strike Newtonia from different directions. There were changes to Blunt's original order, when he was instructed to divide his force of six thousand into two columns. The primary column was to proceed down the Sarcoxie Road towards Newtonia, as Salomon's command had done on September 30, and attack the town from the north. The second column was to move west towards Granby and then strike southwest towards Newtonia from the Granby Road.[74] In the meantime, Cooper had sent the Indian Battalion and the 1st Cherokee

Battalion to Granby to resist any Union advance from Sarcoxie. In the opposite direction, the division under General Totten would move from its camps towards Newtonia via the Jollification Road, hitting the town from the east. Also, on the evening of October 3, Cooper sent Col. Shelby's 5th Missouri Cavalry up the Jollification Road to discover and resist any Union movements from that direction. They were also to buy time for the rest of the Confederate command to evacuate Newtonia.[75]

In the very early hours of October 4, the pickets for the two Indian battalions who had been posted on the road leading to Granby from Sarcoxie, numbering no more than four hundred ranks, encountered and began skirmishing with the Union column advancing from Sarcoxie. Realizing their predicament, Lt. Col. Buster sent a runner to Col. Cooper advising him of the Union movement. As the Union force advanced, Buster and his small battalion pulled out of Granby just as the eastern sky began to lighten. They deployed a line of battle near the Granby to Newtonia Road, but the Union column simply bypassed them and continued towards Newtonia. The two Indian battalions moved down a secondary road until they reached the Neosho Road-Mill Street. There was little activity occurring in Newtonia, so Buster began to organize a defense of the town, first by deploying Howell's battery on the west end of Mill Street and then waiting for the Union Army's approach.[76]

Shelby's Missouri cavalry advanced eastward on the Jollification Road. An advance party conducted reconnaissance along the route until they arrived at the town. The advance troops then attacked and captured some Union pickets who had been sent there the night before. As the sun began to rise in the east, Shelby arrived at Jollification and observed the large Union force under Totten moving quickly in their direction. Shelby wheeled his command around and they quickly retreated back to Newtonia, leaving the captured Union pickets at the blacksmith's shop. When they returned to town, Shelby discovered that Cooper had already sent the supply train back to Camp Coffee, so Lt. Col. Buster had taken the initiative and had begun preparing for the defense of the town. As ranking officer, Shelby took command and began deploying the remaining troops in a defensive posture. Buster and Bryan had already deployed their respective battalions to the left of the battery. Shelby placed Jean's Jackson County Cavalry to the right of the battery. He placed his own regiment on the east part of town to provide a defense from Totten's advance. They had been in position for only a short time when Shelby received word that communications had been cut off with Camp Coffee and the Pineville Road was going to be cutoff by Union troops very quickly. Shelby ordered all of the Confederate troops to evacuate the town towards Camp Coffee via a small, secondary road on the west end of town. As they were starting to withdraw, with Buster on the left wing and Shelby on the right, Union artillery began to rain down upon the town and the retreating Confederates.

Mary Grabill and the other civilians in town had been notified by the Confederates of the upcoming attack by the Union. She stated that the Confederates were already retreating when the Union Army appeared and began its furious bombardment with thirty-six field guns and four 12-pounder mountain howitzers. Mrs. Grabill and the other citizens evacuated the town as best they could. She recalled those moments in a letter to her daughter:

> "About a mile out of town, we saw approaching on the open prairie and immense body of cavalry – thirty thousand in number altogether. They came at double-quick, and we were so terrified we did not know what to do – especially as in their rear their artillery was shelling the Southerners, who were retreating toward the south as rapidly as possible, and these shells went woo-o-shing over our heads just as though aimed at us. And we were in no means sure in our own minds that they were not. We fell down in cover of a zigzag fence, and taking little white things of our bags, waved them as flags of truce!"[77]

Wiley Britton of the 6th Kansas Cavalry recalled years later:

> "…A brisk cannonade was opened upon the enemy for half an hour or upward from the batteries of the three divisions, when at a signal the troops of these divisions unfurled their flags, sounded the march from their bugles, and moved forward to battle, marching in columns of companies and squadrons. As the troops came near town, rail fences were thrown down where they were in the way, and the infantry came into line with the cavalry marching on the flanks. As soon as the rebel pickets and outposts ran in and announced the Federal advance,

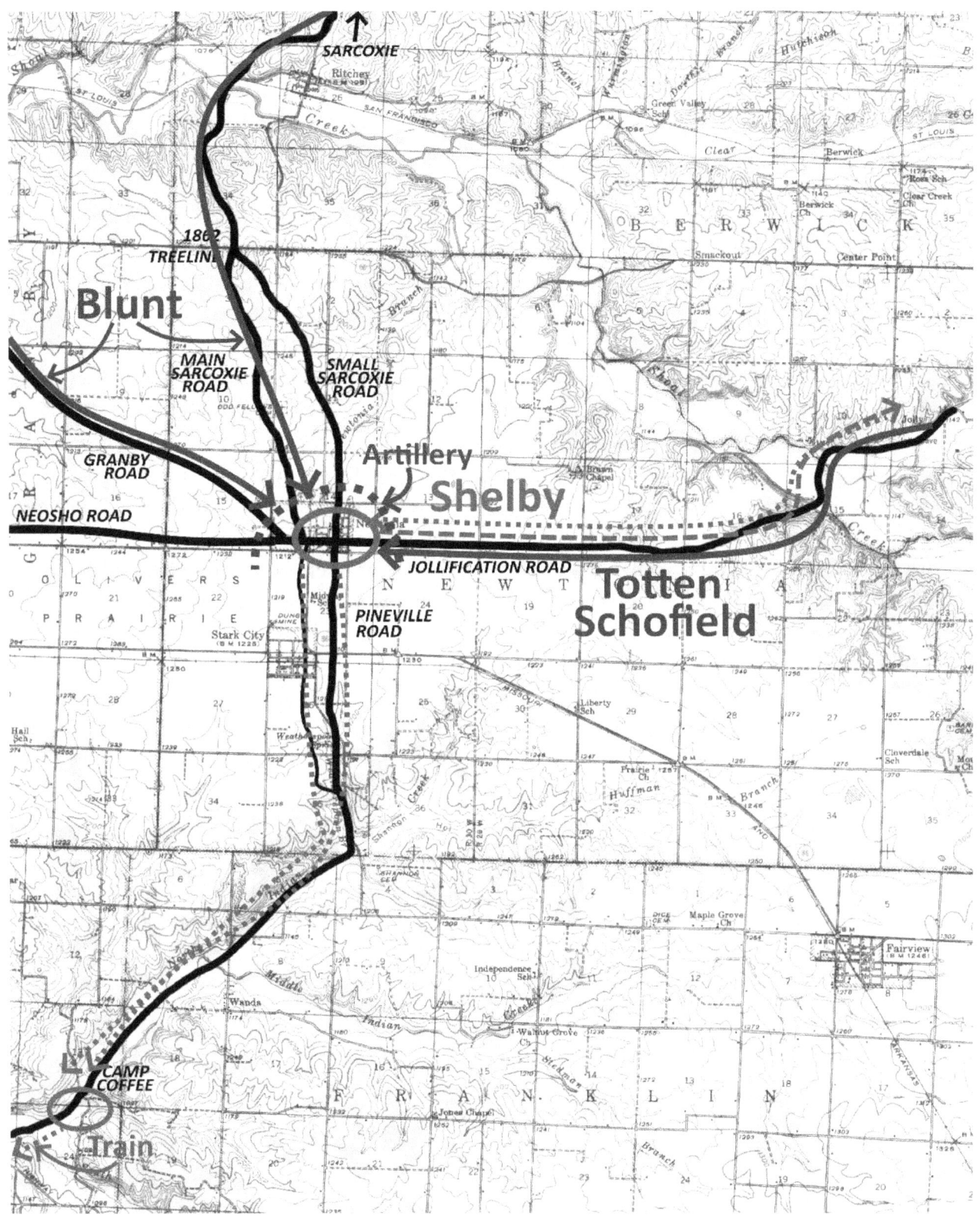

**Planned Union movements and Confederate withdrawal on
October 4, 1862**

Map courtesy of USGS / Additions by Author

General Cooper posted his troops behind the stone fences and other advantageous positions. He then watched the advancing columns as they came steadily on, until he became satisfied that they were going to meet him, man to man, on ground of his own choosing, and then he ordered a retreat, after firing a few scattering shots from his small-arms…"[78]

Back at Camp Coffee, Cooper heard the same information that communications had been cut off with Newtonia. He left Col. Folsom with his 1st Choctaw regiment at the camp to escort the wagon train south towards Pineville. Cooper took his own 1st Choctaw and Chickasaw regiment and Stevens's 22nd Texas Cavalry northward on the road to Newtonia and found that the Pineville Road was still open with only a few Union soldiers in view. When he reached town, he discovered that Shelby had already withdrawn, the town was occupied by Union skirmishers, and there was massive artillery fire upon the town and surrounding areas. Cooper then saw the large columns of Union infantry coming toward the town from the west, north, and east who would soon be upon his command. He rode to his left and took the same small, secondary road that Shelby had used. As Cooper moved southward, he met with Capt. Loring, whom he had sent with his company of the 1st Choctaw and Chickasaw regiment to gather stragglers and check on Col. Folsom and the 1st Choctaw regiment. Loring reported that the wagon train was still at Camp Cooper, but that Folsom and his regiment were gone and that its commander had "quit his post." Cooper maintained his position at the edge of

Oliver's Prairie until he had gathered the remainder of his stragglers.[79]

When the Union commanders saw Cooper and his troops falling back, they sent a large force of Union cavalry in pursuit, and Cooper sent a portion of his 1st Choctaw and Chickasaw regiment to counter-attack and buy some time for the Confederates retreat. The Union pursuit lasted most of the day without any notable success. Blunt was very disturbed by the lack of aggressiveness of Totten's division coming in from the east during the battle. It was actually Schofield who was in command of this portion of the army since he was with Totten. They had agreed on signal guns to announce each other's arrival at their deployment points. Blunt, because his columns had encountered much resistance from the Confederates, was a few minutes behind his assigned deployment time. He fired his signal guns but did not receive a response from Schofield's column. After his column had driven the Confederates from Newtonia, Blunt stated in his remembrances:

> "…After the bird had flown, General Schofield's column could be seen approaching over the prairie from the east. He had five miles less distance to march than I had, did not encounter even a picket, and yet failed to carry out his part of the arrangement, which, had he done as agreed upon, the greater portion of the rebel force could have been captured…"[80]

The retreating troops camped that night at Dog Hollow, about four miles north of Pineville. Over the next few days Cooper and the Confederate Army plodded southward through Pineville and finally into Northwest Arkansas, passing through the Pea Ridge battlefield. Cooper met with General Rains at Mud Hollow (near modern-day Lowell, Arkansas). While they were together, Rains asked Cooper what he thought should be the next move for the Indian Brigade. Cooper suggested that they invade Kansas and capture Fort Scott, including all of the supplies meant for the Kansas Division. He believed that it would force Blunt to move back towards his supply base, which in turn would force Schofield and the remainder of the Union Army of the Frontier to fall back as well because they believed that only Blunt was aggressive enough to attack. Rains agreed, but

then detached all of the Texas units from Cooper's command and ordered him back to Indian Territory.[81] Cooper insisted that he at least be able to keep Buster's Missouri Battalion along with his brigade. Rains relented and they agreed on October 22 as the day the Kansas operation would begin, to coincide with a Confederacy-wide Fall offensive. Cooper and the Confederate Indian troops, including Capt. Howell's Missouri battery, arrived back at Old Fort Wayne via Maysville, Arkansas, over the next few days to re-group, re-supply, and recruit.[82]

[1] Lincoln, Abraham. *Abraham Lincoln papers: Series 1. General Correspondence. 1833 to 1916: John Ross to Abraham Lincoln, Tuesday, Relations between the U.S. and Cherokee Nation*. 1862. Manuscript/Mixed Material. Retrieved from the Library of Congress

[2] "Collected Works of Abraham Lincoln. Volume 5 [Oct. 24, 1861-Dec. 12, 1862]." In the digital collection *Collected Works of Abraham Lincoln*.

[3] Lincoln, Abraham, Letter, October 10, 1862, pp. 723.

[4] Lincoln to Curtis, *OR,* Ser. I, Vol. XIII, pp. 723.

[5] Curtis to Lincoln, *OR,* Ser. I, Vol. XIII, pp. 723.

[6] Pike, Albert. Letters, *OR* July 15 & 20, 1862, Ibid: 856-874.

[7] Pike to Chiefs, *OR,* Ser. I, Vol. XIII, pp. 869-871.

[8] Pike was liable to the penalties prescribed by section 29 of the act of Congress regulating intercourse with the Indians and to preserve peace on the frontiers, approved April 8, 1862, as follows:
*If any person shall send, make, or carry, or deliver any talk, speech, message, or letter to any Indian nation, tribe, band, chief, or individual, with intent to * * * make such nation, tribe, band, chief, or Indian dissatisfied with their relations with the Confederate States or uneasy or discontented, the person so offending shall, on conviction, be punished by fine not exceeding $10,000 nor less than $2,000, and by imprisonment not less than two nor more than ten years, and the intent above*

mentioned shall be conclusively inferred from knowledge of the contents of any such talk, speech, message, or letter in writing.

[9] Hindman to Cooper, *OR,* Ser. I, Vol. XIII, pp. 41.

[10] Hindman, *OR,* Ser. I, Vol. XIII, pp. 928.

[11] Pike to Holmes, *OR,* Ser. I, Vol. XIII

[12] Abel, Annie Heloise. The American Indian as Participant in the Civil War, Cleveland, Ohio: A.H. Clark Company, 1919. pp. 201.

[13] Monaghan, Jay. Civil War on the Western Border, 1854-1865. Lincoln and London: University of Nebraska Press. 1955. pp. 256.

[14] Public Laws of the Confederate States of America, Passed at the First Session of the First Congress; 1862.

[15] "The Partisan Ranger Act of 1862; Article, Wikipedia

[16] Spencer, John D. The American Civil War in the Indian Territory, New York and Oxford: Osprey Publishing, Ltd. 2006. pp. 14-19.

[17] Ibid. pp. 24-41.

[18] Charles D. Collins, Jr. *Battlefield Atlas of Price's Missouri Expedition of 1864.* Fort Leavenworth, Kan.: Combat Studies Institute Press, 2016, p. 21.

[19] "The First Battle of Independence," Historical Marker, Missouri Department of Natural Resources.

[20] Eakin, Joanne Chiles, *Battle of Independence, August 11, 1862,* Two Trails Publishing, 2000, pp. 4.

[21] Eakin, Joanne Chiles, *Battle of Lone Jack, August 16, 1862*, Two Trails Publishing, 2001, pp. 147-236

[22] Matthews, Matt and Lindberg, Kip, "Shot All to Pieces, the Battle of Lone Jack, Missouri, August 16, 1862", *North and South*, Vol. 7, No. 1, January, 2004, pp. 61.

[23] Ibid. pp. 71.

[24] Stevens, Walter Barlow. Centennial History of Missouri S.J. Clarke Publishing. 1921. pp. 629.

[25] Blunt to Schofield, *OR*, Ser. I, Vol. XIII. pp. 235-236.

[26] Matthews, Matt and Lindberg, Kip, pp. 66.

[27] Britton, Wiley, The Civil War on the Border, Volume I, New York and London: G.P. Putnam's Sons, 1899. pp. 348.

[28] Britton, *Brigade*, pp. 89.

[29] Shea, William L. Fields of Blood: The Prairie Grove Campaign, Chapel Hill: The University of North Carolina Press, 2009. pp. 16-17.

[30] General Orders No. 4, *OR*, Ser. I, Vol. XIII. pp. 595.

[31] Totten to Blunt, *OR*, Ser. I, Vol. XIII, pp. 610.

[32] Weer to Blunt, *OR*, Ser. I, Vol. XIII, pp. 665-666.

[33] Cooper to Rains. *OR*, Ser. I, Vol. XIII, pp. 296-297.

[34] Oates, Steven B. Confederate Cavalry West of the River, Austin: University of Texas Press, 1961. pp. 35.

[35] Wood, Larry. The Two Civil War Battles of Newtonia, Charleston & London: The History Press, 2010. pp. 50-51.

[36] Special Orders No. 28, *OR*, Ser. I, Vol. XIII. pp. 630.

[37] Schofield to Totten, *OR*, Ser. I, Vol. XIII. pp. 633.

[38] USGS Topo Map: Carl Junction Quad, Sec. 24.

[39] Ritchie to Blunt, *OR*, Ser. I, Vol. XIII, pp. 278.

[40] Britton, Wiley, *Border 1*, pp. 352.

[41] Weer to Blunt, *OR*, Ser. I, Vol. XIII, pp. 665-666.

[42] Britton, Wiley, *Border 1*, pp. 355.

[43] National Park Service, Newtonia Battlefields Special Resource Study, Newtonia, Missouri, January 2013

[44] USGS Topo Map: Ritchey Quad: Sec. 13, T24N; R31W

[45] These Granby lead ore mines were first discovered in 1850 and were a large source of income for the area until the mines began to close in the late 1950s.

[46] Cooper to Rains, *OR*, Ser. I, Vol. XIII, pp. 296-297.

[47] Lynde to Salomon, *OR*, Ser. I, Vol. XIII, pp. 291-293. Wood, Larry, pp. 55-58.

[48] Salomon to Schofield, *OR*, Ser. I, Vol. XIII, pp. 286-288; Lynde to Salomon, *OR*, Ser. I, Vol. XIII, pp. 291-292; Jacobi to Salomon, *OR*, Ser. I, Vol. XIII, pp. 291-292.

[49] Blunt to Salomon, *OR*, Ser. I, Vol. XIII, pp. 692.

[50] Bryan to Cooper, *OR*, Ser. I, Vol. XIII, pp. 301.

[51] National Park Service, Newtonia Study

[52] Cooper to Rains, *OR*, Ser. I, Vol. XIII, pp. 296-297.

[53] Hardee's Rifle and Light Infantry Tactics, School of the Battalion, Title Fourth, Article Fourteenth, Dispositions against Cavalry.

[54] Lynde to Salomon, *OR*, Ser. I, Vol. XIII, pp. 292

[55] Hawpe to Cooper, *OR*, Ser. I, Vol. XIII, pp. 305-306.

[56] Grabill, Mary. Letter entitled "To My Daughters." Transcribed copy on file at Wilson's Creek National Battlefield, Republic, Missouri. Wood, Larry. *Newtonia*, pp. 66.

[57] Lynde to Salomon, *OR*, Ser. I, Vol. XIII, pp. 292.

[58] Wood, Larry. *Newtonia*, pp. 67.

[59] Hadley to Stockton; Stockton to Blocki, *OR*, Ser. I, Vol. XIII, pp. 295.

[60] Wood, Larry. *Newtonia*, pp. 71.

[61] Britton, Wiley, *Border 1*, pp. 357.

[62] Judson to Salomon, *OR*, Ser. I, Vol. XIII, pp. 291.

[63] Wood, Larry. *Newtonia*, pp. 79-80.

[64] Cooper to Rains, *OR*, Ser. I, Vol. XIII, pp. 298-299.

[65] Wood, Larry. *Newtonia*, pp. 80.

[66] Grabill, Mary. Letter

[67] Britton, Wiley, *Border 1*, pp. 359.

[68] Weer to Blunt, *OR*, Ser. I, Vol. XIII, pp. 288. Judson to Salomon, *OR*, Ser. I, Vol. XIII, pp. 291. Cooper to Rains, *OR*, Ser. I, Vol. XIII, pp. 298-299.

[69] Stockton to Blocki, *OR*, Ser. I, Vol. XIII, pp. 296.

[70] Hall to Brown, *OR*, Ser. I, Vol. XIII, pp. 289-290.

[71] Britton, Wiley, *Border 1*, pp. 360-361.

[72] Hall to Brown, *OR*, Ser. I, Vol. XIII, pp. 289-290.

[73] Cooper to Rains, *OR*, Ser. I, Vol. XIII, pp. 299.

[74] Schofield, Special Field Order #12; *OR*, Ser. I, Vol. XIII, pp. 706-707.

[75] Cooper to Rains, *OR*, Ser. I, Vol. XIII, pp. 299.

[76] Wood, Larry. *Newtonia*, pp. 93-94.

[77] Wood, Larry. *Newtonia*, pp. 94-95.

[78] Britton, Wiley, *Border 1*, pp. 362.

[79] Wood, Larry. *Newtonia*, pp. 96.

[80] Blunt, *Experiences*, pp. 227.

[81] Shea, William L. pp. 34.

[82] Cooper to Hindman, *OR*, Ser. I, Vol. XIII, pp. 331-333.

Chapter 9
Blunt vs. Cooper:
The Battle of Old Fort Wayne

As the Summer of 1862 passed into Fall, the question of loyalties was firmly set. One was either pro-Southern or pro-Union. Riding the fence was no longer an option as the war dragged on, with greater and greater battles and their accompanying high casualty counts. The Confederate forces in the Western Theater were having a difficult time with losses at Shiloh, Pea Ridge, Island No. 10, the capture of New Orleans in April, and the loss of the critical rail junction at Corinth without a shot in May. Later, the Confederate Army had some small successes such as at Richmond, Kentucky, but the Battle of Perryville, Kentucky, in October ended General Braxton Bragg's Heartland Offensive. In the Eastern Theater, the Confederates were having much more success. They saw the failure of General McClellan's attack on Richmond via the Virginia Peninsula, the defeat of General John Pope at Second Manassas, and General Thomas "Stonewall" Jackson's victories in the Shenandoah Valley as clear signs that the Confederacy was still a powerful force to be reckoned with. Their only setback during this period was the Battle of Antietam in September which, although a tactical draw, was a strategic Union victory. This was the victory President Lincoln needed to issue the Emancipation Proclamation that would free the enslaved people who were living in the states deemed to be in "rebellion" against the United States. Although this included all of the states currently in the Confederate States, it did not apply to the

Border States of Missouri, Kentucky, Maryland, Delaware, and the District of Columbia, since slavery was still constitutionally protected in those areas.

In August 1862, the Second Session of the Confederate Congress was called to order. On October 2, Elias Boudinot first arrived and was seated as the official delegate to the Confederate Congress for the Cherokee Nation on October 9. In the following months, Robert M. Jones, a Choctaw, who served as the Chickasaw and Choctaw representative, and Samuel Benton Callahan discharged the same duties for the Creek and Seminole tribes. These positions were created by virtue of the treaties that Albert Pike had negotiated between the Five Civilized Tribes and the Confederate Government in 1861. According to the treaties the Seminoles and Creeks together were awarded one delegate; the Chickasaw and Choctaws received one delegate; and the Cherokees were given one delegate. These delegates were chosen by ballots managed by the tribal governments and overseen by the Confederate Indian agents. But Albert Pike had promised the Indian tribes the same status as other Confederate states. Jefferson Davis pushed the House of Representatives to not allow the Indian Nations delegates to vote or to introduce legislation. They also were only able to place their delegates in the House of Representatives and not in the Confederate Senate. It took two years to actually outline the duties and privileges of the Indian delegates. Most of the legislators did not believe that the Five Civilized Tribes would actually send representatives to Congress. So, the arrival of the first delegates took them by surprise although the legislators had passed the treaties themselves. There was also a feeling of distrust between the white members and the Indian delegates. Many of the Confederate legislators were the same politicians who forced the tribes to move west twenty-five years before. Most of the business the Indian delegates contended with was responding to tribal members complaints and attempting to bridge the gap between what their treaties stated and what the Confederacy was actually willing to adhere to. Although Boudinot was able to get the Cherokee Relief Act passed in early 1864, they always had difficulty in getting their legislation introduced since another full-voting delegate was needed to bring it to the floor. Whatever measures they succeeded in enacting or issues they tried to correct, they did not relieve their

constituents' problems and this could be attributed to wholly inadequate financing of the Confederacy as the Civil War years dragged on. This was another broken promise by the Confederacy towards the Five Civilized Tribes.[1]

Back in August 1862, Brig. Gen. Blunt had again pushed for another Indian Expedition, under his command, that would return the Indian refugees back to the Cherokee and Creek Nations. But by this time, as shown, the Union Indian Brigade had been broken up among the three brigades of Blunt's First Division. The higher Union commanders had set their sights on other goals that kept the Indian Territory low on the priority list. On September 19, Indian Agent Carruth wrote to Commissioner Coffin in Washington that he was enroute to Fort Scott after he had received an order from General Blunt to assist him in the removal of the Indian refugees back to their homes.[2] Unfortunately, the events along the Missouri and Kansas border including the battles at Independence, Lone Jack, and Newtonia changed all of the plans for the second expedition. At this time the conditions for the Indian refugees was starting to get critical. Roving bands of guerillas, including Missourian William Quantrill, were raiding the refugee camps in Kansas and depriving the Indians of even the most basic essentials. A Shawnee group known as Black Bob's Band, had been assigned a settlement in Johnson County. This one camp was raided by Quantrill and his guerilla band in August and again in the first week of September. This band was finally removed to the far western part of the Indian reserve. The problem became worse as time went on since other small tribes, such as the Seneca, Wyandots,

War-time Photograph of a Union Army Camp within the Indian Territory
Location and Units are Unknown
Photo courtesy of Oklahoma Historical Society

and Quapaws, were being similarly attacked along the border and were forced to move up towards the refugee camps near Fort Scott. Conflict between the Army and the Indian Affairs representatives again came to a head, with each claiming the other was profiteering from Congressional appropriations for the refugees. In July, Congress suspended the annuity payments to the tribes, including the Five Civilized Tribes, using those monies instead for assistance to the Indian refugees in Kansas. This was another opportunity for greed and mismanagement of government funds intended to help the Indians. Although the funds were distributed to the Indian

agents in the same way as before, there were no clear Indian chieftains to assist in the proper distribution of funds. Regardless, the Indian refugees were in much better physical condition than they were when they had arrived in Kansas the previous winter. In October, the number of refugees stood at 7,500 persons, and the Indian agents requested $69,000 to support them that quarter. It is unclear if the total refugees included those serving in the three Indian Home Guard regiments, which, if not, would put the Indian population at over ten thousand.[3]

View from Fort Wayne towards Maysville, route taken by Blunt and the Union Army
Photo by Author

View of Confederate Battery during 2nd Kansas Attack
Photo by Author

Looking southwest across prairie where the 2nd Kansas Cavalry attacked
Photos by Author

Spavinaw Creek & Crossing
Photo by Author

Arkansas State Marker naming the Old Fort Wayne Battle as the Battle of Maysville
Photo by Author

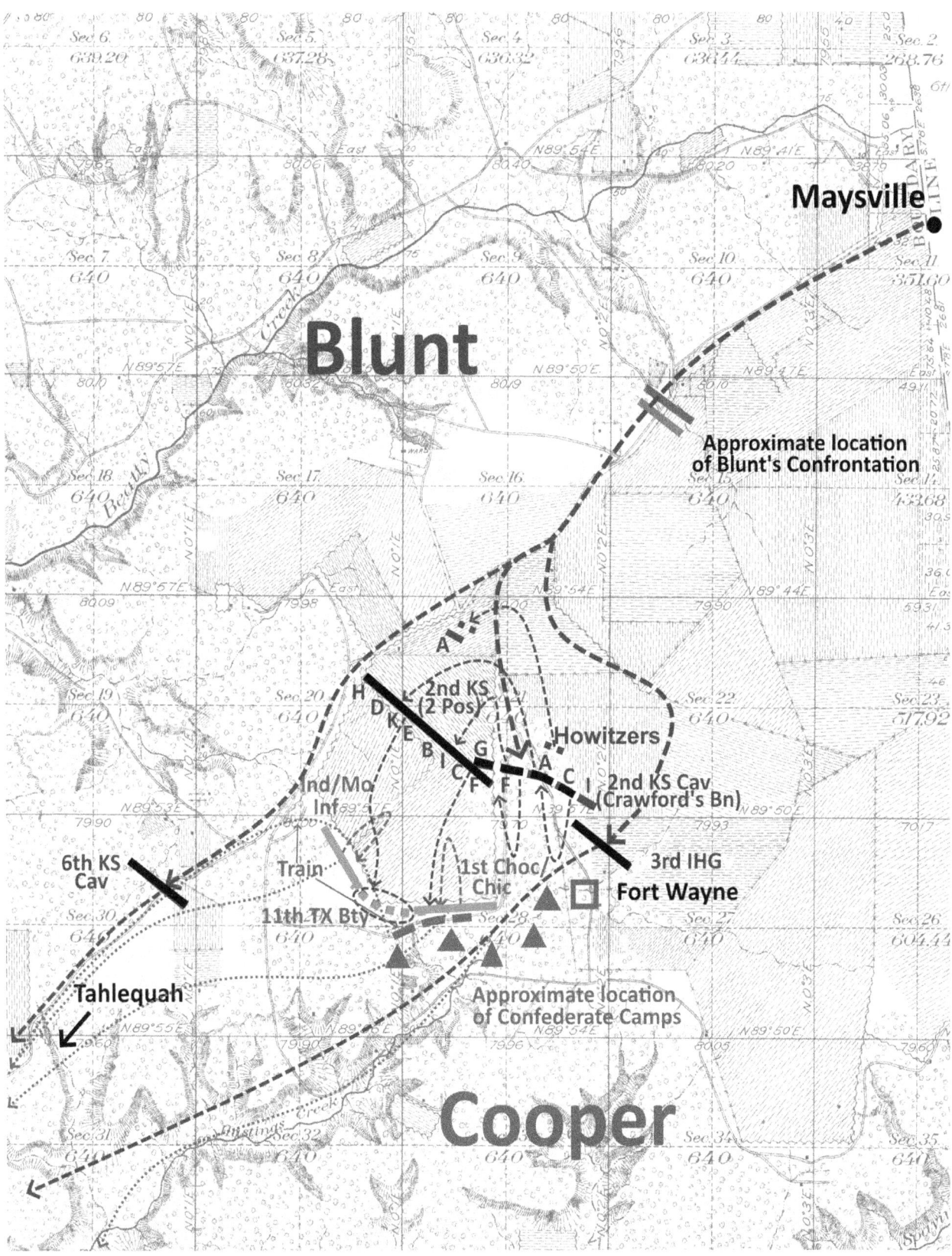

Area of Operations for the Battle of Old Fort Wayne (or Beattie's Prairie, or Maryville)
October 22, 1862

Map courtesy of BLM- General Land Office / Additions by Author

Meanwhile life in most of the Indian Territory remained relatively peaceful except for the severe divisions within the Cherokee Nation. The Choctaw, Chickasaw, Creek, and Seminole Nations remained, thus far, mostly untouched by the war. They did have to contend with Confederate-leaning refugees moving south from primarily the Cherokee and Creek Nations. Although the drought of 1862 also affected this portion of Indian Territory, they were still able to plant and harvest crops during the summer and fall and provide foodstuffs and horses to the Confederate Army. All of the military posts that had been garrisoned by the U.S. Army prior to the beginning of the war were now staffed by large numbers of Texas troops. Along with the Confederate Indian agents, they took responsibility for maintaining peace between the civilized and wild Indian tribes, especially out in the western Leased Lands. The Cherokee Nation stands out as the largest casualty of the war at this time. The pro-Union and pro-Confederate bands were still roaming the remote roads, farms, and settlements. Confederate Col. John Drew's 1ˢᵗ Cherokee Mounted Rifles had been given the mission of tracking down and capturing the rival bushwacking gangs, but that ended when most of Drew's regiment changed sides to the Union Army. Stand Watie and his regiment operated out of the Spavinaw Creek area and had no interest in protecting the majority Union-leaning population. In all probability many of those attacking the Ross-faction or Union-supporting Cherokees were soldiers of Watie's regiment. Francis Elizabeth Kemp was a young child in the Chickasaw Nation and observed the Cherokee refugees as they fled their war-torn lands:

"…The refugees from the Cherokee Nation came in bunches and settled near us during the War. They were without food, and I have often seen them gathering the render leaves from Mulberry trees and cooking for greens. Father would kill beef and hogs and divide out among them; also, let them have corn to make bread. They would dig Briar Root, which was sweet and brittle like potatoes, and mix with the meal when they didn't have enough meal for bread…"[4]

The Creek Indians had their share of troubles after the flight of Opothleyahola to Kansas. Those followers of Opothleyahola that had not fled north tried to keep to themselves. But with many of the men gone with the Confederate Creek regiments, those older men and women, along with the wives and children who were left behind, were also subject to harassment and violence. Malucy Bear, who was a child growing up in the Creek Nation when the Civil War erupted remembered:

"…Anything that had been left was stolen and the house burned and destroyed by the raiding parties who even took things from the homes that were still occupied by the women and children, The pleas of the women were disregarded and they were helpless and powerless to do anything about keeping their belongings. It was a fearful time so that most of the women who had been left without the protection of the men were scared to say anything as the raiders often had no mercy. Most of the neighboring country was laid in waste by the raiders and the small skirmishes that took place by the different hostile parties. We would often take to the bushes and thickets where we would stay in hiding during the day and

return to what was left of our homes during the night…"[5]

Although there were some points of discontent in the Choctaw, Chickasaw, and Seminole Nations, due to those tribes almost fully supporting the Confederacy, they did not have the divisions present in the Cherokee or Creek Nations. And in a larger sense it appears that the Union was only concerned about those areas north and west of the Arkansas and Grand Rivers.

The U.S. Army command in Washington was disturbed by the series of defeats of Union forces in Missouri. As a result, the Union Army in Missouri and Kansas went through another re-organization in mid-late September 1862. The Departments of Kansas and Missouri were merged into a new Department of the Missouri with Maj. Gen. Samuel Curtis in command out of St. Louis. It also acquired the Territories of Nebraska and Colorado within its control. The field commands of General Blunt in Kansas and General John M. Schofield in southwestern Missouri were combined into the new "Army of the Frontier" with Schofield as senior officer. General Blunt was to command the First Division (Kansas Division) which included the Indian Brigade and other troops in Kansas. General Totten would command the Second Division (Missouri Division), composed of a mixture of Missouri, Iowa, and Illinois regiments. The Third Division

(Missouri Division) was created from the joining together of regiments from Missouri, Indiana, Wisconsin, Illinois, and Indiana. This division would be commanded by Brig. Gen. Francis Herron. This was a good management decision since it gave more flexibility to the troops operating in the region without being constrained by department or district boundaries.

The three Union divisions spent the next couple of weeks poking and probing into far Southwestern Missouri and Northwestern Arkansas for reconnaissance and to gather intelligence concerning Confederate movements and strengths. The information they were gathering was, as usual, wildly inaccurate, with reported Confederate troop strengths between fifteen and thirty thousand. They were rumored to be everywhere, but nowhere. The stated purpose of the Union movements was to push the Confederates south across the Boston Mountains and away from Missouri and Kansas.[6]

Confederate Brigadier General Cooper soon began issuing orders to his Indian regimental commanders to immediately gather at the abandoned former Fort Wayne near the border of Arkansas and Indian Territory, just about three and a half miles south-southwest of the town of Maysville, Arkansas. This was in preparation for a planned Confederate invasion of Kansas and capture of Fort Scott as part of a Confederacy-wide Fall offensive. Unfortunately for General Cooper, only a few of his subordinate units responded to the orders to gather at Fort Wayne. Some of the units failed to receive the order while, again, others simply disregarded it.[7] The 1st Choctaw regiment, the same one that had disappeared from Camp Coffee, remained about thirty miles south of Fort Wayne. The 1st Creek

Battalion, under the command of Lt. Col. Chilly McIntosh, quickly responded to the call, but the 1st Creek Mounted Rifle regiment, although camped close by, did not immediately respond to Cooper's orders. Since Col. John Drew's 1st Cherokee Mounted Rifles was now defunct, Col. Stand Watie's regiment was given that title. Unfortunately, only a limited number from that regiment were available, most being on other assignments. The Seminole Battalion had been given special permission to remain in the Seminole Nation under Chief/Lt. Col John Jumper as provost guard or military police. It is also believed that Cooper did not trust John Jumper since he was such a close friend of Albert Pike. Due to the lack of response from the orders he had issued to the other commands within the Confederate Indian Brigade, Cooper was forced to wait for his forces to arrive before he could take any offensive action against Fort Scott, as Rains had ordered.[8]

Unexpectedly, Cooper received one honor from Richmond that he hoped for:

Special Orders
Adjt. And Inspector General's Office
No. 227
Richmond, Virginia,
September 29, 1862

XVIII. Brig. Gen. D.H Cooper is assigned to duty as Superintendent of Indian Affairs by virtue of act of Congress permitting such assignment.
By command of the Secretary of War:

JNO. WITHERS,
Assistant Adjutant-General

Cooper was now confident that his personal action plan was working in his favor. On the other hand, orders from the Trans-Mississippi Department in Little Rock dated the previous day assigned command of "all troops in the Indian country" to Brig. Gen. J.S. Roane. Unfortunately, along with Cooper's notification of his promotion to brigadier general, Secretary of War Randolph informed General Holmes that Capt. T.J. Mackey of the Provisional Army Engineer Corp had charged the newly promoted general with "habitual intoxication and notorious drunkenness." Although Cooper was cleared to exercise his role as Indian Superintendent, the Secretary of War did not want him officially in command of troops until the charges were examined and a court of inquiry was conducted.[9] Therefore Brig. Gen. Roane was to fill in temporarily. The new commander was a Tennessean who had relocated to Arkansas in the 1840s, where he served as a state legislator and was governor from 1849-1852. In another Arkansas political twist, during the Mexican-American War, Roane served as lieutenant colonel of a regiment of Arkansas volunteers. At the Battle of Buena Vista, he took command of the regiment after the death of its colonel. A young Capt. Albert Pike challenged Roane to a duel over the conduct of the regiment during the battle. The duel was fought but neither was wounded.[10] At least Cooper and Roane had one thing in common, they both disliked Albert Pike. This situation between Pike and Roane may have been why General Hindman had asked for Roane to command the Indian Territory troops. General Holmes had a different opinion. He believed Roane was an unsavory character who was absolutely worthless as a commander and was sent to care for the Indians just to get rid of him.[11] In operational terms, Cooper remained the de-facto commander of the Indian Territory. Unfortunately, his available Confederate Indians were not truly soldier material. They commonly disobeyed orders, responded only when it suited them, and had to be bribed with promises of plunder to get them to fight. In many ways they acted in a similar manner to untrained American militia resisting Regular Army officers. Regardless, although Cooper was fairly popular with the Indians, he was unable to spur them into action.

From Newtonia, the First Division of the Army of the Frontier under Union General Blunt followed the retreating Confederates for the next few days, finally camping at the town of Keetsville, Missouri. Schofield and the Second and Third Divisions meanwhile, were spending time making maps of roads and terrain features rather than pursuing the Confederates. On October 15 Blunt moved his First Division forward and camped several days on the old battlefield at Pea Ridge. Apparently, collecting battlefield souvenirs is not a recent concept. Wiley Britton commented on their time there:

"…The signs of the great battle fought in March, between the Federal forces under General Curtis and the Confederate forces under Generals Van Dorn, Price, and McCulloch, were still visible on every hand, and many of the soldiers availed themselves of the opportunity of going over the field which had recently been the scene of blood and strife, for the purpose of picking up something, such as a piece of shell, or bullet, or buckle, as a souvenir to be sent north to relatives or friends…"[12]

Blunt was anxious to get moving and stated:

"…Our arrival at Pea Ridge was about the 15th of October, and the time since leaving Newtonia had been spent by General Schofield in making a survey of the country and mapping out roads in our rear, while the enemy kept just far enough in our advance to avoid danger and gather from the surrounding country the supplies that we should have appropriated to the use of our command. At Pea Ridge, where we lay in camp for a week, the same farce was reenacted, and during this time the rebel forces, which we had driven out of Newtonia on the 4th of October, were encamped at Elm Springs, twenty-five miles south of us… On the morning of the 20th of October, information was received that the rebel forces had divided at Elm Springs, Cooper and Stand Watie, with six thousand men moving west to Maysville… General Schofield then came to my headquarters… asked me if I had any suggestions to make relative to future movements. I proposed that, with his permission, I would take the second and third brigades of my division and move against Cooper and Stand Watie at Maysville…"[13]

Later, on October 20, Brig. Gen. Schofield issued General Orders #6, which set the Army of the Frontier in motion to force the Confederates either into battle or back across the Boston Mountains. The new Second and Third Divisions were sent southward towards Evansville, Arkansas, and Brig. Gen. Blunt's First Division was ordered to pursue Cooper and the Confederate Indian Brigade into the Indian Territory. Cooper was thought to be in the vicinity of Maysville. The distance between the First Division's encampment at Pea Ridge and Maysville, Arkansas, was about 45 miles over fairly rough ground, with many hills, hollows, and streams to cross. Blunt advanced with the Second (Col. Weer) and Third Brigades (Col. Cloud) of the First Division, Army of the Frontier. These brigades consisted of the 2nd and 6th Kansas Cavalry regiments, the 10th and 11th Kansas Infantry regiments, the 1st and 3rd Indian Home Guards, and were supported by the 1st Kansas and 2nd Indiana Batteries and four 12-pounder mountain howitzers attached to the cavalry regiments. General Salomon and his First Brigade remained behind with General Schofield back in Missouri. They were expected to protect Blunt's rear and maintain his supply line.[14]

Blunt sent out scouting parties to find any concentration of Confederate troops in the area. His scouts reported the gathering of Cooper's and Watie's troops at Fort Wayne, again overestimated to be between five and seven thousand men. Blunt immediately set out towards that location, passing through Bentonville on October 21. He left his supply train to camp three miles beyond Bentonville with orders to follow at daylight. Col. Cloud and the Third Brigade took the advance and moved towards Maysville, Arkansas. They entered a wide-open prairie of about five miles just shy of the town. Ordering a forced march, at about 2 a.m. the column halted and Blunt rode forward to discover why they had stopped. Cloud informed Blunt that he had known they were approaching the Confederate forces and wanted to give the column, especially the infantry, a chance to

close up. Blunt took this opportunity to move forward with a small advance party of Capt. Russell and two troopers from the 2nd Kansas Cavalry. As they closed in on Maysville, they observed a large house and farm. Blunt disguised himself as a Confederate straggler and was able to convince the farmer's wife, whose husband was in Cooper's camp, to provide him with the exact location of the Confederate stronghold by stating that he was lost and looking for his regiment. She advised him where to find the picket outposts and further informed him that Cooper's force had been reinforced by two Texas regiments. Blunt returned to the advance party and ordered Companies B and I, 2nd Kansas Cavalry, under Capt. Hopkins, to move around the town and cut off and capture the Confederate pickets manning the outpost. Unfortunately, the Confederate pickets had already heard the Union Army's advance and had awakened the townspeople and then sent runners back toward the camps near Fort Wayne. At this time, nearly 5 a.m., Blunt discovered that only one other company of the 2nd Kansas Cavalry had followed the advance. He had to send Col. Cloud, who had accompanied the advance party, back to Lt. Col. Bassett to discover what had happened to the missing troops. Bassett sent Maj. Fisk back to stir up the remainder of his brigade. The remainder of the entire command was seven miles to the rear, mostly sleeping along the road.[15]

Meeting an escaped enslaved African, or as Blunt termed him, an "intelligent contraband," whose master was in the Confederate camp, he recruited him as a guide by promising him his freedom for his assistance. With this help Blunt was able to find the approaches to the camp. Passing

Maysville, Blunt and the three companies of cavalry, moved into the Cherokee Nation, Indian Territory. The three and a half miles between Maysville and Fort Wayne was fairly level and open prairie with a few small groves of timber. Because Blunt was basically following the Maysville to Tahlequah Road as it moved southwest, Blunt's small force was able to quickly move toward the Confederates. An interesting occurrence happened as they were leaving Maysville. A small Confederate cavalry detachment was observed just to the southwest of town. As Blunt approached with only Capt. Crawford, Sergeant Cooper, the contraband riding a stolen Confederate horse, and two troopers of the General's bodyguard, the Confederates fled towards their camp. Wiley Britton described the scene:

"…General Blunt and Captain Crawford, being mounted on fleet horses, were gaining rapidly on the rebel detachment when, suddenly coming to a sharp curve in the road, they found themselves face to face with the enemy's grand guard of perhaps not less than eighty men, mounted and in line. The General and Captain Crawford checked their horses suddenly, but Sergeant Cooper's horse could not be stopped, and carried him right through to the Confederate line, and he was captured… then one of the rebel officers ordered a charge, and at the same instant one of his men fired a shot which passed directly over the heads of General Blunt and his party. Captain Crawford, who was carrying his Colt's army revolver in his hand,

immediately brought it in line with his eye and fired two shots into the rebel guard, which seemed to disconcert the rebel officer who had just ordered the charge, for the charge was not made…"[16]

Fortunately for the careless Blunt and Crawford, Companies B and I, 2nd Kansas Cavalry appeared, which convinced the Confederates to turn back for their own lines. An interesting story regarding this incident was recorded by some participants. According to the story, Sgt. Cooper had told others he was a nephew of Col. Douglas Cooper, the commander of the Confederate Indian Brigade, but most believed that he was only joking because of the common last name. Regardless, Sgt. Cooper was with the Confederates either as a prisoner of war or as a welcomed guest.[17] By this time the remainder of the 2nd Kansas Cavalry arrived, under the command of Lt. Col. Bassett, along with the two attached mountain howitzers.

The Confederate camps were located within the tree line of Hastings or Hog Eye Creek, (when the name was changed is unknown) north, northwest, and southwest of the old structures of Fort Wayne. As Blunt advanced, he was able to capture many of Cooper's pickets out in the open prairie. At Fort Wayne, Cooper had been waiting for the remainder of the Confederate Indian Brigade to respond to his orders to rendezvous at his headquarters. His invasion of Kansas and taking Fort Scott was supposed to occur on the morning of October 22. He had been informed of the Union advance on the evening of October 21 and he sent runners to Col. Stand Watie, who was camped south along the Tahlequah Road with his approximately 500 ranks, more than likely on Spavinaw Creek near the crossing or near his mills, about a half mile east of the crossing. Cooper was torn between staying to fight or retreating towards Tahlequah. Cooper ordered Watie and his 1st Cherokee Mounted Rifles to his headquarters after the pickets reported Blunt's troops advancing on his camp in the early morning hours of October 22. He also sent messengers to Col. D.N. McIntosh of the 1st Creek regiment who was also camped along Spavinaw Creek.[18]

Since Col. Cooper reported that he was sick, Lt. Col. Buster of the Indian (Missouri) Battalion was given field command of the Confederate Indian Brigade.[19] Cooper himself withdrew from Fort Wayne in an ambulance. Buster knew an attack was coming and sent out Capt. J. Henry Minhart and his company to reconnoiter the area. (This may be the grand guard that Blunt's entourage ran into) He also ordered Capt. Israel Vore, the brigade's assistant quartermaster, to get the supply train ready to move south. Buster's intention was not to fight at this location but to move everything three miles to the southwest along the Tahlequah Road to the Spavinaw Creek crossing and challenge the Union to a fight as they attempted to cross. Unfortunately, Vore was unable to get the train moving quickly enough. In addition, they had to move the wagons a half mile on a small road to the west to reach the main road. He also did not have many troops to call upon. He had Capt. Howell's four-gun battery, one hundred and forty-six infantrymen, four companies of the 1st Choctaw and Chickasaw Mounted Rifles, Capt. Minhart's and Capt. John Scanlon's squadrons, and Maj. Bryan's 1st Cherokee Battalion. Buster estimated he had one thousand effective fighters available.

None of the other regiments ordered up by Col. Cooper had yet responded. He ordered Minhart to take charge of covering the retreat and placed the Choctaw and Chickasaw companies along the fence on the north side of the road leading to the main road. Buster reported:

> "…I then gave orders to retreat, with the artillery and infantry, moving off. When about halfway from camp to the main road, Lieutenant-Colonel Folsom was attacked by greatly superior numbers. The rear of the train still in sight, I was forced into battle. I ordered the Artillery into battery, supported by my infantry on the left and Lieutenant-Colonel Folsom on the right, a more spirited engagement I never witnessed. There being such odds against us we were forced to give way. Most of the Artillery horses were either killed or wounded. I succeeded in bringing off 2 caissons…"[20]

Blunt initially ordered Bassett to dismount his command and advance and skirmish into the timber on the north side to determine the actual location of the Confederate forces. Bassett was unable to find Cooper's main camp, so Blunt ordered them out of the woods and had them remount. Observing a quarter-mile wide opening in the timber, Blunt reported:

> "…Advancing through an opening in the timber, about a quarter of a mile in width, I discovered the enemy in force, their line extending across the open ground in front and occupying

the road. Between the point I occupied (reconnoitering their position and movements) and their line was a pasture of open ground, some 200 yards across, and two fences intervening. Believing that the enemy were contemplating a retreat, I determined to lose no time in trying the effect of a few shells upon their ranks from the two little mountain howitzers. The Second Kansas was accordingly moved forward in line to the first fence, and the two howitzers, under the command of Lieutenant E. S. Stover, supported by Company A, of the Second Kansas, under Lieutenant Johnston, were ordered to advance through the fence to within 200 yards of the enemy's battery, from which position Lieutenant Stover opened upon them with shell and with much animation…"[21]

In response to Blunt's orders, Bassett then re-formed his regiment into a line of battle (Left to Right: Companies G, F, C, I, B, E, K, D, H, with Lt. Stover's mountain howitzers behind Co. A) and advanced towards the half-mile Confederate line. Lt. Col. Bassett stated:

"…Lieutenant Stover, with his howitzers, led the column, and, arriving at the first field, I formed a line of battle near the fence, selecting for Lieutenant Stover a commanding position inside the field, where he immediately opened fire with good effect. The enemy had formed line of battle in the road south of the field and had advanced even to the center

of the upper field. I gave the order to prepare to fight on foot. The men dismounted with celerity and passed over the fence, pouring volley after volley into the ranks of the enemy, driving him completely from the first field within five minutes, and followed up the advantage thus gained by moving forward to the division fence…"[22]

The 2nd Kansas Cavalry was fortunate in that they were armed with Springfield rifles and were organized more like mounted infantry. These rifles were accurate up to three hundred yards. Capt. Crawford also described the scene as he deployed his battalion of five companies of the 2nd Kansas Cavalry:

"…General Cooper had formed his line of battle across the field at the south end of the prairie, in front of a heavy body of timber, and artillery on a slight elevation near the center. General Blunt stationed Stover's howitzers on the right, protected by Company A. On the left were companies C, I, F, and G, leaving a wide open space between the two wing[s]. I came up with my battalion at a gallop in front of the rebel center and directly in front of the rebel battery…"[23]

The Confederate gunners changed their shot from shell to canister but were shooting over the heads of the advancing Federals. At this point, Capt. Crawford had his dismounted troops open fire on the Confederate line. Blunt realized that his small force of five companies would be overwhelmed without significant reinforcements. He saw that the 2nd Kansas Cavalry battalion was being flanked on the left and right under the cover of the timber. At this time the 6th Kansas Cavalry, under Col. Judson, and the 3rd Indian Home Guards, under Col. Phillips, arrived. Judson was ordered to sweep to the right flank and Phillips to the left. Realizing that his flanks were now protected, Crawford ordered Lt. Horace Moore to take the battery. Lt. Moore stood in front of his company and started whirling his sword over his head and yelled; "Forward, D Company!" Crawford reported:

> "…I came up with my battalion at a gallop in front of the rebel center and directly in front of the rebel battery, in the open field between our two flanks. While we were dismounting the rebel battery was turned onto my battalion, but it seemed to be shooting at the stars. The shells flew high over our heads. …the line advanced at the quickstep to and over a rail fence within 50 yards of the rebel battery, which was belching in our faces. Over the fence the battalion leveled one volley at the battery and the rebel line of support, and then dashed forward, driving everything before them. When we had captured the battery I ordered Captain [Henry] Hopkins to turn the guns on the enemy…"[24]

Following Moore's company, the remainder of Crawford's battalion, Companies B, E, H, and K swept forward across fifty yards, knocked over a rail fence, and captured Howell's Confederate battery. Hopkins attempted but was unable to turn the guns on

the former possessors since he could not locate any caps. One Union officer stated: "The men went at it with a yell, and never halted for an instant, until they had surrounded the guns."[25]

Buster states that at this time the "brave Colonel Stand Watie" arrived on his left for support. This contradicts Watie's own report which has him not going forward until the battle was nearly over. He also stated that Lt. Col. Chilly McIntosh also arrived along with Watie. In the haze of battle, Buster may have misidentified the responding Indian brigade units since Cooper reported that Chilly McIntosh and his Creek Battalion had reported to the Fort Wayne encampment.

Capt. Howell of the 11th Texas Battery later reported:

> "…Nearly all the Batteries horses were shot in consequence of which we were unable to remove the battery from the field and the 4 guns, 3 six-pounders and one 12-pounder howitzer and 2 caissons were captured by the enemy. The other 2 caissons were brought off the field…"[26]

Hopkins's missing artillery caps were probably in one of the caissons that Howell was able to remove from the field. Lt. Col. Buster continued:

> "…Capt. Howell's Texas Artillery company, under command of Lieutenant Ruth, fought bravely, never forsaking the battery until the enemy were within twenty paces. Lieutenant Ruth being the last to leave the battery… My little handful of infantry

stood to the last, which I cannot say of the mounted men…"[27]

The confusion was not yet over for the retreating Confederate Indian Brigade. Capt. Rabb's 2nd Indiana Battery, consisting of four 10-pounder James rifles, and two 6-pounder guns, arrived and began to blast shot and shell into the disorganized and retreating Confederate Indians. Once the train was over the Spavinaw Creek crossing, Buster had to request the tardy Col. D.N. McIntosh and his Creek regiment to help guard the train and cover the retreat since the fleeing troops were raiding the wagons. It became so difficult and dangerous that Buster, fearing that he would be flanked while passing Hildebrand's Mill, sent the brigade's important papers and funds to Fort Gibson by way of the old Military Road instead of on the Tahlequah Road. Buster was able to get the train safely to the vicinity of Tahlequah.[28]

The Union infantry regiments came up too late to participate in the attack and capture of the battery. Capt. Edmund Ross of Company E. 11th Kansas Infantry later wrote:

"…The 2nd Kansas, being cavalry, got ahead of us and attacked the enemy's guns, we started on a double-quick, running the 3 miles in about 25 minutes, just five minutes too late to take part in the charge made on the enemy cannon, but still just in time to prevent them from making a charge to take it back again. They could not stand the sight of our fixed bayonets in heavy column, coming over the hill on a full run and they immediately broke and run like the devil, pell mell over hills and woods…"[29]

1st Lt. McAfee shared an interesting bit of information in his writings. It seems that one of General Blunt's bodyguards decided at the high point of the battle for the artillery to suddenly desert to the defending Confederates. Seeing this take place, the 11th Kansas Infantry reacted to "which our fire was directed to him." It is unclear if this is the same event mentioned at the very beginning of the battle or if this indeed happened twice during the course of the battle. Since the 11th Kansas Infantry was still east of Mayfield when the first event took place, they would not have witnessed the first take place. Yet, the post-war years are always filled with embellished stories from those actually present, or from those who can weave a good tale from rumor.[30]

Stand Watie's Cherokee regiment moved a half mile from their camp toward the battle but stopped and fell back to Spavinaw Creek. Watie reported:

"…In the morning, having received information of the near approach of a considerable body of the enemy, I immediately ordered the different companies (amounting in the aggregate to 500 men) to be in readiness to move, which order was promptly obeyed. While in line I received your order [to repair with my regiment to your headquarters] but having proceeded about half a mile I received intelligence from scouts I had sent out that a detachment of the enemy was flanking our force on the left. Without waiting for orders I counter-marched and took position on the

Tahlequah road to meet this movement…"[31]

In bureaucratic terms, Watie is saying that he probably thought better of it and refused to move forward. And again, he moved back to Spavinaw Creek to guard the crossing, but it is also where he had his milling businesses. Also, it is not clear what Union units Watie's scouts might have identified. It wasn't until late in the battle that any Union troops had been able to penetrate that far, although Blunt indicated that after the Confederate lines were broken, he did send cavalry south in pursuit of the retreating Confederates. 1st Lt. Josiah McAfee, Company I, 11th Kansas Infantry, recalled that it was "…The two Indian regiments, officered by white men, [who] did efficient service in pursuing and capturing the routed enemy…"[32] Watie reported the actions he took:

"… My men were dismounted, leaving their horses hitched some 300 yards in the rear. They here awaited the appearance of the enemy, who in a short time discovered our position and commenced an attack. Most of the cannonading and firing in other parts of the field had at this time ceased, and leaving Maj. Joseph Thompson in charge of the line, I advanced and made a rapid personal reconnaissance of the enemy, whom I discovered now moving with a considerable force of cavalry and infantry (about 3,000 strong) on the left, as before, and advancing on the road to Tahlequah. The regiment had then engaged the enemy with spirit, but I judged it best under the

circumstances to order the men to retire to their horses and fall back to the Spavina."[33]

There are many questions about this small battle that have no satisfactory answer. Why did it take so long for Watie to get his regiment active? He had been notified that a large Union force was heading their way at least three hours before the attack. They were camped only three or so miles from the main Confederate camp at Fort Wayne. Plus, since he had been assigned to lead the Confederate Indian Brigade on its Kansas and Fort Scott mission on that day, Watie and his regiment should have been far more prepared than they were. And what of the 1st Creek Mounted Rifles? Cooper stated that they were camped along Spavinaw Creek as well, and Col. McIntosh was also advised of the attack at the same time as Col. Watie. They appeared even later than Watie's troops, just in time to join the chaotic retreat. Cooper had also reported that Lt. Col. Chilly McIntosh's 1st Creek Battalion had responded but this unit is not mentioned in Cooper's report, although it is mentioned in passing in Lt. Col. Buster's report. Cooper stated only that the Creek Battalion responded promptly and, in conjunction with Maj. Bryan's 1st Cherokee Battalion, had rendered "important service by checking the advance of the Federal cavalry at Spavina [sic] Creek, thus enabling the train to escape."[34] Folsom's 1st Choctaw regiment had been ordered to Fort Wayne as well, but Cooper probably wrote them off as unreliable as demonstrated by their desertion of the supply train at Camp Coffee during the Newtonia campaign. As for Cooper being sick during the battle, one Union soldier from Company A, 2nd Kansas Cavalry, by the name of Vincent Osborne, wrote in his diary that Cooper was "intoxicated" and "managed the battle unskillfully." Osborne also wrote that the Confederate artillery was consistently missing the Union batteries (probably the 12-pounder howitzers with the cavalry) and that when the other Union guns arrived (probably the 1st Kansas Battery and the 2nd Indiana Battery), the Confederates simply ran.[35] Although the fact that Cooper was an alcoholic is clearly established, it is not known how this Union soldier came by this information unless passed to him by prisoners. Even so, Blunt defeated this large Confederate force with two understrength Kansas cavalry regiments supported by four 12-pounder howitzers. Blunt's quick attack also prevented a fairly large-scale invasion (for the Trans-Mississippi) of Kansas and the potential loss of Baxter Springs Depot and Fort Scott.

The Battle of Beattie's Prairie or Old Fort Wayne was a resounding Union victory that resulted in the capture of four pieces of artillery (three iron six-pounders and one 12-pounder mountain howitzer) and part of a large supply train.[36] Capt. Hopkins and his company were awarded the four artillery pieces and later became officially the 3rd Kansas Battery, or the "Trophy Battery." Colonel Cooper and Colonel Watie retreated in confusion. Watie's force retreated down the Tahlequah Road, followed the Long Prairie, and took refuge the first night at the Monrovian Mission (present-day Oaks, Oklahoma). Over the next few days, the Confederate Indian Brigade eventually fell back to Tahlequah, Fort Gibson, Fort Davis, and finally to Scullyville in the Choctaw Nation. Cooper blamed the "disaster" on his superiors who withdrew the four Texas

regiments from his command prior to the battle, and also on his Indian troops whom he felt had let him down by failing to follow orders to gather at Fort Wayne.[37] The Union had five dead and five wounded. Cooper reported six killed, thirty wounded, and twenty-six missing.

The Federals, who had been force-marched for two nights, returned to the Confederate encampments and feasted on the breakfasts that had been cooking on the fire that morning before the attack. They also confiscated camp equipment, food stores, and other items that were essential to the Confederate units. The white Union troops were also shocked to see the level of hatred between Cherokees on either side. Sgt. Eli Gregg of the 10th Kansas Infantry stated that they held a severe grudge between them. He wrote that "They hate the rebel Cherokees with a hatred that can only be appeased by blood. Stand Waitee's [sic] band is particularly obnoxious to them and they kill every one they get a hold of."[38]

General Hindman's response to the loss at Old Fort Wayne was probably the best to simplify what happened:

> "…it is difficult to understand how Colonel Cooper could have been anywhere near Maysville on the day given, and still more so to believe that any sane man could have been surprised in that open country."[39]

After the defeat Hindman asked Col. Emmett MacDonald, who was serving as the Provost Marshal for Missouri, to investigate the actions of the Confederate Indian Brigade and Cooper's leadership. He reported back on October 27, 1862:

"...I am in-formed that the Indian forces were, at the time of the attack, encamped some as from five miles from the battle ground. The Federals making the attack got between the Indian forces and the body attacked. Our troops did but little fighting. Col Buster was in command and his Reg't was supporting the battery. Almost at the very onset the artillerymen abandoned their guns and the Reg't gave back in disorder and confusion. Many of them throwing down their arms and fleeing into the woods..."[40]

MacDonald's report was taken as being unbiased since he was a Missourian, and Buster's unit was the only white Missouri battalion with the Indian Brigade. Cooper wrote to Hindman through Capt. Newton and explained the difficult problems he was encountering with the Indian Brigade:

"...I was at the time of the attack, and for some days previous and am now extremely ill, and can hardly dictate a line... In regard to my making a move up into Kansas this season, I think it doubtful. The Indians are in a destitute condition – barefooted and nearly naked. They feel they have been abandoned by their white brethren – and some regiments are almost demoralized... you need not depend on them..."[41]

An interesting bit of evidence of the severity of the Confederate Indian's retreat was observed approximately six months after the battle when Wagon Boss Peck took a supply train past Fort Wayne on the Old Military Road. He commented:

"...In the vicinity of the Fort Wayne fight, as we passed along the road, we saw plenty of signs of the battle of six months previous; such as dead horses and mules, broken down wagons, pieces of tents and tent poles, camp kettles, broken muskets and old bayonets, belts and cartridge boxes, etc. And such rubbish was strung along the road for several miles beyond the battle ground, showing that the enemy had left there in a hurry and somewhat demoralized..."[42]

On October 25 Brig. Gen. Schofield sent a congratulatory message to Gen. Blunt regarding his victory over Cooper and the Confederate Indian Brigade. Schofield stated that the victory "was greeted by the Missouri division with rousing cheers for General Blunt and the Kansas division. I heartily congratulate you on your success." He also gave preparatory order for an upcoming operation with the approval of General Curtis. The order stated in part:

"...I propose to send your entire division along the border of the Indian country as far south as may be necessary to rid the Territory of Cooper and his band, your Indian regiments being sent to their homes to re-establish themselves and their families there, your other troops being kept where they can be called quickly in the vicinity of Fayetteville, if necessary..."[43]

Chapter 9: Blunt vs. Cooper: The Battle of Old Fort Wayne

Blunt responded quickly on the same date. He wrote to Schofield:

> "...I have information from scouts returned from Indian Territory that Cooper and Stand Watie have fled by way of Fort Gibson across the Arkansas River to Fort Davis—Cooper's old camp of last summer. They retreated in great haste, their advance reaching Fort Gibson, 70 miles distant, in thirty hours after the battle here. It will be impossible for them to remain there long for want of subsistence and forage; they must either retire into Texas or go down the valley of the Arkansas in the direction of Fort Smith..."[44]

Regarding the Indian Territory, squabbles among the Union commanders in Missouri made this mission very difficult to accomplish since Blunt was the only Union general to believe that the Indian Territory was worthy of attention.[45] He also stressed the importance of the occupation of the Indian country as far south as the Arkansas River, and further towards the Canadian River and the eventual capture of Fort Smith. Blunt firmly believed that the Confederates Indians were alarmed at the advance of the Union army. He reasoned that since the rout of Cooper at Fort Wayne, many of the Indians in the Confederate Army had more than likely discarded their firearms and had gone home with the intention of either joining the northern forces or just staying out of the fight. Blunt's reported purpose in this action had three parts. First, he hoped to send the three Indian regiments to fulfill their original purpose of occupying this section of Indian Territory to secure it from reoccupation by Confederate forces. They were also to protect the inhabitants from raids by their inter-tribal adversaries and bushwhackers. Second, the 3rd Indian Home Guard was made up largely of Cherokees who understandably wanted to remain near their homes. The final reason was to reoccupy the abandoned salt works in the Indian Territory. Blunt was extremely disappointed in Schofield's response to allowing the First Division to pursue Cooper's retreating forces deeper into Indian Territory. Blunt recalled:

> "...I now urged Schofield to permit me to move forward with my division, but instead of obtaining such permission, I received an order 'to fall back to the vicinity of Pea Ridge, to be within supporting distance of the other two divisions.' Where the danger was, to the second and third divisions, requiring this support, I have never yet been able to learn..."[46]

After the Battle of Old Fort Wayne, there was little danger from the Confederate Indian Brigade threatening the right flank of the Army of the Frontier. Because of this, Gen. Schofield decided he did not want to waste an entire division of his army chasing Cooper and his Indian Brigade around the Indian Territory. It would take Blunt until the Summer of 1863 before he could bring the entire First Division into the Indian territory.

The Union Indian Home Guard regiments were still dealing with severe desertion problems. On November 5, Maj. Ellithorpe, who was commanding the 1st Indian Home Guard regiment, wrote to his brigade commander, Col. William Cloud of the Third Brigade, First Division, that he had been informed that two companies of Creek soldiers belonging to his regiment were at Leroy, Kansas, and absent without leave. He wanted to communicate with the Indian Superintendent William Coffin to request that he entice Opothleyahola to convince these troops to return to their regiment. Apparently, Opothleyahola had been influencing Creek Indian soldiers to disobey the commands of the Army and in doing so was providing aid and comfort to the enemy. Ellithorpe wrote to Coffin and informed him that his desire was that every man who had enlisted as a soldier in the Indian Home Guard to return to his regiment via Fort Scott. He also stated that he would try to get them back pay due to the bad counsel they had received from the "old Chief." To complicate matters, Lt. Col. Wattles, commander of the 1st Indian Home Guard, managed to abscond with four thousand dollars belonging to the regiment's troops and departed on a leave of absence before the theft was discovered. Ellithorpe swore out an arrest warrant, and Blunt immediately sent troops after the wayward regimental commander. [47]

In the aftermath of the Fort Wayne battle, Col. William Phillips, with the best mounted men of the 1st and 3rd Indian Home Guard regiments, left camp on November 6 to reconnoiter the area around Fort Gibson, Park Hill, and Tahlequah. Since it had been reported that Stand Watie had crossed to the north side of the Arkansas River with six hundred men, Blunt wanted to know for sure. Phillips arrived at Fort Gibson on November 9 and reported back that Cooper and Watie were still south of the Arkansas River. Although Phillips's primary mission was to

determine if any of Cooper's force were north of the Arkansas River, he was to use part of his command to process some salt for the Union Army. As mentioned earlier, most of the best salt works were located in the Cherokee Nation, especially the Mackey's Salt Works, a large, well-established operation started by Samuel Mackey in the early 1820s. It consisted of over one hundred salt kettles and was located on the Illinois River near the Fort Gibson-Fort Smith Military Road. As stressed before, salt was desperately needed by both sides, and Blunt believed that he could produce enough to supply the Army of the Frontier as well as Union supporters in the region. [48]

Meanwhile, Blunt sent detachments out to the Jones and Hildebrand mills to process wheat into flour for his command. This was a common activity for armies serving in the Trans-Mississippi West. Many times the campaigns in the region were fought simply to gain access to the established grist mills as well as any salt works that could be activated. A "special correspondent" wrote of the activities of the First Division during this period after the Fort Wayne battle while working at Brown's Mill, Arkansas:

"…The Kansas troops form the First Division of the Army of the Frontier, and occupy the extreme right. It is under command of Gen. Blunt. We are now camped about eight miles south of Bentonville, on Prairie Creek. A detachment of three companies of the Second and two companies of the Eleventh are running Brown's Mill, about ten miles further south. Detachments from the First and Second Brigades, are running mills at other points near this. So far we have had no difficulty in getting wheat, but still the country has been pretty well foraged, and a large army would find it impossible to subsist here any length of time…"[49]

As we have already discussed in the previous chapter, grist mills and salt works were very important for an army, which is why the Confederates held on to Newtonia, with its steam-powered mill, so desperately. Many of the actions taking place along the Indian Territory and Arkansas boundaries in the next few weeks would center around these grist mills and salt works. Grist mills were buildings in which grain was ground into flour. In some areas they are also known as corn or flour mills. The mills served as a valuable resource for both armies as they could produce flour and then other food locally instead of transporting the supplies by wagon. Their ability to produce food locally also made them a prime target for both the military and any bushwhackers passing through the area. The destruction of the mill would prevent its use by enemy soldiers. An example would be Rhea's Mill in Arkansas, located approximately five miles northwest of Prairie Grove. During the Civil War, William Rhea, a prosperous merchant and miller, whose mill was taken over by both Confederate and Union troops–control of the mill changed hands seven times during the war. Rhea was decidedly pro-Confederate as he charged Confederates two and a half cents per pound for flour, while Union soldiers were charged twice that. A correspondent with the Chicago Evening Journal who was accompanying the First Division wrote:

"…I am sure you will approve of our industry. We are in the vicinity of four flouring mills, with large quantities of wheat and corn in the vicinity. We have taken possession of the mills and gone into the milling business. We are now turning out about twenty thousand pounds of flour and meal daily…"[50]

This gave an incentive to both the Union and Confederate commands to keep these natural resources within their control and to prevent the enemy from possessing them. As for wheat and corn for milling, it was a challenging year for harvesting due to the severe drought. This made wheat and corn for milling to wheat flour and cornmeal a scarce and valuable commodity. All armies depended on bread as an absolute staple of a soldier's diet. The Union Army paid Union-supporting farmers for their wheat. Confederate supporters simply had their farm goods confiscated. At the start of the war, both followed the same ration guide that existed before secession. These rations allotted just over a pound of meat, likely beef or pork, just under a pound of "hard bread," and a small collection of dried vegetables. Any meat products required salt for preservation and taste. The most common form of hard bread was called "hard tack," a basic wheat biscuit that did not easily decay and could survive a rough march. It was extremely hard, and was often soaked in water, coffee, or in meat fat to soften it enough to eat. Other items, such as beans, peas, rice, coffee, sugar, or salt, were also issued, but not on a daily basis. Many of these items could be grown locally such as wheat, corn, and barley. But beans, rice, coffee, and sugar were only grown elsewhere

Chimney Stack from Rhea's Mill at Prairie Grove State Battlefield Park
Photo by Author

Original Millstone from Rhea's Mill at the foot of the Chimney Stack
Photo by Author

and needed to be transported in by whatever logistics system worked for the particular army.

Maintaining an army's subsistence needs was a paramount concern.[51] Due to the lack of major rivers and railheads in the Southwest Missouri, Northwest Arkansas, and the Indian Territory, these goods needed to travel by wagon across extremely rough roads that crossed mountains, rivers, and broken terrain. Railroads had not reached closer to the Indian Territory or the other areas of operation than Rolla, Missouri. During the war both the Union and Confederate armies experienced the same difficulties with supply logistics and distribution as merchants did pre-war. In the Ozarks and in the Indian Territory, goods were transported by wagons to supply soldiers with food, clothing and arms. As soldiers progressed further into the countryside the subsistence and equipment supply trains had a difficult time keeping pace, especially with mounted units. Soldiers often

needed to forage the countryside for food because the troops had outpaced their supply line. Farms and gristmills were prime targets for soldiers and bushwhackers as they often stored large quantities of food.[52] Many times commanders would not destroy a mill or salt work, even if they were to lose it, based on the experience that they would more than likely need it in the future.

Unfortunately for Blunt and the First Division of the Army of the Frontier, events across the Arkansas state line were going to dictate their activities for the next month or so, as Hindman was rebuilding his newly conscripted forces into the new "Trans-Mississippi Army." With this newly constituted force, the Confederates were planning on driving the Union Army out of Arkansas and Southwest Missouri. Phillips was recalled from the Indian Territory again.

---

[1] Wilson, Terry Paul, "Delegates of the Five Civilized Tribes to the Confederate Congress;" *Chronicles of Oklahoma*, Volume 53, Number 3, Fall 1975. pp. 353-366.
[2] Carruth to Coffin, 1862 Report, pp. 166.

[3] Able, Annie H. *Participant*. pp. 206-209.

[4] Kemp, Francis Elizabeth. Interview, 1938. Indian-Pioneer Papers, Western History Collections, University of Oklahoma, Norman, Oklahoma.

[5] Bear, Malucy. Interview, October 25, 1937. Indian-Pioneer Papers, Western History Collections, University of Oklahoma, Norman, Oklahoma.

[6] Curtis to Halleck, *OR*, Ser. I, Vol. XIII, pp. 729-730.

[7] Cooper to Newton, *OR*, Ser. I, Vol. XIII, pp. 332.

[8] Cooper to Hindman, *OR*, Ser. I, Vol. XIII, pp. 332-335.

[9] Randolph to Holmes, *OR*, Ser. I, Vol. XIII, pp. 906.

[10] Warner, Ezra. --Generals in Gray: Lives of Confederate Commanders, Baton Rouge and London: Louisiana State University Press. 1987. pp. 257-258.

[11] Able, Annie H. *Participant*. pp. 199-200.

[12] Britton, Wiley, *Border*, Vol. 1. pp. 366.

[13] Blunt, *Experiences*, pp. 227-228.

[14] Schofield, General Order #6, *OR*, Ser. I, Vol. XIII, pp. 754-755.

[15] Blunt to Schofield, *OR*, Ser. I, Vol. XIII, pp. 325-326.

[16] Britton, Wiley, *Border*, Vol. 1. pp. 370-372.

[17] Collins, Robert. General James G. Blunt: Tarnished Glory, Gretna, Louisiana: Pelican Publishing Company, 2005. pp. 80-82.

[18] Cooper to Hindman, *OR*, Ser I, Vol. XIII, pp. 334.

[19] Cooper to Hindman, *OR*, Ser I, Vol. XIII, pp. 335.

[20] Buster to Cooper, Supplement to the Official records of the Union and Confederate Armies. Jane Hewitt, ed. Wilmington: Broadfoot Pub. Co. 2001. Vol. 3, Pt. 1. pp. 65-67.

[21] Blunt to Schofield, *OR*, Ser I, Vol. XIII, pp. 327.

[22] Bassett to Hill, *OR*, Ser I, Vol. XIII, pp. 330.

[23] Blunt to Schofield, *OR*, Ser I, Vol. XIII, pp. 328.

[24] Crawford, Samuel J. *Kansas in the Sixties* (Chicago: A.C. McClurg and Company, 1911), 59-61.

[25] Ibid.

[26] Edwards, Whit. The Prairie was on Fire: Eyewitness Accounts of the Civil War in Indian Territory. Oklahoma City: Oklahoma Historical Society, 2001. pp. 31.

[27] Ibid

[28] Buster to Cooper; *Supplement OR*, Vol. 3, Pt. 1: 59-62.

[29] Ross, Edmund. Letter, Edmund G. Ross Collection, n.d. Kansas State Historical Society: Topeka

[30] *Leavenworth Daily Conservative,* November 13, 1862. Newspapers.com

[31] Watie to Cooper, *OR*, Ser I, Vol. XIII, pp. 336-337.

[32] Edwards, Whit. pp. 33.

[33] Watie to Cooper, *OR*, Ser I, Vol. XIII, pp. 336-337.

[34] Cooper to Hindman, *OR*, Ser I, Vol. XIII, pp. 335.

[35] Collins, Robert. pp. 82.

[36] Blunt to Schofield, *OR*, Ser I, Vol. XIII, pp. 325-328.

[37] Cooper, Douglas H. Report, October 25, 1862, Ibid: 331-332.

[38] Eli H. Gregg. *Muscatine Daily Journal*, November 18, 1862; Shea, William L. Fields of Blood: The Prairie Grove Campaign, Chapel Hill: The University of North Carolina Press, 2009. pp. 42.

[39] Hindman to Holmes. *Supplement OR*, Vol. 3, Pt. 1: pp. 63.

[40] MacDonald to Hindman October 27, 1862. Peter Wellington Alexander Papers, Box 1, Rare Book & Manuscript Library at Columbia University in the City of New York. Missouri Digital Heritage Website.

[41] Cooper to Newton, October 28, 1862. Peter Wellington Alexander Papers, Box 1, Rare Book & Manuscript Library at Columbia University in the City of New York. Missouri Digital Heritage Website.

[42] Peck, R.M. *Wagon Boss*

[43] Schofield to Blunt, *OR*, Ser. I, Vol. XIII, pp. 763-764.

[44] Blunt to Schofield, *OR*, Ser. I, Vol. XIII, pp. 765.

[45] Ware, James W. "Indian Territory", *The Western Territories in the Civil War*. Fischer, LeRoy H. ed. Manhattan, Kansas: Sunflower University Press. 1977: 107.

[46] Blunt, *Experiences*, pp. 228.

[47] Johansson, M. Jane, *Ellithorpe*. pp. 46-50.

[48] Wyant, Sharon Dixon. "Colonel William A. Phillips and the Civil War in Indian Territory," unpublished Master of Arts Thesis, Oklahoma State University, 1964. pp. 19.

[49] *The Leavenworth Daily Times*. November 15, 1862. Newspapers.com

[50] Johansson, *Ellithorpe*, pp. 80. *The Chicago Evening Journal*: Published November 25, 1862.

[51] State of Florida, Division of Historic Resources Website: https://dos.fl.gov/historical

[52] Rhea's Mill Ledger, The Impact of the Civil War in the Ozarks, ozarkcivilwar.org; Dorothy Johnston, Ph.D, "History of Rhea Community", in *Flashback*, Vol. 37, No. 3, 1987, pg 18.

Two unidentified members of the Union Indian Home Guard Regiments
Photos courtesy of the Library of Congress

Clement Vann   **D.N. McIntosh**
Officers of the Creek Mounted Rifles Regiments

A photograph of a younger
Colonel Stand Watie

Chapter 10
The Arkansas Border War – Phase 1
The Battle of Cane Hill

As illustrated in the previous chapters, the Indian Territory was never a high priority for either the Union or Confederate governments. In the Trans-Mississippi West, the United States was concerned about protecting Kansas, and the Confederate States wanted to protect Arkansas and Texas. Both wanted to protect and occupy Missouri due to its rich natural resources and large population. Arkansas was in greater peril than Kansas, although the bushwhackers and Partisan Rangers from Missouri were a continuous threat to Kansans and the relocated Union Indians. The governor and legislature of Arkansas were continually asking the Confederate government for more troops and better protection from the Federal armies. At one point Governor Henry Rector told Richmond that if he did not get the protection he needed from the Confederacy, he was going to hold a convention to secede Arkansas from the Confederate States. The Union Army under Maj. Gen. Samuel Curtis had the Army of the Frontier operating out of Springfield, Missouri, which was mostly in control of Northwest Arkansas. There was also a large army under Brig. Gen. Fredrick Steele that was operating out of Helena, Arkansas, and this force was continuing to threaten Little Rock, the state capital. Maj, General Theophilus Holmes was the commander of the District of Arkansas of the Confederate Trans-Mississippi Department. He was being pressured by both the Confederate government in Richmond and

Kidd's Mill, Cane Hill
General Marmaduke's Headquarters
Photo by Author

the state government in Little Rock to take actions to prevent the Union from capturing the capital. Holmes, as he usually did, recalled Brig. Gen. Thomas Hindman to Little Rock from his headquarters at Fort Smith to assist in planning the defense of Arkansas's capital city. Hindman did not believe that the greatest Union threat was from Helena but from the northwest where he was attempting to regain the territories lost after the losses at Pea Ridge, Newtonia, and Old Fort Wayne. Finally on October 15, General Hindman returned to Fort Smith after having spent the previous month in Little Rock helping to organize a defense of the capital. Unfortunately, he spent most of his time trying to convince General Holmes to let him return to his command in Northwest Arkansas. When he arrived, he was angry that his troops had, for the most part, abandoned Southwest Missouri, and the "rich granaries of Missouri," as he called them, along with its lead mines, and the rich food producing areas of Northwest Arkansas to the Union. Hindman later forced Gen. Rains to resign or face court-martial. Several other ranking officers were demoted or dismissed from the Confederate Army. Hindman soon established a large camp on Massard Prairie, south of Fort Smith, as a gathering and training area.[1] He was planning on making full use of the new Confederate Conscription Act of 1862 to build a new army using conscripts. Unfortunately, the new law allowed conscripts to be drafted only into existing state regiments. This created a legal problem since, when General Van Dorn left the state, he took nearly all of the existing Arkansas state regiments with him. Hindman, with Holmes's permission, bypassed the law and began drafting men into new Arkansas regiments. This did not sit well with the Confederate government. Here is what the Hon. Henry S. Foote of the Confederate Congress had to say about General Hindman in his book, *Casket of Reminiscences*:

> "…Perhaps the most cruel and atrocious conduct perpetrated by any of President Davis's military servitors during the war was that practice by his especial favorite, General Hindman, in the State of Arkansas he said, that the very comprehensive provision of the conscription law were not quite comprehensive enough to suit his purposes, deliberately amplified them by proclamation; declared martial law throughout Arkansas and the northern portion of Texas, and demanded the services of all whom he had thus lawlessly embraced in his wide-sweeping conscription list. All who refused to obey his mandate, as

he in terms confesses, were apprehended, subjected to trial by military court, appointed by Hindman himself… he had them all executed…"[2]

To encourage volunteers who wished to avoid conscription, Hindman announced that volunteer companies raised before a certain date would be allowed to elect their own officers, as was the custom, while conscript companies would have their officers appointed. Through rigorous enforcement of new Confederate conscription laws, Hindman was able to raise a new army in Arkansas in the summer of 1862. Maj. Ellithorpe of the 1st Indian Home Guard (IHG), acting as an anonymous correspondent for the *Chicago Evening Journal*, remarked in an article:

> "The fearful ravages of the rebel conscript law are everywhere more apparent. Fathers, husbands and brothers, who would have remained quietly at home, have been forced into the rebel ranks – their families left to battle with want and the inevitable abuses tha[t] follow in the wake of an army. It is hard to witness the fears of those who are not responsible for this great calamity…"[3]

On the Federal side it was now obvious that General Schofield had little stomach for aggressive action. He failed to bring the Second and Third Divisions close to the Confederates at Huntsville or to bring on a decisive battle with Hindman, beyond a small strike on their rear guard on the morning of October 28. This rear guard was commanded by Col. Thomas Bass and his

Texas Brigade. Neither Bass nor Hindman were able to determine where the Union forces were heading or their size. Bass had been struck by the advance of the Third Division under Brig. Gen. Herron, consisting of the 1st Iowa Cavalry and the 7th Missouri State Militia (MSM) Cavalry, at Cross Hollows but was soon overwhelmed by the size of the Union advance. He also lost all of his camp equipage and a few wagons. Hindman initially believed that Bass was simply not aggressive enough and sent Col. William Bradfute to take command of the rear-guard action and "resist for as long as possible." They also received a report that the Union forces were moving towards both Huntsville and Carrollton roads. Hindman ordered Col. Shelby to divide his brigade and advance to both locations to support Bradfute and Bass as well as to hold the line until the Confederates could get over the mountains. Meanwhile, Hindman began to order the remainder of their forces southward towards Ozark and Clarksville. Although Hindman had close to ten thousand ranks, at least four thousand, almost half of his total command, were unarmed. Hindman sent numerous requests for arms back across the Mississippi River. Many weapons were transferred to the Trans-Mississippi District from Vicksburg in what became known as the "Fairplay Affair." A shipment of 18,000 arms were dispatched to Pine Bluff, Arkansas, from Vicksburg, Mississippi, by way of Monroe, Louisiana, but 5,000 of those 18,000 were captured on the steamer *Fair Play* by Union forces, and 2,500 arms were redirected to Major General Richard Taylor's army in Louisiana. Only 11,000 arms made it to Pine Bluff. These weapons had come from the arsenal of eastern Confederate states that had been returned to the state arsenals as the

Confederates in the Eastern and Western Theaters had re-equipped themselves with the better captured Union arms. Most of the guns were castoffs and unusable weapons from the various state armories which had been returned to the armories after the Confederate armies east of the Mississippi had been re-equipped from the "Battlefield Quarter-master" of Seven Days Battles, the Battle of Second Manassas and the capture of Harper's Ferry.[4] Hindman continually pleaded with General Holmes for supplies and arms since he believed he would not have access to forage or subsistence if he had to sustain his troops in the mountains. Gen. Schofield, for his part, refused to follow the Confederate army any farther since he only had five days' rations available.[5]

It was during this time that another sad event took place in late October 1862 in the far western portion of the Indian Territory. The Confederate Indian Department had continued operating the Wichita Agency near Fort Cobb. From here Agent Matthew Leeper maintained his control over the 300-390 Confederate-supporting Tonkawa Indians. They had antagonized the other Plains Indians in Texas by supporting and supplying scouts for the Texas Rangers in their campaigns against these other tribes, especially the Comanches. They had been moved up from Texas in 1859 for their own protection. Using this weakness and lack of Confederate forces, and coupled with the distraction of the war, a combined group of warriors, made up mostly of Union-leaning Caddos and Osages, who were believed to be supported by Shawnee, Delaware, Comanche, Kickapoo, Kiowa, Wichita and Seminole members decided to strike. This renegade group attacked the Agency on October 23 and

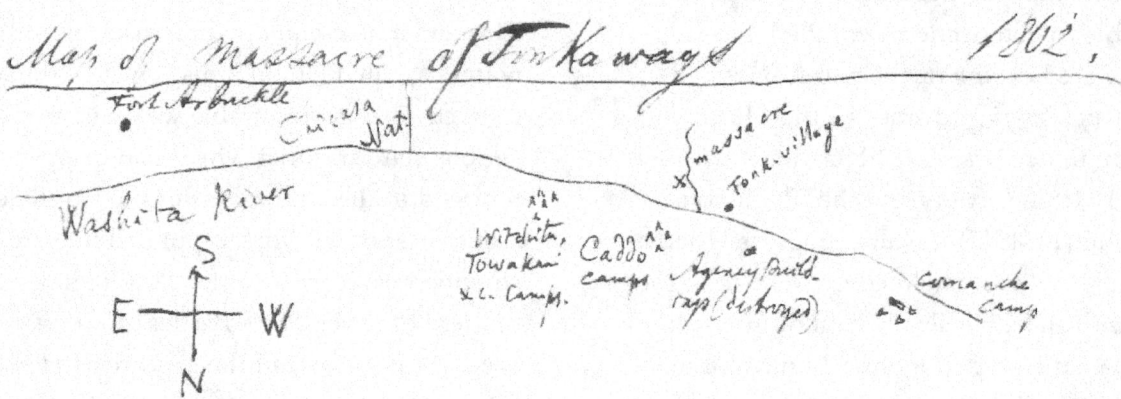

MAP 2.—Albert Gatschet's sketch map of the location of the Tonkawa Massacre in 1862. (Smithsonian Office of Anthropology Archives.

killed Agent Leeper, who was hated by most of the region's tribes, as well as four other agency employees. They also burned all of the Agency's buildings. The Tonkawa's, seeing the danger they were in, fled the area towards Fort Arbuckle where the Confederate Chickasaw Battalion was posted. But on October 24 Kiowa and Comanche scouts discovered the Tonkawa's camp and notified the other renegade warriors. This force caught up with the Tonkawa's on the south side of the Washita River, downriver from Fort Cobb. In what became known as the "Tonkawa Massacre," the estimates of the Tonkawa's killed were 137-240 men, women and children, among them Chief Placido. The remainder of the Tonkawa's drifted into Fort Arbuckle over the next few days. Today, historians differ on the reasons for the near massacre of the Tonkawa tribe. Some believe it was due to rumors of the Tonkawa's participating in cannibalism, of which there is very little evidence of ever taking place. Others state that there was no clear reason for the attack beyond their feelings toward Agent Leeper. It's also possible that they were attacking them due to their support of the Confederacy, although the Plains Indians

didn't really have strong feelings for either side, unlike the Five Civilized Tribes in the eastern Indian Territory. The most plausible reason was as stated earlier: the Tonkawa's providing aid to the Texas Rangers for their campaigns against the Plains Indians created an animosity that could be avenged under the ravages of the Civil War.[6]

Back in Arkansas, following orders, Brig. Gen. James Blunt brought his division back towards Bentonville and went into camp (Camp Ewing) four miles south of that town, twelve miles in advance of Schofield's headquarters, until November 10. By this time Schofield had pulled the Second and Third Divisions back to Springfield, Missouri. He failed to send any orders or instructions to Blunt during this period. Blunt stated:

"…Here I remained until about the 10th of November, and receiving no instructions from Schofield, but learning unofficially that he had abandoned the country, and with the second and third divisions moved back towards Springfield, the question naturally arose in my mind, what I should do. Not yet having had

272

much experience in military affairs, I did not know but that it was a part of West Point tactics for a superior officer to abandon his subordinate, and leave him in the face of the enemy, with an inferior force, without any order or instructions, but I was not well enough versed in the science of war to appreciate the 'strategy' of such a movement. I was now well convinced that I had been abandoned to my fate, and must act upon my own responsibility…"[7]

In Schofield's defense, he had been ordered by General Curtis to bring his troops back towards Springfield in case they were needed to deploy to Southeastern Missouri to counter any moves against the garrison at Pilot Knob or Gen. Fredrick Steel's command in Helena, Arkansas. There were also reports of large Confederate concentrations near Yellville, Arkansas.[8] Nothing in the Official Records indicates that this information was passed on to Blunt, so he was correct in being in the dark about decisions being made without his knowledge. Schofield did instruct Blunt on October 27 to move from Fort Wayne to a location along the Arkansas state line directly west of Fayetteville. He also informed Blunt that a new regiment, the 13th Kansas Infantry, was enroute to his location with twenty-five wagons of rations and that a main supply train from Fort Scott was due in two days.[9] Regardless, Blunt failed to move. He informed Schofield that the road from the Fort Wayne and Maysville area to Fayetteville was a much better road than the "neighborhood road" that would be available if he moved south. But by November 10, the forage in the area around Maysville was

becoming scarce, and Blunt moved his division south along the old military road to its crossing of Flint Creek and encamped on Lindsley's Prairie. From this point he sent out scouts and spies to Van Buren and Fort Smith to get a reading on Confederate intentions, actions, and strengths.[10] Maj. Ellithorpe described the campsite of the 1st Indian Home Guard as "a fine little bottom & well supplied with corn standing in the field of good quality. My camp is in a turnip patch & we find this esculent a great rarity." This camp was designated "Camp Babcock."[11]

On November 2, 1862, General Curtis issued General Orders #11, which divided up the Department of the Missouri into twelve districts "for convenience of police regulations." Blunt was placed in command of the 8th District of Western Arkansas; the 9th District of Indian Territory; and the 10th District of Kansas. Commanders were authorized to cross district boundaries with their troops and to cooperate with the commanders of the various districts. The order did not address the three-division organization of the Army of the Frontier, although it stated that all communications from within the various districts to the department commander would be made through the district commander. Apparently, it did not affect the structure of Schofield's Army of the Frontier since later communications still used the divisional organization titles.[12]

It was during this period Blunt received information that must have had him shaking his head:

"…I received a copy of the St. Louis Democrat containing a letter from Schofield's 'army correspondent,' and dated at his (Schofield's) headquarters, saying that 'the Army of the Frontier' had fulfilled its mission, and had gone into winter quarters near Springfield, and that General Schofield was about to leave for St. Louis to recruit his health, which had been shattered by long and arduous duties in the field… This newspaper letter afforded me the only information as to the whereabouts of the second and third divisions that I had been able to obtain since -- in compliance with Schofield's order -- I had moved from Maysville back to the vicinity of Pea Ridge, to 'support him."[13]

While camped on Lindsley's Prairie, or Camp Babcock, Blunt's spies and scouts returned from Fort Smith and Van Buren. They informed him that Gen. Hindman was at Van Buren and had Gen. Mosby Parson's Missouri troops moving to that location from Yellville, which by chance had ended the threat to Springfield from that location. Hindman planned on gathering all of the Confederate troops available at Van Buren for a final thrust into Southwest Missouri. The Confederate commander also had General John S. Marmaduke, a Missouri West Pointer, to take his cavalry division across the Boston Mountains, then swing to the northwest and take position at Cane Hill and Rhea's Mill. This was an excellent place for the Confederates to gather provisions. Wiley Britton noted:

"…The region around Cane Hill and Rhea's Mills was the best agricultural part of Washington county, besides nearly every family had an orchard

that produced an abundance of fine apples. The season had been especially favorable for their maturing and ripening, and they were delicious. The people were quite willing to exchange them with the Union soldiers for sugar, coffee and tea, which the citizens had difficulty in getting in that section…"[14]

In addition to their potential offensive actions, they were also to operate the mills to provide provisions for the Confederate Army. General Hindman believed that with the two other divisions of the Army of the Frontier in Missouri, this was a great opportunity to attack and destroy the First Division with overwhelming strength. Blunt also received information that Marmaduke planned to attack the First Division since the other two divisions had been withdrawn. Blunt had intended to attack Marmaduke's force on November 17 at Cane Hill, but a reconnaissance by the 6th Kansas Cavalry under Col. Jewell discovered that the Confederate cavalry commander had withdrawn his forces from Cane Hill on November 14, retrograding back across the Boston Mountains. Jewell was able to transit the Cove Creek Road through Cane Hill southward until they ran into Confederate pickets near the road's junction with Lee's Creek. After a short skirmish Jewell's troops captured a number of Confederate prisoners.

It was believed by most of the leadership of the First Division that the Confederates had amassed an army of twenty thousand by combining the commands of Hindman, Cooper, Marmaduke, and Parsons, with its first primary mission to be the

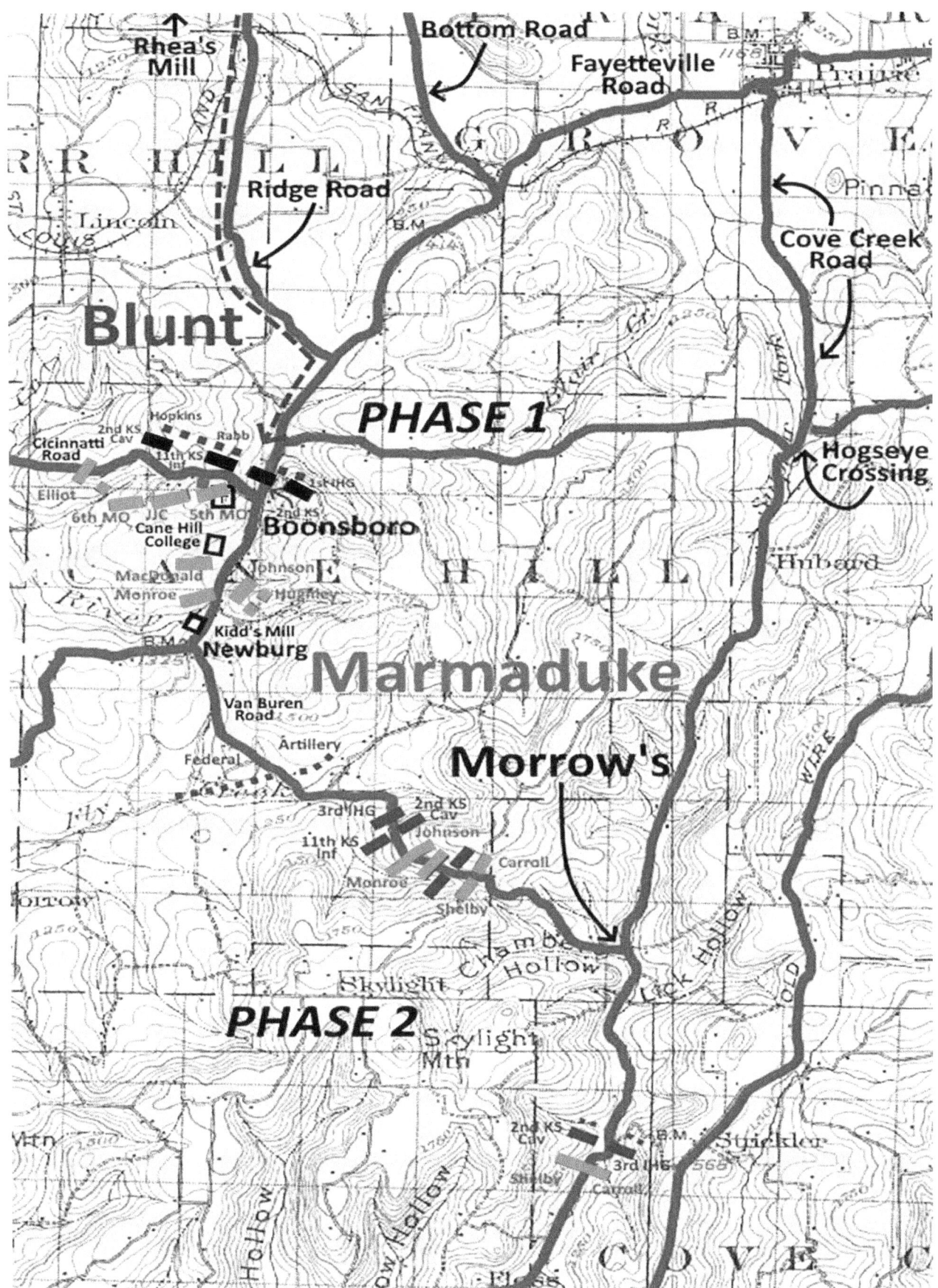

destruction of the First Division. The second mission was to drive the Union Army out of Northwest Arkansas and Southwest Missouri. The First Division had approximately eight thousand troops and consisted of the three brigades along with four artillery batteries and several 12-pounder mountain howitzers. Another reconnaissance by Jewell's 6th Kansas Cavalry found that Marmaduke was assembling in large numbers at Cane Hill, the advance of Hindman's army. Blunt wrote:

"…On the 26th of November, I learned that Marmaduke had again ordered Companies B and I, 2nd Kansas Cavalry, under Capt. Hopkins, to move around the town and cut off and capture the Confederate pickets manning the outpost. Unfortunately, the Confederate pickets had already heard the Union Army's advance and had awakened the townspeople and then sent runners back toward the camps near Fort Wayne. At this time, nearly 5 a.m., Blunt discovered that only one other company of the 2nd Kansas Cavalry advanced to Cane Hill with eight thousand mounted men, and eight pieces of artillery, and that Hindman, with over twenty thousand infantry and artillery, then on the south side of the mountains, would join him by the 30th, when they intended moving against me in force, and crush me before I could receive assistance…"[15]

Blunt decided that he would attack Marmaduke at Cane Hill before Marmaduke could be reinforced by Hindman's column coming up from Van Buren. He also recalled Phillips and the Indian detachment back from the Indian Territory to assist in the upcoming offensive operations. Blunt delayed his strike at Marmaduke for twenty-four hours in order to insure the safe arrival of a train of two hundred wagons from his supply depot at Fort Scott, Kansas. The immense commissary train, two miles in length, rumbled into the Union camp at sunset on November 26, much to everyone's relief.[16] Blunt was determined to prevent the joining of the two Confederate columns occurring on the evening of November 28, as his intelligence sources had indicated. Blunt ordered his supply and subsistence trains to Camp Babcock. He designated the 9th Wisconsin Infantry to provide security for the camp and trains under the direction of Gen. Salomon. On the morning of November 27, the Union Army of five thousand troops with thirty pieces of artillery, packing four days of cooked rations, and forty rounds of ammunition for each soldier, began their march towards the Confederates who were approximately thirty-five miles south of the Union camp. As was his custom, Blunt rode at the head of the column in a carriage accompanied by a small escort and preceded by a bevy of scouts. Instead of continuing a few miles farther to the junction with Cincinnati Road, the usual route to Cane Hill, Blunt turned east on a primitive track that crossed the southern end of Wedington Mountain. "Road quite rough and country broken," observed an Ohio artilleryman. After another five miles, the Federals halted for the night on Moore's Creek a short distance west of Rhea's Mill. Blunt's command had covered almost twenty-five miles that day. This left another seven to ten miles over mountainous terrain left to traverse in order to strike the

Confederates at Cane Hill. The First Division reached Rhea's Mill at sunrise and turned south on an "obscure and unfrequented" lane called Ridge Road. A short time later the Federals encountered the "ridge" that gives the road its name, but which is actually the northern escarpment of Cane Hill. Progress slowed to a crawl as men and animals struggled up the steep incline. Cane Hill is not a hill at all but a broad, rolling plateau fronted by steep escarpments to the east and north. An army entering the Cane Hill area via the Fayetteville Road from the north would encounter a narrow defile with high ground on either side, perfect for a defensive position.[17] The Fayetteville Road traversed Cane Hill southward through the narrow valley cut by Jordan Creek. It was a much-broken country cut by deep ravines, and very steep hillsides. In addition, the banks of Jordan Creek were extremely steep, all of this steep terrain made crossing very difficult. Blunt's army would need to squeeze through this area to strike at the Confederates.

Confederate Gen. Marmaduke had set up his headquarters at Kidd's Mill, on the road between Boonsboro and Newburg (present day Canehill and Clyde). This mill was originally known as Truesdale's Mill, but during the war it was owned by Tandy Kidd, and known to both armies as Kidd's Mill. Marmaduke ran this mill and four others in the area in an effort to build up a substantial supply of provisions to last the Confederate Army through the winter. He initially planned an attack on the Union positions at Elkhorn Tavern since the hated 1st Arkansas Cavalry (USA), that consisted of Union-supporting Arkansans called "Mountain Feds," was stationed there. His scouts informed him early on the morning of November 28 that Blunt's

forces were advancing up the Line Road, so the Elkhorn Tavern movement was cancelled. He ordered his supply wagons to safety by crossing the mountains and moving south along the Cove Creek Road towards Lee Creek crossing. Marmaduke decided to fight a delaying action to buy time for the wagons and for Hindman's main force to come up from the south. He was faced with many disadvantages and would not be able to completely stop Blunt's advance with the force he currently had. Although Blunt's intelligence stated that Marmaduke had a force of five thousand, this advance Confederate force only had two thousand effectives (soldiers fully equipped and armed). While Blunt's men had plentiful ammunition, Marmaduke's Confederates were short on both ammunition and percussion caps. Marmaduke only had six cannons to Blunt's sixteen. Furthermore, most of the Confederate cannons were obsolete and nearing the end of their effective life. In addition, the Union cavalry's horses were in better condition than those of the Confederates, and the Confederates were inadequately uniformed, a problem because the weather had turned cold. Many Confederates were without warm clothing, and many were barefoot.[18] When Marmaduke received the news of the Federals approach down the Line or Old Military Road, he assumed that Blunt would strike the Confederates coming up the Cincinnati Road, the main route between the two towns. The Confederate commander decided to deploy his small command across the Cincinnati Road with only pickets out on the Fayetteville Road.

Marmaduke placed Col. Jo Shelby and his brigade in command of this sector of the

Maj, General Theophilus Holmes
Trans-Mississippi Department
Photo courtesy of the Library of Congress

Brig. Gen. John Schofield, USA
Army of the Frontier
Photo courtesy of the Library of Congress

Colonel John S. Marmaduke
Division Commander
Photo courtesy of the Library of Congress

Corner where Blunt waited in his carriage at start of the battle. *Photo by Author*

Union Approach to Confederate Defenses at Cane Hill. *Photo by Author*

Initial Position of Bledsoe's Missouri Battery in Cane Hill Cemetery. *Photo by Author*

Location of Rabb's Battery in the Cane Hill Cemetery. *Photo by Author*

Van Buren Road Junction with Summit of Reed's Mountain. *Photo by Author*

Junction of the Van Buren and Cove Creek Roads at Morrow's *Photo by Author*

Looking South on Cove Creek Road. 6th Kansas Charged Confederate Rear Guard. *Photo by Author*

Probable location of Confederate Ambush of 6th Kansas Cavalry. *Photo by Author*

Post-War Photograph of Rhea's Mill, Arkansas
Photo courtesy of The Shiloh Museum of Ozark History / Washington County Historical Society Collection

Cane Hill College
Photo by Author

line. Shelby in turn assigned Major Benjamin Elliott's small Missouri battalion in advance on the Cincinnati Road. Shelby's main battle line rested with his right flank on the Booneville Cemetery and his left on a high piece of ground about a half mile to the west. He placed Col. Gideon Thompson's 6th Missouri Cavalry (CSA) on the left flank, Col. Beal Jean's Jackson County Cavalry in the center, and Lt. Col. B. Frank Gordon's 5th Missouri Cavalry (CSA) on the right flank at the cemetery. Capt. Bledsoe's Missouri Battery of two Mexican War-era iron six-pounder cannons were placed along Shelby's line. He placed one of the guns to cover the Cincinnati Road and the second to cover the Fayetteville Road. Shelby also had a company of Quantrill's Partisan Rangers, under the command of Lt. William Gregg, who would make up the reserve. Although the battle line assignments had been made, Shelby was surprised by the quickness of the Union attack and still had many of his troops back in camp to the rear. The time was about 10 a.m.[19]

As shown earlier, Blunt opted to take the Ridge Road, which had been pointed out to him by Col. Cloud, who had used the road a few weeks earlier. Unfortunately for the Federals, the Ridge Road was extremely steep in certain sections, which made the travel hard on the troops, especially the infantry. The 11th Kansas Infantry probably suffered the most since they were placed between two cavalry regiments on the line of march, and since this was a small local road, there was no room to pass. They were also carrying their newly issued heavy, M1818 Prussian, .72 caliber, long muskets that fired "buck and ball," meaning their rounds consisted of one standard musket ball and three buckshot. The Kansas infantrymen were forced to march at double-quick step most of the seven miles from Rhea's Mill to Cane Hill. This slow movement by the infantry affected the entire column including the following cavalry units, the artillery batteries, and the ambulances carrying ammunition and medical supplies. Maj. Albert Ellithorpe of the 1st Indian Home Guard wrote to *The Chicago Evening Journal*:

> "…The column moved as rapidly as possible over the mountain roads; indeed one of the mountains was so precipitous that the men had to lay hold of the guns and assist the jaded animals to make the ascent. These difficulties did not deter the men or officers; silently as possible we pressed forward…"[20]

Blunt had ridden ahead believing that the column was right behind him, but only Maj. James Fisk's battalion of the 2nd Kansas Cavalry, with its two 12-pounder mountain howitzers commanded by Lt. Stover, and Capt. Rabb's 2nd Indiana Battery, were actually with him.[21]

The Ridge Road ends at the Fayetteville Road about a mile northwest of Boonsboro. Blunt and his column reached the junction without being detected, so he abandoned his carriage and mounted his horse. He ordered Col. Cloud to push up through the narrow defile with his Third Brigade, not realizing the rest of the brigade was at least a mile in the rear. Cloud obeyed the order with the 2nd Kansas Cavalry and its two howitzers, providing approximately three hundred soldiers. Blunt promised Cloud that he would send support as soon as they arrived. The 2nd Indiana Battery remained with Blunt at the junction. Gunfire was heard from Cane Hill indicating Cloud had encountered the Confederate picket. Blunt sent a staff officer, Maj. Verplanck Van Antwerp, up the Ridge Road to speed up their march. At the worst possible time, Col. Ewing's entire 11th Kansas Infantry had collapsed on the narrow road, and they refused to get out of the way of the second battalion of the 2nd Kansas Cavalry so they could respond to Blunt's call for reinforcements. Lt. Col. Bassett, in command of the second battalion of the 2nd Kansas Cavalry, tried in vain to get Col. Ewing to clear a path for the cavalrymen. Maj. Van Antwerp told Ewing in no uncertain terms that they needed to get out of the way.[22] He then told Bassett to basically ride over them if they failed to get out of the way, which was done. Capt. Crawford recalled:

> "…But they did not move until the Rebel batteries opened fire, when a staff officer came dashing back with orders for the Second Kansas to the front. I took the six companies we had, passed the Eleventh, and reached the field in a few minutes. On arriving General Blunt directed me to leave one company with him, and take the other five and move rapidly to the enemy's left and, if possible, roll up his flank…"[23]

The 1st Indian Home Guard (IHG) commanded by the accused thief, Lt. Col. Wattles, was next in line and did the same thing, single file at a gallop. They were followed by Capt. Henry Hopkin's Kansas Battery (consisting of the guns his company had taken from the Confederate Indian Brigade at the Fort Wayne battle). Meanwhile, Maj. Fisk and his battalion had advanced up the road leading to Boonsboro about a half mile when they encountered Shelby's

Confederate battleline, which he had hastily deployed on the road after his pickets came running back with the Federals on their heels. Cloud ordered Fisk to place Stover's howitzers on the west side of the Jordan Creek valley and to engage with Bledsoe's gun, which was approximately four hundred yards away near the cemetery. Fisk complied and, while placing the guns, was wounded in the head by a piece of shrapnel from one of Bledsoe's guns. Shelby then brought the second of Bledsoe's guns to the cemetery, along with a 12-pounder mountain howitzer from Capt. John Shoupe's Arkansas Battery.[24] If Shelby had known how small this Union advance party was, only about 300 effectives, he could have easily driven them back up the Fayetteville Road. Unfortunately, with the narrowness of the valley and the limited sightline due to hills and forest, it was reasonable to overestimate the number of Federals he was facing, especially with the amount of artillery support involved.

As soon as Blunt heard the artillery exchange, he sent Rabb's six-gun battery, consisting of four 6-pounder James Rifles, and two 6-pounder bronze smoothbore guns, up the road to Col, Cloud, but he was having difficulty finding room in the narrow valley to utilize all of the guns he had available. Bassett later wrote:

> "…The enemy was drawn up in force on a hill to the right front of Major Fisk's line, and with three pieces of artillery opened fire. Lieutenant Stover immediately turned his howitzers to bear upon them. Rabb's Second Indiana Battery coming up, Major Fisk moved a portion of his force to the rear, over the brow of the hill,

dismounted it, and ordered the men to lie down and hold themselves as a support to Rabb's battery…"[25]

Cloud placed Fisk's battalion in the center across the main road, supported by Stover's howitzers. He sent the 1st Indian Home Guard (IHG) regiment to the left flank adjoining Fisk. Ellithorpe continued with *The Chicago Evening Journal*:

> "…Gen. Blunt was not to be caught in this kind of trap. The column was at once moved from the main road up the steep hillside and through the thick brush, completely out of sight of the enemy. A position was gained upon the top of a hill overlooking the town and the enemy. Three mountain howitzers, put in position, at once commenced the battle…"[26]

Rabb's Indiana Battery was placed on the right side of the road facing southwest towards the Confederate guns on the cemetery hill. He ended up sending Hopkins and his Kansas battery, three 6-pounder bronze guns and one 12-pounder howitzer, to the top of the hill west of the valley, approximately one mile northwest of the town. Hopkins reported:

> "…Finding that the enemy had changed their position farther south, I was ordered to occupy a point one-half mile to the front. From this position I proceeded to shell the woods below and in my front, where a body of the enemy's cavalry was moving. Immediately to the left a rebel battery was discovered posted on a high hill. I directed the fire of my guns upon it, dislodging and forcing it to retire. It appeared shortly afterward in the main road, passing through the town. Again changing the direction of fire, shell were thrown with evident effect, the enemy retreating behind the hills to the left of Boonsborough…"[27]

Ewing, having been embarrassed by being dressed down in front of his troops, had his 11th Kansas Infantry drop their packs and push forward at double-quick step for the final mile and a half. Cloud had them move to the right and fill in the space between Rabb's and Hopkins's batteries. He sent the final two battalions of the 2nd Kansas Cavalry to the far right to support Hopkins and protect the right flank. At one point the Union artillery struck one of the Confederate 12-pounder howitzers and destroyed its carriage. The artillerymen were observed carrying the howitzer tube by hand to the rear. The artillery duel lasted more than an hour without any clear results.[28] Unfortunately for the Confederate artillerymen, the Union had far superior weapons and better trained troops. Wiley Britton remembered:

> "…The Confederates would have had some advantage in the contest by their tactics of retiring and choosing strong positions had not General Blunt been well supplied with splendid field artillery. They could not operate their battery of six guns in one position more than a few minutes when it would be located by the Federal artillery officers, who in another moment would be pouring such a

storm of shot and shell into it as to force it out of action…"[29]

Marmaduke informed Shelby that he was content with simply delaying the Federals until the commissary wagon trains with the flour and other foodstuffs had crossed the mountain between Newburg and the Cove Creek Road. Once Shelby was told that the train was safe, he began to withdraw down the main road. Knowing that the one gun commanded by Capt. Bledsoe himself was in danger on the left flank since it was facing advancing Union infantry and cavalry with significant artillery support, Shelby ordered it to limber up and head south. He wanted to keep the second gun that was being commanded by Lt. Conners to remain as long as possible covering the Fayetteville Road. Shelby noted in his report:

"…General Marmaduke, after surveying the position, and I having notified him that a heavy body of infantry was endeavoring to flank me on the left, I received orders to fall back, which I did, by ordering Colonel Jeans to mount his men and directing Bledsoe to withdraw his piece, at the same time ordering Lieutenant [B. A.] Collins, who was in charge of the piece that commanded the Fayetteville road, to keep a steady fire on the enemy until I could mount and form all my regiments,.. I then ordered Colonel Thompson to mount his regiment, which was done in the best order, moving the piece under Bledsoe by the right to the rear; Thompson's regiment followed, after which came Jeans, the Collins gun

following, covered by Gordon's regiment…"[30]

The Confederates were falling back just in time as over fifteen hundred Federal infantry with bayonets mounted and cavalry with sabers drawn suddenly swept down the Jordan Creek Valley towards their lines. Shelby began using this tactic of a "fighting retreat" from this point on during the war. As the Federals advanced, the Confederates would go back into a line of battle and, after firing a few rounds, would retreat to the next designated location. He organizationally broke down his brigade into its thirty component companies. He placed fifteen companies in column on each side of the road. The forward company would engage the Federals until they began drawing casualties and would fall back behind the rear company. This meant the Union troops engaged fresh troops during every interval. This continued as the Union advanced towards, then past Newburg, happening at least three times during the first phase of the Cane Hill battle.[31] Ellithorpe continued with *The Chicago Evening Journal*:

"…The regiments comprising the 1st brigade rapidly advanced, covered by the artillery. Defining shouts went up from out lines as they pressed forward. The rebels could not stand it any longer, and now the skedaddle commenced. From one hill to another, through deep ravines, up and down mountains, and through the woods, they fled, occasionally making a stand in some masked place, until charged and shelled out. Thus the battle continued, the retreat and pursuit, from 10 in the morning until dark…"[32]

Unfortunately, the Union artillery fire upon the retreating Confederate forces resulted in many buildings being damaged in Booneville and Newburg. Probably the greatest loss was the Cane Hill College, a Presbyterian men's school, which was hit several times by Union artillery.

Due to the advance of the Union forces, Gen. Marmaduke was required to abandon his headquarters at Kidd's Mill, which he did while Shelby's Brigade filed past. Col. MacDonald's Missouri Cavalry Brigade, numbering only around two hundred ranks, was deployed astride the main road just south of Booneville in an area that spread out wide after the narrow valley. They realized that they could not stop the quick advance of Blunt's forces, so Marmaduke came forward and ordered the brigade to the rear. Col. Charles Carroll's Arkansas Cavalry Brigade had been deployed adjacent to the mill facing north from bluffs on either side of the Fayetteville Road. In his After-Action Report, Carroll stated he only had 389 men available in his brigade, others being detailed to picket duty or without serviceable horses or arms. Lt. Col. James Johnson's Arkansas Cavalry regiment was posted on a substantial bluff on the east side of the road and Lt. Col. James Monroe's Arkansas Cavalry regiment was deployed on lower ground on the west side of the main road. Lt. William Hughey's small battery of two 12-pounder mountain howitzers was located on top of a large hill on the right flank of Carroll's brigade. Just as MacDonald's force left the area to rejoin Shelby's brigade, Union artillery under Capt. Hopkins opened fire on Hughey's battery with their three 6-pounder guns and one 12-pounder howitzer, from a hill on the other side of the Jordan River Valley. Realizing their danger, the

Confederate artillerymen pulled their guns from the hill. One of the prairie carriages for the howitzer became damaged, from either Hopkin's fire, or just due to the rough terrain. Again, the artillerymen abandoned the carriage and carried the howitzer tube away by hand. Marmaduke again came forward and ordered Carroll to fall back and set up a second defense along the Van Buren Road, which branches off of the Fayetteville Road just south of the town. Carroll placed his remaining mountain howitzers on a small hill that overlooked the main road coming south from Booneville and Newburg towards Fly Creek. He deployed Lt. Col. James Johnson's Arkansas Cavalry regiment on the right side perpendicular to the Van Buren Road, and placed Lt. Col. James Monroe's Arkansas Cavalry regiment on the left side. As he was deploying his small brigade, the final units of Shelby's withdrawing brigade began filtering through his line as they began their climb over Reed's Mountain (modern-day Skylight Mountain), a four-hundred-foot sandstone obstacle that separated their location and the Cove Creek Valley. Carrol had received an order to deploy Johnson's regiment forward as skirmishers when that order was countermanded by new orders. His new instructions were to fall back through the Fly Creek Valley and up over the Van Buren Road over Reed's Mountain to Morrow's. He was to set up a defensive line at the highest point on the Van Buren Road and delay the Federal's advance as long as possible. He was to join with Shelby's brigade so the entire division would be available to counter the Union advance.

General Marmaduke began to worry about how hard and fast Blunt and the Union Army were pursuing his force, and he became concerned that his retreating troops would press up against the southbound commissary train. The first Confederate line of defense up Reed's Mountain was set in place approximately a third of the way up. Scattered units belonging to Shelby's brigade formed up across the road in an effort to slow the Federals. These cavalrymen were dismounted and placed into the heavy woods on either side of the road. Carroll's brigade proceeded to the top of the mountain's saddle where the road crossed and formed behind a fence and a cornfield. Bledsoe's two-gun battery was also unlimbered and began exchanging rounds with the fourteen guns and howitzers of the Union artillery that had deployed along Fly Creek at the bottom of the mountain. Bledsoe soon ran out of ammunition, limbered up, and began to go down the mountain towards the Cove Creek Road. When Blunt saw the Confederate defense up on the mountain, he remarked that the hill was a "most admirable position for defense." Believing that the only way the Confederate line was going to be broken was by storming it, Blunt pushed forward the first troops to reach the bottom of the mountain. This force was a mixture of Second and Third Brigade units including the 11th Kansas Infantry, who with their heavy Prussian muskets (many in jest called these guns "light artillery"), took up the center across the road; the 2nd Kansas Cavalry that swung over to the left flank and advanced; and finally, Col. Phillips and his 3rd IHG regiment, which was considered by most commanders as the best of the three Indian regiments, came up on the right flank of the 11th Kansas Infantry, and the line moved forward. Silas Marple of the 11th Kansas Infantry wrote home:

"…I supposed… that the rebels were in full sight, but when I brought up my gun to fire I saw no one to aim at. So, I took it down again. In a moment more the Regiment fired their second volley. I looked in vain to see what they were firing at, but could see nothing… so I shot about the height of a man's breast… I may have shot some Rebel, and I may not…"[33]

All moved up the mountain under the cover of the huge Federal artillery barrage falling on the top of Reed's Mountain. Blunt stated in his after-action report:

"…I accordingly ordered up the Second Kansas and dismounted them. They charged up the steep acclivity in the advance… The resistance of the rebels was stubborn and determined. The storm of lead and iron hail that came down the side of the mountain, both from their small-arms and artillery, was terrific; yet most of it went over our heads without doing us much damage. The regiments just named, with a wild shout rushed up the steep acclivity, contesting every inch of ground, and steadily pushing the enemy before them, until the crest was reached, when the rebels again fled in disorder…"[34]

Jo Shelby did his best to try and stem the Federal advance up the mountain. By this time most of the retreating units were all mixed up and organization had fallen apart. Shelby wrote:

"…I had ordered Lieutenant Gregg at that point over to the right, but finding the enemy were making a move still to his right, I withdrew him, and had him to form back on the main road to await further orders. Immediately on top of the mountain I had a part of Colonel Thompson's command, under Major [M. W.] Smith, formed to receive the enemy, and a little to the rear of Smith, on the right, I had one company of Elliott's scouts, commanded by Captain Martin. Smith and Martin calmly awaited the coming of the enemy, and as they came charging up the hill in solid columns, they poured a deadly fire on them… The enemy pushing us about this time with all the force he could urge on, and the ground being of such a nature as not to allow us to form by regiments or squadrons, I was compelled to detach companies and form them on both sides of the road, receive and fire on the enemy, load, form, and reform, using in that manner every company in the regiments of this brigade. We fought them in this manner about three hours, never once allowing them to reach our rear in sufficient numbers to capture any of the men."[35]

Shelby and Carroll quickly pulled their troops from the summit and headed them to Morrow Junction, where the Van Buren Road met the Cove Creek Road. They continued their fighting retreat all the way to Morrow Junction.

A cheer rose from the Federals remaining in the Fly Creek Valley when they saw the Union troops at the summit wave the Stars and Stripes. Along with the 3rd IHG, 11th Kansas Infantry, and the 2nd Kansas Cavalry, Blunt brought Rabb's battery and four howitzers up to the summit. He then moved forward with the 11th Kansas Infantry on the left of the road and the 3rd IHG on the right. The 2nd Kansas Cavalry was joined by the 6th Kansas Cavalry and, along with Rabb's battery, they moved downhill towards Morrow Junction on the Van Buren Road behind the 3rd IHG and 11th Kansas Infantry. These two units were the ones responsible for engaging the Confederate rear guard. Blunt had been riding with the advance, and when they reached Morrow Junction, he observed the quickly retreating Confederate batteries just a little over half a mile south of the junction. Blunt stated:

"…Down the valley, in front of us, the ground appeared adapted to the use of cavalry to good advantage, and I determined to make an effort to capture their artillery, of which they had six pieces. A large force of their best cavalry was acting as rear guard, with a portion of their artillery just in front of them. Waiting for my cavalry to come up, I called for volunteers to make a charge. Three companies of the Sixth Kansas, nearest at hand, responded promptly to the call…"[36]

Wiley Britton, a member of the 6th Kansas Cavalry, described the scene:

"…Colonel Jewell responded to the call, offering to lead the charge, and every man of his command present volunteered to follow their leader, and in a moment the bugle sounded the charge, and the Colonel, at the head of his men, hardly exceeding two hundred and fifty, with drawn sabers, dashed forward down the valley and soon came upon the enemy filling the road and commenced sabering them right and left, some of his men putting up their sabers and using their Colts revolvers more effectively…"[37]

Blunt continued his description:

"… [The 6th Kansas Cavalry] dashed on to the rear of the rebel column, cutting and shooting them down with sabers, carbines, and revolvers. The charge continued for about half a mile down the valley, to a point where it converged in a funnel shape, terminating in a narrow defile. At this point a large body of the enemy were in ambush in front and upon the flanks…"[38]

In his haste to catch the artillery, Blunt had run into a clever ambush set up by Shelby and Carroll where they let the riders in front go past then opened up on the main body of troops. During this ambush many men were killed or wounded. Blunt and the survivors quickly turned around left the kill zone. Blunt further commented:

"…As soon as the party we were pursuing had passed through the defile, they opened upon us a most destructive fire, which, for the moment, caused my men to recoil and give back, in spite of my own efforts and those of other officers to rally

troops. During this ambush many men were killed or wounded. Blunt and the survivors them; whereas, if they had, after receiving the enemy's fire, passed on 200 or 300 yards, we could have secured, in a moment more, what we so much coveted— the enemy's artillery…"[39]

As the Union cavalry fell back, they were met by a four-gun battery of mountain howitzers and accompanied by the 3rd IHG. Wiley Britton continued:

"…the Colonel [Jewell] was just entering one of the narrow gorges described, to the left of which was a flat elevation ten to fifteen feet above the road at the proximal end of a ravine, made by the soil washing down from the mountain, and received a volley from the rifles of a company stationed on the elevation within two or three rods and was struck in the region of the hip with a ball and mortally wounded, together with several of his men."[40]

All formed into a line of battle across the road just as they became aware of an attempted Confederate counterattack. Blunt further reported:

"…Emboldened by their success in defending the defile and checking our advance, they raised a wild yell and advanced toward us. With the aid of Colonel Judson, Major Campbell, and Captains [H. S.] Greeno and [D.] Mefford, I succeeded in rallying the three companies of the Sixth Kansas, who had suffered severely in the charge, and formed them across the valley, and the four howitzers, coming up at the same time and opening on the enemy with shell, soon forced them to retire…"[41]

Seeing the artillery-supported battle line, the Confederate horsemen stopped short of the line and slowly backed away, giving their artillery time to continue southward. Blunt, at this point, discovered that the terrain beyond the defile opened into a wide-open valley, and he was determined to drive through the gorge if he could bring up his infantry in time. He ordered the four howitzers, one section of Rabb's battery loaded with canister, and the three companies of the 6th Kansas Cavalry to act as support behind the artillery. The Union line advanced to about one hundred yards from the enemy. Just as he was about to execute the assault, a Confederate rider with a white flag came forward with a request by General Marmaduke for a truce to gather the wounded and dead of each side. Blunt was correct in assuming this was a ploy to give the Confederates time to continue to move their artillery and commissary train southward out of danger. But by this time, it had grown dark, as it does in late November, and there was no further pursuit. Blunt was also concerned about Lt. Col. Jewell and the other troops from the 6th Kansas Cavalry who had fallen during the ambush. Unfortunately. Lt. Col. Jewell died from his wounds on the night of November 29. Jo Shelby wrote in his report:

"…About sunset the enemy made the last and desperate charge, led by Colonel [L. B.] Jewell, in person. Colonels Thompson's and Jeans' men received him with a fire the effect of which will ever be remembered by Jewell's regiment. In that charge Jewell fell, mortally wounded. Upon the fall of Jewell, Colonel Gordon, with a portion of his regiment and a portion of Colonel Jeans', under Captain Jarrett, charged the Federals hotly and fiercely, sending them back in perfect confusion, and thus ending a hard day's fight…"[42]

Darkness ended the day's fighting. The wounded were collected and cared for by the surgeons. The dead were collected by each respective side and buried wherever convenient. For the amount of lead expended and artillery fire each side endured, there were surprisingly few casualties. Blunt reported eight killed and thirty-six wounded. Marmaduke on the other hand reported ten killed and around seventy wounded. Marmaduke's weary troops continued down the Cove Creek Road and stopped at Lee Creek crossing for the night. There the weary Confederates fell into restless sleep. In contrast, the Federals who had already marched upwards of eight miles before reaching the battlefield, now had to climb back over Reed's Mountain to get back to Cane Hill. Residents along the way were subject to severe foraging by the hungry Union soldiers. Since this was enemy territory, they felt no obligation to pay or ask for what they took.

That night Blunt wrote to Schofield and stated that he had attacked Marmaduke and "thrashed him out of his boots and britches and fought him for ten miles over the Boston Mountains in his retreat until night closed the conflict." Blunt then told Curtis that the Confederates were "badly whipped and worse chased." Blunt didn't mention his small command being ambushed by Shelby and Carroll in the narrow defile. This was one of the primary complaints of Blunt's subordinates. He sometimes forgot that he was a general and was not supposed to be leading charges against the enemy. That was the job of the field and company officers. Still, Blunt performed well and could put two wins in his column in a little over a month. Department commander Samuel Curtis was impressed enough with Blunt's actions to create a new District of Western Arkansas and place it under his jurisdiction. Blunt and his commanders decided to make Cane Hill a base of operations with the Second and Third Brigades operating out of new camps in the area. The division's trains were moved from Lindsey's Prairie down to the new operating area. The First Brigade was detailed to Rhea's Mill to provide security for the supply trains and routes and operate the millworks to provide flour and grain for the First Division (Kansas Division). Brig. Gen. Salomon still did not have Blunt's confidence due to his lackluster performance at Newtonia in September and his mutinous actions back in the Indian Territory with Col. Weer. He and his brigade were usually kept away from the other brigades.[43] Maj. Ellithorpe, of the 1st Indian Home Guard, was detailed to bring Kidd's Mill back into operation. He stated in his journal dated December 4, 1862:

> "This morning I had the mill ready for operation & got up steam[.] I upon trial [sic] that I can turn out about 12,000 lbs of flour every 24 hours. I have got it in fine operation & making bread for the command very rapidly[.] how long we shall run it is uncertain.

> For the enemy are within 12 miles of us in force 20,000 strong, & may march upon us at any hour. I shall run the mill untill [sic] the bulets [sic] begin to whistle[.]..."[44]

While camped at Cane Hill, Maj. Preston Plumb of the 11th Kansas Infantry found an old printing press that had been abandoned by Evan Jones, the missionary to the Cherokees. This printing press included both English and Cherokee type fonts, and Plumb, a printer by trade, decided to print a newspaper for his regiment. He titled the newspaper the *"Buck and Ball,"* which referred to the ammunition used by the 11th Kansas Infantry's Prussian rifles. His initial run of fifteen hundred copies carried the "news" that Marmaduke and Hindman were marching in their direction with twenty-five thousand Confederates. These enemy troop levels were always exaggerations based on rumor, poor mathematics, and a misunderstanding of the enemy's levels of popular commitment.[45] As for the rest of Blunt's forces, they spent the next few days constructing defenses, barricading roads, patrolling, and attempting to drill on the uneven ground.

Blunt was intent on holding his position at Cane Hill. Using portions of all three of his brigades, he began to cover roads and avenues of approach to the Union lines. He put troops to work building defenses along Newburg Heights, a set of high ground southeast of Newburg that overlooked Fly Creek Valley and the Van Buren Road coming off of Reed's Mountain. He placed detachments on Cove Creek Road, on the Telegraph Road, on the road to Evansville, and on Hogeye Road. Many contemporaries questioned why Blunt did not simply move his entire division across the narrow Cove

Creek Valley. However, this would have severely bunched up the advancing Trans-Mississippi Army in those narrow defiles such as the one he had impetuously charged into on the evening of November 28. Perhaps he worried that Hindman would cross over to the

Telegraph Road to bypass his defensive position. Instead, Blunt knew that the rest of the Army of the Frontier was enroute, and if he could divert a portion of Hindman's force over Reed's Mountain, then Herron's divisions could attack the Confederates in the flank down the northern end of the Cove Creek Road.

[1] Prairie Grove Battlefield Handbook, pp. 8.

[2] Henry Stuart Foote, Casket of Rememberance, Chronicle Publishing Company, 1874

[3] Johansson, M. Jane, Editor Albert C. Ellithorpe, The First Indian Home Guards, and the Civil War on the Trans-Mississippi Frontier. Baton Rouge: Louisiana State University Press, 2016. pp. 55; The Chicago Evening Journal: Published Nov 25, 1862.

[4] Shea, William L. Fields of Blood: The Prairie Grove Campaign, Chapel Hill: The University of North Carolina Press, 2009. pp. 81.

[5] Hindman to Holmes; Supplement OR, Vol. 3, Pt. 1: 59-62.

[6] Joseph Connole, A Terrible Truth: The Tonkawa Massacre of 1862. Chronicles of Oklahoma, Volume 97, Number 4, Winter 2019, pp. 450-467

[7] Blunt, Experiences, pp. 228-229.

[8] Curtis to Halleck, OR, Ser. I, Vol. XIII, pp. 779.

[9] Schofield to Blunt, OR, Ser. I, Vol. XIII, pp. 766.

[10] Britton, Wiley, Border, Vol. 1. pp. 380.

[11] Johansson, Ellithorpe, pp. 56.

[12] General Orders #11, OR, Ser. I, Vol. XIII, pp. 777.

[13] Blunt, Experiences, pp. 229.

[14] Britton, Wiley. Brigade, pp. 107.

[15] Blunt, Experiences, pp. 229.

[16] Lindberg, Kip and Matt Matthews, eds., "The Eagle of the 11th Kansas: Wartime Reminiscences of Colonel Thomas Moonlight", Arkansas Historical Quarterly, 62 (2003): pp. 23.

[17] "Nassau" letter in St. Louis Daily Missouri Democrat, Dec. 11, 1862; Blunt, Experiences, pp. 230; Wiley Britton, pp. 108; Nov. 27 and 29, 1862, entries in Robert T. McMahan Diary, University of Missouri-Columbia; Nov. 27, 1862, entry in Sherman Bodwell Diary, Kansas State Historical Society, Topeka.

[18] Shea, Prairie Grove. pp. 93.

[19] Shelby to Marmaduke, OR, Ser. I, Vol. XXII (P1). pp. 56.

[20] Johansson, Ellithorpe, pp. 80. The Chicago Evening Journal: Published December 11, 1862.

[21] Crawford, pp. 69.

[22] Shea, Prairie Grove, pp. 95-97.

[23] Crawford, pp. 69.

[24] Cloud to Blunt, OR, Ser. I, Vol. XXII (P1). pp. 45-46.

[25] Bassett to Cloud, OR, Ser. I, Vol. XXII (P1). pp. 47.

[26] Johansson, Ellithorpe, pp. 80. The Chicago Evening Journal: Published December 11, 1862.

[27] Hopkins to Cross, OR, Ser. I, Vol. XXII (P1). pp. 51.

[28] Ewing to Cloud, OR, Ser. I, Vol. XXII (P1). pp. 52.

[29] Britton, Border, Vol 1, pp. 385.

[30] Shelby to Marmaduke, OR, Ser. I, Vol. XXII (P1). pp. 56.

[31] Collins, Robert. pp. 88.

[32] Johansson, Ellithorpe, pp. 80. The Chicago Evening Journal: Published December 11, 1862.

[33] Marple to wife, December 4, 1862. Marple, Silas Hough. Burlingame, Kansas, Transcribed Letters, 1855-1862. University of Arkansas, Fayetteville, Arkansas. Box: 66, Folder: 15; Shea, Prairie Grove, pp. 100.

[34] Blunt to Curtis. OR, Ser. I, Vol. XXII (P1). pp. 45.

[35] Shelby to Williams. OR, Ser. I, Vol. XXII (P1). pp. 57.

[36] Blunt to Curtis. OR, Ser. I, Vol. XXII (P1). pp. 46

[37] Britton, Brigade, pp. 114.

[38] Blunt to Curtis. OR, Ser. I, Vol. XXII (P1). pp. 45

[39] Ibid.

[40] Britton, Brigade, pp. 114.

[41] Blunt to Curtis. OR, Ser. I, Vol. XXII (P1). pp. 46

[42] Shelby to Marmaduke, OR, Ser. I, Vol. XXII (P1). pp. 58.

[43] Blunt to Curtis. OR, Ser. I, Vol. XXII (P1). pp. 41-46; Blunt to Schofield. November 28, 1862, Schofield Papers, Library of Congress; Shea, Prairie Grove, pp. 105.

[44] Johansson, Ellithorpe, pp. 84-85.

[45] Monaghan, pp. 260.

Chapter 11
The Arkansas Border War - Phase 2: The Battle of Prairie Grove

General Marmaduke and his forces spent the second night after the Cane Hill battle at Dripping Springs as they continued southward towards Van Buren. At this point he joined up with Hindman's forces after the latter had crossed the Arkansas River from Fort Smith on December 3, 1862. The Confederates did not see the battle at Cane Hill as a defeat, only as a minor setback in their plans. They were determined to reclaim the fertile agricultural area around Prairie Grove, Cane Hill, and Rhea's Mill from Blunt and the Union Army. Both Hindman and Marmaduke mistakenly believed at this time that Blunt's division was isolated and that there was no support for him closer than Springfield, Missouri. The Confederates' intention was to get their eleven thousand-man army between Blunt and the Missouri divisions and defeat them individually. Hindman knew that if the entire Union Army of the Frontier was able to reunite, his Confederate Army would be vastly outnumbered. This merged Confederate "Trans-Mississippi Army," consisting of nine thousand infantry and two thousand cavalrymen, supported by twenty-two pieces of artillery, moved slowly northward back up the Telegraph Road to Lee Creek Crossing. Hindman was forced to leave almost two thousand men behind at the training ground near Fort Smith since he did not have weapons or other equipment for them.

Hindman had a difficult time convincing his department commander, Maj. Gen. T.H. Holmes, to let him make this advance. Holmes ran hot and cold on the attack as he wavered in his support. One message was: "You must save the country if you can. Do so without risk of being destroyed." The next message read: "You must not think of advancing in your present condition. You will lose your army." He then said: "If your army is destroyed or demoralized ruin to us will follow." Holmes final message simply said, "Use your discretion and good luck to you."[1] Hindman had the permission he needed.

Col. Jo Shelby's division took the point, followed by Col. David M. Frost's division and Col. Shoup's division. A small wagon train followed the column on Telegraph Road. The trek up Cove Creek Road was difficult for the exhausted Confederates since it was mostly uphill and crossed the creek's cold waters many times. It took Hindman's army three days to travel forty-five miles. Knowing they outnumbered Blunt's Federals, Hindman planned to sweep up and over Reed's Mountain, which he knew he had to have to control the battlefield, and then attack Blunt's forces at Cane Hill. The Confederates first contacted Blunt's army along a narrow stretch of Cove Creek Road. Two battalions of the 2nd Kansas Cavalry under Lt. Col. Bassett had been assigned to the Morrow junction to slow the Confederate advance. All three battalions of the regiment were supposed to be part of the defense, but a miscommunication between Blunt and Col. Cloud left the remaining battalion in camp. Bassett and one of the battalions moved southward on Cove Creek Road intending to discover the location of the Confederate advance. They ran into Company F of Gordon's Cavalry regiment, who made up the point of Jo Shelby's brigade, a force approximately twelve hundred strong, adjacent to Morrow's place at the junction of the Van Buren Road and the Cove Creek Road. Bassett fought a retreating skirmish while sending messengers to Blunt warning him of the Confederate advance. This was one of the areas fought over just a few days prior. As expected, the Federals observed the Confederate column turn left on Van Buren Road towards Reed's Mountain. The Confederate cavalry followed the Federal cavalry up the Van Buren Road and finally tried unsuccessfully to drive the Federals from the summit of Reed's Mountain. Undeterred, Hindman had his troops camp around Morrow's place and junction.

To prevent Blunt from having the ability to fall back to the west towards Cincinnati or Evansville, two small towns between Prairie Grove and the Indian Territory. Hindman ordered Cooper and the Confederate Indian Brigade to proceed to Evansville and prevent Blunt from retreating into the Indian Territory. Col. Stand Watie reported:

"General, On Wednesday morning, December 3, I received an order from you inclosing instructions from Major-General Hindman to proceed to the neighborhood of Evansville, and, if possible, open a communication with the pickets of the Confederate army on the Line road... Early next morning (Saturday, the 6th) sent a scout to Evansville, in order to communicate with the pickets of our army; found none, but- on entering the town discovered a Federal scout going out I learned from the citizens that our pickets had not been there since the Monday... On Sunday morning I sent a scout to the Line road, but found no pickets on that road; same day cannonading was heard at a distance... On the day of the battle at Prairie Grove the enemy sent his trains on a different route from the Dutch Mills... Monday morning took possession of Dutch Mills and notified General Hindman of the fact..."[2]

Blunt had been informed as early as December 2, by scouts and informants, as well as by Curtis, that Hindman's Confederates were moving north to engage with his command. Blunt immediately contacted Curtis in St. Louis and requested assistance from Brig. Gen. Herron, who was in command of the Second and Third Divisions of the Army of the Frontier during Totten's and Schofield's absences. Curtis in turn notified Herron at 9 a.m. on December 3, and by 12 p.m. the Second and Third Divisions were leaving their encampment at Wilsons Creek. The two divisions quickly marched enroute to Blunt's assistance, a distance of almost 125 miles on the Telegraph Road to Cane Hill. Herron and Col. Dudley Wickersham's cavalry brigade, consisting of the 10th Illinois Cavalry, 1st Iowa Cavalry, a battalion of the 2nd Wisconsin Cavalry, and the 8th Missouri Cavalry (USA), arrived at Elkhorn Tavern on the evening of December 5, when they received a message from Blunt requesting to send as much cavalry his way as possible. The Second Division arrived at Cross Hollows on the evening of December 6, rested until midnight, and then resumed their march. The Second and Third Divisions were traveling at a rate of thirty-five miles per day, a remarkable feat for infantry (This march far outdistances those of Thomas "Stonewall" Jackson's "foot cavalry"). Herron did order his infantrymen to place their heavy packs in wagons to lighten their loads while marching.[3] However, the quick pace of the Army's movements did result in some injuries and deaths. At one point a runaway wagon killed one soldier and severely injured two others. Fatigue and stress were affecting the fast-moving column.[4] The remaining two divisions of the Army of the Frontier arrived at Fayetteville, Arkansas, late on December 6 or early December 7.

On December 6, Hindman again attempted to drive the Union forces from Reed's Mountain with Col. Monroe's cavalry brigade and with the addition of Brig. Gen. Parsons's Missouri infantry brigade. Blunt initially had no intention to hold the summit. In fact, the 2nd Kansas Cavalry had been pulled from the mountain's summit and had camped the night of December 5 down along Fly Creek. Late in the afternoon of December 6, he changed his mind and decided to slow the approaching Confederates by sending the 2nd Kansas Cavalry back to the summit. Bassett placed the First Battalion under Capt. Crawford in a battleline along the summit, centered on the Van Buren Road. He placed the Second Battalion in a line of battle about a half mile west of the summit, also centered on the road. Pvt. Vincent Osborne of the 2nd Kansas Cavalry recalled:

> "…We dismounted and went up as skirmishers sheltering ourselves as much as possible behind trees and arrived at the top with out discovering any enemy then kept on about thirty rods when we saw about a dozen fired on them and they retreated one of them had a flag he got behind a tree and waved it at us and then put spurs to his horse and was out of sight in a moment We now halted and in a few minutes fell back to the top of the mountain and formed an ambush expecting the enemy to soon return…"[5]

Lt. A.F. Bicking of the 1st Indian Home Guard (IHG) arrived with one hundred Union Indians to assist. Lt. Stoval's howitzer section was also deployed along the summit. Companies C and G, 2nd Kansas Cavalry, had been sent down the east side of the mountain along the road early the previous night and were to take station as near the Morrow junction as possible. Monroe's Confederates appeared that morning and, after a small skirmish, pushed them back to the foot of the mountain. Capt. John Gardner of the 2nd Kansas Cavalry reported:

> "About 6 a.m. this morning, the enemy advanced with considerable force, attacking me in front and on both flanks, bringing my men under a very heavy fire. I retired slowly, keeping up a skirmishing fire for 2 miles, when I formed line and drove the enemy back. I maintained this position for half an hour or more, and then retired to the foot of the mountain, and awaited the arrival of re-enforcements…"[6]

The two companies later returned to the top of the mountain and joined the defenses being prepared along the summit. Soon Bassett's cavalrymen observed dismounted Confederate cavalrymen, armed with shotguns, coming up the slope from the east side of the mountain. As soon as the Confederates were within carbine-range, the Kansans opened fire on Monroe's advancing troops. Seeing the Confederate advance line falter, Monroe requested help, and Parsons deployed his Missouri infantry into a line of battle and advanced up the mountain, guiding off of the Van Buren Road. Bassett realized that his small force could not hold the mountain summit without help. Blunt

received the request for reinforcements and dispatched a battalion of the 11th Kansas Infantry under Capt. Joel Huntoon to Reed's Mountain. Later Lt. Col. Wattles arrived with two hundred Union Indians of the 1st IHG, who were placed into a defensive line. Wanting to take the summit with his own brigade alone, Monroe again made a drive towards the Federal line at the top and nearly broke the Kansans' line but were again driven back. This time Parsons sent Maj. Lebbeus Pindall's Missouri Sharpshooter Battalion to assist Monroe's highly driven cavalrymen. Monroe and Parson's finally deployed their entire detached force against the Kansans, flanking them on the right. The Kansans began to fall back down the west side of the mountain along the Van Buren Road. The Confederates did not pursue. The 2nd Kansas Cavalry battalions and the battalion of the 11th Kansas Infantry fell back to the Union defenses along the Newburg Heights. The Confederates immediately began felling trees and gathering rocks to establish breastworks at the top of the mountain, indicating that they intended to hold it.

On the evening of December 6, Hindman received word from scout patrols he had sent towards Fayetteville on the Telegraph Road. They informed him that there were large numbers of Union troops in and coming into Fayetteville from the direction of Elkhorn Tavern. Hindman quickly realized that these were the Missouri divisions he had believed were still near Springfield. He quickly organized an officers call with his senior commanders for a Council of War. This situation left him with two choices and neither one was an especially good one. He could get between the two Union forces by continuing over Reed's

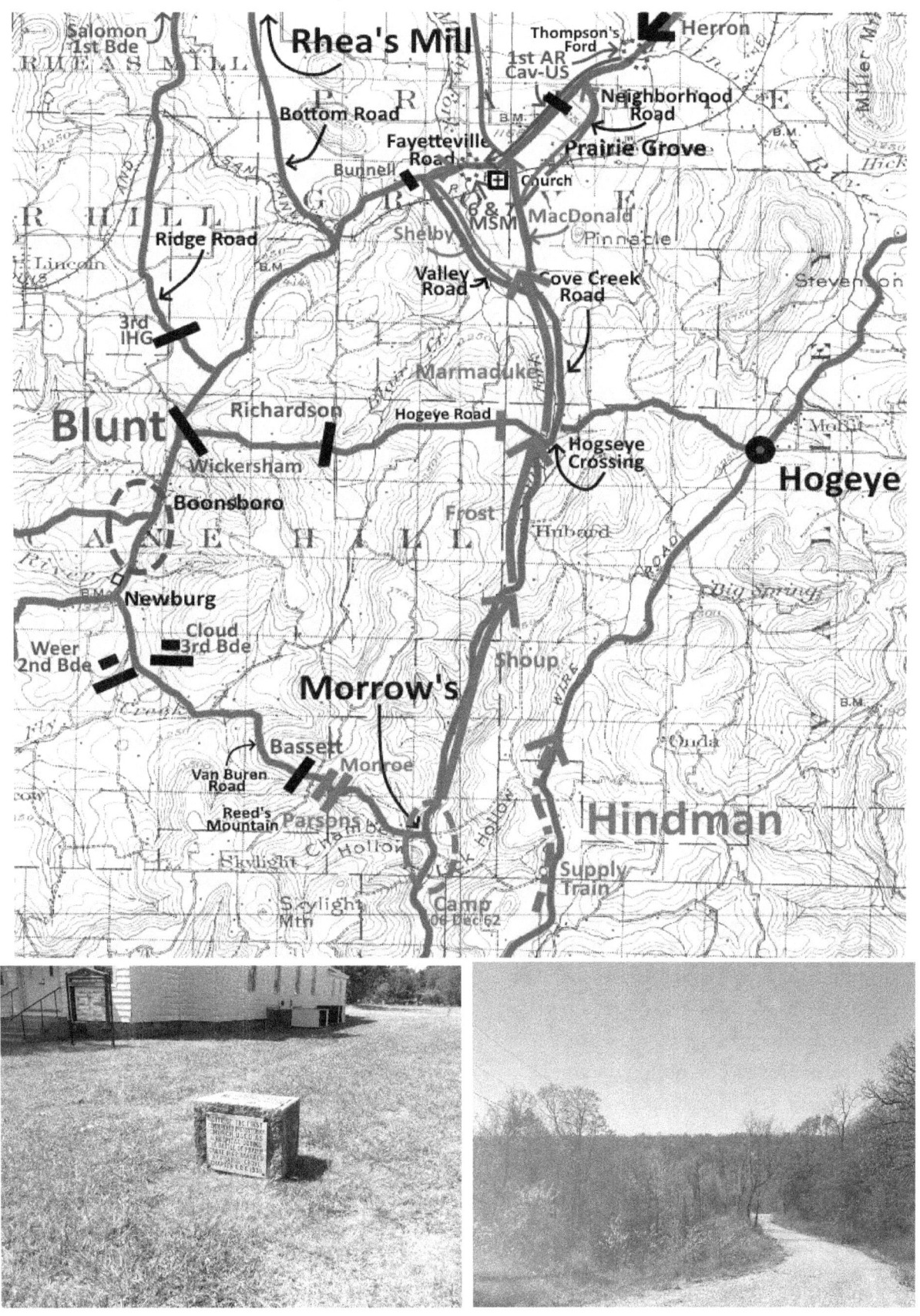

Location of Prairie Grove Church
Photo by Author

Summit of Reed's Mountain
Photo by Author

Mountain to attack and hopefully defeat Blunt before he could join up with the Missouri divisions. They would then turn around and deal with those same divisions. Or he could keep Blunt in place with a feint on his lines while he moved towards Fayetteville to engage the Missouri divisions. Hindman chose the latter. He sent Col. Monroe and his brigade of five hundred troops up on the now undefended Reed's Mountain. They were to start campfires all along the summit and keep them going all night long in an effort to fool the Federals into believing that Hindman's entire Trans-Mississippi Army was encamped on the top of Reed's Mountain, poised to strike at first light. In the morning Monroe's troops advanced in skirmish formation down the mountain to imply that they were the advance of a larger force. To their credit, it mostly worked. Those fires burning and some Confederate soldiers milling around the top was just enough to convince Blunt that the threat was real. The Federals did not discover the ruse until later in the morning of Sunday, December 7. By that time the entire situation had changed, and Blunt was going to be late to the fight.

Blunt also fell victim to one of his commanders, Col. J.M. Richardson of the 14th Missouri State Militia (MSM-US), for failing to keep him informed of enemy movements. Richardson had been assigned to proceed with his regiment and one hundred extra soldiers to the junction of Hogeye Road and Cove Creek Road, known locally as Hogeye Crossing. They were to set up a simple defense of the junction, send pickets down the Cove Creek Road, and report any activity to Blunt. Not only did Richardson not go to the junction, but he also never got within two miles of it. He sent a small picket to the

junction, and they were immediately driven back by the Confederate advance force. The Union commander insisted that he did not hear from Richardson until 10 a.m. on December 7 when he reported that he observed the entire Trans-Mississippi Army pass by the junction. Blunt was furious that the entire Confederate army had passed, and Richardson had failed to report any of it. In his official report, Richardson states that he sent a message to headquarters at 9 p.m. on December 6, detailing the advance of the Confederate force.[7] Blunt did not believe him and stated that Richardson was "deserving of the severest censure."[8] In turn, Blunt sent Col. Judson and his 6th Kansas Cavalry across on Hogeye Road to harass the rear elements of the Confederate force as they moved north on Cove Creek Road.[9]

After Blunt had contacted Curtis and Herron on December 3, he also ordered the 1st Arkansas Cavalry (USA) under the command of Colonel M. La Rue Harrison, and the 7th Missouri State Militia (USA), both of which had units posted at Elkhorn Tavern, to begin patrolling the Telegraph Road between that location and Fayetteville. They were to keep the road clear for Herron's divisions to advance as quickly as possible without any delays. On the night of December 6, they were relieved of this duty once Herron's divisions began arriving at Fayetteville. They were ordered forward to join up with Blunt's command near Cane Hill. The 1st Arkansas Cavalry (USA), consisting of 485 troops, had been originally ordered by Blunt that evening to proceed to Boonsboro, but Harrison insisted his horses were in such poor condition that they really couldn't go much farther. Harrison instead requested permission from Capt. Harris Greeno, one of

Blunt's staff officers who was responsible for directing the arriving units leaving Fayetteville, to stay in Fayetteville and rest and re-shoe his almost five hundred horses. Greeno insisted that Harrison had to move towards Cane Hill as directed. His command made it about eight miles before the horses began to give out. Harrison camped along the Fayetteville Road at Thompson's Ford on the Illinois River, about a mile northeast of the small hamlet of Prairie Grove. He sent Lt. James Roseman to inform Blunt about the condition of his horses. Blunt realized that the unshod and rundown 1st Arkansas Cavalry (US) would be of little use. Instead, Blunt ordered Harrison to proceed to Rhea's Mill to assist in the defense of the mill and the supply train because the First Brigade had been ordered to join the Union force at Cane Hill. The Arkansans were also expected to begin operating the mill since they were not available for a combat role. Roseman upon returning reported encountering only Union troops along the route.[10] The "Mountain Feds" were eating supper at Thompson's Ford when Col. Dudley Wickersham's cavalry brigade, consisting of the 10th Illinois Cavalry, 1st Iowa Cavalry, a battalion of the 2nd Wisconsin Cavalry, and the 8th Missouri State Militia (MSM-USA), stopped to water their horses at the same Illinois River crossing. Wickersham's force did not linger and continued its ride on the Cane Hill-Fayetteville Road to report to Blunt at Cane Hill. The 7th MSM (USA), and two companies of the 6th MSM (USA), under the command of Major Bredett, with approximately 650 troops, had been directed by Capt. Greeno to proceed to Cane Hill and join Blunt as well. Bredett and his command opted to stop for the night and camp at the Farmington Branch, a few

miles outside of Fayetteville. After a few hours rest, they were moving again by 2 a.m. At daylight they passed the 1st Arkansas Cavalry camp at Thompson's Ford and continued on for another two and a half miles, stopping at the Muddy Fork crossing of the Cane Hill-Fayetteville Road. Maj. Bredett had his troops water, graze, and rest their horses, losing bridles and girths, and cook some breakfast before starting their final four mile stretch to Cane Hill.[11]

As Marmaduke's Confederate division moved northward on the Cove Creek Road, they discovered a small cut-off road known locally as Valley Road that proceeded northwest of the Cove Creek Road, followed the Muddy Fork, and dead-ended at the Fayetteville-Cane Hill Road southwest of the crossing. Jo Shelby was tasked with leading his brigade up Valley Road. Shelby was glad to be on the trail although his troops were tired, hungry, and cold. He wrote in his standard report in poetic terms:

> "Upon the eventful morning of the 7th, long before the full round moon had died in the lap of the dawn; long before the watching stars had grown dim with age, my brigade was saddled, formed, and their steeds champing frosted bits in the cold, keen air of a December morning, ready and eager for the march. After advancing rapidly and without intermission for several hours, I struck their trail, hot with the passage of many feet, reeking with the foot-prints of the invader. It needed no command now to close up. There was no lagging, no break in serried ranks, no straggling from the line, but each man grasped his gun with the

strong, firm grasp and the strange, wild looks of heroes and born invincibles [sic]…"[12]

Shelby's scouts observed the camp and fires from the Muddy Fork camp of the 6th and 7th MSM and they quickened their pace. Shelby continues:

"…After riding hard for about an hour, my advance came full upon the foe, and, with the mad, fierce whoop of men who have wrongs to right and blood to avenge, they dashed on and away at the pas de charge. Rapidly and in splendid style Colonel Jeans, by my command, rushed on to follow up the attack, while Colonels Thompson's and Gordon's regiments were dismounted and formed in the dry bed of a creek, and so stationed that they could resist an attack either from the east or west…"

Tramp, tramp, along the land they ride, Splash, splash, along the lea; The scourge is red, the spur drops blood, The flashing pebbles flee!…"[13]

At a little past dawn, Company M, 7th MSM (US), under Lt. Lafayette Bunner, had been assigned as the advance company and moved ahead past the junction of the Valley Road and the Cane Hill Road. Also, at nearly the same time, a company of the 8th MSM that had become separated from Col. Wickersham's brigade overnight, passed through Bredett's resting regiment. As they advanced towards the Valley Road and Cane Hill Road junction beyond the crossing, they were fired on by unseen troops, shielded by the underbrush and cornfield. They then observed that the troops who were engaging them were wearing blue Union uniforms. Believing this was part of the 1st Arkansas Cavalry (US) and simply a friendly-fire situation, the MSM troops stopped and held their fire until it became obvious that these were Confederate troops. In fact, they were Lt. Gregg's company of Quantrill's Raiders, wearing captured Union uniforms. This guerilla group included both future outlaws Frank James and Cole Younger, who were discussed in an earlier chapter. Just the knowledge that Quantrill's bandits were present was enough to create a panic among the Union Missouri troops. Upon hearing the gunfire, Lt. Bunner and his company turned around and began to head back towards their regiment at the crossing. They also saw the blue-clad horsemen and believing them to be the Mountain Feds, concluded again to be a friendly-fire incident due to the early morning darkness. They, then, approached the blue-clad guerillas the same way that the 8th MSM troops did on the other side with Lt. Bunner calling out "Cease firing; we are friends!" Finally realizing these were Confederates once butternut and gray infantry came in view, Lt. Bunner attempted to advance and cut his way through the enemy line. Unable to do so, and fearing he was being surrounded, he split his company into squads and assigned them to observer status. He wanted them to collect as much information as they could on the enemy's strength and location and provide a report to Gen. Blunt.[14]

The blue-clad leading elements of Shelby's brigade happened to arrive at the junction of the Valley Road and the Fayetteville Road between the advance company of the 7th MSM and the left-behind company of the 8th MSM. It was a very cold morning, and it would have been difficult to convince the Confederates wrapped in their heavy blue Union Army overcoats to take them off before the fight. Regardless, this gave them an upper hand that somehow happens in times of war. As Shelby's advanced regiments, assisted by Capt. Bledsoe's remaining 6-pounder iron gun, pushed the company of the 8th MSM back onto the Muddy Fork crossing, causing confusion among the 6th & 7th MSM regiments having breakfast. They had let their horses graze in adjacent cornfields and when surprised by the attack, had difficulty regaining control of their steeds. Maj. Bredett was able to get a portion of his command mounted and attempted to put up a defense. Capt. Milton Brawner reported:

> "…While feeding (with bridles off and girths loosened), a cavalry troop (part of the Eighth Missouri Cavalry) passed through the lane in which we were feeding. When nearly through they were assailed by a volley of small-arms, fired by an unseen foe (concealed by thick underbrush and cornfields), which threw them into great disorder, they retreating through our column, causing great confusion. Major Bredett at this juncture behaved with great coolness and bravery, using his utmost exertion to rally and form the men. The line was twice formed, but the enemy pouring in heavy volleys on our front, left, and rear, the retreat was sounded…"[15]

Unfortunately, the exhausted Federals were beyond the point of rallying. They began to break to the rear towards Prairie Grove and they soon encountered another problem.

At 7 a.m. the 1st Arkansas Cavalry (US) had mounted up and began their move to Rhea's Mill. They planned on traveling approximately four miles on the Fayetteville Road to Bottom Road, where they would turn north for the march to the mill. As they proceeded towards Prairie Grove in a column of fours, they encountered the first elements of the fleeing MSM troops from Muddy Branch. Officers attempted to stop the panicked Missourians but were unable to do so. In addition, the Fayetteville Road at this point had fencing on either side so the cavalrymen were funneled on the road. Once again, misfortune visited the Union cavalry. Gen. Marmaduke had intended to go directly to Prairie Grove on the Cove Creek Road but then opted to send Col. MacDonald's brigade up a small lane known locally as the Neighborhood Road. This road left Cove Creek Road on the right, moving northeast and ending at Fayetteville Road between Prairie Grove and Thompson's Ford. This placed MacDonald's Confederates almost behind the 1st Arkansas Cavalry (US). MacDonald divided his force into two columns, one under Lt. Col. R.P. Crump to attack the Federal's flank, and one under Col. M.L. Young to attack from the rear. Caught between two Confederate brigades and trapped by the funnel effect of the fence lines, the Union Mountain Feds broke to the rear and began to stampede back towards Fayetteville. Many Union troops were killed, and many simply surrendered. Maj. Bredett of the 8th MSM was killed moments after decapitating a Confederate cavalryman with his saber. Some Federals attempted to set up a dismounted defense along some of the fence lines, but they were soon overrun as well. Confederate Col. Shelby actually became a prisoner during the melee when he was surrounded by Union troops but was soon released when a force of Confederates surrounded the Federal soldiers holding Shelby. At the end of this first action of the day the Federals had already lost over three hundred killed, wounded, or captured. The 1st Arkansas Cavalry (USA) also lost its entire 21-wagon supply train to the Confederates. When the regiment's quartermaster tried to save them by turning them around, the wagons created a roadblock with turned over wagons and dead and wounded horses.[16]

General Herron and his two Missouri Divisions of the Army of the Frontier had been enroute along the Fayetteville Road towards Cane Hill when they encountered the fleeing cavalrymen of the Missouri State Militia and 1st Arkansas Cavalry. This was about six miles west of Fayetteville and just a couple of miles short of Thompson's Ford near the small town of Walnut Grove. Herron recalled:

> "It was with the greatest of difficulty that we got them checked, and prevented a general stampede of the battery horses; but after some hard talking, and me shooting one cowardly whelp off of his horse, they halted…"[17]

Realizing that the Confederates were between his forces and those of Blunt, Herron quickened the step of his exhausted infantrymen until they encountered Shelby's cavalry brigade in line of battle approximately one and a half miles from the Illinois River crossing. By having his troops cut roads through the woods, Herron was able to bring

eighteen artillery pieces into use to support his deployed infantry. After seeing the Union artillery deploying to counter the battle line, Shelby's Confederates were soon pulled back across Thompson's Ford. The 94th Illinois Infantry was in the lead and was not challenged as they crossed the Illinois River at Thompson's Ford at approximately 9 a.m. What they did observe were the three brigades of Marmaduke's division on the high ridge of Prairie Grove straddling the Cane Hill Road, supported by three batteries of artillery. It took Herron until almost 1 p.m. to get the first of his Third Division brigades across the river and deployed in a line of battle. Herron directed Brig. Gen. Daniel Huston, Jr., who was acting division commander in Totten's absence, upstream to Taylor's Ford to deploy along the high ground north of Prairie Grove. During this same time Hindman was able to deploy Brig. Gen. Francis Shoup's division (an Indiana-native who somehow ended up in the Confederate Army) to the west of Marmaduke, extending his line along the Prairie Grove ridge, giving his artillery far reaching fields of fire to protect the lines from infantry attack or to conduct counter-battery operations.

At this time Blunt was still unaware that the movement by Col. Monroe's Confederate brigade on Reed's Mountain was a feint until Monroe deployed his troops in a half-mile line of battle. The Union officers realized that this small force of mostly dismounted cavalry was not a main attack. It was at about this time that Col. Richardson of the 14th MSM finally returned with his overdue report of the main Confederate army moving northward on the Cove Creek Road. Blunt hoped to connect with Herron's divisions before Hindman was able to get between them. Upon hearing the artillery fire

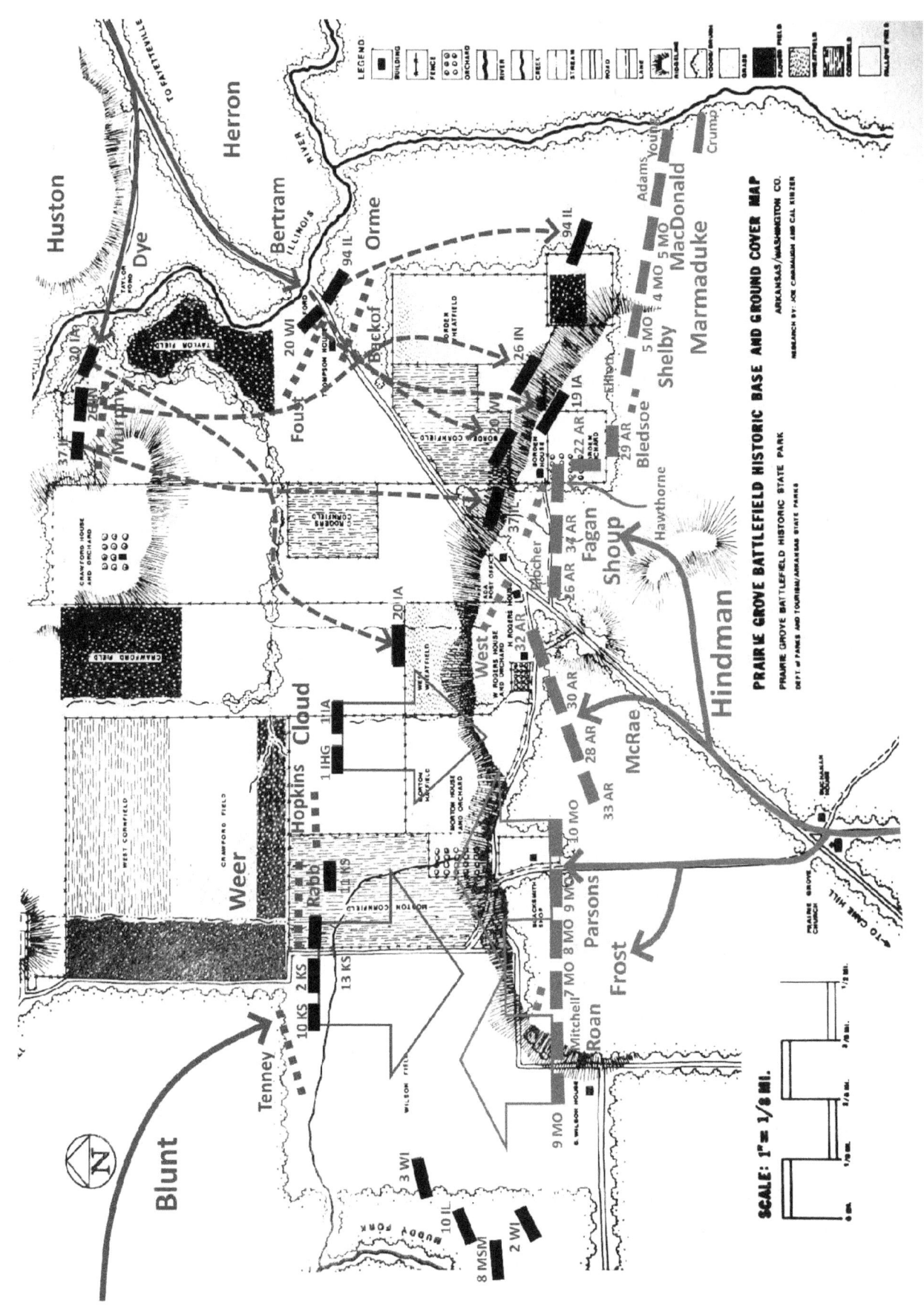

PRAIRIE GROVE BATTLEFIELD HISTORIC BASE AND GROUND COVER MAP

ARKANSAS/WASHINGTON CO.

PRAIRIE GROVE BATTLEFIELD HISTORIC STATE PARK

RESEARCH BY: JOE CARVAUGH AND CAL KINZER

DEPT. of PARKS AND TOURISM/ARKANSAS STATE PARKS

SCALE: 1" = 1/8 MI.

from the northeast, he sent Col. Wickersham and his cavalry brigade up the Fayetteville Road to rendezvous with any elements of Herron's force. He also sent messengers up the other roads for the same purpose, but they returned reporting that they encountered Marmaduke's forces blocking the roads. Knowing that the Fayetteville Road was obstructed, and intending to break through to Herron, Blunt directed the Second and Third Brigades to follow Wickersham's cavalry up the Fayetteville Road. For some reason, however, Wickersham turned his cavalry northward towards Rhea's Mill. Blunt, desiring to keep the two brigades together, and hoping to find a clear path to the east that could be used to join up with Herron's two divisions, followed the lead elements toward Rhea's Mill.

Over the next several hours Herron and Huston advanced their divisions over Crawford's Prairie against Hindman's divisions deployed along the wooded Prairie Grove ridge. The heaviest fighting took place around the Archibald Borden house and orchard, on the ridge near the center of the battlefield. The Union line of battle was deployed approximately two hundred yards west of Thompson's Ford. Battery E (Lt. Joseph Foust), Battery F (Capt. David Murphy), Battery L (Capt. Frank Backof), 1st Missouri Artillery, and Battery A (Lt. Herman Borris), 2nd Illinois Artillery, with their combined twenty-four 10-pounder Parrott rifles, advanced toward the Confederate line up on the ridge. The Confederates only mounted fourteen guns and all but four were bronze 6-pounder smoothbores with limited range. The Federal gunners were taking aim at individual Confederate guns and when that gun was disabled and its gunners out of action, they moved to the next gun. By 1 p.m. all of the guns on the eastern side of the ridge

had been silenced or withdrawn. Capt. Blocher's Arkansas Battery (CS) of four 6-pounder guns were located in the yard east of the Borden house. They had been abandoned by their gunners due to heavy fire from the Union batteries. They could not pull their guns back because their artillery horses had been killed in their harnesses. The artillerymen took shelter on the south side of the orchard. A sergeant with Blocher's artillery later wrote:

> "The enemy advanced upon us with their artillery under cover of their infantry, until within range of our battery when they opened a most disastrous fire on us from both arms. Hail from Heaven never fell thicker than the shot, shell, and minie balls did for minutes. Having no support, Captain Blocher ordered our men to fall back and save themselves…"[18]

When Gen. Herron observed the abandoned guns, he sent the 19th Iowa Infantry and the 20th Wisconsin Infantry regiments forward to capture the guns. The two completely exhausted regiments easily captured the four guns, but when they attempted to sweep through the orchard, they encountered the Confederate infantry from Col. Fagan's brigade, cavalry from Col. Shelby's brigade, and Capt. Bledsoe's two-gun battery. The two Union regiments found themselves in a crossfire, and although they did their best to hold onto their position, eventually they were overrun, losing almost half of their strength. They fell back until they were under the protection of the Federal artillery. Lt. Will S. Brooks of the 19th Iowa Infantry recalled years later:

"…The fight was most determined and the slaughter immense. I was struck… while we were being driven back from a too far advanced position. We were outflanked and had to run 300 yards over open ground and exposed to murderous fire from the right, left, centre, or rear. Here we lost Lieutenant Colonel [Samuel] McFarland. We lost one-half of our regiment, and in Company D more than half our effective men…"[19]

The 20th Wisconsin Infantry sustained the greatest loss of any Federal regiment at Prairie Grove with fifty killed and one hundred fifty-four wounded. Lt. Col. Henry Bertram of the 20th Wisconsin Infantry stated:

"…I observed a battery of the enemy, supported by infantry, trying to get into position in my front. I immediately ordered the Twentieth Wisconsin to charge the battery, which was done in gallant style, Major [H. A.] Starr leading. After taking the battery, the regiment advanced under a heavy fire to the brow of the hill, where they met a heavy force of the enemy's infantry, some four or five regiments, advancing, which poured a terrific fire into the Twentieth Regiment Wisconsin Infantry, and obliged them to fall back, which they did in good order, destroying what they could while falling back of the battery taken before. The Twentieth fell back in good style across an open field to a fence, where they reformed…"[20]

The Confederate counterattack pushed the Union infantry until they were one hundred yards from the Union guns. At this point, the twenty-four Union guns opened on the Confederate line with double canister and drove the butternut and gray soldiers fleeing back to the ridge and orchard. Earlier the 26th Indiana Infantry and the 37th Illinois Infantry had been ordered to advance and support the 19th Iowa Infantry and the 20th Wisconsin Infantry regiments in their attack on the abandoned battery. Due to the heavy smoke and lack of good communications, the Indiana and Illinois regiments did not know the other two regiments had been driven off or receive a counter-order to return to the Union lines. They charged up the same hill and though the same orchard as the Iowa and Wisconsin regiments. They recaptured the battery and swept the entire way through the orchard. On the other side of the orchard, they encountered not only Col. Fagan's and Col. Shelby's brigades, but also Col. Dandridge McRae's brigade that had been brought over from the left part of the Confederate line. Once again, the Federal bluecoats were driven out of the orchard, past the abandoned battery, and back to the protection of the Union artillery at the north edge of the Borden cornfield. Colonel John Clark of the 26th Indiana Infantry recalled the horrific struggle he and his regiment experienced:

"…by rapid marching, reached the field of action (about 10 miles distant) at 1 p. m… The regiment was at that time ordered on the left of the Thirty-seventh Illinois Volunteers, which formed the extreme right of the line of battle. Soon after, the Thirty-seventh

Muddy Fork Ford where Battle of Prairie Grove began. *Photo by Author*

Thompson's Ford on the Fayetteville – Cane Hill Road. *Photo by Author*

Union Army Approach from Thompson's Ford towards Borden Hill. *Photo by Author*

Union Army Approach from Confederate lines on Borden Hill. *Photo by Author*

Area of Blunt's Approach and Attack from near the Morton House
Photo by Author

Painting of the Union Defense of the Borden House Line
Painting courtesy of Prairie Grove Battlefield State Park / Photo by Author

Re-Built Borden House on Prairie Grove Battlefield. *Photo by Author*

Confederate Line between Morton House and Roger's House on shelf. *Photos by Author*

Morton House and Cellar
Photos courtesy of Prairie Grove Battlefield State Park

Illinois and the Twenty-sixth Indiana were ordered forward, and moved to the left of the line, where they were ordered to charge the enemy, who were strongly posted on a hill covered with timber. My regiment succeeded in reaching a point some 75 yards beyond the crest of the hill, but was overpowered by being outnumbered two to one, and driven back in considerable disorder, then rallied before they were beyond the reach of the fire of the enemy. The regiment was then ordered to fall back…"[21]

Lieutenant Colonel John Black of the 37th Illinois Infantry attempted to bring his regiment and the 26th Indiana Infantry regiments back down off of the ridge to a fence line at the base of the hill due to the threat of a Confederate counterattack. Lt. Col. Black's brother described the scene in a letter to their mother:

"…we had the rebels now just where we always wanted them, on level clear ground, and we felt now was an hour of vengeance. The regiment rose as one and poured in a volley… which stopped the rebel pursuit. Our [Colt] revolving rifles kept playing and one and another fell, and one and another fled into the woods whence they came…"[22]

Behind the lines Hindman was having difficulties of his own. An entire regiment of Arkansas conscript infantry in Col. Shaver's brigade, under the command of Col. Adams, deserted en-masse to the rear. Hindman himself had to go over to Frost's division and get the brigades organized again.[23] Even with

the confusion and the two different strong Union attacks, the Confederate soldiers were holding their own during the battle. Col. Shelby in his poetic prose reported:

"…All along the lines the near fire of the infantry rose, crash upon crash, the dense smoke filling the air and the wild powder gloom getting darker and darker. This terrible fire soon rippled out in one vast, mighty wave of bullets, that circled and roared like a storm at sea, varied incessantly by the thunder of impatient cannon and the yell of exultant and furious combatants. On the right, four regiments of Federal infantry formed in the open field, and came up in splendid order, with flaunting banners and waving pennons, the light of battle on their faces and their steps proud with the thoughts of an easy victory. My skirmishers were steadily driven in, and down to meet them like an avalanche our own infantry swept. They met, the shock was terrible, but, broken and rent, our boys drove them back and followed at the charge. Again and again, they returned to the fight, and again and again were they repulsed with great slaughter…"[24]

Hindman had Frost's division with Shaver's, Parson's, and Roane's independent cavalry brigades available to the west. They remained in this position in order to counter Blunt coming from Cane Hill. This gave the Confederates four thousand fresh troops to counter any more attacks as well as being used to counterattack the Union Army. Hindman and his commanders felt that victory was at hand. Since their left flank overlapped Herron's and Huston's divisions, they planned on attacking the Federal's right flank to roll up the Union forces and drive them back across the Illinois River hoping for a Bull Run-style retreat.

Blunt, still down at Cane Hill, heard Herron's artillery began pummeling the Prairie Grove ridge at about 10:30 a.m. This point confirmed his suspicion that he had been deceived and began to make his move towards Herron's divisions via Rhea's Mill. There was a road that connected the mill town to Fayetteville that would give his commissary and supply wagons a clear path for retreat in case the Confederates tried to take Rhea's Mill. Hindman was expecting Blunt to be very late and coming from the southwest. Instead, by using the roundabout route through Rhea's Mill, Blunt was approaching the battlefield from the northwest. At approximately 3 p.m., as Hindman massed his forces in preparation to attack Herron's right flank, two shells fell in front of the massed troops. This artillery fire announced Blunt's arrival on the battlefield. He deployed his three artillery batteries, 2nd Indiana Battery (Capt. John Rabb), 1st Kansas Battery (Capt. Marcus Tenney), and the 3rd Kansas Battery (Capt. Henry Hopkins), and total of sixteen powerful guns, across the fields adjacent to the William Morton house and blacksmith's shop. They immediately went to work dropping shells among the gathered Confederates. There were about twenty civilians in the cellar of the William Morton house during the battle. At the time of the battle Nancy Morton lived with her parents in the Morton house. Nancy recalled in 1896:

"...Early in the day the battle commenced on the Borden farm east of the grove, lasting until sunset, winding up on the Morton farm one mile west. The families were ordered west to the first cellar, which was Morton's. Those in the cellar during the battle were N. J. and J. M. Morton, William Morton, William D. Rogers, wife and three children, A. Borden, wife and five children, Eliza Borden, Dr. Rogers, wife and two children. We all remained in the cellar until dark, but I went into the house several times to get victuals and some bedclothes and wraps for the children. They fought through and around the house, the shots flying like hail in every direction, only a few cannon balls striking close, Mrs. Borden's pony stood hitched close to the cook room, saddled and was not hurt, and after the firing ceased, she and her three children mounted the pony; passed the guards, and rode to Mrs. Mock's in safety..."[25]

Also, in the cellar that day was Caldonia Ann Bordon, who was nine years old at the time of the battle. She recalled the battle in her memoir in 1937:

"...One early morning [December 7, 1862] Pa told us to move out as there was to be a battle very soon on our hill. We went to a neighbor's a mile away, taking what we could carry and some food. The battle started on the hill where our house was. We could hear the cannons and see their heads rise up to fire. We hadn't had any breakfast, we were too excited to be hungry. We stayed out and watched the battle until two o'clock in the afternoon. The cellar was full of apples, potatoes, and barrels of other foods. I had to sit on a big barrel of vinegar and hold my little sister… After dark it got quiet and we came out of the cellar. There was a dead man across the cellar door, wounded and dying men all around. I can still hear them yelling "help-help-help." The men worked through the night helping the wounded. Yankees and Rebels all got the same care. Four died that night. One soldier's leg was just hanging by the skin and the doctor cut it off and threw it outside. It was scary and pitiful. Some of us got sick."[26]

Blunt then deployed Col. Weer and his Second Brigade in a line of battle with the 10th Wisconsin Infantry, 2nd Kansas Cavalry, 11th Kansas Infantry, and the 13th Kansas Infantry from right to left. A small detachment of forty-four Union Cherokees from the 3rd Indian Home Guard under Lt. Gallaher, protected the right flank of the 10th Kansas Infantry. They quickly advanced and pushed into the thickets separating them from the Confederates. They met Roane's and Parson's brigades of Frost's division on what was known locally as "the bench," ground that was lower than the ridge but higher than the fields to the north. The heaviest fighting took place in a ten-acre area around the Morton house and outbuildings. The fighting seesawed back and forth for more than an hour. Eventually, Frost's two brigades were able to push Blunt's Federals off the bench and back onto the corn and wheat fields. The

two Confederate brigades counterattacked but were driven back by the powerful Union artillery. The 1st Texas Partisan Rangers made an attempt to strike Weer's brigade from the west but were beaten off handily. Blunt sent Col. Wickersham and his brigade out to the west to keep any roving Confederate cavalry from surprising his flanks. Col. Cloud and the Third Brigade arrived, and Blunt placed them between Weer's left flank and Herron's furthest right units. The 1st Indian Home Guard of Cloud's brigade and the 20th Iowa Infantry of Col. William Dye's Second Division brigade joined up and made a desperate attack on McRae's brigade near the William Rogers house, near the center of the Confederate battle line. They were also supported by Lt. Stover and his two 12-pounder mountain howitzer battery, who took station in the yard. Major William Thompson of the 20th Iowa Infantry later recalled:

> "…At every step we took, our brave lads fell wounded or killed. I am happy to say but few of our regiment were killed but many were wounded. My horse, a noble animal, was hit twice but I escaped without a scratch… But just at sundown and the very last shot they shot at us… I was hit… The ball struck me in the side of the hip… hurting the leaders and nerves so that I cannot have the free use of my leg for some time."[27]

The 1st IHG was made up of two hundred and fifty Creeks and Seminoles. They advanced across the open Morton hayfield and West wheatfield in what appeared to be a heavy skirmish line. To fight in close order drill was still a mystery to the Union Indians.

Lt. Col. Stephen Wattles reported:

"…On arriving at the battle-ground, we dismounted and entered the wood on the left of the center, with the Eleventh Kansas Volunteers on our right and an Iowa regiment on our left, and rapidly penetrated to the line of battle of the enemy, which gave way on our approach. At this time the Iowa regiment gave our left the partial effect of a volley."[28]

They went up the bench and passed the William Rogers house on the west. They began to receive fire from behind. Inadvertently, they had drifted in front of the 20th Iowa Infantry. This resulted in a five minute panic among the Union Indians. Lt. Col. Joseph Leake of the 20th Iowa Infantry reported:

"…General Blunt having taken position on our right in the middle of the field, I was ordered to move forward in support of his Indians. Skirmishing, I moved forward in line of battle rapidly across the field, obliquing to the left, across the orchard fence, at the foot of the hill; drove the enemy's skirmishers through the orchard, and advanced beyond the fence, through the wood, a short distance. The left wing being more severely engaged, the right had passed farther in advance, through the wood, where some of the Indians came running back through the wood to the right, gesticulating violently and pointing toward the direction whence they came. At this moment an officer shouted to me that we were firing on friends…"[29]

At one point snipers were observed in the W. Rogers house, so Blunt had Rabb's battery open fire on the house, setting it on fire. Again, Hindman's superior numbers came into play and the Iowans and Union Indians were driven back from the bench. The 1st IHG fell back to a small hill north of the bench.

As darkness began to fall early on this December day, Hindman wanted to conduct a full-scale counterattack on the Second and Third Brigades of Blunt's First Division. Parsons's and Shaver's brigades drove northward from between the Morton and W. Rogers houses. Parsons shouted to his Missouri troops: "My brave soldiers, these cut-throats stand between you and your outraged homes – cut them down and stamp them to earth. Give them cold steel: charge bayonet!" The Federal 10th, 11th, and 13th Kansas Infantry regiments and the 2nd Kansas Cavalry held their own on the bench around the Morton orchard but were overwhelmed by the massed Confederate forces of Frost's two brigades along with support from the 7th Missouri Infantry (CSA) from Roane's brigade. The Union regiments were forced back until they were under the protection of their artillery. A very sad event took place at about this time. Col. Cloud ordered Capt. Rabb to fire incendiary shells into the haystacks that were scattered around the battlefield to give the Union gunners better light in the growing darkness. Capt. Rabb stated in his report:

"…by your order, I threw several shell into a straw pile, near the edge of the timber, around which large bodies of the enemy swarmed. The straw was soon ignited, and again we opened with canister for about fifteen minutes. My guns were worked rapidly, making sad havoc in the ranks of the enemy, who retreated to the wood. I gave them a few shell as a parting salute..."[30]

Unknown to the gunners, the haystacks were filled with wounded soldiers who had crawled into the stacks to keep warm. Many wounded soldiers from both sides were burned to death.

In a final action, Parsons's Missouri brigade (CS), consisting of the 7th, 8th, 9th, and 10th Missouri Infantry regiments, with support from and Clark's Missouri Infantry and Mitchell's Missouri Infantry regiments, and aided by West's and Reid's artillery, charged past the burning haystacks to strike at Blunt's right flank. The attack was centered on Weer's Second Brigade that had just been driven off of the Morton house area. Blunt ordered Capt. Tenney to bring his six 10-pounder Parrott rifles farther to the right in the fading light to enfilade the Confederate lines. Blunt also dismounted the 9th Kansas Cavalry and the 44 Cherokees of the 3rd Indian Home Guard, to move along with Tenney's battery in the flanking movement. Confederate Brig. Gen. Mosby Parsons recalled:

"...At this time the enemy made furious assaults upon my lines with a galling fire from three batteries, and a concentration of their musketry on my center with the evident intention breaking it. This fire was more terrible than any I had experienced during the present war. About this time... the enemy were massing their columns in great force on my extreme left. I immediately ordered my reserve regiment... to advance in that direction.... In the meantime... General John S. Roane, having ascertained the critical condition of my left, ordered down Col. John B. Clark's Missouri regiment to support me. This it did in the gallant style characteristic of that officer and his gallant command... The enemy was badly beaten... and the sun was getting low... I rode to Colonel Steen and remarked to him that the contest must be closed and that I had determined to charge the enemy with the bayonet..."[31]

Col. Alexander Steen was killed during this final attack. Once again, the crossfire and the Union gunners decimated the advancing Confederates with double canister, sending them reeling back to their lines on the bench. At this point, the battle of Prairie Grove was over.

In the succeeding hours a thirty-six hour truce was arranged between the two armies to gather and treat the wounded and to gather and bury the dead of each side. The three divisions of the Army of the Frontier reported 175 killed, 813 wounded, and 263 missing. The Confederate Trans-Mississippi Army claimed 164 killed, 817 wounded, and 336 missing. The death toll probably rose significantly over the next few weeks as the wounded succumbed to their injuries through infection or loss of blood. Many witnesses remarked on finding the burned bodies of those wounded who sought shelter in the haystacks that were set afire. This may account for the large number of missing on each side. Samuel Pittman was a local resident who was serving in the 34th Arkansas Infantry of Brig. Gen. Fagan's brigade who received permission to check on his home and family. He observed many persons he recognized who had been wounded during the battle. He later recalled encountering one of these acquaintances:

"…It was midnight and up from the orchard came cries and moans, and men calling for water. I remembered seeing as we returned from the last charge in the evening, a young officer whom I had known from childhood, horribly mutilated by a cannon shot, lying by a fence. He was alive when I passed him and begged piteously for a drink. Not being able to respond to his appeal, I only bent over him long enough to see that his wound was fatal and passed on…"[32]

In the early morning light Lieutenant C.W. Huff of the 19th Iowa looked out on the bloody fields of Prairie Grove and recorded in his diary, "I have been an eye witness to the battle fields of Wilson Creek and Pea Ridge but the devastation and destruction visible are nothing in comparison to… Prairie Grove…" The young officer also observed local residents, many of whose family members and friends were serving in Hindman's army, moving among the dead and wounded in a desperate search for loved ones. He told a story of a woman, holding onto two small children, who was looking for her husband and brother. Huff described her actions:

"…At length a smothered groan fell upon my ear & the exclamation – O my brother – escaped her lips. She pushes back to gory locks and low words of grief murmered [sic] over him… A few yards distance lay another brother who she was ignorant of being in the fight. She kneeled down over him, her agony was only a few wild words such as – O God, is this not enough… her children following her not comprehending anything of the scenes around them… The object of her mission on this bloody field was soon accomplished, for only a few rods from where her brothers lay she found her husband… His face showed no agony and I rejoiced that her in her loved one she had been spared the disfiguration which so many had suffered… A wild unearthly shriek and the wife had encircled her arms around the form of her husband… The suffering of that woman none but God can know…"[33]

The morning dawned cold and crisp around Prairie Grove. The battle had been a tactical draw since losses were fairly equal and both armies were still in place. Blunt held the upper hand though. That morning Brig. Gen. Salomon arrived with the First Brigade from Rhea's Mill, which gave the Union nearly three thousand fresh troops with which to renew the conflict. Hindman for his part knew he was going to have to fall back since his ammunition train was nearly thirty miles south of Prairie Grove. Hindman and his generals had met after the firing stopped and, together, they decided to withdraw to Van Buren. The Confederate retrograde began just after midnight. By the time the Union and Confederate commanders had met that morning to discuss the details of the truce, Hindman admitted that most of his army had moved south with only Marmaduke's division still on the field. And these began to withdraw as soon as the leaders' conference was over. The two cavalry regiments that were designated to bury the Confederate dead were caught simply collecting firearms from the battlefield. Notified of this, General Herron advised the two Confederate colonels in command that if their soldiers were found doing anything beyond burying the dead and helping their wounded, they would be taken as prisoners of war. Hindman followed his army as they re-crossed the Boston Mountains back to Van Buren, then crossed the Arkansas River to Fort Smith.

Federal soldiers examining the former Confederate lines discovered a substantial number of unfired minie and musket balls littered over their part of the field. It was learned from Confederate prisoners that the Arkansas conscripts had been loading their weapons with paper and powder, disposing of the ball, and just firing blanks at the Union soldiers. Maj. Ellithorpe, in his role as the anonymous journalist for the Chicago Evening Journal, wrote:

> "…From prisoners and deserters we learn that the Arkansas conscripts were forced to the front during the battle, and compelled to fight – guided and forced on by the butternut rebels of Missouri and their regiments of regulars from Texas, Thus the partially loyal men of Arkansas, who would have been glad to remain quietly at home, have been forcibly conscripted,

led to the front ranks and slaughtered like so many sheep. General Hindman is guilty of these murders…"[34]

An interesting note that Ellithorpe noted was when the Union prisoners were exchanged, they had been stripped of their heavy overcoats, accoutrements, and in many cases their boots, especially cavalrymen's boots. It seems the Confederacy's largest supplier of arms and equipment was the Union Army.

Blunt and Herron received information from Schofield that he was returning to field service and ordered them to withdraw back to Missouri. Blunt wrote in his post-war memoir:

> "…I received a telegram dated at Wellsville (between Springfield and Rolla) from General Schofield, who had recovered his health, or in other words had failed to secure the promotion to major general, that he went to St. Louis for, and was returning to the command that he had, two months before, deserted… I considered that a decidedly cool proposition to come from an officer who had deserted his command in the face of the enemy, and immediately replied to him that 'I was in command of the Army of the Frontier, and that until a superior officer arrived there and assumed command by general order, I should direct its movements, and that I should commence moving on the enemy at Van Buren at daylight the next morning'…"[35]

Both Union commanders believed

that Hindman was weakened and the opportunity to destroy his Trans-Mississippi Army was at hand. Blunt later recalled:

> "…On the 25th of December, I learned through my scouts and spies, that Hindman had been reinforced at Fort Smith, with nine thousand infantry from Little Rock and that he contemplated moving against me again and risking another battle, and I at once determined to 'beard him in his own den.' Hindman's forces were on the south side of the Arkansas river, and knowing the facilities he had for ferrying them across at Van Buren, I was convinced that he could not have more than half his force on the north side before I could reach that point; and although the proposition was dissented to by all my subordinate commanders, I determined to move on him rapidly, surprise and attack him in detail, or in other words, while the river divided his force, to defeat those on the north side…"[36]

Disregarding Schofield's orders, because they believed that the commanding general would waste the opportunity, the combined Army of the Frontier pushed south across the Boston Mountains. By effectively using his Union cavalry and artillery along the shores of the Arkansas River, they entered Van Buren on December 28, capturing hundreds of Confederate soldiers and burning military stores, four steamboats, and a ferry. Blunt gained information that Frost's division was camped about five miles below Fort Smith on the south side of the river. Blunt sent some artillery batteries to a location on the north side of the river and began to lob

shells into the Confederate camp, causing them to evacuate the area. Hindman correctly believed that his small remaining force was not in any condition to stop the Federals. He abandoned Fort Smith, destroyed all of his military stores he did not have transportation for, along with two steamboats, the *Eva* and *Arkansas*, filled with supplies, and moved his remaining forces downriver towards Arkadelphia.[37] Blunt and Herron, failing to get Hindman to fight, decided to not cross the Arkansas River and began to withdraw northward towards the Cane Hill area. When they met with and reported their activities to Schofield at Dripping Springs, the commanding general showed his true colors by court-martialing both. General Schofield was deeply offended by having two aggressive subordinate commanders winning battles and gaining headlines while he was in bed being nursed for some malady.[38] Wiley Britton later wrote:

> "…why continue to fall back and give up the country we have gained at the cost of so many lives and of so much toil and suffering? Is it because the present Commanding General did not direct the movements of our army in gaining the splendid victories that we have won? The jealousies of military rivals have already in other instances been a curse to our arms…"[39]

While all of this was being played out back in Arkansas, on December 23 Blunt sent the 1st Indian Home Guard, 2nd Indian Home Guard, and detachments of the 3rd Indian Home Guard and 6th Kansas Cavalry, approximately twelve hundred troops under the command of Col. William Phillips, from their camps near Cane Hill and sent them into the Cherokee and Creek Nations. Maj. Ellithorpe wrote in the Chicago Evening Journal:

> "…The object was to penetrate these countries, and drive the enemy out, and at the same time assist the suffering Union families to remove to some place of security. The command passed through the entire Cherokee country by the way of Park Hill and Taliqua [sic] to Fort Gibson, surprising and capturing a small party of Col. McIntosh's command at the Fort…"[40]

On December 27 Phillips moved his command down to the Arkansas River and crossed at Frozen Rock, and drove into the lightly defended Fort Davis (General Pike's million dollar mistake). The Union Indians set fire to all of the fort's buildings in the evening and watched it burn into the night. Col. Phillips reported on December 28:

> "…I only got in half an hour ago from the Creek Agency. I drove the enemy toward the Canadian and Red Rivers; crossed Arkansas River, with my whole force, at the Frozen Rock Ford; took and burned Fort Davis, reducing all the barrack and commissary buildings and the whole establishment to ashes…"[41]

Phillips later wrote to Schofield:

> "…I burned Fort Davis, to root the rebel army out of "house and home," on the south bank of the Arkansas River, and as an exhibit of power to affect the Indians. It was no mere

wanton destruction. I treated the private property of even rebels scrupulously, so as to pave the way for negotiations, and spared the house of Colonel McIntosh, near Fort Davis, although sold by him to the rebel Government, through Albert Pike. In a note to him I told him I spared the house as private property, not regarding the sale…"[42]

Neither Cooper nor Watie were in any condition to make any effort to oppose Phillips. Maj. Ellithorpe continued his statement:

> "…We then crossed the Arkansas river near the junction of the Grand [Neosho] river, captured and burned Fort Davis, and destroyed what stores were there. The buildings were fired just at sunset. The dense columns of smoke from over fifteen buildings, and the lurid glare of the flames, were truly a grand sight. The vast prairie country rendered this beacon of destruction visible for many miles… I then tooke [sic] a party of 200 men & made a dash of five miles up the river to the Old Creeke [sic] Agency driving the forces of McIntosh & burning all the Rebel Govt property I could find…"[43]

Phillips also believed that with this action he was going to be able to bring the Creek Nation over to the Union side. He reported to Blunt:

> "…I have entered into negotiation with Colonel McIntosh, and am to

meet him to-morrow. I expect to disarm or bring over the whole Creek Nation. I sent messengers to the Choctaw Nation, and was in hopes of opening the gates to Texas through friends… Your order breaks off my negotiations, and I start for the place you ordered me to in one hour…"[44]

Unfortunately, Blunt recalled the entire Indian Territory detachment due to Schofield's orders to reform at Prairie Grove before Phillips could finalize any agreements. Schofield had no plan or vision for any further advancements. He clearly wanted all three divisions together for the simple reason to keep the Army of the Frontier under his direct control. Blunt was relieved of his command, and it was passed down to Col.

Weer. But the Union Indian forces did make a reappearance in the Indian Territory and showed it could move in and out of the Indian Nations at will.

A further disappointment for Phillips was that Col. McIntosh was not sincerely negotiating for a surrender of the Creek Nation. The Confederate Trans-Mississippi Department knew about the offers that Phillips had made because McIntosh was providing all of his contact information and documentation with his commanders. Perhaps McIntosh was hoping that by knowing the Union was making peace offers, he could use this information to spur the Confederate government to provide a better deal for the Creeks.[45]

---

[1] Shea. *Prairie Grove,* pp. 111-113.
[2] Watie to Cooper. *OR,* Ser. I, Vol. XXII (P1). pp. 66-67.
[3] Britton. *Brigade.* pp. 118-120.
[4] Shea. *Prairie Grove*, pp. 130.
[5] Farlow, Joyce and Louise Barry, eds., "Vincent B. Osborne's Civil War Experiences," *Kansas Historical Quarterly*, 20 (1951) pp. 204.
[6] Gardner to Cross. *OR,* Ser. I, Vol. XXII (P1). pp. 62.
[7] Richardson to Wood. *OR,* Ser. I, Vol. XXII (P1). pp. 87.
[8] There is disagreement about this timeline, however. Blunt wrote in his report that he sent Richardson's command to Hogeye Crossing on the evening of December 6. Other historians state that Richardson was not sent to Hogeye Crossing until around 6 a.m. on December 7, (Shea. *Prairie Grove,* pp. 202-203) In his official report Richardson indicates he was out on Hogeye Road that night when he encountered an element of the 9th Kansas Cavalry who had been driven from Hogeye Crossing.
[9] Blunt to Curtis. *OR.* Ser. I, Vol. XXII (P1). pp. 72-73.
[10] Blunt to Curtis. *OR,* Ser. I, Vol. XXII (P1). pp. 72.
[11] Shea. *Prairie Grove*, pp. 138-139.

[12] Shelby to Marmaduke. *OR,* Ser. I, Vol. XXII (P1). pp. 149.
[13] Shelby to Marmaduke. *OR,* Ser. I, Vol. XXII (P1). pp. 150.
[14] Bunner to Brawner. *OR,* Ser. I, Vol. XXII (P1). pp. 114.
[15] Brawner to Clark. *OR,* Ser. I, Vol. XXII (P1). pp. 113.
[16] Shea. *Prairie Grove*, pp. 137-143.
[17] Herron to Curtis, *OR,* Ser. I, Vol. XXII (P1). pp. 102-103.
[18] Sallee, Scott E. "The Battle of Prairie Grove: War in the Ozarks, April '62 – January '63.' Blue & Gray Magazine. David Roth, Ed. Volume XXI, Issue 5. Fall 2004. pp. 21-22.
[19] Prairie Grove Battlefield State Park: Interpretive Marker #6
[20] Bertram to Herron; *OR,* Ser. I, Vol. XXII (P1). pp. 128.
[21] Clark to Huston. *OR,* Ser. I, Vol. XXII (P1). pp. 111.
[22] Prairie Grove Battlefield State Park: Interpretive Marker #9
[23] The Battle of Prairie Grove: Park Booklet. pp. 20.
[24] Shelby to Marmaduke. *OR,* Ser. I, Vol. XXII (P1). pp. 151.

[25] Nancy Jane Morton Staples Civil War reminiscences, Arkansas State Archives, Little Rock, Arkansas.

[26] Caldonia Ann Borden reminiscences, Arkansas State Archives, Little Rock, Arkansas

[27] Prairie Grove Battlefield State Park: Interpretive Marker #11.

[28] Wattles to Cloud, *OR*, Ser. I, Vol. XXII (P1). pp. 93-94.

[29] Leake to Lake, *OR*, Ser. I, Vol. XXII (P1). pp. 121.

[30] Rabb to Cloud. *OR*, Ser. I, Vol. XXII (P1). pp. 100.

[31] Prairie Grove Battlefield State Park: Interpretive Marker: West Overlook

[32] Sallee. pp. 46.

[33] Sallee. pp. 47.

[34] Johansson, *Ellithorpe*, pp. 97.

[35] Blunt, *Experiences*, pp. 235.

[36] Blunt, *Experiences*, pp. 234.

[37] Britton. *Brigade*. pp. 162-163.

[38] Sallee. pp. 46-47.

[39] Wiley Britton, *Memoirs of the Border, 1863*. Chicago: Cushing & Thomas, 1882, pp. 80.

[40] Johansson, *Ellithorpe*, pp.107.

[41] Phillips to Blunt, *OR*, Ser. I, Vol. XXII (P1). pp. 881.

[42] Phillips to Schofield, *OR*, Ser. I, Vol. XXII (P2). pp. 62.

[43] Johansson, *Ellithorpe*, pp. 105-107.

[44] Phillips to Blunt, *OR*, Ser. I, Vol. XXII (P1). pp. 881.

[45] Cooper to Hindman, *OR*, Ser. I, Vol. XXII (P2). pp. 770.

Chapter 12
The Second Federal Invasion of the Indian Territory

The dawn of the new year of 1863 brought about a great many changes to the forces making up the militaries of the Trans-Mississippi West. In the Western Theater on New Year's Day, 1863, Maj. Gen. William Rosecrans was successful in pushing back Confederate General Braxton Bragg's Army of Tennessee at the battles of Stones River and later the Tullahoma Campaign in the Spring, resulting in Bragg abandoning Tennessee. And although Maj. Gen. Ulysses Grant was successful in defeating Maj. Gen. John Pemberton and his Army of Mississippi in a series of battles in central Mississippi, he found himself bogged down in a siege around Vicksburg after Pemberton pulled his troops into the substantial defenses in and around the city. In the Eastern Theater, after the victory at Antietam, President Lincoln finally fired Maj. Gen. George McClellen for inactivity and insubordination. This was followed by the disastrous defeat of the Army of the Potomac under its new commander, Maj. Gen. Ambrose Burnside, at Fredericksburg, Virginia, in early December 1862. This was followed by the equally embarrassing Mud March to the North Anna River later in the month where Burnside attempted a not well thought out winter campaign against the Army of Northern Virginia. On a positive note, the anticipated long-range effects of the Emancipation Proclamation, which became law on January 1, 1863, were soon becoming realized. The newly freed men could now be used for the

Fort Leavenworth, Kansas
Photo courtesy of the Frontier Army Museum

Union cause in any number of ways, including as soldiers to fill the ranks of the Federal Army. But the end of the war was still over two years away.

After the December victories for the Union Army of the Frontier at Cane Hill, Prairie Grove, Van Buren, and the destruction of Fort Davis, Brig. Gen. John Schofield wrote to Maj. Gen. Samuel Curtis, undermining Brigadier Generals James Blunt, Francis Herron, and James Totten. He sticks the knife especially deep into Blunt's back by trying to have him reassigned out of the Army of the Frontier. Since these victories were achieved during Schofield's "sick leave" in St. Louis, his petty nature got the best of him. He stated to Curtis:

> Fayetteville, January 1, 1863.
> Major-General Curtis :
> General Blunt desires to go to
> Leavenworth to attend to business

connected with his district. I do not feel at liberty to withhold my consent, since, as district commander, he is independent of me. He can, doubtless, be spared better now than a month hence. If General Blunt is to retain command of his division, it seems to me that it will be necessary to place the Kansas district under some other officer, or, if he is to retain command of the district, he should be relieved from that of the division in the field. The latter would, I believe, be the wiser arrangement of the two. The operations of the army, since I left it, have been a series of blunders, from which it narrowly escaped disaster where it should have met with complete success. At Prairie Grove Blunt and Herron were badly beaten in detail and owed their escape to a false report of my arrival with re-enforcements. I state this simply as a fact which it is my duty to let you know, without intending to pass censure upon any officer. This it [sic] would be improper for me to do without seeing their official reports, which I have not.

J. M. SCHOFIELD,
Brigadier-General[1]

The report that the Confederates believed Schofield was coming with reinforcements was obviously fabricated to simply enhance his position because there is nothing in the records that indicates that the Confederates believed that Schofield was enroute with reinforcements. In fact, Curtis rebuked Schofield for submitting his comments on the conduct of the battle before the official reports had been filed, especially since he had not been present at any of the engagements. In the following days, further exchanges between Schofield and Curtis became strained since Schofield obviously wanted Blunt out of his command.[2] Blunt was removed from command of the First Division on January 3, 1863, and proceeded to Fort Leavenworth to oversee the District of Kansas from that location. Blunt never really reported on why he quickly accepted the new organization of the Army of the Frontier. In later years he wrote:

> "...the arrival of Schofield defeated all further plans, and on the third of January I left the "Army of the Frontier" and proceeded to Fort Leavenworth to attend to the administration of affairs in my district, that had been much neglected in my absence. My geographical district now comprised Kansas, the Indian Territory and western Arkansas... Before leaving Arkansas, I made application to General Schofield for troops to hold the conquered territory then embraced in my district, and for which I was responsible..."[3]

Wiley Britton of the 6th Kansas Cavalry later wrote about his feelings regarding Blunt's departure from the Army of the Frontier:

> "...Gen. Blunt has been relieved, and bade his troops farewell to-day, and, with his staff and escort, started to Forts Scott and Leavenworth. On account of his personal bravery and the brilliant achievements of his

campaign, he has greatly endeared himself to his troops. I speak from personal knowledge of his bravery. He was to the front all day during the battle of Cane Hill and was only a few yards from Col. Jewell when he fell mortally wounded. At Prairie Grove too, he was on the field all the afternoon in dangerous positions, directing the movements of his troops. And at Dripping Springs he was at the front with us when we charged the enemy's camp, and rode with the advance squadrons when we dashed into Van Buren..."[4]

Major Ellithorpe of the 1st Indian Home Guard was also perplexed at the removal of Blunt from command of the First Division. He wrote in the *Chicago Evening Journal*:

"...The reasons for this change have not transpired to the army here. History will accord to Gen. Blunt the praise of planning and perfecting one of the most successful campaigns of the present war. He has surmounted the thousand little obstacles that, to many Generals, would have been sufficient excuses for delay. Quick to perceive, prompt to execute, and possessed with an indomitable energy, he has led his little army to glorious victories in every important engagement. All the country north of the Arkansas river is now free from any organized rebel forces..."[5]

In response to the new challenges, in the new organization Schofield placed Col.

William Weer in command in Blunt's place. Brig. Gen. Fredrick Salomon was probably next in line for that position but, at some point between the end of the Prairie Grove campaign and the first part of January 1863, he was reassigned to the First Brigade, 13th Division, XIII Army Corps, Army of the Tennessee, in the Helena, Arkansas, area as a brigade commander. He probably understood that his career was going to be limited within the Army of the Frontier.[6] Schofield also reorganized the regiments and batteries of the First Division. The new Third Brigade became the "Indian Brigade," and was comprised of the three Indian Home Guard regiments, a battalion of the 6th Kansas Cavalry, and Hopkins's 3rd Kansas Artillery. Col. Phillips was given command of the new brigade but was to serve under Blunt's command. This brigade's mission was to protect the Union sympathizers along the Arkansas and Indian Territory border and within the Territory itself. They were to provide supplies to these sympathizers and attempt to bring other Confederate tribes, especially the Creeks and Choctaws, into the Union fold. They were also to return the relocated Union Indians in Kansas back to their homelands as soon as possible. Unfortunately, squabbles among the Union commanders in Missouri and Arkansas made this mission difficult to accomplish since Blunt was the only Union general to believe that the Indian Territory was worthy of attention.[7] But, for the time being, Blunt needed to contend with another rash of border issues in Kansas from a group known as the "Red Legs." Blunt described this group:

"...During my absence in the field, matters left in charge of subordinates had been running rather loosely in the district. Among other things, an organization had sprung into existence known as 'Red Legs,' and whatever had been the primary object and purpose of those identified with it, its operations had certainly become fraught with danger to the peace and security of society. The organization embraced many of the most desperate characters in the country… A reign of terror was inaugurated, and no man's property was safe, nor was his life worth much if he opposed them in their schemes of plunder and robbery. In this condition of things I considered it my duty to interfere for the protection of honest and peaceable citizens, and to a great extent was successful…"[8]

After the new Union Indian Brigade was established with Phillips in command, he wrote a very open and honest assessment of the three Indian Home Guard regiments under his command. His first-hand knowledge of the operations and inner workings of these unusual regiments is important information for the understanding of how they were able to perform within a white-man's army. His description is presented here in its entirety:

HDQRS.3d Brig., 1ST DIV., ARMY OF THE FRONTIER, Camp Curtis, January 19, 1863.
Maj.-Gen. CURTIS, Cmdg.
Department of the Missouri:

SIR: I desire to report the peculiar features, character, and present condition of the three Indian regiments. My close connection with them in active service during the past nine months has given me opportunities to judge, and I submit a report as brief as it can be made, believing it is necessary to give the Government a clear idea of the nature and wants of this branch of the service.

1st. The First Indian Regt. is of Creeks, mustered at Leroy. The only white officers at first were field officers. The regiment did some service in June and July; it became badly demoralized for want of sufficient and competent officers; partially broke up in August; was collected in October, and had white first lieutenants mustered, under Gen. Blunt's order. Some 300 or 400 of the regiment, who had gone to Leroy in August, and who had refused to leave it, got down with the train just at the time the Army of the Frontier was rebrigaded.[sic] The regiment had drilled very little; are indifferently informed as to their duties. These Creeks are about equal in scale of intelligence to the Delawares of Kansas; they are inferior to the Cherokees. They are in bad shape, get out their details slowly, sometimes desert a post, or a party when sent on duty; yet I would be lacking in my duty to them or the Government if I failed to say that, with one or two good field officers, military men, and two, or even three, company officers, they could be made very effective. No party of them should be sent without a competent officer. Their own officers are, with few exceptions, useless, but there are one or two men of influence amongst the captains, brave fighters in the field, and of influence not be overlooked. This Creek regiment gives me much more concern than either of the others.

2d. The Second Regt. originally consisted of Osages, Quapaws, &c., and, when it got into the Cherokee Nation, finally of Cherokees. The Osages, who were neither more nor less than savages and thieves, who brought the whole Indian command into disgrace, were finally mustered out during one of their periodic desertions, which fortunately happened at pay time. So of the Quapaws and other broken fragments of tribes that were little better. Under Gen. Blunt's orders, I recruited for the Second Indian Regiment, and its numbers have been brought up to its present status (see reports) from Cherokees, half-breeds, and whites. Last summer the regiment drilled but little; lately it has improved in that respect. It still lacks necessary officers, but is in a fair way to make a useful force.

3d. The Third Indian Regiment, which was my own, rejoined after its organization, was literally taken from the enemy, and was the heaviest blow dealt in the Southwest last summer. Profiting by the experience of the first two regiment[s], it was organized by

Gen. Blunt's orders, at my suggestion, with first lieutenant and orderly sergeants picked out of the white regiments in the field. I endeavored to secure active, intelligent men, conversant with their duties as soldiers, or non-commissioned officers, and just so far as I succeeded in this the result has been favorable. Unless when on actual march, the regiment had dress parade every evening, and drill and officers' schools every day. The result is that it is as well drilled as many white regiments that have been a longer time in the service. The regiment has done a great deal of active service, besides innumerable scouts and skirmishes. They were for two hours and forty minutes under his musketry and finally artillery fire at Newtonia. They participated at Fort Wayne, Cane Hill, Dutch Mills, Prairie Grove, and other engagements. This is the only Indian regiment that is really so far, although the Second undoubtedly will be, but there are several errors in its organization, and some few of this command and also the Third absent themselves without leave, which is a chronic Indian weakness. The error in all the Indian regiments has been in not mustering the captains or white officers to be fully responsible for property, and to see orders carried out. I take the liberty of suggesting that the necessary officers for an Indian company are, the captain (first lieutenant might be an Indian) and second lieutenant white men; or, better yet, the captain a white man, first lieutenant a white man, second lieutenant an Indian, an orderly sergeant a white man. The white men to be selected from the volunteer army, or from men who thoroughly understand military duties, and who will work hard. It is a blunder to put men of poor ability in an Indian regiment. It requires character, so that the Indians will respect him, and a thorough knowledge of military duties. In a white company, if the captain and lieutenants are ignorant, perhaps some privates in the company can run it, but an Indian company improperly officered is in a frightful mess. The officers in an Indian regiment have to work very hard to get things in shape. The besetting sin of Indian is laziness. They are brave as death, active to fight, but lazy. They ought invariably to be mounted; they make poor infantry, but first-class mounted riflemen. The Third Regiment, most of the Second, and half of the First entered the service with their own horses; were paid as infantry, but foraged and shod by department order of Gen. Blunt. Their horses have nearly all been used up in the service. At this time the stock is very poor. The Third Indian Regt. is of twelve companies mounted riflemen, and has two howitzers attached. They are only paid as infantry, but used as mounted men. About 100 of them are on foot, as their horses have died in the service. To be efficient they ought to be mounted on Government horses in the spring. The Third is armed with

Mississippi and Prussian rifles. The Second, Prussian rifles and muskets, and the First with hunting rifles, and have to mold their bullets. Nothing but active steps to supply necessary orders can save the First Indian Regt. from utter demoralization. My orders to drill are disregarded. As I compel the regiments to draw on consolidated provision returns, I have difficulty in getting reports from them. I am much embarrassed, as arresting all the officers of a regiment is not to be thought of, and permitting it to run loose has a bad effect on the rest. I earnestly desire instructions and necessary authority to myself or some others. In the mean time I shall do the best I can.

With great respect,
WM. A. PHILLIPS,
Col., Cmdg. Third Brigade.[9]

It should be understood that the addition of white officers from the volunteer regiments was not necessarily going to be any better. Few would have had any real military experience, and they would still have language issues in providing instruction or in giving orders to the non-English speaking Indians. But with this plan, the Indian regiments did improve, and they began to train and drill in the same manner as the white regiments. Yet, even Col. Phillips had his detractors, especially Maj. Ellithorpe of the 1st Indian Home Guard who wrote in his journal:

"...-- Col Phillips, the former Major of this Rgt & now Col of the 3rd Cherokee – is put in command of the brigade. He has not got this position because he has earned it for I have done much more labour [sic] in the Indian service than he has[.]

He takes the prefferance [sic] simply because he is a Kansas man, his abilities as a military man are of a very low order – his acquirements as a business man, are only medium. But every favored Kansas politician, of the Jim Lane school, & who will pledge a future support to Lane for U.S. Senator, can obtain promotions irrespective of personal quali-fications…"[10]

Ellithorpe especially took offense to Phillip's assessment of the 1st Indian Home Guard stating it was deficient and un-disciplined to which Ellithorpe disagreed. Although Ellithorpe's assessment of Col. Phillips was meant to be private, it is also plainly obvious that he was more than likely jealous of not being selected as the com-mander of a regiment or the new brigade. Even though Ellithorpe seemed to be in command of the 1st IHG for longer periods of time than Lt. Col. Wattles, he was not selected for advancement to that position. He reported "Lieutenant-Colonel Wattles's resignation is returned as informal. Although serious charges are preferred against him, he is yet in command…"[11] In fact, Wattles not only avoided any repercussions for the theft of the money he owed his soldiers, but he also remained in command of the 1st Indian Home Guard until the end of the war, even reaching the coveted rank of Colonel. As discussed in an earlier chapter, Ellithorpe had agreed to drop the charges against Wattles if he repaid the soldiers for the money he stole and

resigned from the Army. Wattles repaid the missing funds but reneged on resigning from the Army, and Brig. Gen. Blunt did not push the matter.[12] Still yet, Ellithorpe had a positive attitude about the potential of the new brigade:

> "…This brigade, well equipped and filled, will hold the Indian counties, and I am of the firm opinion that the Indians can be used in no other locality to, so good an advantage; in fact, I believe that to divert them to any other field of operations than the Indian counties will tend to demoralize them to dissolution…"[13]

Very little military action took place in the Indian Territory during the late winter and early spring of 1863. In an effort to suppress the bushwacking activities in the Cherokee Nation within the Honey Hills and near Spavinaw Creek, Col. Phillips moved the Union Indian Brigade (this will be the designation of the Third Brigade for the remainder of the book) to Maysville, Arkansas, near the site of Old Fort Wayne. From this position he could monitor activities in the Indian Territory and also be within reach of the other brigades of the First Division. Phillips sent out patrols to monitor any Confederate activity along the Arkansas and Indian Territory border. On January 15 a patrol from the 3rd Indian Home Guard under Capt. H.S. Anderson ran into a 200-man detachment of Confederate Partisan Rangers under Col. Thomas Livingston near Spavinaw Creek, twelve miles south of the brigade's encampment at Maysville. The IHG unit separated into three columns and attacked from different directions, pushing the

Confederate forces from one column to another until the guerillas finally broke and scattered.[14] Wiley Britton later wrote:

> "...Captain Anderson, having ascertained the position of the enemy, attacked them with such energy that they soon broke up into small parties, and were pursued until they were lost in the woods and hills of that region... In the action and in the pursuit. Captain Anderson reported one of his own men killed, and eight of the bandits killed, including Captain Smith (CS)..."[15]

The newly minted Lieutenant Colonel Albert Ellithorpe was a part of this action and wrote this report:

> "...I proceeded the advance of our brigade from Elm Springs to this place in command of 500 mounted men. The distance of 25 miles I made during the night, and surprised a party of Livingston's gang. I killed 9 and captured 13 of the gentry. We have sent the prisoners on to Ft. Scott. The country is still full of marauding devils. This light snow for the past 2 days enables us to track them to a charm. One of the devils that I took had the scalp of one of our soldiers in his pocket..."[16]

With most of the Confederate Indian Brigade located south of the Arkansas River, and a large number of them even further south beyond the Canadian River, movement of Union forces in the Indian Territory resumed with little opposition. Since the area was fairly safe for the time being, it was determined by the Unionist Cherokees that it was time to conduct a meeting of the National Council to determine the future of the Cherokee Nation. Phillips reported this development to Curtis on January 29:

> "...I deem it proper to communicate to you that there is a proposed meeting of the Cherokee council and committee. In the Second and Third Indian Regiments are a quorum of these bodies, and other loyal citizens not in the army are near or with it. I think it desirable that the representative bodies of the Nation should meet. The acting chief, Captain Pegg, is here. I understand they propose rescinding the ordinance of secession, that was forced on them, and of other actions of a similar nature, and loyal demonstrations to the Government will likely follow..."[17]

Beginning on February 3, the Unionist Cherokees gathered in council at the Cowskin Prairie in the far northeastern Cherokee Nation. Back in August 1862, Chief John Ross had left for exile in Washington, D.C., and he faced the daunting task of convincing the United States that the Cherokees had in fact remained loyal to the Union.

Complicating this position was the existence of a rival Confederate Cherokee government led by Stand Watie. Watie's government not only reaffirmed the nation's treaty with the Confederacy, but it also declared Ross a traitor. More problematic, Watie's forces controlled some parts of the Cherokee Nation, but due to the limited size of the 1st Cherokee Mounted Rifles, they spent most of their time raiding and harassing the Unionist Cherokees or raiding the smaller towns in Kansas and Missouri. The Cherokee National Council's meeting at Cowskin Prairie was an attempt to strengthen their extremely difficult political situation. The council was composed primarily of Cherokee soldiers serving in the Union's Indian Home Guard regiments, many of whom were probably members of the Keetoowah Society. Not surprisingly, many were also members of the same National Council who had voted to join the Confederacy back in October 1861. The council officially elected Thomas Pegg as the acting chief of the Cherokees until John Ross's return, disavowed the treaty with the Confederate government, declared Stand Watie and his followers as outlaws to the Cherokee Nation, and ordered the confiscation of their property. Pegg argued that the Confederate treaty had been coerced, and that the tribe had always remained loyal to the Union. The council's other primary decision was to abolish slavery in the Cherokee Nation, which would bring the Cherokee government in line with the United States government and its implementation of the Emancipation Proclamation. Between February 18-23, 1863, the Cherokee National Council passed the following legislation:

An act revoking the alliance with the Confederate States and re-asserting allegiance to the United States.

An act deposing all officers of any rank or character whatsoever, inclusive of legislative, executive, judicial, who were serving in capacities disloyal to the United States and to the Cherokee Nation.

An Act Providing for the Abolition of Slavery in the Cherokee Nation

An Act Emancipating the Slaves in the Cherokee Nation[18]

The Acts were to become effective on June 25, 1863. Unfortunately, the council members simultaneously made it clear that freed slaves had no right to citizenship and were expected to immediately leave the nation. (This requirement or restriction became a significant legal issue since the Cherokees insisted that only they had the right to determine who was a tribal member or not. This question was not resolved until the 21st century.) Throughout the meeting, the council affirmed its ties with the Union while simultaneously asserting their right to govern the Cherokee Nation. The council primarily directed its actions toward the United States, leaving no doubt of its opposition to Stand Watie's Confederate government. These acts by the council established two separate governments of the Cherokees. In order for the Union-backed government to take precedence, the Union army would have to establish itself as the dominant force and drive Col. Watie and his forces completely out of the Cherokee Nation.[19]

Colonel Stand Watie did attempt to disrupt the February meeting by sending a force under Maj. J.M. Bryan's 1st Cherokee Cavalry battalion (CSA) to cross the Arkansas River at Fort Gibson, and one force under his own command to cross at Webber's Falls. They then received word of a patrol led by Lt. Col. Lewis Downing of the 3rd IHG in the vicinity of Fort Gibson and quickly abandoned their plans to disrupt the meeting.

Since there was an official quorum present at the council meeting, the United States accepted the provisions outlined in the official acts. The Cherokee Nation was again officially aligned with the Federal Union.[20]

The Confederate command of Indian Territory changed again during the early part of 1863. In January, the Indian Territory was detached from the District of Arkansas and was placed directly under the control of the Trans-Mississippi Department as an independent command. Maj. Gen. Thomas Hindman was transferred to a division command at Vicksburg. Brigadier General William Steele, a native New Yorker who had married into a wealthy Texas family, was given command of the Indian Territory District, with his headquarters at Fort Smith. He was essentially taking the position previously held by Albert Pike. He diligently tried to exercise control over his widespread command and provide for the needs of the units in the field. His field commander in the Indian Territory continued to be Brig. Gen. Douglas Cooper. Cooper, for his part, attempted to forestall any Union invasion by sending his scattered units, as well as Partisan Rangers, to harass the Union forces, hoping to buy time for General Steele to gather more troops from Texas and Arkansas for duty in the Indian Territory. But Cooper was again in full retreat after Blunt's Federals captured Van Buren. Private Dallas Bowman of the 1st Creek regiment wrote:

"…All the Federals came to 4 miles of Ft. Smith, it was reported they were on to Scullyville and Gen. Cooper not having enough sufficient force to fight them we retreated to this camp about 80 miles west of Scullyville. Col.

[Walter P.] Lanes regiment was engaged when the Federals came down they had several heavy skirmishes with the Pins. Lewis Attaway was killed and Martin Craig was taken prisoner…"[21]

An interesting point that Steele made in a January 13, 1863, letter to the Confederate governor of Arizona was that he was now the commandant of Indian Territory and the "*ex officio*" Superintendent of Indian Affairs. It is not clear if he was assuming that role from Cooper or if he simply assumed it came with his position. In a separate turn of events, Steele wrote to Cooper on January 18 complaining:

"…Since the retirement of General Hindman from this vicinity, the intermediate country between this and Dardanelle [Arkansas], and the south of this in the direction of Waldron, has been infested with lawless bands of robbers and murderers. These bands are composed chiefly of Union men and deserters from General Hindman's army… Several of the most respectable citizens of the valley of the Arkansas have been murdered, and numerous robberies committed by these outlaws. Having received no intelligence from headquarters for some time, I am induced to believe that my communication in that direction has been cut off. I have sent in pursuit all the cavalry at my command…"[22]

In other words, the Confederates were dealing with the same type of bushwacking activities as they were conducting against the Federals up in Indian Territory, Kansas, and Missouri.

The Confederate high command in Richmond once again directed department commanders to aggressively move north, especially in the west, to attempt to relieve pressure on the besieged city of Vicksburg. These actions were to coincide with General Robert E. Lee's planned invasion of Pennsylvania.[23] Unfortunately, the Trans-Mississippi Army was in poor condition and not in a position to do anything but defend the areas they currently occupied. Steele wrote after his arrival at Fort Smith: "…Thus impressed, I ordered the main body of the troops in the Territory to encamp as near Red River as was convenient…" The supplies directed to Holmes and Steele's commands were routinely diverted to other Confederate commands at Vicksburg, Port Gibson, or Louisiana.

Federal Colonel Phillips had stationed himself at Camp John Ross, adjacent to Park Hill in the Cherokee Nation. From this location he attempted to relieve the food shortages crippling the Cherokee people. On February 4, the same day the Cherokee National Council began at Cowskin Prairie, he wrote to Maj. Gen. Curtis:

"…I left a post at Maysville, near Camp Curtis, of about 200 men to guard my connection to the Arkansas and to Fayetteville, and to run a small mill that otherwise would feed the rebel guerrillas.

I sent another train of provisions down toward Fort Gibson, to relieve destitute and starving citizens. I have a distributing agent at Park Hill, and one at Hildebrand's Mill, about the center of the Nation, where I have a company running the mill. There is no grain there, and I have to supply it from above and east of this place.

The extreme want of the people below here steadily assumes a more serious cast. My movements are much embarrassed for want of transportation… Have sent two trains of flour and meal into the Indian Nation, and have subsisted about 1,000 starving refugees, principally women and children, round my camp…"[24]

The Confederates did try to take advantage of a lull in the fighting, in the belief that Springfield, Missouri, was only lightly defended by Missouri State Militia. Maj. Gen. Holmes believed that they could severely disrupt the supply chain between the railroad terminus at Rolla and Springfield, weakening the Federal forces occupying Southwest Missouri and Northwest Arkansas. Gen. Marmaduke was given command of the raid but was instructed to not incite any major engagements since manpower and supplies were in short supply. Marmaduke acted quickly by moving his three thousand cavalrymen in two separate columns northward out of Arkansas during the first week of January 1863. After crossing into Missouri, Marmaduke's force attacked and destroyed the Union outpost at Ozark, Missouri. He then received word that confirmed Springfield was lightly defended by approximately twenty-one hundred Missouri State Militia under the command of Brig. Gen. Egbert Brown. The Union commander learned of Marmaduke's approach only on January 7 and frantically began to improve the defenses surrounding Springfield. Not wanting to waste his opportunity, Marmaduke disregarded his orders not to incite any major engagements and attacked Springfield on January 9. After an all-day battle, Brown's Union forces held their ground. Marmaduke retreated that night, believing that Springfield would soon be reinforced. The Union victory resulted in estimated casualties of 163 Union and 240 Confederate.[25]

The second column of Marmaduke's raid was led by Col. Joseph Porter who was leading his Missouri Cavalry Brigade (CS). This column moved north from Pocahontas, Arkansas, and on January 9 they eventually attacked and captured a Union outpost at Hartsville, Missouri. On January 10, some of Porter's men raided some other Union outposts in the area before catching up with Marmaduke's retreating column east of Marshfield, a small town on the road between Springfield and Rolla. Marmaduke had received reports of Union troops approaching to surround him and prepared for a confrontation. Col. Samuel Merrill, commander of the 21st Iowa Infantry, with a 700-man Union column, arrived in Hartville, discovered that the garrison had already surrendered, and set out after the Confederates. A few minutes later, the fight began. Marmaduke feared being cut off from his retreat route back to Arkansas, so he pushed Merrill's force back to Hartville, where the Union established a defense line. A four-hour battle began in which the Confederates suffered many casualties but forced the Federals to retreat. Although they won the battle, the Confederates were forced to abandon the raid and return to Arkansas.

The Confederate victory resulted in 78 Union casualties and 329 Confederate.[26] Marmaduke invaded Missouri a second time in April 1863. This raid was unsuccessful as well after a severe loss near Cape Girardeau, Missouri, to Union forces under Gen. John McNeil.

As the Spring progressed Cooper and Watie began to bring their Confederate Indian forces, supplemented with some Texas troops, back up into the Indian Territory. Their goals were to recruit new members into their ranks, to continue to harass any Union forces, and attempt to reassert some dominance in the regions north of the Arkansas River. In February 1863, Bryan's Battalion was consolidated with additional companies from the 1st Cherokee Mounted Volunteers to form the 2nd Cherokee Mounted Volunteers commanded by Col. William Penn Adair, who had been paroled, exchanged, and released from a Union prisoner of war camp after he was taken prisoner during the first invasion in 1862. This action was designed to bolster the number of troops available to Cooper as well as provide some organization to the idle companies. Most actions between the warring sides during this period were small and had little effect on the outcome of the upcoming campaigns. Wiley Britton, who was serving as a sergeant with Company K of the 6th Kansas Cavalry, describes one of these small affairs that took place on March 30:

"…A detachment of this division just arrived from Park Hill, Cherokee Nation, reports that seven of our Indians, known as Pins, were killed at that place a few days ago by a party of rebels wearing the federal uniform. By this deception and dastardly act the enemy were permitted to approach within a few yards of the Indians, and, by a well-directed fire, shot them down before they had time to offer any resistance. This is not the only instance during the past year of small detachments of our troops having been entrapped by the enemy who were dressed in the federal uniform…"[27]

As the fields and prairies had become green due to the heavy spring rains, it was finally decided to reoccupy Fort Gibson permanently. Col. Phillips sent out various reconnaissance patrols along the Old Military Road, the Texas Road, and other trails between Arkansas, Missouri, Kansas, and the Indian Territory. Captain N. B. Lucas of the 6th Kansas Cavalry was sent on a scout of several days in the direction of Fort Smith with one hundred men of the Indian Brigade, traveling from Illinois Creek to Cincinnati, a small town on the Arkansas line, then to Dutch Mills and finally to Park Hill, which was only seven miles from Tahlequah. Col. Phillips also sent Maj. John Foreman, with three hundred mounted men of the 3rd IHG, in the direction of Webber's Falls, on the Arkansas River approximately twenty-three miles downstream from Fort Gibson, with instructions to thoroughly scout that section and to rejoin the main command at Park Hill. Although they encountered and skirmished with a formation of Confederate Indians near Webber's Falls, killing six, they also discovered that there were no significant numbers of Confederates north of Arkansas River. On April 9 a refugee train from Neosho, accompanied by two hundred Union Indian soldiers of the 3rd Indian Home Guard

under the command of Capt. Alexander Spillman, arrived at Camp John Ross. Wiley Britton recalled the actions of the Indian soldiers:

> "…When it had become definitely known to the Indian soldiers that they were on the march to Fort Gibson, every morning on starting out they put on their faces their brightest warpaint, and their war-whoops, which commenced at the head of the column and ran back to the rear several times in succession, had more animation than usual, for the realization of their hopes appeared to them near at hand…"[28]

Finally, on April 13, Col. Phillips marched his combined command, along with all baggage and supply trains and the artillery, from Park Hill into Fort Gibson. They immediately set to work on establishing a camp and building fortifications around the fort site. The quartermaster and commissary buildings had earlier been constructed of stone and were in good repair. The barracks and other structures were in poor condition, except for the remaining officers' quarters. Lt. Col. Ellithorpe remarked in another *Chicago Evening Journal* article:

> "…Col. Phillips is having this point fortified, so that in a few days we will not be afraid of twenty-five thousand of the enemy, should that many make their appearance, which is not at all probable, as that is more than they can possibly concentrate at that point. It is one of the finest places in the country to fortify. It commands all of the

surrounding country and has plenty of good water, and roads leading out in every direction, which make it a central point…"[29]

The Indian refugees immediately began to plow and cultivate grain and corn seeds provided by the Department of the Interior, while others went out in search of loose cattle to round up and bring closer to the fort as a food supply. Teams were sent out onto the prairies with scythes to cut and collect as much forage as possible from the new prairie grasses. Wiley Britton recorded in his journal some thoughts on Fort Gibson:

> "…This is quite an old post. It was established as a military post by the United States before the Cherokees left their Tennessee and Georgia homes and emigrated to this Territory. There are now two or three persons living here who say that they have a distinct recollection of Jefferson Davis, a Lieutenant of Dragoons, when he was stationed at this post as far back as 1832…"[30]

As mentioned earlier in this account, one of the most significant aspects of this entire Indian Territory affair was the multi-cultural composition of the troops. Up to this point, Union and Confederate Indians had fought each other alongside white troops belonging to their respective sides. The Emancipation Proclamation and the Cherokee Emancipation Acts freed the enslaved persons in areas considered to be in rebellion. This provided the Federal government with a new source of manpower. When the war began, President Lincoln had been approached by many abolitionists throughout the North to

recruit these free black men and contraband from the South into the U.S. Army. Lincoln deferred at the time since he was struggling to hold on to the border states of Missouri, Kentucky, Maryland, and Delaware, as well as the District of Columbia. Freeing the slaves in those areas at the start of the war would have driven those states into the awaiting arms of the Confederacy. There was also a constitutional issue because slavery was allowed, if not clearly expressed, and protected by rulings of the Supreme Court under the U.S. Constitution. Lincoln felt that freeing the enslaved persons in the rebellious areas during a state of war could be done by executive order.

Maj. Gen. David Hunter, formerly of the Department of the Missouri, had been reassigned to command the Department of the South where he began enlisting available black men into the Union 1st South Carolina Volunteers Regiment in May 1862. They were organized, equipped, and issued firearms. Since Hunter was unable to secure enough volunteers for his black regiment, he resorted to conscription of any able-bodied black male. This did not garner him any good graces among the enslaved population or the plantation owners. Unfortunately for Hunter, the hue and cry of the political class in Washington brought an end to this experiment and the unit was disbanded in August without ever having taken to the field.[31]

Yet, a new force was beginning to take shape in Kansas in July 1862. An act of Congress on July 17, 1862, authorized the president to accept persons of "African descent" into the military service of the United States. But the new law required the president to authorize it, and Lincoln had not yet done so. Unsurprisingly, Senator James

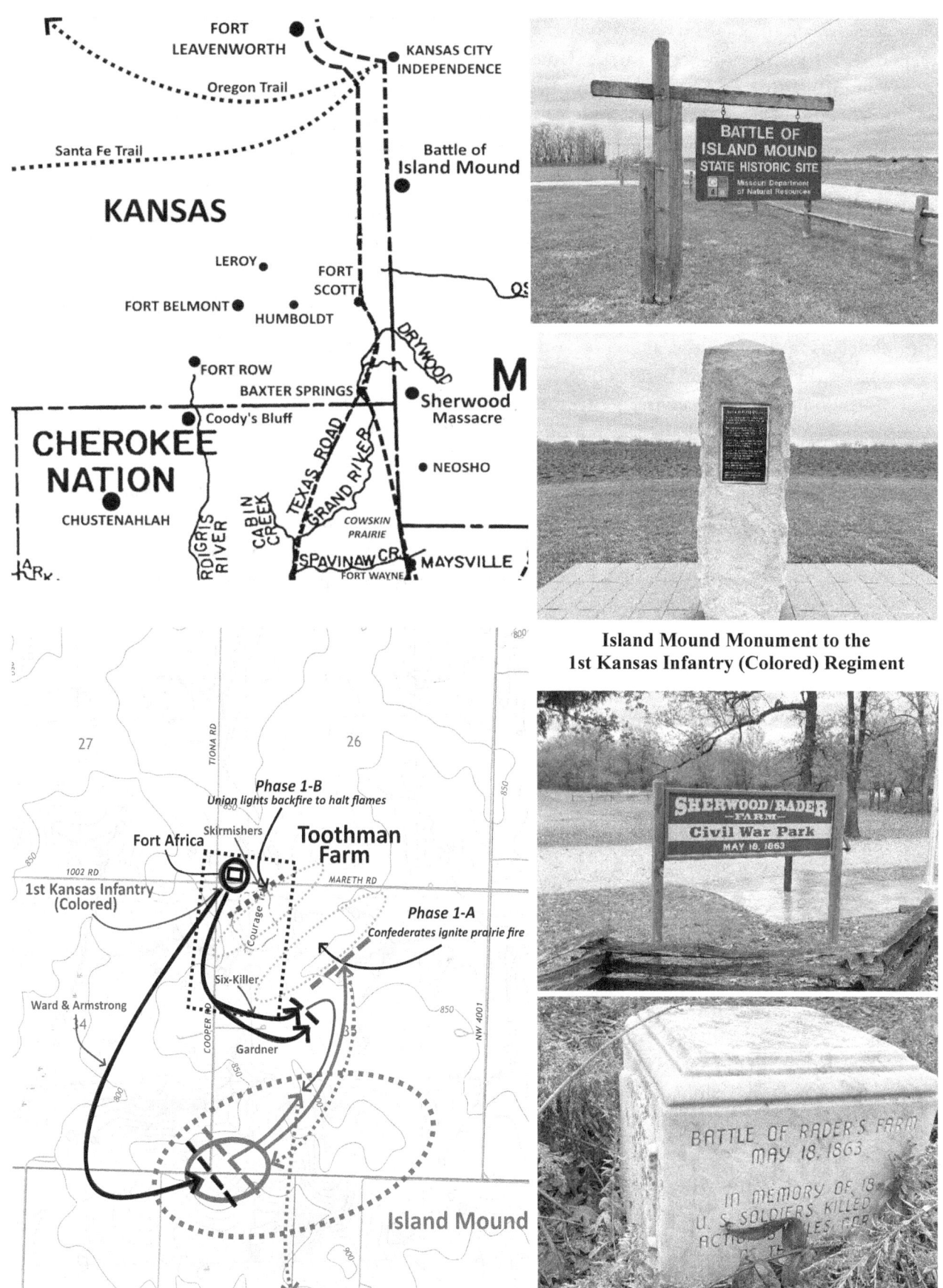

Island Mound Monument to the
1st Kansas Infantry (Colored) Regiment

Lane, once again disregarding the War Department's instructions, appointed a Union recruiting officer for a new regiment, but Lane did not specify the race of this new unit. On August 4, 1862, Lane appointed Capt. James Williams and Capt. Henry Seaman to enlist both blacks and whites into service, and issued General Orders No. 2, an unauthorized order that allowed enslaved men of African descent to enlist in the service of the United States. This action created much controversy for politicians both in Kansas and Washington. Lane was again given a direct order forbidding him from enlisting men of African descent, but again Lane ignored the message. These actions gave Senator Lane an opportunity to advance the Radical Republicans' agenda for freedmen's civil rights, and he was not going to miss the opportunity to prevail. Under Lane's watchful eye, free blacks and escaped slaves in Kansas were being recruited and formed into the 1st Kansas Colored Infantry regiment. Although most of the troops were recruited in mid-to-late 1862, because they had not been accepted into United States service yet, (that would not come until January 13, 1863) they were enrolled as Kansas Militia, and equipped with uniforms, accoutrements, and the same Austrian and Prussian muskets that had been issued to the Kansas regiments. Yet, because Senator Lane was a powerful politician in Washington and Union sympathy in Kansas was, for the most part, abolitionist in nature, they had no great issue with recruiting black soldiers. Moreover, anything happening in far-off Kansas was far removed from the lobbying and back-scratching in Washington.[32]

James M. Williams was given the assignment of raising the regiment, and officer positions were given to white men who wished to serve in the regiment. Williams was a lawyer from New York who somehow found his way out to Kansas when the war began and was able to secure a captain's commission in the 5th Kansas Cavalry. He caught the eye of Senator Lane who had him resign that position to take the helm of the new black regiment and be awarded a commission as Colonel. Yet, although most Kansans approved of abolition, many did fear the arming of black troops. Along with this fear was the presence of many pro-Southern citizens in Kansas who tried to disrupt the formation of the regiment. Williams and his new recruits were hampered many times by local officials and found it difficult to properly supply the new regiment.[33] The initial enrollment of the 1st Kansas (Colored) was to be six companies of infantry, which were based at Fort Scott throughout the summer of 1862. In October it was reported that a guerilla/Partisan Ranger group that had previously been led by the currently imprisoned Enoch John Toothman, remained in the area around his homestead in Bates County, Missouri, located about nine miles on the other side of the Kansas-Missouri border. Kansas scouts identified this large band currently led by local Confederate guerrillas Bill Truman and Dick Hancock. These guerilla forces were holed up on Hog Island in the Osage River, near the Toothman homestead. Captain Richard G. Ward's 170-man battalion and Captain Henry C. Seaman's 70-man battalion of the 1st Kansas Colored Infantry were ordered by Maj. Benjamin S. Henning, commanding Fort Scott, to proceed toward the Missouri River. The two hundred and forty men of the 1st Kansas (Colored), accompanied by six members of the 5th Kansas Cavalry serving as scouts, among them some Cherokees, advanced and occupied the Enoch John Toothman farm on October 27. The black regiment began to build breastworks for defense and designated it "Fort Africa." At about the same time, the Confederate guerillas were joined by a group of mounted Missouri State Guard (CSA) under the command of Col. Jeremiah Cockrell. The guerillas and Missouri guardsmen now outnumbered the Kansas regiment. Skirmishing between the two sides was continuous for most of two days. On October 29, the food situation within the Kansas force was becoming a concern, and runners were dispatched back to Fort Scott for more supplies. In an effort to cover the movement of the runners, the Kansas soldiers began advancing and firing volley after volley from their long range rifles, which far outmatched the shotguns and pistols of the Confederate force. After the foraging party returned, the skirmishers fell back to their fort to eat. While they were eating, the Confederates set fire to the prairie south of Fort Africa hoping to burn out the Kansas troops. Capt. Seaman in turn set a backfire to inhibit the flames coming in their direction. Pvt. John Six-Killer, a Cherokee who had enlisted into the Kansas forces along with his slaves, with seven men, was directed by Capt. Seaman to scout around the southwest edge of the fire to determine the location of the enemy force. He was ordered to stay in view of the main force. Six-Killer and his small group went around the left flank of the Confederates, began skirmishing with them, and then disappeared. Seaman sent Lt. Joseph Gardner with sixteen men to support Six-Killer and bring his group back. This group also disappeared from sight. Capt. Ward and fifty men of Lt. Armstrong's detachment went directly up the small mound that the guerilla force had occupied and forced Cockrell's Confederates back up and over the mound. In

doing so they ran directly into Gardner and the remnants of Six-Killer's detachment. The mounted Confederates had struck this small group hard. Screaming his war whoops, Six-Killer had dispatched a number of Confederates before he fell with six bullet holes in his body. After a further time of intense melee, the Confederates broke and fell back through a severe crossfire by the Kansas forces.[34] Casualties were somewhat heavy for the number of combatants involved. The 1st Kansas (Colored) reported eight killed and eleven wounded out of approximately two hundred-fifty soldiers involved. The casualties for the approximately four hundred guerilla and Missouri State Guardsmen is harder to report. Their loss was estimated to be eighteen killed and twenty-five wounded.[35]

This small battle or skirmish was significant as it was the first time African American troops engaged Confederate forces during the Civil War. At the time it registered so little interest that it was not mentioned in the Official Records of the War of the Rebellion. Yet, the significance of the battle was reported in *The New York Times* by a correspondent who had accompanied the Kansas unit. The heroic action of the African Americans was headlined as "desperate bravery." On a related note, guerilla leader Bill Truman told supporters in Butler that the blacks had fought "like tigers." This is not surprising since the black troops were fighting for their freedom. They wanted to ensure they would never go back to slavery, and they understood that the guerrillas would give them no quarter, having promised to kill blacks rather than take them prisoner. This attitude lasted throughout the war.[36]

Senator Lane had six companies completed by January 1863, and all ten companies were completed by May. There were times when recruits became hard to find. In this case Lane, just as Hunter had done in South Carolina, authorized raids into Missouri to capture enslaved black men and impress them into Federal service.[37] Unfortunately, the two black commissioned officers in the 1st Kansas, Capt. William Matthews and Lt. Patrick Minor, were forced to give up their commissions upon entry into federal service since the officers in the black regiments were required to be white. But the 1st Kansas (Colored) did receive orders for its first mission. Fearing that the regiment would be harmed if left near the Kansas population centers, on May 2, 1863, Gen. Blunt ordered the ten companies of the 1st Kansas Colored Infantry to march from Fort Scott to the Union supply depot at Baxter Springs just outside Indian Territory. During the black regiment's time at Baxter Springs, it was involved in patrolling the military road in the region to assist in keeping the lines of communication open between Forts Gibson and Scott.[38]

A significant event helped Colonel Phillips accomplish his new mission out of Fort Gibson. He received information about a proclamation by the Confederate Cherokee Council, which stated it would meet at Webber's Falls on April 25, 1863. Webber's Falls was home to numerous relatives of Stand Watie and yet had a notorious reputation as the site of the 1842 slave revolt by the enslaved Africans against their Indian owners. On November 15 of that year, more than two dozen enslaved Africans in Webber's Falls decided that they had enough. At the Joseph Vann plantation, the enslaved Africans waited until the slaveholders and overseers were asleep. The enslaved black

men, women, and children took all kinds of things that would help execute their rebellion: horses and mules, guns and ammunition, food, and other supplies. At sunrise, they headed out on their journey, stopping to pick up more enslaved people in the Muscogee (Creek) Nation. They were determined to reach Mexico, where they believed they would be free from their pursuers. Two days after the escape, the Cherokee National Council, afraid that word of their rebellion would spread and influence other enslaved people to revolt, sent a Cherokee militia of eighty-seven men to capture the fugitives and bring them back to the Indian nations. The runaways eventually lost their way and were caught by Col. Thomas Livingston's militia near the Red River on November 28, after nearly two weeks on the run. All were returned to their owners, less the few who were killed by pursuers during their escape. They continued in their enslaved state until the end of the Civil War.[39]

The Confederate Cherokee Council had initially held a preliminary meeting at Scullyville, the Choctaw National capital located southwest of Fort Smith. This group selected Stand Watie to be the chairman of the assembly to take place at Webber's Falls, where the council would select a permanent principal chief (for the Confederacy), as well as discuss the military situation in the Indian Territory. The council members gathered on the evening of April 24, before the scheduled meeting on the next morning. Watie was disheartened by the situation that the Confederate Cherokees had found themselves in. He spoke "with a heavy heart, for evil times had come upon the country." He further declared that "disaster upon disaster has followed the Confederate arms in the

Cherokee country." He was hopeful that the continuing successes of the Confederate Army in the Eastern Theater in the winter and spring of 1863 would bring further assistance to the desperate Confederate forces in the Indian Territory. He stated, "that the time is near at hand when the tide of success is due to be returned to the Confederate arms when we shall be able to drive the federal forces out of our country." They went to their rest that night hopeful for a constructive meeting where many issues would be discussed and resolved.[40] The first gathering of the assembly was to be in the afternoon of April 25. The council was to be protected by four companies of the 1st Cherokee Mounted Rifles regiment.[41]

After being informed of the gathering of the Confederate Cherokee Council, Col. Phillips put together a force of Union troops on the night of April 24. This raiding party would consist of six hundred mounted troops consisting of details from all three Indian Home Guard regiments and a battalion of the 6th Kansas Cavalry. The command made a night march that crossed the Arkansas River a few miles south of Fort Gibson and, after a thirty mile night march, they attacked Colonel Watie's Confederate force at Webber's Falls on the morning of the 25th, soundly defeating them and preventing the Confederate council from meeting. Phillips reported:

"...Crossed the Arkansas River on the night of the 24th, and marched 30 miles in the night, and at daylight struck the rebels of Stand Watie's command near Webber's Falls; routed and broke them up, killing a number and taking prisoners 5 took the equipage, &c., that they had. Lost 2

killed... By a proclamation issued, the rebel Cherokee Legislature was to meet on the 25th, at Webber's Falls. Prevented, and dispersed with the rebel forces..."[42]

Watie's Confederate Indians, most of whom were not dressed and many of whom were still asleep, were completely surprised by the dawn attack on their camp. They fired a few token rounds at their attackers and fled, not only leaving their camp equipment and supplies behind, but failing to protect the council members. All soldiers and council members were dispersed and retreated towards Fort Smith, Arkansas, and North Fork Town, where Col. Cooper had been gathering his Confederate forces. One sad loss for the Union Army was the killing of Dr. Gilpatrick, who had been with the Union Indian Brigade from almost the beginning. Wiley Britton described what happened:

"...But our most serious loss was the killing, or rather assassination of Dr. Gilpatrick, a special agent of the Government, who accompanied us on this reconnoitering expedition. After the skirmish was over, he was called upon by a rebel woman to dress the wound of a rebel soldier, who had fallen a hundred yards or so from where we halted. While performing this duty of mercy for a fallen foe, he was shot by a rebel from a concealed position, and he died immediately afterwards. We all felt indignant that he should have been thus basely entrapped. We brought him back with us, and he is to be buried on Sunday with military honors..."[43]

Lt. Col. Lewis Downing, USA
3rd Indian Home Guard
Photo courtesy of Library of Congress

Maj. Gen. William Steele, CSA
Indian Territory District
Photo courtesy of Library of Congress

Colonel James Williams, USA
1st Kansas Infantry (Colored)
Photo courtesy of Kansas Historical Society

Enoch John Toothman Farm/Fort Africa
Photo by Author

Island Mound (Location is on Private Property)
Photo by Author

Site of the Rader Homestead near Sherwood, MO
Photo by Author

Route of Union Troops retreat
Photo by Author

Osage Warriors.

Osage Chief Hard Rope
*Photo courtesy of the Kansas
Historical Society*

COLONEL WARNER LEWIS.

Area near Okay, Oklahoma where Confederates attacked the supply train enroute to Fort Gibson
Photos by Author

Remnants of the Earthworks constructed by Col. William Phillips at Fort Gibson / Blunt
Photos by Author

Dr. David Hitchcock who was serving as the Assistant Surgeon for the 2nd Indian Home Guard remarked on the loss of Dr. Gilpatrick:

> "…The doctor's body was brought up to Gibson about noon. He had been called upon to attend a wounded Secesh, Bill Pettit, some little distance from the main body of the army. Return Foreman, John Johnson, and David Vann rode up and fired on him while in the house. He asked what they meant by firing upon an uniformed surgeon; they told him to come out and surrender and he should not be hurt. He advanced a few paces, when the four fired simultaneously and lodged 18 buckshot in his body; he was killed instantly…"[44]

Dr. Gilpatrick was also Maj. Gen. Blunt's uncle and was married to the general's sister-in-law. Angry Union soldiers took out revenge by setting fire to the houses in Webber's Fall. Few structures remained when Col. Phillips and his detachment turned back north towards Fort Gibson. Phillips returned to his headquarters at Fort Gibson after the battle and, in May 1863, renamed the installation Fort Blunt, in honor of the district commander who had recently been promoted to major general. (Although Phillips made this name change, the name will remain Fort Gibson in this work to avoid confusion.) Phillips built a series of fortifications around the open-grounded Fort Gibson. He described it as:

> "…Under your orders, demolished the works at Hildebrand's Mill, and have concentrated the force at Gibson. Here I have a strong work which cannot be taken by any force or artillery the enemy can bring to bear on it. It is on a commanding hill, with rear bluffs, on Grand River; water from river within lines; incloses 15 acres; defensible now, but needs much more work—a line of works, with angles and facings, over 1 mile in length, built by Indian soldiers. My rear is up Grand River Valley…"[45]

The 1st Kansas Infantry (Colored) departed Fort Scott on May 4 enroute to Baxter Springs as directed by Brig. Gen. Blunt, to provide security between Fort Scott and Fort Gibson. On May 5 while enroute along the Military/Line Road, the regiment was contacted by a sixty-man detachment of the 2nd Kansas Cavalry, commanded by the regiment's adjutant, M.M. Ehle. He had been sent by Blunt to investigate a report of a gathering of guerillas encamped on Centre Creek near the town of Sherwood, Missouri. Ehle's command found the guerilla's encampment approximately twelve miles from their current location. The size of the bush-whacker group was estimated to be around two hundred and fifty men, which far outnumbered his small detachment. Adjutant Ehle requested assistance from Col. Williams and his Kansas infantry which was quickly approved. Williams provided two companies of his black soldiers and one cannon from the 2nd Kansas Battery which was accompanying the regiment to Baxter Springs. At daybreak on May 6 the combined Kansas infantry, cavalry, and artillery attacked the guerilla encampment. Maj. Charles Blair reported:

> "…With this added force, he attacked

the enemy at daybreak, carrying the camp in gallant style and dispersing the rebels in every direction. He subsequently attacked and took another camp nearer the town and dispersed its occupants. Some few prisoners were taken, and about 50 head of young horses and mules, part of which, with the prisoners, were delivered over to Colonel Williams…"[46]

The failure to get the guerillas into a face-to-face battle at this time highlighted the shortcomings of infantry units when encountering irregular forces. Most of the partisan ranger and guerilla members were mounted and were not interested in conducting regular combat operations. If a report of guerilla activity came in, by the time the infantry would arrive the perpetrators would be gone. Since they were so loosely organized, the guerillas would just fade into the woods or towns and would rendezvous at a later time with the other members, usually at a pre-planned site (This is the same tactic employed by Iraqi and Afghan militias against modern U.S. forces in those hotly contested areas). The infantry could not effectively pursue these highly mobile forces. The Missouri-Kansas-Arkansas-Indian Territory region was much too large for one regiment of infantry to control. The best use of these troops was to escort the slow-moving wagon trains along the supply roads. The members of the 1st Kansas Infantry (Colored) were also fearful of being caught out in the open or away from any support or reinforcements.[47]

Major Thomas Livingston of the Missouri Partisan Rangers led a raid that began at the Creek Agency along the Arkansas River on May 6. He intended on raiding targets of opportunity while enroute to Missouri from the Indian Territory. His forces crossed the Arkansas River at this point and then moved northward, crossed the Verdigris River at Sandtown, and continued moving north on the Texas Road toward the Cabin Creek crossing. On May 8, near the house of Capt. Joe Martin at Cabin Creek, this Confederate guerilla force encountered a group of Creek Indian stragglers from the 1st Indian Home Guard who were being marched from Fort Scott to Fort Gibson. The Union Indians took cover around the houses near the small settlement and fought a short, hot battle that resulted in a small number of casualties. A company of Union cavalry from Fort Gibson came into view, so Livingston opted to cross Cabin Creek and continue heading north, passing the Union-held saltworks near Spavinaw. Blunt reported to Schofield that Livingston and his guerillas had captured twelve Union Creek soldiers at Cabin Creek and had murdered them all. The Confederate guerillas later crossed over into Missouri without any further contact with Union forces until they reached Centre Creek, Missouri. On May 14, at the nearby lead mines, they were confronted by a detachment of 8th Missouri State Militia (MSM), numbering about 125 soldiers. The Missouri militiamen attacked Livingston's guerillas, who were posted behind thick brush and an old log shop building at the mine. The Confederate bushwhackers were able to drive off the MSM troops after a short, 15-minute fight. The accounts differ on casualties, with the Union commander, Major Edward Eno, 8th Missouri State Militia, recording four dead and two wounded. Livingston reported only two slightly wounded. Lt. Col. Thomas

Crittendon, who commanded the Union post at Newtonia, reported to Col. William Cloud that the guerillas buried fifteen bodies. Either way, it was a small encounter with limited results.

On May 18 a twenty-five man foraging party of the 1st Kansas (Colored) with five mule-team wagons set out from Baxter Springs to gather corn near Sherwood, Missouri. They did not have a cavalry escort with them on this day that could scout the area for guerillas and bushwhackers. Wiley Britton of the 6th Kansas Cavalry later remarked, "There was at this time among the white soldiers a decided feeling against serving with colored troops, and this was perhaps the reason he had no cavalry attached to his command…"[48] Blunt also had a severe lack of cavalry to spare for escort duty. Instead, Williams mounted fifteen white soldiers of the 2nd Kansas Battery to escort the foraging detachment. Confederate scouts observed the Union soldiers moving toward Sherwood and contacted Maj. Livingston, who quickly deployed sixty-seven of his best Partisan Rangers to attack the Union detachment. They found the Union foragers at the Rader homestead that belonged to one of Livingston's guerillas. The black infantrymen had stacked their weapons while they searched the Rader house for hidden caches of corn. At this point the Confederate guerillas attacked. Hugh Thompson of the 3rd Wisconsin Cavalry recalled after the war:

> "…shooting began in the rear, and almost instantly all around us. In surrounding us the enemy had cut off the 6 men on picket, 20 unarmed men loading corn and 5 teamsters, also unarmed, leaving but 29 men hemmed

in a short, narrow lane with a gate at one end, with arms to fight our way out. Without knowing it we had rode into a trap…"[49]

The white officers and the artillerymen, being mounted, were mostly able to escape the surprise attack, but the black infantrymen were quickly overrun since their weapons were stacked and probably unloaded. Capt. John R. Graton of Company C of the 1st Kansas Infantry (Colored) described the scene in a letter home to his wife:

> "…A party of us boys got badly whipped on the afternoon of the 18th, a party of our men of about twenty five, and some 15 of the artillery boys mounted on their horses with 5 six mule wagons went to Sherwood Jasper County, Mo, about 15 miles from this place, to get corn. The officers in command were Maj Ward Cap Armstrong & Lieut Edgerton. They had just arrived at the place, and four out of five were engaged in loading corn, and of course were away from their guns when they were surprised by a force of Bushwackers [sic] numbering some one hundred and fifty or two hundred men, coming from some timber opposite the house, and came upon them before many of them could get to their guns, and as a matter of course our men were used up. The Artill [sic] boys and our officers being mounted were able to get out of way, but the black boys being on foot had to take it and most of them were killed, they chased the mounted men some six miles, some of

the Artilery [sic] horses giving out, two of the Artillery men killed…"[50]

When the survivors made it back to Fort Scott, Col. Williams gathered five companies of his regiment and conducted a forced night-march back to the Rader homestead. They found the body of one of the artillerymen in the road as they were marching. When they arrived at the house, they found that the Confederate guerillas had killed thirteen black soldiers whose bodies lay around the curtilage of the house. They could not bury the soldiers since they failed to bring any shovels, so they placed the thirteen bodies in the Rader house (Mrs. Rader had abandoned the house the previous night, taking her personal property). The command also executed a recently paroled local Confederate bushwhacker named John Bishop, who was discovered wearing Federal boots and had blood on his clothing. He was accused of participating in the attack in violation of his parole. His body was also placed into the house which became a funeral pyre after the soldiers set it ablaze. In a further action, Williams in retaliation ordered his soldiers to destroy everything within a five mile radius of the Rader homestead. The soldiers destroyed thirteen homesteads and the entire town of Sherwood, which was never rebuilt.[51]

Maj. Livingston soon contacted Fort Scott stating he had five prisoners, three white soldiers and two black. He said he was willing to exchange the three white soldiers for any three Confederates held in custody. Livingston also stated that he was unwilling to recognize the two black men as soldiers, so they were being confiscated as contraband of war and being sent back into slavery. (It is

believed that the two black soldiers were killed while in Confederate custody)[52]

Many authors have attempted to claim that Colonel Phillips and the Federal forces at Fort Gibson were under siege by General Cooper's Confederates. The evidence shows that although there were raids against Fort Gibson, the Union forces were able to move about freely and, when confronted by Confederate forces, were almost always successful in driving them off. The supply line between Fort Scott and Fort Gibson was indeed tenuous, but the vast majority of supplies from Fort Scott made it to Fort Gibson unmolested. In military strategy, raids are easy to conduct against small portions of the enemy's lines of communication or logistics. Small raids such as these did not produce much in the way of results, unlike those that Jeb Stuart conducted in Virginia and Nathan Beford Forrest did in Tennessee. But there was a period of about two months after the Webber's Falls engagement in which Watie tried to disturb the operations of the Union Indian Brigade and Fort Gibson on a somewhat regular basis. This was accomplished by a series of raids, mostly upon working parties gathering forage, or the theft of livestock and horses. At times the Confederates were successful, at other times they were not. If nothing else, it kept the Union forces on alert.

In May 1863 an unusual, confusing, and much-debated incident took place in the Osage Nation, in what is now southern Kansas, between modern day Coffeyville and Independence. A February 1910 article appeared in *The Osage Magazine 2*, an industrial periodical dealing with Oklahoma mineral and agricultural topics, describing a battle between a Confederate party and a group of Osage

Indians. According to the various accounts of the conflict, on May 16 (the dates do not agree in the various accounts) a band of 18-20 Confederate officers were dispatched from Jasper County, Missouri, led by Confederate Colonel Charles Harrison. Harrison was a comrade of Missouri guerrilla William Quantrill, who was a cardsharp, former gambling hall operator, and infamous gun-slinger who had rioted with pro-secessionist miners in Denver, Colorado, in 1861. Harrison was accompanied on the mission by Colonel Warner Lewis, a St. Louis lawyer who joined the Confederate Army in 1861. Their mission was to make an expedition across southern Kansas and eventually into Colorado and New Mexico territories. The expedition had two objectives: to recruit men from the westward prairies to join the Confederate Army, especially those who had fled Missouri to avoid the conflict, and to incite the western Indians against the Kansas settlers, a majority of whom held free state or anti-slavery views. The party departed Jasper County, Missouri near the town of soon-to-be-destroyed Sherwood at the point where Center Creek crosses the line into Kansas/Cherokee Unassigned Lands. They boldly started the trek through the reserve in broad daylight and stopped their first night at the site of New Town, which had been a trading post until burned in 1861 by Federal troops (later rebuilt and renamed "Chetopa"). The Confederate party, many dressed in stolen or captured Union blue uniforms, continued westward the next day and quickly encountered a group of ten Osage Indians of the Old Hill clan east of modern day Independence, who were preparing for the summer buffalo hunt. The Osage hunters, led by Chief Hard Rope (We-He-Sa-Ki) inquired

as to their identity. The Indians stated that the U.S. government held them responsible for what occurred in their country, The Confederate commander stated that they were Union soldiers out of Humbolt, Kansas, on patrol. At the time, Union Capt. Willoughby Doudna, Company G, 9th Kansas Cavalry, commanded a post of Federals at Humboldt. The post commander had a policy of keeping the Union-friendly Osages familiar with post activity, especially the personnel. The Osage leaders were familiar with most of the Union soldiers stationed at Humbolt but did not recognize these as being a part of that unit. The Osage, wanting to stay on the good side of the United States and fearing this was just another bushwhacker or partisan group, told the party that they would have to accompany them to the Union post at Humbolt and verify their identities. The Confederates refused and shot and killed an Osage Indian. Because they were outnumbered, the Osage retreated back to their village on Big Hill Creek. Quickly, a force of as many as two hundred warriors of the Big Hill Osage under chiefs Hard Rope and Little Beaver, including dozens of nearby Big Hill Osage led by Shaba Shinka, were gathered. Hard Rope and nine "Heart-Stays" scouts (a special group of the Osage scouts) picked up the unmistakable trail of the Confederates and confronted them five miles east of the Verdigris River near Drum Creek. The angry warriors proceeded to drive the intruders across the tallgrass prairie as if conducting a buffalo hunt. The Osage pursued and drove the Confederate officers west to a loop of the Verdigris River near its confluence with Elk River. Skilled Osage riders flanked the Confederates and blocked any of their attempts to turn, maneuvering their harried enemies straight toward the steep, unfordable bluffs along the Verdigris River. The Confederates were soon surrounded and, since the banks were too steep to cross the Verdigris, they were backed into a sandbar. Eventually, all but two were killed.[53]

The two survivors hid in the brush and made their way upstream from the battle site. After a very difficult journey over many days, they were able to make it back to Missouri. One of the survivors was Col. Warner Lewis. His recollection of the encounter was significantly different from the one described in the February 1910 issue above. Lewis's *The Osage Magazine 2* May 1910 issue was a response to the February issue. The now elderly Colonel wrote:

"…We had begun saddling up to renew our journey when we discovered a body of men coming on our trail at full gallop. By the time we were all mounted they were in hailing distance, and proved to be a body of about 150 Indian warriors. To avoid a conflict we moved off at a brisk walk, and they followed us. We had not gone far until some of them fired and killed one of our men, Douglas Huffman. We then charged them vigorously and drove them back for some distance… The Indians kept gathering strength from others coming up. We had a running fight for eight or ten miles, frequently hurling back their advance onto the main body or with loss… The Indians here got all around us at gunshot range, and kept up an incessant fire. We had only side arms and pistols and were out of range… I tried to get the men

to halt and give them a fire so as to let him get into the timber but did not succeed. We could not cross the stream with our horses, owing to the steepness of the banks on both sides. I went down to get a drink and heard the Indians coming to the bank below us. John Rafferty stood on the bank above me, and I said to him, "Follow me.""[54]

Chief Hard Rope immediately sent a messenger to Humbolt and Capt. Doudna and a hundred troops responded to the location within a couple of days. In the aftermath of the battle, the Osages stripped, scalped, sliced, and decapitated the eighteen Confederate soldiers. Papers were found among the dead troops that had orders indicating what the mission was. Which of the two accounts is the correct one, we may never know. Each author attempted to show that they were the initial injured party and had been significantly outnumbered during the chase.[55] There is also a dispute between the locations where the battle took place. One site is the loop in the Verdigris River, immediately northeast of Independence. A second location is along the banks of a smaller creek known as "Rebel Creek," northeast of the river loop. As yet no formal archeological studies have been completed, and metal detector operators have yet to find anything of consequence at either location. And, since the Bureau of Indian Affairs still considered the Osage to be "wild," it's not surprising that the Osage held a war dance in celebration of their victory on the Verdigris River with Capt. Doudna and his command present.[56] Southern Superintendent of Indian Affairs William Coffin stated in his 1863 Annual Report:

"…Had this party of rebels reached the wild tribes of Indians on the plains, restless and warlike as they are, and organized and led them, a vast amount of damage would have resulted to the emigration and supply trains destined for the military posts in New Mexico, Colorado, and Dakota, and might have cost the government millions of dollars to have them crushed out. So important was this service deemed by me that I immediately called the entire Osage nation in grand council at Convill's trading post to thank them, on behalf of their Great Father…"[57]

Back in the Indian Territory, Cooper was trying to keep his troops occupied by conducting various raids around Fort Gibson. He sent Col. Watie and his Cherokees up into the northeast portion of the Cherokee Nation to disrupt any milling or salt works operations depended upon by the Union forces. Col. Phillips for his part worried about his tenuous supply line with Fort Scott and committed a substantial number of his available troops to protecting that logistics chain. Phillips also knew that Cooper, and the Confederate Indian Brigade and the attached Texas regiments outnumbered him and feared that the enemy would combine his forces and sweep over Fort Gibson and drive the Union Army back to Kansas. He countered by attempting small offensive operations against the Confederates to keep them from assembling into a larger force that could achieve that goal. Phillips found that the prairie pastures around Fort Gibson would always attract raids on Union livestock,

especially horses and mules.

One plan to attack Fort Gibson and its livestock herds ended up being recalled in a story in *The National Tribune* in 1886, in sort of a light-hearted way regarding two young warriors who were supposed to be surveilling Fort Gibson:

"…Gens. Watie and Cooper were anxious to find out how their neighbors [Fort Gibson] were getting along without the trouble of paying them an actual visit. Besides, they had not been invited. It was agreed between Cooper and Watie to send two youngsters of Watie's command with an old telescope to a high hill that overlooked Fort Gibson and prairie, some four miles off. These scouts spent two days alternately looking through the tube and encouraging a lass whom they found living under it… But they observed that many hundreds of horses, mules and ponies were sent out early in the morning under slight guard to graze on the prairie all day, from one to three miles off. This fact was duly reported to Gen. Watie, and it was about all that was reported of any value. Little as it was, it was enough to give Watie an idea of some importance…"[58]

The story does not give any date for the raid but one such raid took place on May 14, 1863. Col. Watie and his Cherokee regiment crossed the Arkansas River, attacked the herders and the military guards, made off with a large number of horses and mules, and destroyed a few Army wagons. Diversions had

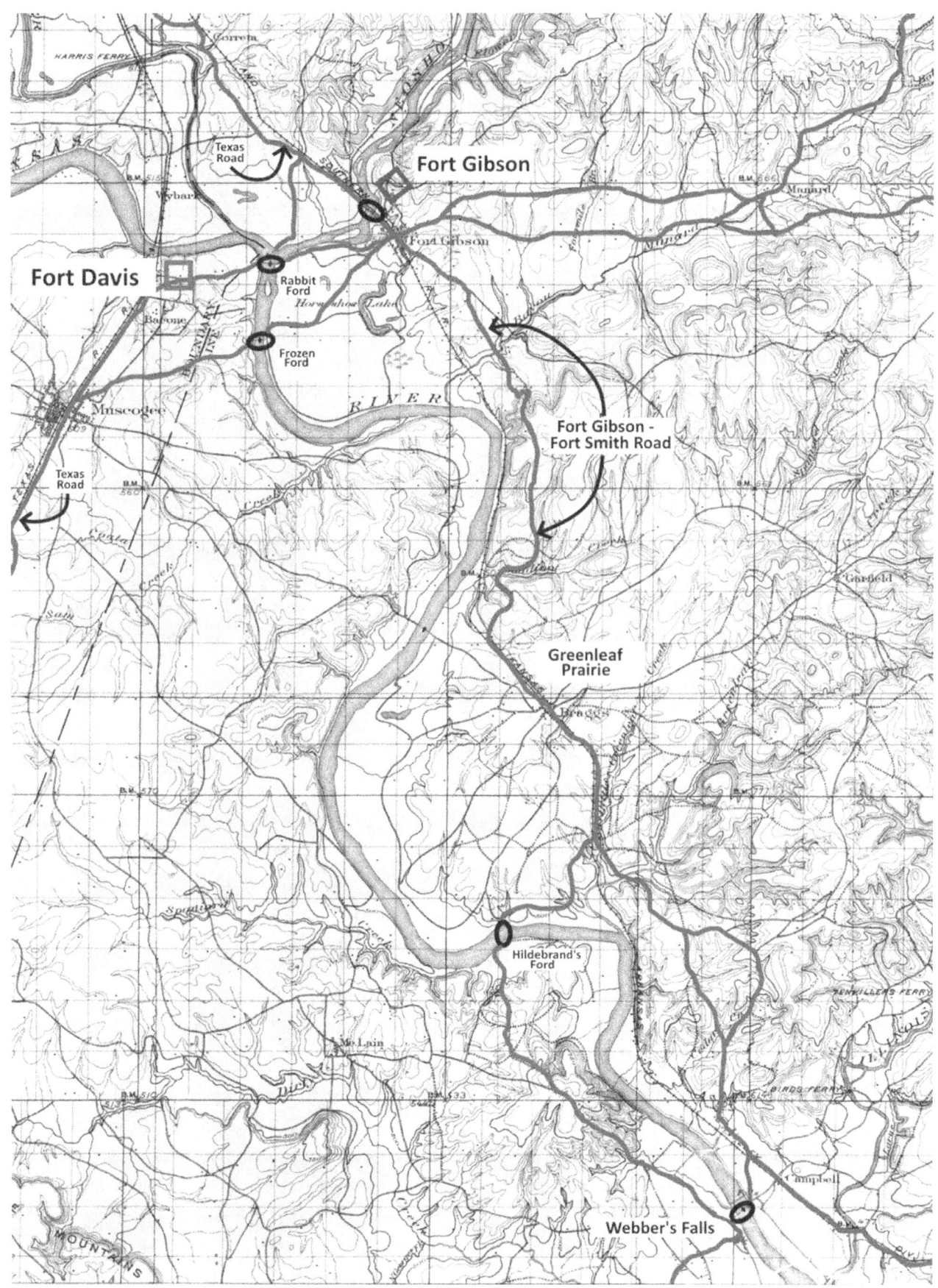

been set up in which the pickets on either side of the river were exchanging rifle fire with the intention of distracting the Union pickets from the Confederates crossing the river at an unguarded ford. Lt. Bradford Bassett of the 3rd Kansas Artillery deployed from Fort Gibson with a section of artillery and drove off the Confederate raiders with a few rounds of canister and spherical case shot. Pvt. Robert Elder Horn of Company K, 5th Texas Partisan Rangers recalled:

> "…We had a little skirmish near Ft. Gibson and frequently had picket fighting across the river. Corporal Finney had a little argument with the enemy when we tried to capture a train of men and wagons near the fort. After we destroyed some wagons and captured some mules and horses, the reinforced army came out and we had to skedaddle to other parts, Finney was in the rear. The Federals ordered him to halt but there was no halting this corporal. He put his spurs to his mare, and as the water was not deep reached the other side in safety. They shot at him several times close…"[59]

The records do not indicate which of the Arkansas River fords Watie used, although the most probable would have been Hildebrand's Ford opposite Greenleaf Prairie.

After Federal scouts had reported that the area around Fort Gibson was void of Confederates for at least twenty miles, on May 20 Col. Phillips ordered that the Army stock be released to graze on the prairies around the post during daylight hours. Unfortunately, the scouts failed to check one road leading to the Fort from the southeast, probably the Fort

Smith-Fort Gibson Military Road. Just before 9:00 am Col. Watie crossed the Arkansas River on an unguarded ford and with several hundred men and swept down upon the vulnerable livestock and herders. The post's garrison was down by eight hundred troops since they were deployed up the Texas Road to protect the incoming supply train from Fort Scott. Phillips reported:

> "…the stock was, therefore, being sent out to graze, when the enemy pounced upon it. Sending all the mounted men I could raise, the larger part of the stock was taken from them. The Creek regiment refused to charge, or it could all have been saved. I sent forward Majors [J. A.] Foreman, Wright, and Pomeroy, with all the present available force, and as rapidly as possible moved everything within the works. The enemy, being strongly posted 5 miles distant, drove back Major Foreman and the others for some distance, although the ground was hotly contested. Captain [K. B.] Lucas, of the Sixth Kansas, was nearly surrounded, as was Captain [Henry S.] Anderson, of the Third Indian, but they gallantly cut their way through…"[60]

It's not recorded why the 1st Indian Home Guard regiment would not charge when ordered to do so, although it could be indicative of the regiment's issues that Phillips referred to in his descriptions of the three Indian Home Guard regiments. Col. Phillips called on two battalions of Indian infantry, probably the 2nd IHG, and a section of the 3rd Kansas Battery and swept into the fray, pushing the Confederates back over the mountain separating the Bayou Manard and Greenleaf Prairie. He reported that Watie and his force retreated down river to Webbers Falls where they recrossed at the ford. Apparently, some of the 1st Indian Home Guard did not stand by and watch. Sgt. Legus Perryman of Company I (Future Principal Chief of the Creek Nation) later wrote:

"…The rebels run up on the prairie here and got into the heard [sic] of horses and mules and killed fifteen herders and drove off about 200 mules and we attack the main body of the enemy at the edge of the wood, and had a good fight. About 10 rebels were killed there, among whom was George West. We pursued them across the river and slew a great many more…"[61]

Perryman's description was embellished since the Federals did not pursue the Confederates across the Arkansas River. But the Federals were not the only ones to embellish their stories. Lt. Col. James Bell of the 2nd Cherokee Mounted Rifles reported:

"…Our 1st and 2nd Cherokee and some Creeks went over and completely surprised them killed between 60 and 100 men and captured at least 1500 head of stock if they could have got them away but they could not get them all off and fight too. George West was killed and a man by the name of Bean was also killed…"[62]

Maj. Elias Boudinot of the 1st Cherokee Mounted Rifles provided some interesting observations of the battle/raid. He agreed that only about 500 head of stock was captured and admitted to about twenty killed and wounded on the Confederate Side. He reported in the *Washington (Arkansas) Telegraph* issued on June 3, 1863:

"…The enemy used his artillery, and were much superior to our men, both in number and quality of their arms. I had a view of the whole battle ground: it was as interesting an affair as I have ever witnessed. Many feats of individual prowess are related…"

Private Black Fox of the 1st Cherokee Mounted Rifles provides an interesting perspective of the battle from an enlisted man's perspective. He later reported in an undated and unidentified newspaper article:

"…I happen to be one of the thieving renegade Indians in that raid, consequently have some little knowledge of the facts… Our [command] was halted in the Bayou [Manard] bottom under cover. A number of our cowboys with the best horses were selected and made a dash across the prairie, taking the herders by surprise and shooting at every man in sight. They ran the heard [sic] together and when they got them started in our direction, our command covered the retreat… The nearest they came to getting me was when they got my little gray cap. We crossed the Arkansas at Hildebrand Ford. The Federals did not have enough horses to follow us and a small squad that had horses came out as far as Bayou Mountain…"[63]

A few days before this raid, a Confederate officer had appeared with a flag of truce on the Union side of Rabbit Ford on the Arkansas River, almost adjacent to Fort Gibson. He stated to the Federal corporal of the guard that he was a staff officer from General Cooper with a message to Col. Phillips regarding the exchange of prisoners between the two sides. The corporal failed to take any precautions in taking the Confederate to Phillips's headquarters through the Union lines such as blindfolding him. Phillips was a nervous and somewhat anxious man, and he suspected the Confederate officer was on a scouting mission to report the size of the Union garrison and the defenses, "although the rebel officer claimed to be as verdant as the Corporal, and perfectly innocent of any intention of spying on us." Phillips decided to confine the officer in a single room in his headquarters but still treated him well. It was just a few days after this officer was blindfolded and returned to the Confederate side that the raid occurred. Robert Peck, the civilian wagon master of the 2nd Indian Home Guard supply train, recalled the confusion of the raid:

"…One morning (May 20, 1863), shortly after this flag of truce bearer had returned to his command, a mounted scouting party had just returned from a reconnaissance on the Fort Smith Road, and the officer in charge was making his report in the Adjutant's office that there was no sign of rebels on that route, when just at that moment a breathless runner came rushing in to tell that the prairie out along the Fort Smith Road was swarming with rebels, who were killing all the pickets and herders and driving off the herds of mules and horses. At the same time a scattering rattle of firearms in that direction seemed to confirm the messenger's statement, although the astonished officer of the patrol insisted that there were no signs of an enemy near the road as he came in… In a moment all was excitement. Colonel Phillips was about the most excited person I saw. He was sure that Cooper's whole rebel force was coming in to take the place. The Indian regiments were quickly formed and moved out in the direction of the firing, but when they reached the grazing grounds- our Indians being afoot and the rebel force well mounted- it was only in time to see the enemy retiring into the timber on the Fort Smith Road, as they had come, driving the herds of horses and mules before them…"[64]

Phillips remarked in his official report that three of his Indian picket stations had deserted their posts and left his flanks exposed. This again was a common occurrence among the Indian troops whose loyalties lay with their tribes, and not with the United States. Both Albert Pike and Douglas Cooper had experienced the same issues. Phillips initially deployed his infantry, cavalry, and artillery, approximately one mile east of the fort. Mounted skirmishers were deployed to the northeast, and a large force of Confederates were discovered in some dense brush about a half mile east of their initial deployment line. The two artillery pieces began to drop shells into that forested area, which appeared to create a significant amount of movement. Other mounted troops were assigned to gather the remaining livestock and drive them back to the fort. The infantry began to move toward the Confederate line and the artillery followed, pausing every few minutes to drop some more shells on the Confederate position. The mounted Union troops deployed on the left of the infantry, a bugler sounded the charge, and the cavalrymen swept forward into the dense thicket. The Confederates in that defensive position wisely began to evacuate the site and headed towards the Fort Smith Road that lead over the mountain to the south of Fort Gibson. Meanwhile, when Phillips discovered that the Texans were attempting to cross the river at Rabbit Ford, he sent whatever mounted men he had after the raiders up the mountain towards the Greenleaf Prairie. These troops were able to recapture some of the horses and mules that had been taken by Watie's force but had been left behind. Phillips took his infantry and two cannons down to the ford to counter the crossing. Since the Arkansas River is quite wide at Rabbit Ford, both sides ended up just lobbing shells at one another across the river for the rest of the day.[65] Adjutant Leslie DeMorse of the 29th Texas Cavalry, also the son of regimental commander Colonel Charles DeMorse, reported that their unit was to make a demonstration of crossing the Rabbit Ford opposite Fort Gibson to distract the Union pickets at that point. This action was to give Stand Watie and the raiding party, consisting of members of the Cherokee and Creek regiments, the opportunity to use a different ford, probably the Hildebrand Ford,

with which to make their attack. The diversion obviously worked since there were no Federal troops keeping an eye on whichever ford Watie and his command used to gain access to the Fort Smith-Fort Gibson Military Road, although the Arkansas River was reported to be low at the time.[66]

Phillips reported his losses at upwards of twenty to twenty-six killed and between ten and thirteen wounded. In addition, he had lost much, if not most, of his livestock in the raid. The 6th Kansas Cavalry lost half of its horses. Fresh horses and mules were absolutely necessary for both the Union and Confederate armies fighting in the Indian Territory. In fact, during the previous week, a force of about four hundred men of the 6th Kansas Cavalry crossed the river and proceeded upriver towards the Old Creek Agency where they captured about sixty head of horses and mules of Cooper's command. Meanwhile, the first part of the expected supply train arrived at Fort Gibson without incident.

Due to the actions of the Confederates sending patrols to the fords to capture Union pickets, Col. Phillips ordered all of his troops on the southern and eastern slopes of the hill to break down their camps and move them into the new fortifications that had been built around Fort Gibson. Sgt. Britton later recalled:

> "…We took our tents down and packed everything up, and in less than two hours were inside the fortifications. Some few of the soldiers thought it useless, while most of them were perfectly satisfied to trust to the judgment of our commanding officer, as he was in a position to know very nearly the exact

situation. The thought of being somewhat crowded, it is true, was not a pleasant feature; but nearly every one was willing to forego a little freedom of movement for the sake of greater safety. So this morning when the sun had climbed the mountains, which, from our more elevated position, looked lovely fringed with green, the parapets were bristling with the guns of Captain Hopkins battery…"[67]

Knowing there was another supply train due to arrive from Fort Scott, Colonel Phillips decided that a demonstration across the Arkansas River at Rabbit Ford would keep the Confederates closer to their camps rather than attempting to capture the supply train. In the late afternoon of May 24, Phillips gathered approximately five hundred mounted and dismounted soldiers and two 12-pounder howitzers and marched them from the fortification to the ford. Due to the heavy brush on both sides of the ford, the Confederates were unable to see the cannons being deployed. As rifle and musket shots were being exchanged between the two shores, suddenly the artillery opened fire, dropping shells among the Confederate defenders, most of who fell back away from the shoreline and onto the small rises in their rear. Suddenly, Phillips and his mounted troops advanced across a firm sandbar and moved into the river channel as if they were going to cross in force. The infantry came in behind, deployed in a battleline on the sandbar, and began providing volleys of suppressive fire on the defenders on the western shore. At this point the Confederates began to feed reinforcements into the area

around the ford. Phillips and his mounted troops reached the center of the channel and stopped when the horses began to lose their footing. Since Phillips had no intention of crossing to the other side of the river, they slowly backed off and fell back to the eastern side of the ford. The infantry and artillery remained in place and kept the Confederates occupied for approximately three hours. With the Confederates attention diverted to Rabbit Ford, Phillips sent the 3rd Indian Home Guard and a section of Hopkins Battery, under the command of Lt. Col. Frederick Schaurte, up the Texas Road to reinforce the incoming supply train.

General Cooper saw Phillip's attack as an opportunity. Cooper had in fact been planning to attack the incoming supply train and had Maj. Livingston monitoring its progress. Yet when Phillips started to cross, Cooper came to believe that this attack was the main thrust. Cooper believed this because Phillips had already demonstrated that he was not averse to crossing and attacking. He hoped that he could catch Phillips and the Union attackers with their backs to the river, so he diverted as many troops as possible to the ford. Col. DeMorse of the 29th Texas Cavalry, whose command was responsible for protecting Rabbit Ford, had earlier been ordered to mount five companies of his command and proceed towards the Old Creek Agency, passing the destroyed Fort Davis while enroute. Cooper believed that Phillips had observed these companies leaving their camps which is why the Union commander had decided to attack. Col. DeMorse reported:

"…Four hours after we left, the enemy who from some of their lookouts had a sight of our men moving off, and thought our camp deserted—brought down their artillery, a body of Cavalry, and of Infantry, to the Rabbit ford, a half mile above our camp, and attempted a crossing. They commenced shelling, feeling about for the camp, the exact locality of which they did not know and threw two or three shells near it which hastened the departure of the Artizans [sic], the Invalids, etc., with the camp equipage. Capt. Elliott's Co. (I) was on duty at the ford, and sent down word for reinforcements which they soon got… The enemy soon commenced firing from the bank by the Infantry and under cover of this the Cavalry attempted a passage, and came midway of the stream, but became confused there and turned back. The rattle of grape and musketry was rapid and sharp. Maj. Carroll who has been in three or four actions previous, including that at Elkhorn, says the hail of balls was thicker than he has ever seen before… The infantry and Artillery withdrew at dark. The bed of the river was so thoroughly enveloped in smoke, that the effect produced by us could not be seen; but some horses without riders were seen to go up the opposite bank."[68]

Cooper finally realized that the Union attack was just a feint, so he went back to his original plan to attack the Federal supply train. Unfortunately, he did not discover that Phillips had sent reinforcements to assist in protecting the supply train from a Confederate attack until it was too late.

Phillips had the right idea. Very early in the morning of May 28 (there is some discrepancy in regards to the date, either the 25th or 28th), the Confederates did attempt to attack the supply train as it moved southward along the Texas Road. Lt. Col. Schaurte and the reinforcements made contact with the train's escort ten to twelve miles from Fort Gibson, approximately the south-side of the modern-day town of Wagoner, Oklahoma. The train consisted of about two hundred wagons and was stretched out for at least a mile. Sgt. Wiley Britton accompanied the train and reported:

> "…We marched with a detachment of about one hundred cavalrymen, say a quarter of a mile in advance of the escort just in front of the train, with detachments of cavalry at convenient distances from each other on both flanks, and with a strong rear guard. Skirmishers were also kept out a half mile on each side of the road, with instructions to keep up with the advance guard…"[69]

They got close to the Grand River timber near the J.D. Wilson place, southeast of Will Roger's cow pen, about an hour before sunrise. This is a distance of about five miles northwest of the fort, near the present town of Okay, Oklahoma. The Confederate raiders swept up from a slight ridge in the prairie and opened fire on the escort. By this time Col. Phillips himself had joined the escort, and he took command of the train's defense. He ordered the wagons to come in line, two-abreast, and tighten up their follow distance. He also firmly ordered the wagon teamsters to remain with their assigned wagons and teams. The Confederates swarmed towards the supply train and its escort from the south, east, and west, with an estimated number of close to a thousand raiders. Sgt. Britton described the defensive measures employed:

> "…Colonel Phillips formed his troops into a kind of oblong square, which inclosed [sic] the train. The two short sides of the square were made quite strong, and when the enemy made an effort to break either of the long and weak sides, we cross-fired him, and all his efforts were fruitless. Nearly half of our troops fought dismounted, which enabled them to fire with greater precision. We held the enemy in check in this manner for upwards of an hour, and until towards daylight, repulsing him in every attack, when Colonel Phillips determined to take the offensive, and at the decisive moment ordered the bugle sounded and led his troops to the charge. We moved forward with a shout, and in a few moments completely routed the enemy all along the line…"[70]

The Federals claimed they chased the Confederate raiders nearly to the Verdigris River crossings to the west. The Confederates claimed to have captured fifty wagons but to have lost them in the counter-attacks.[71]

The previous evening, Cooper had sent Col. D.H. McIntosh with his 1st Creek regiment, and Col. Leonidas Martin and his 5th Texas Partisan Rangers to make the attack on the Federal supply train. Cooper indicated that Maj. Livingston and his Partisan Rangers were following the wagon train, but if they participated in the attack, they are not mentioned in any of the reports or witness recollections. Private John Thomas Howard of Company F, 5th Texas Partisan Rangers later recalled:

> "…We were ordered on a scout across the Arkansas River to intercept a supply train from Fort Smith [Scott] to Fort Gibson. Col. D.H. McIntosh was in command. Late in the evening our scouts discovered the train. McIntosh was afraid that the whites and Indians could not distinguish each other and it was decided to wait till morning. Just before daylight we were discovered and they pulled out and we made a move to cut them from Ft, Gibson and struck the road just after they had passed. As the Yankees ran in and thought they were safe, leaving 2 men, some of our straggling Indians found these men and stripped them of their clothing and as we came up we found the Indians murdering one of the men and we got there in time to save the other. And after he was dead they "gobbled' over him…We were ordered to charge after the Yankees to within about 2 miles of Ft. Gibson, capturing a portion of the train. The Yankees ran out artillery and infantry and the Indians ran and there were not enough white men to hold them. Lost the wagons which had been captured. We retreated south of the river…"[72]

Wagon Master Peck definitely had ideas of how and where the attack would take place. When the reinforcements under Col.

Lt. Col Frederick Schaurte
2nd Indian Home Guard
Photo courtesy of Kansas State Historical Society

MAJOR JOHN A. FOREMAN

Photo courtesy Of Kansas State Historical Society

Frederick "Shorty" Schaurte arrived, they brought intelligence that Phillips had received information from a spy in Cooper's camp that there was going to be an attack on the wagon train. The attack was to occur before the train reached the timber, about four miles northwest of Fort Gibson. Up until that time the wagon train and its escort were not aware of the pending attack. Peck wrote in *The National Tribune*:

> "…This train had the usual cavalry escort of several companies of the 3d Wisconsin Cavalry from Fort Scott, in addition to which it had been met at Cabin Creek, 50 miles north of Fort Gibson, by a reinforcement of about 200 Indian Infantry from Fort Gibson, and all were coming along without any suspicion that the rebels across the Arkansas had made all

arrangements to divert the outfit from Fort Gibson to Fort Davis by cutting them off at a point about four or five miles our from Fort Gibson, on the road, and running them across the Verdigris (pronounced Verdigree) and Arkansas Rivers, right into the rebel camp…"

Although Sgt. Wiley Britton gave a good account of the Union troops making up the escort, Wagon Master Peck had a different perspective that did not show the 3rd Wisconsin Cavalry in the best light:

> "…Part of the 3d Wisconsin Cavalry constituted the advance guard. The rest of them, under command of Major Stout, 3rd Wisconsin Cavalry, were riding alongside the train near the point where the rebels cut across

the road. At the first charge and fire of the rebs the valiant Major and his cavalry bolted for Gibson at good speed, and never stopped until they were safe in town. Major Stout afterwards gave as an excuse for running away from the fight that he had 'brought the train safely inside of Colonel Phillips' pickets' (which was not true, as we had no pickets so far out) and he thought the Indian Brigade ought to take care of it then."[73]

Regardless of the differing eyewitness accounts, the Union supply train made it safely to Fort Gibson, and the Federal garrison was supplied for another period. An added bonus was that the Army paymaster was accompanying the wagon train to distribute the pay for the Union Indian Brigade.

Later on May 28, the waters on all of the major rivers began to rise significantly, probably due to the late spring snowmelt from the Rocky Mountains finally reaching the lower plains. Regardless of the reason, it made crossing at most of the fords impossible, although some of the remaining ferries were able to operate. Col. Phillips wanted to ensure that the wagon train headed back to Baxter Springs was safe, so he sent Maj. John Foreman and his battalion from the 3rd Indian Home Guard to provide security. Although he correctly believed that Col. Watie would not waste valuable resources to attack an empty train, Phillips did worry that they would find a way to damage or destroy the wagons or capture and kill the mule teams. Watie was in fact moving up through the Cherokee Nation and foraging to support his troops. Once again, the rumor mill had Watie everywhere but nowhere.

There were small skirmishes with portions of Confederate commands at various locations within the Cherokee and Creek Nations during this time including at Grand Saline, the Union Salt Works, and one near Cabin Creek. None of these minor skirmishes resulted in more than four or five casualties on either side. Maj. Foreman attempted to pursue Watie through Park Hill and Tahlequah, but his horses were worn out so he limped back to Fort Gibson. Meanwhile, not knowing that Foreman had abandoned the chase, Phillips had ordered Lt. Col. Wattles to take four hundred mounted troops from the 1st Indian Home Guard and a section of Hopkin's 3rd Kansas Artillery to move down the north side of the Arkansas River and try to cut off Watie from the crossings. Wattles was soon notified that Foreman had abandoned his pursuit of Watie and that the Confederate Cherokee leader had been reinforced by the 20th Texas Cavalry under Col. Thomas Bass giving the Confederates a two-to-one advantage. If Foremen, unaware of this fact, had attacked Watie and his reinforced force, he more than likely would have been severely defeated. Wattles was also informed that the 29th Texas Cavalry had crossed Hildebrand's Ford near Greenleaf Prairie and was now between Wattles and Fort Gibson. He had missed the crossing even though he was close by. Col. Phillips reported to Brig. Gen. Blunt:

> "… Colonel Wattles in the night passed close to where the rebels were crossing the river without discovering them; but two scouts, who were watching Greenleaf Prairie, notified

him at daylight that the rebels were forming in his rear. His force returned 6 miles, and engaged the enemy on Greenleaf Prairie, 18 miles distant from the fort... I mounted everything, on mules, horses, &c., and sent out Colonel Schaurte with 500 men and one piece of artillery (part of his force with the gun was infantry), and pushed him forward to aid Colonel Wattles..."[74]

General Cooper had ordered Col. DeMorse on Saturday, June 13 to move eight companies of the 29th Texas Cavalry regiment, along with eight companies of Col. Tandy Walker's 1st Choctaw and Chickasaw regiment, a total of approximately 760 soldiers, over the Arkansas River and onto Greenleaf Prairie. By noon on June 14, DeMorse had begun to move the regiment across the river at Hildebrand's Ford. The first 120 troops to get over the river were ordered to scout the area. Captain Brown with Co. H, was pushed northward to the edge of Greenleaf Prairie, and Capt. Vann with 12 men, was directed to scout the areas he felt were necessary. Due to the condition of the river at the ford, it took until the night of Monday, June 15 to get the sixteen companies across. The entire command moved inland about five and a half miles northwest of the ford and camped near Chalybeate Springs, the source of modern day Salt Creek, west of Greenleaf Prairie. Small detachments were sent out to determine if any Union troops were in the area. DeMorse moved his command to the eastern edge of the Greenleaf Prairie to some high ground where they could observe the Hildebrand's Ford Road and the Fort Smith-Fort Gibson Military Road. Some Union troops were observed on the Hildebrand Road. Lt. Heiston was sent on a reconnaissance with thirty men and reported back that the Union force was only a company-sized unit. DeMorse ordered two squadrons from the 29th Texas Cavalry, Companies D & K under Capt. Hook, and Companies F & E, under Capt. Olivers, to proceed to the location to intercept and capture the Federals. He also posted Company B, commanded by Lt. A. G. Bone to the left of the road to Gibson, and near the Brewer's farmstead. Company A, commanded by Capt. T. W. Daugherty, was sent to observe the telegraph road.[75]

During the night of June 15, Col. Wattles's command advanced south along the Fort Smith-Fort Gibson Military Road were through Greenleaf Prairie just before dawn on June 16. They had advanced about two more miles and crossed Greenleaf Creek when scouts reported that the Confederates were observed behind them on Greenleaf Prairie. Wattles immediately countermarched his command and sent one hundred troops under Lt. Robert Thompson as an advance guard back towards the prairie to "hold the enemy in check until I arrived with the rest of the forces." They approached to within one mile of the Confederate line when Wattles sent Capt. Bowlegs and seventy-five men around the right to sweep up their left flank. With his remaining mounted troops, Wattles charged Maj. Caroll's Confederate line, leaving the infantry to support the artillery. The surprise charge caught the Confederates off-guard and they fell back nearly three-quarters of a mile. Bowleg's attack on their left flank crushed Company K of the 29th Texas Cavalry, but Company D rallied to fill the open spot in the line.[76] Battalion Commander Maj. Carroll reported:

Greenleaf Prairie Battlefield, Camp Gruber, Oklahoma Army National Guard Base
Photos by Author

"…I ordered Capt. Oliver's Squadron into line and dismounted it, which was promptly done, and just at that time Co. K Lieut. Littlejohn com'dg was thrown into confusion losing all organization and not being controlled by its officers, broke through Capt. Oliver's line, scattered his horses and confused his men…"[77]

At this point, the Confederates had fallen back to the main line that Col. DeMorse had set up on a small rise in the middle of the western portion of the prairie, approximately where modern-day Camp Gruber is located. With the remainder of his command, including the Choctaw and Chickasaw regiment, DeMorse counter-attacked the Union line and drove it back until they reached the infantry line and the artillery.[78] Maj. Carroll continued:

"…They rallied upon their howitzer, about 400 strong. I ordered company C, Capt. Harmon com'dg who I found on the left of the Choctaws forward with a view of capturing the howitzer. Capt. Harmon dismounted his men

within seventy five yards of the Gun, and with the support of about 200 Choctaws drove the enemy about thirty yards from it, when the Choctaws were thrown into confusion and began slowly to retire. I attempted with the assistance of Lieut. Col. Parks of the 1st Cherokee to rally them in which we failed, and noticing the enemy were beginning to close in on our right, I retired slowly and in good order; feeling satisfied that with one more company I could have taken and held the howitzer…"[79]

Capt. Solomon Kaufman opened fire with his howitzer, which brought the attack to a halt. At this point, DeMorse was informed that Federal reinforcements were coming into the field from the direction of Fort Gibson. This was Lt. Col. Schaurte's relief column coming up behind the Confederates. Wattle was able to move his command around the flank of the Confederate line intending to meet Schaurte's command. DeMorse saw Wattles's movement and attempted to cut them off but was unsuccessful. At this point DeMorse knew he would be severely outnumbered and ordered his troops back across the Arkansas River but was unable to find an unprotected ford:

> "…I reluctantly ordered a movement down the Ft. Smith road, to take in a piquet[sic] of 30 men, said to be at Flakes or Lacy Milsom's. We found none at either place, though usually kept their [sic] heretofore; and we hurried through a defile in the mountains towards Mackey's Saline, and encamped at night within a quarter of a mile of the Illinois river. Our horses were much jaded, and foot sore, from the rocky route through the mountains. In the morning we started for Webber's falls; there being no road to Hildebrands, where I desired to go…"[80]

At Webber's Falls they encountered one of Watie's men who stated that the crossing there was still too high. Rather than take the very rocky and steep road for fifteen miles along the river to Hildebrand's Ford, DeMorse opted to take the easier and smoother road south to the Canadian River where they were able to get across the Arkansas River.[81]

Colonel Phillips reported four killed and seven wounded. Col. DeMorse reported ten killed and eight wounded. The casualties seemed very low when compared to the nearly two thousand troops involved. Sgt. Britton probably summed up the Battle of Greenleaf Prairie the best when he wrote: "There was very little dash displayed on either side."[82]

[1] Schofield to Curtis, *OR,* Ser. I, Vol. XXII (P2) pp. 6.

[2] Collins, *Tarnished Glory*, pp. 116-119.

[3] Blunt, *Experiences.* pp. 238.

[4] Wiley Britton, *Memoirs of the Border, 1863.* Volume 2. Chicago: Cushing & Thomas, 1882. pp. 77-78.

[5] Johannson, *Ellithorpe*, pp. 109.

[6] Returns of the Army of the Tennessee, 13th Division, XIII Army Corps, February 28, 1863. *OR,* Ser. I, Vol. XXII (P2) pp. 134.

[7] Ware, James W. "Indian Territory", *The Western Territories in the Civil War*. Fischer, LeRoy H. ed. Manhattan, Kansas: Sunflower University Press. 1977: pp. 107.

[8] Blunt, *Experiences.* pp. 239.

[9] Phillips to Curtis, *OR,* Ser. I, Vol. XXII (P2) pp. 56-58.

[10] Johansson, *Ellithorpe*, pp. 114.

[11] Ibid.

[12] Johansson, *Ellithorpe*, pp. 94.

[13] Ellithorpe to Chipman, *OR,* Ser. I, Vol. XXII (P2) pp. 49.

[14] Phillips to Schofield, *OR,* Ser. I, Vol. XXII (P1). pp. 219.

[15] Britton, *Memoirs*, pp. 21.

[16] Johansson, *Ellithorpe*, pp. 127.

[17] Phillips to Curtis, *OR,* Ser. I, Vol. XXII (P2). pp. 85.

[18] Abel, *Participant*, pp. 256-257.

[19] Trevor M. Jones, "Cowskin Prairie Council (1863)," *The Encyclopedia of Oklahoma History and Culture*, https://www.okhistory.org

[20] Phillips to Curtis, *OR,* Ser. I, Vol. XXII (P2) pp. 85.

[21] Edwards, pp. 38-39.

[22] Steele to Cooper, *OR,* Ser. I, Vol. XXII (P2) pp. 775.

[23] Josephy, Alvin M. "War on the Frontier." (The Civil War Series), New York: Time-Life Books, 1986: pp. 150.

[24] Phillips to Curtis, *OR,* Ser. I, Vol. XXII (P2) pp. 97.

[25] Marmaduke to Newton, *OR,* Ser. I, Vol. XXII (P1). pp. 194-198.

[26] Warren to Chipman, *OR,* Ser. I, Vol. XXII (P1). pp. 189-190.

[27] Britton, *Memoirs*, pp. 198.

[28] Britton, *Border, Vol. 2.* pp. 33.

[29] Johansson, *Ellithorpe*, pp. 160.

[30] Britton, *Memoirs*, pp. 207.

[31] Rampp, Lary C. "Negro Troop Activity in Indian Territory." Kepis & Turkey Calls: An Anthology of the War Between the States in Indian Territory: pp. 189.

[32] Ian Michael Spurgeon. *Soldiers in the Army of Freedom: The 1st Kansas Colored, the Civil War's*

First African American Combat Unit, Norman: University of Oklahoma Press, 2014. Pp. 53-55.

[33] Rampp, pp. 191.

[34] Spurgeon, pp. 80-97.

[35] Terry Beckenbaugh, "The Battle of Island Mound," U. S. Air Force Command and Staff College

[36] Ibid.

[37] James Williams, Regimental History 1st Kansas Colored Late "79th U.S." 1 Jan 1866. Muster Rolls and Payrolls, 79th United States Colored Infantry, AR 117, Kansas State Historical Society, 167.

[38] Rampp, pp. 191.

[39] Linda Mayes Miller, "Webber's Falls," *The Encyclopedia of Oklahoma History and Culture*, Online, Oklahoma Historical Society

[40] Franks, *Stand Watie*. pp. 136-137.

[41] Ibid

[42] Phillips to Curtis, *OR*, Ser. I, Vol. XXII (P1). pp. 314-315.

[43] Britton, *Memoirs*. pp. 226.

[44] Howard, "Frivolous History of Fort Gibson," T.L. Ballenger Collection, NSU/UA. pp. 260. Edwards, pp. 42.

[45] Phillips to Curtis, *OR*, Ser. I, Vol. XXII (P1). pp. 315.

[46] Blair to Blunt, *OR*, Ser. I, Vol. XXII (P1) pp. 320.

[47] Spurgeon, pp. 136-137.

[48] Britton, *Border, Vol. 2*. pp. 77.

[49] Hugh Thompson, Interpretive Panel, Sherwood/Rader Farm Civil War Park, Joplin, Missouri

[50] John R. Graton to Dear Wife, May 22, 1863. John R. Graton Collection, Kansas State Historical Society

[51] Livingston to Price, Ser. I, Vol. XXII (P1) pp. 321-322.

[52] Spurgeon, pp. 144.

[53] Elder to Blunt, *OR*, Ser. I, Vol. XXII (P2) pp. 286.

[54] "Massacre of the Confederates by the Osage." The Osage Magazine 2. February 1910

[55] "The Only Survivor's Story of the Tragedy." The Osage Magazine 2. May 1910.

[56] CSA vs. Osage in S.E. Kansas: treasurenet.com

[57] Southern Superintendent of Indian Affairs William Coffin. *Annual report of the Commissioner of Indian Affairs*, for the year 1863. pp. 173.

[58] "A Horse Raid" *The National Tribune*, February 18, 1886. pp. 2.

[59] Robert Cannon Horn, The Annals of Elder Horn: Early Life in the Southwest. Ed. John Wilson Bower and Claude Harrison Thurman. New York: R.R. Smith, Inc. 1930. pp. 41-42; Edwards, 45-46.

[60] Phillips to Blunt, *OR*, Ser. I, Vol. XXII (P1) pp. 337.

[61] Interview with Legus Perryman, in Grant Foreman "Indian-Pioneer History," Oklahoma Historical Society; Edwards, pp. 47.

[62] James Bell, Letter, May 25, 1863, E.E. Dale Collection, Box 112, F10, University of Oklahoma Western History Collection: Edwards, pp. 48.

[63] Edwards, pp. 48.

[64] Peck, R.W. "Wagon Boss and Mule Mechanic," *The National Tribune*

[65] Phillips to Blunt, *OR*, Ser. I, Vol. XXII (P1) pp. 336-337.

[66] Leslie DeMorse, *Supplemental OR*, Vol. 68, pp. 242; Edwards, pp. 49.

[67] Britton, *Memoirs*, pp. 262.

[68] *Clarksville (Texas) Standard*, June 16, 1863

[69] Britton, *Memoirs*, pp. 267-269.

[70] Ibid

[71] Ibid

[72] Mamie Yeary. Reminiscences of the Boys in Gray, 1861-1865. Non-Published Work: Dallas. 1912. pp. 351-352; Edwards, pp. 50-51.

[73] Peck, R.W.

[74] Phillips to Blunt, *OR*, Ser. I, Vol. XXII (P1) pp. 369.

[75] DeMorse to Cooper, *Clarksville (Texas) Standard*, June 21, 1863

[76] Wattles to Phillips. *OR*, Ser. I, Vol. XXII (P1) pp. 351.

[77] Carroll to DeMorse. *Clarksville (Texas) Standard*, June 16, 1863

[78] John C. Grady & Bradford K. Felmly. Suffering to Silence: 29th Texas Cavalry, CSA, Regimental History. Quannah, Texas: Nortex Press. 1975.

[79] Carroll to DeMorse. *Clarksville (Texas) Standard*, June 16, 1863

[80] DeMorse to Cooper, *Clarksville (Texas) Standard*, June 21, 1863

[81] Ibid

[82] Britton, *Memoirs*, pp. 294.

Chapter 13
The Honey Springs Campaign

With all of the activity happening down in the Cherokee Nation it is easy to forget that there were four other Indian Nations involved in the war. The reason is a significant lack of reliable information about the activities of the Creeks, Choctaws, Chickasaws, and Seminoles under the Confederacy. There were few chroniclers of tribal activities within the Indian Territory during the conflict. Since the ties between the United States and the Cherokee Nation had always been tighter than the ties with the others, it is no surprise that more information about the Cherokees is available. Southern Superintendent of Indian Affairs William Coffin reported in 1863 that the Creeks who remained after Opothleyahola and his followers had been driven from the Creek Nation were, for the most part, destitute and in need of provisions that could not be provided by the Confederate States. Many Creeks took refuge down in the Chickasaw Nation around the Wapanucka ["Eastern Land People"] Academy, located about twelve miles northwest of the Confederate supply post at Boggy Depot. This was a girl's school built in 1852 by the Chickasaw Nation and operated by the General Assembly of the Presbyterian Church. It was a large, three-story stone building with a two-story extension at one end that was designed to house up to one hundred Chickasaw young women. Due to the hostilities, it was decided to close the school and sell whatever property was owned by the school. During the War, the stone building was used as a Confederate

1st Kansas Colored Volunteer Infantry Monument @ Cabin Creek Battlefield
Photo by Author

hospital, with some of the rooms at one time being barricaded for a guard house or prison, as well as a refugee collection point.[1] Whatever caused the destitution of the Creek Nation's citizens, it could not necessarily be blamed on the Union Army since the Federals had not entered the Creek Nation in any significant way. Although they too had contended with the drought of 1862, the rains came later in the summer, which still left time for a quick crop to be planted and harvested. The probable cause was the population, which had been reduced by at least half with the evacuation of the Unionist Creeks. Superintendent Coffin reported 3,200 Unionist Creek refugees remained in Kansas awaiting return to the Creek Nation but were prevented from doing so because of the danger posed by the Confederate Creeks. Another cause was that most of the men who had remained behind were serving in one of the Confederate Creek regiments commanded by the McIntoshs' and their close associates. It was probably a combination of a lack of manpower and the probable impressment of food and livestock by the Confederate Army that caused the Creeks to be destitute. Even with these difficulties and the defection of many Creek soldiers to the Union Army, the Creek Nation still operated. They held councils, passed laws, conducted elections, and attempted to support their two Confederate regiments in the field.[2]

The Choctaws and Chickasaws remained almost universally behind their allegiance with the Confederacy. Since they were further south and closer to their supply system in Texas, they did not seem to encounter the same subsistence issues that plagued the Creeks and the Cherokees until later in the war. And since the Union Army made few forays into these two nations, they were not affected by the war in any significant way. Few members of either of these tribes renewed their allegiance to the United States during the war. Only the Seminoles, who had reluctantly sided with the Confederacy, for the most part stayed out of the war and provided only a small battalion of men to the Confederate Army. As mentioned earlier, the Chickasaw Nation did provide a refuge for members of the Five Civilized Tribes, and a few others, especially after Texas instituted a head-tax for any Indian refugee who crossed

the Red River into that state. The Confederate refugee depots provided beef, flour, and soap to 4,823 Creeks on the Washita and at the Wapanucka Academy, 2,906 Cherokees at Tishomingo, and 574 Seminoles at Oil Springs, west of Fort Washita. Later in the war they provided aid to many Osages and Choctaws as the Federal armies moved south.[3] Unfortunately, most of the destruction that occurred within the Indian Territory during the war can be attributed to the actions of guerilla bands made up of Texans, Kansans, Missourians, Arkansans, and many local tribal members. These bands brought havoc upon and created desperation in the population of the various Indian tribes, many of whom were simply trying to stay out of the way of the war. These bandits took advantage of the lack of manpower since many of the men were serving in the Confederate or Union armies. Moreover, many homes, farms, and businesses had been abandoned when their occupants had relocated to Texas or other areas of the South to escape the danger. By the end of the war the Indian nations had been stripped clean of anything of value. Most of the livestock was gone and many of the freshwater wells and springs had been contaminated or intentionally poisoned. The Indian Territory was a terrible place to live during the Civil War.

Near the end of June 1863, a 200-wagon supply train was to be sent from Fort Scott to Fort Blunt with a large quantity of food and munitions. The train was to be escorted by a battalion of the 2nd Colorado Infantry and a section of the 2nd Kansas Battery. Colonel Williams was to take command of the supply train, and his First Kansas Colored Infantry regiment was to accompany it from the Baxter Springs Depot.

Area of Operations for Cabin Creek and Honey Springs Campaigns

Colonel Phillips at Fort Blunt, fearing that the supply train would be attacked by Colonel Watie's Confederates, dispatched six hundred men and a howitzer under Major John A. Foreman, the acting commander of the Third Indian Home Guard. The supply train moved south from Fort Scott and joined with Colonel William's and Major Foreman's troops on June 24 near Baxter Springs.

On June 23 Confederate Brig. Gen. Douglas Cooper was notified by scouts, probably Maj. Livingston's guerilla band who were monitoring Baxter Springs, that a large

Union supply train was preparing to depart for Fort Gibson. He then notified Maj. Gen. William Steele that he had ordered one thousand mounted Confederate troops up the west side of the Grand River, bypassing Fort Gibson, with the intention of attacking and capturing the incoming supply train. He sent Col. Stand Watie with his five hundred men in advance to the Cabin Creek crossing of the Texas or Military Road. Once Steele had confirmed that the wagon train was heading south from Baxter Springs, he ordered Brig. Gen. William Cabell to move his forces from their camps in Arkansas and immediately proceed to Grand Saline on the Grand River to meet up with Cooper. Together they were expected to jointly attack the Federal train. Unfortunately for the Confederates under Cooper and Cabell, recent heavy rains had swollen the Grand River to the point that it was impossible to cross, leaving Cabell's force on the wrong side of the river. This also affected the remaining column of Cooper's command that was moving up the west side of the Grand River because the feeder streams and creeks that needed to be crossed were swollen as well. The Cabin Creek crossing was also too high to cross. Steele was disappointed when he learned that Cabell's force was not able to cross the Grand River to support Cooper because he had little confidence in the Confederate Indian Brigade. Steele wrote of his apprehensions to Maj. W.B. Blair, Lt. Gen. T.H. Holmes's adjutant, on July 1 stating: "The Indian brigade, under General Cooper, will dissolve in a great measure if forced to fall back."[4]

The train and its escort moved out of Baxter Springs and into the Indian Territory on June 26. The wagon train escort included five companies of the 2nd Colorado Infantry which had been formed in February 1862 in Canon City, Colorado Territory, and was commanded by Colonel James Hobart Ford. They spent most of their existence at Fort Garland, Colorado Territory, until they were ordered first to Fort Riley via the Santa Fe Trail, then Fort Scott, and finally to act as an escort for the supply train bound for Fort Gibson/Blunt. The battalion present at Cabin Creek, consisting of Companies B, E, G, H, and I, was under the command of Maj. J. Nelson Smith. The Federal column was also slowed by the high waters on the Grand River

and its feeder streams. They were forced to hold up at the Hudson's Creek crossing until June 29 due to the high water. They were able to advance to the Horse Creek crossing where the command camped on the night of June 29. On June 30 scouts reported to Maj. Foreman that they had found a large trail that had been used by a large number of troops. Foreman reported:

> "…I immediately detached Lieutenant [Luke F.] Parsons, of the Third Indian Regiment, with 20 Cherokees, to ascertain what had made the trail, as it was fresh. Parsons followed the trail about 4 miles, when he found 30 of the enemy, who proved to be Stand Watie's advanced picket. He gallantly attacked and defeated them, taking 3 prisoners and killing 4…"[5]

This was the trail that was discovered by Maj. Foreman and his Indian scouts. They knew that the horses that made the side trail were Confederate since there were only three horseshoe nails in each side of the horse's hooves. Due to the lack of horseshoe nails in the Confederacy, they only used three nails on each side while the Union Army used four. Since the Federal advance party had moved to the point where they were no longer under the observation of the Confederates on Timbered Hill, this is where Foreman sent Lt. Parsons to follow the trail. Parsons and his small command followed the obvious trail and discovered the Confederate observation point, first making contact with the scout holding the horses. The scout fired a shot, got on his horse, and rode off across the prairie in the direction of Cabin Creek. The sudden gunshot scared off the other horses and, although it warned the other scouts, it was too late, and they were either killed or captured by Lt. Parson's small force.[6] An interesting story was related by Wagon Master R.M. Peck in which he tells that one of the Confederate scouts captured was a man named Mage Lipe, a mixed-blood Cherokee. After his capture, the Union Cherokees with the detachment wanted to kill Lipe and the other captured Confederates. Lt. Parsons had to step in between to prevent this from happening. Eventually Parsons released the prisoners on parole. Mage Lipe wanted to join the Union Army and took his oath of allegiance to the United States. He refused to join any of the Indian Home Guard regiments, opting instead to enlist in the 14th Kansas Cavalry. He knew that other Union Cherokees in those Indian Home Guard regiments would eventually kill him.[7]

It should be noted that the leadership of the Union wagon train fully expected to have a substantial engagement with the Confederates at some point before they reached Fort Gibson/Blunt. It was also going to be a baptism of fire for many of the Union troops who were marching with the supply column. The African American troops of the 1st Kansas Infantry (Colored) were nervous but anxious to fight. Wiley Britton described what he saw and knew of the black soldiers:

> "…Soldiers of the colored regiment expressed themselves as eager for a fight; it would be the first fight in which they had taken part as a regimental unit since their organization; they knew that there was prejudice among the white soldiers against them on account of their color, and they determined that if they

became engaged with the enemy, that their performance should be creditable to any military organization, and Colonel Williams, who had drilled them and handled them, had the utmost confidence in their steadiness under fire…"[8]

The Federal supply train approached the Cabin Creek crossing of the Texas Road at about noon on July 1. At this point, the Foreman's Union advance ran into parts of Colonel Watie's Cherokee regiment and Colonel D.N. McIntosh's Creek regiment at the ford. They first encountered a small picket force on the north side of the crossing which quickly fell back across Cabin Creek after losing three men. The main Confederate battleline was hidden by the thickets and fallen logs and heights on the south side of the crossing where they had a commanding view of all of the approaches to the ford. Foreman deployed his command into a line of battle on both sides of the road leading to the crossing. He placed Lt. David Painter of the 2nd Indian Home Guard (IHG) and Lt. Luke Parsons of the 3rd IHG on the right of the road. He deployed Lt. Fredrick Crafts of the 1st IHG and Lt. Benjamin Whitlow of the 3rd IHG to the left side of the road. He placed Lt. Jule Cayot of the 3rd IHG in the center of the approach road along with one of the 12-pounder mountain howitzers. As Foreman's battalion advanced, the Confederates opened fire on the advancing Federals. Shots were exchanged in a "brisk fire" for over thirty minutes until eventually the return fire began to drop off. Under the protection of the cannister loads from the mountain howitzer, which were keeping the Confederates heads down, the Union force advanced to the ford and attempted to cross and make contact with their foe. They were unable to do so when the soundings of the ford indicated the water was too deep for the troops or the wagon train. Instead, Col. James Williams of the 1st Kansas Infantry (Colored) ordered the wagon train to pull back approximately two miles from the crossing, corral in the open prairie, and go into camp while waiting for the water levels to recede. The escort troops were ordered to surround and protect the supply train while also monitoring the Confederate activities.[9]

Watie and McIntosh were in desperate need of Gen. Cabell's 1,500 Arkansas troops and the four artillery pieces he had with him, although he had little artillery ammunition supply with him. Gen. Steele in Fort Smith had been pleading with the Confederate Ordinance Service to no avail stating that his units had severe shortages for the few artillery pieces available to his command. Cabell was expected to arrive at Cabin Creek on July 2 after crossing the Grand River at Grand Saline, so Watie and McIntosh were simply hoping to hold the crossing until Cabell's reinforcements arrived. In the interim Watie's Cherokees and Col. McIntosh with his 1st Creek Mounted Volunteers were reinforced by the combined six hundred men of the 5th Texas Partisan Rangers under Col. Leonidas Martin, and the 29th Texas Cavalry under Col. DeMorse. Watie probably had nearly a thousand troops under his command. They were spread out about a mile up and down stream with the ford in the middle. They could see the Union signalmen in the trees reporting the Confederate movements along the southern shore. For some reason Watie did not order the troops to build defenses beyond hiding behind the thickets, trees, and logs. Unfortunately, Indian commanders had not learned how to dig trenches or build breastworks since that style of warfare was not practiced by the Indian tribes. They also had no artillery to counter the Federal guns or to use in challenging the crossing at the ford.

When Watie came north with his command, he was accompanied by a military engineer who had determined that the Cabin Creek crossing was the best place to defend against a Union attack or to simply hold up a supply column from the ford. Cabin Creek had a strip of timber and brush nearly two miles wide that ranged up and down stream for most of the region. In addition, Mustang Creek, which entered Cabin Creek from just north of the ford, provided a funnel effect for the approaches to the crossing. There were heights on both sides of the ford that would provide coverage for defense, although the heights on the south side would be a disadvantage to the Union force since they had to drop down to the water level to cross. If Cabell and his fifteen hundred Arkansas soldiers and its artillery arrived in time, the Cabin Creek ford would be almost impregnable.[10]

As instructed, Cabell and his reluctant Arkansas volunteers did arrive at Grand Saline on the Grand River on July 2 but were prevented from crossing due to the extremely high water and strong current. Instead of waiting for the water to recede, Cabell turned his force around and by July 8 he was back in Van Buren. In hindsight, although Cabell was at the Grand River crossing on July 2, he would not have been able to help the Confederate Indian Brigade hold on to the Cabin Creek crossing. By the time he would have been able to fully cross his Arkansas troops and the artillery, the Union force would have already punched through Watie's force. The Confederate Indian Brigade and the Texas detachments were on their own.

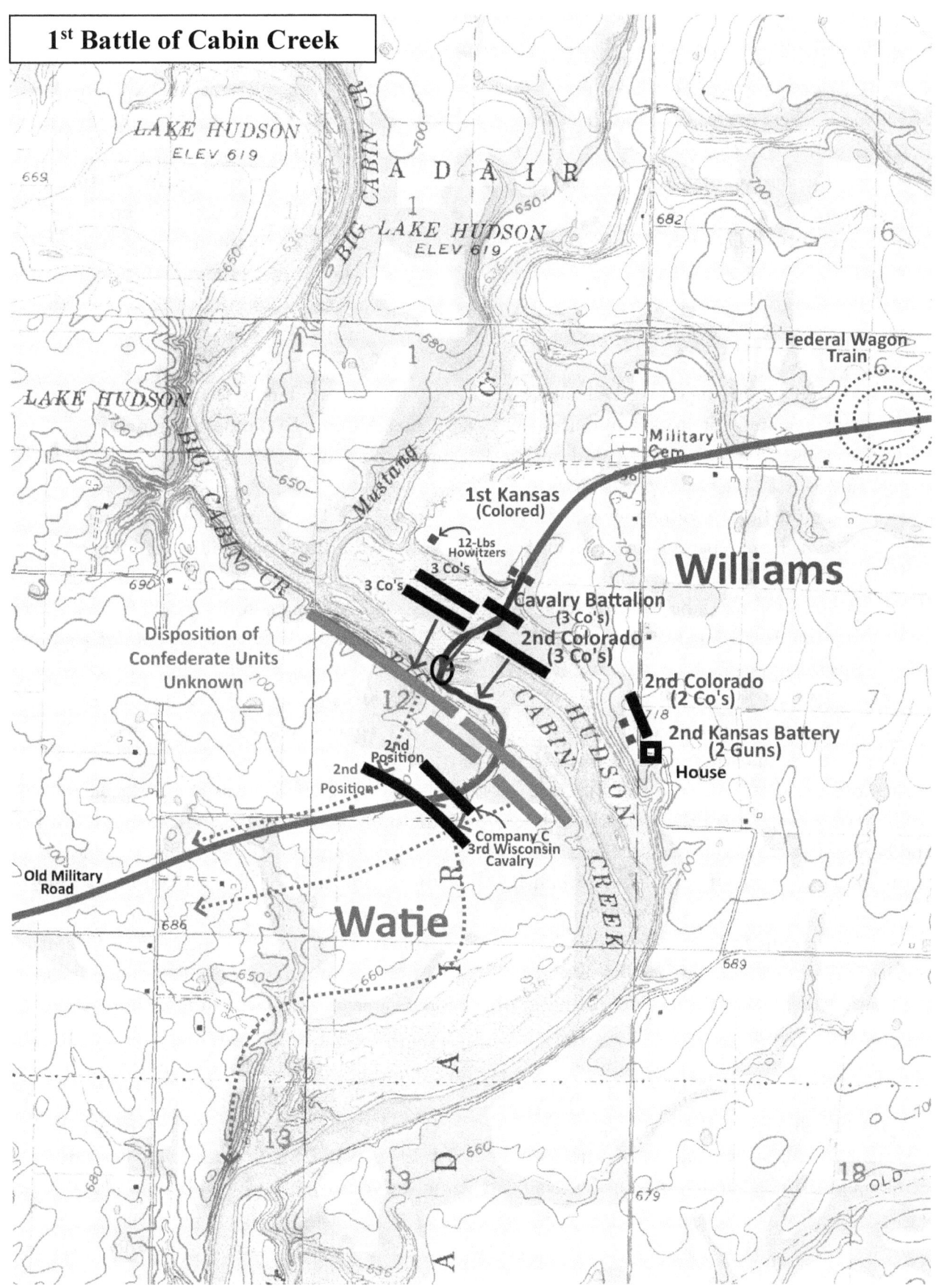

1st Battle of Cabin Creek

When they observed the Union cavalry advancing onto the ford, the Confederates sent a line of infantry forward out of the brush and thickets to challenge the crossing. Pvt. G.S. Killgore of Company H, 3rd Wisconsin Cavalry, remarked in an article in the *National Tribune* in 1883 where he discusses what he believed the Confederate plan was:

> "…The small force which opposed our crossing at the creek was to hold us there by digging down the bank so that we could not cross with our wagons, and then hiding behind trees and logs on the opposite side and doing the best shooting they could to keep us from repairing the road, but, as luck would have it, Grand River was too full for Cabel [sic] to cross, and our cavalry made a charge which put the opposing force to rout…"[11]

At this point the Union batteries began a forty-minute cannonade. The artillery scattered the Confederates on the south side of the ford, causing them to dive under logs and behind trees. After the cannons had fallen silent and after finding that the creek level had dropped during the night, Major Foreman led the single company of Indian Home Guard cavalry across the creek and viciously engaged the Confederates on the other side. As they reached the opposite shore, the hidden Confederates opened fire on the advancing Federals. This volley brought the advance troops to a halt, especially when Maj. Foreman was hit by two rounds. Private Christopher Kimball of the 9th Kansas Cavalry wrote:

> "…In the attempt, he [Major Foreman] was severely wounded in the back and neck. Col. Williams took command of the column, at the head of the troops, dashed into the stream. They got across with little loss, and charged on the rebel position. They fled from the center precipitously when the Negroes and Colorado boys charged, leaving arms and accouterments scattered as they went…."[12]

Private James McCombs of the 1st Creek Mounted Volunteer Regiment remembered,

> "…The men were in the water. Then the shooting commenced. I turned around and shot the Federal, Major For[e]man, in the arm. His men took him back. Everybody commenced shooting and it was hard to tell what was going on. We went out to the prairie where our horses were…"[13]

As Foreman was being evacuated to the surgeon wagon, the 1st Kansas Infantry (Colored) under Colonel Williams deployed three companies of his regiment into a firing line to the right side of the crossing. The artillery was ordered to open fire again, and for twenty minutes the artillery rained shells and canister onto the southern shore while the Kansas infantrymen fired volley after volley into the thick brush and thickets on the other side. Lt. Luther Dickerson of Company B, 1st Kansas (Colored), saw one of his soldiers, a Pvt. Smith [there were two Smiths in the company] get hit with a round:

> "…while in the act of loading his gun,

[Smith] was struck by a bullet which passed between the gun and his hand tearing the flesh from the inside of his fingers... Standing close beside him at the time he turned to me and asked me to load his gun for him which I proceeded to do..."[14]

As Dickerson raised his left arm to draw the ramrod, he was struck in his upper arm, and the bullet came to rest in his shoulder. He carried that bullet in his shoulder until the end of the war.[15] Over on the Confederate side of Cabin Creek, Pvt. John Thomas Howard of Company F, 5th Texas Partisan Rangers, was on the receiving end of the cannonade. He later wrote:

"...While the hardest of the fight was going on and while grape and canister were flying thick... It hardly looked like twenty men could stand under fire as we did, but everyone had a tree and had it not been for that we could not have lived, for the artillery was not over a hundred yards from us. I could hear them every time they would swab their guns..."[16]

At this point the 1st Kansas Infantry (Colored) was ordered to advance and drove into the ford. The Kansas infantry took the right side of the ford, and the 2nd Colorado Infantry battalion crossed on the left side. The water was up to the infantrymen's armpits, and they held their rifles and cartridge boxes over their heads as they crossed. When they reached the opposite shore, they quickly deployed into a line of battle, and Williams ordered them to advance. The charge broke the Confederate line, and they quickly

retreated out of the heavy vegetation of Cabin Creek. The Union infantry quickly drove off what few stragglers remained on the southern shore of the ford. The Federals could see Col. Stand Watie riding through his retreating Texas and Indian troops and attempting to get them to form up and hold the line. He was finally able to secure a weak battleline at the edge of the prairie approximately a quarter mile south of the crossing. Williams ordered Company B, 3rd Wisconsin Cavalry, and Company B, 14th Kansas Cavalry, under the command of Lt. John Stewart, to the right flank to protect the troops from that direction. Company C, 9th Kansas Cavalry, under the command of Lt. R.C. Philbrick, was ordered to take position at the front of the attacking columns, replacing the Indian Home Guard company. Company C filed through the Union infantry and pursued the retreating Confederates. They quickly came upon the new Confederate line, formed into a line of battle, drew sabers, and the buglers sounded the command: "Charge!" Lt. Philbrick's Kansans charged directly into Watie's small remaining force of Texans and Indians, which quickly broke, and Watie's command disintegrated, his men scattering to the east, west, and south. Company C attempted to pursue them, but they failed to catch the retreating Confederates. Some sources say a few drowned after they attempted to swim across the Grand River in their efforts to get away.[17] Even before Company C had returned from their pursuit, the wagon train had already begun to ford Cabin Creek.

On the Confederate side, there was considerable finger pointing as to who was responsible for the defeat and failure to capture the Federal supply train. Although there is no official record from the

Confederate commanders, Col. Watie wrote to his wife Sarah on July 12, 1863, stating:

> "My Dear Wife, I returned yesterday [from] my trip up the country… I had a hard trip of it, while I was gone had one of the severiest [sic] fights that has been fought in this country on the bank of Cabbin [sic] Creek fought for two days, the Feds forced and drove our men away from the ford, the second day by a severe cannonade, had two pieces the first day the 2nd had four, I lost but few men. I am safe you must not be uneasy about me I will take care of myself…"[18]

Watie placed the bulk of the blame on his lack of artillery support and the failure of the Creek regiment to hold its ground. He went into some greater detail into his reasons for failing to hold onto the Cabin Creek ford in a letter on the same date to "Sally," which may have been a pet name for his son Saladin, but this is just conjecture:

> "…Cannon not having arrived. I had to fight under great disadvantage. No cannon against a superior force with 4 pieces of cannon, I fought them about six hours at the crossing of Cabin Creek at Joe Martin's place if the Creeks had stood I would have fought them longer but they would not do it. Was a 2 day fight about 4 hours the first day and 2 hours the next. I am proud of my men they withstood cannonade as well, I lost but few men only 4 killed and few wounded. My whole loss in killed, wounded and prisoners does not exceed 15. The loss

of the Federals must be heavy…"[19]

Despite the fighting, the supply train and its escorts arrived safely at Fort Gibson/Blunt in the evening of July 5 and began the tedious task of ferrying the wagons and teams across the swollen Grand River. During the day of July 6, the wagons were unloaded at Fort Gibson/Blunt. On July 7, after another day of ferrying the wagons and teams back across the Grand, they began the slow trek back north to Fort Scott. Due to the threat from bushwhackers and Partisan Rangers, a battalion of the 6th Kansas Cavalry was assigned to protect the train as it moved northward. There also was the possibility that Gen. Cabell had finally been able to cross the Grand River at Grand Saline and could get between the train and Fort Scott. As a precaution they moved north with parts of the 2nd Colorado Infantry, a detachment of Indian Home Guard, and a section of the 2nd Kansas Battery accompanying them until at least they reached Cabin Creek at which point they would return to Fort Gibson/Blunt. They camped the first night on the return journey at Flat Rock Creek crossing, the same location in which the previous year's Indian Expedition had camped. On July 9 they met a Union column headed by Maj. Gen. James Blunt, who was accompanied by at least six hundred soldiers consisting of a different battalion of the 6th Kansas Cavalry with two 12-pounder mountain howitzers, commanded by Col. William R. Judson; a battalion of the 3rd Wisconsin Cavalry, commanded by Capt. E.R. Stevens, and a section of the 2nd Kansas Battery consisting of two 6-pounder field pieces, commanded by Capt. E.A. Smith. Blunt had gathered this large detachment under the command of Col. Judson as soon as

Texas/Military Road Crossing @ Cabin Creek Battlefield, Oklahoma
Photo by Author

Texas/Military Road approach to Cabin Creek ford from the North
Photo by Author

Cabin Creek floodplain across where the Union units drove the Confederates from left to right
Photo by Author

Location of 2nd Kansas Artillery
Photo by Author

View of Confederate 2nd Line
Photo by Author

Grand River Ferry & Ford near Fort Gibson, Cherokee Nation
Photo courtesy of the Oklahoma Historical Society

he heard of the battle at Cabin Creek. They had traveled 120 miles in less than three days. As a surprise, Blunt had some welcome news for the troops from Fort Gibson/Blunt. He told them of the decisive victory of the Union Army of the Potomac against General Robert E. Lee and his Army of Northern Virginia at Gettysburg in Pennsylvania; of General Ulysses Grant's capture of the Confederate stronghold of Vicksburg, Mississippi, the last major enemy stronghold on the Mississippi River; and the defeat of the Confederates under Gen. T.H. Holmes at Helena, Arkansas. The tides of the war were beginning to turn.[20] Satisfied that the situation was working in the Union's favor, Blunt and his command pressed onward to Fort Gibson/Blunt, arriving there on July 11. Gen. Blunt was not pleased with what he discovered at the post. He recalled:

> "...I reached Fort Gibson on the morning of the 11th, where I found that the administration of military affairs had been very badly conducted. Detachments of the enemy had been allowed to cross the Arkansas river at pleasure, and amuse themselves by capturing all stock sent out to graze, and in every other way annoy our troops, who were kept close to the fortifications, while rebel spies were inside of the garrison in the full confidence of the commanding officer, and acting as his military advisers, and in this way they (the rebels) were enabled to 'play both hands,' and it is not to be wondered at that they always 'took the tricks'..."[21]

There is much evidence that the lines between the Union forces north of the Arkansas River and the Confederates on the south side were very porous. This was illustrated by a comment in a Clarksville, Texas, newspaper:

> "...The Feds have four stores near Gibson; sell Coffee at 25 cents per pound, and Calico at 25 cents per yard. Many of the Indians from this side have been over to trade with them.--They are said to be interesting..."[22]

Blunt was also responding to the continual demands of Gen. Schofield to do more with fewer men. Prior to departing Fort Scott, Blunt had received instructions from Schofield that:

> "...I should take the field in person and if possible maintain the line I then held," which was the Arkansas river. This was what I desired and intended to do if I could be provided with troops..."[23]

Blunt had also heard that the Confederates in Fort Smith had hanged eleven Unionists. Blunt wrote, "I have now learned where the game is, and ache to get across the river. I shall make them pay dearly for their barbarity."[24] Blunt had decided that he had dealt with the Confederate Indian Brigade long enough and was intent on destroying Gen. Cooper's army. It was to be a monumental undertaking since he had only three thousand men to take on the Confederates who were fielding at least six thousand in the Indian Territory, consisting of the Confederate Indians and Texas cavalry. As soon as he had arrived at Fort Gibson/Blunt on July 11, he began making plans to cross the swollen Arkansas River and advance to and destroy the Confederate forces at Cooper's new supply depot at Honey Springs. Before Blunt could start any serious preparations, the next night he was invited to a reception in his honor due to his exertions to keep the "bread-line" open between the two posts. Sgt. Britton recorded the event:

"…Colonel Phillips presided at the reception, and in introducing the General stated that that feature of the program was not necessary, for the Indian people and soldiers present were familiar with the brilliant achievements of the General in closing the campaign of last year, which had to his credit Newtonia, Fort Wayne, Cane Hill and Prairie Grove, ending in the capture of Van Buren and destruction of the enemys [sic] steamboats and supplies on the Arkansas River, and that his presence was a guarantee that while he was in the field the enemy would not find it convenient to pitch their camps as near that post, as they had been doing during the spring when the Indian command was holding the most advanced position of any Federal force west of the Mississippi River…"[25]

In response, Gen. Blunt thanked everyone for their patience and support and presented the news of the Union victories in the Eastern and Western Theaters. He also outlined the general plans for the Department of Missouri and the District of Kansas and Indian Territory. He was optimistic that the troops diverted from the Trans-Mississippi to support Gen. Grant's Vicksburg campaign would soon be brought back to expand the operations in Kansas, Missouri, Arkansas, and the Indian Territory.

In preparation for another ill-advised movement to push the Union Army out of the Indian Territory and drive into Kansas, by early summer of 1863 Gen. Cooper had established a small supply depot at the old settlement of Honey Springs in the Creek Nation, about 25 miles south of Fort Gibson/Blunt. Honey Springs lay near the Elk Creek crossing of the Texas Road and had been established many years before as a stage stop and watering hole for persons traveling through. The Confederate supply depot consisted of at least one frame building, a log hospital, hastily built wooden arbors, abandoned houses and barns from the mostly deserted settlement. The Texas Road crossing of Elk Creek was located approximately three miles north of Honey Springs and was served by a series of shallow fords and a toll bridge at the main crossing.[26] There was a skirt of timber and brush about a half mile wide on either side of the Elk Creek crossing. Beyond these were big and little bluestem prairies with patches of scrub brush that ranged in size from waist high to about a chest high. It was a good place to hide your defenses. The prairies provided a wide and long view towards the north. The banks of Elk Creek were steep, usually around 15 feet deep, which made crossing difficult, hence the exclusive use of the fords and toll bridge. The toll bridge was a wooden structure owned and operated by William McIntosh, the brother of D.N. McIntosh of the 1st Creek Mounted Rifles. The bridge was not collecting tolls during this period since most civilians had left the area earlier in the war.

Colonel Douglas Cooper's General Order #25 describing his plan for defending the Honey Springs Depot

General Orders,)
 No. 25   )

Hdqrs. First Brig., Indian Troops,
Elk Creek, July 14, 1863.

I. The First and Second Cherokee Regiments will constitute the right wing of the brigade, Col. Stand Watie, senior colonel, commanding.

II, The left wing will be composed of First and Second Creek Regiments, Col. D. N. McIntosh commanding.

III. The center will consist of Twentieth Texas dismounted cavalry, Twenty-ninth Texas Cavalry, Fifth Texas Partisan Rangers, and Lee's light battery, Col. Thomas C. Bass, senior colonel, commanding.

IV. Seanland's squadron, [L. E.] Gillett's squadron, and First Choctaw and Chickasaw Regiment, Col. Tandy Walker commanding, will be attached to headquarters and constitute the reserve, to which such other troops belonging to this brigade as may report will be added until further orders. Captain [John] Scanland will fall back to a position which will be assigned him near headquarters, Honey Springs.

The right wing will encamp convenient to the two lower crossings on Elk Creek the center near or at such places as may be convenient to the middle ford, and the left wing at or near the upper ford; the reserve near headquarters, Honey Springs Depot. Commandants of each wing will see that necessary ways are opened along the front and near Elk Creek to enable the troops to move with facility from point to point, and also that proper roads from the camps perpendicular to the way along the bank of the creek are opened. Each regiment will occupy a front at least equal to the number of files, minus one-fifth. For example: If the total of a regiment be 1,000 men, or 500 files, the front will be 400 yards. The proper intervals between squadrons and regiments will be observed, and kept free from obstruction, to allow the passage of the troops. These intervals may be increased where the ground is obstructed, and in timbered places the line may be extended. In case of attack there should be an advance party thrown out to and along the skirt of the prairie in front (north side of the creek), with adequate supports formed near the creek. The enemy must, if possible, be prevented from gaining the cover of the timber on the north side. Commandants will examine the ground in front of them, and especially creeks, bayous, or wooded ways leading from the prairie north and west of camp down southward and connecting with the main bottom of Elk Creek. These smaller creeks will be used in case of attack by the enemy to penetrate to Elk Creek, and thus flank the different positions near the fords. These can be used by our troops to advantage in gaining a position in advance of the general line of the prairie to flank the columns of the enemy while advancing on the roads leading to the fords. It is necessary that commanding officers should examine and understand the ground in front of their own positions, and also those occupied by other corps.

By order of Brig. Gen. D. H. Cooper:
THORNTON B. HEISTON

Colonel Cooper's General Order #25 execution plan

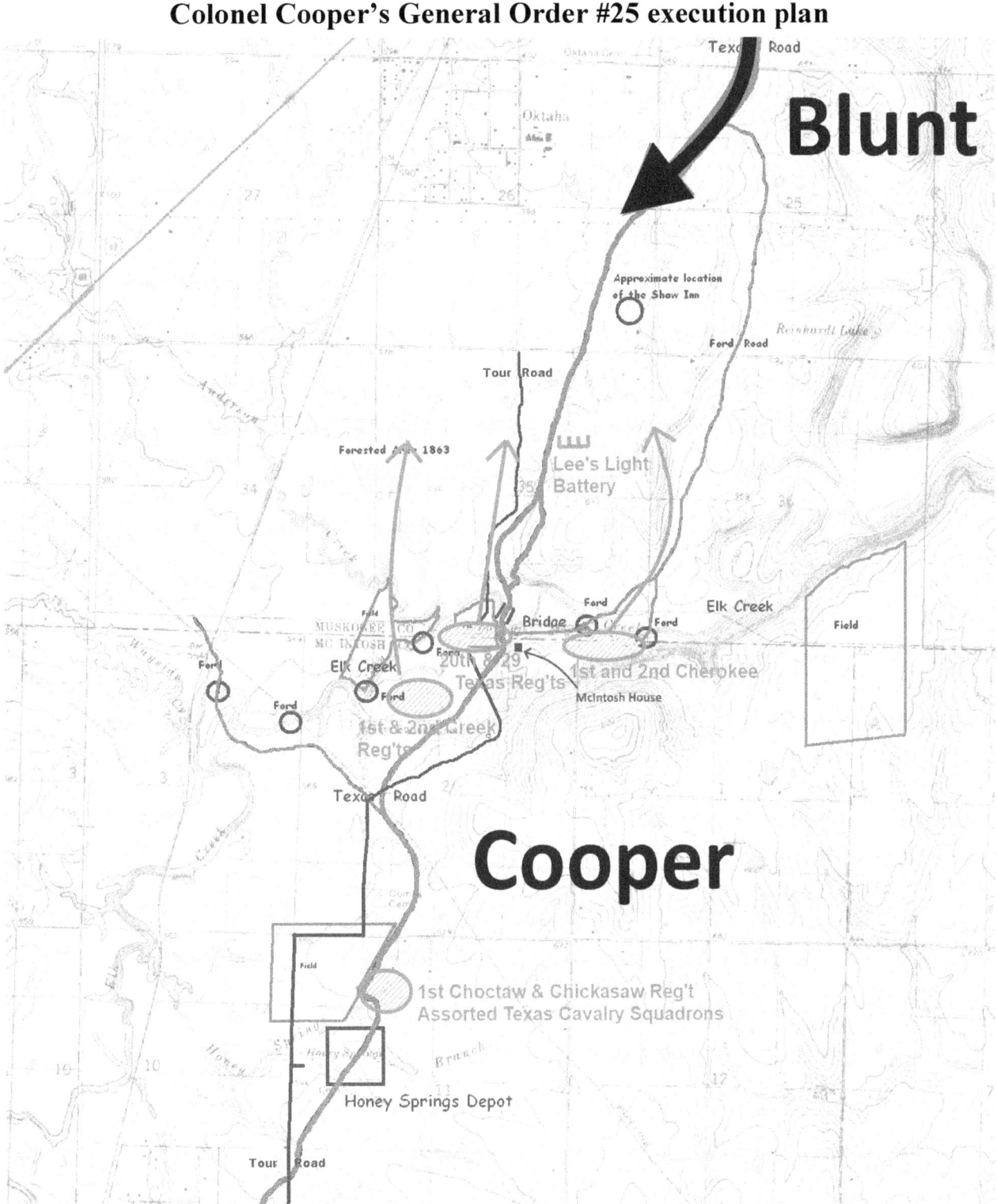

Map courtesy of USGS / Additions by Author

In the aftermath of the severe defeat that the Confederates had suffered at Cabin Creek, a council of war was held at the Honey Springs Depot with all of the commanders who had participated in the battle. Together they discussed the battle and determined the main issues to be addressed to correct those errors. They also discussed further movements of the Confederate forces in Indian Territory, with the usual push from Richmond to advance into Kansas. Believing that a Union attack from Fort Gibson/Blunt to prevent his moving northward was imminent, Cooper and his commanders developed a defensive plan for the Honey Springs Depot. On July 14, 1863 Cooper issued General Order #25, which detailed his plan for defending the Elk Creek line. Cooper would station his commands along Elk Creek centered on the toll bridge. The orders also divided his small brigade into three wings. The right wing, consisting of the 1st and 2nd Cherokee Mounted Rifle regiments with Col. Stand Watie commanding, would encamp near and cover the two lower fords. The center wing, consisting of the 20th Texas Cavalry (Dismounted), 29th Texas Cavalry, 5th Texas Partisan Rangers, and Capt. Lee's light battery (Cooper's only artillery), Col. Thomas Bass of the 20th Texas in command, was to cover the middle crossings including the bridge over Elk Creek. Lee's battery consisted of three 12-pounder howitzers mounted on prairie carriages and one rifled bronze field piece. This bronze gun was an experimental mountain rifle with a bore diameter of two and a quarter inch and was designed and intended to support cavalry operations. According to historian Wayne A. Stark, this was one of only 20 such cannons made by the Tredegar Iron Works in Richmond under an order placed in November 1861. It is unknown how this unusual gun made its way out to the Indian Territory. The left wing of Cooper's force consisted of the 1st and 2nd Creek regiments, commanded by Col. D.N. McIntosh, and was to encamp and cover the upper fords. The 1st Chickasaw and Choctaw regiment, Scanland's Texas Squadron, and L.E. Gillett's Texas Squadron were attached to headquarters and were to constitute the brigade's reserve. The troops were dressed in the "butternut" brown uniforms common with Confederate armies, especially in the West. Some wore common clothes due to a lack of uniforms and, like the Union Indians, the Confederate Indians tended to accent their uniforms with traditional items. They were armed with a wide variety of weapons since no standard weapon had been issued to them.[27] Cooper's plan for defense was well thought out and clear in its execution.

To head off any advance toward his positions along Elk Creek, Cooper ordered that all fords of the Arkansas River from west of Fort Gibson/Blunt downstream to Webber's Falls have picket posts established. Cooper's entire force numbered just over five thousand troops. Once again, Gen. Cabell of Arkansas was ordered to move from Fort Smith to Honey Springs to support Gen. Cooper with his brigade of Arkansas conscripts. Once again, he would be too late. On July 10 Gen. Steele issued this order to Gen. Cabell:

Hdqrs. Department of the Indian Territory,
Fort Smith, Ark., July 10, 1863.
Brig. Gen. W. L. Cabell,
Commanding Cabell's Brigade

General: The general commanding directs me to say that you will move forward, with the least possible delay, with that portion of your command with which you have just returned from Northwestern Arkansas, to some suitable camping place between Webber's Falls and Fort Gibson. For the greater convenience of obtaining water and grass, it would be advisable to move your command in detachments.

I have the honor to be, &c.,

B. G. DUVAL,

Lieutenant and Assistant Adjutant-General. [28]

General Blunt immediately began plans for an offensive movement against Gen. Cooper's stronghold at Honey Springs. Blunt divided his limited troops into two brigades. The First Brigade was commanded by Colonel Judson and consisted of the 1st Kansas Infantry (Colored) under Col. Williams, the 2nd Indian Home Guards (dismounted) under Col. Fred W. Schaurte, a battalion of the 6th Kansas Cavalry with two 12-pounder mountain howitzers under Lt. Col. William T. Campbell, a battalion of the 3rd Wisconsin Cavalry with two mountain howitzers commanded by Capt. E.R. Stevens, and two sections (four guns) of the 2nd Kansas Battery under Capt. E.A. Smith. [29]

The Second Brigade was under the command of Col. Phillips and was comprised of six companies of the 2nd Colorado Infantry under Lt. Col. T.H. Dodd, the 1st Indian Home Guards (dismounted) under Col. S.A. Wattles, and the 3rd Kansas Battery (four guns in two sections) under the command of Capt. Henry Hopkins. An interesting note is that Hopkins' battery was

the one formed from Company B, 2nd Kansas Cavalry using the artillery pieces captured from Stand Watie's force at the Battle of Old Fort Wayne. The Union troops were dressed in the familiar dark blue wool blouse with gold buttons and light blue kersey wool trousers topped with a blue kepi, although the Indian troops tended to accent their uniforms with various types of Indian dress, such as feathers and beads. Most infantry soldiers were armed with the Enfield rifle or the standard model 1861 Springfield rifle, .58 caliber. The cavalry troops were armed with a variety of shorter carbines including Spencers and Sharps. Even though they were well armed, Blunt still had only approximately 3,000 troops for the upcoming conflict. [30] The 3rd Indian Home Guard remained behind to provide security for Fort Gibson/Blunt.

Unfortunately for the Union force, the Arkansas River was swollen and unfordable in the immediate vicinity of Fort Gibson/Blunt. Gen. Blunt, fearing that Cabell would arrive on July 17 to reinforce Cooper on the Elk Creek line, quickly ordered the construction of flat boats to ferry the troops, artillery, horses, and supplies across the river. According to Wagon Master Peck, each flatboat could hold one company of soldiers, or one six-mule team and wagon, or one artillery cannon. On July 15 scouts sent out from Fort Gibson/Blunt reported that the Arkansas River level had begun to fall and was fordable upstream of the mouth of the Verdigris River, approximately 13 miles west-northwest of the fort. At approximately midnight of the 15th, Blunt took the battalion of the 6th Kansas Cavalry with its two howitzers and a section of the 2nd Kansas Battery, approximately 250 men, and crossed the ford of the Arkansas River near the Old

Creek Agency. They also brought along a flat boat to be used as a ferry for sensitive items. When they approached the crossing, they ran into Confederate pickets in rifle pits and after a quick skirmish, the Confederates retreated across the river and Blunt's troops hurried across. The Union force moved swiftly downstream in an attempt to get behind and capture the Confederate picket posts at the mouth of the Grand River, opposite Fort Gibson/Blunt. The picket posts became aware of the Union advance and abandoned their defenses and retreated south towards Honey Springs before Blunt could reach their location. A somewhat sad event happened during the night crossing of the Grand River. Wagon Master Peck recalled:

> "…It was found that a narrow and difficult ford on Grand River just above the mouth could be used by the cavalry and some teams, and then an easy ford of the Arkansas above the mouth of Grand would expedite the crossing of both troops and teams. While crossing at the Grand River ford a team belonging to the 1st Kansas Colored was washed away down into deep water, and the six mules, with wagon, driver and two colored soldiers, who had crept into the wagon, were drowned. The black teamster never left his saddle mule, and as he sunk out of sight was still jerking the lead-line to try to steer his team to the shore. The two men in the wagon were caught like rats in a trap- the cover being tied down all round- and drowned before they found out where they were "at"…"[31]

By 10:00 p.m. of July 16 the Union forces had completed the crossing of the Arkansas River by using the flat boats and began the long march south along the Texas Road to meet Cooper and his forces on Elk Creek.[32]

General Cooper received word from his picket posts that the Union Army was crossing the Arkansas River, but he was unsure if it was just a large scout or the main body of troops. He sent Colonel Tandy Walker and his 1st Chickasaw and Choctaw Regiment along with Captain Gillett's Texas Squadron out to the vicinity of Chimney Mountain where the Texas Road intersects with the road leading to the Creek Agency, near the mouth of the Verdigris River. This force was to send out patrols on both roads to find the direction of approach of the Union force. Walker was also directed to take along Captain Lee's light artillery battery but failed to do so, owing to a misunderstanding of Cooper's orders.

During the early morning hours of July 17, 1863, the advance companies of the 6th Kansas Cavalry made contact with Walker's force on a small rise near the crossroads. Company F, 6th Kansas Cavalry, under the command of Capt. William Gordon, attacked Walker's Confederate force but soon found itself in trouble. The other four companies of the 6th Kansas arrived, observing Gordon's situation, charged forward and together they drove Walker's force back down the Texas Road. Col. Thomas Moonlight, Blunt's Chief of Staff, described the night's action:

> "…On the 16th the command moved and by dark that night the enemy's entrenched pickets had been driven

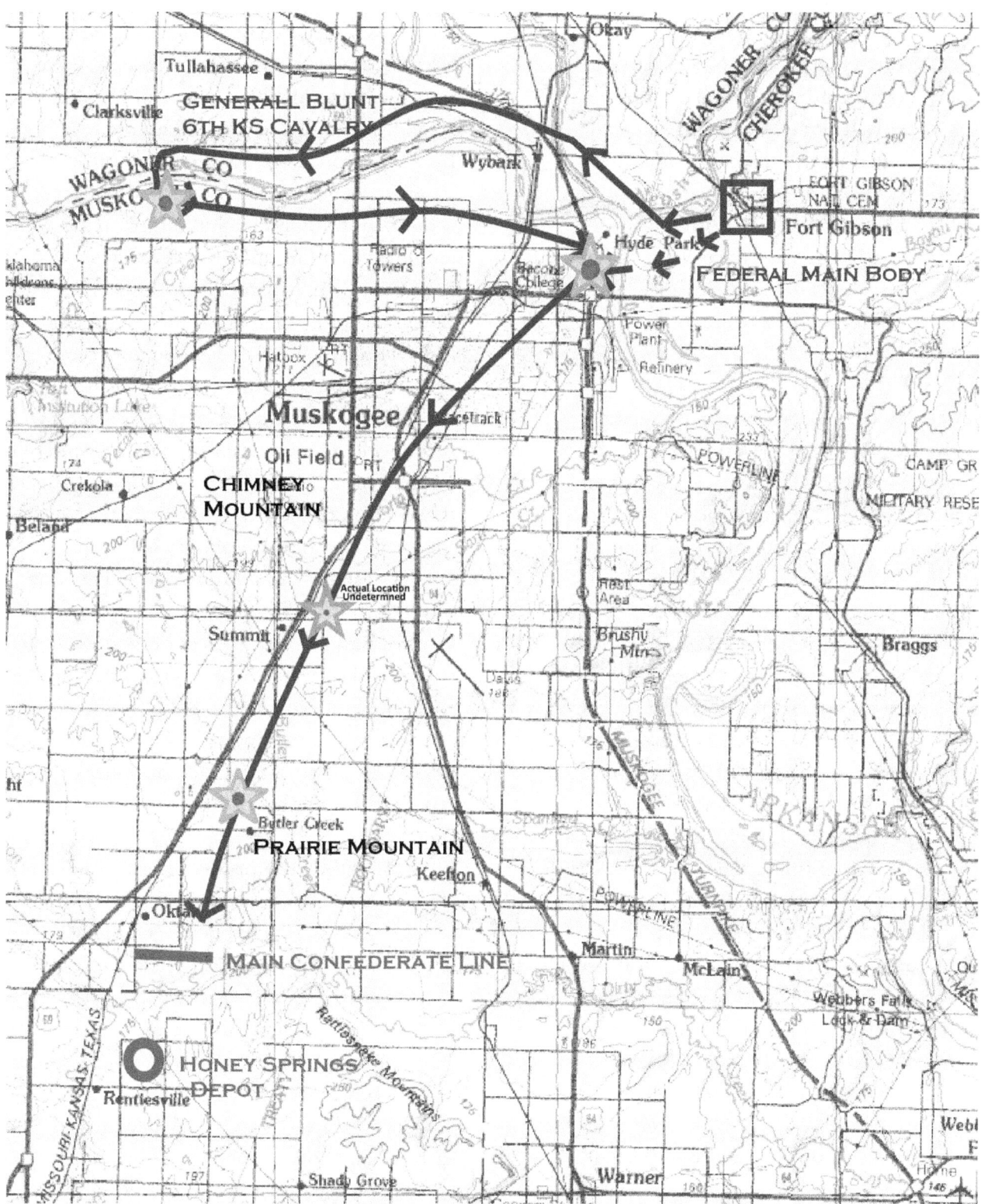

Area of Operations for Brig. Gen. James Blunt's advance on Honey Springs from Fort Gibson on the night of July 16-17, 1863

from the southern bank of the of the river by a small cavalry and artillery force and under cover of darkness the command was landed on the enemy's side. The march was immediately taken up at early break of day. The crack of rifles could be heard in the extreme advance. With their customary dash the cavalry chased and routed the enemy's advance, composed of a regiment of cavalry…"[33]

After the brief fire fight Walker withdrew his troops back towards the Elk Creek camp. On the way back, Walker left Lieutenant T.B. Heiston, Cooper's aide-de-camp who had accompanied the scout, in command of a number of troops at Prairie Mountain, approximately three miles north of Elk Creek along the Texas Road approach to the crossing. Heiston was to deploy his troops to make it appear to be a larger force. The ruse worked because when General Blunt's Union force was fired upon by Heiston's troops, they quickly deployed in a line of battle giving Lieutenant Heiston an opportunity to estimate the size of the Union force. He estimated and reported to Cooper that the Union Army consisted of approximately 4,500 soldiers. Heiston also reported that the day was dawning cloudy and rainy, which caused the powder of the Confederate troops, which had been purchased in Mexico, to quickly absorb moisture and turn to paste. This rendered the ammunition completely worthless. After the brief contact with the Union force, Heiston withdrew to General Cooper's headquarters.[34]

Unfortunately for the Union force, Blunt had become sick just prior to the invasion and was attempting to coordinate his army's activities from the back of an army ambulance.[35] After the brief encounter at Prairie Mountain with Lieutenant Heiston's small force, Blunt regathered his deployed troops and once again marched toward Cooper's defenses. After passing Prairie Mountain, Blunt pulled himself from the ambulance and went forward with his staff escorted by a company of the 6th Kansas Cavalry. From a high point on a small ridge just east of the Texas Road, General Blunt was able to discover the location of the Confederate defense line. He estimated it to be one and a half miles in length and centered on the Texas Road. He was unable to discover the location of the Confederate battery due to the heavy brush and timber in which Cooper's line was established. During this activity one of General Blunt's escort, Pvt. White, was shot. Blunt returned to meet his approaching troops and ordered them to gather behind the small ridge out of view of the Confederate line to rest and eat since they were weary after the forced night march. Col. Moonlight describes Blunt's actions:

> "…Gen. Blunt was laboring under a severe fever, scarcely able to hold his head up, and while he laid himself down in a fence corner to await the arrival of the infantry and artillery before making the attack…"[36]

Blunt sent the battalion of 6th Kansas Cavalry forward within a half mile of the Confederate line and had them deploy to provide security for the Union brigades.[37] In the vicinity of Blunt's makeshift bivouac site was an inn operated by a Cherokee named Shaw. Local tradition states that during this rest period Gen. Blunt and his officers had their meals served at the Shaw Inn, although there is very little documentation to support this beyond a mention of a house by Wagon Master Peck. He recorded:

> "…About the middle of the plain that lay between the hills where our command had halted and eaten our snack, and the timber, stood a farm house- the only one in sight. After our line had passed this house some distance it was halted…"[38]

Blunt's officers were probably busy preparing themselves and their troops for the upcoming battle. Fortunately for the Union troops, a heavy rainfall just prior to the stop gave the weary and thirsty soldiers a chance to fill their mostly empty canteens.[39]

Blunt used the break to finalize his plans for the upcoming battle. He wanted to deceive Cooper into believing he had a larger army than he truly had. Lt. Col. Moonlight described the scene:

> "…Genl Cooper had taken up a strong position in the northern edge of the timber of Elk creek, while his headquarters & stores were some 3 miles in his rear at a place called Honey Springs, the name given to the battle. I here desire to state that the enemy never for a moment supposed that we were anything more than a cavalry and artillery force which had driven him from his entrenchments on the river the day before, and [with this in mind] had so arranged his plan of battle and stationed his regiments and batteries as to commanding

officers to accomplish this work in just 30 minutes. This came from the lips of the commanding officer of the 20th Texas…"[40]

He planned to have the entire command move the final one half mile or so toward the Confederate line with the infantry in close columns of companies, the cavalry in close columns of platoons, and the artillery by sections. The First Brigade would be on the right side of the Texas Road and the Second Brigade on the left. This would appear to the enemy to be a large block of troops, difficult to estimate in size. At about 10:00 a.m., after resting the troops about two hours, Blunt formed his troops for the final movement towards the Confederate lines. They marched down the left and right sides of the Texas Road until they were approximately one quarter mile from the Elk Creek timber where Cooper's troops were deployed. Suddenly, the massed Union troops deployed in a wheel movement left and right into line of battle which stretched close to one mile in length. The Union batteries were unlimbered on both sides of the road and prepared to fire.[41] On the right of the road, the Union First Brigade formed with the 1st Kansas Infantry (Colored) lined with its left flank on the edge of the road. Lt. Col John Bowles reported to Col. Judson:

> "…Previous to forming a line of battle, Colonel [James M.] Williams was informed that his regiment would occupy the right and support Captain Smith's battery. Colonel Williams then called "attention," and said to the men, "I want you all to keep cool, and not fire until you receive the com-

mand and in all cases aim deliberately and below the waist. I want every man to do his whole duty, and obey strictly the orders of his officers." We then moved in column, by company, to the position assigned us, and formed in line of battle, when the engagement was opened by the battery…"[42]

The 2nd Indian Home Guards formed on the right of the 1st Kansas Colored, and the battalion of the 3rd Wisconsin Cavalry with its howitzers swung far to the right to protect the brigade's flank. Capt. Edward Smith moved his 2nd Kansas Battery, consisting of two 12-pounder Napoleons and two 6-pounder iron rifled cannons, about 600 yards west from the road and unlimbered.[43] Capt. Smith reported:

> "…I was ordered forward, preceded by the Third Wisconsin Cavalry, and supported by the First Regiment Kansas Colored Volunteers and the Second Regiment Indian Home Guards. Changing direction to the right of the road, I continued in that direction about 600 yards, when I wheeled the battery into line, and moved down upon the left of the enemy's line, which could be faintly discerned through the timber and brush. At this moment the rebel batteries on their right opened upon Captain Hopkins's battery, in Colonel Phillips's brigade…"[44]

The Second Brigade deployed with the six companies of the 2nd Colorado Infantry holding its right flank on the edge of the Texas Road. Lt. James Burell, the regimental quartermaster of the 2nd Colorado, recalled in an April 4, 1877, article in the *Colorado Transcript*:

> "…Our little battalion of the Second bore a prominent part in the engagement, being drawn up in line of battle with the First Kansas Colored on their right, and an Indian regiment on their left, in front of a rebel battery, that was pouring its deadly missiles into their ranks, supported by a Texas regiment…"[45]

The 1st Indian Home Guards formed to the left of the 2nd Colorado Infantry. At the last minute, Blunt ordered the 6th Kansas Cavalry to Colonel Phillips' brigade to be assigned to the far left to prevent a flanking movement by Cooper's Confederates. Colonel Judson retained control of the 6th's howitzers. Capt. Henry Hopkins's 3rd Kansas Battery of two 12-pounder Napoleons and two six-pounder iron rifled guns was placed on the left supported by the 2nd Colorado Infantry and the 1st Indian Home Guards. Capt. Hopkins described his batteries' actions in the first moments of the battle:

> "…Moving up in line of battle to within 300 yards of the enemy's position, we were ordered by yourself to commence firing and shell the woods in the immediate front, which continued for one hour and a quarter. Immediately after our fire opened, the enemy's battery was discovered occupying a position to our right and front, which opened fire upon us with shot, shell, and canister…"[46]

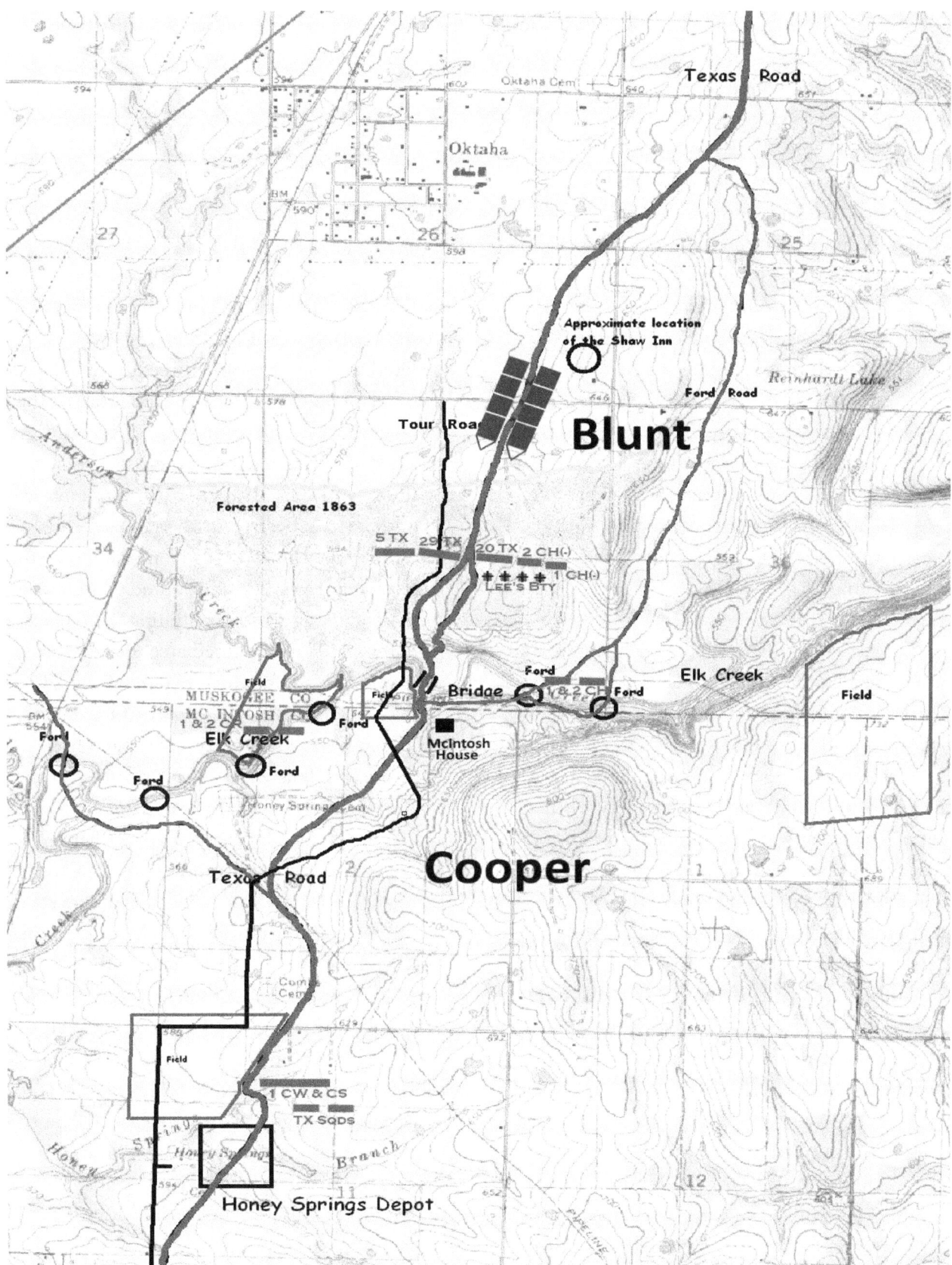

Map courtesy of USGS / Additions by Author

The entire battleline moved forward in unison to within 300 yards of the Confederate battleline and the Union artillery opened fire.[47] Wagon Master Peck observed:

"…It was a very hot day, and our soldiers had stripped themselves of everything in the way of clothing and equipments [sic] that could be dispensed with, pilling the surplus stuff in heaps in rear of their line. I noticed that the men of a battery near me (Hopkin's 2d Kansas) had stripped to their undershirts and pants, and the 1st Kansas Colored-Colonel Williams,- about the center of the line, had even taken off their shirts, and their black skins glistened in the sun…"[48]

Although General Cooper had issued orders detailing his defense plan for the Elk Creek line, the plan went awry as soon as the Union force deployed before him. Captain Lee's battery was posted on a small rise near the line of timber about 100 yards east of the Texas Road. Captain Lee had deployed the battery after being left behind earlier by Colonel Walker who was to take the battery with him on his scout. Cooper sent Colonel Bass and his 20th Texas Cavalry (dismounted) forward to support Lee's battery. Lt. Col. Otis Welch of the 29th Texas Cavalry, who commanded the left of the regiment, describes his initial actions:

"…In obedience to orders from you on the morning of the 17th inst. this regiment was promptly mounted and marched across Elk Creek to its north fork, when it was dismounted under cover of the timber, and proceeded rapidly on foot across the skirt of timber into the prairie, where we were under your directions posted in line of battle…"[49]

Welch directed Capt. Harmon's Squadron, [Co's C and A] and Capt. Mat Daughtery's company "E" to be thrown forward as skirmishers. They remained in this position some half-hour when they were ordered to the right some three hundred yards and closed on the left of the remainder of the 29th Texas. Capt. Harmon's squadron was retained as skirmishers in front of the new position, and Capt. Daugherty's company was sent up a small bushy ravine that extended into the prairie a few yards from the left to the front. Some 15 or 20 dismounted Choctaws came along from the rear hunting for a place to fight, and Welch immediately placed the warriors on his left, which extended the 29th Texas's line nearly to the center of the ravine, up which Capt. Daugherty's company had been posted. This was Welch's position at the commencement of the fight.[50]

Capt. Scanlon's and Capt. Gillett's Texas troops were ordered to support Col. D.N. McIntosh and the Creek regiments at the upper crossings of Elk Creek after Cooper observed the Union cavalrymen of the 3rd Wisconsin moving that way. Col. Walker was ordered to remain in the vicinity of his camp near the Honey Springs Depot and act as a mobile reserve while sending pickets out onto the Prairie Springs Road to the east to ensure that no Union movement would come from that direction. Once again, Walker misunderstood one of Cooper's orders and moved his entire command out on the road instead of just pickets. This left Cooper

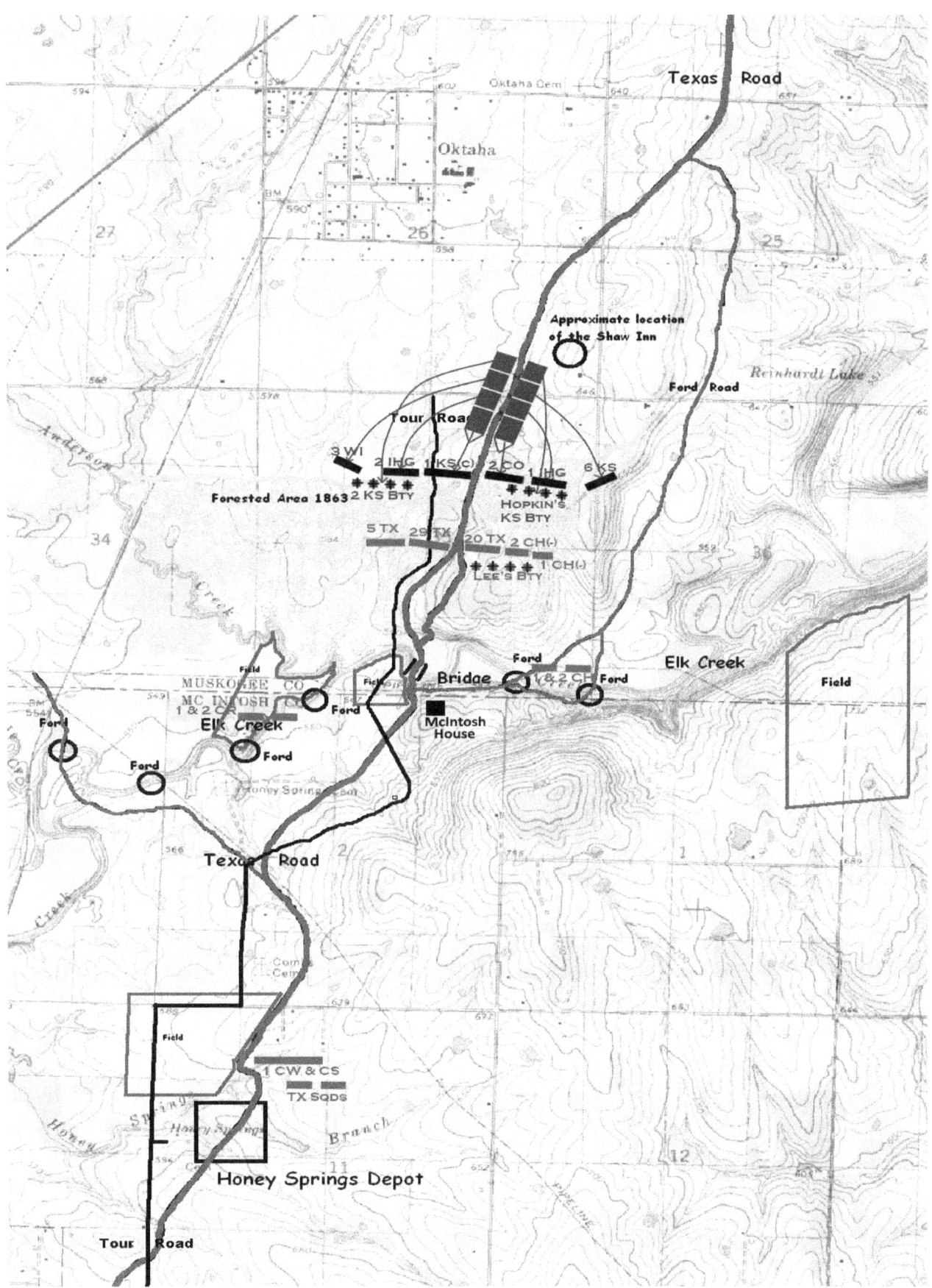

without a mobile reserve. Further right, Cooper found the 20th Texas and part of the 2nd Cherokee regiment supporting Capt. Lee's battery. A portion of the 1st Cherokee regiment under Capt. Hugh Tinnin was deployed as skirmishers to the right of the 2nd Cherokee. The remainder of the Cherokee regiments were still posted near the lower crossings of the creek. Cooper ordered half of the 20th Texas to the right to support the 1st Cherokee skirmishers and ordered the remainder of the Cherokees, who were still eating breakfast, up and posted them near the edge of the prairie on the right of the 20th Texas.[51] Colonel Charles DeMorse of the 29th Texas Cavalry had moved his regiment northward to the edge of the prairie as was stated in the general orders. He dismounted his troops and deployed three companies as skirmishers. DeMorse soon received orders from Cooper to move to his right about 300 yards and link up with the left flank of the 20th Texas near the Texas Road.[52] This may be the reason Lt. Col. Welch, in command of the left of the 29th Texas, continued to attempt to move to his right and meet up with the right of the regiment. At some point early in the battle Col. DeMorse was wounded and taken to the rear. Nothing in the official reports or later news reports indicated whether or not Welch had been notified of DeMorse's wound, which would place him in command of the 29th Texas. Welch reported in the news account:

"…. I sent Major Carroll to the right to ascertain the position of your regiment, in order to close upon it, and to rally our right. I remained with the left. The Major not returning I sent to the right myself, to ascertain

the cause, when I found that all on our right had given away, and that the enemy were passing rapidly to our rear, on the right…"[53]

Cooper also sent orders to the Creek regiments and the 5th Texas Partisan Rangers to move downstream and join up with the 29th Texas regiment. Apparently, only Col. DeMorse and Col. Martin's Texas Partisans responded to Cooper's orders. With Col. Walker's Chickasaw and Choctaw regiment out on the road to Prairie Springs, and the Creek regiments under Col. McIntosh not responding to orders, as they also failed to do at Old Fort Wayne, Gen. Cooper's defenses were at their maximum possible condition to meet Blunt's Union army.[54] Capt. George Washington Grayson of the 2nd Creek Mounted Rifles described his experience while waiting for the battle to begin:

"…With my own men and a picked body of others from the 2nd Creek regiment, with our colonel Chilly McIntosh in immediate command, we were ordered to occupy a certain position in a dense bottom of the Creek (Elk), and to remain out of sight under cover of the foliage of the trees until ordered to take active part in the day's work. Here we remained listening breathlessly at the rattle of small arms and an occasional exulting whoop that was one of the characteristics of the fighters of the southern forces, and anxious to take a hand in the affray, but could do nothing without orders…"[55]

Grayson later recalled the words

spoken by Colonel Chilly McIntosh in the Creek language before the battle:

> "…When you first saw the light, it was said of you 'a man child is born.' You must prove today whether or not this saying of you was true. The sun that hangs over our heads has no death, no end of days. It will continue indefinitely to rise and to set; but with you it is different. Man must die sometime, and since he must die, he can find no noble death that which overtakes him while fighting for his home, his fires and his country…"[56]

Colonel Chilly McIntosh was the eldest son of William McIntosh and the older brother of Daniel N. McIntosh of the 1st Creek Mounted Rifles. He was one of the Lower Creeks who had signed the removal treaty so was an enemy to Opothleyahola, leader of the Upper Creeks.

In his official report, Cooper does not mention the role of Stand Watie beyond being in command of the Right Wing. A post-battle account states that Cooper had sent Stand Watie on "detached service," a diversionary movement at Webber's Falls with a portion of his command.[57] Apparently this diversion was to make Blunt believe that Confederate forces were going to move on Fort Gibson/Blunt if he moved south towards Honey Springs. Unfortunately, no contemporary documentation exists to substantiate this assignment. It seems unlikely that Gen. Cooper would send his most experienced commander on such a movement before what he knew would be the largest engagement up to that time. It is more likely that Cooper did not know where Watie was at that time. Watie had several relatives in Webber's Falls and spent much of his time there. In addition, since being elected the Principal Chief of the Cherokee Nation, Watie may have felt that he was on an equal footing, if not a superior one, to Cooper in rank and believed he could come and go as he chose. In fact, neither of the Cherokee regimental commanders were present at the battle. Cooper sent a supplemental Official Report on September 19, 1863, stating:

> "…In your introduction to the publication of the Official Report, of the affair at Elk Creek, on 17th July, it is stated that the first, and second Cherokee regiments were commanded respectively by Colonels Watie, and Adair, and in justification to these officers, it is proper to say, that the first Cherokee regiment was commanded by major Thompson, Colonel Watie being at the time on detached services at Webber's Falls; and the second Cherokee regiment by Lieutenant Colonel Bell, Colonel Adair being absent—sick…"[58]

The most important mission facing Blunt was the discovery of the Confederate artillery battery. He had been unable to find it during his earlier scouting mission. But as the Union army deployed left and right before the Confederate line, Capt. Lee's artillery opened fire on Capt. Hopkins' unlimbering 3rd Kansas Battery. In doing so they gave away their hidden location to the other Union batteries.[59] The small mountain rifle of Lee's battery fired a shot that barely missed Capt. Smith of the 2nd Kansas Battery, which was in support of the 1st Kansas Infantry (Colored).[60]

Capt. Hopkins located the general location of the Confederate battery to their front and right and exchanged fire for about an hour and a quarter with shot, shell, and canister. During the exchange, one of Hopkins' 12-pounder Napoleons was hit, mortally wounding a sergeant and killing four horses. Hopkins further reported:

> "...Discovering one of their guns occupying an open space in the woods, an order was given to direct the fire of two guns upon it, and, if possible, dismount it, which was soon effected. By the explosion of one of our shells, the cannoneers belonging to that piece and all their horses were killed or wounded..."[61]

Capt. Smith's battery of four pieces had turned to their left to fire upon Cooper's lone battery when Blunt ordered Smith to move his section of 12-pounder Napoleons back to the left to the edge of the Texas Road and move forward. Capt. Smith complied with this order, moving the section forward through the 2nd Colorado Infantry about 100 yards, almost to the edge of the timber. The fire from this artillery section caused Capt. Lee to withdraw his three remaining pieces about 300 yards where he went back into active battery.[62]

Concerned about the flanks of his advancing force, Blunt ordered the cavalry battalions on his right and left to sweep into the timber on either side of the Confederate line. The 3rd Wisconsin Cavalry, with two mountain howitzers, on Blunt's right flank advanced into the woods. They encountered Confederate skirmishers who quickly fell back through the woods and crossed the North

Fork of Elk Creek. The Union cavalrymen approached Elk Creek itself where they discovered a line of Confederate troops, most likely McIntosh's Creek regiments, formed behind a rail fence bordering a cornfield on the other side of the creek. Capt. Stevens, commanding the 3rd Wisconsin Cavalry, brought up the two howitzers and opened fire on the Confederate position. Capt. Stevens sent a portion of his command farther to the right in an effort to flank the Confederates. Further, he dismounted the troops in front of the Confederate line and had them advance, supported by the howitzers. The forward and flanking movements caused the Confederate troops to fall back in confusion towards Honey Springs.[63]

On the Union far left flank, Lt. Col. William T. Campbell of the 6th Kansas Cavalry observed some Confederates, most likely troops of the Cherokee regiments, attempting to out flank the Union line. He dismounted Companies C, F, and H and sent them forward into the woods as skirmishers. He retained Company A as a mounted reserve. After about an hour and a half, Campbell's troops were successful in pushing the Confederates across the two lower crossings of Elk Creek. At one point the Cherokee Confederates attempted to slip around the right flank of the 6th Kansas Cavalry but were stopped by a charge of the 1st Indian Home Guards during their forward movement.[64] Col. Steven Wattles of the 1st Indian Home Guard described the action:

> "...My command was ordered to the left, in support of Hopkins battery, and then ordered to charge the enemy out of the timber. I advanced, under a destructive fire from the enemy, after hard

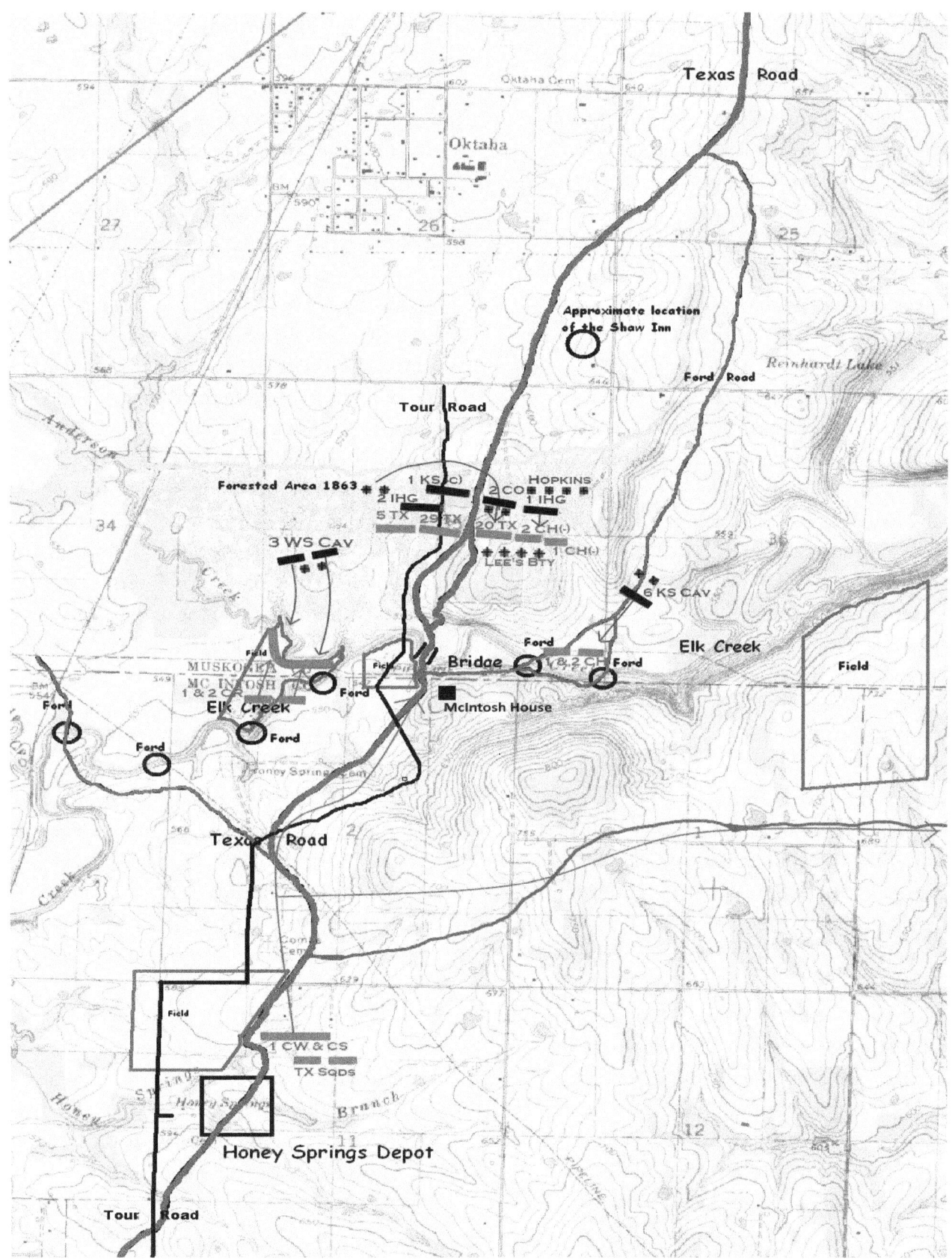

Map courtesy of USGS / Additions by Author

fighting, gained a position in the timber, and finally drove them across the stream, on the left of the bridge, the enemy forming several times, and desperately contesting every foot of ground…"[65]

Although General Cooper had meticulously planned his defense of the Elk Creek line, designating both primary and secondary defensive positions, when General Blunt and his forces began to push at his primary defense line, most of the defeated troops just kept retreating past the secondary defense line. Cooper found out too late that Col. Walker and his 1st Chickasaw and Choctaw Regiment had moved off towards Prairie Springs when he needed support for the Texas regiments that were supporting the battery in the center. As the Confederate troops fell back towards Honey Springs, Cooper established another defensive line north of the depot. Using Major Carroll's right wing of the 29th Texas Cavalry to cover the retreat, Cooper placed Capt. Gillett's squadron of two companies and Capt. Scanlan's Texas squadron of two companies across the Texas Road leading from the toll bridge to Honey Springs. At about this time, Col. Walker and his regiment arrived back at Honey Springs and were immediately placed to the right of the Texas Squadrons. Col. McIntosh's Creek regiments were placed on the left as they began to drift back from their primary positions. Gen. Cooper hoped to hold off Blunt's Union forces until Gen. Cabell and his Arkansas brigade could arrive.[66] This was only wishful thinking on Cooper's part. Even if Cabell, who was enroute from Briartown on the Canadian River, was to arrive, Cooper's forces were so widely scattered by this time that a successful counterattack was impossible.

General Blunt, although incapacitated by illness, still smelled victory in the air. Capt. Hopkins' battery was sent across the creek and down the road towards Honey Springs, one section on the right and the other on the left of the road. The battery was supported by the 2nd Colorado Infantry and was soon joined by the 6th Kansas Cavalry after they had replenished their ammunition supply. The 1st Kansas (Colored) was returned to its original line of battle to retrieve the equipment and uniforms left behind before being ordered by Lt. Col. Moonlight to move forward another three miles to Honey Springs, skirmishing with remnants of Cooper's Confederates along the western high ground south of the bridge. The remaining Union forces continued skirmishing in the woods with the remnants of the Cooper's Confederates and eventually went into camp.[67]

The 2nd Colorado and 6th Kansas Cavalry pushed Major Carroll's Texans back upon the new Confederate defense line. Hopkins's battery periodically unlimbered and fired rounds into the massing Confederates. Eventually, the 6th Kansas made a gallant cavalry charge in the face of the Confederate artillery, finally breaking the defensive line and sending the remaining Confederate soldiers scurrying down the Texas Road past the Honey Springs Depot.[68] But the delaying action of Maj. Carroll's Texans and the final defensive line gave General Cooper enough time to load the supplies he could into a wagon train and move it off on the Prairie Springs Road to the east. He felt that sending the supply train on that route would prevent the Union Army from capturing it since he believed they would look for it to move south on the Texas Road. Cooper then had the rear guard of his army set fire to the remaining supply of salt, sugar, and flour since there was no room left in the supply train. One Confederate ambulance driver sacrificed his ambulance by placing it across the road in an effort to slow the Union advance. Cpl. W.K. Makemson, of Cooper's headquarters staff, recalled in a 1910 letter:

"…The Federals were within 200 yards of myself and detail while we were setting fire to the commissary and Quartermaster stores. Col. Martin's regiment and Capt. Scantlins [sic] Squadron were the only commands that left the field in formation. We left our dead and wounded on the field… I was ordered by General Cooper to assist in hearding [sic] up the stragglers and turn them in a south-easterly direction. No heard [sic] of Texas cattle was ever more thoroughly scattered or demoralized than our Indian forces. Every Indian was running his dead level best…"[69]

Cooper moved his forces to the east and south and eventually joined up with General Cabell on the Briartown Road. By 2:00 p.m. the battle was over.[70] Private Dallas Bowman of the 1st Choctaw and Chickasaw Mounted Rifles lamented:

"…We fought over 3 hours under heavy fire all the time but at last their firing began to get too heavy for us, at the same time dismounting one of our pieces of cannon. We were compelled to fall back and our men began to

scatter which caused confusion and we had a general stampede…"[71]

As mentioned earlier, the Union center consisted of the 1st Kansas Colored Infantry and the 2nd Indian Home Guards on the right, and the 2nd Colorado Infantry and the 1st Indian Home Guards made up the left. The Texas Road generally divided the left and right centers. During the first hour or so of the engagement, the center made no firm forward movement. Instead, they kept up a strong volley of rifle fire on the mostly hidden Confederate line. At approximately noon Gen. Blunt ordered the entire line forward.[72] Blunt came to Col. Williams prior to the main assault and stated:

"…After the lapse of ten minutes, during which time the fire from the battery was incessant, General Blunt came in person to Colonel Williams, and said, 'I wish you to move your regiment to the front and support this battery (which was already in motion); I wish you to keep an eye to those guns of the enemy, and take them at the point of the bayonet, if an opportunity offers'…"[73]

The 1st Indian Home Guards, initially supporting Hopkins' 3rd Kansas Battery, quickly moved into the timber and began to push back the mixture of Confederate Cherokees and Texans that Cooper had sent to the right. Hopkins' battery moved forward, occupied and unlimbered on the original position of the Confederate battery. Colonel Wattles' 1st Indian Home Guards continued to push the Confederate forces back. The Confederates made several stands and

Remnants of the actual spring at Honey Springs
Photos by Author

Tredegar Iron Mountain Rifle

Chimney Mountain from the South
(The dip is from 19ᵗʰ & 20ᵗʰ Century quarrying)
Photo by Author

Prairie Mountain from the route of the Texas Road
Photo by Author

Pumpkin Mountain to the southeast of Honey Springs
Photo by Author

Prairie Mountain from the Texas Road north of the Battlefield
Photo by Author

**Remnants of the Texas Road from the main line
of battle**
Photo by Author

**Location of the initial line of battle for General
Blunt's army looking to the southeast.**
Photo by Author

**Capt. Lee's Texas Battery was located near the
structure at the center of the photograph**
Photo by Author

**Remnants of the Texas Road headed south towards
the tree line of Elk Creek**
Photo by Author

Texas Road approach to the Elk Creek Bridge
Photo by Author

Remnants of the Elk Creek Bridge
Photo by Author

attempted to get between Wattles' regiment and the 6th Kansas Cavalry on the far left. As mentioned earlier, Wattles's troops charged into the Cherokee and Texan troops and relieved the flank of the 6th Kansas Cavalry. They continued to push the Confederates back, finally crossing Elk Creek on the left of the toll bridge.[74]

The 1st Kansas Infantry (Colored) under Col. Williams had the most to prove and also the most to lose. Prior to their departure from Fort Scott, they received word from the Confederate command stating their revulsion to the Union concept of arming the former black slaves. They further stated that any black soldiers, or their officers, used by the Union army, if captured, would receive no quarter (mercy) at the hands of the Confederate Army. Prior to departing the resting area before the beginning of the battle, Colonel Williams spoke to his nervous troops stating:

> "…Show the enemy this day that you are not asking for quarter, and that you are eager to fight for your freedom and finally, keep cool and do not fire until you receive the order, and then aim deliberately below the waist belt. The people of the whole country will read the reports of your conduct in this engagement; let it be that of brave, disciplined men…"[75]

When the order came from General Blunt to move forward and attempt to capture the Confederate battery to their front and left, they were ready.[76]

Colonel Williams took personal command of the left wing of the 1st Kansas (Colored) and Lt. Col. John Bowles took command of the right wing. The regiment moved forward and stopped on the right of the section of 12-pounder Napoleons belonging to the 2nd Kansas Battery, which had earlier moved up to within 300 yards of the Confederate line. The 2nd Colorado Infantry, on the left of the 1st Kansas Colored, moved up towards the Confederate line. The 1st Kansas (Colored) moved forward to line up with the 2nd Colorado, but the left wing had to halt when three companies of the 2nd Colorado encountered a deep ravine and had to file to the right, directly in front of the 1st Kansas (Colored). Both regiments then moved smartly up to approximately 40 paces from the concealed Confederate line.[77] Pvt. R. McDermott of Company A, 20th Texas, recalled being in that position:

> "…We was ordered to lay down and not fire till the enemy came in forty yards. Our lines were 2 deep and the enemy was 4 and 8 men deep. We lay down and they marched on firing as they came and killed our men before they came in range of our guns. When they came near enough, we let them have it and killed lots of them, though they made no halt, and the Texas boys had to fly. The bullets came faster than we ever saw before…"[78]

Colonel Williams of the 1st Kansas gave the command "Ready, aim, fire" to his black regiment and two long lines of smoke and fire erupted from his troops. The Confederates stood up and immediately returned fire, wounding Col. Williams in the right breast, face, and hands. Col. Williams later described his intentions if he hadn't been wounded:

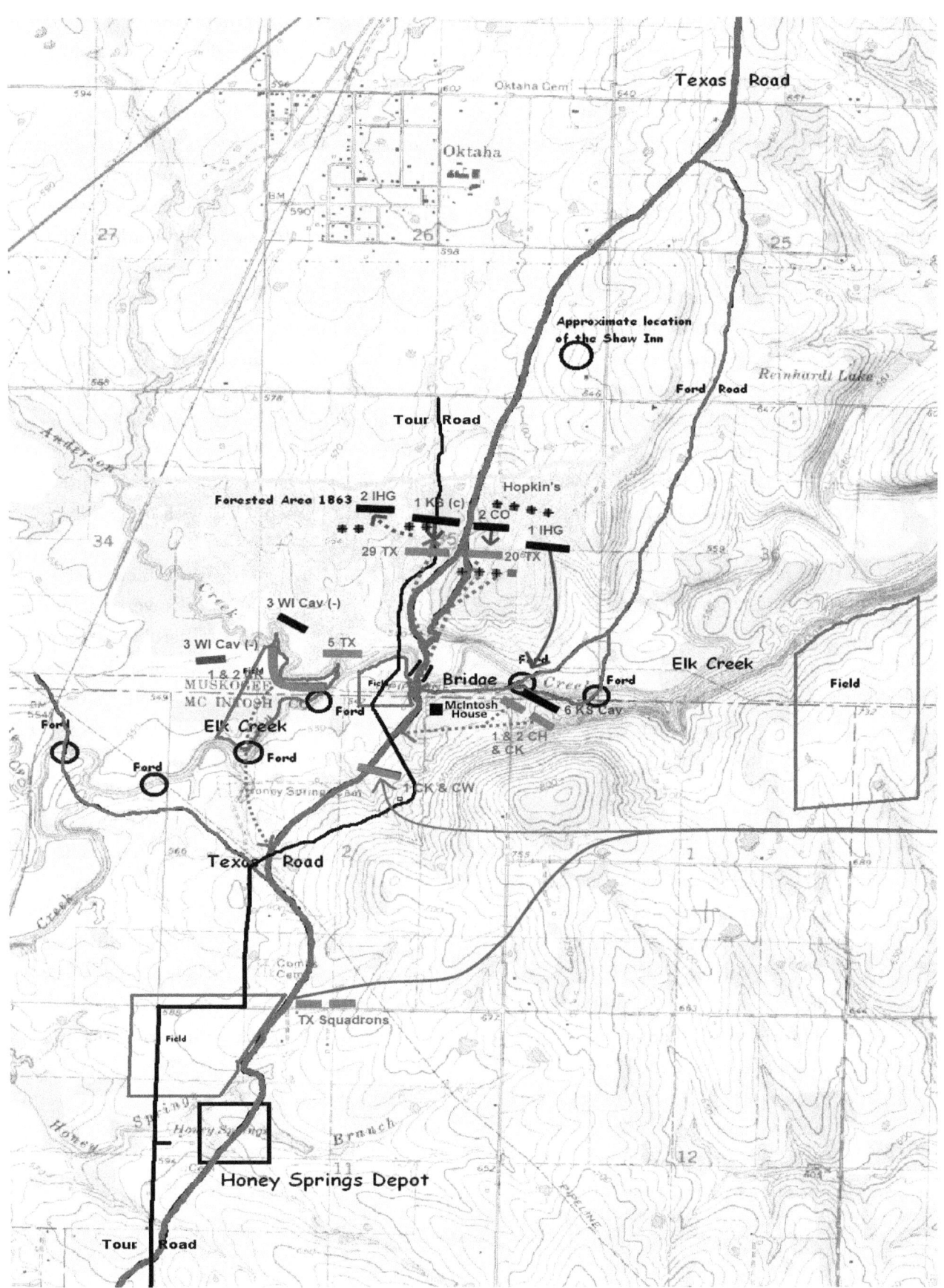

"…My intention was after delivering this volley to charge their line and take their battery, which the effects of my volley had placed completely at my disposal.…"[79]

Wagon Master Peck observed the Colonel being wounded:

"…Just after the last advance of our line began, Colonel Williams, of the 1st Kansas Colored, was struck by a musket ball and fell; but knowing that if he left the field it would have a depressing effect on his men, although he could not sit on his horse or stand up, he had some of his men to prop him up on a pile of knapsacks in rear of their position, and from this seat directed their movements…"[80]

Command of the regiment fell to Lt. Col. Bowles until Col. Moonlight, ordered by Blunt, could make it to the front. Capt. Benjamin Van Horn, commanding Company I of the 1st Kansas Infantry (Colored) wrote in a letter:

"…The Rebels knew we were coming and came out 2 miles and chose their ground to fight on. They were located in the brush and tall grass on a hillside. We skirmished around some time before we got them fairly located, our Colonel was wounded soon after the fighting began. Tom Moonlight was put in command of our regiment. Our main line marched within 52 yards (I stepped the ground afterward) of where their main line lay concealed in the brush…"[81]

The 2nd Indian Home Guards, under Lieutenant Colonel Fred W. Schaurte, located on the right of the 1st Kansas Colored, were ordered by Colonel Phillips to deploy themselves as skirmishers and clear the timber between the 1st Kansas Colored and the 3rd Wisconsin Cavalry. They entered the woods and fought an uncoordinated battle with various Confederate troops, including portions of Colonel Martin's 5th Texas Partisan Rangers, before they were able to drive the Confederates across Elk Creek in the area of the upper crossings on the right of the toll bridge.[82] Sgt. Maj. Richard Ross of the 1st Creek Mounted Volunteers, a very senior position within the regiment, recalled not really having much to do with the battle:

"…Only a few of our men were engaged. The Creeks were not in the fight only a few Cherokee and some Texans and one Arkansas Regiment and Lees battery were engaged. The Creeks having been placed on the left to keep them from flanking the main force. Our men fought well considering the number they were engaged with, but were compelled to retreat at last…"[83]

The 1st Kansas (Colored) and the 2nd Colorado were approximately 30 yards from the Confederate line comprised of Col. DeMorse's 29th Texas, with portions of the 20th Texas on his right and the 5th Texas Partisans on his left. Lieutenant Ed Berthoud of the 2nd Colorado stated:

"…The battalion of the 2nd was ambushed by 500 secesh, who gave them a volley from the bushes when

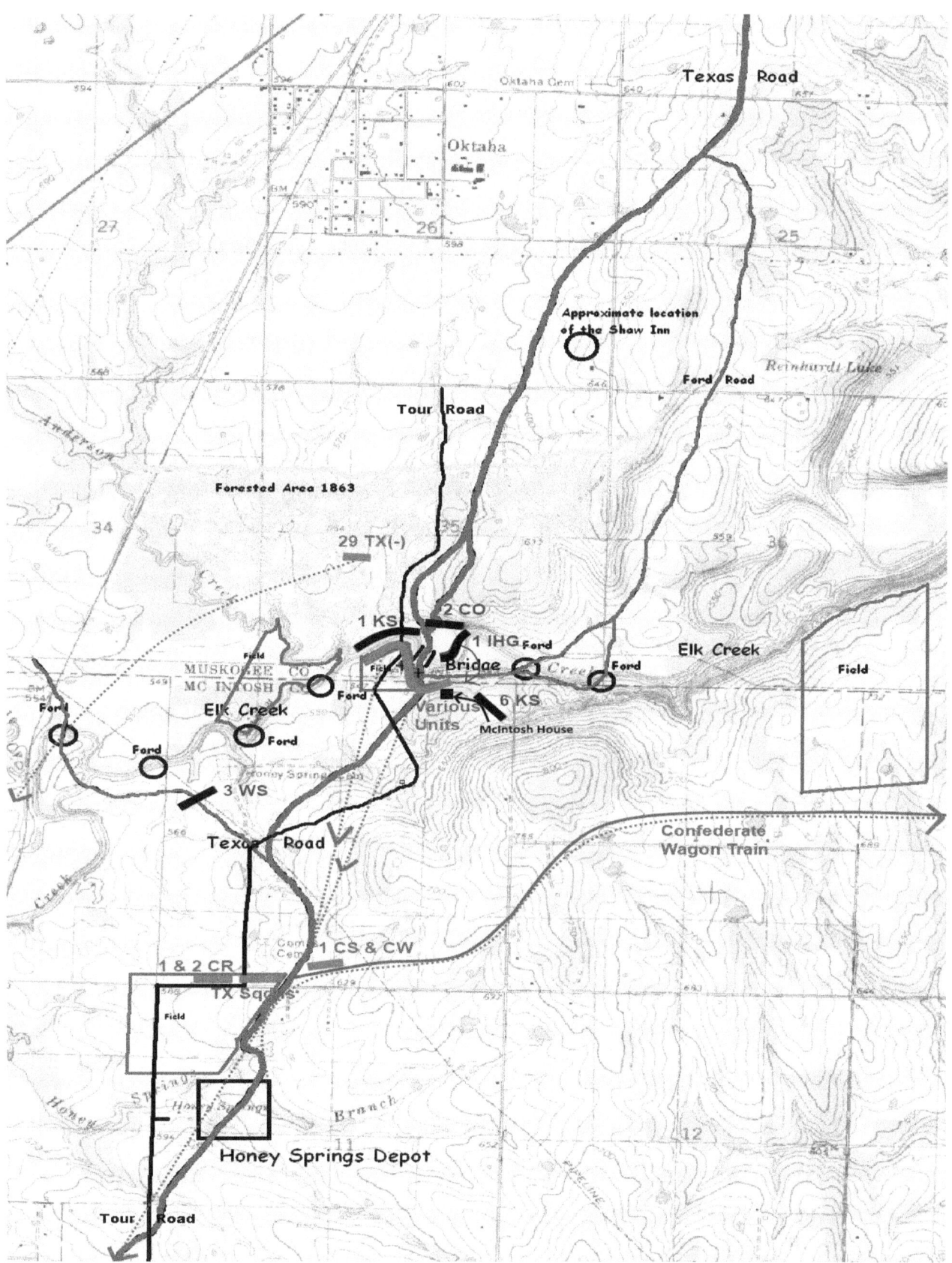

Map courtesy of USGS / Additions by Author

marching in line of battle... A returning volley from our battalion of Colorado boys sent 17 Confederates to their last rest..."[84]

Lt. Col. Bowles was just about to order the black regiment to "charge bayonet" when a mounted portion of the 2nd Indian Home Guards rode in between the right wing of the 1st Kansas (Colored) and the 29th Texas Cavalry. Lt. Col. Bowles rode out and loudly ordered the 2nd Indian troops to fall back. Col. DeMorse of the 29th Texas mistakenly believed that the order had been given to the 1st Kansas (Colored) regiment to retreat and immediately ordered a counterattack. The Confederate troops stood up and advanced to within 25 paces when the 1st Kansas Infantry (Colored) delivered a vicious volley that practically mowed down the entire first rank of the 29th Texas, killing two color bearers and sending the Texans back in confusion. Unfortunately for the 1st Kansas (Colored), the 2nd Indian Home Guards that were falling back picked up the fallen colors of the Texas regiment and later claimed them as their trophy.[85] Wagon Boss Peck described the flag:

"...This flag, which I saw at Blunt's headquarters after the battle, was of three stripes- red at top, white in the middle, red at the bottom, with a blue corner covering the ends of the first two stripes, red and white, and containing 11 stars. This was the style of colors adopted by the rebels at the beginning of the war, but, as they found that at a distance they could scarcely tell their colors from our Stars and Stripes, and often mistook one for the other..."[86]

During the encounter Col. DeMorse was riding behind his troops on his favorite horse, "Selim," when he was struck in the left arm or hand by a minie ball from the black regiment. After attempting to conceal his wound, he released command of the regiment to Lt. Col. Welch.[87] Private McDermott wrote of the exchange at this point by later writing: "...Our officers and men behaved bravely as long as it was prudent to stay and then we run [sic] like hell..." The black troops continued to press the Texans back towards Elk Creek. Peck recalled:

"...Although the rebels had all the advantage of position, and made it hot work, I could see that our men were slowly crawling up on the enemy's position, and it soon became evident that the rebels were falling back in some parts of their line; but, for fear of being drawn into an ambush in the timber, no rush was made by our men- only a gradual creeping up..."[88]

The right wing of the 1st Kansas (Colored) succeeded in pushing the Confederates at their front across the creek and into a cornfield to the right of the toll bridge. The Union left wing drove the Texans back towards the toll bridge in an attempt to capture Capt. Lee's battery. Portions of the 20th Texas and 29th Texas regiments along with Lee's 3-gun battery made a final stand at the bridge but were unable to stop the drive of the 1st Kansas (Colored) and the 2nd Colorado. A mixture of the various Texas regiments held off the onrushing Union forces until Lee was able to pull his battery across the bridge. Gen. Cooper remarked in his official report what he saw at this point:

"…the men of Colonel Bass' regiment stood calmly and fearlessly to their posts in support of Lee's battery until the conflict became a hand-to-hand one, even clubbing their muskets and never giving way until the battery had been withdrawn and, even when defeated and in full retreat, the officers and men of different commands readily obeyed orders, formed, falling back and reforming at several different positions, as ordered, deliberately and coolly. Their steady conduct under these circumstances evidently intimidated the foe, and alone enabled us to save the train and many valuable lives…"[89]

This was a difficult situation since the Texas Road approaching the bridge splits in two about a half mile north of the bridge and both sides were using both roads. Gen. Blunt reported:

"…My men steadily advanced into the edge of the timber, and the fighting was unremitting and terrific for two hours, when the center of the rebel lines, where they had massed their heaviest force, became broken, and they commenced a retreat. In their rout I pushed them vigorously, they making several determined stands, especially at the bridge over Elk Creek, but were each time repulsed…"[90]

Cooper also mentioned this intensive fight for the bridge over Elk Creek:

"…I then rode to where I expected to find the Choctaws, in order to bring them to the support of Colonel Bass' command and the battery, which was engaged with that of the enemy. Colonel Walker, mistaking the order, had moved off on the mountain several miles with his whole force, instead of sending a picket. Messengers were sent after him and he returned promptly, but too late for the defense of the bridge. Riding back near the creek, I discovered our men in small parties giving way. These increased until the retreat became general. Colonel Bass' regiment and Captain Lee's battery, after a most gallant defense of their positions, were compelled to fall back…"[91]

Regardless, the battery was pulled to safety, minus one 12-pounder mountain howitzer, and began to fall back towards Honey Springs, a mile and a half to the rear. Col. Martin's 5th Texas Partisans were ordered by General Cooper to hold the ford above the bridge, but they were unable to hold that location. They found their flanks compromised by the 1st Kansas (Colored) and the 2nd Indian Home Guards. They eventually fell back as well. The entire Confederate line was now in full retreat, and now Blunt's Union forces controlled most of the Elk Creek crossings as well as the toll bridge and were driving forward to Cooper's depot at Honey Springs.[92]

Lt. Col. Welch of the 29th Texas was with the left wing of the regiment when the right wing gave way. Finding Union troops on both of his flanks as well as to his front, he began to fall back to the designated second line of defense near Elk Creek, believing he would be supported by the rest of the brigade at that location. As Welch and his small command reached the creek, they made another stand but found that the rest of the Texans on their right and left had already fled and that they were behind the main Union line. Welch filed his men westward through the Elk Creek ravine and made it out to the prairie. They hiked back south eventually reaching North Fork.[93] Maj. Joseph Carroll, who was left in command of the regiment's right wing, succeeded in getting his troops back to the regiment's horses and assisted in the final defense of the Honey Springs Depot.[94]

Pvt. Wesley Walk Bradly of the 5th Texas Partisan Rangers described what he saw that afternoon:

> "…Since I last wrote to you, we have had another battle with the feds which resulted, as usual, with our defeat…. The enemy succeeded in turning the right wing of our army and a retreat was ordered. Our regiment was placed on the eastern left wing of our army but was not in the engagement. We were very nearly cut off in attempting to get off of the battlefield. We covered the retreat of our army and brought off every thing in good order…"[95]

What makes this letter from Pvt. Bradly significant is that here is a second regiment that had no actual contact with the Federals. It is apparent by the casualty lists that Col. Bass's 20th Texas Cavalry and Col. DeMorse's 29th Texas Cavalry, along with Lee's Light Battery, bore the brunt of the Battle of Honey Springs. Col. Martin's 5th

Texas Partisan Cavalry and the two Creek regiments under the two McIntosh brothers on the left flank failed to take an active part in the battle, although Martin's regiment did cover the retreat. Capt. Grayson of the 2nd Creek recalled:

> "…While awaiting orders for a forward move, here came orders for us to fall back and retire in good order, which reluctantly our men did. The good order of our retirement was perfect since there was no enemy or other thing in sight to cause disorder to our movements. This fiasco I think was the nearest I had to up to that time ever come to being engaged in a battle with the enemy…"[96]

There are only three Confederate official reports, one overall report from Gen. Cooper, which was basically what he saw and did during the battle. The other two reports are from Col. DeMorse and Lt. Col. Welch of the 29th Texas Cavalry, one of which was simply correcting an earlier version of Cooper's report. No other official reports from the Confederate side have been found. Due to the low casualties of either Union or Confederate units on either flank, it is apparent that the Confederate Indians, Creeks on the left flank and Cherokees on the right, simply melted away to the rear when confronted by the Federal cavalry units at the Elk Creek crossings.

An issue confronting historians is what road the Confederate supply train traverse away from Honey Springs Depot. The problem is that no wartime maps of the area show a road to Camp Prairie Springs, which is believed to be near Brushy Mountain

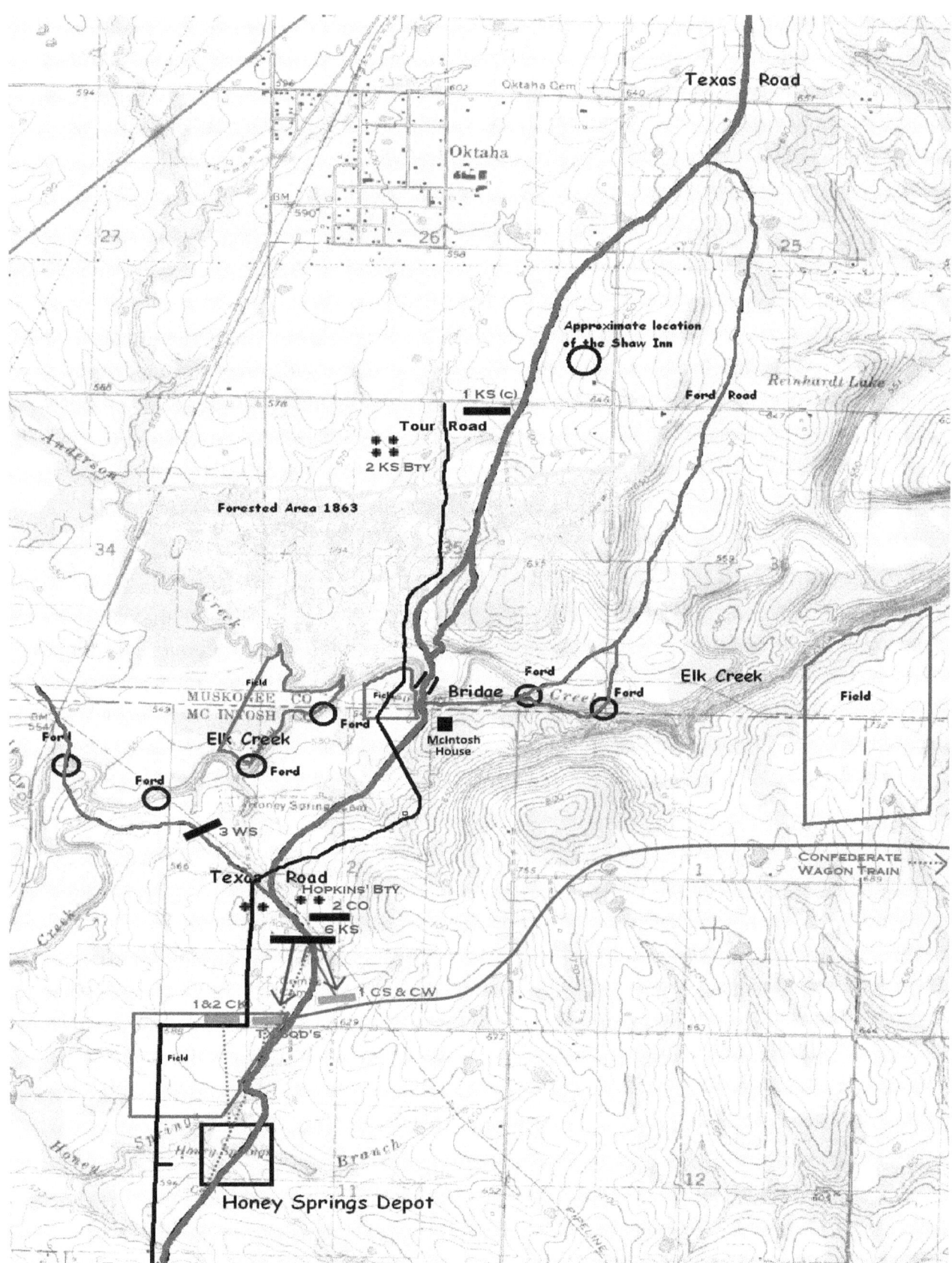

Map courtesy of USGS / Additions by Author

in Muskogee County. Nor do maps show a direct route to Briartown where Cooper states he sent the train to meet up with Gen. Cabell's reinforcements. Cooper remarked that he sent them on the road to the east that goes around the south side of Pumpkin Mountain to confuse Gen. Blunt. Cooper wrote in his report:

> "…thus giving the train time to gain some 6 or 8 miles on the road to Briartown, which had been indicated by yourself as the route by which re-enforcements would be sent… The retreat of the forces under my command eastward instead of south completely deceived the enemy, and created, as I anticipated, the impression that re-enforcements from Fort Smith were close at hand, and that by a detour in rear of the mountain east of Honey Springs our forces might march upon Gibson and destroy it while General Blunt was away…"[97]

This road did exist in the 1898 USGS survey but did not really show a direct route to Prairie Springs. Although the 1898 surveys showed many roads existing between Briartown and Honey Springs, there was no direct route since they had to cross the Rattlesnake Mountains and many creeks in between. The probable solution is that the road to Briartown may have connected with the road to Prairie Springs. Plus, it had been thirty-two years from the time of the battle to the time the area was surveyed, in which case many roads may have been built and others fallen into disrepair. Either way, the Confederate train got away safely from Honey Springs.[98]

After Cooper's retreat, the Union force pursued until their horses and soldiers were exhausted. Gen. Blunt established his headquarters at Honey Springs and began planning his next moves. His troops extinguished the fires set by the retreating Confederates and began caring for the wounded and burying the dead.[99] Peck and his muleskinners became very busy:

> "…I rushed up to the ambulances to go to gathering the wounded and hauling them to the hospital. At the commencement of the battle General Blunt had taken possession of the farm house for a hospital (perhaps the Shaw Inn). I don't know what became of the family, but as very little ceremony was used in such matters in war times, I suppose they were given short notice to get out, and "got" … We soon had the ambulances scampering over the field, picking up the wounded and bringing them in. The attendants were ordered to bring in the wounded only- leaving the dead to be gathered up and buried afterward- our own men first, and the rebel wounded afterward…"[100]

Corporal Makemsum, who returned to Honey Springs days after the battle, remembered that the Federal dead were buried in the corner of a garden adjacent to the McIntosh house, the home used by William McIntosh, who ran the toll bridge, that sat a couple hundred yards south and east of the bridge. This house was also used as a makeshift hospital for the Confederate wounded who were being cared for by Union Army surgeons. Most of the Federal wounded were quickly transported back to Fort Gibson-Blunt. The Confederate dead were gathered and buried in a cornfield on the bottom of the north side of Elk Creek, just above the bridge at a point right where the Confederate army rested during the battle, and which was immediately on the left of the road leading from North Fork Town to Fort Gibson-Blunt.[101] Union soldiers picking through the remains of the Confederate supplies at the Honey Springs Depot discovered approximately three or four hundred sets of handcuffs. Statements made by Confederate officers of the command in later years confirmed that these handcuffs were to be used on the captured black troops that they intended to send back to slavery in the South. That evening the Union Army established a defensive position and went into camp for the night.[102] The Confederates must have been quite thorough in destroying buildings because George Duffield, working as a cowboy on a cattle drive in 1866, left a diary describing his experiences driving cattle up the Texas Road through the Creek Nation. On June 15, 1866, he recorded:

> "…Beautiful warm day makes us all feel thankful for some sunshine. Beeves behaved well. Are camped at the old site of Honey Springs where a great Battle took place between the Confederate Indians & Federals. There is no part of a house left. The other Herd are still out. 50 Beeves are still water bound & time moves slow…"[103]

The next morning, General Blunt prepared to move back up the Texas Road to Fort Gibson/Blunt. He still had Confederate cavalry screening his front but had received reports from scouts that Confederate Generals Cooper and Cabell had joined up and had moved south of the Canadian River. Together this united Confederate Army posed a significant threat to Blunt and his extended line of supply and communications. The Union troops were also close to exhaustion and were nearly out of ammunition. They arrived back at Fort Gibson-Blunt at various intervals between July 18 and 19.[104]

Within a few days of General Blunt's withdrawal to Fort Gibson/Blunt, Gen. Cooper's combined force reoccupied Honey Springs. They found the bodies of the dead buried in shallow graves on the battlefield and most structures had been destroyed. Only the graves of the Union troops were marked, and these were eventually removed to the Fort Gibson National Cemetery after the war. Although the dead Confederates were also buried on the field, the only marked Confederate grave on the field was that of Lieutenant H.H. Molloy who was wounded and carried to the Shaw Inn along the Texas Road along with other Confederate wounded. He soon died and was buried in the Inn's peach orchard. Wagon Master Peck claims that he was with Lt. Molloy when he died after having his leg amputated by a Union surgeon. Molloy told Peck that he knew he was going to die, and his biggest regret was that Colonel Thomas Bass, his commanding officer, was going to get the girl back home that both had been courting. Molloy suspected that Bass had set him up to get killed:

> "…You see, he and I were both courtin' her, an' there was no show for him while I was on the string, so I

think he concluded today in the fight that now was a good time to get rid of me; and so he left us in the cornfield when he retreated with the rest of the regiment, without sayin' a word about retreatin' at all, knowin', as he must, that we would be killed or captured. Well, he's succeeded only too well in puttin' me out of his way; for even if I should live to get well, the girl would hardly choose a one-legged cripple in preference to a whole man."[105]

Even after numerous archeological surveys, the Confederate graves have never been located[106]

Although he once again held Honey Springs, the morale of Cooper's troops was extremely low, and Gen. Cabell's Arkansan troops were deserting at alarming rates. For all intents and purposes, Confederate strength in northeastern Indian Territory was devastated to a point of no return. Blunt reported a loss of 14 killed and 61 wounded in the Battle of Honey Springs while General Cooper reported a loss of 134 killed and wounded and 47 taken prisoner. Cooper blamed the loss on the poor quality of the Mexican-purchased ammunition that turned to paste when exposed to moisture, and he blamed his lack of sufficient artillery to counter Blunt's.[107] The Confederate Commissioner of Indian Affairs S.S. Scott reported this problem to Richmond by stating:

"...The powder is perfectly worthless. The mere charging of the gun grinds it into the finest dust, which is little likely to explode; and, should it do so, its power is scarcely more than sufficient to drive the ball out of the piece. A surgeon was left behind,

after the late skirmish {Honey Springs}, by Gen. Cooper, to take care of his wounded, who states that balls were extracted from the bodies of wounded Yankees, in his presence, which were not even buried in the flesh. The Indians have taken up the idea, which I endeavored to overcome, that the powder (which came from Matamoras) was made at the North, and sent out especially to be sold to our army..."[108]

But Cooper's problems went far beyond his ammunition. A successful military commander, even with a good plan, must have subordinate commanders who will execute those plans and orders efficiently and effectively. Although the Texas regiments under Cooper's command performed their duties honorably, his Indian troops were obviously less than effective. Gen. Cooper later wrote, "Perhaps if they were all furloughed, it would be as well." After the retreat of the Confederate troops from Honey Springs through Perryville and Boggy Depot, the Indian Territory department commander, Gen. William Steele, wrote his opinion of the Indian troops by stating:

"...As I have feared, the Indian troops, with the exception of one regiment of Choctaws, were no service whatever. The greater part of the Cherokees were absent, and the Creeks utterly refused to leave their country after the occupation of their country by the enemy..."[109]

But in many instances, it would be understandable why the Indian troops acted as they did. First, they were treated very poorly by the Confederate government, especially when it came to supplies and equipment. Secondly, although they were considered "civilized", the Indian troops still maintained old-standing traditions of fighting when they felt like fighting and coming and going as they pleased. They also did not care for the white man's military discipline. Given the Indian troops treatment by the Confederate government and their non-white lifestyle, which the white commanders did not understand, it is no wonder that they behaved as they did. Even still, Cooper had 24-hour notice of Blunt's advance and failed to adequately prepare his forces for the upcoming battle.

General Blunt attributed his victory to many factors. First, he was fortunate to have superior artillery firepower, which was successful in turning the tide of battle. He also praised the quality of his troops, especially his Indian and black troops. His three Indian Home Guard regiments were organized, disciplined, and better supplied than their Confederate counterparts, and as a result, they performed better. The Union Indian regiments consisted of a mix of tribes rather than just one tribe. This prevented one tribe from exercising too much influence on regimental operations. The Confederate practice of raising tribal troops serving under tribal officers was not followed in the Union Army. Instead, the Union commanders strove to remove the Indian troops from the influence of their tribal leaders by placing white men or qualified Indians in most leadership positions. Although they had the same tribal companies, they attempted to mix the tribes within the regiments. These factors contributed greatly to their greater performance in the campaign and battle.[110]

General Blunt reserved special praise for the 1st Kansas (Colored) Infantry regiment. He felt this regiment was responsible for the Union carrying the day. Blunt wrote in his official report of the Battle of Honey Springs:

> "…The First Kansas (colored) particularly distinguished itself; **_they fought like veterans_**, and preserved their line unbroken throughout the engagement. Their coolness and bravery I have never seen surpassed; they were in the hottest of the fight, and opposed to Texas troops twice their number, whom they completely routed. …It would be invidious to make particular mention of any one where all did their duty so well…"[111]

Although Blunt was an abolitionist and believed in the black regiment, it is still clear that he was justifiably impressed with their conduct. The charge by the 1st Kansas (Colored) took place just one day before the famous charge of the all-black 54th Massachusetts Infantry at Battery Wagner, South Carolina.

General Blunt moved his forces back to Fort Gibson on July 19. His intention was to gather supplies and reinforcements for a final push to the Red River. General Schofield initially refused reinforcements and instead wanted Blunt to pull his troops back towards Fort Scott. Blunt argued that Schofield had ordered him to "hold the line of the Arkansas River,"[112] and he fully intended to do so. Blunt later stated:

> "…My position now was a delicate and trying one. Prostrated by severe sickness; far in the enemy's country, with but a handful of troops, and in

the face of a foe greatly my superior in numbers, and constantly increasing, I felt that I was purposely abandoned to fate…"[113]

A very enlightening description of the events during the Battle of Honey Springs, as Gen. Blunt had officially named it, was that of Evan Jones, Chaplain of the 1st Indian Home Guard. He wrote in a letter on July 23:

"…We have returned from a fight which we call the battle of Honey Springs. The heaviest fighting took place in the edge of the timber as you approach Big Elk Creek. It was a well selected place for the sesech to defend. But Gen. Blunt moved on them in their concealment, with two brigades formed in line of battle stretching more than a mile in length. The fight began a little before 10. We had two good batteries of four guns each and four howitzers. Chilly McIntosh's rebel regiment did not fight but run. D.N. McIntosh's fought a while and then run. Folsom's and Tandy Walker's Choctaws fought a little more than McIntosh's. Stand

Watie's fought desperately for awhile and then run. The Texas and Arkansas troops fought better and displayed a good deal of bravery. They formed several times for fights but we quickly broke their lines every time. Honey Springs was Cooper's Headquarters. As they fled they set fire to their stores…"[114]

Eventually Blunt convinced Schofield that the movement south was important. Finally, the forces at Fort Gibson were reinforced by a brigade consisting of the 2nd Kansas Cavalry, the 1st Arkansas Infantry (US), the 6th Missouri State Militia Cavalry, a battalion of the 8th Missouri State Militia Cavalry, and two sections of the 2nd Indiana Battery under Colonel W.F. Cloud. Blunt began his movement south of Fort Gibson-Blunt on August 22, 1863, with approximately 4,500 troops in an attempt to drive the Confederate forces south of the Red River and out of the Indian Territory.[115] The campaign was not yet over.

---

[1] Murial H Wright, "Wapanucka Academy, Chickasaw Nation," *Chronicles of Oklahoma*, Volume 12, December 1934. pp. 402-431.

[2] Angie Debo. The Road to Disappearance: A History of the Creek Indians, Norman: The University of Oklahoma Press, 1941. pp. 154-155.

[3] Arrell M. Gibson, The Chickasaws, Norman: University of Oklahoma Press, 1972. pp. 270.

[4] Steele to Blair, *OR*, Ser. I, Vol. XXII (P2) pp. 902.

[5] Foreman to Phillips, *OR*, Ser. I, Vol. XXII (P1) pp. 382.

[6] Robert Morris Peck. "Wagon Boss and Mule Mechanic," *The National Tribune*, 1904.

[7] Ibid.

[8] Britton, *Brigade*, pp. 258

[9] Williams to Phillips, *OR*, Ser. I, Vol. XXII (P1) pp. 381.

[10] Britton, *Brigade*, pp. 260.

[11] G.S. Killgore. "Spirited Actions Not in History." *The National Tribune*, Washington, DC. November 1, 1883, pp.7. Newspapers.com

[12] Frank Moore, ed. The Rebellion Record: A Diary of American Events, with Documents, Narratives, Illustrative Incidents, Poetry, etc. Vol. 7. New York: D. Van Nostrand, 1864, pp. 179-180.

[13] James McCombs, Report, n.d. Section X, Biographies, Oklahoma Historical Society/Archives Division.

[14] Ian Michael Spurgeon. Soldiers in the Army of Freedom: The 1st Kansas Colored, the Civil War's

First African American Combat Unit, Norman: University of Oklahoma Press, 2014. Pp. 152.
[15] Ibid
[16] Yeary, *Reminiscences of the Boys in Gray*; Edwards, pp. 61.
[17] Phillips to Blunt, *OR*, Ser. I, Vol. XXII (P1) pp. 378-381.
[18] Dale and Litton. *Cherokee Cavaliers*. pp. 131; Edwards, pp. 60.
[19] Stand Watie to Sally. Letter, July 12, 1863, J.L. Hargett Collection, Box H-57, OU/WHC (?)
[20] Britton, *Memoirs*, pp. 341-342.
[21] Blunt, Experiences, pp.243-244.
[22] "A Soldier of the 29th," *Clarksville Standard*, May 30, 1863, pp. 2.
[23] Blunt, Experiences, pp.243.
[24] Blunt to Curtis, *OR*, Ser. I, Vol. XXII (P2) pp. 368; Christopher M. Rein, The Second Colorado Cavalry: A Civil War Regiment on the Great Plains. Norman: University of Oklahoma Press, 2020. pp. 91.
[25] Britton, *Brigade*, pp. 269.
[26] Charles D. Cheek, Editor. *Honey Springs: Search for a Confederate Powder House, An Ethnohistorical and Archeological Report*. Series in Anthropology. Oklahoma City: Oklahoma Historical Society, 1976: pp. 8.
[27] Cooper, Douglas H. General Order #25, July 14, 1863, *OR*, Ser. I, Vol. XXII (P1) pp. 461-462.
[28] Duval, B.G. Message, July 10, 1863, *OR*, Ser. I, Vol. XXII (P2) pp. 916-917.
[29] Ibid, 273.
[30] Blunt to Schofield, *OR*, Ser. I, Vol. XXII (P1) pp. 447.
[31] Peck. "Wagon Boss and Mule Mechanic," *The National Tribune*, 1904
[32] Blunt to Schofield, *OR*, Ser. I, Vol. XXII (P1) pp. 447.
[33] Thomas A. Moonlight, "Wartime Reminiscences of Thomas Moonlight," unpublished manuscript. Thomas Moonlight Collection, Yale University Library, New Haven, CT; Edwards, pp. 62.
[34] Blunt to Schofield, *OR*, Ser. I, Vol. XXII (P1) pp. 457-458.
[35] Curtis, H.Z. Message, July 22, 1862, *OR*, Ser. I, Vol. XXII (P2) pp. 392.
[36] Moonlight; Edwards, pp. 62.
[37] Blunt to Schofield, *OR*, Ser. I, Vol. XXII (P1) pp. 447-448.
[38] Peck.
[39] Britton, *Union Indian Brigade*: 275-276.
[40] Lindburg and Matthews, Eagle of the 11th Kansas. pp. 31.
[41] Blunt to Schofield, *OR*, Ser. I, Vol. XXII (P1) pp. 447-448.
[42] Bowles to Judson, *OR*, Ser. I, Vol. XXII (P1) pp. 449.
[43] Blunt to Schofield, *OR*, Ser. I, Vol. XXII (P1) pp. 448-454.
[44] Smith to Judson, *OR*, Ser. I, Vol. XXII (P1) pp. 454.
[45] James Burrell, April 4, 1877, *Colorado Transcript*, Golden, Colorado
[46] Hopkins to Phillips, *OR*, Ser. I, Vol. XXII (P1) pp. 456.
[47] Blunt to Schofield, *OR*, Ser. I, Vol. XXII (P1) pp. 452-453, 455-457.
[48] Peck. "Wagon Boss and Mule Mechanic," *The National Tribune*, 1904
[49] *Clarksville Standard*, September 12, 1863
[50] Welch to DeMorse, *Clarksville (TX) Standard*, September 12, 1863, p. 2, c. 2-3
[51] Blunt to Schofield, *OR*, Ser. I, Vol. XXII (P1) pp. 458-461.
[52] Welch to Bass, OR, Supplement, Vol. 4, Addendum to Vol. 22, pp. 148-149.
[53] Welch to DeMorse, *Clarksville (TX) Standard*, September 12, 1863, p. 2, c. 2-3
[54] Cooper to Steele, *OR*, Ser. I, Vol. XXII (P1) pp. 459.
[55] Baird, W. David, Editor. A Creek Warrior for the Confederacy: The Autobiography of Chief G.W. Grayson, Norman and London: University of Oklahoma Press. 1988. pp. 62.
[56] Ibid; Attributed to Chilly McIntosh, pp. 63.
[57] Edward Everett Dale, & Gaston Litton, eds. Cherokee Cavaliers: Forty Years of Cherokee History as told in the Correspondence of the Ridge-Watie-Boudinot Family, Norman: University of Oklahoma Press, 1939: 140-141.
[58] DeMorse to Cooper, OR, Supplement, Vol. 4, Addendum to Vol. 22, pp. 145.
[59] Hopkins to Phillips, *OR*, Ser. I, Vol. XXII (P1) pp. 456.
[60] Smith to Judson, *OR*, Ser. I, Vol. XXII (P1) pp. 454.
[61] Hopkins to Phillips, *OR*, Ser. I, Vol. XXII (P1) pp. 456.
[62] Smith to Judson, *OR*, Ser. I, Vol. XXII (P1) pp. 454.
[63] Stevens to Judson, *OR*, Ser. I, Vol. XXII (P1) pp. 453.
[64] Campbell to Judson, , *OR*, Ser. I, Vol. XXII (P1) pp. 452-453.
[65] Wattles to Phillips, *OR*, Ser. I, Vol. XXII (P1) pp. 455.

[66] Cooper to Steele, *OR*, Ser. I, Vol. XXII (P1) pp. 461.

[67] Cooper to Steele, *OR*, Ser. I, Vol. XXII (P1) pp. 448, 450, 453, 456-457.

[68] Cooper to Steele, *OR*, Ser. I, Vol. XXII (P1) pp. 453, 455, 457.

[69] W.K. Makemson, Letter, November 19, 1910. Copy in author's possession.

[70] Cooper to Steele, *OR*, Ser. I, Vol. XXII (P1) pp.460-461.

[71] Dallas Bowman, Letter, n.d. Dallas Bowman Collection, Box 96.47. Oklahoma Historical Society/Archives Division.

[72] Blunt to Schofield, *OR*, Ser. I, Vol. XXII (P1) pp. 448.

[73] Bowles to Judson, *OR*, Ser. I, Vol. XXII (P1) pp. 449

[74] Stevens to Judson, *OR*, Ser. I, Vol. XXII (P1) pp. 453; Smith to Phillips, *OR*, Ser. I, Vol. XXII (P1) pp. 455; Hopkins to Phillips, *OR*, Ser. I, Vol. XXII (P1) pp. 456.

[75] Britton, *Brigade*. pp. 277.

[76] Ibid

[77] Bowles to Judson, *OR*, Ser. I, Vol. XXII (P1) pp. 449-450; Britton, *Brigade*. pp. 280-281.

[78] R. McDermott, Letter, July 30, 1863, 20th Texas Collection, Harold B. Simpson Research Center, Hillsboro, Texas.

[79] James M. Williams, July 30, 1863. Williams Collection, Kansas State Historical Society

[80] Peck.

[81] Benjamin F. Van Horn, Letter, July 17, 1863, Benjamin F. Horn Collection, Military History Collection, Manuscript Division, Kansas State Historical Society; Edwards, pp. 64.

[82] Schaurte to Judson, *OR*, Ser. I, Vol. XXII (P1) pp. 451-452.

[83] Richard J. Ross, Letter, July 21, 1863. Richard J. Ross Letters, Private Collection, Patsy Mann, Checotah, Oklahoma; Edwards, pp. 68.

[84] (Denver, Colorado) *Weekly Commonwealth*, September 3, 1863.

[85] Bowles to Judson, *OR*, Ser. I, Vol. XXII (P1) pp. 450-451.

[86] Peck.

[87] Bradford K. Felmly, and John C. Grady, Suffering to Silence: 29th Texas Cavalry, CSA, Regimental History. Texas, pp. 91.

[88] Peck. "Wagon Boss and Mule Mechanic," *The National Tribune*, 1904

[89] Cooper to Steele, *OR*, Ser. I, Vol. XXII (P1) pp.460.

[90] Blunt to Schofield, *OR*, Ser. I, Vol. XXII (P1) pp.

[91] Cooper to Steele, *OR*, Ser. I, Vol. XXII (P1) pp. 459.

[92] Reports, July 1863, *OR*, Vol. 22, (P1) pp. 448, 450, 455, 458.

[93] Welch, OR, Supplement, Vol. 4, Addendum to Vol. 22: pp. 149.

[94] Cooper to Steele, *OR*, Ser. I, Vol. XXII (P1) pp. 460-461.

[95] Wesly Walk Bradly, Letter, July 20, 1863, Wesly Walk Bradly Letters, Private Collection, George Warren Blankenship, Austin, TX.

[96] G.W. Grayson, Baird, Ed. pp. 63.

[97] Cooper to Steele, *OR*, Ser. I, Vol. XXII (P1) pp. 460-461.

[98] USGS Quad Maps, Muskogee, Sansbois, Okmulgee, 1896-1898.

[99] Blunt to Schofield, *OR*, Ser. I, Vol. XXII (P1) pp. 448.

[100] Peck

[101] Makemsum, 1910

[102] Britton, *Brigade,* pp. 282-283.

[103] George C. Duffield, "The Long Drive: Driving Cattle From Texas to Iowa, 1866." Diary entry, Kansas State Historical Society, pp. 91

[104] Britton, *Brigade*, pp. 284-285.

[105] Peck. "Wagon Boss and Mule Mechanic"

[106] Yates, Catherine H., Wyckoff, Don G., Baugh, Timothy G., Harrington, John A. Jr. A Survey of Prehistoric and Historic Sites in the Honey Springs Area, McIntosh and Muskogee Counties, Oklahoma. Oklahoma Archeological Survey, Archeological Resource Survey Report No. 11. Norman, Oklahoma. 1981. pp. 61.

[107] Cooper to Steele, *OR*, Ser. I, Vol. XXII (P1) pp. 460.

[108] Scott to Holmes, *OR*, Ser. I, Vol. XXII (P2) pp. 1097.

[109] Steele to Sneed, *OR*, Ser. I, Vol. XXII (P2) pp. 1013.

[110] Phillips to Curtis, , *OR*, Ser. I, Vol. XXII (P1) pp. 56-62.

[111] Blunt to Schofield, *OR*, Ser. I, Vol. XXII (P1) pp. 448.

[112] Blunt, James G. "Civil War Experiences," *Kansas State Historical Quarterly*, May 1932: 245

[113] Blunt, James G. "Civil War Experiences," *Kansas State Historical Quarterly*, May 1932: 245

[114] Evan Jones, Letter, July 23, 1863. John Ross Papers, Folder 1189, The Gilcrease Museum, Tulsa, Oklahoma; Edwards, pp. 65.

[115] Blunt to Schofield, *OR*, Ser. I, Vol. XXII (P1) pp. 597.

Chapter 14
The Re-Occupation of Fort Smith and the Guerilla War

On August 9, 1863, Stand Watie wrote to the Governor of the Choctaw and Chickasaw Nations in his role as the Confederate Principal Chief of the Cherokee Nation (The United States recognized John Ross or Thomas Peggs as the Principal Chief). In this letter he laments the poor response of the Confederate Government in providing military support to the Indians that were loyal to the South. He also tries to instill the idea of unity among the various Nations in their common goal of driving the United States forces out of Indian Territory, especially the Cherokee Nation. His basic premise was that, if they want to win, they will have to do it without substantial support from Richmond. This dual role of Principal Chief and Confederate Officer at times clouded Watie's decisions and actions. Many times during the course of the war, he is notably absent from significant events, usually being on "detached service" or on a reconnaissance raid. He may have seen his role as political leader of the Confederate Cherokee Nation as being more important than his role as a military officer with the Confederacy. Perhaps when tribal business needed to be taken care of, he would return to Tahlequah or go to where he believed his presence was needed.[1]

After the battle of Honey Springs, the Creek Nation became wide open for bushwhackers and other independent guerilla groups. The two Confederate Creek regiments were serving with Brig. Gen. Cooper's Confederate Indian Brigade along the

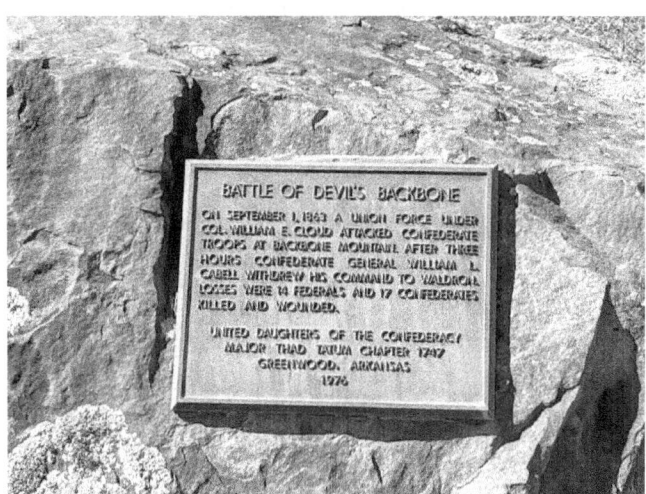

Arkansas River. The civilian population that remained after Opothleyahola's flight in 1861 remained largely pro-Confederate. The Creek or Muscogee Tribal Council also remained firmly committed to the Southern cause. In fact, the Creek national government had passed a series of punitive laws against those tribal members who still held Union sympathies. Some of these laws included the seizure and sale of property of those who supported the Union or Opothleyahola. The proceeds of those sales were to go to the treasury of the Creek Nation. Another law stated that any blacks who were captured as soldiers or in some other way were supporting the Union cause, would be sold back into slavery. After Maj. Gen. Blunt had driven Cooper and the Confederate Army south to the Arkansas River, the civilian population of the Creek Nation began to evacuate their lands. Their destinations included refugee camps in the Choctaw and Chickasaw Nations, or further south into Texas. In the weeks following the Honey Springs battle, a flurry of activity could be witnessed among the civilian populace.[2] Elsie Edwards was a Creek child during the post-battle exodus of the Creek Nation. She recalled:

> "…The neighboring people were so excited that folks were fixing up the wagons while the women folks were busy getting the quilts ready, gathering up pots and other cooking utensils, and loading up the wagons. Ropes were made of cow hides cut up the light articles… I remember we made our camp across the Red River near a high hill. The people made shelter in any way and out of anything that could be used, but most people made their crude houses out of bark which was usually of hickory bark…"[3]

Both Confederate and Union supporters suffered greatly during the war. But things improved with the Unionist Indians in Kansas since money was flowing into their families from the pay of soldiers serving in the Indian Home Guard, allowing them to buy those things the government could not or would not provide. In addition, the United States had a more efficient logistics system so, after a slow start at the beginning of the war, supplies and armaments were rarely in short supply. Not so in the Southern Indian camps. Although there was a supply system set up for the refugees, the instability of the Confederate logistics system in the Trans-Mississippi was a continual problem. Frances (Fanny) Elizabeth Kemp was interviewed several times in 1937. Her father operated the Kemp Ferry on the Red River between the Chickasaw Nation and Texas.

She described seeing the Cherokee refugees in the Chickasaw Nation:

> "…The refugees from the Cherokee Nation came in bunches and settled near us during the War. They were without food, and I have often seen them gathering the render leaves from Mulberry trees and cooking for greens. Father would kill beef and hogs and divide out among them; also, let them have corn to make bread. They would dig Briar Root, which was sweet and brittle like potatoes, and mix with the meal when they didn't have enough meal for bread. I have beaten mortar and made shuck bread to send to the men in camp. The Rebel soldiers would pass our house for days, fifteen and twenty together, and stop for food. Mother would cook a whole hog in the wash-pot; they would eat everything and move on. …"[4]

In fact, it almost seems that the U.S. Government was fighting the war with one hand tied behind its back. Congress was able to pass at least two substantial acts with immense ramifications that indicated a growing country, not one in the midst of a civil war: the Homestead Act of 1862, which allowed citizens and immigrants to claim 160 acres of public land in exchange for living on and improving it, and the Pacific Railway Act of 1862, which provided for the trans-continental railroad in an effort to connect the eastern and western parts of the United States.

The lack of a clear chain of command and defined roles hampered the entire Confederate military operation in the western

Campaign Map of Post-Honey Springs Operations including
Perryville, Fort Smith, Devil's Backbone, and Dardanelle

Trans-Mississippi area. So many times, during the course of the war the question of: "Who is in charge?" hampered Confederate actions in Indian Territory. Many of the problems presented were somewhat common on both sides, Union and Confederate. Some of the problems could be simple, such as bickering over dates of rank. But some issues were serious as when both Confederate Generals T.C. Hindman and Albert Pike were exercising command over the Indian Territory or, as shown during the First Federal Invasion, when both Colonel James Clarkson and Colonel D.H. Cooper were operating in the same area for the same objective but neither knew the other's orders. Another issue confronting the Confederate command were the treaties they had executed with the Indian Nations, which clearly spelled out the conditions for where and when Indian troops could be used and who would command them. Overall, poor lines of authority and communications between commanders and subordinates created serious difficulties for the Confederate Army in the Indian Territory.[5]

The United States Army chain of command was much clearer than that of the Confederate Army simply because of the structure already in place under the pre-war Regular Army, although the Confederates attempted to model their military in the same way. But this arrangement did not guarantee success since there were very few former Regulars or West Point graduates fighting in the Indian Territory on either side, so even the Union forces had to learn some things from scratch. Most senior officers in the Army of the Frontier were friends of Senator James Lane of Kansas, the ardent abolitionist and the reputed leader of the "Jayhawkers,"

guerilla fighters who clashed with pro-Slavery "Border Ruffians" during the period of "Bleeding Kansas." He would pressure President Abraham Lincoln to commission his friends and supporters to ensure the commitment of their cause, the abolishment of slavery.[6] This created an environment of privilege and status that obviously affected the operations of the Union Army of the Frontier. This promotion of military officers whose only qualification was support of a certain politician resulted in wasting soldiers in a battle or campaign. This can be illustrated by Senator Lane's selection of numerous newspaper owners and lawyers to leadership positions while ignoring several former enlisted Regular U.S. Army soldiers who were available in and around Fort Leavenworth. Some did slip through the cracks such as Thomas Moonlight, who had been a first sergeant in the 1st U.S. Cavalry, who was now leading the 2nd Kansas Cavalry and became a primary point of military knowledge for Gen. Blunt. Wagon Master Robert Peck was another intelligent former Regular Army sergeant, as shown by his journal records that have been used in this work, who could have obviously whipped a Kansas or Indian Home Guard company or regiment into a fine military unit. Instead, he saw what was happening and opted to work with the supply trains since that position paid much, much more.

During the Spring session of the U.S. Congress, they passed legislation requiring the Department of the Interior to relocate all Indians, regardless of status as refugee or on titled reservations, to the Indian Territory. In addition, the Bureau of Indian Affairs was to work with the War Department in Kansas to organize these Indians for their relocation as well as form military regiments from the eligible men in those tribes. General Schofield notified Blunt of this congressional action on August 10. He stated to Blunt:

"…It is desired by the Department of the Interior to comply with the recent act of Congress "providing for the removal of the Indians from the State of Kansas," &c. Complete and peaceful possession of the Indian Territory by our forces, however, is necessary to be obtained before this object can be accomplished. I am, therefore, authorized by the War Department to add to your forces by putting into service such Kansas Indians as may be willing to enter the military service of the United States, and thus secure the possession of the Indian Territory. You are hereby authorized to carry out the wishes of the Government by organizing these Indians into regiments and battalions as rapidly as possible…"[7]

Blunt was not happy with these instructions, and he quickly responded with a series of reasons why this was not a workable idea. He replied on August 22 and did not mince his words:

"…There are several reasons why I do not think such a policy practicable or advisable. It would take several months under the most favorable circumstances to organize and put into the field the Indians referred to, even were they ready and willing to enlist, of which fact I am not advised, but presume they would be very slow

to enlist; besides, my experience thus far with Indian soldiers has convinced me that they are of little service to the Government compared with other soldiers… they have become greatly demoralized and nearly worthless as troops. I would earnestly recommend that (as the best policy the Government can pursue with these Indian regiments) they be mustered out of service some time during the coming winter, and put to work raising their subsistence, with a few white troops stationed among them for their protection… I would not exchange one regiment of negro troops for ten regiments of Indians…"[8]

Fortunately for the Union cause, this policy did not begin to be implemented until after the war was over.

After his retreat from Honey Springs down the Texas Road, General Cooper began reforming his troops at Perryville, a small town south of the Canadian River within the Choctaw Nation. He commandeered the town's buildings and began to stockpile supplies for a thrust northward towards Fort Gibson-Blunt. The "A Soldier of the 29th" had earlier written to the *Clarksville Standard* and stated: "…In the morning we passed through Perryville, a very small unattractive village. There was a hospital at Perryville, and a Quartermasters dept, Blacksmith shop etc.…"[9] Cooper also maintained an advanced force at Briartown about half way between Perryville and Honey Springs. Cooper had finally been reinforced by Cabell's brigade, which included the 1st and 7th Arkansas Cavalry Regiments and the 13th Arkansas

View of Perryville from the north including the battlefield
Photos by Author

Battalion. On paper, Cooper had a force of approximately 6,000 although this number is questionable. The Arkansas conscripts were deserting in droves, and many of his disillusioned Indian troops had gone home after the defeat at Honey Springs. The Confederate War Department even still expected this decimated remnant of the Confederate Trans-Mississippi Army to drive north and push Blunt back into Kansas. Before the arrangements could be made, scouts reported Blunt's movement south on the Texas Road. Cooper withdrew his force from Briartown back towards Perryville. There was confusion in North Fork Town when the Confederates withdrew through the area and began to file over the Canadian River. Civilians were told to leave the area as the Confederates attempted to empty the military supplies that had been stored in the town. Reverend Stephen Foreman recalled in a diary entry on August 24 his surprise at the withdrawal of the Confederate Army:

> "…For a few moments I was confused not knowing what to do, or where to begin. It was dark and we had but one small two horse wagon in which to put our stuff. We went to

work, however, and loaded up, putting in some of our most valuable goods, barely leaving room for the children and myself…"[10]

General William Steele had also set up a general military hospital at North Fork Town to treat the wounded from the Honey Springs battle.

As the Confederates withdrew, they took those wounded soldiers they could and left the rest. The hospital quickly fell into Union hands.[11] Capt. Samuel J. Crawford of Company A, 2nd Kansas Cavalry, described what information the Union had:

> "…General Blunt's scouts came in and reported a Confederate train of three hundred wagons at Perryville, forty-five miles away. Within an hour after receiving this report General Blunt directed me to take the Second Kansas Cavalry, a part of the Third Wisconsin Cavalry, and a part of the Fourteenth Kansas Cavalry, with a section of artillery, and swing around Cooper to the west by way of North Fork Town and then make a forced march to Perryville…"[12]

Battle of Perryville, Choctaw Nation, August 26, 1863

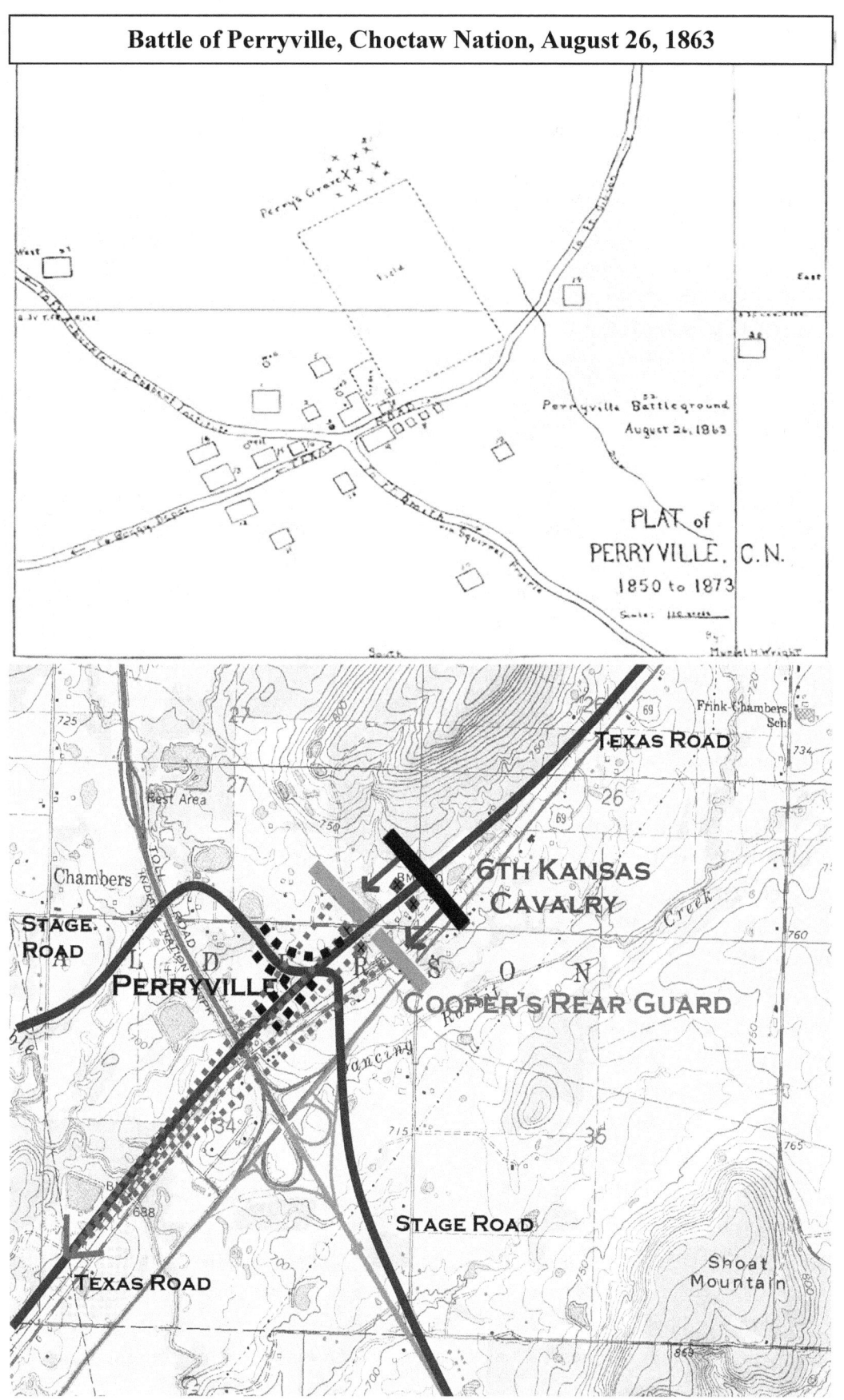

Overall Confederate commander Brig. Gen. William Steele was unsure of Blunt's objective. He sent Gen. Cabell and his increasingly undependable brigade back towards Fort Smith on the Perryville-Fort Smith Road in an attempt to lure Blunt in that direction. Cooper with the rest of the estimated 5,000 Confederates would then either flank him from the south or get between the Federals and Fort Gibson, thereby cutting off their supplies or route of retreat. Steele was anticipating Gen. Smith P. Bankhead and his Texas brigade to arrive by way of the Texas Road to assist him in resisting the Federal advance. Steele also sent Chilly McIntosh and his 2nd Creek regiment west in the direction of the headwaters of the Canadian River to prevent a movement in that direction.[13]

General Blunt refused to take the bait and moved onto the Perryville supply depot, marching 60 miles in 48 hours. He sent only patrols to monitor McIntosh and Cabell's brigades. Since Bankhead had not arrived as he had hoped, Steele did not want to face the reinforced Blunt at Perryville, so he ordered Cooper to evacuate the depot and begin moving southward towards the Red River Valley where they could occupy Fort McCulloch, the earthwork fortification Albert Pike had built in 1862 on Blue Creek. The Confederates left a rear guard to delay the Federals consisting of hastily built breast-works consisting of boxes, wagons, and some shallow earthworks, complemented with naturally heavy timber in the area. Two 12-pounder mountain howitzers were centered on a slight rise on the Texas Road before it entered the town. Steele and Cooper had recently been reinforced by a newly recruited Choctaw regiment, although they were completely green and only about half of the 800 Indian soldiers were armed.[14]

On the August 26 approach to Perryville, Blunt deployed his Indian troops as skirmishers with the cavalry bringing up the rear. Whenever Confederates forces were encountered, the cavalry would be called up to push them back. Capt. Crawford mistakenly reported that they were fighting Confederate General Richard Gano's force (this was Cabell's brigade as Gano was not yet in Indian Territory) to his front and Cooper's from the rear. He reported:

> "…It was quite a mix up. Gano was in my front, Cooper in my rear, and Blunt in Cooper's rear. Had Cooper moved forward promptly, he might have crowded my line out, because no one can fight a force in front and one in the rear. But Cooper did not do this. He pulled off the main road with his whole army and passed around on the open prairie to my left and allowed Blunt to move within supporting distance in my rear. With Cooper off my rear I moved forward in earnest and we had a running fight to Perryville, where the enemy formed on top of the hill at the edge of the village and raked the road with artillery, until we flanked them on the right and left and drove them from their positions…"[15]

At approximately 8 pm, the 6th Kansas Cavalry took the advance and came upon Cooper's main force who fired at them with the two howitzers. Blunt deployed his two howitzers with the advance and dismounted the 6th Kansas, centering them on the Texas

Road. They advanced through the timber upon Cooper's units while the howitzers blasted holes in the simple defensive barricades. The Confederates became disoriented by the darkness as well as the artillery. Their lines disintegrated, and they began to flee from the town although some members of the rear guard attempted to save the precious supplies. Blunt stated in his official report:

> "…On entering the town I learned the force there was trying to remove the commissary stores which they had not time to accomplish. This being a regular military post, nearly every building contained government stores. I directed the burning of the whole place…"[16]

Few, if any, buildings were spared, and the town of Perryville ceased to exist for the remainder of the war. The Confederates continued their retreat toward a new supply base at Middle Boggy Creek, near modern day Atoka, Oklahoma. Losses on both sides were very light. Pvt. William Prior Bates of Company A, 29th Texas Cavalry, later recalled that, "I was wounded at Perryville Choctaw Nation. I was shot in the forearm which broke under the bone and was wounded at the same time in the left side." Six soldiers from the Confederate Indian Brigade were captured. Corporal J. Clyburn of Company L of the 6th Kansas Cavalry wrote that the regiment lost twenty horses in the battle. The Union cavalry continued the pursuit throughout the night before returning to Perryville. Wesly McCoy was an eight year old enslaved child who lived near Perryville. He recalled the battle and the aftermath in 1938:

> "…I was about eight the year Father died. That was in 1864. I remember a battle that occurred that year; I didn't see the actual fighting, but I was near enough to hear the cannons. This battle was near old Perryville. Stand Watie, Sam Cooper, and a Colonel Williams were the commanding officers. The Federals won, and I remember seeing the Southern Army in disorganized retreat; it was composed of about half Indians and half white men, Texans. Some were barefoot, some hatless, and many wounded…"[17]

Private Christian Isley of the 2nd Kansas Cavalry described the event in a letter to his wife Elise:

> "…They made a stand at Perryville and fired into our advance with their Artillery, but they were soon driven out shot & shell sent after them; but it became dark and we could not pursue them any more that night… Our troops were very tired and the horses used up, the roads very dusty, the weather extremely hot, and water poor and scarce, and so we destroyed the town and rebel property, and turned back next morning, took up our march for Ft. Smith…"[18]

The Confederate loss of the supplies at Perryville dealt a severe blow to a force already short on commissary and military stores. For the time being they became completely disorganized and disillusioned and ceased being an effective military force. Only Stand Watie's Cherokee regiment maintained a semblance of military discipline, but few remained. Gen. Steele reported that:

> "…The Creeks, who were encamped above North Fork Town, were ordered to join at Perryville, which they had ample time to do, but failed to do so. I have not heard from them. A Choctaw regiment joined, but about half of its numbers were unarmed. Col. Stand Watie, who was on a scout to Webber's Falls [*Again?*], where the enemy were reported crossing, has not joined. Many of the Cherokees have left to look after their families. Of the two regiments, there are probably not more than 100 in camp…"[19]

At this point Blunt divided his force in half, sending Col. Judson's brigade consisting of the 1st Indian Home Guard, the 2nd Colorado, and the 6th Kansas Cavalry, west to confront McIntosh's Creek regiments. As Judson's force moved west, McIntosh's Creeks began to melt away until only approximately 150 soldiers remained. Many of the Creeks surrendered to the Federals, and most of these signed oaths of allegiance to the United States. In return they were provided with food and protection. Judson's brigade returned to Fort Gibson-Blunt without sustaining a single casualty.[20]

Blunt proceeded with Colonel William F. Cloud's brigade and drove east up the Fort Smith-Fort Arbuckle / California Road (incorrectly reported as the Butterfield Overland Route) after Cabell and towards Fort Smith. On August 31 Blunt encamped his troops approximately three miles from the ford over the Poteau River called McLean's Crossing. This ford was the location of one of the first iron truss bridges on the Beale Wagon Road. The iron bridge was reported by Federal cavalry officer, Lt. W. W. Averell, as having been destroyed, as he rode west from Fort Smith into Indian Territory seeking Lt. Col. W. H. Emory, commander of Indian Territory, on April 27, 1861, at the start of the Civil War. This bridge was destroyed but it is unknown just who destroyed it.[21] Colonel Cloud took a small detachment forward and discovered that Cabell, with about 2,500 troops, had entrenched at the ford crossing. Cabell also had his troops drop trees or otherwise block roads to hinder the Union advance. Blunt planned on assaulting the Confederates on the morning of September 1 but when the Union brigade approached the ford, they discovered that the Confederates had slipped away during the night back towards Fort Smith. Blunt ordered Col. Cloud forward with the 2nd Kansas Cavalry and the 6th Missouri State Militia Cavalry, supported by two sections of the 2nd Indiana Battery, to engage Cabell's retreating forces. After discovering that Cabell, instead of heading towards Fort Smith, moved eastward toward the town of Jenny Lind. Thus began a chase of approximately sixteen miles until Cloud's force caught up with Cabell's rear guard at the Devil's Backbone, a steep, razor-back, east-west ridgeline that bisects the Waldron Road between Fort Smith and Waldron, Arkansas. Cabell had realized he could not hold Fort Smith with the forces he had at his disposal, and he needed to protect a wagon train of supplies that he had sent through the town of Jenny Lind to Waldron. Cabell's remaining forces evacuated Fort Smith, and the rear guard made up of Col. J.C. Monroe's 1st Arkansas Cavalry (CSA) deployed along the

Maps and Photos of the Campaign and Battle of Devil's Backbone

Photos by Author

View Northward from the top of Devil's Backbone Mountain

Fort Smith-Waldron Road across Devil's Backbone Battlefield

Remnants of Confederate Breastworks along Fort Smith-Waldron Road side of the mountain hidden within the heavy brush. Cabell wrote:

> "...I placed Monroe's regiment in ambush at the foot of the mountain, and placed all the different regiments en echelon along the sides of the mountain, near the road; the battery being placed so as to command the whole field of operations..."[22]

Cabell apparently intended to make a full-fledged stand on the Devil's Backbone since he sent only a small force with his wagon train. His forces were concealed to a point they were able to ambush the advance elements of the 2nd Kansas, killing Captain Edward Lines and wounding and mortally wounding many more. Pvt. Christian Isely of the 2nd Kansas wrote of his experience during the ambush at Devil's Backbone to his wife Elise:

> "The scouts and Col. Cloud's bodyguard, Company C, were in the extreme front, then it was our Co., and I was 3rd in number [at the advance] of our Co., I prepared myself for the worst... We stopped to water our horses and a tremendous fire of musketry opened on Co. "C." A regiment of rebs placed themselves on one side of the road in a cornfield and another in the woods & brushes, on the other side and it is very strange that one man of Co. "C" escaped... the Captain and one private mortally wounded. 2 of our Co. were wounded besides 2 others of other companies were mortally wounded..."[23]

In a letter to the widow of Capt. Lines, Dr. Joseph P. Root, the regimental surgeon for the 2nd Kansas Cavalry lent his skills to try and save the Captain and described his actions:

> "...We overtook the enemy and engaged him, strongly posted among the mountains, immediately upon what is called Back-bone. The battle was opened by a volley of musketry, nearly in our faces, from a company ambush. Capt. [Edward] Lines' company was in the advance and received the first shock, at which your husband received a fatal wound from a rifle ball. Capt. Lines fell while gallantly leading his men at the head of our column. I was riding with Col. Cloud close behind, and immediately went to the Captain's assistance. All I could do was alleviate. He lived long enough to be cheered by the knowledge that the enemy had been routed completely..."[24]

In an effort to drive the Confederates from their mountain defenses, Cloud deployed his two sections of the 2nd Indiana Battery to shell the hillsides. This was a difficult situation for the Arkansas troops. Cabell wrote that he was disgusted that at least eight companies from three regiments fled from the battlefield. Only Monroe's regiment had remained somewhat cohesive. But after a sharp three-hour fight on the ridge, one of Col. J.C. Monroe's dismounted Arkansas cavalry battalions broke and ran from the pressure exerted by the Federals as they moved up the mountain. Cabell stated in his official report:

> "...There was nothing to make these

regiments run, except the sound of the cannon. Had they fought as troops fighting for liberty should, I would have captured the whole of the enemy's command, and gone back to Fort Smith, and driven the remainder of the enemy's force off and retaken the place..."[25]

The fleeing soldiers drove through the General's headquarters, flying southward on Waldron Road. At this point Cabell had to break off contact with the Union Army and move south, surrendering the Devil's Backbone to Cloud's forces. In their hasty retreat they lost a number of wagons filled with weapons and other supplies to Cloud. The Confederates also left their dead and wounded on the field. The men of the 2nd Kansas Cavalry felt good about their actions. Private Isely wrote:

"...[the regiment] advanced and dismounted and formed into a line of battle and drove the rebels about a mile... [the rest of the battle] was chiefly done with artillery... Their artillery fired [sic] did not do us much damage..."[26]

Captain Samuel Crawford added:

"...This was a species of warfare to which the Second Kansas never condescended. That regiment fought in the open and was always there at the beginning and the ending, but never once did any soldier of the regiment sneak around in the brush and shoot an enemy in the back..."[27]

The Federals occupied the field and began to process prisoners as well as receive scores of Confederate deserters known as "Mountain Feds," a term describing their supposed true loyalty to the United States instead of to the Confederacy.[28]

Meanwhile, when Gen. Blunt realized that Cabell and his Confederate forces were retreating to the south, he took his personal bodyguard and the 1st Arkansas Infantry and marched triumphantly into Fort Smith, the town as well as the military post. Soon the Stars and Stripes was raised above Fort Smith for the first time since it had been taken down by Capt. Samuel Sturgis's company of the 1st U.S. Cavalry on April 23, 1861. This was another major milestone objective attained by the United States in the Indian Territory. The Union Army now controlled all of the Indian Territory north of the Arkansas River completely and had made substantial inroads to areas south between that river and the Canadian River. Col. Cloud rejoined Gen. Blunt at Fort Smith the next day, assuming temporary command of the post. With just over 2,300 troops, including the 2nd Kansas Infantry (Colored) and the 2nd Colorado Infantry, Blunt pushed out the defenses of Fort Smith with outposts at Clarksville, Rosedale, and Van Buren. On September 9 Blunt ordered Cloud southeast with 200 troopers from the 2nd Kansas Cavalry towards Little Rock by way of the Fort Smith-Little Rock Military Road through Dardanelle, Arkansas. On September 12, as the Federals entered Dardanelle, they were soon joined by around 300 local Unionist volunteers who had organized themselves into six companies. Most of these men had previously served under the command of Cabell at the Devil's Backbone battle and had subsequently deserted the Confederate Army. Cloud's forces then met a brigade of Confederates under Colonel Ras Stirman with four pieces of artillery. The Union force quickly dispatched the Confederate troops after a two or three hour fight resulting in the capture of two hundred head of cattle, several hundred bushels of wheat, flour, and other commissary stores, and about thirty Confederate prisoners of war.[29] Before the Dardanelle battle, Col. Cloud encountered an unusual situation. He wrote:

> "…One gratifying feature of great interest and importance to the cause presented itself in our march, i.e., we were joined by six companies of Union men, about 300 all told, with the Stars and Stripes flying, and cheers for the Union. These men assembled at one day's notice and accompanied me in the attack on the town, and justly share the victory…"[30]

He later notes the novel sight of seeing soldiers in Confederate gray fighting alongside those wearing Union blue.[31]

In early September 1863, Union Major General Fredrick Steele (no relation to Confederate General William Steele) moved up the Arkansas River from the Mississippi River and, after a series of battles, occupied the Arkansas capital city of Little Rock. Colonel Cloud took 100 men of the 2nd Kansas Cavalry and reconnoitered along the Arkansas River to establish communications with General F. Steele, capturing two Confederate steamboats along the way. Once Cloud's force arrived at Little Rock on September 18, the result was that the entire navigable length of the Arkansas River was now under Union control. At this point Blunt recommended to Schofield that the Federal posts, including Fort Smith and Fort Gibson, be supplied by steamboats from the supply posts on the Mississippi River and then transported up the Arkansas River. He insisted that if each boat had an infantry detachment with two howitzers, it was a safer way to supply these posts than by wagon train from Fort Scott and Fort Leavenworth. He was also successful in getting telegraphic service to St. Louis and Little Rock. Edwin Bearss and Arrell Gibson wrote:

> "Blunt's ubiquitous demands were usually reasonable and his assessment of military situations correct; but these facts seemed to only irritate department and division commanders, who perhaps coveted his talent for quickly seeing a fault or flaw and moving to correct it and his promptness in offering a solution."[32]

This jealousy would come to a head in the next few months.

After the loss of Fort Smith, the Confederate Army began to move south towards Texas and Southern Arkansas in an attempt to regroup and resupply.[33] The Confederate command was again reorganized in October 1863, with Indian Territory becoming an independent district within the Trans-Mississippi Department with Gen. W. Steele remaining in charge and Gen. Cooper commanding only the Indian troops. Cooper had been maneuvering for many months to discredit Steele in order to replace him, claiming that the Indian forces had lost confidence in his abilities, as they had with General Albert Pike. Tired of a lack of supplies and support from Richmond, as well as the intrigues of Cooper and others, and with little respect for the quality of the Indian troops, Steele resigned as the commander of Indian Territory in December 1863.[34]

Cooper's plan was derailed when Steele was replaced by Major General Samuel Maxey, one of the few West Pointers to serve in the Trans-Mississippi Theater. In fact, he graduated in 58th place in the Class of 1846 that included George McClellan, Thomas J. Jackson, and George Pickett. He was a Mexican War veteran who had served with the 7th U.S. Infantry and had later left the military to pursue a career in law in Kentucky and later in Texas. He was the first Colonel of the 9th Texas Infantry Regiment early in the war. Before coming to the Indian Territory command he had served in the war's Western Theater, serving under Albert Sydney Johnston in Kentucky and East Tennessee. [35]

General Maxey did not have much to work with either in manpower or supplies. His division consisted of the Confederate Indian Brigade under Cooper with about 2,000 assigned, and the 5th Texas Cavalry Brigade under Brig. Gen. Richard Gano with approximately 1,800 soldiers. Gano had attended Bacon College in Kentucky, Bethany College in Virginia, and the Louisville University Medical School. He practiced medicine in Kentucky and, like his commander Gen. Maxey, eventually drifted from Kentucky to Texas in the years prior to the war. His previous experiences in battle included serving with John Hunt Morgan's cavalry, including his Great Raid in 1863 through Kentucky, Indiana, and Ohio. He also served with Nathan Bedford Forrest at the Battle of Chickamauga in September 1863. Then in October 1863, he was sent to the

Indian Territory.[36]

Maxey wanted to reorganize his division into three over-sized brigades, two made up of the Indian troops and one of all the Texas troops. Cooper, Waite, and Gano would be the three brigade commanders. Unfortunately for Maxey, Cooper pressured his old friend Jefferson Davis to support his idea of a light division made up of Indian troops with the Cherokees brigaded under Stand Watie, the Choctaw and Chickasaw troops brigaded under Colonel Tandy Walker, and the Creeks and Seminoles brigaded under Colonel Daniel McIntosh. Cooper would be made divisional commander with the accompanying rank of Major General, with Watie, Walker, and McIntosh holding Brigadier General commissions. Cooper, with the support of the three Indian delegates to the Confederate Congress, was able to sway President Davis to support the light division proposal, but the approval came without the expected promotions, and the proposal was never implemented. Moreover, with the reorganization came a long period during which only the Indian troop units were stable while the Texas and Arkansas units were moved in and out of the Indian Territory as needed. Before the next re-organization could take place, the Union Army drove southward again out of Fort Gibson.[37]

On the Union side the command of the Indian Territory's troops also changed drastically. Although Gen. Blunt had been successful in defeating Cooper's Confederate forces in each encounter as well as recapturing Fort Smith, a chance encounter with William Quantrill's outlaw raiders brought a quick, but brief, end to his command of the District of the Frontier. After the capture of Fort Smith, Blunt had returned to Fort Scott to organize the 2nd Kansas (Colored) Regiment and the 14th Kansas Infantry Regiment, as well as to prepare to move his district headquarters to Fort Smith. He also needed to address the allegations against him and his district for possible corruption within the Army of the Frontier, including fraud, embezzlement, cattle theft, and other violations. Blunt agreed that the corruption existed, but that it was limited to the Union Indian Brigade for the most part. Blunt tried to explain to his superiors that theft was common and accepted among Indians. Col. William Phillips at Fort Gibson took offense to Blunt's allegations concerning the Indian troops but never filed an official complaint. Most of the financial corruption occurred within the currency operations between the Indian Office and the War Department represent-tatives. There was also a fair amount of financial speculation and land transactions that seemed obviously questionable.[38]

Nevertheless, Blunt tried to get a grip on the situation when, soon upon his return to Fort Scott, he was informed of a possible attack on Fort Smith by Generals Maxey and Cooper. Blunt gathered his staff and an escort totaling approximately 100 men, which included the clerks, the brigade band and eight wagons containing supplies and the military district's archives. Although he was still in poor health, or "burning up with fever," as he reported to Maj. H.Z. Curtis, he was able to travel by his personal buggy. They moved south out of Fort Scott along the Military Road on the morning of October 4, 1863.

Captain William Quantrill and his unit of Partisan Rangers, or outlaws depending on your viewpoint, had been very active along the Kansas and Missouri border that summer.

Western Missouri was his home territory, and he believed that the Union sympathizers in that area, and especially in Kansas, were responsible for the dangerous living conditions in the area. Wagon Master Peck made an interesting observation regarding Quantrill:

> "…While I was a soldier in the Regular Army in Utah, on the Mormon Expedition in 1858, under command of Brevet Brigadier-General Albert Sidney Johnston, I used frequently to see this young man, Charley Hart (his real name), as they called him, who was a hanger-on with the Army at that time, and he was there noted as being one of the most reckless gamblers in our camp. One of his big gambling exploits I saw myself, at Fort Bridger…"[39]

It was during this period that this former schoolteacher realized that most Kansas-based troops were deep in the Indian Territory chasing Cooper and Steele and decided to take advantage of their absence. On August 21, 1863, in one of the most brutal and criminal acts of the entire Civil War, he and approximately 300-500 of his men attacked the town of Lawrence, Kansas, believed by Quantrill to be the seat of the Jayhawker/anti-slavery movement. His men looted then burned the town, and the guerrillas, on Quantrill's orders, killed around 150 to 200 unarmed men and boys who were old enough to carry a rifle. Quantrill was hoping to find Senator James Lane, but he managed to escape through a cornfield in his nightshirt. After taking the cash from the banks, looting stores and homes for valuables,

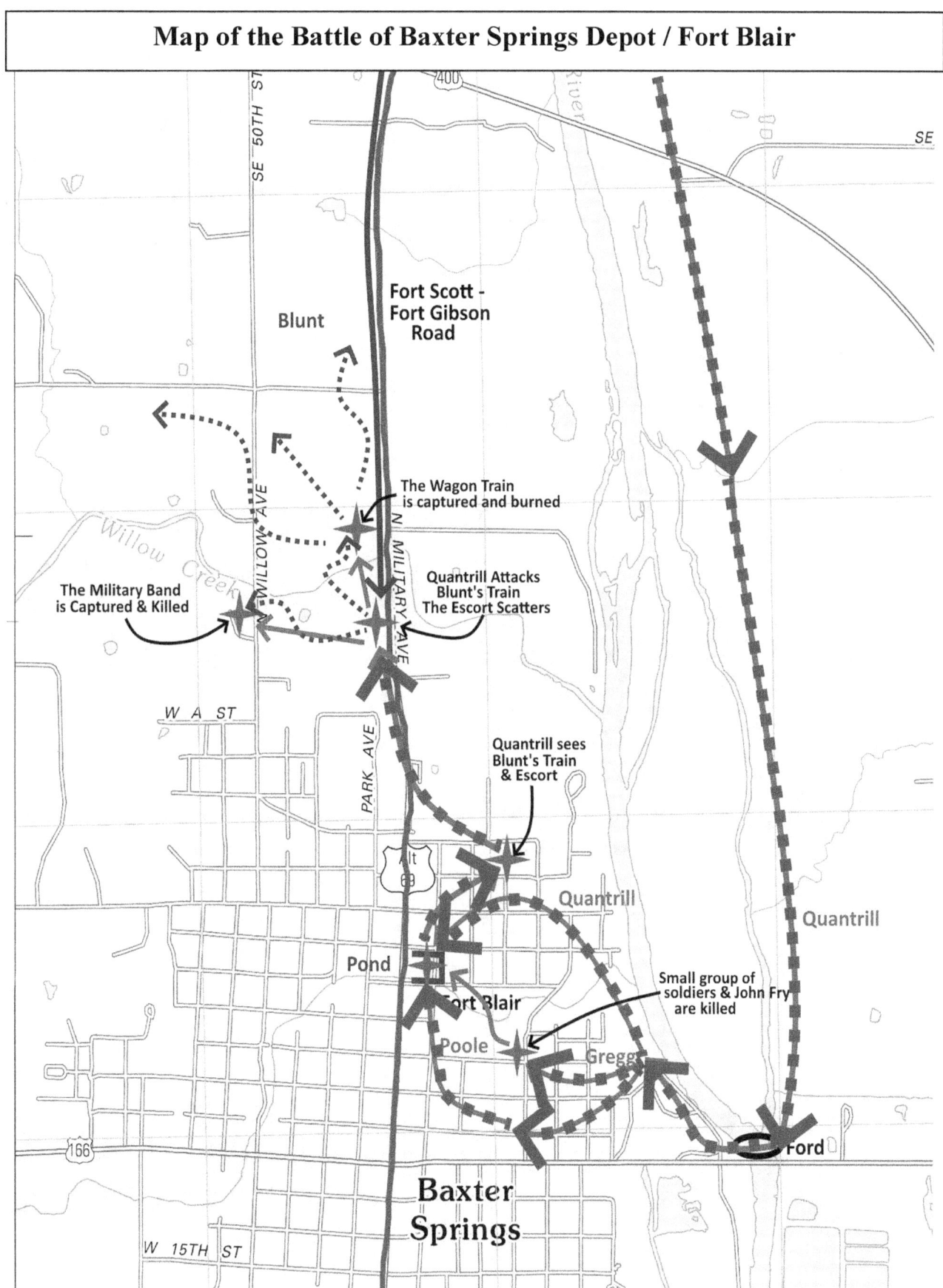

Map of the Battle of Baxter Springs Depot / Fort Blair

**Brig. Gen. William L. Cabell, CSA
Commander, Northwest Arkansas**
Photo courtesy Library of Congress

**Major General Samuel Maxey,
CSA**
Photo courtesy the Texas State Archives

Brig. Gen. Richard Gano, CSA
Photo courtesy of Texas State Archives

Reproduction of Fort Blair at Baxter Springs

Arbors where soldiers were eating mid-day meal

Photos by Author

Ford used by Quantrill to attack Fort Blair

Memorial at site of Wagon Train attack

Congressional Medal of Honor

Brig. Gen. James Blunt & Regimental Band in Kansas prior to Baxter Spring Massacre

Lt. John Pond, USA
3rd Wisconsin Cavalry

Location where Regimental Band was massacred
Photo by Author

General John McNeil
District of the Frontier

Chief John Ross's "Rose Cottage" Antebellum View

Photos courtesy of Oklahoma Historical Society

George Murrell's "Hunter's Home" in Late 19th Century

the outlaws swept back into Western Missouri and melted away before Federal troops could respond. A somewhat sympathetic Charles L. Robinson, the first governor of Kansas and an eyewitness to the raid, characterized the attack as an act of vengeance:

> "…Before this raid the entire border counties of Missouri had experienced more terrible outrages than ever the Quantrill raid at Lawrence… There was no burning of feet and torture by hanging in Lawrence as there was in Missouri, neither were women and children outraged…"

Governor Robinson explained that Quantrill targeted Lawrence because Jayhawkers had attacked Missouri "as soon as war broke out" and Lawrence was "headquarters for the thieves and their plunder."[40] This attack resulted in Union General Thomas Ewing, Jr. issuing General Order 11, which ordered the complete depopulation of three and a half Missouri counties (Bates, Cass, Jackson, and part of Vernon) except those living within one mile of a listed community. It was correctly believed by Federal authorities that most of Quantrill's supporters lived in these areas.[41]

Quantrill and his guerillas laid low for the rest of August and all of September, but his fellow Partisan Ranger, Capt. Thomas R. Livingston, remained active along the Kansas-Missouri borders, performing small raids while avoiding detection by the numerous Union patrols that had fanned out over the region after the Lawrence Massacre. The entire area was suffering from "Quantrill fever." On September 5, 1863, Capt. John Gardner of the 2nd Kansas Cavalry was carrying dispatches from Springfield,

Missouri, to the command staff of that regiment who were enroute from Fort Smith. He was being escorted by 75 troopers of the 1st Arkansas Cavalry (USA) who were stationed at Cassville. During the course of their mission, they twice encountered detachments of Livingston's guerillas. The first contact was with an approximate 65-man force about four miles from Flint Creek. They were able to break away and traveled south about eight miles along the Fort Gibson-Fort Scott Military Road that ran along the state line. Soon the running battle ensued until the Union detachment was confronted by a 300 man force. Fifty of the Union soldiers fled and the remaining 25, including Capt. Gardner, were captured by the guerillas.[42]

On September 7, Col. M. La Rue Harrison of the 1st Arkansas Cavalry (USA) embarked on an expedition into Southwest Missouri, Northwestern Arkansas, and the Indian Territory. His force was comprised of 265 cavalrymen from his 1st Arkansas, a detachment of the 1st Arkansas Light Artillery (USA) with 25 artillerymen, and one company of the 18th Iowa Infantry, with 20 infantrymen, a total force of 310. He moved to Cassville on September 10 and departed two days later after receiving information that Partisan Rangers under Capt. William Brown were gathering near Enterprise, Arkansas. The Union force moved through Elk Mills then southward across Cowskin Prairie. Harrison reported:

> "…Moving forward rapidly, I drove in the enemy's pickets near Elk Mills (killing 1 man), and attacked his skirmishers at 10 a. m., whom I found in line 1 mile west of Enterprise, in a dense thicket. I immediately

dismounted a portion of my command as skirmishers, and at the same time commenced shelling the town, where his reserve was stationed. After the engagement had continued about one hour, my right and rear were attacked by a strong force, said to be Brown's, which was repulsed and scattered in a short time. The enemy ceased to reply to our fire at 12 m., and retreated through the thicket in great disorder…"[43]

The guerilla band again spread out into the emptiness of the Spavinaw Hills of the Indian Territory. Harrison continued his pursuit of the partisans into the Cherokee Nation but was unable to bring the enemy to fight.[44]

In late October Quantrill and his men, seeking warmer weather, had been enroute from Missouri down the Texas Road to spend the coming winter in Texas. The raiders were short on supplies and needed to acquire some before continuing on. On October 6 they learned of the small Union garrison at Baxter Springs, known as Fort Blair. This small, enclosed earth and wood fort was begun in August 1863, and consisted of parts of two previous garrison posts, Camp Joe Hooker and Camp Ben Butler, both built by the 1st Kansas Infantry (Colored) before the Battle of Honey Springs. Fort Blair was commonly called Fort Baxter since it was located at the springs where the original settler, John Baxter, had his home. The fort itself consisted of a log blockhouse, some log cabins, and breastworks made of logs and earth surrounding the camp area. Lt. John Crites was sent down from Fort Scott with parts of Companies C and D of the 3rd Wisconsin

Cavalry to do the construction, including hauling the logs from the previous two posts. On October 4, Lt. James Pond arrived from Fort Scott with the remainders of the two companies of the 3rd Wisconsin Cavalry along with Company A of the new 2nd Kansas Infantry (Colored). Sixty men of the garrison, including all of the cavalry, had been sent out that morning to forage, which left the remaining ninety men to continue work on the fort. Lt. Pond had ordered the west wall to be dismantled so the enclosure could be enlarged.

Quantrill, leading between 400-500 men, with many dressed as Union soldiers, decided to attack Fort Blair and capture its supplies. "Colonel"[45] William Quantrill reported in one of his few known official reports:

"…I started on the morning of October 2, at daybreak, and had an uninterrupted march until night, and encamped on Grand River for three hours; then marched to the Osage. We continued the march from day to day, taking a due southwest course, leaving Carthage 12 miles east, crossing Shoal Creek at the falls, then going due west into the Seneca Nation… On October 6, about 2 p. m., the advance reported a train ahead. I ordered the advance to press on and ascertain the nature of it. Captain Brinker being in command of the advance, he soon discovered an encampment, which he supposed to be the camp of the train…"[46]

The guerillas crossed the Spring River east of the encampment and broke up into three groups. The first group, under David

Poole, moved west to a small hill south of the fort and attacked from that direction. The second group, under William Gregg, moved up the ravine on the east side of the fort area, surprised and killed a small group of soldiers and civilians including John Fry, famous at the time for being the first Pony Express rider, who was serving as a Union scout when he was killed in the ravine. This group attacked from the east. Quantrill himself led the third group, and they circled around the east and attacked the fort from the north. The guerillas surprised the Federals as they were eating their mid-day meal under arbors near the small fortification. Although surprised, outnumbered three-to-one, and now surrounded, the 90 or so Union men quickly formed their defenses and, with the aid of a mountain howitzer, manned singlehanded by Lt. James Pond of the 3rd Wisconsin Cavalry, the post commander, they were able to drive off the bandits after a 30-minute fight. Lt. Pond reported:

> "…The attack was made from the woods east of the camp. It was unexpected, as I had sent my cavalry out not more than an hour previous to reconnoiter on the same road the enemy came in on. My men were at dinner when the attack was made, and most of them were obliged to break through the enemy's lines in order to get their arms, which were in camp. In doing this, 1 of my men were shot down… I saw the camp surrounded on all sides by mounted men two ranks deep. I called what men were near to me to get inside the camps if possible. At the same time I ran through the enemy's ranks myself, and got safely inside, where I found the enemy's men as numerous as my own. In a moment every man was rallied, and we soon succeeded in getting the enemy outside the camp…"[47]

Ten Union soldiers were killed in the melee, including one black soldier who was killed by his former master who had recognized him.[48] Failing in their attack on the Union garrison, Quantrill's guerillas soon discovered Blunt's small force moving southward on the Military Road unaware of the attack when they approached within a mile of the Federal supply depot at Baxter Springs. Quantrill wrote:

> "… I moved with three companies—Captains Todd, Estes, and Garrett, in all 150 men—out on the prairie north of the camp, and discovered a train with 125 men as an escort, which proved to be Major-General [J. G.] Blunt and staff with body guard and headquarters train, moving headquarters from Fort Scott, Kans., to Fort Smith, Ark. I immediately drew up in line of battle, and at this time I heard heavy firing on my left, and on riding out discovered, for the first time, the fort, with at least half of my men engaged there. I ordered them to join me immediately, which they did, on the double-quick…"[49]

Blunt's column had stopped to let the wagons catch up and to allow the General the opportunity to put on his dress uniform. He needed to do this since he did not like uniforms and could usually be found around his posts or with the troops in civilian attire, but he still wanted to make a good impression on the troops he was visiting. While stopped, he observed troops moving towards him. Believing that the soldiers might be a welcoming detachment, the General sent his scout, Mr. Tough, to reconnoiter who the troops were. Tough returned and reported that they were the enemy and that Fort Blair was under attack. Blunt reported:

"…On the 6th we met with a party of guerrillas, numbering six hundred and fifty, under Quantrill, in the vicinity of Baxter's Springs. As they were dressed in blue uniform and carried our flag, they were at first supposed to be federal troops, but a doubt arising as to whether they were friends or enemies, I approached their line, alone, to ascertain their true character, and when within three hundred yards of them, they opened a fire on me. When, upon turning to my escort to signal them to return the fire and charge their line, I discovered that the entire escort (who were new recruits) had broken at the first fire of the enemy and were flying in disorder over the prairie. In vain I endeavored to halt and rally any portion of them until they had continued their stampede for a distance of two miles, when I succeeded in halting a squad of fifteen men, with which I checked the advance of the enemy, and followed them back over the field that was strewn with our dead…."[50]

The escort's officers were able to deploy only a small line of battle consisting of about 40 troops from Company A, 14th

Kansas Cavalry on the right and 25 men of Company I, 3rd Wisconsin Cavalry formed on the left. Twenty soldiers of the escort remained with the train and did not assist in its defense. Major Benjamin Henning recalled:

"…I had ridden forward myself and discovered that the force was large, and reported the same to the general, who then rode forward to reconnoiter for himself. At this time I discovered that the enemy were being re-enforced from the southwest, on a line between us and the camp at Baxter Springs, the main body of the enemy being east of us) and, wishing to ascertain the condition of things in that quarter, I rode forward to the crest of the hill, where I saw that the camp was nearly surrounded by the enemy, and the fighting very brisk…"[51]

Blunt and Curtis were desperately attempting to rally their troops and form some type of defense. Quantrill's men approached the Union defenders while walking and firing, and when the distance between the two forces came within 20 feet, the Union line broke and the men fled north and west. The guerillas swept forward toward the fleeing Federals and killed 76 and wounded another 18. The unarmed military band had attempted to escape in their wagon but, in true Hollywood fashion, it lost a wheel and spilled them out on the prairie. They tried to surrender by waving white handkerchiefs but were murdered on the spot instead. Quantrill's men piled the dead musicians into the wagon and set it on fire. Henning witnessed the attack on the band wagon:

"…the enemy had just got to it as I returned, giving me an opportunity to see every member of the band, Mr. O'Keal, the boy, and the driver shot, and their bodies thrown in or under the wagon and it fired, so that when we went to them, all were more or less burned and [the wagon] almost entirely consumed. The drummer-boy, a very interesting and intelligent lad, was shot and thrown under the wagon, and when the fire reached his clothes it must have brought returned consciousness, as he had crawled a distance of 30 yards, marking the course by bits of burning clothes and scorched grass, and was found dead with all his clothes burned off except the portion between his back and the ground as he lay upon his back…"[52]

Estimates of Quantrill's losses range from one to 50 killed but, in reality, his losses were probably very light, in the neighborhood of ten or so. The official Union reports state that the dead soldiers, musicians, and teamsters had been shot numerous times in the head. Quantrill's men had demanded the surrender of Blunt's escort soldiers but killed them as they surrendered. Sergeant Jack Splane of the 3rd Wisconsin had surrendered to a guerilla who told him: "Tell old God that the last man you saw on earth was Quantrill." He was then shot five times, but feigning death, he was able to survive. Among the dead was included James O'Neill, a special artist for Frank Leslie's newspaper who is believed to be the only war correspondent to be killed in actual combat during the war. Major H.Z. Curtis, the son of Major General Samuel Curtis and who was serving as Blunt's aide-de-camp, was also killed after his horse

was shot while attempting to jump over Willow Creek. Some reports state that he too was killed after surrendering. Blunt was able to escape leading the wife of the quartermaster on horseback while Henning made a mad dash through the attackers and reached Fort Blair. He asked Lt. Pond to assist Blunt's escort. Pond refused stating that he had no troops available to assist. Pond wrote in his report:

"…This I could not furnish him, as every effective man had been sent out in the morning, and all I had was about 25 of my own company (C) and 20 of Company D, Third Wisconsin Cavalry (none of which had serviceable horses), and 50 negroes. The major thought that, under the circumstances, I could do no better than hold my camp…"[53]

As reported, Blunt and his officers were only able to rally 15 or so who were able to fall back from the field in a semblance of order. This small group was able to drive off Quantrill's men. The guerillas looted the wagons for supplies and then burned them, including Blunt's personal buggy. Henning reported the issues that contributed to the disaster:

"…The ground on which the fight took place is rolling prairie, extending west a long distance, covered with grass, and intersected with deep ravines and gulleys, on the banks of which grow willow bushes, sufficient to conceal any difficulty in crossing, but not sufficient to protect from observation; and in retreating, many of our men were overtaken at these ravines, and killed while endeavoring to cross…"[54]

Lt. Pond had held onto the fort and was in a good enough position to refuse a surrender demand from Quantrill:

"…At 2 o'clock a flag of truce approached. The bearer, George Todd, demanded the surrender of the camp, which, being refused, he stated that he demanded in the name of Colonel Quan-trill, of the First Regiment, First Brigade, Army of the South, an exchange of prisoners. I answered that I had taken no prisoners; that I had wounded several of his men, whom I had seen fall from their horses, and would see that they were cared for, provided he would do the same by our men… This, I think, was intended for a blind to find out what I had done, as they had already murdered Major Curtis and all the prisoners…"[55]

After a quick celebration southwest of the fort, using the alcoholic beverages found on the supply wagons, the guerillas continued moving south through Indian Territory along the Texas Road. Quantrill reported his results to Maj. Gen. Sterling Price:

"…We have as trophies two stand of colors, General Blunt's sword, his commission (brigadier-general and major-general), all his official papers, &c., belonging to headquarters. After taking what we wanted from the train; we destroyed it, fearing we could not carry it away in the face of so large a

force… So at 5 p. m. I took up the line of march due south on the old Texas road. We marched 15 miles, and encamped for the night. From this place to the Canadian River we caught about 150 Federal Indians and negroes in the Nation gathering ponies. We brought none of them through. We arrived at General [D. H.] Cooper's camp on the 12th in good health and condition…"[56]

Although Blunt had sent messengers to the commanders of both Fort Smith and Fort Gibson ordering them to intercept Quantrill, the Confederate guerillas were able to reach the Red River without any further definitive encounter with Federal forces. Quantrill erroneously reported that he had killed Blunt, and this rumor spread among the Confederate forces in the Trans-Mississippi.

Blunt and the survivors returned to Fort Scott where the General put together another expedition to Fort Smith with a wagon train of nearly 400 wagons. Although the official reports show that he had been doing his best during the disaster, Blunt was severely criticized for his actions at Baxter Springs and, coupled with accusations of mismanagement and corruption in his command (a very common occurrence in Civil War armies), was quickly removed from command. Scholfield had finally found a reason to rid himself of his only successful commander. Schofield then ordered General John McNeil to take command of the District of the Frontier. Wiley Britton of the 6th Kansas reflected on what most of the soldiers of the Army of the Frontier thought:

"…It is not the thought by a good

many that General Blunt should be relieved at this time. The Baxter Springs disaster, should not, his friends say, be deemed a sufficient cause for his removal. It was more of an accident than a blunder. He is a brave officer and has never before met with defeat..."[57]

Quickly Blunt brought forth an issue that stopped Schofield cold:

"...The day before we were to leave Fort Scott, I received an order from General Schofield, directing that Brig. Gen'l McNeal [McNeil] should relieve me at Fort Smith of the command of the "District of the Frontier," when I was to proceed to Leavenworth and report to him (Schofield) by letter. A few days subsequent, information was received from Washington of the decision of the question of rank between Schofield and myself, which was adverse to Schofield and sustaining me in every point that I had raised, affirming that "Schofield was only a brigadier general... Instead of proceeding to Fort Leavenworth and reporting by letter to Gen'l Schofield, I wrote to the Secretary of War, enclosing a copy of Schofield's order, and telling him that "I should not obey it, or any other order from him (Schofield) or hold any further intercourse with him unless it should be to prefer charges against him for imbecility and cowardice..."[58]

Blunt was later restored to his command of the District of the Frontier, but

by that time the war in Indian Territory had become a stalemate.[59] Lieutenant James B. Pond of the 3rd Wisconsin Cavalry and commander of Fort Blair was awarded the Congressional Medal of Honor for his defense of the Union encampment. His citation reads:

"...For extraordinary heroism on 6 October 1863, while serving with Company C, 3d Wisconsin Cavalry, in action at Baxter Springs, Kansas. While in command of two companies of Cavalry, First Lieutenant Pond was surprised and attacked by several times his own number of guerrillas, but gallantly rallied his men, and after a severe struggle drove the enemy outside the fortifications... First Lieutenant Pond then went outside the works and, alone and unaided, fired a howitzer three times, throwing the enemy into confusion and causing him to retire..."[60]

When examining the results of the Battle of Baxter Springs, one obvious fact was never mentioned. It is quite possible that the Union soldiers in the escort somehow recognized Quantrill's raiders from a distance and, given that the reputation of Quantrill's force giving no quarter was well known, it is not surprising that they fled. These were new recruits who were, for the most part, from Kansas and may have recognized the guerillas from a distance or at least recognized the fairly non-military approach as Gen. Blunt had pointed out. And once the word got out among the escort troops, the fight or flight response spurred the men into panic mode. Since it had only been about six weeks since the Lawrence Raid, that massacre was fresh in everyone's memory and certainly influenced the green recruits when they encountered Quantrill out on the open prairie. The worst part is that their fears were confirmed by witnessing the brutal murder of surrendering soldiers as well as unarmed non-combatants by Quantrill's band. None of the officers who had submitted reports of the battle and massacre mentioned the cause of the panic, so there can only be speculation. In addition, since so few of the soldiers involved in the massacre survived, it is difficult to know exactly what was on their minds when the battle began.

The conflict between Generals Schofield and Blunt continued unabated through the end of 1863. Things heated up when on November 2, 1863, Col. William Weer, commanding the 10th Kansas Infantry, wrote a letter directly to Schofield claiming that the wagon train that Blunt was taking to Fort Gibson and Fort Smith contained contraband of war to be sold to the Confederates. Weer claimed:

> "...Blunt has gone to Fort Smith with a large Government train, 200 wagons, loaded with contraband of war. He is partner. He openly defies you and the Government. Lane has encouraged him. The goods are to be sold to rebels. Allow me to suggest the stoppage of the train, and its search, via Cassville. The report at Fort Scott is that a large amount of buried treasure is at Fort Smith and Van Buren. I make this statement upon my honor as an officer. I believe a treasonable design is on foot. If mistaken, no harm can result from an

examination of the train. McNeil should be warned, as I believe there is a design to overawe him..."[61]

Schofield immediately sent a message to McNeil at Fort Smith telling him to search for the contraband and arrest Blunt if any is present:

> Saint Louis, November 2,1863.
> Brigadier-General McNeil :
> It is officially reported to me that Major-General Blunt has started from Fort Scott for Fort Smith with a large train loaded with goods contraband of war, of which he is part owner; that he openly defies me and the Government. You will at once search the train, and ascertain the truth of this matter. If you find the report true, you will arrest General Blunt in my name, and send him to Saint Louis. If he refuses to obey the order of arrest, or refuses to turn over the command to you and return to Leavenworth, you will arrest him by force and send him to Saint Louis under guard.*
> You will seize all contraband goods and arrest all persons engaged in contraband trade.
> J. M. SCHOFIELD,
> Major-General.
> Headquarters District of the Frontier[62]

General McNeil reported back to Schofield on December 1. He stated:

> "...The train General Blunt brought through from Fort Scott is here— 248 [wagons], by this morning's report... I did not report on McDonald stock, as

the provost marshal found it only an ordinary stock of merchandise. My personal observations confirm this report. The ostensible parties are McDonald and Brooks. General Blunt has not returned to Fort Leavenworth. He is acting under authority of the War Department as commissioner to raise the Eleventh Regiment Colored Troops. He asked my aid in that capacity. Desirous to facilitate the service, and with respect to his rank, I have assigned him an office and such assistance as his duty requires. By General Blunt's train we received 100,000 rations, without flour and little hard bread…"[63]

As McNeil reported, there was no contraband on the wagon train, only commercial sutler goods, nor was any buried treasure ever found or, at least it was never reported. This was apparently just another attempt to discredit Blunt by those who were lurking in the shadows, especially by one whose career was saved by the same general after the disastrous First Invasion of the Indian Territory in 1862. Schofield was still looking for ways to rid himself of Blunt, so he sent Major Champion Vaughn down to Fort Smith to monitor Blunt's activities. Vaughn was a former editor of the *Leavenworth Conservative* newspaper and had served as a staff officer under Blunt earlier but must have had a falling out with his commander. In two private letters to Schofield, Vaughn comes across as a sniveling bootlicker bent on ruining Blunt. Vaughn wrote on December 10:

"…General Blunt is still here, and his immediate friends are urging every effort to promote his interests and advancement. Still, all their efforts are of a subdued nature. We have no more of that noisy insolence so much in vogue when General Blunt was first relieved… The opinion generally prevails that Blunt is hand-in-glove with some of the army speculators hereabouts, but General McNeil tells me that he has failed to find evidence sufficient to warrant action. Of course, Blunt and his friends still claim that he is not under your orders nor subject to your authority... General McNeil is warmly your friend, and is working faithfully and energetically… General Blunt announces that he is going to Kansas in "the course of time." He holds that he had to come here to turn over the command to General McNeil… I must congratulate you on the kindness and cordiality I have invariably heard you spoken of, and the earnest wishes expressed for your final triumph over the 'embattled hosts of darkness,' to all of which do I most heartily cry, 'Amen."[64]

Vaughn wrote to Schofield again on December 12 with the same unmanly manner that his previous letter had. He was particularly offended when Blunt came into Gen. McNeil's new office at Fort Smith and proudly read the letter he had submitted to the Secretary of War in which he outlines his charges of imbecility and cowardice towards Schofield in front of a group of officers and civilians in McNeil's office. His new letter continued:

"…Yesterday morning General Blunt sought me out at General McNeil's headquarters, and, in the presence of the latter officer, Mr. Hutchison, Lieutenant [Joseph T.] Tatum, and some other gentlemen, told me that he knew I was here as your represent-ative… I replied that his offensive manner compelled me to decline any conversation with him. He then said he would like to facilitate my business, and in order to do this would read me a copy of a letter he had lately written and forwarded to Hon. Edwin M. Stanton, Secretary of War… The letter denounces you as a liar, an imbecile, and a coward. He declines obeying your orders, and distinctly affirms that he will not report to you… Without losing my temper, I sought to give him 'a Boland for his Oliver,' [*a comparable or equally skilled opponent*] and to sustain you as an officer, a gentleman, and the commander of this department. I told General Blunt his letter was an outrage on all decency; that he had much better obey your orders… If you desire any action, and wish my services, a telegram will notify me, and I will do my duty… Having about finished my business for the State, I now proffer my services to you. If proper, I should like to have you place me upon your staff on the receipt of this, and I will remain with you at least till we see the end of this Blunt business. If you think it wise to abstain from any action at present, I will assent…"[65]

These are strong words for a Major to address to a Major General. Vaughn probably knew he had Schofield to protect him, even though it had already been established that Blunt was a major general and Schofield was still a brigadier general, at least at this time. Even still, the insubordination of Maj. Vaughn was quite striking. This author opines that the actual conversation only took place in Vaughn's head and then verbally only with the Schofield supporters. Regardless of the opinions expressed by Vaughn, it did not seem to make any difference since no real change took place.

Guerilla war was always a substantial element of the conflict in the Indian Territory, and the Union and Confederate Armies were experiencing how terrible this type of conflict could be. In the early period the internal battles within the tribes ebbed and flowed with whichever side was in control. The guerilla actions were especially violent in the Cherokee Nation, which had the deepest divisions between Union and Confederate supporters. After the Confederate loss of Fort Smith and the withdrawal of most Confederate forces from the Indian Territory, the war drifted into a series of small raids against the Union supply lines. The most successful Confederate raider was the newly promoted Brigadier General Stand Watie and his mounted Cherokee regiments. During his time as a commander, Watie had mastered the art of guerilla warfare. For the final two years of the war this Confederate general would cause havoc to the Federal forces attempting to maintain their supply lines between Forts Smith, Gibson (the name reverted from Blunt after Baxter Springs), and Scott. Since the Confederate Government could not commit any substantial resources to the war in Indian Territory, Watie was given a free hand in conducting guerilla raids throughout the Union-held areas. With both Confederate raiders and Federal patrols covering the same places, livestock virtually disappeared from the areas north of the Arkansas River. Horses were either confiscated or stolen by both sides for use as cavalry mounts. Crop fields were empty, and pastures were bare and overgrown with weeds. The Union Army, being stretched thin by protecting its supply routes, could not adequately protect the Union-supporting Indians. This situation became obvious when, on October 29, 1863, Watie and his command attacked and burned the Cherokee Council House at Park Hill as well as burning down Chief John Ross's home, the beautiful and stately Rose Cottage. They were also successful in breaking up a meeting of the Union tribal council and killing several members. Watie wrote to his wife Sarah on November 12 regarding this raid:

> "…When Medlock went away I was out on a scout. I went to Tahlequah and Park Hill… Killed a few Pins in Tahlequah. They had been holding council. I had the old council house set on fire and burnt down, also John Ross's house. Poor Andy Nave was killed. He refused to surrender and was shot by Dick Fields… Another scout has since been to Tahlequah under Battles. He returned today. They found some negro soldiers at Park Hill, killed two and two white men. They brought in some of Ross's negroes…"[66]

But the situation reversed when Watie withdrew, and the Union supporters took revenge on their Confederate brethren.[67]

On December 17 Col. Watie and his mounted force of about 500 Cherokees and 150 Creeks crossed the Arkansas River below Coodey's Creek, burned the Old Creek Agency, and attacked the pickets about three miles from Fort Gibson. They then proceeded towards Park Hill and the Tahlequah area. and, once again, began exacting revenge on the Union supporters.[68] Watie's aim had originally been to help re-locate Confederate Cherokees from the region around the Spavinaw Creek area south to the Choctaw Nation or Texas. His force crossed the Arkansas River at Webber's Falls and then deep into the Cherokee Nation. They remained in the area and continued to make periodic raids on Union sympathizers and attempting to break up any planned meetings of the Union-supporting Cherokee Tribal Council. Another goal was to either break or destroy the flour and grain supply chain between Rhea's Mill in Arkansas and Fort Gibson. Alerted to this possibility, Col. William Phillips, still in command of Fort Gibson, sent Maj. John A. Foreman and 300 men of the 3rd Indian Home Guard regiment to that location to protect both the mill and the supply trains. They were also expected to operate the mill and provide flour to Fort Gibson. By mid-December 1863 Watie's forces finally moved south from the northern Cherokee Nation down into the Tahlequah and Park Hill area.[69]

Col. Phillips at Fort Gibson was notified of the raid in the early afternoon of December 17. He ordered Capt. Alexander Spilman of the 3rd Indian Home Guard regiment to gather available troops from the 1st, 2nd, and 3rd Indian Home Guard regiments at Fort Gibson. Spilman rounded up 290 infantry soldiers plus a howitzer and crew

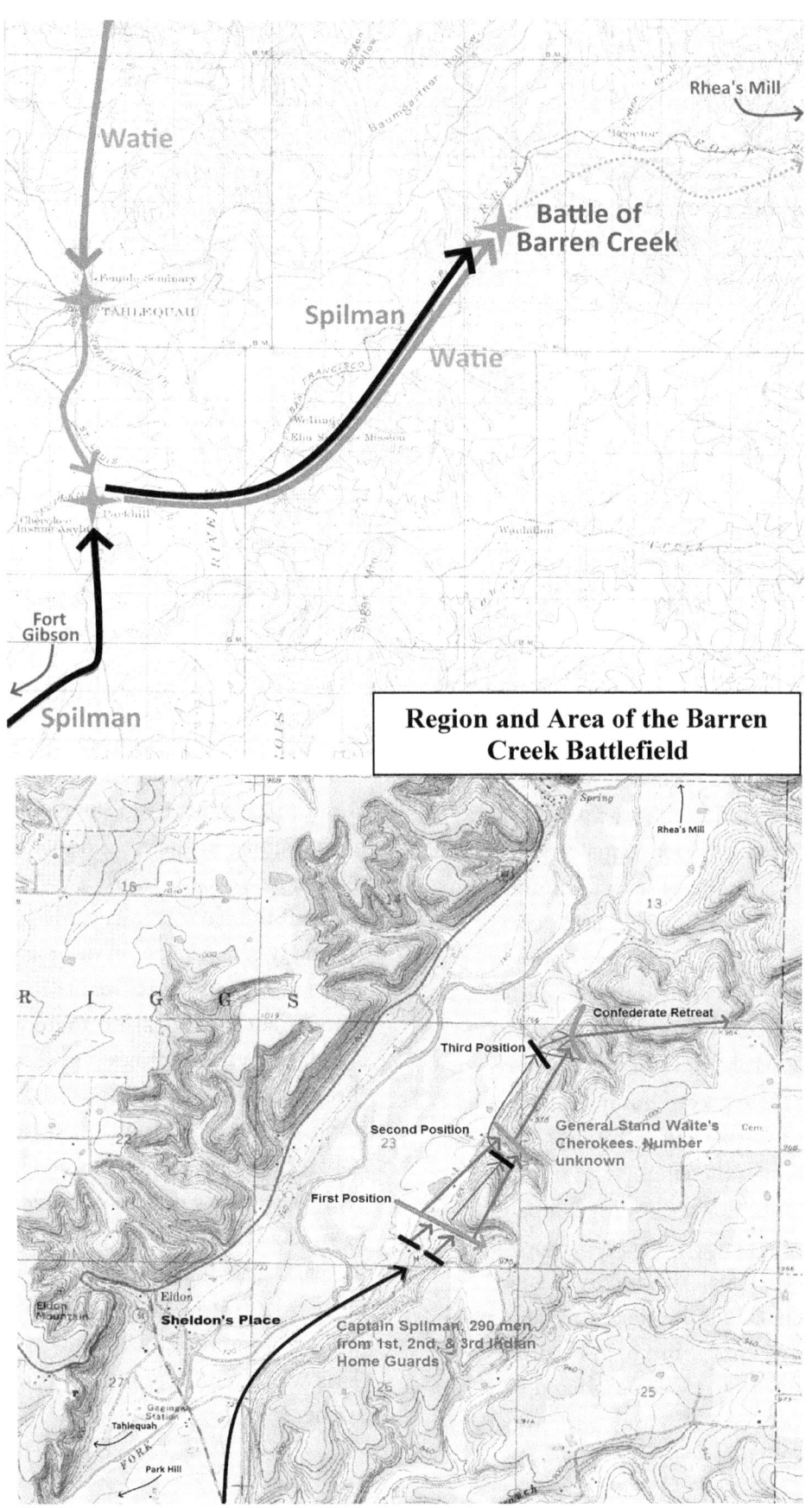

Region and Area of the Barren Creek Battlefield

commanded by Sergeant Hendricks of Company I, 3rd Indian Home Guard, and marched out at approximately 3:30 pm. This Federal reconnaissance-in-force proceeded up the Park Hill Road and, when they reached that place, they were notified that Watie had ridden in with between 500 and 800 troopers the previous day. Watie's force had plundered "Hunter's Home," the residence of George Murrell, a Union supporter and close friend and relative of Chief John Ross. The raiders also burned the remaining slave cabins at John Ross's Rose Cottage. Later they moved out of the area and camped along the Illinois River bottoms. Captain Spilman led his slower moving infantry force to the Illinois River near the Barren Fork Crossing and camped at approximately midnight.[70]

The next day dawned as a cold December day as Spilman began to move his troops across the Illinois River and up the road along the Barren Fork. The valley that Barren Fork flowed through was narrow and rocky, with flat prairie floodplains on one side and forested hills on the other with limited visual range due to the heavy foliage. The creek bed also weaved from side to side through the valley which forced the road to cross the cold stream water many times. Spilman states that he did not know where Watie's Confederates were but must have received some information or found some sign to indicate that the enemy was up the Barren Fork. Since he was leading almost 300 Indian soldiers, it is reasonable to believe that many were able to track the enemy. As he moved northeast along the road, the signs of large numbers of horses and men became evident, and they were twice countered by small Confederate patrols, driving them off in each encounter. The *Leavenworth Daily Times*

correspondent accompanying the Union force reported on December 23:

"…As Capt. Spilman marched his men to the river, a portion of the enemy came down to contest the passage. The bottom of the Illinois was covered, as were the surrounding hills, with dense brush. The stream was rapid and waist deep. In those deep dells and in that thickly bottom lay the enemy. An enemy much more than twice, nearly thrice as strong. To cross the river, to hazard all—to "wade in." It is such steps, that seem unimportant that are pregnant with events. For a moment the thoughtful young Captain wavered on the brink, and then the command swept over…"[71]

The Confederate camp was located about three miles from the crossing. Satisfied he was going in the right direction, Spilman moved his force forward along the road through the narrow valley. He sent a company-sized force forward as an advance guard as well as detachments to the flanks as a security screen to warn if there was a potential attack from those directions. After proceeding three-quarters of a mile past the Sheldon[72] home-stead the Union advanced guard made contact with the main Confederate force. Watie's troops were dismounted and had formed a battle line with their right flank on the road near the water and the line extending to their left into a deep ravine filled with thick timber and underbrush. The Confederate left flank was anchored on a hill at the end of the ravine.[73]

The *Daily Times* correspondent again reported:

Views of the Barren Creek Battlefield
Photos by Author

"…As our command swept slowly through the dense woods near this line all was silent, for the rebels who had been in front went off into the hill precipitately as if beaten. The small advance of mounted men discovered the rebel line and galloped back. A line of battle was formed against it, and swept on until the rebels indiscreetly poured out a volley at too long a range to be effective. Halting his men, and using his howitzers, Capt. Spilman briskly shelled the ravine…"[74]

Unfortunately, most of the information known about this small battle is from the Federal perspective, there only being a brief mention of it in the Confederate records, so the intention or plan of Watie's command is basically unknown.

Captain Spilman ordered the howitzer forward, supported by 95 men of the 1st Indian Home Guard under Capt. Oliver Willets, and placed it on the right side of the road. Spilman then ordered Lt. Parsons's detachment of the 3rd Indian Home Guards, which consisted almost entirely of Cherokees,

to the far right, between the gun and the bottom of the hill. As soon as the howitzer began to shell the Confederate positions their line broke, and they fled out of range of the "shooting wagon." Parsons and his Cherokees charged and drove the Confederates up and over the hill. The Confederates attempted to reform again about one-quarter mile beyond but were again pushed back. They finally regrouped on the crest of the next hill to the east. Spilman called his troops back to the road and moved up a short distance to an area of high ground with several log buildings. From this point the two sides exchanged gunfire for the next two hours, neither side having a distinct advantage. The Federals were protected by the buildings and supported by the howitzer, and the Confederates were hidden in the heavy timber and brush and protected by large rocks. Capt. Spilman eventually fooled Watie's troops into the open by feigning an organized retreat from the cover of the buildings. As soon as the Union force pulled back, Watie's dismounted soldiers thrust forward only to be met by a barrage of Federal gunfire as Spilman's troops quickly returned to their covered places. This final act was too much for the Confederate Indians and they fled back over the hill. Spilman did not order a pursuit.[75] The *Daily Times* correspondent wrote:

> "…Here the rebels made their last desperate effort. Supposing the movement to be a retreat, they formed and rushed forward in pursuit as rapidly as the nature of the ground would permit. Here the rebel Col. Adair had his horse shot from under him. For a moment the two lines of battle swung. Spilman stayed by the guns. Lieut. Parsons, Luke Parsons, one of the old John Brown boys, was on the right, and taking his hat swung it and called his Cherokees to charge. The effect was as ludicrous as irresistible. Mistaking the nature of the demonstration, each Cherokee soldier pulled off his hat likewise, and with a terrific roar of enthusiasm went forward. Talk of a bayonet charge— that was a hat charge, and goes to prove that the enemy who goes in wins. The whole rebel line gave way, and were driven back to the hills in confusion, not to form again that day…"[76]

Reported casualties were very light for both sides. Two Union soldiers were killed, including Capt. O.P. Willets of the 1st Indian Home Guard who was mortally wounded early in the battle. Confederate losses were estimated to be 12 killed and 25 wounded. Spilman's detachment marched forward after the battle for a total of 18 miles that day to continue on their mission to reinforce Maj. John Foreman at Rhea's Mills. Watie's force in turn moved eastward. The primary result of this small affair was that it stopped any further raids by the Confederates for the winter of 1863-64. Watie eventually spent considerable time that winter moving Southern Cherokees down to Texas while at the same time leading his troops in small parties northward to winter near his mill along Spavinaw Creek.[77]

Colonel Watie and his Indian troops had one final hurrah in 1863. On December 26 they captured a supply train enroute from Fort Smith to Fort Gibson along the Military Road. Pvt. Warren Day of Company G, 1st Indian Home Guard, stated that the lieutenant in charge of the train failed to circle the wagons as was protocol, and instead let them scatter. Watie's Indians got the train, and the Union escort had one man killed and between six and eight were captured.[78]

The year of 1863 had ended on a high note across the various theaters. With the Union victories at Gettysburg and Vicksburg, the substantial Union loss at the Battle of Chickamauga was revenged by the combined efforts of Generals Sherman, Thomas, and Hooker under the leadership of General Grant. Together they broke the siege of Chattanooga with the victories of Lookout Mountain and Missionary Ridge on November 24 & 25, 1863. This year was also the high tide of the Civil War in the Indian Territory, but the war was not yet over. The year saw the pinnacle of troop strength for both sides as well as the heaviest actions, especially the Battle of Honey Springs. The Federals now controlled most of the Cherokee and Creek Nations north of the Arkansas River. The Union also controlled the entire navigable length of that river from the Mississippi River. The Union Indian Brigade had shown itself to be a reliable and fairly formidable military force, while the

Confederate Indian Brigade had failed miserably in every one of its encounters with the Federal army. From their actions at the Honey Springs battle it showed that the Texas regiments fought the brunt of the battle but were overwhelmed by the better organized and led Union Army. The year also showed the dependability of the Union regiments of African-decent when they stood firm at Island Mound, Cabin Creek, Fort Blair, and especially, Honey Springs where they broke the Texas lines. Any doubts of their ability to stand and fight were put to rest. General Blunt had again shown himself as the most aggressive and successful commander in the Trans-Mississippi Theater. Blunt had severely defeated Confederate General Cooper's

Confederate Indian Brigade at Perryville, captured Fort Smith, and destroyed three Confederate supply depots. If not for the Baxter Springs disaster, Blunt may have been on his way to command the entire Trans-Mississippi since President Abraham Lincoln was always in search of aggressive commanders who wanted to fight and win. Unfortunately, in this case, Lincoln was deceived that General Schofield was a good commander. In fact, Schofield did his best to downplay and filter Blunt's successes from the War Department and the President, many times taking the credit for himself. The Civil War in the Indian Territory would substantially change in the next year.

[1] Watie, Stand. Correspondence, August 9, 1863, *OR*, Vol. XXII, (P2) pp. 961-962.

[2] Ward, *When the Wolf Came*, pp.179-180.

[3] Angie Debo. The Road to Disappearance: A History of the Creek Indians. Norman: University of Oklahoma Press, 1941. pp. 156.

[4] Frances Elizabeth Kemp, Interview, Indian-Pioneer Papers, Western History Collections, University of Oklahoma, Norman, Oklahoma.

[5] Manning, They Fought Like Veterans: The Civil War in the Indian Territory, Blue & Gray Magazine, David Roth, Ed.

[6] Ware, James W. "Indian Territory", *The Western Territories in the Civil War*. Fischer, LeRoy H. ed. Manhattan, Kansas: Sunflower University Press, 1977: 106.

[7] Schofield to Blunt, *OR*, Ser. I, Vol. XXII (P2) pp. 440.

[8] Blunt to Schofield, *OR,* Ser. I, Vol. XXII (P2) pp. 465.

[9] "A Soldier of the 29th," *Clarksville Standard (Texas),* May 30, 1863, p. 2, c. 1-2

[10] Entry of August 24, 1863, Stephen Foreman Diary, Western History Collections, University of Oklahoma, Norman, Oklahoma.

[11] Abel, American Indian in the Civil War, 1862-1865, pp. 296.

[12] Samuel J. Crawford. Kansas in the Sixties. Chicago: A.C. McClurg and Company, 1911. pp. 97.

[13] Steele to Bankhead, *OR*, Ser. I, Vol. XXII (P2) pp. 980-981; Steele to Boggs, *OR*, Ser. I, Vol. XXII (P2) pp. 984.

[14] Steele to Sneed, *OR*, Ser. I, Vol. XXII (P1) pp. 599-600.

[15] Crawford, pp. 98-99.

[16] Blunt to Schofield, *OR*, Ser. I, Vol. XXII (P2) pp. 597-598.

[17] Wesley McCoy, Interview, March 21, 1938. Indian-Pioneer Histories, University of Oklahoma Western History Collection.

[18] Ken Spurgeon. A Kansas Soldier at War: The Civil War Letters of Christian & Elise Dubach Isley. Charleston: The History Press, 2013. pp. 105.

[19] Steele to Sneed, *OR*, Ser. I, Vol. XXII (P2) pp.600.

[20] Blunt, *"Civil War Experiences:"* pp. 247.

[21] Messer, Carroll. Beale's Wagon Road to the Pacific Coast: Western Camel Road and Eastern Iron Bridge Road. Professional Paper. Self-Published, 2021. pp, 45.

[22] Cabell to Duval, *OR*, Ser. I, Vol. XXII (P1) pp. 606.

[23] Spurgeon, *Kansas Soldier*, pp. 105-107.

[24] Charles B. Lines, Memorial of Edward C.D. Lines, Late Captain of Company C, Second Regiment, Kansas Cavalry. New Haven: Tuttle, Morehouse and Taylor, 1867. pp. 34; Address by Rev. S.W.S. Dutton, D. D., at the Funeral of Capt. E. C. D. Lines. pp. 11-12. The Kansas Collection, KanCall.org.

[25] Cabell to Duval, *OR*, Ser. I, Vol. XXII (P1) pp. 607.

[26] Spurgeon, *Kansas Soldier*, pp. 105-106.

[27] Crawford, pp. 100.

[28] Cloud to Schofield, *OR*, Ser. I, Vol. XXII (P1) pp. 599.

[29] Central Arkansas Library System: Encyclopedia of Arkansas: "Skirmish at Dardanelle, 1863."

[30] Britton, Border, Vol.2, pp. 159.

[31] Cloud to McNeal, *OR*, Ser. I, Vol. XXII (P1) pp. 603-604.

[32] Bearrs and Gibson, *Fort Smith*, pp. 272-273.

[33] Cloud to McNeal, *OR*, Ser. I, Vol. XXII (P1) pp. 602-604.

[34] Steele, William. Report, February 15, 1864, Vol. 22, Part 1: 28-36.

[35] Warner, Ezra J. *Generals in Gray: Lives of Confederate Commanders*, Baton Rouge and London: Louisiana State University Press. 1987: 216.

[36] John C. Waugh. Sam Bell Maxey and the Confederate Indians, Abilene, Texas: McWhiney Foundation Press. 1995. Pp. 72.

[37] Cantrell, M.L. and Harris, Mac. Editors. *Kepis & Turkey Calls: An Anthology of the War Between the States in Indian Territory*. Oklahoma City: Western Heritage Books. 1982, Editor's Comments: 131

[38] Edwards, pp. 80.

[39] Peck, "Wagon Boss and Mule Mechanic," *National Tribune*, 1904.

[40] "Governor Robinson's Speech". *Lawrence Daily Journal and Evening Tribune*. August 23, 1892.

[41] General Order 11, *OR*, Ser. I, Vol. XXII, (P2) pp. 494.

[42] Gardner to Laurant, *OR*, Ser. I, Vol. XXII, (P1) pp. 612.

[43] Harrison to McNeil, *OR*, Ser. I, Vol. XXII, (P1) pp. 614-615.

[44] Ibid.

[45] Quantrill claimed to have received a promotion to Colonel from President Davis. No documentation exists to support his claim. On his War Department headstone he is listed as a "Captain."

[46] Quantrill to Price, *OR*, Ser. I, Vol. XXII, (P1) pp. 700.

[47] Pond to Blair, *OR*, Ser. I, Vol. XXII, (P1) pp. 698-699.

[48] Collins, *Tarnished Glory*, pp. 153.

[49] Quantrill to Price, *OR*, Ser. I, Vol. XXII, (P1) pp. 700-701.

[50] Blunt, *"Civil War Experiences:"* 247-248.

[51] Henning to Greene, *OR*, Ser. I, Vol. XXII (P1) pp. 695-696.

[52] Ibid

[53] Pond to Blair, *OR*, Ser. I, Vol. XXII, (P1) pp. 699

[54] Henning to Greene, Ser. I, *OR*, Ser. I, Vol. XXII (P1) pp. 696.

[55] Pond to Blair, *OR*, Ser. I, Vol. XXII, (P1) pp. 699.

[56] Quantrill to Price, *OR*, Ser. I, Vol. XXII, (P1) pp. 701.

[57] Britton, *Memoirs*, pp. 424.

[58] Blunt, *"Civil War Experiences:"* pp. 248-249.

[59] Blair to Greene, Ser. I, *OR*, Vol. XXII, (P1) pp. 690-693.

[60] Congressional Medal of Honor Society

[61] Weer to Schofield, *OR*, Ser. I, Vol. XXII, (P2) pp. 689-690.

[62] Schofield to McNeil, Ser. I, *OR*, Vol. XXII, (P2) pp. 690.

[63] McNeil to Schofield, *OR*, Ser. I, Vol. XXII, (P2) pp. 727.

[64] Vaughn to Schofield, December 10, 1863. *OR*, Ser. I, Vol. XXII, (P2) pp. 738.

[65] Vaughn to Schofield, December 12, 1863. *OR*, Ser. I, Vol. XXII, (P2) pp. 742.

[66] Stand Watie to Sarah C. Watie, Letter, November 12, 1863, from Camp on North Fork; Dale and Litton, Cherokee Cavaliers, pp. 144-145.

[67] Rampp, Lary C. "Civil War Battle of Barren Creek, Indian Territory, 1863," *Kepi's and Turkey Calls*: 123-124.

[68] Anderson to Cooper, *Supplement OR*, Vol. 4, (P1) pp. 165-166.

[69] Britton, Wiley. *The Union Indian Brigade in the Civil War*, Kansas City, Missouri: Franklin Hudson Publishing Company, 1922: 334.

[70] Rampp: 123-130.

[71] "Incidents of the Late Fight in the Indian Nation," *Daily Times [Leavenworth, Kansas]*, January 23, 1864, p. 2, c. 2-3.

[72] "Sheldon Homeplace" is probably a disambiguation of the name "Eldon" which is located in Sec. 27, T17N-R23E.

[73] Spilman to Phillips, *OR*, Vol. XXII, (P1) pp. 781-783.

[74] "Incidents of the Late Fight in the Indian Nation," *Daily Times [Leavenworth, Kansas]*, January 23, 1864, p. 2, c. 2-3.

[75] Spilman to Phillips, *OR*, Vol. XXII, (P1) pp. 781-783.

[76] "Incidents of the Late Fight in the Indian Nation," *Daily Times [Leavenworth, Kansas]*, January 23, 1864, p. 2, c. 2-3.

[77] Rampp: 123-130.

[78] Edwards, pp. 88.

Chapter 15
The 1864 Phillips Expedition

By the mid-winter of 1864, the crisis for the Confederate-supporting Indian population was becoming desperate. As the Federal forces moved south and west through the Cherokee Nation and into the Creek Nation, refugees fearing a Union backlash fled deep into the neighboring Choctaw and Chickasaw Nations as well as over the Red River into Texas. The promised annuities and provisions failed to arrive due to the inadequacy of the Confederate Commissary System. The Confederate Army's commissary officers were successful keeping the Army of Northern Virginia well supplied but failed miserably when it came to supplying the western Army of Tennessee and the Trans-Mississippi Department. It became even worse as the war continued. The refugees placed a severe strain on the Confederate and tribal authorities who were hoping to drive the Union Army back into Kansas. Although the Indian refugees who fled north into Kansas had periods of want, especially early in the war, they never experienced the kind of despair that the South-supporting Indians experienced.[1]

During the winter of 1863-1864, the bonds between the Confederate Five Civilized Tribes and the Confederate States of America were becoming strained due to the lack of attention from Richmond. Tribal represent-tatives held a council at Armstrong Academy, the designated Choctaw national capital, in mid-November 1863. When the conference was opened, Colonel Stand Watie, chief of the Confederate Cherokee faction, was called

Misplaced Middle Boggy Battle Marker Near Atoka, Oklahoma (Battlefield is 60 miles West of the Marker)
Photo by Author

upon to address the members. He arose and said in substance:

"...Chiefs and leaders of the Warrior Tribes: It is with a heavy heart that I rise to perform the solemn duty you have imposed upon me. Evil times have fallen upon us and our people. The troops of the enemy have been occupying the country of my people for a year, and they are now occupying the greater part of your country, and Fort Smith and Western Arkansas, and I do not see any present prospect of arresting their progress south to Red River. You know they have come like a whirlwind and swept our forces from Elk Creek, Northfork and Perryville, and captured and destroyed our supply depots at those places, leaving us neither arms, ammunition, food nor clothing, which we had been collecting with the greatest difficulty. Our people have become discouraged and indifferent. Many of our warriors have returned to their homes and are hiding out, or gone over to the enemy. When by treaty stipulations we pledged our allegiance to the South, the agents and officers of the Confederate Government told us that we would organize our men for the Confederate service, we would be furnished with arms, ammunition, clothing, food supplies and equipments, and that white troops would co-operate with us in defending our country and preventing its invasion by the enemy; yet you know that our warriors have been poorly armed, with few exceptions, and furnished with worthless ammunition, and have received but little clothing, being part of the time almost naked, without hats, shoes, and their clothing in rags. We have had to depend upon the country most of the time for our food supplies, and the promised protection from invasion of our country by our enemies has not been made good; nor have our people suffering for the common comforts of life been regularly paid their annuities, as under the Old Government with which we gave up our treaty obligations; that in the spring a force of hostile Indians and negroes and a battalion of white troops from Kansas, numbering about two thousand, took possession of Fort Gibson and have since held and fortified the place without any effort being made to dislodge them, when it

was known we had three times as many troops as they, who were two hundred miles from their base of supplies… Chiefs and leaders of the warrior tribes: You know we have lived up strictly to our treaty stipulations. We have performed our part; we have even gone out of our country to defend it; the people of my country devoted to the South have been robbed of nearly everything they possessed. They are scattered over the Choctaw and Creek Nations and in Texas, and are utterly destitute. Though the commanding General, Steele, has been strengthened by infantry and artillery, the same lethargy and procrastination prevails, and our prospects are more gloomy than ever. The movements of our troops have been around and about, but never against a much inferior foe, and this has produced universal dissatisfaction and despondency. I would therefore propose that we lay the whole distressing situation as I have depicted it, before General Smith, commanding the Trans-Mississippi Department, and the President of the Southern Confederacy at Richmond to determine whether we are to be abandoned or must continue the struggle with our own resources, unaided by our white allies. You know the Confederate Government has done practically nothing to relieve the destitution among our people…"[2]

Regardless of the tone and depressing attitude of the representatives of the Indian nations, they again swore their complete and continued allegiance and support for the Confederacy. By passing resolutions in a general council, they had formed a type of Indian confederacy that was speaking as one voice. Many of the issues that were outlined by Stand Watie, were formed into a letter to President Jefferson Davis in December.[3] Meanwhile, the Choctaw and Chickasaw regiments went into winter quarters around Doaksville, Choctaw Nation at the Spencer Academy. The Creeks holed up at Boggy Depot, and Stand Watie's Cherokees set up their winter quarters at Preston, Texas.

The United States War Department wanted to take advantage of the desperate condition in the Indian Territory as part of its overall strategy for the Union armies in the spring of 1864. It was hoped that Union forces would drive all Confederate forces out of the Indian Nations and deep into north Texas. This would disrupt Confederate supply and reinforcement in the region before General Nathanial Bank's scheduled campaign up the Red River through Louisiana.

On January 1, 1864, the Department of Kansas, which included Nebraska, Colorado, and the Indian Territory, along with Fort Smith, was separated from the Department of the Missouri and placed under the command of Major General Samuel Curtis. He was to be headquartered at Fort Leavenworth. At the same time, Schofield was replaced as commander of the Department of the Missouri upon his request for a field command. Maj. Gen. Schofield can be commended for his later services during the Atlanta Campaign under Gen. Sherman and his defense of the line at Franklin, Tennessee, against the bloody charge of Confederate General John Bell Hood. Unfortunately, his

time in the Trans-Mississippi West can be remembered only as a time of lies, intrigue, backroom politics, and self-promotion. Schofield was replaced by Maj. Gen. William S. Rosecrans, who had fallen into disfavor after the disastrous Union loss at the Battle of Chickamauga in Northern Georgia in September 1863. He was nearly captured by General James Longstreet's corps when they broke through a hole in the Union line on the second day of the battle.

As mentioned in the previous chapter, Major General James Blunt had been removed from command of his District of the Frontier on October 19, 1863, because of the Baxter Springs Massacre and the unfounded accusations of selling contraband of war. He had been replaced by the untried Brigadier General John McNeil who had been stroking department commander Maj. Gen. John Schofield's ego.[4] Blunt had remained at Fort Smith ostensibly to recruit for the 11th U.S. Colored Infantry regiment, basically ignoring McNeil and continuing to exercise command over the District of the Frontier. A new issue was created when the new Department of Arkansas was established on January 6, 1864, which included the entire state except for the Fort Smith military post. This new department was to be commanded by Maj. Gen. Fredrick Steele, who was one of the few West Point graduates and career Regular Army commanders to serve in the Trans-Mississippi. He had led the battalion of Regulars at Wilson's Creek in 1861 before becoming Colonel of the 8th Iowa Infantry. He also fought under Maj. Gen. William T. Sherman at Vicksburg and led the attack and occupation of Helena, Arkansas. He later commanded the Federal drive up the Arkansas River to capture Little Rock. Blunt recalled what transpired in this period:

"…Early in January, 1864, and after the organization of this regiment [11th U.S. Colored Infantry] had progressed so far that my personal attention with it was no longer required, I made application to the Secretary of War for assignment to other duty, in answer to which I received a telegram from the President to proceed to Washington, where I arrived on the 27th of January, and there learned that the object for which I had been called to Washington was for… the view to a campaign, early in the spring, into Texas, through the Indian country. Before leaving Washington to return to the West, I was assured, by Mr. Lincoln, that I should have every facility afforded me for the organization of this Texas expedition that I desired… I had not left Washington twenty-four hours when General Halleck, with his chronic hatred of Kansas, had determined to defeat the contemplated Texas expedition… which had had the sanction and approval of the President and Secretary of War"[5]

Prior to this point, an issue had developed related to the outposts that Blunt had established that were protecting the approaches to Fort Smith and were garrisoned by Army of the Frontier units. Under the new command structure, all of these units would now fall under the new Department of Arkansas because they were not part of the garrison of Fort Smith. When Curtis and Blunt objected to the General-In-Chief Henry

Halleck, he approached Schofield for his opinion. As true to his nature, Schofield recommended attaching the Department of the Indian Territory to Steele's command and re-establishing the independent Department of Kansas and placing Blunt there to get him out of the way. In typical "Old Brains" Halleck manner, he took the scheming Schofield's advice, going so far as convincing a skeptical President Lincoln to sign on to the plan. Blunt wrote of Halleck:

"…he had been in collusion with General Steele in robbing my district of all available troops before I could arrive there. Being satisfied that so long as General Halleck was commander in chief, I was to be the special object of his malice…"[6]

On February 22 the tepid Brig. Gen. McNeil was replaced by Brig. Gen. John Thayer, a Massachusetts lawyer who had relocated to Nebraska before the war. He was selected as the Colonel of the 1st Nebraska Infantry and served as a brigade commander at the Fort Donelson and Shiloh battles. Brig. Gen. Blunt proceeded back to Fort Leavenworth and took command of the smaller Department of Kansas. His responsibilities were now limited to keeping the nomadic Plains Indians in check and ensuring the overland trails remained open and protected. These were all parts of a difficult, confusing, and poorly executed set of decisions that would have been humorous if they were not so damaging. It is obvious that these changes were initiated strictly for the eventual removal of the politically connected Maj. Gen. Blunt.

Colonel Phillips was retained as

commander of the Federal Army of the Frontier troops within the Indian Territory and established his headquarters at Fort Gibson. He had not been idle during the winter as he worked on keeping the garrison of Fort Gibson supplied. Since most of the white troops in the Indian Territory had been sent to Fort Smith, he had only the Union Indian Brigade available for duty. And of these, only about one-third were mounted. Phillips wanted to fully mount the 2nd and 3rd Indian Home Guard regiments as mounted riflemen for service along the line of the Arkansas River and the military roads between the posts. The army quartermasters were hesitant to spend $125-$150 on cavalry horses for the Indian Brigade. This was especially true of Gen. Curtis who believed that the Indians were too hard on the standard cavalry mounts. Phillips instead suggested purchasing Indian ponies for the Indian Brigade because they were substantially more durable on the plains than regular horses. In addition, they could be purchased for as little as $25-$30 a head. Curtis agreed and the ponies were purchased during the winter.[7] Phillips began to send mounted picket squads to each of the fords up and down the river from Fort Gibson. These pickets would watch and report any crossings of Confederate or partisan groups that were moving north towards Southern Kansas or Southwest Missouri. A runner would be dispatched from the squad to Fort Gibson, and a message would be relayed to Fort Smith by courier. At Fort Smith they would contact the department commander in St. Louis by telegraph who in turn would notify the Southwest Missouri and Kansas military commands to intercept the intruders. This entire enterprise stretched the Indian Brigade, which had less

than 2,100 men available. And Phillips was lucky to have an Indian Brigade. Apparently, after the Battle of Honey Springs, Gen. Blunt began to push for the disband-ment of the Indian regiments, claiming that they were "worthless." According to Wiley Britton, it was believed at the time that Blunt was jealous of Phillips's success in turning the Indian Brigade into a viable force. Blunt, as a staunch abolitionist, probably wanted all of these Indian regiments replaced with African American regiments. By his actions, it was sometimes clear that he preferred the black regiments over the Indian Brigade. Despite all this, nothing came of this although it did create enmity between Blunt and former subordinate.[8]

During the winter period, after the Confederate losses around Fort Smith, their forces fell back deeper into the Choctaw and Chickasaw Nations, attempting to keep some distance between themselves and the Union Army while they tried to re-supply, re-arm, and recruit for the upcoming campaign season. Apart from Col. Watie's periodic raiding in the Cherokee Nation, most of the Confederate Indian Brigade remained stationary. Even Watie's Cherokees were fairly quiet after their defeat at the Barren Fork battle. The Confederate Indian Territory command tried to set up a defense line along what became known as the Red River Line. Considering that the vast majority of Confederate Indian troops had gone home on furlough, authorized or not, there were only a limited number of troops available for defense. The Confederate Red River defense line in Indian Territory ran from Fort Arbuckle eastward to Fort Washita, then along the north side of Red River through Boggy Depot, to Fort Towson and eastward to where the Indian Territory-Arkansas-Texas state lines meet on the Red River. That is where Confederate Brig. Gen. Richard Gano and his newly recruited Texas Brigade were wintering on what became known as Gano Island. The line west from Gano's command to Fort Arbuckle was the responsibility of the Chickasaw and Choctaw regiments and their attached Texas units. The line from Fort Arbuckle to the Seminole Agency was the responsibility of the Seminole Battalion. The Confederate Creek regiments under the McIntosh's were responsible for the defense line from Cochran's Store on Clear Boggy Creek eastward to Perryville, at which point the Cherokee regiments under Stand Watie took responsibility along the Arkansas River to Fort Smith. To counter any Union advances into the Indian Nations, Gen. Maxey ordered the various roads and crossings of the main trails to be monitored for any Federal incursions.[9]

On February 1, 1864, a Grand Council of all Confederate Indians that had been called by the tribal councils in November met at Armstrong Academy in the Choctaw Nation. The purpose of this meeting was to discuss inter-tribal issues of peace and harmony and continued Indian support for the Confederacy. They also wanted to plan an invasion of southern Kansas, in particular Humboldt along the Neosho River. [Humboldt must have been symbolic since it held no strategic value] The attack was planned to be a diversion for the forces south of the Arkansas River. Israel G. Vore, from the Confederate Indian Office, wrote to inform Gen. Maxey of the meeting. He stated:

"…General: In the Grand Council of the United Nations, which meets

February 1, a plan will be proposed to effect a peace between our Indians and all those of the prairies extending as far north as it is possible to communicate with them between this and spring, the time for holding the Grand Peace Council of the Prairies. This council has two objects in view; one is to establish peace and friendship between all the Indian tribes, to unite all and win them on our side; the other is, after the peace is effected, to make a raid into Kansas, and, if practicable, attack Humboldt, producing a diversion from our front. This matter is confined to the Indian tribes…"[10]

General Maxey had requested permission from Lt. Gen. E. Kirby Smith to attend the Indian council, and his commander responded:

"…The lieutenant-general commanding approves of your plan of visiting the Indian council, and earnestly desires you to do all in your power to cheer and encourage them. He has on the east bank of the Mississippi upward of 20,000 stand of arms, some of which he will send immediately to you on their arrival here…"[11]

General Maxey became one of the keynote speakers and promised that with the commitment of the Five Civilized Tribes, he would drive the Federals out of Indian Territory and return refugee Indians back to their homes. Here is a portion of his address:

"…Chiefs and Leaders of the Indian Nations: I have been assigned by the Confederate Government to the command of the Confederate forces in the District of the Indian Territory; I salute you as your friend and brother in the desperate struggle in which we are mutually engaged. I beg to assure you that I shall do everything in my power for the happiness and welfare of your people, and to protect and restore them to their homes. But we must first prepare ourselves for the work before us and drive out the invaders who have driven your noncombatant people from their homes and taken or destroyed their property… We must not retire another step; we must advance and make the enemy retire. Let us advance, and let advance be our watch word all along the line; we can do it; we must do it…"[12]

He also advised all non-warriors to return to their homes and raise crops for the Confederate forces as well as for their families. Although Maxey's words failed to rouse the tired and lethargic Indians, he did secure a continued commitment from all Confederate-leaning Five Civilized Tribes to the Southern Confederacy. Although a plan for the Kansas movement was developed, it never had a chance to be implemented.[13] Instead, the Confederate command almost immediately fell back on a somewhat static area defense with the Indian and Texas troops stationed along roads and possible areas of Union advancement from either Fort Gibson or Fort Smith. They were completely surprised by what would happen next.

Colonel Phillips had begun to gather a force of 1,500 at Fort Gibson in January 1863 for a planned thrust down into the Choctaw and Chickasaw Nations and a push eventually to the Red River. His primary mission was to try and bring the remainder of Indian Territory under Federal control and perhaps make a strike towards North Texas. The Colonel was to carry with him copies of the "President's Proclamation of Amnesty and Reconstruction" that, although not particularly written exclusively for the Indian Nations but for all persons involved in the "rebellion," it was modified to their particular needs. It was hoped that the long suffering Confederate Indians would submit again to Federal authority and in doing so, retain their previous treaty agreements. Phillips had developed a deep respect for the Indian Nations, so he had the President's Proclamation translated into the various Indian languages. On the next page is a partial transcript of the message that had been translated into Cherokee. In an effort to spark a sense of motivation on his troops, Phillips also issued a circular to his forces not only outlining the expedition's purposes but also hinting that no prisoners would be taken.[14] The circular stated:

> Circular Hdqrs. First Brig.,
> Army of the Frontier Fort Gibson,
> January 30, 1864:
>
> Soldiers! I take you with me to clean out the Indian Nation south of the river and drive away and destroy the rebels. Let me say a few words to you that you are not to forget. Do not begin firing in battle until you are ordered. When you fire, aim low,

Made Public by the President of the United States

It is true, that the government's constitution allows the President to pardon those who have wronged the U. S. Government. If they have been arrested by the use of paper, then it is against the law to pardon them. It is true, a war against our central government is now here. That's the reason the representatives of our states and the work of the states' government have been turned over [in turmoil] for a long time. There are more people who are against our government and those who are with me are obligated to punish them. During this war between states laws have been passed against those who fought us. Those who own workers [slaves] are to let them go. The President made it evident that this was handed to him and it started at once [law] though it was some time later when it was published. What the judges have decided makes sense. The President has the power to pardon. He has published that all slaves are to be freed. All who were members in the war between states and those who were in the war against the central government need to take new stand. They are to take part in our own government. The reason is:

I, Abraham Lincoln, the President of the United States, have told them and I speak to all of those who were in the war against states. They are the only ones who worked, they are known as followers. They must not hate the rest of them, as individuals, they must pledge to this. Those who pledge must sign their names. From then on, love must be in their hearts. That is the way to be for those who pledge on the long paper, that is to be preserved [filed away]. [and this what they say when they pledge:] repeated later.

I, Abraham Lincoln, President of the United States, am making known by speech to all who took part in the War against the states. Those who worked can be identified as well as those who followed. Don't let what is written below be unpleasant. For each and everyone I have given pardon. What was theirs in the past, I have made it theirs the second time. If those individuals who will pledge or those who have pledged will write their names, from then on they must live in friendship and love. That's what the pledge says in the long paper that is preserved and this is what the pledge or oath is:

"The beloved oath I now take, witnessed by God whose knowledge and strength is unbound. That from now on, without deceit I'm to be a part of it [the nation], aid it, and be included by the United States Government laws, and must obey those in the command of the laws. In the same way and without deceit, I'll obey all the laws of this nation that were passed since the war [that which was against slavery] until they [laws] are repealed, or changed, or made void [useless] by the council or the delegates, and what the Supreme court judges."

Now that this has been published or made known, you are not to dislike it. Neither are the past workers that I pardoned from all of the districts, nor the delegates from the other districts where the law is known; and the freed judges of the United States Government, those who are against the nation who fought the war and all who were officers. Those who are on the outside and are against the government judges, those who caused the war against the United States Government, I speak to them. [Also I include those that were the bravest who fought in water and those who fought on land.] Also I include those who aided in the war against the government, the blacks and of the whites who had slaves.

Now that we have published, spoken and have made it known what has been wrong in the different districts or states…

I used my own hand to write this, here in the town of Washington, now this 8th day of December in the year of 1863, and since the forming of the United States Government 88 years.

I, Bill SuWati   Abraham Lincoln

Chief Ass't to the President[1]

about the knee, or at the lower part of a man's body, if on horseback. Never fire in the air. Fire slowly and never until you see something to shout at that you may hit. Do not waste your ammunition. Do not straggle or go only that leave their comrades in the face of the enemy; nearly all the men we get killed are stragglers. Keep with me close and obey orders and we will soon have peace. Those who are still in arms are rebels, who ought to die. Do not kill a prisoner after he has surrendered. But I do not ask you to take prisoners. I ask you to make your footsteps severe and terrible. Muscogees! the time has now come when you are to remember the authors of all your sufferings; those who started a needless and wicked war, who drove you from your homes, who robbed you of your property. Stand by me faithfully and we will soon have peace. Watch over each other to keep each other right, and be ready to strike a terrible blow on those who murdered your wives and little ones by the Red Fork along the Verdigris or by Dave Farm Cowpens. Do not be afraid. We have always beaten them. We will surely win. May God go with us.

> Wm. A. Phillips
> Colonel, Commanding[15]

Also on February 1, the same date the Indian Grand Council was beginning, Col. Phillip's 1,500-man invasion force began to move southward. The Expedition consisted of elements of the 1st & 3rd Indian Home Guard regiments (the 2nd Indian Home Guard

regiment was to garrison Fort Gibson and also operate Rhea's Mill in Arkansas), a small battalion from the 14th Kansas Cavalry (Companies B, L, & M), under Major Charles Willets, and a section of howitzers under Captain Kaufman. It began its southern movement by crossing the Arkansas River then moving southwest along the Texas Road, past the Honey Springs battlefield, and then deep into the Creek Nation. Phillips reportedly had 450 mounted troops in total, the remainder infantry or support. Phillips had arranged for the remaining nine companies of Col. Thomas Moonlight's 14th Kansas Cavalry to join the expedition at North Fork Town by way of the Beale Wagon Road. This was the trail first blazed in 1858-59 by Edward F. Beale, which traveled on the south side of the Arkansas River until it crossed the Canadian River near its confluence with that river. The road crossed at a ford into North Fork Town and then continued west on the north side of the Canadian.[16] Phillips was continually worried because he was planning on moving deep into the Creek, Choctaw, Chickasaw, and Seminole Nations, and was counting on having this reinforcement of the remaining 500 men of the 14th Kansas Cavalry from Fort Smith. When Phillips arrived at North Fork Town, Moonlight's force had not yet arrived. Phillips left a small detachment of the 14th Kansas at North Fork Town with forage and corn to supply the newcomers and then lead the reinforcements to the main body. Phillips had brought only six-days' worth of provisions as it was expected that the expedition would be supplied by living off the country, raiding and taking what they needed from the Indian settlements. The expedition left the Texas Road and struck out west on Beale's Wagon

Road. With this combined force he intended to eventually make a southward movement, driving the Confederate Indians or Texans before him. Unknown to Phillips, a small demonstration by Confederates in the Fort Smith region caused Moonlight's orders to be changed with the other nine companies remaining at that post. Phillips ended up wasting many man-hours searching for and waiting on the reinforcements that would never come. It was probably for the best since Maj. Gen. Samuel Curtis, commander of the Department of Kansas, advised him to not venture too far south even with Colonel Moonlight's force since they had no clear intelligence of the strength of the Confederate forces that may oppose the expedition. Curtis suggested establishing advanced posts in the areas to impose the government's will on the Confederate Indians, which would be a better alternative. Obviously, Curtis did not completely appreciate the supply situation in the Indian Territory and the difficulties of supplying Forts Gibson and Smith without having to supply distant outposts. On the fifth day out, the Federals halted and established Camp Willetts at the Creek Hillabee settlement, near modern-day Hanna, Oklahoma, while the expedition waited for Moonlight's reinforcements. Phillips sent out advanced parties further into the Choctaw and Chickasaw Nations as well as forces up the Canadian River into the Seminole Nation. He reported that over the next three days these advanced parties killed upward of 50 enemy combatants, although that number is unconfirmed. Nevertheless, the small number of Confederate Creek soldiers who had not evacuated with their families, remained at their posts, and tried to slow the Federal advance, were consistently driven back. Col. Chilly McIntosh and Lt. Col. Pink Hawkins of the 2nd Creek Mounted Rifles finally pulled

"Site of Edward's Trading Post and Texas Trail Established 1835. Dedicated by David Dallas, Aug. 9, 1938 in honor of the Texans" Also known as Little Rivertown
Photo courtesy Oklahoma Historical Society

their remaining troops back to the southwest in the direction of Fort Arbuckle.

As for the expected reinforcements, on February 11 a battalion of the 14th Kansas Cavalry under Major J.G. Brown reported that it had been enroute from Fort Smith to meet with Phillips's command but stopped about 15 miles east of North Fork Town and about two miles south of the Canadian River. Major Brown reported that he sent a scout forward to the town but they did not meet the detachment left behind to meet them. He further stated that his horses were completely worn and were starving due to a lack of forage or grain.[17] Unfortunately, this battalion was almost a week late for its rendezvous, as it is likely that the scout did not exercise any due

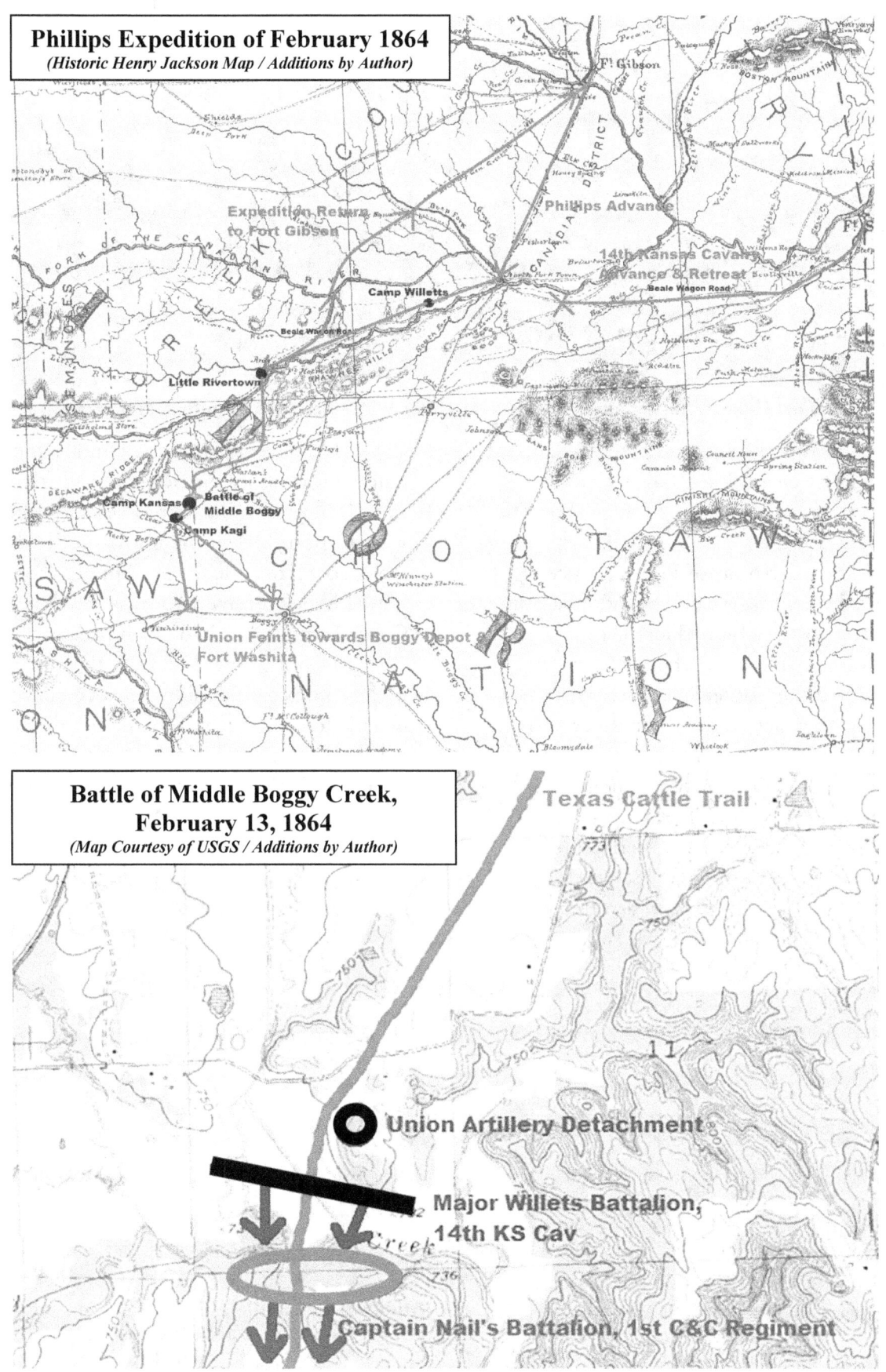

Phillips Expedition of February 1864
(Historic Henry Jackson Map / Additions by Author)

**Battle of Middle Boggy Creek,
February 13, 1864**
(Map Courtesy of USGS / Additions by Author)

Fort Holmes Marker and Location near Holdenville in Hughes County, Oklahoma,

diligence in attempting to find the detachment left behind at North Fork Town, especially since they had not found the forage and corn left behind for them.

The main body of Phillip's force continued west past old and abandoned Fort Holmes; a post established in 1834 by Lt. T.H. Holmes (the same officer who now commanded the Confederate forces in Arkansas) then across the Little River to the Edward's Store settlement known as Little Rivertown. Originally the log home of Thomas Edwards, the site became important as a stop on the Butterfield Overland Mail Route, which operated from 1858 to 1861 between Tipton, Missouri and San Francisco,

California. The area that the Union expedition entered was under the nominal control of John Jumper and the Seminoles since they were responsible for the north-south line from Fort Arbuckle to the Seminole Agency. Here Phillips received information that the Confederate forces had fallen back to the Boggy Depot supply point. By this time the Federal commander didn't believe he would be able to bring them to battle before they reached "Pike's Ditches," a derogatory name for Confederate Fort McCulloch further south on the Blue River. Major Charles Willetts and his command of the small battalion of the 14th Kansas and the howitzers, along with the mounted portion of the 1st

Dragoon Crossing of Middle Boggy Creek at Cochran's Store
Photo by Author

Middle Boggy Battlefield south of Allen in Pontopoc County, Oklahoma
Photo by Author

Indian Home Guard, were the southbound advance units. The 3rd Indian Home Guard, consisting mostly of infantry, formed the rear guard for the column. They crossed the Canadian River at Little Rivertown and moved southwest on the California Road past Shawnee Town. Further on, Willetts's advance turned south on the 1834 Dragoon Trail, which followed the Little Sandy River on its east side. (After the war the 1834 Dragoon Trail became known as the Texas Cattle Trail which remained active until it was replaced by the better known Chisholm Trail in the 1870s). The infantry and wagon train remained at the Little Rivertown encampment.

On February 13 Willetts discovered a small Confederate force on the south side of the Middle Boggy Crossing of the Dragoon Trail. Phillips states in his report that the enemy force consisted of the Chickasaws, Seminoles and Texans. It has been widely believed that the Texas unit involved was the 20th Texas Cavalry although there is not any documentation to support this, Nevertheless, there were Texans serving in the Indian regiments, especially in the recently formed Chickasaw Battalion. It is established that the Texas regiment was being used as a garrison unit at Boggy Depot to protect the army's property. In any case the Confederate force probably numbered around ninety men under the command of Captain Adam Nail of the Chickasaw and Seminole battalion, a newly minted patchwork force. Willetts began the action by placing Captain Kaufman's howitzers on a small hill that overlooks the Middle Boggy Crossing from the north and began shelling the Confederates on the other side. Willets then attacked the Confederate defenders with his entire force of about 350 troopers. After a stiff fight of about a half-

hour, the Confederate units fell back toward the camp of the Seminole Battalion, a few miles back down the road towards the Dragoon Crossing of the Clear Boggy Creek. This unit was under the command of Lt. Colonel John Jumper, the Principal Chief of the Seminole Nation. After the rendezvous of Nail and Jumper, they moved cautiously forward to the battle site and found that the Union force had moved back north. They also found 49 dead Confederates, about half of Nail's command. Per the January 30 circular by Colonel Phillips, the Federals took no prisoners and left no wounded. This brutal, if not murderous, action at Middle Boggy was to be the largest of the engagements during the Phillips Expedition.[18]

Since the planned for and expected, reinforcements had failed to arrive, the invasion soon transformed into a large scale raid. From the Little Rivertown camp, Phillips sent the bulk of his force and infantry, along with provisions, forage, and an ox train loaded with captured supplies, back to Fort Gibson under Col. Wattles of the 1st Indian Home Guard. Meanwhile, Willetts moved forward with 450 mounted men of the 14th Kansas Infantry and the 1st Indian Home Guard to the Middle Boggy battle site and established Camp Kansas on the evening of February 13. The next day Phillips's column moved south and west from this location along the Dragoon Trail another 20 or so miles and, after crossing at the stone-bottomed Dragoon Ford, established Camp Kagi on Clear Boggy Creek near the Cochran's Store and Trading Post settlement. Robert L. Cochran, a Georgian, had built a trading post at this point on the south side of Clear Boggy Creek. The trading post consisted of a large frame building with a ware room on the north side of the store extending the entire length of the building.[19] The Cochran's Store settlement contained both government buildings and schools of the Chickasaw Nation. It was also the home of their current governor or principal chief, Winchester Colbert, who had a home two miles west of Cochran's Store. It was also home to the Colbert Institute, a Chickasaw Nation school, located approximately one mile southwest of Cochran's Store. It is believed that these school buildings were used as winter quarters for Confederate troops in the area, making them subject to destruction. The Fort Arbuckle-Fort Smith Military Road continued southwest past Cochran's Store to Fort Arbuckle, and smaller roads led to Boggy Depot and Fort Washita. The Confederates later reported that Phillips had burned down Governor Colbert's house, and there is some evidence to suggest that Phillips's men attempted to completely burn down the Colbert Institute but were unable to completely do so due to two days of heavy rain.

After learning of the large Federal expedition into the Choctaw and Chickasaw Nations, Cooper and Maxey frantically attempted to gather troops at Fort Washita or Boggy Depot to counter the Federals. Col. D.N. McIntosh of the 1st Creek regiment wrote in a letter:

> "…The raid by the Federals was made at the mouth of the Little River. I was on my way to the Grand Council when the intelligence of which caused me to return from Boggy Depot. When I reached my command, they had left that place and had been gone too long for me to over take them. I

had gotten information that the Federals had returned in the direction of Ft. Gibson. They took no prisoners but killed all without mercy, what number I am not able to learn, but shall learn when we return. How brutal the actions of the enemy! The much savage tribe of Indian who never heard of Civilization would shudder at such barbarity…"[20]

What is unique and telling about this information provided by McIntosh is that the letter was written on February 9, 1864, fully four days before the Middle Boggy battle. Whatever deprivations he is referring to must have happened earlier while the Phillips Expedition was marching through the Creek Nation. 2nd Lt. Riley Perryman of Company H, 1st Creek Mounted Rifles had apparently been informed of the same issue:

> "…They came out in a force of about two thousand as far as Mill Creek in the Chickasaw Nation and went back from there taking all the women and children who were back on the Canadian back with them. It has not yet been ascertained how many Creeks they killed as they were scattered when the raid was made, but the general supposition is that there is thirty to fourty [sic] killed. Ab Lott and Colbert Lowe were killed near Wewoka and their families taken back to Fort Gibson. There are several others that are supposed killed who have not yet been heard of…"[21]

Unfortunately for the Confederate Indian Brigade, they were scattered throughout the region south and east of Boggy Depot. Only the Seminole Battalion, the new Chickasaw Battalion, and perhaps the 20th Texas Cavalry were positioned to slow or stop Phillips's Union force. But after the defeat of these forces under Capt. Nail and Lt. Col. Jumper at Middle Boggy, the remaining Confederate forces melted back towards Boggy Depot, Fort Arbuckle, or Nail's Crossing on the road to Fort Washita.

As General Curtis had warned, Phillips realized that even if he was to push forward through Boggy Depot, he would not be able to move past Boggy Depot or Fort Washita into north Texas without substantial reinforcements. Instead, he resorted to splitting his command to feign an advance on Fort Washita as well as on Boggy Depot. At this point Phillips was able to distribute the President's proclamation while also adding personal notes to the principal chiefs of the Choctaw, Chickasaw, and Seminole Nations. Phillips was confident enough to write a personal letter to Chickasaw Principal Chief Winchester Colbert who lived near the Cochran's Store settlement, but the Chief was not at home to welcome him. He was probably enroute home from the Grand Council at Armstrong Academy. Still, Phillips chides Governor Colbert:

"…Governor: When I passed your house I could not find you. Were you a fugitive from fear, or did you flee as a man who wants to be an enemy? Sad you come to me frankly you would have found a friend. The Government has not believed that you really desired to fight it, but your conduct leaves the matter in great doubt and will expose your people. Why did you send for

soldiers to keep the troops of the United States out of the Chickasaw Nation? Your treaties require you to admit their presence. Are we to understand that you now want formally to break these treaties? These questions must be answered, and answered soon. Your power as head man was not given you to gratify your prejudice or pride. You are responsible to your people, and have no right to expose them to ruin when the Government offers them mercy…"[22]

The unnamed person to whom Phillips gave the letters and proclamations, and presented them to the Confederate authorities, was arrested as a spy. His disposition is unknown as well. Regardless, on February 15, Phillips turned his forces back to the north and retraced his steps to the Little Rivertown settlement.[23] The entire force was back at Little Rivertown by the evening of February 16.[24] His column eventually caught up and rejoined the remainder of the command on its march back to Fort Gibson on the 1834 Dragoon Trail by way of the Council Hill in the Creek Nation. Sometime during the return march, as a warning they burned down the Seminole Nation's Oak Ridge Mission near old Fort Holmes. The Union Army was back at Fort Gibson by February 23, 1864, with wagons full of provisions. Private Warren Day, detached from Company C, 3rd Kansas Cavalry to Company E, 1st IHG, reported to his family:

"…We brought out of the Creek Nation about 1000 head of cattle about 250 ponies, 30 yokes of work cattle and about 800 bushels of corn, between 130 and 200 refugee Indians and about 30 or 40 Negros and killed 90 bushwhackers. I think that the Rebels will clear out for Texas…."[25]

Although Phillips was unable to draw the Confederate Army out for a decisive action, he was able to demonstrate that the Union Army could basically move anywhere throughout Indian Territory with little opposition.[26] Although Phillips did a very good job of keeping Maj. Gen. Curtis informed of what he was accomplishing, he did receive a somewhat firm rebuke from his superior about going on the expedition without notifying headquarters. Curtis wrote:

"…I am glad to hear a good account of your expedition south, for seeing a poor chance for support with cavalry I felt that you were not exactly arranged for the success which might have been realized by a complete equipment with all arms of the service and entire harmony on your left flank. Of course all expeditions making general movements by commanders of districts and posts and brigades must be submitted for proper arrangements and support to department headquarters for approval, as you. will see prescribed by all orders, the proper exception being carefully designated. The shifts and changes of your superiors about the time of your movement made the delay of a movement from Fort Smith to join you as you expected…"[27]

Since Phillips' expedition/raid did not encounter a significant Confederate force, the campaign's impact was more psychological on the Indians than due to military achievements. The council at Armstrong Academy did not completely eliminate Indian suspicions, and Phillips' propaganda encouraged those leaders inclined towards the Union. Another council, consisting of seven delegates from each tribe, met on March 16 at Tishomingo, near Fort Washita, to discuss the matter. Confederate Indian Brigade commander D.H. Cooper again worked towards securing a decision favorable to the Confederacy, but there were signs of discouragement as the council decided to enlist men for the defense of the Red River area instead of reclaiming the Cherokee Nation. Unlike at Armstrong Academy, opposition at Tishomingo could not be reconciled, leading a small group of Choctaws to meet at Scullyville and seek peace for the Choctaw Nation. Union officials saw this as a potential precursor to a larger scale defection from the Confederate alliance.[28]

General Maxey developed a deeper comprehension of the issues facing the Indian tribes through a series of tribal meetings and direct interactions with numerous Indian leaders. Maxey wrote to Lt. Gen. E. Kirby Smith pleading for white troops to be sent to the Indian Territory to bolster and firm up the Indian troops who seemed to drift away whenever Federal troops came their way.[29] From this point onward, the Confederates made a concerted effort not to confront the Union forces directly ,and the war in Indian Territory soon drifted into a guerilla war of raids and ambushes. It was in these actions that Confederate Cherokee leader Stand Watie finally blossomed into a successful cavalry commander if not a distinguished military officer. On the Union side, command of the

Armstrong Academy / Choctaw Nation Council House
Photo courtesy of the Huntington Digital Library

District of the Frontier fell to Brigadier General John Milton Thayer on February 22, 1864.[30]

It is not known if the Phillip's Expedition was intended to have a direct effect on Union General Nathanial Banks's combined Army and Navy Red River Campaign which began in March 1864. One of the expedition's purposes might have been to prevent Confederate manpower and supplies from Indian Territory and North Texas being used against Banks's advance up through Louisiana. By returning quickly to Fort Gibson, Phillips was not in position to divert the Indian Territory Confederates when the Union advance towards Shreveport began on March 10, 1864. By the end of that campaign, almost 4,000 Indian and Texas Confederates had been sent to Arkansas and helped repulse General William Steele's Federal army in what became known as the Camden Expedition. Meanwhile, the Union Indian Brigade spent most of the rest of the Spring of 1864 in routine guard and escort duties. After returning to Fort Gibson from his expedition into the southern part of the

Indian Territory, Col. Phillips directed Maj. Willetts, who was commanding a battalion of the 14th Kansas Cavalry, to relieve Maj. M. B. C. Wright, 3rd Indian Home Guard, at Rhea's Mill, Arkansas. Wright had been at that place for the past month with a battalion of his regiment, collecting wheat and corn to make into flour and meal for the Indian command at Fort Gibson. A part of the 2nd Indian Home Guard, under Col. John Ritchie, was stationed at Mackey's Salt Works, in the Cherokee Nation, about thirty miles northwest of Fort Smith on the north side of the Arkansas River. They spent most of the Spring manufacturing salt for the Indian Brigade and watching the fords of the Arkansas River in that vicinity, both to prevent the enemy from crossing to the north side and to protect the Cherokee people in raising a crop. His position was not far from Webber's Falls, and his force was intended to guard the river above and below that point for some distance. In the first part of April, Col. Watie was able to cross the Arkansas River near Briartown with about 300 men. Watie sent Col. Adair northward toward Park Hill and Maysville. Watie himself crossed and

Original Salt Kettle from Mackey's Salt Works located on the grounds of Bacone College, Muskogee, Oklahoma *Photo by Author*

proceeded towards Spavinaw Creek. Col. Phillips at Fort Gibson was, as usual, short of men and serviceable horses. The Confederate Indians did not seem to have any special mission beyond simply forcing the Federal troops to chase them. One 100-man mounted detachment of the 3rd Indian Home Guard under Capt. Anderson received information about the location of Col. Adair's 2nd Cherokee regiment. The Union Indians found the main body of the Confederate Indians at Huff's Mills, ten miles west of Maysville. They pushed forward and attacked them vigorously on May 8, routing them in a few minutes, killing six men and wounding as many more.

On the night of May 13, Major Milton Burch, leading twenty men of the 8th Missouri State Militia Cavalry (USA), from Neosho, Missouri, found a camp of thirty men of Adair's Confederate Indians on Spavinaw Creek, south of Maysville. The Missouri cavalrymen surprised and attacked the Confederate Indians, killing two men and capturing their horses, arms, and equipment. (These are just some examples of the type of actions that kept both the Confederate and Union Indian Brigades occupied during this time.) [31]

The Indian Territory's role in the war was about to expand far to the East.

[1] McMurry, Richard M. *Two Great Rebel Armies: An Essay on Confederate Military History*, Chapel Hill: University of North Carolina Press, 1989.

[2] Britton, Wiley. *The Union Indian Brigade in the Civil War*, Kansas City, Missouri: Franklin Hudson Publishing Company, 1922, pp. 338.

[3] Abel, *Participant*, pp. 317.

[4] McNeil to Schofield, *OR*, Ser. I, Vol. XXII, (P2) pp. 535.

[5] Blunt, *Experiences*, pp. 249-250

[6] Ibid.

[7] Curtis to Stanton, *OR*, Ser. I, Vol. XXXIV, (P2) pp. 462.

[8] Britton, *Brigade*, pp. 379-381.

[9] Carroll Messer, "Phillips Expedition of 1864 in Indian Territory," Self-Published, 2014.

[10] Vore to Maxey, *OR*, Ser. I, Vol. XXXIV, (P2) pp. 928.

[11] Smith to Maxey, *OR*, Ser. I, Vol. XXXIV, (P2) pp. 819.

[12] Britton, *Brigade*, pp. 343.

[13] Phillips to Curtis, *OR*, Ser. I, Vol. XXXIV, (P1) pp. 106-108.

[14] Phillips to Curtis, *OR*, Ser. I, Vol. XXXIV, (P1) pp. 106-108.

[15] William Phillips, Circular, *OR*, Ser. I, Vol. XXXIV, (P2) pp. 190.

[16] Foreman, Grant. "Survey of a Wagon Road from Fort Smith to the Colorado River." *Chronicles of Oklahoma*, Volume 12, Number 1, Oklahoma Historical Society, March 1934: 74-96.

[17] National Archives and Records Administration, AGO RG 94.

[18] Carroll Messer, "Phillips Expedition of 1864 in Indian Territory," Self-Published, 2014.

[19] George W. Burris, "Reminisces of Old Stonewall," Chronicles of Oklahoma Volume 20, No. 2, June 1942, Oklahoma Historical Society, pp. 152.

[20] D.N. McIntosh, Letter, February 9, 1864, MS378, Microfilm Division, University of Arkansas Library, Fayetteville; Edwards, pp. 91.

[21] Riley Perryman, Letter, February 15, 1864, Richard J. Ross Letters, Private Collection, Patsy Mann, Checotah, Oklahoma; Edwards, pp. 91.

[22] Phillips to Colbert, *OR*, Ser. I, Vol. XXXIV, (P1) pp. 109-110.

[23] Carroll Messer, "Phillips Expedition of 1864 in Indian Territory," Self-Published, 2014.

[24] Carroll Messer, "Phillips Expedition of 1864 in Indian Territory," Self-Published, 2014.

[25] Warren Day, Letter, March 3, 1864, Alice Robertson Collection, OHS/AD.

[26] Ibid.

[27] Phillips to Curtis, *OR*, Vol. XXXIV, (P2), pp. 537

[28] Hood, Fred. "Twilight of the Confederacy in Indian Territory," Chronicles of Oklahoma, Winter 1963; Oklahoma City, Oklahoma. pp. 433.

[29] Maxey to Smith, *OR*, Vol. XXXIV, (P2), pp. 994-997.

[30] Warner, Ezra J. Generals In Blue: Lives of Union Commanders, Baton Rouge and London: Louisiana State University Press. 1987: 499.

[31] Britton, *Border*, Vol 2, pp. 239-242.

Chapter 16
The Camden Expedition

The Union successes in all theaters during 1863 convinced the Lincoln Administration that the end of the war was near, so it promoted an across-the-board advance of the Union Armies. Ulysses S. Grant was promoted to General-in Chief of the United States Army on March 2, 1864, and took command on March 8. Lincoln and the War Department had been making the plans for a spring offensive across the majority of the war's theaters. Together they planned five coordinated Union offensives to prevent Confederate armies from shifting troops along interior lines. Maj. Gen. George Meade would drive into Virginia to confront General Robert E. Lee's Army of Northern Virginia. Maj. Gen. William T. Sherman would move south out of Chattanooga, Tennessee, directing their actions toward Atlanta, Georgia, and taking on Lt. Gen. Joseph E. Johnson's Army of Tennessee. Maj. Gen. Benjamin Butler was to drive up the James River in Virginia towards Richmond. Maj. Gen. Franz Sigel would move down the Shenandoah Valley to capture the rail lines while capturing the granaries that were providing the eastern Confederate forces with provisions. Originally, Grant wanted Maj. Gen. Nathaniel Banks to attack Mobile, Alabama and shut down that active port.

Successes in the Trans-Mississippi West, especially the control of the Mississippi and Arkansas Rivers, led the War Department under Stanton and Hallack, to continue a series of upper cuts by way of the various rivers and other waterways deep into the

Confederacy. Interestingly, the operation that would become the Red River and Camden Campaigns had their start in the desires of northern investors to invade and occupy Texas. These investors wanted this done in order to establish a free-soil, cotton-growing colony in Texas and Louisiana to supply northeastern textile manufacturers with raw materials, the lack of which was hampering production, since most of their cotton had been coming from the Southern states. Their only other option was to purchase and transport the cotton from Egypt or India for exorbitant prices. There was also the possibility that Napoleon III of France would attempt to interfere with the domestic issues of Mexico and having a substantial U.S. Army and Navy expeditionary force in the area would help ensure that the Monroe Doctrine could be enforced. In fact, Banks was issued these instructions prior to the start of the Expedition:

Department of State,
Washington, D. C., March 14, 1864.
Maj. Gen. N. P. Banks,
New Orleans, La.:

General: I give you herewith an extract from a dispatch which was addressed to the Department by Mr. Corwin, the minister of the United States in Mexico, on the 26th ultimo. In view of the representation thus made by Mr. Corwin, the President thinks it necessary that you should be specially charged to do whatever is practicable, consistent with the national safety and dignity, to avoid any collision between the forces under your command and either of the belligerents in Mexico, and even to guard so far as may be possible against suffering any occasion to arise for dispute or controversy between your command or the authorities of Texas and either or both of these parties.
I am, general, your obedient servant,
WILLIAM H. SEWARD.
[Indorsement.]
War Department,
March 15, 1864.
**The foregoing instructions having been submitted to this Department are approved, and General Banks will conform to them.
EDWIN M. STANTON,
Secretary of War.[1]

President Lincoln, who needed the support of the nation's manufacturing sector since they were basically bankrolling the war, acquiesced to their demands and turned over the issue to the War Department. The Union developed four goals at the start of the campaign. The first was the capture of Shreveport, the state capital and headquarters of the Trans-Mississippi Department. The second was the destruction of the Confederate forces in the District of West Louisiana commanded by General Richard Taylor. The third as to confiscate as much as a hundred thousand bales of cotton from the plantations along Red River. The fourth was to organize "pro-Union" state governments in both Louisiana and Texas.[2]

The plan to take Mobile was put on hold, and Maj. Gen. Nathanial Banks was reassigned as the overall commander of the Red River Expedition. He was a Massachusetts politician who had wrangled his general's commission from Lincoln due to his influence and money from that state. His military performance during the period prior to this assignment was very poor. Other than mounting a stiff resistance at the Cedar Mountain battle, he had been driven from the Shenandoah Valley by Confederate General Thomas "Stonewall" Jackson in 1862 and had led his troops into a disastrous advance at the Battle of Port Hudson. His father had been the superintendent of a cotton mill in Massachusetts, and Banks had worked there during his youth beginning as a bobbin boy. His experience with cotton and the influence of Massachusetts textile manufacturing officials most likely affected his selection to lead this expedition. Although Banks was selected to lead the expedition, he did not plan it. President Lincoln, Secretary of War Edwin Stanton and the War Department set the goals and resources available.[3]

The Red River Expedition's plan involved three simultaneous thrusts towards

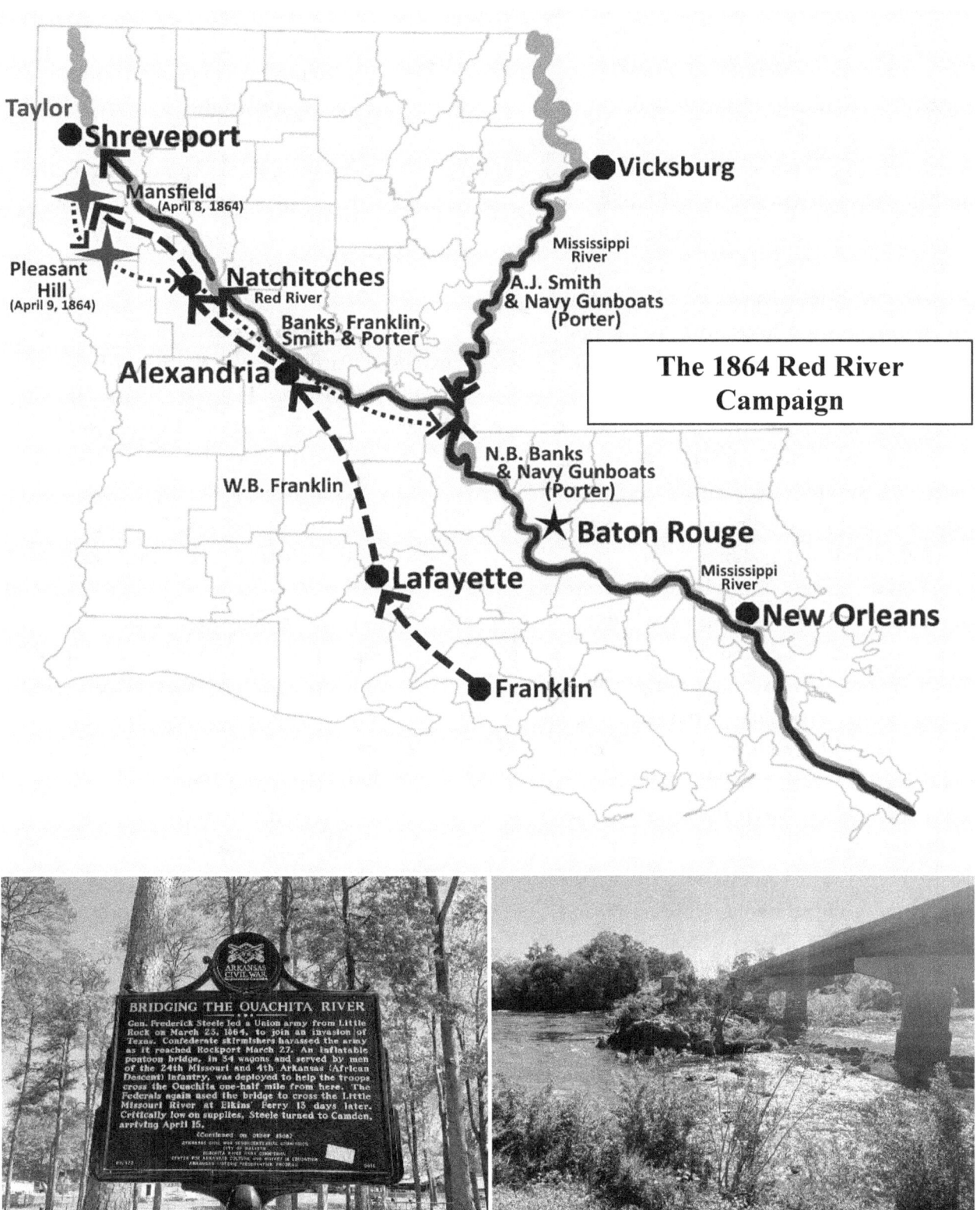

The 1864 Red River Campaign

Rockpoint Crossing of the Ouachita River

Shreveport. The southern column, consisting of 17,000 troops under Maj. Gen. William B. Franklin, commander of the Union XIX Corps, was to move up the Red River to Alexandria, Louisiana. An eastern drive from Vicksburg by 10,000 men under Maj. Gen. A.J. Smith was to rendezvous with Franklin at the same place. The U.S. Naval Flotilla, Mississippi Squadron, commanded by Rear-Admiral David Porter had at its mission to provide river gunboats and supplies by water. The northern pincer movement to the Red River Campaign was assigned to Maj. Gen. Frederick Steele, the commander of the Union Department of Arkansas. Steele was ordered to cooperate with Banks during the latter's northward movements. Steele firmly believed that the expedition was a suicide mission, and he dragged his feet in preparation. Steele claimed he was reluctant to move his forces due to the poor and wet condition of the roads and the lack of forage for the animals so early in the season; which would present a problem of traveling through an area that had been heavily foraged over the last three years, a problem not yet experienced in the Gulf region. Steele wrote to Maj. Gen. Hallack on March 12:

> "…General Banks, with 17,000 and 10,000 of Sherman's, will be at Alexandria on the 17th instant. This is more than an equal for everything Kirby Smith can bring against them. Smith will run. By holding the line of the Arkansas secure I can soon free this State of armed rebels. Sherman insists upon my moving upon Shreveport to co-operate with the above-mentioned force with all my effective force. I have prepared to do

so, against my own judgment and that of the best-informed people here. The roads are most if not quite impracticable; the country is destitute of provision on the route we should be obliged to take. I made proposition to General Banks to threaten the enemy's flank and rear with all my cavalry and to make a feint with infantry on the Washington road…"[4]

These issues were coupled with the extensive Partisan Ranger activities in Southern Arkansas. Halleck, still issuing orders until Grant was up to speed, replied:

> Washington,
> March 13, 1864—12.30 p. m.
> Major-General Steele, Little Rock, Ark.:
> --I advise that you proceed to co-operate in the movement of Banks and Sherman on Shreveport, unless General Grant orders differently. I send to him the substance of your telegram of the 12th.
> H. W. HALLECK,
> General- in-Chief.

Lt. Gen. Grant backed up Halleck and Sherman on the next day:

> Louisville, Ky., March 14, 1864-2 a.m.
> Maj. Gen. F. Steele, Little Rock:
> --Maj. Gen. W. T. Sherman is now commander of the Military Division of the Mississippi. You will therefore treat his request in regard to your co-operation with Maj. Gen. N. P. Banks accordingly.
> U. S. GRANT, Lieutenant-General.[5]

Although he had vehemently opposed the Red River operation, Grant issued these orders because the plans had been made before he had been placed in overall command and it was too late to alter them.[6] Realizing that the expedition was going to happen, Steele ordered Brig. Gen. Thayer to rendezvous with him at Arkadelphia on April 1 with his 3,600 troops from Fort Smith. Thayer began his movement on the 170-mile march on March 21, bringing most of the Frontier Division with him. The Union Indian Brigade, which was still in the Department of Kansas, was essentially on its own and was garrisoning Fort Gibson, grinding flour at Rhea's Mill, and making salt at a few of the salt works in the area.

Thayer's advancing force consisted of the 1st and 2nd Arkansas Infantry, 18th Iowa Infantry, 1st and 2nd Kansas Infantry (Colored), the 12th Kansas Infantry, the 2nd, 6th, and 14th Kansas Cavalry regiments. Thayer also brought along the 2nd Indiana Battery (4 guns), and the 1st Arkansas Battery (4 guns).[7] Most of these units were veterans of service in the Indian Territory. Low water on the Arkansas River coupled with the early part of the year, prevented Thayer's Frontier Division from moving with any real supply of provisions or forage for the journey. They took what little rations they could and hoped they could find supplies along the way.

General Steele finally marched out of Little Rock on March 23 with 6,800 troops including Brig. Gen. Frederick Salomon's 3rd Division of VII Corps, and two brigades of cavalry under Brig. Gen. Eugene Carr. He left Brig. Gen. Nathan Kimball in charge of the Federal arsenal. A veteran of the march recalled:

"…at 9:40 a.m., with rations packed, knapsacks slung, forty rounds of ammunition in the cartridge boxes, with all of the paraphernalia of a long and dangerous march, to the old, accustomed tune of Yankee Doodle, we marched out of Little Rock…"[8]

Steele's division arrived in Arkadelphia on March 29. Arkadelphia was an attractive village of white frame houses that had thus far remained largely untouched by the ravages of war. Steele had failed to account for food supplies and almost immediately placed the troops on half-rations. The Union soldiers did forage for supplies in town but surprised the inhabitants by offering to pay for almost everything they took. Unfortunately, many refused payment because they believed that if they were found with U.S. Dollars, they would be suspected of collaborating with the enemy. The people in the town did remark that the Union troops treated them better than their own troops.[9] After waiting three days for Thayer to arrive, Steele continued by way of the Old Military Road to Washington, Arkansas, where the Confederate Arkansas State Capitol had been relocated into the Hempstead County Courthouse after the fall of Little Rock. Washington was also the military headquarters for the Confederate District of Arkansas. It had a large army supply depot and hospitals for sick and wounded soldiers. After learning of a large Confederate supply of provisions at Camden, Steele changed directions toward that location. Unfortunately, his scouts informed him that Camden was heavily fortified by Confederate troops and would be difficult to capture. Steele again decided to change his destination back towards Washington. Steele was hoping that he could draw out the Confederates from their defenses in Camden and that he could swing behind them and capture the town and its provisions.[10]

Lieutenant General E. Kirby Smith, Confederate commander of the Trans Mississippi Department, anticipated federal movements into Arkansas, Louisiana, and Texas. He had his district commanders establish supply points at various locations throughout the department since so much of the area was now destitute of any usable resources. Smith replaced the aging Maj. Gen. Theophilus Holmes with Maj. Gen. Sterling Price as commander of the District of Arkansas. Price had two small infantry divisions under Brig. Gen. Thomas Churchill's Arkansas Division, and Brig. Gen. Mosby's Parsons Missouri Division, and two cavalry divisions under Brig. Gen. John Sappington's Missouri Division and Brig. Gen. James Fagan's Arkansas Division. Smith also notified Price that Maj. Gen. Maxey, commander of the District of Indian Territory, would be available to assist his Arkansas District. Maxey led Gano's Texas Brigade and Tandy Walker's Choctaw Brigade to join up with Price at Prairie D'Ane. The two Cherokee regiments, two Creek regiments, and the Seminole Battalion would remain in Indian Territory under Cooper's command, to counter any moves that Col. Phillips might make from Fort Gibson. E. Kirby Smith had also determined that the Union movements would be targeting Camden in Arkansas due to its prime location for an advance on Shreveport. He placed Maj. Gen. Richard Taylor in command of the

District of West Louisiana, whose mission would be to stop the joint Banks and A.J. Smith's Union invaders moving up the Red River from Alexandria towards Shreveport. Smith's goal was to defeat all three Union advances in piecemeal fashion.

In an effort to divert Confederate attention from his movement from Little Rock, Gen. Steele ordered Col. Powell Clayton, 5th Kansas Cavalry, commanding the Post at Pine Bluff, to continue to patrol the lower Arkansas River. On March 29 and 30, Powell conducted a feint by moving his forces southeast along the west side of Arkansas River. He moved his troops to the Mount Elba crossing of the Saline River, constructed a small pontoon bridge, crossed and headed in the direction of Camden as a demonstration to confuse the Confederate commanders as to the actual target. Clayton then advanced to the Marks' Mill area where he divided his cavalry forces. He sent one cavalry detachment up the Camden Road and another in the direction of Princeton. Clayton picked one hundred soldiers who were dispatched to destroy the Confederate pontoon bridge over the Sabine River at Long View. While doing so they surprised and captured 250 Confederates at the bridge site, 30 wagons of supplies, plus 300 horses and mules that were heading for Monticello, Arkansas. They burned the bridge and the wagons and brought the animals back to Pine Bluff. Clayton's force engaged in another small skirmish at the Mount Elba crossing, driving off the Confederates. His force was back in Pine Bluff by the evening of April 1. This was the only successful Union operation during the Camden Campaign.[11] As soon as Price was able to determine the intentions of Steele, he ordered Brig. Gen. Cabell's brigade to Tates Bluff near the mouth

The Camden Expedition
Map courtesy of Arkansas.gov / Additions by Author

Battle of Elkin's Ferry Marker and Ferry location on the Battlefield
Photos by Author

of the Little Missouri River. He ordered Confederate Brig. Gen. John Marmaduke to send one brigade to Tates Bluff as well, and another brigade to harass and slow Union Major General Steele's advancing division. The first contact was at Rockport, Arkansas on March 27 where the Union Army needed to build a floating pontoon bridge over the Ouachita River. Confederate skirmishers harassed the 24th Missouri Infantry and 4th Arkansas (African Descent) Infantry who were responsible for constructing the bridge. The bridge parts were carried by a 34 wagon train and, once Steele's army crossed, the bridge was taken up and moved with the troops.

Steele's force did not encounter any organized Confederate resistance until April 2 when Brig. Gen. Joe Shelby's brigade attacked the Union supply train guarded by Brig. Gen. Samuel Rice's brigade at Terre Noire Creek as it proceeded from Spoonville. The Union force consisted of the 50th Indiana, 9th Wisconsin, and 29th Iowa Infantry Regiments, and Capt. Martin Voegle's Wisconsin Battery, which was manned by cross-trained infantrymen from Company F of the 9th Wisconsin. Shelby then attacked Steele's division again at Okolona, Arkansas. After hours of intense fighting during a thunderstorm, Steele was able to drive off Shelby, who fell back to the town of Antoine. A post-war memoir by Shelby's adjutant stated that the only reason they retreated was that a Union shell hit a beehive which attacked the Confederate troops. On April 1-2, 1864, Confederate Generals Shelby, Marmaduke, and Cabell set up a defense on the south side of the Elkin's Ferry crossing of the Little Missouri River. At this point they waited for Steele's division to arrive. They

knew the Federals would need to use Elkin's Ferry because all of the bridges over the Little Missouri River had been destroyed. They did not have long to wait, and the Union force arrived at Elkin's Ferry late on April 2. Steele tasked Brig. Gen. Salomon with the mission to take and hold the ford. He assigned Col. William McLean and the 43rd Indiana to the objective, supported by elements of the 36th Iowa Infantry and the 2nd Missouri Light Artillery. After a forced march they arrived at the Little Missouri River after dark. A squadron of the 1st Iowa Cavalry preceded the infantry and crossed first, establishing a picket line for the night. The Federals were surprised when no attack came at dawn on the 3rd, a Sunday. It was so quiet that the infantry regiments sent out foraging parties to look for whatever meat could be found. Marmaduke only skirmished with the Union troops on April 3. Col. McLean of the 43rd Indiana Infantry reported:

"…Early on the morning of the third instant, I ordered Major Norris, of the 43rd Indiana, to proceed with four companies of that regiment to the front, to reconnoiter the position of the enemy, deploy the men as skirmishers, and support the cavalry pickets. He soon succeeded in discovering the position of the advance pickets and skirmishers of the enemy, drove them back for some distance… On the same evening… I ordered Lieutenant Colonel Drake, 36th Iowa, to proceed with three companies from that regiment, and three companies from the 43rd Indiana, to a position on the main road leading from the ford

Battle of Prairie D'Ane and small portion of the remaining Battlefield Prairie
Photos by Author

immediately in our front, to deploy his men on the right and left of the road, to watch the movements of the enemy, and to resist their approach as long as was prudent…"[12]

On the morning of April 4 Marmaduke attacked the Union bridgehead with 1,600 cavalrymen and succeeded in driving the Federals back to the Little Missouri.[13] Col. McLean continued:

"…At six o'clock on the morning of the 4th the enemy approached in force, and commenced an attack on the advance companies of Lieutenant Colonel Drake, who resisted them gallantly for nearly two hours, being well supported by the artillery of Lieutenant Peetz… The enemy… charged with a yell upon our left, for the purpose of flanking us and capturing our battery. Their approach from the cover of the timber was met gallantly by two or three well directed volleys from the 36th Iowa. Immediately after the charge and

repulse of the enemy, the reinforcements sent for by me arrived, consisting of the 29th Iowa Infantry and the 9th Wisconsin Infantry, of Brigadier General Rice's brigade. But before they were put in position by him the enemy withdrew…"[14]

Salomon's counterattack with his First Brigade of infantry troops from Iowa and Wisconsin, pushed and drove off Marmaduke. The Confederates fell back 16 miles to the breastwork defenses that had been constructed by slave labor at Prairie D'Ane (*Fr.* Donkey Meadow).[15]

Sterling Price's Confederate Army held the strong defensive log and earth breastwork on the western and northern edge of Prairie D'Ane. In 1864, Prairie De Ann was a circular body of open prairie land surrounded by forest, 25 to 30 miles square. On the prairie was a crossroads that intersected with the Old Military Road and with roads between Washington and Camden. Price's main objective at this point was to take his time advancing from Elkin's Ferry as he re-installed the pontoon bridge over the Little

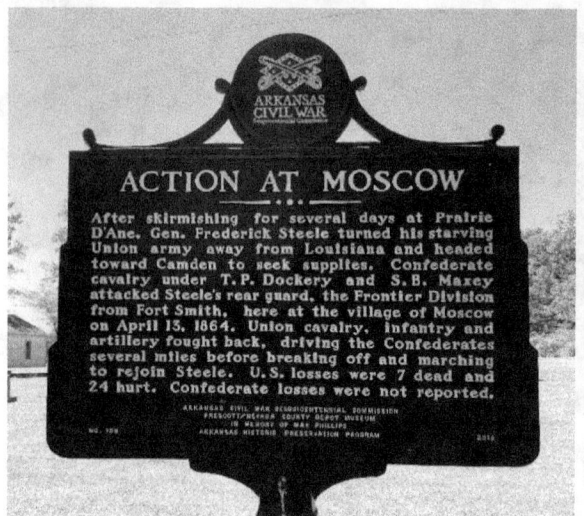 

Battle of Moscow Marker and Historic Moscow Church and Cemetery
Photos by Author

Missouri River to cross the wagons and protect the state capital at Washington. Steele waited for Gen. Thayer to catch up. Thayer and the Frontier Division, slowed by the steep Ouachita Mountains, poor roads, and wet weather, did not arrive until April 9. With their arrival, Steele's Federal Army had now grown to 13,000 men. Pvt. Sperry of the 33rd Iowa described the arrival of the Frontier Division:

> "…While we lay here, the long-looked-for and much-talked-of, reinforcement of "Thayer's command" arrived from Fort Smith. A non-descript style of reinforcement it was too, numbering almost every kind of soldiers, including Indians, and accompanied by a multitudinous vehicles of all descriptions, which had been picked up along the way…"[16]

But with the higher number came the issue of diminishing rations for the troops, since Thayer's division was just as short of provisions as was Steele's. Steele was not pleased about Thayer's tardiness:

> "…Yesterday a messenger sent by Major Green brought us the first intelligence of Thayer. Instead of taking the Caddo Gap road, as agreed upon, he went to Hot Springs, having turned off his road above Mount Ida. It is expected that he will join us tomorrow. He is entirely out of rations, and our delay has caused a consumption of the supplies which might have lasted us to Shreveport…"[17]

On April 10 they began marching south from Elkin's Ferry. The Federal army marched from their camp at 6 a.m. with Carr's cavalry division in the lead, followed by Salomon's 3rd Division, the supply and pontoon train, and the Frontier Division acting as the rear guard. Steele's combined force was approaching the Confederate defenses at Prairie D'Ane when they encountered Price's advanced line of battle and his first line of breastworks defended by Dockery's and Shelby's brigades at Gum Grove. The Federals quickly deployed into a line of battle and attacked with artillery, cavalry and infantry skirmishers, and in a rare night attack they drove the Confederate line back about a mile before being checked by Price at the second line of breastworks. At this point the two sides began exchanging artillery fire. The skirmishing ended at about midnight. The skirmishing then resumed in the morning and continued throughout April 11 until one final Union advance was ordered. The cavalry division advanced forward, and Salomon's infantry division deployed into a stunning line of battle that many witnesses recorded. The Prairie D'Ane was one of the few places that was open enough for anyone to see an entire infantry division deployed from end to end. By the time the 3rd Division reached the second line of breastworks, they found that Price's Confederate forces had abandoned Prairie D'Ane to the south. Pvt. A.J. Sperry of the 33rd Iowa Infantry described what they discovered:

> "…Over a mile of very passable breast-works, alternated with places for cannon, of such a range that they could have literally mowed us down in a direct assault, were now … in our possession, with hardly any loss; and we were much pleased at this result of a flanking movement…"[18]

Price still believed that Steele was intending on capturing Washington and had his forces deploy in a third prepared line of breastworks across the road leading to that town. Steele, realizing that Price expected him to move toward Washington, diverted his command down the road leading to Camden and its supply point of provisions. In a diversionary move, Steele ordered Thayer's Frontier Division to make a feint toward Washington, thereby drawing the enemy into a fight south of the prairie. Steele withdrew from Prairie D'Ane and began marching on Camden on April 12, 1864.[19]

The Federal rear guard was the Frontier Division, consisting of about 5,000 men commanded by Brig. Gen. Thayer. The Frontier Division was camped near the village of Moscow, on the edge of Prairie D'Ane, on April 13, 1864. Steele's army was making slow progress on its march to Camden, so Thayer's division was still in camp at noon as the army's rear-most supply and pontoon wagons began to move off. When it became apparent that Steele was not advancing on Washington, Price went on the offensive. He dispatched Richard Gano's Texas Brigade, Tandy Walker's Choctaw Brigade, under Maj. Gen. Maxey, and Dockery's Brigade who recrossed the prairie and assailed Thayer's troops as they were leaving Prairie De Ann on the afternoon of the 13th. Price sent most of the cavalry from Fagan's division as well as the parts of the Indian Territory division on another road paralleling Steele's route to get ahead of the Union forces and contest their advance on Camden. Using both sides of the road the Confederates attacked the rear guard at about 1 p.m. Thayer realized that this was a significant attack and rushed to build a line of defense. Wiley Britton recalled:

> "…When this charging force came up within range, the Federal infantry poured several volleys of musketry into their ranks, with fatal precision, and the twelve guns in the batteries of Thayer's division also swept their ranks and the field in front with a storm of [fire], causing them to retire

hastily with a heavy loss of killed and wounded…"[20]

The Confederate's attack was fierce and the charge finally broke the initial Union line, capturing a section of artillery. Troops had almost reached the supply and pontoon wagons when a counterattack led by the 18th Iowa Infantry and a collection of other broken regiments drove off the Confederates and recaptured the lost artillery pieces. The Union cavalry ended up pushing the Confederates for several miles back over Prairie D'Ane. Thayer's rear-guard remained in place until the wagons had cleared the Terre Rouge Creek bottoms.[21]

Steele's advance force pushed aside a small defense force and marched into Camden on April 15. After the experience of the rough march between the Prairie D'Ane and Camden, Steele was devastated to realize that the Confederates had emptied the food warehouses they were desperately depending on. This forced the hungry Union soldiers to begin demanding and confiscating food from the citizens. Steele realized this would create undue hardships on the residents, so he put some of his forces into the fortifications around Camden and moved the rest of the army to a camp outside of town. Forage around Camden was scarce since a "destroying mania had seized the rebels … [and] by night the whole heavens were illuminated by reflections of the devouring flames." The civilian inhabitants of Camden were starving; therefore, no relief could be expected from that quarter. To the contrary, the Federal commissary was forced to provide some food for the destitute population. The pontoon bridge was deployed over the Ouachita River to connect Camden with the road to Pine Bluff.[22]

By this time rumors began to swirl that Maj. Gen. Banks had been defeated and that he was withdrawing from the Red River. The rumors turned out to be true because when Steele received information from one of his spies on April 17, he had already decided that the campaign was over, and he would return to Little Rock. In fact, Banks had initially had some success in moving up the Red River from Alexandria by capturing Grand Ecore and Natchitoches. Confederate General Richard Taylor, son of former U.S. President Zachery Taylor, decided to make his final stand at Sabine Cross Roads, near the small town of Mansfield, Louisiana. Although the Confederate Army facing Banks was heavily outnumbered, the Federals had become stretched out along the long narrow road leading to the crossroads. This enabled Taylor to attack a much smaller force to his front. Taylor deployed his 10,500 troops in a line of battle. Banks could bring only 6,400 soldiers into line, and they formed on a small ridge behind a rail fence. The Confederates attacked and attacked until they overwhelmed the Union forces. Although Banks was able to get another 2,000 Federals into a new line of battle about a mile south of the first, these, too, were routed by Taylor's Confederates. A third Federal line of 6,500 was formed at Chapman's Bayou, and they held the Confederate advance in check. During the night the defeated Federals fell back to Pleasant Hill, approximately 16 miles southeast of Sabine Cross Roads. On April 9 the battle of Pleasant Hill took place with both sides attacking and counterattacking throughout most of the day with no clear victor. By this time Banks had lost all of his nerve and ordered his forces back to Natchitoches where they met with the U.S. Navy squadron and began the long retreat to Alexandria and New Orleans. Banks placed much of the blame on Steele for not diverting more Confederates to counter his moves in Arkansas. And now Confederate General E.K. Smith had more troops available to move from Louisiana into Arkansas and did so, sending three infantry divisions to assist Price in defeating Steele.[23]

A Confederate riverboat loaded with corn was captured on the Ouachita River below Camden, providing some relief to the hungry Federals. In the never ending quest to obtain provisions for his army, on April 17 Steele ordered a foraging party of 198 wagons west on the Upper Washington Road to collect corn or any other foodstuffs that they could find. Union foragers ranged north and south of the road, taking clothing, jewelry, silverware, pots, pans, and household items, as well as food and forage. So thorough were the foragers that their 198 wagons were filled to overflowing. The wagon train was accompanied by the 1st Kansas Infantry (Colored) with 500 infantrymen, 75 troopers of the 2nd Kansas Cavalry, 50 from the 6th Kansas Cavalry, and 70 from the 14th Kansas Cavalry, and one section of the 2nd Indiana Battery, for a total of 695 men and two pieces of artillery, all under the command of Col. James Williams of the 1st Kansas Infantry (Colored). Williams reported:

"…I proceeded westerly on the Washington road a distance of 18 miles, where I halted the train and dispatched parts of it in different directions to load, 100 wagons, with a large part of the command under

Major Ward, being sent 6 miles beyond the camp. These wagons returned to camp at midnight, nearly all loaded with corn. At sunrise on the 18th, the command started on the return, loading the balance of the train as it proceeded. There being but few wagon loads of corn to be found at any one place, I was obliged to detach portions of the command in different directions to load the wagons, until nearly my whole available force was so employed…"[24]

Union couriers informed Gen. Steele of the train's situation and location and fearing an attack on it, he immediately dispatched reinforcements. These troops included 375 men from the 18th Iowa Infantry, 25 from the 6th Kansas Cavalry, 45 men of the 2nd Kansas Cavalry with two mountain howitzers, and 20 troops from the 14th Kansas Cavalry. The reinforcement was commanded by Capt. William M. Duncan of the 18th Iowa. This raised the wagon train's escort to 875 infantry, 285 cavalrymen, and a four gun battery of artillery.[25] The reinforcements met with the Union supply train at a location known as Cross-Roads, approximately four miles east of the previous night's camping ground on White Oak Creek. After another mile advancing back towards Camden, the Federals encountered a Confederate picket line. The Union troops quickly drove them back at least one mile.

On the night of April 17, Confederate scouts had discovered the wagon train encamped on White Oak Creek, 16 miles west of Camden and the Confederates began planning an attack. The route selected by the forage train was under continuous surveillance of Confederate pickets from Col. Colton Greene's brigade of Marmaduke's Division. Marmaduke believed that his small cavalry force was substantially outnumbered by the Union escort, so he wrote to Price recommending a strong concentration of troops to cut the Federals off from Camden. At sunrise on the 18th, Marmaduke marched his men northward to the Upper Washington-Camden Road, near Lee's Plantation, a locally known landmark, and Poison Springs. Marmaduke ordered his mounted troops, except Greene's brigade, to dismount. Col. Crawford's brigade was placed to the right, while Cabell's men both blocked the road and occupied the left. Marmaduke's four artillery pieces were placed on a knoll a short distance south of the road. While placing them across the road at Poison Spring, the reinforcements sent by Price arrived: Maxey's Indian Territory division consisting of Gano's Texas Brigade and Walker's Choctaw Brigade. One observer noted upon their arrival: "… [they were] mounted on ponies, dressed in all sorts of clothing, including buckskin with feathers in hats…"[26] Finally, Wood's 14th Missouri Battalion also arrived. The Confederate plan of attack at Poison Spring called for Maxey to deliver the initial assault on the exposed Union right flank. Once the Federals changed front to meet Maxey, Marmaduke would hit straight ahead up the road, roll up the Union left flank, and trap the remaining enemy troops in a crossfire. The balance of the Confederates would block the road to Camden.[27] Marmaduke recorded:

> "…At this juncture Brigadier-General Maxey arrived with his division (a Texas and Indian brigade, some 1,200 or 1,500 men, with a Texas battery of

four small pieces of artillery). As General Maxey was my senior in rank I reported to him for orders. He replied that as I had put on foot the expedition and knew the position of affairs I would make the disposition of the troops and the fight. I then suggested that his whole force be dismounted and placed on the left, his division forming a line nearly at a right angle with my line, which was perpendicular to and across the main road to Camden. Maxey's division was put in position accordingly, his Texas brigade on his right with his battery of artillery…"[28]

Marmaduke's scouts had heavily overestimated the strength of the Union escort force. His total force exceeded 3,100, giving him a 3:1 ratio over the Federals. The only advantage that the Union soldiers had over their counterparts was their firearms. The Federals had fairly modern Springfield or Enfield rifles, which were issued to infantry soldiers that had an effective range exceeding 300 yards. The Confederates, on the other hand, were mostly cavalry and were armed with a mixed variety of short-range carbines, shotguns, and pistols, all of which had limited effective range.

The wagon train could no longer move forward. Col. Williams ordered the wagons parked two and three abreast, as close together as possible, as was standard procedure in wooded areas, unlike the circle used on the prairie. This would have made the line of wagons approximately 2,100 feet long, just less than a half mile. This was a lot of line to manage with only a little over a thousand

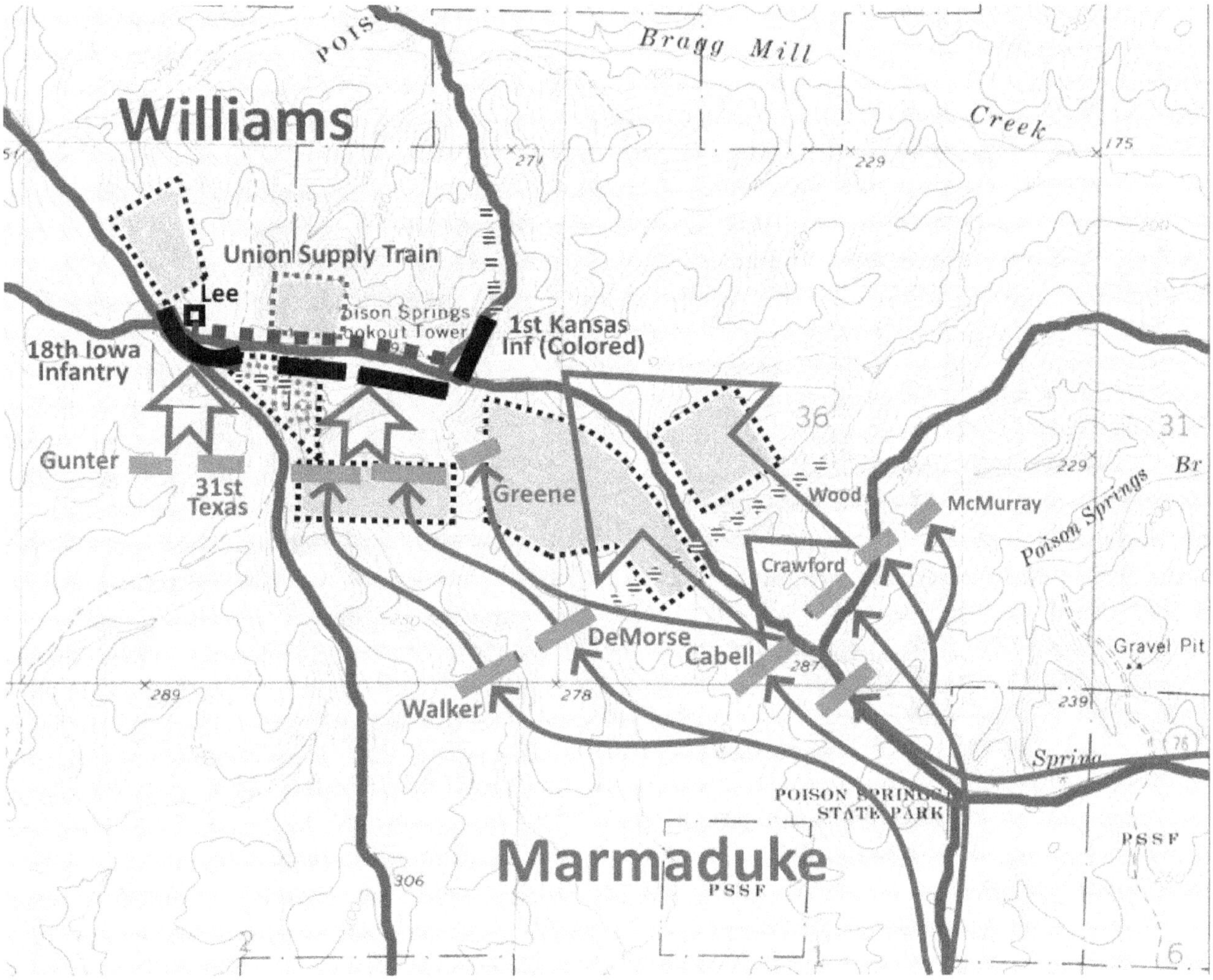

The Battle of Poison Springs, Arkansas
April 18, 1864
Map courtesy of USGS / Additions by Author

Battle of Poison Springs Marker and portion of the Battlefield Park
Photos by Author

soldiers, especially when being attacked from multiple sides. Maj. Richard Ward, commanding the troops of the 1st Kansas Infantry (Colored), ordered his troops to the front forming a semi-circle line with Companies C & I on the north side of the road facing north and east, and Companies D & F on the south side of the road and facing south and east. Ward placed the cavalry detachments on his flanks. Williams then immediately sent word for Capt. Duncan to disperse his troops to protect the rear.

Maxey's Indians and Texans began moving on his right flank and Williams sent orders for Duncan to send help to the front. The courier returned and reported that the Confederates were advancing against Duncan, and he could spare no men. Duncan later wrote:

> "…We first formed a line in the road with the howitzer on our left. Soon after a heavy column of infantry was discovered moving on our right flank. We then changed front and formed in the orchard on the south side of the road, throwing out two companies as skirmishers—one in our front and one on our right. We here were attacked in front and on the right flank. We held this position until the Second Indiana Battery came back in retreat, when I was ordered to form on the north side of the road to protect the battery. Here we held our position, under a heavy fire from the front and left flank, until a portion of the battery had passed into the woods in our rear, when we were ordered to fall back through an open field to the woods…"[29]

The Confederates opened with their artillery, and advanced in front and flank. The left side of Marmaduke's line advanced double-quick through fields and thickets, opening fire as they neared the Union line. Under pressure along his entire line, Williams ordered Duncan to redeploy and cover the retreat of the 1st Kansas. The Federals tried several times to counterattack or make a stand but were continually driven behind the immobile wagon train then forced into the woods north of the road. After a short and hotly contested engagement at close range, the black troops broke ranks, retreating in chaos. The Confederates cut them down right and left in a ravaging fire as they tried to escape. A final charge by Col. Tandy Walker's Choctaw Indian Brigade across the field and wooded ravine, with their screaming and war whoops in usual Indian fashion, caused most of the Union defenders to break to the north in a complete rout. In doing so they abandoned their artillery, wagons and spoils in an effort to escape the howling Indians and westerners. Col. Walker recalled:

> "…The enemy formed next at his wagon train, drawn up on the road which ran along the brow of a wooded hill, but was pressed so closely by this brigade that he soon fled across the road and in a direction up the road to the left, when the train fell into our hands… I feared here that the train and its contents would prove a temptation too strong for these hungry, half-clothed Choctaws, but had no trouble in pressing them forward, for there was that in front and to the left more inviting to them than food or clothing—the blood of

their despised enemy. They had met and routed the forces of General Thayer, the ravagers of their country, the despoilers of their homes, and the murderers of their women and children…"[30]

First Sergeant Charles Pidcocke of Company B, 30th Texas Cavalry, described his observations of the Confederate Choctaws at Poison Springs:

"…Niggers were taken but afterwards hung by the Choctaws. The Indians were turned loose just about the time our Brigade had them whipped. Gentlemen of the colored race suffered. The Choctaws killed every nigger & stripped everything both black and white. Some of these Choctaws can carry bigger loads than their ponies. One man says he saw a Choctaw with his arm broke carrying off the effects of seven dead niggers, guns and ammo… These niggers would enter a house, steal everything cuff & abuse the children, cursing them for d____d secesh rebels end them with stripping them of every vestige of clothing…"[31]

The Confederates captured four complete artillery batteries with guns, caissons, and limbers. Marmaduke's soldiers also captured 175 wagons containing five thousand bushels of corn, bacon, bed quilts, women's and children's clothing, hogs, geese, and other personal property. One hundred and forty five of the wagons were led by prized six-mule teams, a loss the Federals could not easily replace. The Union reported 301 soldiers killed or missing, and Price reported 114 men killed, wounded, or missing. Even though the battle had ended, the killing did not as the Confederates committed atrocity after atrocity against the black soldiers of the 1st Kansas. Further, some of the Choctaw Indians of Walker's Brigade began to scalp the dead white and black soldiers, much in the same way as happened at Pea Ridge two years before. The war had hardened the hearts on both sides, and nothing about the scalpings was said by the Confederate command. With the loss of the wagons, provisions, and mule-teams, Steele and the Union Army were in much deeper trouble than before.[32]

The Confederate command had been successful so far hitting the Union Army at its most vulnerable point: its supply trains. Much to the relief of Steele and his army, a supply train arrived from Pine Bluff arrived on April 20 with 10 days of half-rations. Their arrival somewhat softened the blow of losing the entire forage train at Poison Springs. The Union command, suspecting that General E.K. Smith, freed from the responsibility of pushing Gen. Banks back down the Red River, would send troops to assist Price in driving Steele away from Camden. In fact, Smith had already ordered Taylor to send the three divisions under Generals Churchill, Parsons, and Walker from Shreveport towards Camden on three separate approach avenues. Walker's division also possessed a pontoon bridge intended for the Ouachita River. Smith and Price also knew that they would not be able to push Steele's army out of Camden by a direct assault since the Federals had occupied the extensive fortifications that the Confederate Army had built. These defensive works consisted of nine fortifications surrounding the town that had been constructed in 1863-64 by Confederate soldiers and slaves under the command of Camden lawyer, Col. Alexander Hawthorn. Smith and Price decided to sever Steele's supply and communications lines from Camden to Little Rock and Pine Bluff.[33] Although he was still mostly undecided as to continue his movement southward, Steele began in earnest to prepare for the return trip to Little Rock. His army was secure behind the fortifications, but food rations would still quickly run low. Steele wrote to his immediate superior on April 22:

> "…Although I believe we can beat Price, I do not expect to meet successfully the whole force which Kirby Smith could send against me, if Banks should let him go…"[34]

Steele decided to send the supply train back to Pine Bluff to procure another load of rations. The train consisted of 240 army wagons and some private vehicles. It was protected by three infantry regiments, the 77th Ohio, 43rd Indiana, and the 36th Iowa, 240 troopers of the 2nd Indiana Cavalry and 7th Missouri Cavalry, and Artillery, under the command of Lt. Col. Francis Drake of the 36th Iowa. Drake was an interesting individual. He had a reputation as an Indian fighter. In fact, in 1852, at the age of 19, he led a wagon train from Blakesburg, Iowa, to Sacramento, California, and, while crossing the Nebraska prairie, his train was attacked by an estimated 300 Pawnee warriors. Drake organized and led a spirited defense of his train and, although greatly outnumbered, he and seven companions beat the attackers off, reportedly after Drake personally killed their leader with his knife.[35] Drake's current force totaled approximately 1,600 men. Not included in this total were a wide variety of civilians, including camp followers, sutlers, cotton speculators, and about 300 former enslaved Africans looking for protection. The wagon train, two sections of Battery E, 2nd Missouri Light departed the east side of the Ouachita River on the morning of April 23, traveled 18 miles and camped. On April 24 the wagon train was having difficulty with the wet, almost impassible roads. In addition, the civilian teamsters tagging along behind were growing argumentative and hard to handle due to the slow pace. In response, Drake organized a pioneer corps of 75 recruits from the slave followers to improve the roads by corduroying them with logs. The train camped at the edge of the Moro Creek Bottoms after traveling another 17 miles. The pioneer corps continued working throughout the night of April 24-25. When Major Wesley Norris tried to warn Drake of movement in the woods to his front during the night, Drake laughed off his concerns and told Norris—a combat veteran of the Mexican War—that he "got scared too easily." At dawn the entire train began to move on the Pine Bluff–Camden Road towards the Mount Elba crossing of the Sabine River.[36]

Fagan caught up with the Union column on April 25. He decided to use the same basic tactic as was used at Poison Spring. The Confederates would block the road, forcing the Union advance to stop. As the Confederate's approached the road on which the Federal train was traveling, Fagan ordered Shelby's and Crawford's mounted brigades to the northeast to place themselves between the Union troops and the Saline River.[37] The Confederate forces then blocked

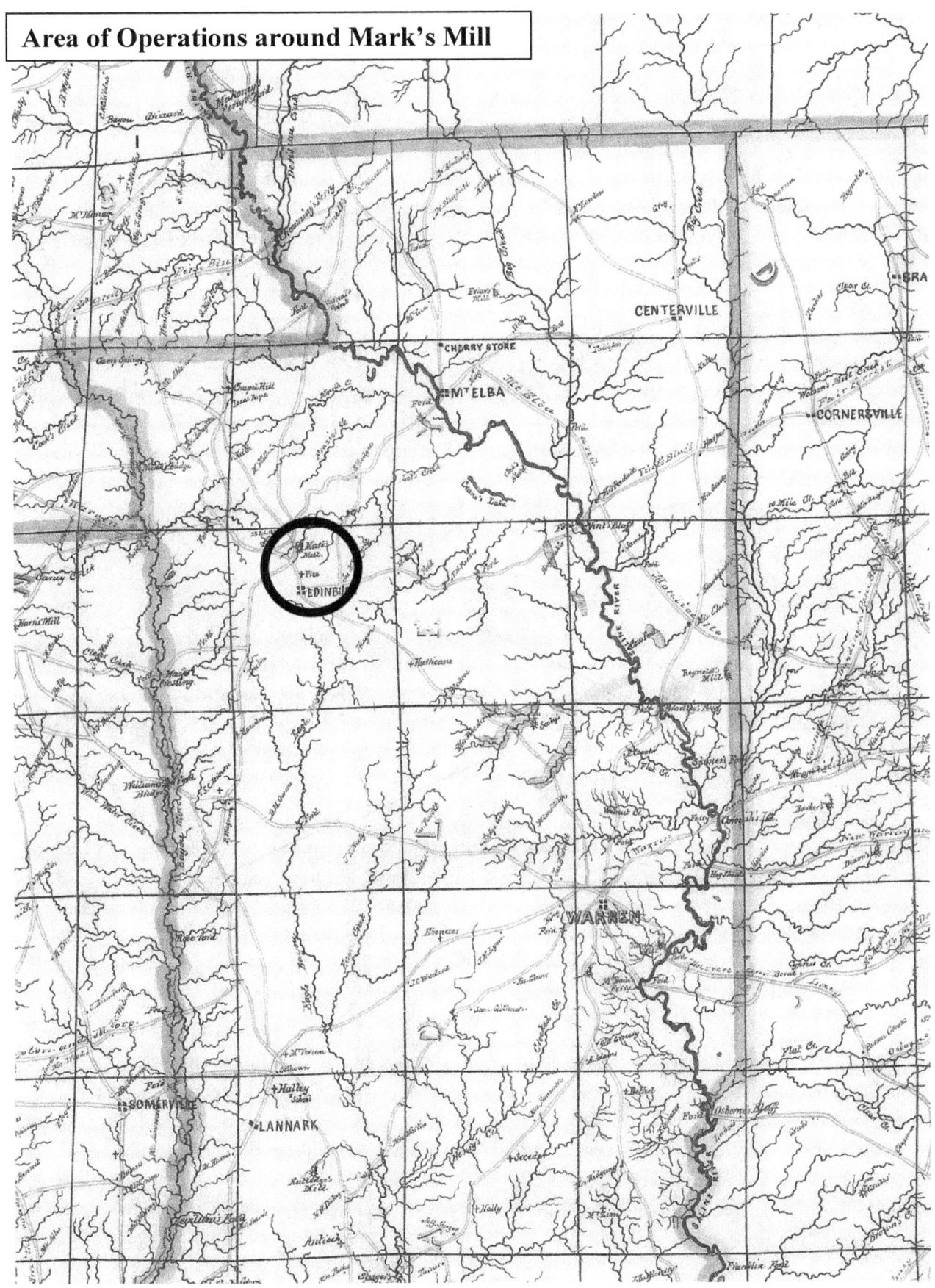

Area of Operations around Mark's Mill

the road leading to a small clearing at Marks' Mill, dismounted and, in piecemeal fashion, they attacked the flanks and rear of the Union wagon train as it moved into the clearing around the crossroads. The Union wagons needed to make a hard left turn at the crossroads, which put them at a severe disadvantage. Sending one brigade into combat before the second was positioned, Cabell's troops easily captured the wagons but found themselves in an intense counterfire. Dockery's brigade was supposed to attack at the same time as Cabell's, but his scouts found a supply of animal forage at a local farm and the commander ordered his troops to stop and feed their horses. Martha Marks, John Marks's daughter, was home at the time of the battle. She wrote:

> "…I saw them shoot down the driver of the first wagon in front of our house. Our home was a temporary hospital and I can see now the wounded and the dying lying on our porches and in the house…"[38]

The 43rd Indiana and one section of the 2nd Missouri Light Artillery were assigned to the advance, and Maj. McCauley and his detachments of the 2nd Iowa Cavalry and the 7th Missouri Cavalry were sent forward on the roads toward Marks' Mill, Warren, Mount Elba and Moro Bay.

On April 23 the Confederates initiated their plans to interrupt the Union supply line. Brig. Gen. James Fagan was given command of four cavalry brigades, Cabell, Dockery, Crawford, and Shelby's, a total of about 4,000 cavalrymen. Fagan's command rendezvoused with Shelby's brigade guarding the pontoon bridge at El Dorado Landing. At this time,

Shelby informed Fagan that his scouts had learned that a large, heavily guarded Union supply train had left Camden heading for Pine Bluff. The Confederate raiders departed El Dorado Landing on the morning of April 24. After a forced march of 52 miles, they crossed the Ouachita River at Moro Bay and moved up the road towards Mount Elba. Fagan divided his four brigade command into two divisions under Shelby and Cabell and began a difficult 45-mile march paralleling the Camden-Pine Bluff Road in order to intercept the Union train before it reached the Mount Elba crossing of the Saline River. They stopped at midnight approximately 8 miles south of Marks' Mill, a local landmark at the junction of the Camden-Mount Elba-Pine Bluff roads. This is the location Fagan decided to spring his trap. John A. Marks and his family lived in this area at the junction of the Moro Bay and Camden-Pine Bluff Roads. They had arrived in the 1830s and developed large landholdings, a cotton gin, grist and flour mills, a sawmill, blacksmith shop, and a brick kiln. To run these various operations, Marks had many enslaved Africans.

Although under attack from three directions and outnumbered, the 43rd Indiana and the 36th Iowa regiments took advantage of Cabell's error and slowed the Confederate attack until Union reinforcements helped stop his advance. Cabell's division had lost its momentum and found it difficult to regain it. The Federals drove off the first enemy assault before being hit on their right flank by a second wave of Confederates. After the Confederates pinned the Federals down near a few log cabins in the clearing, Shelby's cavalry slammed into the Union left flank. As Shelby approached the battlefield from the north, the sounds of artillery and small-arms fire were heard, and the Confederates

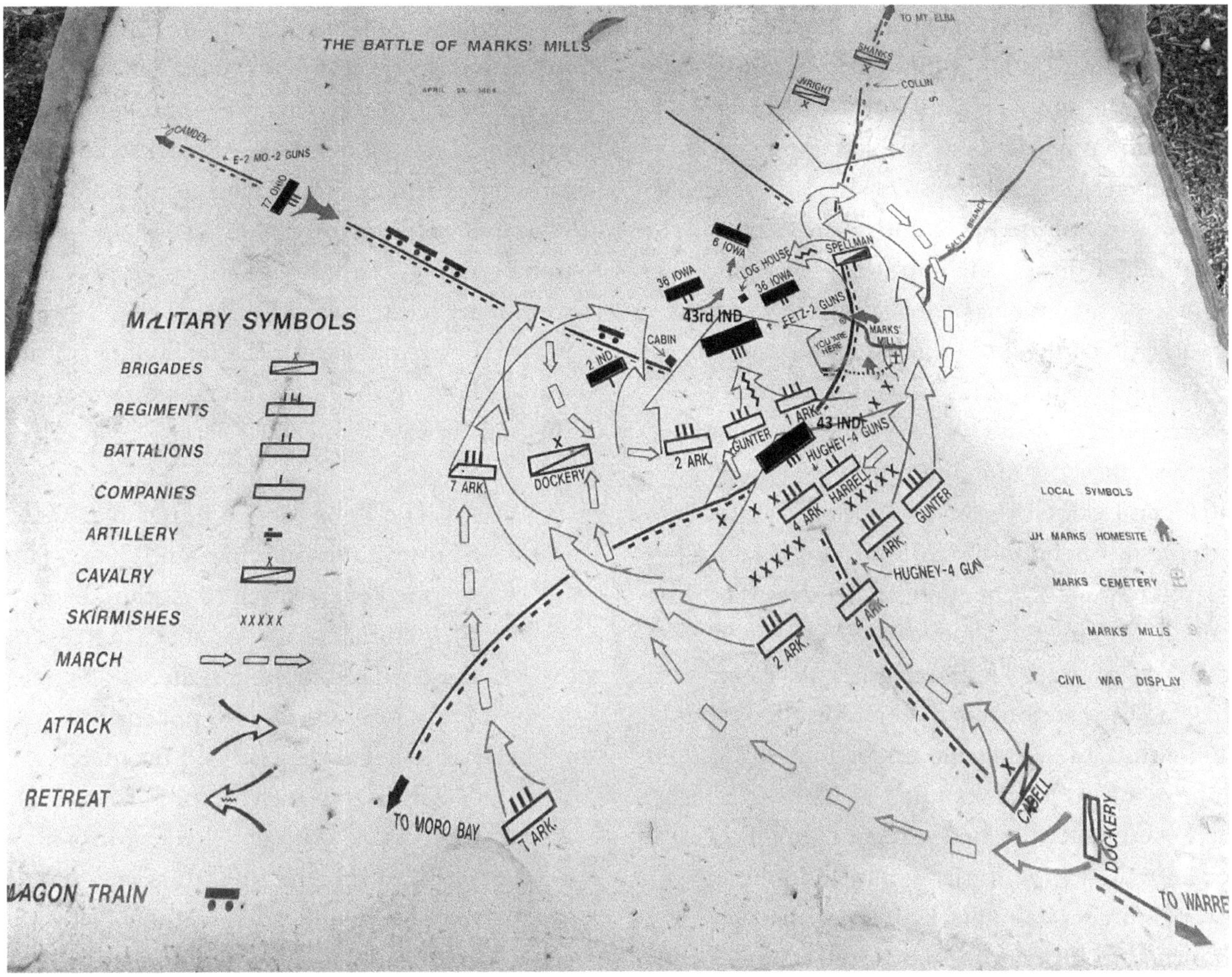

Battle of Marks' Mill Map Marker showing the movements of troops involved
Photo by Author

Marks' Mill Battle Marker at the Salty Branch where Anita Knowles in 1937 stated that "So many horses and soldiers were killed that Salty Branch ran red with blood..."
Photos by Author

encountered routed stragglers and wagons fleeing towards the Mount Elba crossing. Shelby's division deployed into battle lines and swept down on the Federals' left and rear. Nonetheless, the Union units slowly began to lose control of the battle due to the over-whelming Confederate numbers. Over the course of five hours, Cabell's brigade was able to drive back the 43rd Indiana, who were covering the southern flanks of the Union force. The 36th Iowa was covering the advance of the Union column, but they were savagely attacked by Shelby's division. The 77th Ohio regiment was far in the rear and had to quick-march to support the train's defenders. The 1st and 2nd Arkansas regiments (CSA) stormed the guns of the 2nd Missouri Light Artillery, forcing the cannoneers to abandon their posts and guns and take shelter under a nearby log cabin, only to be surrounded by Confederates and eventually surrender. At some point, Lt. Col. Drake was seriously wounded in the hip by a Confederate minie' ball. Drake had returned to his command post to continue managing the battle when he collapsed due to loss of blood. Despite valiant efforts by the 77th Ohio to prevent the ensuing encircle-ment, the Federals found themselves surrounded in the clearing and fighting for their lives. According to Sgt. John Moss of the 43rd Indiana's Co. G, the regiment did not give up *en masse*; rather, continual charges by the Southerners resulted in the capture of small numbers of men each time, until only about 50 of the 43rd remained who had not been killed, wounded, run off or captured.[39]

The Confederate victory at Marks' Mills was nearly absolute. The Confederates reported only 293 casualties (41 killed, 108 wounded, 144 missing). Union losses were staggering. Of the roughly 1,300 Federals engaged in the main battle about 100 were killed and nearly all the rest were captured. Drake later reported that the Confederates captured "a large number" of blacks and pro-Union Arkansans accompanying the column who they subsequently "inhumanly butchered."[40] The 1,300 or so Union prisoners were marched to Camp Ford in Tyler, Texas, where they remained for the remainder of the war. Following the battle, a Federal soldier in the 36th Iowa commented that:

> "…The Rebs robbed nearly every man of us even to our chaplain. They stripped every stitch of clothes, even their shirts, boots and socks, and left the dead unburied and the woods on fire. Clothing was also pulled from the wounded as they begged for mercy. No respect was given for persons rank or age. Old Captain Charles Moss of the 43rd Indiana Infantry was marched bareheaded with his bald head and white locks and beard in the burning sun…"[41]

Fagan found that he had captured a large number of ambulances, hundreds of small arms, 150 contraband slaves, four James rifled guns, and more than 300 wagons, many of which had been partially burnt. Unfortunately, the lack of coordination among Cabell's, Dockery's, and Shelby's divisions severely reduced the effectiveness of the otherwise successful Confederate attack.[42] Many items that the Union troops had looted from farms along the road were found in the captured wagons. The Confederate soldiers also delighted in stripping the sutler wagons of commodities and delicacies they had not seen in many months. Fagan's troops also captured $175,000 in cash that was being transported for the paymaster. After the surrender, a Confederate surgeon examined Lt. Col. Drake and determined his wound to be mortal. He and the remaining Union wounded were taken back to Camden under a flag of truce. Drake surprised everyone by waking up and was able to write his report of the battle by the next day. As for the dead, they were buried on the battlefield. Local tradition states that most of the Union dead were buried in the Marks' family orchard and that the Confederate dead were taken from the field and buried in the local cemeteries or sent home.[43]

An interesting side note is that 520 men of the 1st Iowa Cavalry, commanded by Lt. Col. Joseph Caldwell, were marching dismounted from Camden to Pine Bluff for their veteran's 30-day furlough back in Iowa. The 1st Iowa's surgeon, Charles McLeod, recalled in 1890:

"…The march was being rapidly made for men unaccustomed to marching as infantry, when about noon the booming of artillery was heard in our front. Believing that the train had been attacked, the march was quickened for the purpose of joining in the fight. A halt for a few moments was ordered at a bridge over Moro creek or river—a small, deep, miry stream. thirty-eight miles distant northeast from Camden, near the little town of Moro, in Calhoun county, and five or six miles distant from the place of the engagement with Lieutenant Colonel Drake's command. The halt had scarcely been made, when a most demoralized

crowd of cotton speculators, sutlers, refugees, teamsters, etc., mounted on mules and horses dashed past."[44]

They then heard the sound of the battle being waged. A unit of Confederate cavalry soon appeared in hot pursuit of the fleeting fugitives. Caldwell had his Iowa troopers deploy in a line of battle and threw out skirmishers on the far side of the creek. A well-directed volley from the Iowans stopped the Confederate advance cold, although they did capture a few of the skirmishers. Lt. Silas Nugan was one of those captured. He informed the Confederate commander that they were the "advance of General Steele's army." This misinformation caused the Confederates to halt any further movement towards Camden.[45]

Many authors consider the Battle of Marks' Mill the greatest loss for the Union Army in the Trans-Mississippi West. Steele's loss of almost two thousand men, 440 wagons, and countless animals in the space of a week brought to an end whatever ambitions he may have had of continuing the Camden Expedition. There was a small stream called Salty Branch that crossed the Camden-Pine Bluff Road near the mill site. Anita Knowles' grandfather took part in the battle and was responsible for killing the lead horses of the first wagon. She wrote in 1937: "…So many horses and soldiers were killed that Salty Branch ran red with blood…"[46]

The Confederate artillery had begun shelling the Union forces around Camden on April 22. These were part of the advance units of the three Confederate infantry divisions led by Lt. Gen. E. Kirby Smith and Maj. Gen. Sterling Price who planned on investing the town and forcing Steele to surrender. Somehow, by the evening of April 25, somebody had quickly traveled the thirty-eight miles to Camden notified Gen. Steele of the disaster at Marks' Mills. Late that night Steele held a council of war with Generals Thayer, Salomon, Rice, and Carr. Together they decided that evacuating Camden and retreating to Little Rock was their best chance to save the army. With the loss of the potential supply line to Pine Bluff, they found themselves stranded in Confederate territory with a half-starved army, with little hope of a successful resupply. Quietly, in the dark early morning of April 26, Steele's army began to slip out of town, across the pontoon bridge over the Ouachita River, and northward on the Camden Trail, one of a series of pre-1836 "trunk roads," which crossed the Saline River at Jenkin's Ferry. As they were leaving the Union soldiers were issued two pieces of hardtack, coffee, and a half-pint of cornmeal. This meager ration was to last them for the duration of the trip to Little Rock. Brig. Gen. Eugene Carr and his cavalry division led the Union column onto the Camden Trail. Carr was a native of New York and an 1850 graduate of West Point. He was assigned first to the Regiment of Mounted Riflemen and later to the 2nd U.S. Cavalry under the command of Robert E. Lee. He had spent his pre-war years on the frontier and, once the war started, fought mainly in the Trans-Mississippi. He commanded units at Wilsons Creek, Pea Ridge, and Vicksburg. He earned the Congressional Medal of Honor for his actions at Pea Ridge. On April 26 his division numbered 1,500 troops and six mountain howitzers divided into three brigades. His advance took him 16 miles to the hamlet of Freedo where they found good springs, grazing, and discovered 1,500 bushels of corn. Steele and the rest of the army arrived at this location in the evening. Carr reported that he had not encountered any Confederates along their route.

Not knowing that the Union Army had given them the slip, Smith and Price continued to circle and invest the town of Camden while maintaining the artillery bombardment. They discovered on the evening of April 26 that the town had been abandoned and, fearing an ambush, they waited until the next morning to enter Camden. Once the town was secure, Smith ordered Brig. Gen. Fagan and his cavalry to pursue Steele's army, which had a 24-hour head start. Smith and Price hoped to cut off the Union retrograde movement before the Federals reached Jenkin's Ferry. They hoped to catch and completely destroy Steele's army. But first, Smith had to build a raft bridge across the swollen Ouachita River to begin the pursuit, which added additional time for Steele's army to separate the two sides.

At this time Smith made a questionable decision in which he sent Maj. Gen. Maxey and his two brigades of Choctaws and Texans back to the Indian Territory. This was at a time in the campaign when Smith and Price needed every man they had. Apparently, Smith received information that Maj. Gen. James Blunt was planning a large scale invasion of the non-Federal occupied areas of the Indian Territory. On April 28 Smith ordered Maxey back to the Choctaw Nation to meet up with Brig. Gen. Cooper and Stand Watie to counter the invasion.

> Special Orders No. 1.
> Headquarters Army of Arkansas, Camden, Ark., April 28, 1864.
>
> II. Brig. Gen. S. B. Maxey, with Walker's Choctaw and Gano's brigades, is relieved from duty in the

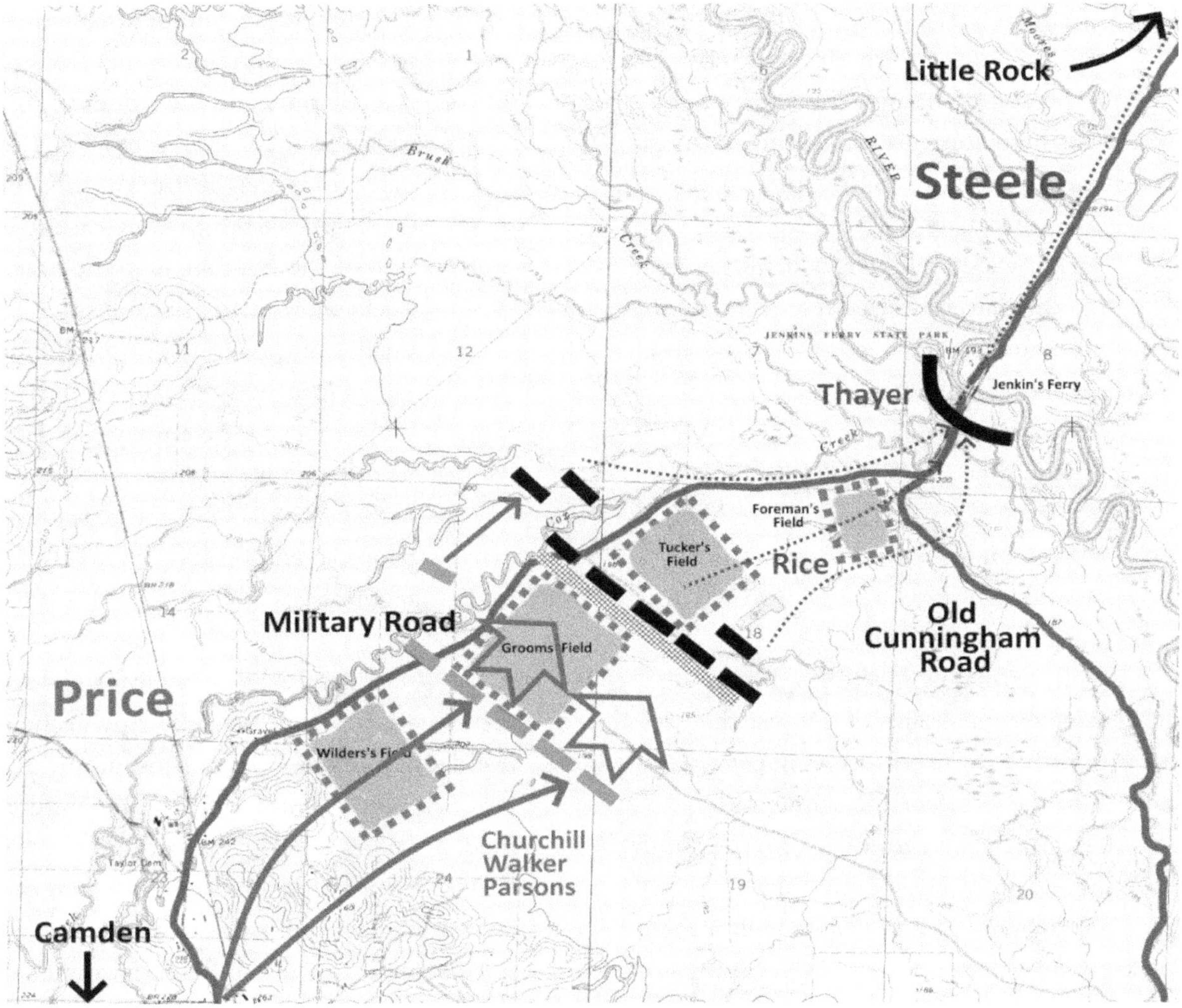

Area of Operations for the Battle of Jenkins Ferry, April 30, 1864
Map courtesy of USGS / Additions by Author

Jenkins Ferry
Confederate Battle Monument
Photo by Author

Remnants of Jenkins Ferry
Photo by Author

Army of Arkansas and will return without delay to the Indian Territory. The commanding general expresses his high approbation of the gallantry of Brig. Gen. S. B. Maxey and his command, and takes pleasure in acknowledging their valuable services in the battle of Poison Spring and in the operations of Prairie D'Ane and around Camden.
By command of
Lieut. Gen. E. Kirby Smith:
GEO. WILLIAMSON,
Major and Assistant Adjutant-General.[47]

Unfortunately for Smith, Blunt's invasion never materialized, but just the suggestion that Blunt was moving caused concern to the Confederate command.

Another problem developed for both sides: three days of steady rain turned the roads into quagmires and flooded all of the creeks, swamps, and sloughs in Central Arkansas. The Federals were hampered by their remaining heavy supply wagons being pulled by exhausted animals along the muddy roads. Instead of making Steele's army the focus of his movement, Fagan took off on independent operations and did not fulfill his orders. First, he failed in his first objective to destroy the Federal supply depot at Pine Bluff, probably because he could not cross the swollen Saline River. Secondly, Fagan failed to occupy a position across Steele's supply and communication lines between Camden and Little Rock, as Price had ordered. Fagan instead went looking for food and forage for his own brigade. Pvt. John West of the 10th Arkansas Cavalry wrote home:

"…General Fagan look his whole force to within Seven miles of Arkadelphia and did not even have a scout on the Camden road not even a man in the neighborhood of Camden. Gen Price & Smith were on the west side of the river shelling the enemy while we were sent austensably [sic] for the purpose of shutting them up but as it turned out for the purpos [sic] of letting the enemy out. I think Gen Fagan is very much to blame. He is no General. I expect he is a very good Seven up player from all accounts…"[48]

On April 29 General Marmaduke's men caught up with Steele's army in the flooded white oak bottoms just south of Jenkin's Ferry. By the time the remainder of the Confederate Army caught up with the Federals, Fagan was not in immediate communication with Price and was not in a position to know about Steele's movements towards Jenkins Ferry. Price soon realized at the time of contact on April 29 that Fagan was not blocking the federal retreat.

Brig. Gen. Carr and his cavalry arrived at Jenkins Ferry in the afternoon of April 29 and discovered the waters were too high to ford the river. Capt. Junius Wheeler, the Chief Engineer for Steele's army, determined that they would need to erect the pontoon bridge to cross the Saline. He ordered up the 34-wagon train that carried the bridging equipment and supervised the placement of the bridge. Since the Federals needed time to complete the pontoon bridge at Jenkins Ferry, Brig. Gen. Salomon's division was given responsibility as the rear guard, with Maj. Gen. Thayer's Frontier Division held in reserve. Salomon should have personally commanded the rearguard action against the pursuing Confederates, but he left the task to Brig. Gen. Samuel Rice and 4,000 Federal infantrymen. Rice had placed the Union infantry behind breastworks, abatis, and rifle pits approximately two miles southwest of the ferry crossing. It was a good location since Cox Creek protected the Union right flank and a soggy wetland forest protected the left flank. This gave the defenders only about a 400 yard front to cover. During the night of April 29-30 Steele was able to get most of his cavalry across, but the conditions were too dangerous for the heavy supply wagons and infantry to cross in the dark.

Before dawn on April 30, 1864, Marmaduke's Confederate cavalry troopers arrived near Jenkins Ferry, dismounted and skirmished with Steele's rear guard infantry force about two miles from the Saline River crossing. Battles raged all day over a series of four farm fields along the Military Road with the Confederates mounting three major attacks against the Union rear guard troops at the breastworks. Pvt. William Braly of the 34th Arkansas Infantry (CSA) described the conditions at Jenkins Ferry:

"… on the morning of the 30 we overtaken them at Jenkins Ferry on Saline river about fifty miles from here The battle was a dredful [sic] one. it rained nearly all that day and day previous. The battle field was a swamp flooded with watter [sic] in places waste deep. Artilery [sic] could not be used with any advantage but for 5 hours there was a continued roar of musketry which was far ahead thing of the kind I ever heard before, we finally drove them off the field and across the river and only lacked our cavalry (which by some misunderstanding did not get there until after the fight) to have made our victory all it could have been wished…"[49]

These attacks consistently failed to break the line or prevent the Union army from crossing the Saline River. Pvt. Milton Chambers of the 29th Iowa Infantry described things from the Union side:

"…morning their main force had come up and came on to us early about 6 o'clock and general engagement commenced and lasted til [sic] noon, they came up very close and fought brave, they tried to drive us three times but we would not drive worth a cent. They came up in 50-75 and 100 yds of us but they had to give back with heavy loss every time, they also brought 3 pieces of artillery up in 125 yds of us but did not get to fire but 4 shots until we [had] every horse killed but every man drove away from it and it fell into [our] hands and was taken back to our rear in less than no time, there was not more than half a dozen Shots of artillery in the whole engagement…"[50]

By 3 p.m. Steele had all of his wagons and artillery across the Saline and up on the high ground. The Federal artillery was unlimbered from this advantageous location and was able to cover the remaining infantry as they pulled back across the bridge. When all Federal troops were across, the bridge was cut and burned to prevent the Confederates from using it. Unneeded wagons were burned or otherwise destroyed.

> "…We privates were not so much interested in the wagons just then, but the officers had all their fine clothes in them; so there came a sudden change of garments, to save the best from burning; and men who had laid down ragged and dirty at dark were see at daylight finely dressed in glossy coats with shining buttons; but hungry and tired as ever. Mess-chests, company boxes, etc. made excellent fuel; and by their blaze the coffee was boiled and the poor pretense of breakfast eaten…"[51]

The Confederate pursuit ended at this point since they had no way to get across the flooded river in time to prevent Steele's army from reaching Little Rock. Kirby Smith's last hope to destroy Steele's army outside of his well-fortified base at Little Rock was dashed as a result of the mismanaged, disjointed, and piecemeal attacks at Jenkins' Ferry.[52]

The Confederates officially reported 86 men killed, 356 wounded, and one missing, for a total of 443 casualties, but Walker's Texas Division did not submit a report of its losses. Federal casualties reported were 63 killed, 413 wounded, and 45 missing, a total of 521 casualties, but like Walker's division, Maj. Gen Thayer failed to make an official report of his losses. In an unfortunate incident, before leaving the field, some black soldiers of the 2nd Kansas Infantry (Colored) regiment shot Confederate wounded near Rice's line in retaliation for the shooting of black soldiers who were trying to surrender at Poison Spring

and the killing of wounded black soldiers at Marks' Mill.

Steele and his badly demoralized and half-starved army finally pulled into their fortifications at Little Rock Arsenal on May 3. Cpl. Christian Isely, Company F, 2nd Kansas Cavalry, gave his opinion from the perspective of an enlisted soldier:

> "…The fatigues of our unhappy campaign seems [sic] just now to overcome me. I feel sore all over, my head & whole body aches… foul play of Gen Steele. The whole army is very indignant about him… [Steele] is a Conservative or Copperhead and is afraid of hurting the rebels…"[53]

The Union troops had travelled 275 miles and had little to show for the effort. The Federal army under Steele had suffered an estimated 2,750 casualties and the loss of 635 wagons, 2,500 animals, eight artillery pieces, and two steamships. When joining the Camden Expedition's numbers with Banks's

Although this Marker indicates that many soldiers of the 1st Kansas (Colored) were killed after they were wounded, it fails to mention the 1st Choctaw Regiment scalping and committing other atrocities on the Black troops.

statistics, the sad totals of Union losses during the campaign surpassed 8,100 men, fifty-seven artillery pieces, 822 wagons, nine naval vessels, and a staggering 3,700 mounts during the Red River Campaign. In addition, it was not until the end of May that the United States Navy flotilla was able to get free from the low water levels on the Red River near Alexandria, Louisiana, using a series of dams to back up the river to refloat the gunboats. Yet neither Union army was destroyed, so the Confederate victories were of little real value. The results of the Red River Campaign clearly showed that Maj. Gen. Banks did not have the leadership qualities needed to push his expedition to victory. If Lt. Gen. Grant had been in command, the Army of the Gulf may have fallen back after the Battle of Mansfield as far as Pleasant Hill, but no further. Grant would have reorganized and resupplied his army and continued the drive towards Shreveport. As for Generals Steele and Thayer, both knew that their forces were in no condition to undertake the Camden Expedition. Every critical issue that Steele had indicated to the War Department came to fruition: a weak and vulnerable supply line, the lack of provisions on hand for both soldiers and animals, wet and muddy roads, and a lack of forage and consumables in the region that had been heavily depleted by Confederate forces.

[1] Seward to Banks, *OR*, Vol. XXXIV, (P2), pp. 595.
[2] United States, Congress, "Joint Committee on the Conduct of the War." Report of the Joint Committee on the Conduct of the War. Washington: Government Printing Office, 1865. 38th Congress, 2nd session.
[3] Ludwell H. Johnson, Red River Campaign: Politics & Cotton in the Civil War. Kent, Ohio: The Kent State University Press, 1993. pp. 19-20.
[4] Steele to Hallack, *OR*, Vol. XXXIV, (P2), pp. 576.
[5] Grant to Steele, *OR*, Vol. XXXIV, (P2), pp. 603.
[6] Ulysses S. Grant, Personal Memoirs of U.S. Grant. New York: Dover Publications, 1995. pp. 281.
[7] Joe Walker. Harvest of Death: The Battle of Jenkin's Ferry, Arkansas. Self-Published, 2011. pp. 180-181.
[8] A.F. Sperry, History of the 33rd Iowa Infantry Volunteer Regiment. Des Moines: Mills & Company, 1866.
[9] William D. Baker, The Camden Expedition of 1864. Little Rock: Arkansas Historic Preservation Program, pp. 4-5.
[10] Walker, pp. 25.
[11] Michael J. Forsyth. The Camden Expedition of 1864: and the Opportunity Lost by the Confederacy to Change the Civil War. Jefferson, North Carolina: McFarland & Company, 2003. pp. 78-81.
[12] McLean to Blocki, *OR*, Ser. 1, Vol. XXXIV, (P1) pp. 705-706.
[13] Forsyth, pp. 92.
[14] McLean to Blocki, *OR*, Ser. 1, Vol. XXXIV, (P1) pp. 707.
[15] Forsyth, pp. 92.
[16] Sperry, pp. 79.
[17] McLean to Kimbal, *OR*, Ser. 1, Vol. XXXIV, (P3) pp. 78.
[18] Sperry, pp. 79.
[19] Forsyth, pp. 91-103.
[20] Britton, *Memoirs*, Vol 2, pp. 273-274.
[21] Baker, pp. 8-9.
[22] Forsyth, pp. 100-101; Richards, Ira Don. "The Battle of Poison Spring." *The Arkansas Historical Quarterly* 18, no. 4 (1959): 340.
[23] Mansfield State Historic Site, Information Panel, Louisiana State Parks.
[24] Williams to Witten, *OR*, Ser. 1, Vol. XXXIV, (P1) pp. 743.
[25] Historic Marker Panel #2 & #3, Poison Springs State Park, Arkansas State Parks
[26] Forsyth, pp. 112.
[27] Richards, pp. 345.
[28] Marmaduke to Belton, *OR*, Ser. 1, Vol. XXXIV, (P1) pp. 819.
[29] Duncan to Williams, *OR*, Ser. 1, Vol. XXXIV, (P1) pp. 750-751.
[30] Walker to Ochiltree, *OR*, Ser. 1, Vol. XXXIV, (P1) pp. 849.
[31] Charles Pidcocke, Letter, May 15, 1864, Butler

Center for for Arkansas Studies, Central Arkansas Library System, Little Rock.

[32] Baker, pp. 10-14.

[33] Forsyth, pp. 122-123.

[34] Steele to Sherman, *OR*, Ser. 1, Vol. XXXIV, (P1), pp. 663.

[35] "Battle of Marks' Mill," Wikipedia Article

[36] Drake to Blocki, *OR*, Ser. 1, Vol. XXXIV, (P1) pp. 714-715.

[37] Baker, pp. 14-15.

[38] Interpretive Panel, Marks' Mill State Park.

[39] William E. McLean. Forty-Third Regiment of Indiana Volunteers: An Historic Sketch of its Career and Services. Terre Haute: C.W. Brown Printer and Binder, 1903.

[40] Drake to Blocki, *OR*, Ser. 1, Vol. XXXIV, (P1) pp. 716.

[41] McLean, pp. 50.

[42] Baker, pp. 19.

[43] Historic Marker, John Marks Home, New Edinburg, Arkansas.

[44] Charles H Lothrop, M.D. A History of the First Regiment Iowa Cavalry Veteran Volunteers: From Its Organization in 1861 to Its Muster Out of United States Service in 1866. Lyons, Iowa: Beers & Eaton, Printers, Mirror Office, 1890. pp. 163.

[45] Forsyth, pp. 132-133.

[46] Interpretive Panel, Marks' Mill State Park.

[47] Special Orders #1, *OR*, Ser. 1, Vol. XXXIV, (P1) pp. 845-846.

[48] John R. West, Letter, June 7, 1864, George W. M. Rubenstein Rare Book & Manuscript Library, Duke University, North Carolina.

[49] William C. Braly, Letter, May 7, 1864. Amanda Malvina Fitzallen Braly Papers in Collections, University of Arkansas Libraries, Fayetteville

[50] Milton Chambers, Letter, May 8, 1864, Special Collections, University of Arkansas Libraries, Fayetteville

[51] Walker, pp. 160.

[52] "Battle of Jenkins Ferry," Wikipedia Article

[53] Spurgeon, pp. 226.

Chapter 17

The Arkansas Raids and an Indian Territory Naval Battle

Major General James Blunt had more than his share of ups and downs during the Civil War. He was obviously the most aggressive commander to serve in the border area of Kansas, Missouri, Arkansas, and the Indian Territory, Union or Confederate. Just a rumor of his move into Indian Territory with an expedition force prompted Confederate Lt. Gen. E. Kirby Smith to quickly send a much needed cavalry brigade under Maj. Gen. Samuel Maxey back to the Indian Territory during the Camden Expedition. Yet, Blunt's aggressive nature provided him with more than his share of distractors. Many of his subordinates, including those whose mistakes he had covered up, had no issue with spreading misinformation, and even substantial lies, to destroy Blunt's reputation. Some of these were the massacre at Baxter Springs in which Blunt was heavily criticized for leading his escort into an ambush by Capt. William Quantrill's Partisan Ranger force. Newspapers and other opponents claimed he failed to adequately scout the route he was taking to Fort Blair at Baxter Springs. Unfortunately, "yellow journalism" was very prevalent in the Kansas border region. Yet most reports show that he had scouts, but that the outlaws were wearing blue uniforms and carrying a U.S. flag. Those soldiers that did survive the battle stated that Blunt and his officers had done their best in forming a defense. There was also the accusation from

Col. William Weer, who should have been cashiered from the Army for drunkenness after his disastrous First Invasion of Indian Territory but, in the end, was saved by Blunt. Weer had accused Blunt of hauling contraband of war from Fort Scott to Fort Smith intending to sell the property to Southern sympathizers. None of this was true. Blunt's relationship with his superiors was also strained by the misrepresentation of his actions by Maj. Gen. John Schofield, Union commander of the Department of the Missouri. Schofield's superior, Maj. Gen. Henry Halleck, who until to March 1864 was the General-In-Chief of the United States Army, did have a severe dislike for anything having to do with the Kansas border areas, especially officers who had been given rank due to their political ties to Senator James Lane. Both Schofield and Halleck were motivated by intrigue and deception to further their own careers. Schofield was kept at his position simply for his ability to pacify the two Union-supporting elements of Missouri politicians, not for any battle-related reasons. Most of the successes in the Trans-Mississippi West could be attributed to Blunt and Maj. Gen. Samuel Curtis. After Ulysses S. Grant was made a Lieutenant General and given command of all of the Union armies, replacing Halleck, who was relegated to the Chief-of-Staff position, one would believe that he would appreciate an aggressive commander like Blunt. Instead, Grant foolishly listened to both Schofield and Halleck, took the Indian Territory away from Blunt and Curtis, and assigned it to Maj. Gen. William Steele's Department of Arkansas. Grant should have recognized that the information he was being provided by Halleck was tainted, especially since Halleck had done the same to Grant after the Battle of Shiloh. In the Shiloh situation, Halleck, who was commander of the Department of the Missouri and Grant's immediate superior, had reported to Washington that Grant had bumbled the battle and had stopped communicating with his headquarters. In fact, however. a Confederate-sympathizing telegraph operator was destroying the messages between the two commanders. In either case, Halleck got permission to replace Grant with himself in command. Grant stated in his memoirs:

> "…A few days afterwards General Halleck moved his headquarters to Pittsburg Landing [Shiloh] and assumed command of the troops in the field. Although next to him in rank, and nominally in command of my old district and army, I was ignored as much as if I had been at the most distant point of territory within my jurisdiction: and although I was in command of all the troops engaged at Shiloh I was not permitted to see one of the reports of General Buell or his subordinates in that battle…"[1]

Blunt was being treated in the same ignominious manner as Grant had been after Shiloh. Although much of the dispute pre-dated Grant taking command of the overall Union Army, it should have still placed a seed of doubt into his mind of anything coming from Halleck or Schofield. Blunt had reason to continually suspect behind-the-scenes manipulations of his command:

> "…I received a telegram from the President to proceed to Washington, where I arrived on the 27th of January, and there learned that the object for which I had been called to Washington was for consultation in reference to the condition of affairs in the Indian territories, and with the view to a campaign, early in the spring, into Texas, through the Indian country. Before leaving Washington to return to the West, I was assured, by

Mr. Lincoln, that I should have every facility afforded me for the organization of this Texas expedition that I desired, to insure its success… I was not at all disappointed, as I had already learned that I had not left Washington twenty-four hours when General Halleck, with his chronic hatred of Kansas, had determined to defeat the contemplated Texas expedition, which had had the sanction and approval of the President and Secretary of War… Being satisfied that so long as General Halleck was commander in chief, I was to be the special object of his malice…"[2]

Regardless, Blunt was sent back to Fort Leavenworth where he was given command of the District of the North Arkansas River in Western Kansas and Nebraska Territory. The most successful commander of the Trans-Mississippi was assigned to chase the Arapahos and Cheyenne across the Upper Plains.

There was much change in the Indian Territory status within the framework of both the Federal and Confederate command structures. Starting back on May 7, 1864, a new higher Union headquarters known as the Military Division of West Mississippi was established with Maj, Gen. E.R.S. Canby in command. It included the Department of Arkansas (Indian Territory was in this department) and the Department of the Gulf. The new division was headquartered at Natchez, Mississippi. On November 29, 1864, Maj. Gen. Joseph J. Reynolds replaced Maj. Gen. Frederick Steele as commander of the Department of Arkansas. Maj. Gen. Grenville Dodge replaced Maj. Gen. William Rosecrans

as commander of the Department of the Missouri in early December. On January 30, 1865, the Department of Kansas was merged into the Department of the Missouri and, on the same day, another new Military Division of the Missouri was created. This division was to oversee the Departments of the Missouri and Northwest and be commanded by Maj. Gen. John Pope, with headquarters in St. Louis, Missouri.

The Confederate command also had changes in late 1864 into early 1865. In August Maj. Gen. J.B. Magruder replaced Maj. Gen. Sterling Price as commander of the District of Arkansas so Price could plan and execute his Missouri raid. On February 14, 1865, Brig. Gen. Stand Watie replaced Brig. Gen. Douglas Cooper as commander of the Confederate Indian Division, which comprised all Indian troops in the Indian Territory. Cooper was redesignated as the Superintendent of Indian Affairs for the Indian Territory. There were further changes later in the spring, but by that time most Confederate forces in the Trans-Mississippi were simply skeleton units. [3]

In the Indian Territory, the consideration of retaking Forts Gibson and Smith remained a focal point for the Confederates throughout May 1864. Despite Maj. Gen. Maxey's dissatisfaction with the inadequate transportation facilities, he informed the Trans-Mississippi Department that while the two forts could be captured with relative ease, maintaining control over them would require securing the Arkansas River area south of Fort Smith. Maxey spoke of this prospect in January of 1864:

"…The wonderful importance of so strengthening this army as to enable it to regain Fort Smith and Fort Gibson and to expel the enemy from this country has never been realized [and likewise] . . . its bearing on northern Texas, and the absolute necessity of the grain, beef, salt, and iron of the country to the Trans-Mississippi Department... The forces at Fort Gibson and Fort Smith must now depend to a great extent upon wagon trains. When the Arkansas rises, if the navigation of that river is left uninterrupted, they can get everything they want and lay in spring supplies."[4]

Although there were plans for a coordinated Confederate offensive aimed at reclaiming northern Arkansas and the Cherokee Nation, such an initiative was never effectively executed. In June, Cooper's preparations to attack Fort Gibson were hindered by Maxey, who believed his forces were too weak to attack the fortifications, which temporarily dampened Confederate aspirations of recovering the fort.[5]

As stated previously, the Confederate command structure changed substantially after the Phillips Expedition. The Army of Tennessee veteran Maj. Gen. Maxey remained in overall command of the Indian Territory with Brig. Gen. Gano in command of the Texans and Brig. Gen. Cooper in command of the Indians. Maxey also desired to separate Watie's brigade out for independent operations against the Federal supply lines. In fact, Confederate Senators Chad B. Mitchell and R.W. Johnson asked Secretary of War Seddon on April 29, 1864, to urge President Jefferson Davis to promote Stand Watie to brigadier general. They believed that, although Watie's successes had been somewhat limited,

"nothing could be more encouraging to our faithful Indian allies." Davis agreed and made the recommendation to the Confederate Senate. After the Senate confirmation, Watie was officially promoted on May 6, the only Native American to achieve a general's rank during the Civil War.[6]

During these final two years of the guerilla war, Stand Watie and his mounted Cherokee Rifles conducted numerous small raids into Union-held territory. Two of these raids were substantial operations against the Union supply lines that were extremely costly to the Federal Army. The first of these raids took place on June 15, 1864, when Watie's force captured the Federal supply steamboat *J.R. Williams* on the Arkansas River at Pheasant Bluff (or Pleasant Bluff, as both names are used in the Official Records) near Tamaha on the Choctaw-Cherokee Nations border.

During the spring of 1864, Watie's scouts had been watching the Arkansas River as well as Fort Gibson, the headquarters of the Union Army in Indian Territory. They discovered that the Federals in the area were beginning to run short of supplies and were anticipating a resupply soon, either by wagon train from Fort Scott or by water from Fort Smith. The water on the Arkansas usually ran too shallow for a steamboat to get over the small falls near Webbers Falls, but on June 10 Watie received information that the river was rising rapidly. He correctly believed that the Federals would send a boat up the Arkansas River. Confederate forces had found it relatively straightforward to gather intelligence on Union movements in the area due to the presence of numerous Southern sympathizers along the Arkansas River. Some of these individuals acted as informers, providing the

Confederates with updates on Federal activities. Many worked on the steamboats operating out of Fort Smith as captains, laborers or rivermen. Both Federal and Confederate forces routinely dispatched scouts and spies to the opposing sides to gather important information. Colonel William Penn Adair, commanding officer of the Confederate 2nd Cherokee Mounted Rifles, had been sent on a raid through the Cherokee Nation. This raid went as far north as Maysville, Arkansas, and the Cowskin Prairie. Adair's actions caused panic among the Cherokees, especially the returned refugees, as well as the civilian populations of Southern Kansas, Southwest Missouri, and Northwest Arkansas. Adair also attempted to contact Capt. Quantrill but was unable to find him. This was probably for the best since the Confederate government had officially disavowed Quantrill and his outlaws after a Christmas 1863 incident in which he and his followers destroyed extensive amounts of private and public property in Sherman, Texas. At the end of March 1864, he was arrested by Confederate officers in Bonham, Texas, but escaped their custody by overpowering the guards. He and his group immediately joined Capt. George Todd's and Capt. Tom Todd's (not related) guerilla groups as an underling. The two groups eventually merged with Capt. "Bloody Bill" Anderson's group. Quantrill eventually broke away with some of his followers, conducted some limited raids in Missouri, moved east of the Mississippi River, and staged raids into Kentucky in the spring of 1865, where he was ambushed and received a gunshot wound to the chest on May 10 near Taylorsville. Quantrill was transported to a military prison hospital in Louisville and died on June 6.

Quantrill was a highly notorious and enigmatic figure during wartime, and many of the narratives or mythologies about him were self-constructed.[7]

Adair's raid provided Watie with information about the *J.R. Williams*, a veteran of both the Vicksburg and Mississippi River campaigns. This vessel could carry 100 tons of cargo, the same amount that would need 100 wagons to transport. Colonel William A. Phillips, commanding Federal Indian troops at Fort Gibson, viewed the river as unreliable due to hostile Confederate Indians between Fort Smith and Fort Gibson. He believed escaping would be difficult and costly if the ship was attacked from the banks.

After receiving news from Adair about the planned trip for the *J.R. Williams*, Watie prepared to capture it along with its precious cargo. The vessels used on the Arkansas River typically featured stern-mounted paddle wheels for propulsion, powered by steam generated from wood-fired boilers. Personal narrative evidence suggests that wood was commonly used as fuel for inland steamboats. (Following the Civil War, coal and oil gained prominence as fuels due to the depletion of American forests in proximity to waterways.)

On June 15, 1864, the *J.R. Williams* was steaming up the river after being loaded at Fort Smith. The vessel was enroute to Fort Gibson, carrying a full complement of commissary supplies, quartermaster stores, and sutlers' goods for contractors McDonald and Fuller. The cargo included yards of cloth, many pounds of cotton yarn, blankets, shawls, skirts, harnesses, boots of various types and sizes, 1,000 barrels of flour, fifteen tons of bacon shoulders, and thousands of yards of linen. According to another account, the cargo also included men's dress clothing such as top hats, dinner jackets with tails, fancy trousers, and spats. Allegedly, Watie's men wore these as their uniforms after the attack. The cargo also contained 400 Sharps rifles and 600 new revolvers, a definite need of the Confederate forces in the Indian Territory. The total merchandise, valued at $120,000 (some accounts state as high as $500,000), was intended not only for the Federal troops stationed at Fort Gibson but also for the 10,000 refugee Indians camped in the vicinity. Fort Gibson was also due to receive another almost 5,000 more Indian refugees from the reserves up in Kansas during the next few months. Lt. George W. Huston, regimental quartermaster for the 14th Kansas Cavalry, was a member of the crew of the steamboat and responsible for overseeing the cargo. He had received instructions from the chief quartermaster's office in Fort Smith to ensure the proper unloading of the steamboat and perform an inspection of the quartermaster's depot at Fort Gibson. Additionally, Huston was directed to report to Captain Greene Durbin, the assistant quartermaster at Fort Smith, regarding the condition of the *J.R. Williams* and the management of the quartermaster department at Fort Gibson. Furthermore, on the return trip, Huston was ordered to load the steamboat with lime and salt at the mouth of the Illinois River near Mackey's Salt Works, where Colonel John Ritchie, commander of the 2nd Indian Home Guard regiment, was in charge.

Watie's plan was to attack the steamboat from the southern shore of the Arkansas River near Pheasant (or Pleasant) Bluff, a considerable height on the south side of the waterway, located approximately five miles downstream of the mouth of the Canadian River and near the present-day town of Tamaha. He and about 300 men of the Cherokee regiments with artillery deployed along the heights above the river. Howell's three-gun battery of howitzers was placed in the brush, with each piece about a hundred yards apart. Questions arise as to who actually placed the guns at these locations. Gen. Watie states in his official record that Lt. Forrester was responsible for their placement.[8] But Captain George W. Grayson of the 2nd Creek Mounted Rifles, who in the post-war years would become principal chief of the Creek Nation, claimed in his autobiography that he was in charge of the artillery and had made the decision regarding emplacements. Grayson recalled:

> "…I was furnished with a detachment of 300 men and a battery of light artillery and ordered to go and picket a certain point on the south bank of the Arkansas River known as Pleasant bluff, to occupy the place at once and so dispose my guns on the banks as to be concealed from, while commanding any possible passing craft as had been known lately to have passed loaded with supplies from Ft. Smith to Ft. Gibson. When a craft was within easy reach of our artillery, a harmless shot was fired across its bow, which I was   informed was a rule of war to warn the enemy of our belligerent intentions…"[9]

The conflicting stories can be probably attributed to the ongoing inter-tribal disputes between the Cherokees and the Creeks, along with post-war chest-beating. As the steamboat passed under the muzzles of Watie's artillery, some well-aimed shots hit the *J.R. Williams* in vulnerable spots. The first

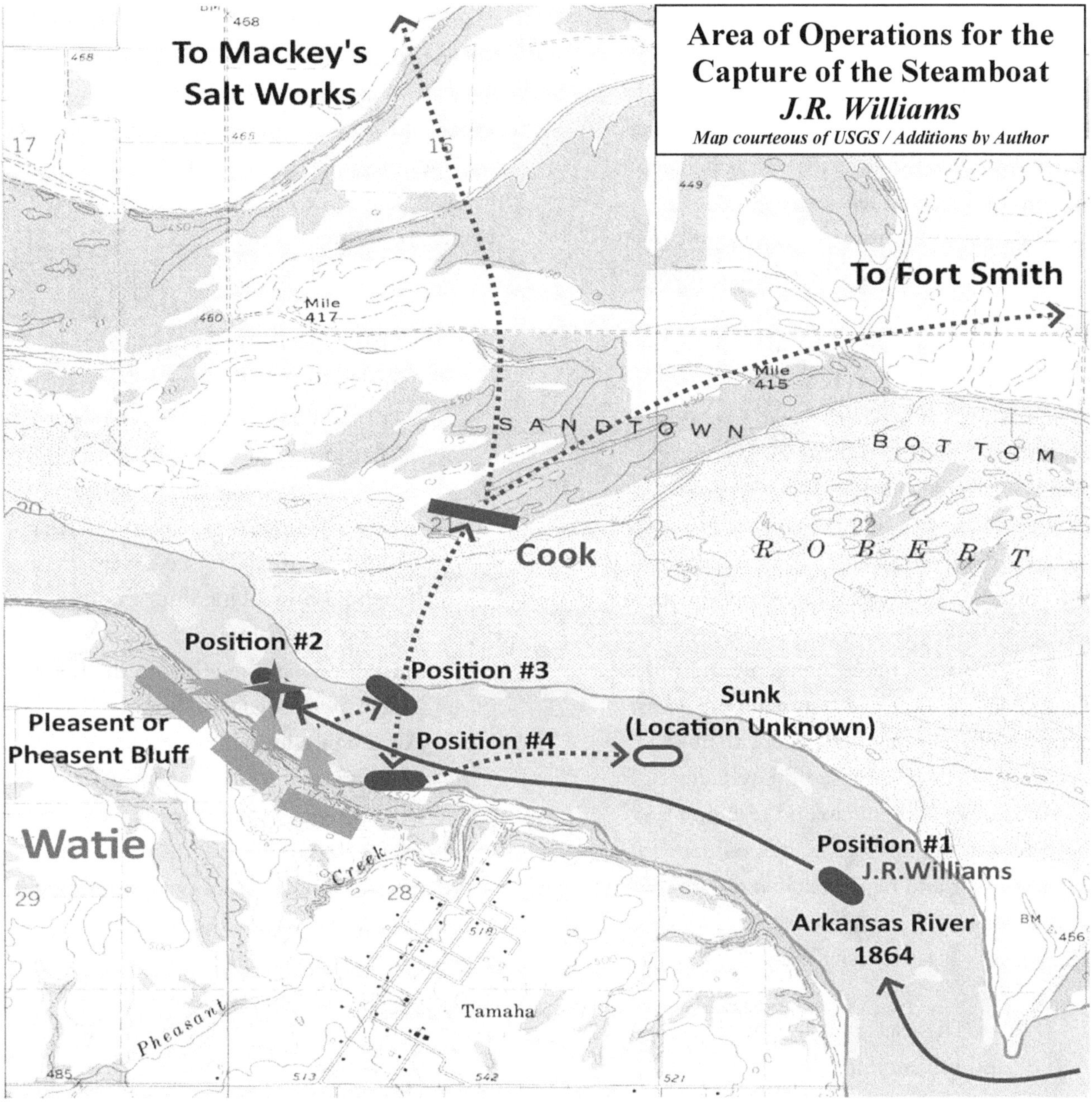
Area of Operations for the
Capture of the Steamboat
J.R. Williams
Map courteous of USGS / Additions by Author
To Mackey's
Salt Works
To Fort Smith
Mile
417
Mile
415
SANDTOWN
BOTTOM
Cook
ROBERT
Position #2
Position #3
Sunk
(Location Unknown)
Pleasent or
Pheasent Bluff
Position #4
Watie
Position #1
J.R.Williams
Arkansas River
1864
Creek
Tamaha
Pheasant

Mural of the Steamboat *J.R. Williams* on the side of a building in Stigler, Oklahoma

J. R.
Williams

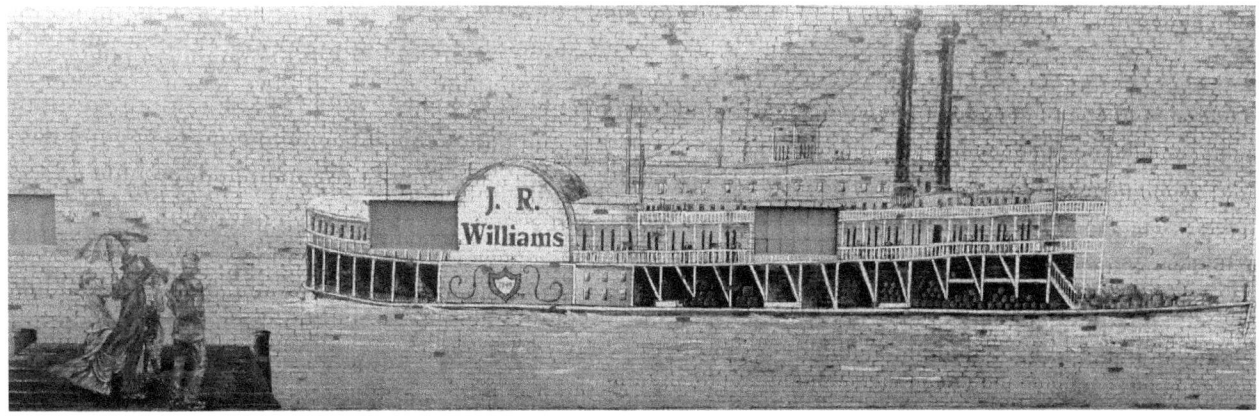

shell struck the chimney four feet above the deck. The second shell hit the pilothouse significantly damaging it. A third shell ruptured some steam pipes, releasing white steam that obscured the pilot's view. The steamboat lost momentum, drifted downstream, and grounded on a sandbar on the north bank.[10] With the engineer and fireman dead, the remaining crew, mainly Southern sympathizers, abandoned the boat, swam to the southern shore, and joined the Confederate force. Union quartermaster Lt. Huston and the captain then used the vessel's small boat to flee to join Watie's force. Huston took with him the written dispatches for Col. Philips from Maj. Gen. Thayer at Fort Smith. The 26-man escort (one sergeant, and 24 privates) from Company K, 12th Kansas Infantry, under the command of Lt. Horace Cook, had been enjoying the warm summer afternoon when the incident occurred. They had passed the remains of Ft. Coffee earlier and anticipated reaching Ft. Gibson without any issues, but suddenly had to take cover and began exchanging fire with Watie's troops. Private Henry Strong of the 12th Kansas Infantry recalled,

> "…The escort got behind the freight and returned the fire to the best advantage we could. When the Rebels opened fire on the boat I was sitting up on the hurricane deck, but got below as soon as possible and got my gun. As soon as the Lt. saw it would be folly to stay on the boat longer, as we would all be killed, he ordered us to leave the boat…"[11]

Realizing their precarious position of trying to mount a defense from within the clouds of steam from the punctured boilers and steam lines, the Union escort waded the 40 yards to the northern shore and then fled across 400 yards of the open sandbar, under fire, and into the timber on the north side of the river. Capt. Grayson of the Creek regiment continued:

> "…We opened up on him in fine style and soon filled the air with the roar of cannons and rattle of small arms. An escort of twenty-five men that was aboard fired one volley without effect, and jumping out of the now disabled boat into the water nearly waist deep, fled to the brush on the other side. The boat now being out of commission and helpless, the sign of surrender was waved from the pilot house when we ceased our attack… Although crippled, the engine and machinery was sufficiently intact to float it over on the side we were…"[12]

Lt. Cook intended to wait until dark, recapture the steamer, and set it on fire to prevent the supplies from falling into Confederate hands. Instead, the captain and pilot of the *J.R. Williams* used the small boat to return and were able to free the steamer from the north bank and drive it into the south bank, right into Watie's hands. After taking some casualties, Cook decided the best option was to lead his remaining force back to Fort Smith. It seems unlikely that Cook's small unit could have held out very long when the Confederates had artillery on the high ground, backed by 300 soldiers. A small group of Cook's command somehow got separated and instead of heading for Fort Smith, they made for Mackey's Salt Works, a distance of approximately ten miles. In the 1930s a 90 year old Judge Michial Ghormley reported to a Works Progress Administration interviewer that he had been a courier with Col. Watie's command as a young man at Pheasant Bluff. He stated that the day after the steamboat was captured, two companies of the "second Cherokee federal regiment" under Colonel John Ritchie arrived from Mackey's Salt Works and appeared on the north side of the Arkansas. These troops began firing on the remaining Confederate Indians guarding the vessel.[13] During this difficult situation, the three or four men of Lt. Cook's detachment had arrived at Col. Ritchie's station at the mouth of the Illinois River. Upon learning about the abandonment of the steamboat, Ritchie promptly assembled approximately 200 of his Indian troops (other sources say 40) and set out toward the location where the steamboat had been left. They traveled throughout the night and reached their destination around 10:00 a.m. the following morning. Col. Ritchie cautiously assessed the area and the positioning of the Confederate forces. At around 12:00 p.m., he instructed his Indian soldiers to fire randomly at the Confederates on the opposite bank of the Arkansas River and to shoot anyone attempting to remove the supplies that had been offloaded from the *J.R. Williams*. Watie's men retaliated, but due to the high river levels, neither side was able to ford the river to launch an attack. Consequently, it was impossible for Col. Ritchie to cross and dislodge Watie, despite having a larger force. Although mostly ineffective due to distance and the inability to cross the swollen river, he was able to temporarily drive the Confederate Indians from the steamboat and make it dangerous to try to carry the captured goods

up the steep slope.[14] They ended up just stacking the captured goods on the sandbar in order to move them up the slope once darkness fell. Gen. Watie requested Lt. Grayson of the Creek regiment to supply sentinels to keep watch over the camp and the captured supplies. Grayson later recalled:

> "…With the men in this state of alarm and perturbation and the well known lack of discipline in our ranks, I was fully aware that I had no ghost of a chance to succeed in obtaining from among them such sentinels as were required of me. I made the effort, however, and, as I had expected, I failed to get a single man to remain, I was sorry and ashamed of my Creek soldiers…"[15]

That night the Arkansas River rose and swept away much of the supplies left on the sandbar. In any case, Watie's blessing also became his curse. Having captured the much needed supplies worth almost a half-million dollars, his men had swarmed over the abandoned steamboat loading up with booty. While the Confederates had secured the essential cargo needed for their survival, Watie soon realized that his victory was accompanied by complications. Although his men diligently unloaded 150 barrels of flour, 16,000 pounds of bacon, and a significant quantity of store goods onto a sand bar, Watie had not arranged a wagon train to transport the spoils of his victory. Additionally, his troops were insufficiently equipped to withstand a substantial Federal force. Furthermore, upon viewing the plentiful merchandise, Watie's men became overly excited and preoccupied with their own needs and those of their families. Consequently, a large portion of the Creeks and Seminoles immediately departed, taking home as many goods as they could carry on their horses and in their arms. Some sixty of the Indians loaded their horses with tinware items such as wash tubs, stew kettles, coffee pots, wash pans, tin plates, cups, and dippers. Noting his suspicion and dislike of those Indians who were not Cherokee, Watie complained that the Indians, especially the Creeks and Seminoles, immediately took what treasures they could carry and fled for their homes, leaving the brigade too short of men to defend the remaining supplies from a Union counterattack. Watie recorded:

> "…I am left here with few men. The enemy is now on the opposite side of the river. Commenced firing on us about 12 yesterday. We have only a portion of flour and bacon brought up on the bluff. The river rose [a] great deal last night and washed off several barrels of flour. If I can get wagons I would move the flour and bacon to Kribb's, otherwise I will be compelled to leave it. The roads are in a wretched condition…"[16]

Cooper dispatched the remainder of the Choctaw and Chickasaw Regiment, but he did not have nearly a sufficient number of wagons to transport the flour, bacon, and the remaining stores to a Confederate supply depot.[17] Despite its symbolic significance, Watie's success in the Indian Territory did not affect the Civil War's outcome. While it boosted Confederate morale and provided supplies locally, for the Union, it highlighted the dangers of river transport and the need to control roadways. Local Confederates eagerly celebrated this event amid 1864's bleak war news.[18]

On June 16, the defeated Union escort arrived back at Fort Smith and reported the ambush. Gen. Thayer, commanding the Federal District of the Frontier, saw their actions a little differently. He stated:

> "…The escort on the boat were, in my judgment, fully able to have prevented the enemy from reaching the boat. If Lieutenant Cook had posted his men in proper positions on the bank he could have prevented the enemy from reaching her until he would have had time to send to me for assistance. I regard the conduct of Lieutenant Cook as a most unjustifiable and criminal abandonment of his post of duty… I therefore respectfully request that he be dismissed from the service…"[19]

Colonel S.J. Crawford was dispatched from that location with about seven hundred Union troops from the 2nd Kansas Infantry (Colored) and a section of artillery. At this point Watie had no choice but to burn the *J.R. Williams* with its remaining captured supplies because he did not have sufficient men to defend it or wagons to transport the captured goods. General Thayer confirms that approximately 40 troops of the 2nd Indian Home Guard did in fact engage the Confederates. Watie retreated approximately 12 miles and met the reinforcements that Cooper had sent. Watie reported:

> "…In this condition I learned that a detachment of Federals of superior

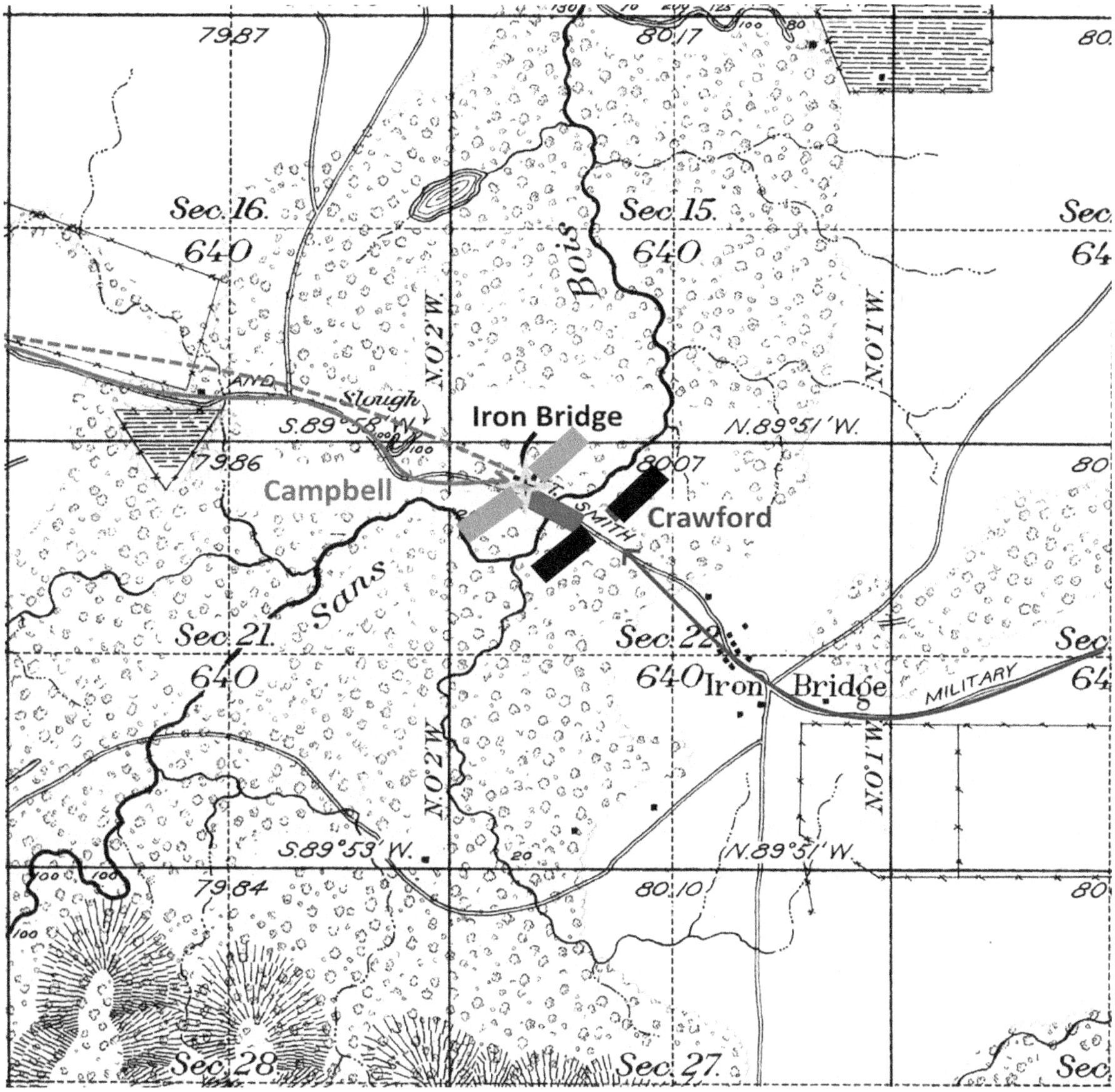

Battle of San Bois or the Iron Bridge
June 18, 1864

Beale Wagon Road approach to the San Bois Iron Bridge. Union troops deployed in the distance.

Former location of the San Bois Iron Bridge
Photos by Author

strength was approaching up the Arkansas on the south side, and I was compelled to burn the commissary stores captured, as I could not defend them successfully with the force I had, and the Canadian River being so high re-enforcements from Colonel Adair was impossible. After retreating 12 miles I met the Chickasaws, who had been ordered to support me. I ordered a party of 150, under Major Campbell, to the iron bridge on San Bois, which they reached about daylight or a little after… The Federals soon made their appearance and a skirmish ensued. The enemy brought up and commenced using his artillery, when the detachment fell back. The skirmish served to check the enemy, who precipitately retreated from this point toward Fort Smith, as was learned by a scout afterward…"[20]

As Crawford's Federal force drew near, they encountered the 150 men under Major Campbell that Watie had sent to hold the iron bridge over the Sans Bois Creek, one of the six iron bridges of the Overland Mail Route. At daylight on June 18, the Federal 2nd Kansas Infantry (Colored) approached the bridge, deployed in a double line of battle, and advanced in precision order. Col. Crawford recalled:

"…On arrival at this crossing I found the bridge flooring torn up, and Texas cavalry dismounted and partially fortified on the other side. The river at this point was narrow and deep, and the Texas troops behind their hastily prepared breastworks were within easy reach of our Endfield rifles. I immediately threw the regiment forward into a line, ran up a section of artillery, and opened fire with both at the same time…"[21]

They opened with their artillery, which drove off Campbell's small force. Fortunately for the Confederate Indians, Crawford hesitated moving forward across the bridge since he had no information about the size or disposition of Watie's force. He also determined it would be foolhardy to chase mounted cavalry with exhausted infantry troops. Crawford continued:

"…For a while the Texans hugged the ground like lizards, and fired as though they were shooting at birds in the trees. We had them to where they could either lie still, nor retreat with safety. Pretty soon our artillery got its bearing and began dismantling the fort. One shell went whizzing across the river and struck the bridge flooring, piled in front of a bunch of Rebels, and sent them whirling back to Dixie… I hastily repaired the bridge, crossed over, and followed them until dark, when they disappeared…"[22]

After resting near the Arkansas River, the Union soldiers turned around and headed back to Fort Smith. Watie had succeeded in capturing a large Union vessel but was unable to exploit it or the riches it contained due to the breakdown of military discipline among the Indian troops.[23]

For the Federal forces as well as the Indian refugees around Fort Gibson, it would be a trying time. Not only were they already short on rations and other needed supplies but, as mentioned earlier, another 5,000 or so refugees were being sent to the region from their camps up in Kansas. The Presbyterian missionary doctor Dwight D. Hitchcock, who was serving as a chaplain for the Union Army in the Fort Gibson area, could not believe that the Army and Interior Department were going to continue with their plan to bring the refugees to Fort Gibson. He wrote to his relatives:

> "…I hear with amasement [sic] that the refugee Indians are even now on their way back to this desolate country, to the number of several thousands. I do not know what the Government means. How supplies are to reach them, and a dozen other questions, are beyond my wisdom to answer. I suppose some gigantic speculation is at the bottom of the movement; and though the Government may for the present meet the expenses, in the end the Indians will have to suffer. Perhaps I do not sufficiently understand the programme [sic] to express myself so confidently as I have above; but at this juncture I do think the removal unwise and cruel…"[24]

The Union garrison at Fort Gibson would need to go back to moving their supplies by wagon from Fort Smith and Fort Scott.

After the severe defeats during the Camden Campaign in southern Arkansas, the Federal forces in the Indian Territory tried to maintain the status quo, not planning many offensive actions, instead focusing on keeping their supply routes open and protected. There were also significant command changes that changed the face of the Union forces in the Territory. On July 30, 1864, Colonel Phillips was relieved of command of Fort Gibson and of the Indian Brigade and was assigned to court-martial duty at Fort Smith. It is believed that Phillips had angered both General Blunt and Senator Lane by interfering in some profiteering undetermined schemes with Union suppliers who were overcharging the Federals for basic supplies in the Indian Territory.[25] In any case, one of the best Union commanders in the Trans-Mississippi was out of field operations.

The summer of 1864 continued to bring severe challenges to the Federal forces in the Indian Territory and in Arkansas, particularly in the vicinity of Fort Smith. The area around this post was dotted with various Union camps containing the troops defending and garrisoning the fort.

They were spread out to provide grazing for the cavalry horses as well as the cattle herds needed to feed the garrison. The location allowed their horses to graze and provided security for the Union garrison at Fort Smith; however, these isolated camps were vulnerable to attack. Confederate scouts identified two isolated Union camps as possible targets; the first was near "Caldwell's" along the Jenny Lind Road, and the second was camped around the Picnic Grove on the southern end of Massard Prairie.

Recognizing the weakness of the Union situation, Brig. Gen. Douglas H. Cooper, commander of the Confederate Indian Brigade, instructed Brig. Gen. Richard M. Gano to launch an assault on the Union camp located on the southern end of Massard Prairie. On July 26, Gano assembled 600 Texas and Confederate Indian cavalrymen at Page's Ferry on the Poteau River, ten miles southwest of the Union camp. As the force gathered before sunset, Gano realized it was insufficient for Cooper's somewhat complex attack plan. Cooper's plan of attack was outlined in his "Special Orders No.86."

> "…The plan, as shown by Special Orders, No. 86 (marked A),* was for Col. S. N. Folsom, commanding detachment from Indian division, to attack the camp of Federals at Caldwell's, on the Jenny Lind road, capture or destroy it if possible, and if pursued by other troops on Massard Prairie or from Fort Smith to retreat by the Fort Towson road over the Devil's Backbone, where McCurtain lay in ambush; the detachment from Gano's brigade to remain concealed near Page's, on Cedar Prairie, until the Federals should pass in pursuit of Folsom, and then attack them in rear, while Folsom and McCurtain should turn upon them at the Backbone…" [*Not Found][26]

Cooper wanted the main attack to be against the camp hoping to pull the other Union forces from the works around Fort Smith to ambush them as they showed themselves. But when the troops gathered on the late evening of July 26, Gano realized he did not have enough troops to fully carry out the mission. When the Second Indian Brigade under Col. S.N. Folsom and the Choctaw Battalion under Lt. Col. John W. Wells arrived at the rendezvous point, they did not have as many mounted troops as Gano had expected. Consequently, he decided to lead the entire force against the Union troops, less the Choctaws under Lt. Col. Jack McCurtain who were already posted on the Fort Towson Road on Devil's Backbone. Cooper remained behind at Page's Ferry on the Poteau River. The Confederates traveled through the night, reaching the vicinity of Massard Prairie just before daybreak on July 27. The Union cavalrymen were encamped in a grove of trees on the southern side of the prairie. Most of the unit's horses had been grazing since daylight, and by approximately 6:00 a.m., the herd was three-quarters of a mile southwest of the camp. After a quick two-hour rest, at dawn on July 27 Brig. Gen. Gano and his 600 white and Indian Confederates swept down onto the Massard Prairie, believing this camp was a better target than the one at Caldwell's on the Jenny Lind Road. They were actually fortunate to find any of the Union's camps, as a correspondent of the *Galveston Weekly News* wrote in August:

> "…Through the incompetency of our guides and the multiplicity of Indian trails, the brigade got lost. We blundered about until about midnight, finally found the road, crossed the Poteau river, and marched to within four miles, as our guides informed us, of the position of the enemy. It was the intention of Gen. Gano to attack them at daylight…"[27]

The camp they found consisted of Companies B, D, E, & H of the 6th Kansas Cavalry under the command of Maj. David Mefford. The four company battalion, consisting of 200 troops, had been at this location since they had returned from the Camden Expedition in May. The locals called the grove of trees "Picnic" or "Round" Grove, and it provided substantial shade for the soldiers and their mounts. They had used rocks to dam the intermittent creek bisecting the camp to create a pool of cool water. The company camps were set up to align with the intermittent creek. Company B's camp was on the western end of the grove, and Companies D, E, and H stretched eastward in a line. Each company had a small parade ground for company drill. The total length of the encampment was approximately 500 feet. Confederate soldiers reported that there were Arkansas "Mountain Feds" in the camp, but they are not mentioned in the Union official reports.[28] A correspondent with the *Houston {Texas) Telegraph* reported:

> "…We traveled nearly all night, halting about two hours, just before day, to arrange plans, and give the men a little rest, soon after sunrise we drove in the enemy's pickets, passed over a high mountain, and came down into Mazzard [sic] prairie, four miles from Fort Smith. At the far end of the prairie, some one and a half miles from the foot of the mountain, we discovered a beautiful island of timber, known as the Diamond or Picnic Grove, at the north end of which we spotted our game, making hasty preparations to give us a warm reception…"[29]

A report by Sgt. Tubbs of Company D, on guard duty at picket station #1, alerted Union Maj. Mefford and his officers. Within five minutes of the initial alarm, the dismounted 6th Kansas Cavalry had formed a defensive line. Orders were given to retrieve the horses, secure the camp's flanks, and send messengers to Fort Smith to report the attack. Gano disrupted these efforts by charging the Union troops from the front and on both flanks, causing the herd to stampede and forcing the Union cavalry to fight on foot. Fifteen year old James Robert Barnes, who was visiting his uncle, James Barnes, whose house was on the southern end of the prairie, witnessed the initial attack. He recalled:

> "…I saw a string of Rebels coming down the hill on the [South] side of the field. I ran to my uncle and said, 'I see a bunch of Rebels coming yonder.' The men made a break for the brush. Jonathan Glenn ran up the road to the West to cut into the brush and, as he did not see some of the Rebels, they got him…"[30]

The Union troops soon found themselves surrounded in the grove of trees. Lt. William M. Burgoyne reported,

> "…The horses were driven in as soon as possible, but too late. The men had not time to saddle. The yells of the enemy and the firing stampeded the horses. Almost all of them started across the prairie in the direction of Fort Smith. The men fell in on their company parades and moved out on the prairie with the intention of gaining the timber on the north side of the prairie, having given up all hope of saving the camp…"[31]

With two companies on the left and

Dwight D. Hitchcock,
Presbyterian Missionary Doctor
Chaplain for the Union Army at
Fort Gibson
Photo courtesy of Oklahoma Historical
Society

Colonel Samuel J. Crawford
2nd Kansas Infantry (Colored)
Photo courtesy of the Kansas State
Historical Society

Evan Jones
Baptist Missionary to the
Cherokees
Photo courtesy of Oklahoma Historical
Society

Massard Prairie Battlefield Park and Monument
Photos by Author

Remnants of Earthworks surrounding Fort Smith

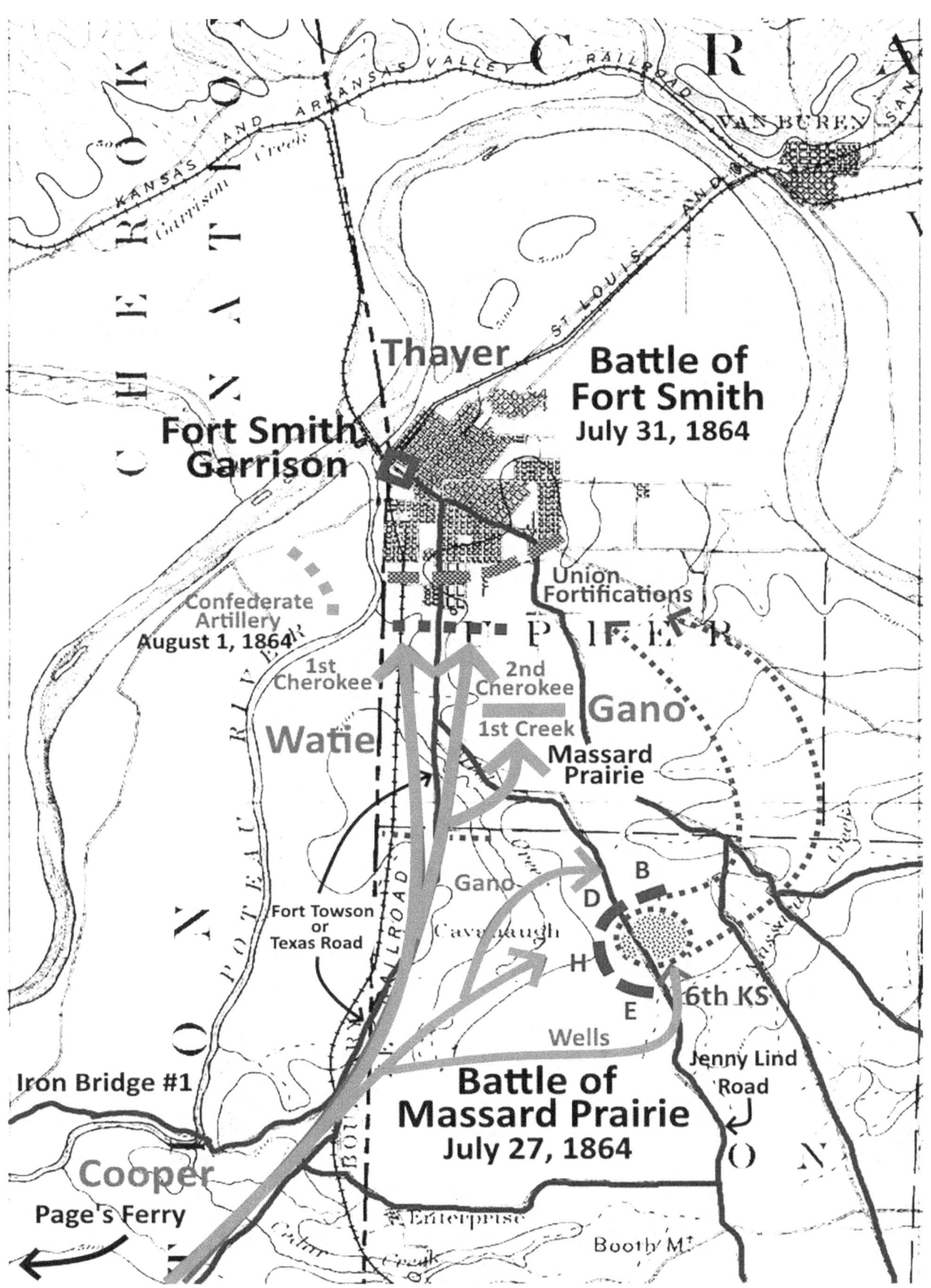

two on the right, armed with superior breech-loading Sharps carbines, Mefford's command initially held its position, repelling all attacks. However, the Confederates significantly outnumbered them and possessed greater mobility. In the open prairie, the mounted Confederates could flank the Union troops at will, advancing, firing, retreating, and advancing again. The Houston (Texas) Telegraph continued:

> "…Co. Fulsom and Lt. Col. Wells were immediately ordered to encircle the grove on the right, while Gen Gano, with his brave 500 at a sweeping gallop dashed around to the left. No sooner had the head of our column come within striking distance than the enemy opened upon us with their Sharpe's rifles. It was but the work of a moment for the general to form his men, and with a Texas yell they dashed forward…"[32]

1st Lt. Jacob Morehead, Company B, believed his first priority was to protect the horse herd, which was grazing less than a mile southwest of the camp. He sent troopers to retrieve the herd and formed a skirmish line to the southwest of the camp. He reported:

> "…I formed my men on the right of camp to protect my herd as it came in and until it could be secured, but before the horses could be brought up the enemy charged on us, which stampeded the herd and left the men on foot to fight as best they could. We drove the enemy back, and as I had received no orders from the commanding officer, I ordered my

men to fall back until they could form on the right of the other companies…"[33]

The four companies of the 6th Kansas Cavalry attempted to form a cohesive battle-line to counter the Confederate charge. The Texas Confederates under Maj. Carroll and Captains Welch and Hard moved through the center of the grove and attacked the thin Union line from across the stream running through the prairie. Lt. Burgoyne stated:

> "…It was not five minutes after the first alarm before firing commenced on the right of camp, Company B being camped on the right of our line. At the same time the enemy were discovered on our left and in front, coming through the timber… When the force was discovered on our left I ordered Sergeant Goss, who had about ten men mounted (of Company D), to go to the rise of ground on our left and if possible check the enemy…"[34]

Eventually, the Union cavalrymen began to falter. Those who managed to secure horses had already retreated to Fort Smith, while the remaining Union troops commenced a fighting withdrawal towards the north across the prairie. Despite several successful charges, the Confederates continued to press their advantage, capturing prisoners throughout the engagement. Ultimately, near a house one mile north of the camp, the last Union contingent was captured. Lt. Morehead wrote:

> "…When I had fallen back to the left

of my company's parade ground I came in speaking distance of Major Mefford, when I received orders to form my company on the right to protect the camp. I immediately took the position assigned me, with Company D on my left. We held our position, repulsing three distinct charges of the enemy. At this time I saw that Major Mefford had, with Companies E and H, been driven from their position on the left of the line and had begun to fall back across the prairie. I knew that I could not hold my ground much longer with what men I had, so, without receiving orders from Major Mefford, commenced falling back toward him. As we fell back I had several men captured by the enemy that was advancing through the timber in the center of our camp. We fought and retreated in good order until we came within half a mile of the house on the prairie, when the enemy closed in on all sides, taking many more of our men prisoners…"[35]

Gano swiftly secured the spoils of victory. They also captured all of the Kansas battalion's camp equipage as well as 200 Sharps carbines and 400 Colt revolvers. The Confederates reported capturing nearly 127 prisoners, along with much-needed carbines, pistols, camp equipment, and other supplies. Union casualties amounted to ten dead and fifteen wounded, while Confederate casualties numbered seven dead and twenty-six wounded. After burning captured items that could not be carried, Gano's troops, along with their prisoners and spoils, returned to camp within twenty-four hours of their departure. An ineffective Union pursuit did not hinder the Confederates. It was one of the very few complete Confederate victories in the Indian Territory operational region. The Union prisoners were double-timed to the rear in the summer sun and heat and were eventually marched to Camp Ford, Texas.[36]

Maj. Mefford and the company officers, and ultimately Col. Judson, were to blame for this disaster. This four company battalion had been in the same camp for four weeks. As exposed as this battalion was to the open prairie and as far away from supporting units as they were, this camp should have been heavily fortified. There was no excuse for not constructing breastworks and rifle pits to protect their camp. The troops' duty only consisted of guard duty and monitoring the herds as they grazed. This left a substantial amount of downtime that could have been put to good use by developing some defensive works around the camp.

From his headquarters in the Choctaw Nation, D.H. Cooper believed that the success of Gano's force at Massard Prairie could be duplicated if they acted quickly. Cooper decided to test Union defenses at Fort Smith with a significantly larger force. Cooper also aimed to escort pro-Confederate families from Sebastian County, which had limited success. He gathered the brigades of Brig. Gen. Stand Watie, Brig. Gen. Richard Gano, and other units, arriving near Fort Smith at sunrise on July 31, 1864. Cooper positioned a diversionary force in the Poteau River bottoms west of Fort Smith to snipe at the post, harassing the garrison with rifle fire all day. Cooper tried to execute his plan:

"…General Watie's command,

Colonel Folsom's command, Wells' battalion, and Howell's section advanced under my personal direction on the main Fort Smith road. General Watie was sent forward with guides to drive in the enemy's pickets on the main road leading directly to one of the principal works in front of Fort Smith, at Negro Hill, and also on the Line road to the garrison..."[37]

The main Confederate force advanced northward up the Fort Towson Road on the south side of Fort Smith. As Cooper's force of Gano and Brig. Gen. Watie's units swarmed out of the Indian Territory and across the prairie south of Fort Smith, their plans went awry. At 11:00 a.m., Watie's brigade led the advance, driving back a picket of the 6th Kansas Cavalry. Lt. Levi Stewart of Company I, 6th Kansas Cavalry, commanding that picket, recalled:

"...Being stationed on outpost duty on the Texas [Fort Towson] road about four miles and a half from Fort Smith, Ark., about 11 [a.m.] on the morning of July 31, 1864, hearing my picket firing, I immediately mounted my men, numbering thirty-five in all, and started to learn the cause of the firing. After proceeding about half mile I met my pickets coming toward one on a run and a number of the enemy following them, at which I halted and formed a line, and after exchanging shots with the enemy I found they were too strong for the number of men under my command and I [was] forced to fall back toward Fort Smith, Ark..."[38]

This picket alerted the Union garrison of the Confederate advance, and the picket exchanged fire with Watie's cavalry as they withdrew. Cooper continued his advance:

"...General Watie executed the order given him with his accustomed gallantry and promptness, sending Colonel Bell, with First Cherokee Regiment, on the main road and Colonel Adair on the road to the left known as the Line road, both detachments charging with the gallant impetuosity for which they and their men are noted. He not only routed the Federal pickets, but ran them up to the line of their intrenchments near Fort Smith...

The enemy having rallied soon began to show themselves on the road, and some sharp skirmishing ensued. The First Creek Regiment was ordered forward to support Colonel Bell, which they obeyed with an alacrity and enthusiasm highly creditable to both officers and men..."[39]

Unfortunately for the attackers, the Federals had pulled most of their outlying camps within the fortifications surrounding the city of Fort Smith, which had grown up near the fort and was a wealthy, as well as a rough and tumble, frontier community. Although again taken by surprise, the Union forces quickly formed into line of battle and brought their artillery into action. Union forces started to appear on the Fort Towson Road and engaged with the advancing Confederates, prompting Cooper to gather several of his scattered commands near the road. Gano's brigade was deployed on the flanks, supported by Watie's dismounted Cherokees. Capt. John T. Humphrey's artillery section, consisting of two light mountain howitzers, advanced and fired upon the Union forces positioned about 600 to 800 yards away. This action led the Union troops to retreat into their fortifications at Fort No. 2 on the Fort Towson Road. For a while, Confederate howitzers bombarded Fort No. 2 and its surrounding trench line. Thayer began to prepare for a response to the Confederates. As the Confederate cannonade continued, the Union forces launched a counterattack with a small cavalry contingent, some infantry, and two cannons from the 2nd Kansas Battery advancing down the Fort Towson road in front of Fort No. 2. Cooper described the action:

"…Soon after General Watie reported the enemy advancing in force, having driven in the Cherokee pickets, but was feeling his way very cautiously. Leaving the reserve (Howell's section and First and Second Choctaw) at Mickles', under Colonel Folsom, Captain Humphreys was ordered forward to General Watie's position and General Gano to cross over from Massard Prairie and join me on the main road to Fort Smith. Having arrived at the camp lately occupied by the enemy I found Brigadier-General Watie with his command in position on the hill south of the spring. General Gano soon arrived, and Captain Humphreys, with his light battery, was advanced and opened on the enemy… The enemy soon

brought up a four-gun battery (Rabb's, I suppose) and commenced a furious cannonade upon our light howitzers, the shot and shell passing harmlessly over our heads for some time. Captain Humphreys, being so unequally matched, was ordered to withdraw…"[40]

Two companies of the First Kansas Infantry (Colored) tangled with Watie's Confederates from the rifle pits that extended from the Towson road to the Poteau River. The Confederate light battery of mountain howitzers was brought up and was just gaining the upper hand when two sections of the 2nd Kansas Battery, armed with 10 pounder Parrot guns drove off the Rebel battery in 15 minutes of firing. Private Henry Strong, of Co A, 12th Kansas Volunteer Infantry described the action "The 2nd Kansas Battery were soon playing on them pretty lively and forced them to fall back…considerable firing with artillery". The Union cannons, equipped with rifled barrels for spinning projectiles, had superior accuracy and range compared to Humphrey's howitzers. Supported by two companies from the 1st Kansas Infantry (Colored), the Union artillery deployed and commenced firing. The Union guns soon dominated the field. When Cooper realized his howitzers were outmatched, he ordered their withdrawal. Although the Union troops had been driven back onto the ridge under the guns of Fort #2, a few quick shots from the cannons drove the Confederates from the field. Capt. E.A. Smith of the 2nd Kansas Battery reported:

"…On the morning of the 31st, Rebels drove in our pickets at Ft. Smith, and

Capt. Smith was ordered out with Lt. [Andrew G.] Clark's section to reinforce Col. [William] Judson, who, with a small cavalry force was skirmishing with the enemy's advance. Taking position about a mile west beyond Fort #2, on the Texas [Fort Towson] road, he opened with such effect as to silence the rebel battery at the first shot. The fire was then directed on a heavy rebel cavalry line, and in fifteen minutes they were driven from the field in great confusion…"[41]

The Confederates were not so confused in that they captured both a herd of beef cattle and the soldiers guarding them. Cooper's force fell back as far as Cedar Prairie in the Indian Territory. The Federals calculated their losses at approximately $130,000. Both sides reported only light casualties.[42] There was only one mention of the battles in the Union Official Records:

Springfield, Mo., August 5,1864.
Maj. O. D. Greene, Assistant Adjutant-General:
The operator at Fayetteville telegraphs to Colonel Harrison, now here, that Generals Cooper, Gano, and Stand Watie had attacked our troops at Fort Smith on last Tuesday, and continued it for three days, with 6,000 men and twelve pieces of artillery, and that our losses are 120 killed, 112 wounded, and 119 prisoners from the Sixth Kansas, and some money and stock, and that the enemy is now posted so as to cut off re-enforcements from Fort Gibson and Little Rock. This

information is from citizens that got through to Fayetteville to-day.

JOHN B. SANBORN,
Brigadier-General.

At dawn on August 1st, the Confederate detachment situated between the two rivers commenced shelling the garrison, maintaining intermittent fire throughout the day. The Confederates engaged in a strategic maneuver, firing at the garrison and then relocating to another position before the Federal artillery could establish their location. This engagement resulted in no damage or casualties on either side. By nightfall, the Confederate forces had withdrawn. Gen. Thayer did not follow his enemy, stating that he did not have enough healthy horses to conduct a successful pursuit, especially with the loss of the 6th Kansas Cavalry's horses on Massard Prairie. On August 2nd, Company K of the 12th Kansas Infantry, including Private Henry Strong, was dispatched across the Poteau River to clear trees and create an open field of fire. The events of the preceding six days had significantly impacted the Federal garrison at Fort Smith. A substantial number of refugees sought protection within the fortification lines, thereby placing additional strain on the outpost at Fort Smith. Meanwhile, the Confederate Indian and Texas Brigades withdrew back into the Choctaw Nation and established camps around Scullyville.

Maxey's Indian and Texas forces continued to monitor any activity around Fort Smith. In fact, at dawn on August 24, Brig. Gen. Watie crossed the Arkansas River with a 500-man detachment from his command above Fort Smith and attacked a Union hay camp on Gunther's Prairie that was being worked by the 11th United States Colored Troops. The camp was located approximately 12 miles northwest of Fort Smith, somewhere near the town of Long, Oklahoma. His intention was to destroy as much hay and capture as much stock as possible. Watie reported that the Federals had about 350 infantry and 70 cavalry in the camp, a larger contingent than he had anticipated. He stated that he killed 20, captured 14, captured 150 mules and horses, and burned a large quantity of hay. The Federal report lists the deaths of one Army surgeon and four soldiers of the U.S. Colored Troops. The Cherokees got away with only one killed and several wounded.[43]

[1] Grant, Ulysses S. Personal Memoirs of U.S. Grant. New York: Dover Publications, 1995. pp. 143

[2] Blunt, *Experiences*, pp. 249-250.

[3] Cantrell, M.L. and Harris, Mac. Editors. Kepis & Turkey Calls: An Anthology of the War Between the States in Indian Territory: Shirk, George H. "The Place of Indian Territory in the Command Structure of the Civil War." pp. 5-6.

[4] Maxey to Cooper, OR, Ser. I, Vol. XXXIV, (P2) pp. 858.

[5] Hood, pp. 435.

[6] Franks, *Stand Watie*, pp. 159.

[7] Carl W. Breihan. Quantrill and His Civil War Guerillas. New York: Promontory Press, 1959. Pp. 130-150.

[8] Watie to Cooper, OR, Ser. 1, Vol. XXXIV, (P1) pp. 1012-1013.

[9] Baird, W. David, Editor. A Creek Warrior for the Confederacy: The Autobiography of Chief G.W. Grayson, Norman and London: University of Oklahoma Press. 1988: 81-83.

[10] Bearss and Gibson, *Fort Smith*, pp. 285.

[11] Diary of Henry Strong, June 15, 1864, Fort Scott National Historic Site, Fort Scott, Kansas, National Park Service.

[12] Baird, ed. *A Creek Warrior*, pp. 83.

[13] Ghormley, Michial O. *Historical Story of M.O. Ghormley*, Indian-Pioneer History Statement, February 2, 1936. Works Progress Administration. University of Oklahoma Historical Collections Online.

[14] Rampp, Lary C. "Negro Troop Activity in Indian Territory." Cantrell, M.L. and Harris, Mac. Editors. Kepis & Turkey Calls: An Anthology of the War Between the States in Indian Territory. Oklahoma City: Western Heritage Books. 1982. 204.

[15] Baird, ed. *A Creek Warrior*, pp. 85.

[16] Watie to Cooper, OR, Ser. I, Vol. XXXIV, (P1) pp. 1013.

[17] Cooper to Scott, *OR*, Ser. I, Vol. XXXIV, (P1) pp. 1012-1013.

[18] Keun Sang Lee, "The Capture of the J.R. Williams." Chronicles of Oklahoma, 60(1), Oklahoma Historical Society, 1982, pp. 22-33.

[19] Thayer to F. Steele, *OR*, Ser. I, Vol. XXXIV, (P4) pp. 503-504.

[20] Watie to Cooper, *OR*, Ser. I, Vol. XXXIV, (P1) pp. 1013.

[21] Crawford, Kansas in the Sixties, pp. 136-137.

[22] Ibid.

[23] Watie to Cooper, *OR*, Ser. I, Vol. XXXIV, (P1) pp. 1013.

[24] Dwight D. Hitchcock to "My Dear Friends," June 2, 1864. Alice Robertson Collection, Special Collections. McFarlin Library, University of Tulsa, Tulsa, Oklahoma; Warde, pp. 205.

[25] Anderson, T.J. General Orders, July 30, 1864, OR, Vol. ILIII, (P2) pp. 476.

[26] Cooper to Scott, *OR*, Vol. XLI, (P1) pp. 31.

[27] *Galveston* (Texas) *Weekly News*, August 17, 1864

[28] Dale Cox. The Battle of Massard Prairie: The 1864 Confederate Attacks on Fort Smith, Arkansas. Bascom, Florida: William Cox, Publisher. 2008. pp. 34-50.

[29] *Houston (Texas) Telegraph*, August 16, 1864. Pp. 85.

[30] James Robert Barnes, n.d. Accounts of the original Choctaw enrollees

[31] Burgoyne to Judson, *OR*, Ser. I. Vol LIII, Supplement, pp. 481.

[32] *Houston (Texas) Telegraph*, August 16, 1864. Pp. 85.

[33] Morehead to Judson, *OR*, Vol. XLI, (P1) pp. 25.

[34] Burgoyne to Judson, *OR*, Ser. I. Vol LIII, Supplement, pp. 481.

[35] Morehead to Judson, *OR*, Vol. XLI, (P1) pp. 25

[36] Cox, pp. 34

[37] Cooper to Scott, *OR*, Vol. XLI, (P1) pp. 32-33.

[38] Stewart to Judson, *OR*, Vol. XLI, (P1) pp. 25-26.

[39] Cooper to Scott, *OR*, Vol. XLI, (P1) pp. 33.

[40] Ibid, pp. 33-34.

[41] Kansas, *Report of Adjutant General 1861-1865*.

Pp. 265-266.

[42] Maxey, S.B. *OR*, Vol. XLI, (P1) pp. 29-30; Cooper, pp. 31-36.

[43] Maxey to Anderson, *OR*, Vol. XLI, (P1) pp. 279.

Chapter 18
The Final Confederate Raids and the Last Surrender

At the end of August 1864, Confederate General Sterling Price began his historic raid into Missouri from Camden, Arkansas. Price was formerly commander of the Missouri State Guard but had risen to the rank of major general in the Confederate Army and was now in command of the Department of Arkansas. It had always been his intention to bring Missouri into the Confederacy and wanted to try at least one more time. Lt. Gen. Edmund Kirby Smith, Confederate commander of the Confederate Trans-Mississippi Department, also desired to retake Missouri to lift the spirit of the Southern cause. In planning the raid, Kirby Smith ordered Gen. Maxey to cause as much havoc in the Indian Territory as he could to draw attention and forces away from Price's cavalry force. Although Maxey had less than twenty-five hundred troops available, he was determined to follow his orders to the best of his ability. In mid-August, Stand Watie had submitted a plan for a major cavalry raid into the Federal-held areas north of the Arkansas River with both his and Brig. Gen. Gano's brigades. Maxey submitted the plan to Kirby Smith who now approved it and directed that it be conducted in conjunction with Price's Raid.[1] Brig. Gen. Cooper described the mission in a letter to Capt. Thomas Scott, Maxey's adjutant general:

> "…Generals Watie and Gano have been ordered across the river and will

cross above Gibson to-morrow night, sweep around by William Alburty's, twenty miles above Gibson at Grand River, destroy a large hay camp there, take in mules, &c., herded; perhaps run into a train now expected from Fort Scott; return by Mackey's Salt-Works and the camp on Sallisaw and back, recrossing Arkansas at the mouth of Canadian or Webber's Falls.

If Gibson is weak enough they may look in there…"[2]

Watie reported that he would have to postpone the beginning of his operation until sufficient healthy horses could be procured. Both Watie and Gano had earlier received their commissions as brigadier generals, but Gano's date of rank preceded Watie's. Watie graciously accepted Gano as the commander of the upcoming raid. The raiding force would consist of Gano's 5th Texas Cavalry Brigade, which included the 29th, 30th, and 31st Texas Cavalry regiments, Martin's Cavalry Battalion, the Gano Guards, and Howell's Battery of six guns, about 1,200 men. The Gano Guards consisted of remnants of his original two-company Texas Battalion that had been assigned to the Western Theater and was later merged into the 7th Kentucky Cavalry (CSA). This unit was involved in the battles of Perryville and Lexington, Kentucky before participating in Gen. John Hunt Morgan's Raid through Kentucky, Ohio and West Virginia. After suffering debilitating losses, Gano brought those Texans that remained back to the Trans-Mississippi.[3]

Watie commanded the First Indian Brigade which was made up of portions of the 1st and 2nd Cherokee regiments, the 1st and 2nd Creek regiments, and Col. John Jumper's Seminole Battalion, a total of about 800 soldiers. They left out of Camp Pike, located on the south side of the Canadian River opposite Briartown and near the current city of Whitfield, Oklahoma. They camped at Camp Prairie Springs south of Fort Gibson on September 14. The Confederate force spent most of the 15th crossing the swollen Arkansas River west of Fort Gibson near the Old Creek Agency, and then the Verdigris

River at Vann's Ford, near the small hamlet of Sandtown. Local tradition states that Confederate scouts led by Bill McCraken rode to the top of Blue Mound, west of present-day Wagoner, and at a distance observed a large Federal hay camp. This is a reputable record since Blue Mound is a significant height from which much of the prairie region can be observed. In fact, Gen. Gano remarks on this activity in his official report:

> "…General Watie and staff with my staff accompanied me to the top of a mountain while the command was halted below, and from our elevated position we could view their camps, and with spy glasses could see them at work making hay, little dreaming that the rebels were watching them…"[4]

Col. DeMorse of the 29th Texas Cavalry also notes this activity:

> "…whilst the remainder of the command moved to the foot of a large mound in the open prairie, and halted to give time to the two former regiments to get into position. From the top of this mound, with the aid of the glass, could be seen the working party, mowing hay as if in perfect security. Sufficient time having elapsed for the first detachment to get into position, and move simultaneous with our own movement, the command moved rapidly forward, to another mound within a mile and a half of the encampment…"

Gano sent the 30th Texas Cavalry and the 1st Cherokee Mounted Rifles, under the command of Capt. Strayhorn, to the far right to block the Texas Road between the camp and Fort Gibson. The rest of the Confederate force moved closer to the hay camp, and Gano began to deploy his regiments for the upcoming attack.

At this point in the war, the greatest issue facing the Federals in Indian Territory was keeping Fort Gibson resupplied. It took a significant amount of provisions and forage to maintain both the troops garrisoning the post as well as the refugee Loyalist Indians who had returned to Indian Territory with the Union Army. Unfortunately, the guerilla war that plagued the Cherokee Nation took its toll in destroyed food crops, livestock, and forage. This lack of forage severely impacted any cavalry operations. Troops were forced to graze their horses at long distances from the post or depend on various hay stations where they could gather then transport the grass to the military commands. One such hay station was located on a small creek south of the larger Flat Rock Creek, approximately 15 miles northwest of Fort Gibson. It was garrisoned by 125 men of Company K, 1st Kansas Infantry (Colored) and Company C, 2nd Kansas Cavalry, with detachments from Companies G & L, all under the command of Captain Edgar A. Barker. The camp's location was described by Wiley Britton as:

> "…nearly two miles from the Grand River timber, along a prairie branch along which, every hundred yards or so, there were pools or lagoons from a few yards to fifty yards long, and in places perhaps two feet deep, connected by narrow threads of water. The low banks of the lagoons were generally precipitous or caving, with

overhanging boughs of small willows. In some of them there were numerous water lilies, with large palm-like leaves floating on the surface…"[5]

Working a haying station was one of the hardest and worst jobs that enlisted men had to perform, ranking up there with digging trenches, mines, and latrines. This when coupled with the hot September weather that was still occurring in the region, made the job almost unbearable. At this haying camp, the 37 black soldiers were the primary ones who did the cutting and loading of the hay while the white troops provided camp guards, mounted patrols and scouts at the river crossings to maintain a watch for Confederates or bushwhackers.

On September 16 Barker's scouts monitoring the Verdigris River crossings noted the advance of a large body of Confederates moving toward the Union camp. Unfortunately, they only reported a force of approximately 200, not realizing this was only the Confederate advance party. Believing his force was sufficient to defend the camp, Barker ordered his troops into line of battle in the ravine behind the camp, the best available position in a prairie environment. Barker and a squad of mounted troopers advanced approximately two miles and encountered Watie's and Gano's entire force on a ridge southwest of the camp. Barker quickly realized his mistake, and he and his troopers quickly retreated back to the ravine and dismounted with the infantry. Barker reported:

> "…I met them about two miles from my camp, 1,000 or 1,500 strong, with six pieces of artillery. I immediately

fell back, skirmishing with their advance, which made several unsuccessful attempts to cut me off from my camp; after reaching which I dismounted my men and placed them in the ravine with the others, which was no sooner accomplished than the main body of the enemy appeared and attacked…"[6]

When the Confederates came within a mile of the hay camp, Gano deployed two Texas regiments, 29th and 30th Texas Cavalry, under Lieutenant Colonel Otis G. Welch to the right and the Indian Brigade, the 2nd Cherokee Mounted Rifles, 1st Creek Mounted Rifles, 2nd Creek Mounted Rifles, and the 1st Seminole Battalion, under Watie to the left. General Gano led the remaining forces in the center for the attack. His force consisted of Martin's Texas Cavalry Battalion, the Gano Guards, and Howell's artillery battery of six guns. Gano described his plan of attack:

> "…From thence we moved to within one mile of their camp unperceived, and I sent Lieutenant-Colonel Welch to the right with a column composed of the Twenty-ninth and Thirty-first Texas Cavalry (De Morse's and Hardeman's), while General Watie conducted the Indian column to the left, while I carried forward the center, with Howell's battery supported by Martin's regiment, the Gano Guards, under Captain Welch, and Head's and Glass' detachment of companies. I could distinctly see Captain Strayhorn formed in the enemy's rear. The clouds looked somber and the V-shape procession grand as we moved forward in the work of death…"

They advanced to approximately 200 yards from the Union force and attacked from five different points. The small Federal detachment was able to fend off the Confederate attack, including three cavalry charges, for about thirty minutes before they were surrounded and casualties began to take their toll. But for the barefooted Confederates, the attack was somewhat difficult as Pvt. W.T. Sheppard with the 5th Texas Partisan Rangers later wrote:

> "…While we were making a charge on the enemy's camp on foot, we passed over some ground where beef cattle had been corralled while the ground was wet, which left it in a rough condition, but at this time was dry and full of holes…"[7]

Capt. Barker mounted up those white Kansas troopers he could find and ordered the remaining dismounted cavalry as well as the black infantrymen to retreat to the timber along the Grand River, approximately one mile to the east. Barker led his mounted troopers in a successful charge through a weak point of the Confederate line and was able to escape along with 15 others. The remainder of Barker's 50 or so white troopers were killed or captured. The black infantry troops continued their fight under Lt. Thomas B. Sutherland for almost two hours from the ravine and banks of the creek. This is a remarkable feat since they were outnumbered nearly 12 to 1. Eventually Gano brought up his artillery and began to fire grapeshot into the black Union line, which soon disintegrated. Gano stated in his report:

> "… I sent Major Stackpole with a captured Federal lieutenant under flag

Confederate Camp Pike near Whitfield, Oklahoma

Battle Creek looking East from Battlefield North of Wagoner, Oklahoma

Battle Creek looking West from Battlefield North of Waggoner, Oklahoma of Wagoner, Oklahoma

Thickets and water-filled pools on the Union Hay Camp Battlefield
Photos by Author

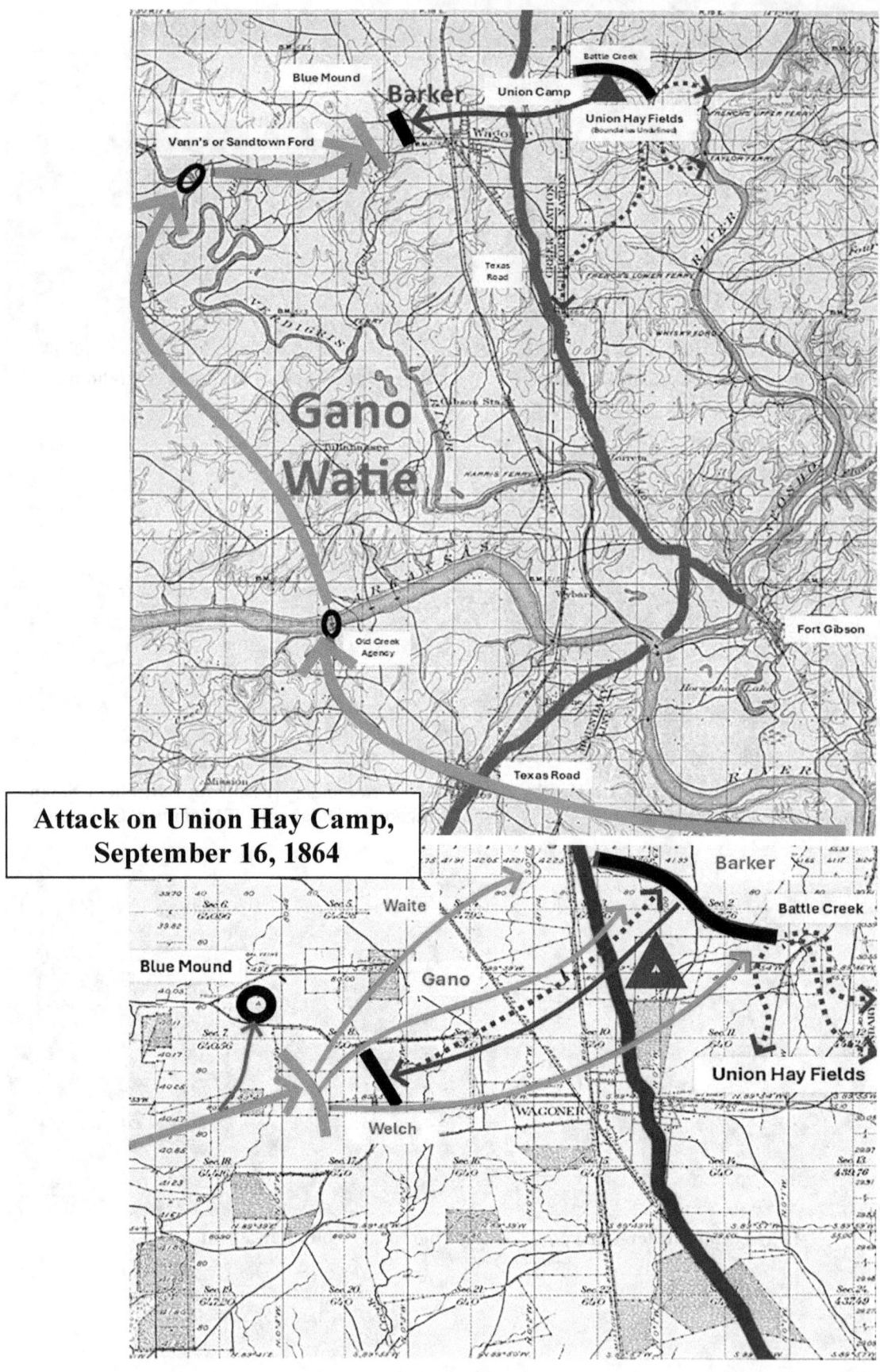

Attack on Union Hay Camp,
September 16, 1864

of truce to demand surrender, but they fired upon my flag and then commenced the work of death in earnest. The sun witnessed our complete success, and its last lingering rays rested upon a field of blood…"[8]

The 1st Kansas Infantry (Colored) probably recalled the treatment they received by the Trans-Mississippi Confederates after their surrender at Poison Spring earlier in the year. The survivors successfully evaded capture by making it to the Grand River timber or by concealing themselves in the pools or prairie shrubbery. The Confederates offered no quarter to the black troops. Capt. Grayson of the Creek Regiment later stated in his autobiography:

"…The men proceeded to hunt them out much as sportsmen do quails. Some of the black soldiers realizing that they were about to be discovered would spring up from the brush and cry out. O Master! Spare me! But the men were in no spirit to spare the wretched unfortunates and shot them down without mercy…"[9]

Some of the black soldiers attempted to hide under the water with only their noses in the air to breathe. These men were soon discovered and executed, except for one who was able to maintain his nose under a water lily leaf until the Confederates tired of hunting them down. Unfortunately, in the end, 33 of the 37 black soldiers were killed in the ravine. In fact, Private Jefferson P. Baze of the 13th Texas Partisan Rangers later wrote:
"…The Negroes were nearly all killed in a little creek in which they had

jumped when our army came up. The water was red with the blood of the dead Negroes. The few Indians who were along with the army dragged the dead bodies from the river and took all that was any value from them…"[10]

(Note: This creek was officially called "Nigger Creek" on maps until the middle of the 20th century. It then remained unnamed on maps until a college student from Wagoner, Oklahoma, petitioned the Oklahoma Geological Survey to rename the waterway as "Battle Creek" to honor the soldiers killed during the battle.)[11] Capt. Grayson continued:

"…I confess this was sickening to me, but the men were like wild beasts and I was powerless to stop them from this unnecessary butchery. Toward the end one of our men found a young white man and brought him to where I sat on my horse and asked me, 'Should we not kill him too?' or words to that effect. I told him not to kill him but to take and turn him in with the bunch of prisoners already taken, that it was negroes we were killing now and not white men…"[12]

Lt. Barker reported a total of 40 killed, wounded, and missing along with 66 being captured, including Lt. Sutherland. He also reported the loss of 1,000 tons of hay, all of the haying equipment, all camp equipment, and many horses, mules, and wagons. The Confederate commanders also learned from Union prisoners that a large wagon train was expected from Fort Scott at any time.[13]

The Confederate raiders camped on

the Hay Camp battlefield on the night of September 16. On the morning of September 17, they proceeded northward along the Old Military Road at Flat Rock Creek. Receiving no information about the wagon train throughout the day, Gano became concerned that it might have taken the road to the east of the Grand River, thus circumventing his position. On that day, Gano also sent a party to burn the hay supply at another camp at the Hickey Place, located about seven miles east of the Flat Rock crossing. They found this camp to be much better prepared and defended, and the Confederate raiders were driven off after reinforcements arrived from Fort Gibson.[14] That night, the Confederates established their camp at Wolf Creek, equidistant between both roads. The next morning, Gano instructed Watie and his brigade to conduct reconnaissance on the east side of the Grand River. By doing so Gano learned that the Union wagon train had not traversed either road. The same night a scout from Fort Gibson met with the Federal wagon train approximately 30 miles north of the Confederate camp on Wolf Creek. The scout informed Major Hopkins that a large Confederate force was enroute towards his position, intent on attacking his train in the next day or two. The Cabin Creek Station had a small defensive stockade on the south side of the crossing, which overlooked the prairie in the direction of Fort Gibson. Gen. Gano described the work as "a strong fortification constructed of logs set up in the earth." The stockade was open on the side towards the creek. The station was garrisoned by 170 Cherokees of the 2nd Indian Home Guard under the command of Lt. Benjamin Whitlow.

The Union supply train had departed Fort Scott on its southward journey on September 12, 1864. The train's escort was a mixture of the 2nd, 6th and 14th Kansas Cavalry Regiments totaling approximately 260 soldiers, less than half of them mounted. This detachment was under the command of Major Henry Hopkins of the 2nd Kansas Cavalry. The train itself consisted of 205 U.S. Government wagons, 91 sutler wagons, and four standard ambulances. The value of the supplies exceeded $1 million. Major Hopkins was also responsible for $500,000 in currency to be used as pay for the troops. The train also contained herds of horses and mules to be used at Fort Gibson as well as replacements for the wagons. The Federals had anticipated an attack on this train and planned to increase the size of the escort as it moved south on the Military or Texas Road into Indian Territory. Major Hopkins had sent word to all stations along the road to send out scouting parties to ensure the passage was clear. The escort was increased by the addition of 100 Union Cherokees from the 2nd and 3rd Indian Home Guard when it arrived at Baxter Springs. Major Hopkins was then notified by Col. C.W. Blair at Fort Scott that Maj. Gen. Sterling Price had crossed the Arkansas River near Dardanelle, Arkansas, and was heading north. The Federal commanders realized that there was a real possibility that the Confederates would conduct a diversionary raid in the Indian Territory to prevent Union troops from being pulled to assist Missouri and Kansas forces in stopping Price. Hopkins immediately dispatched riders to forward the information to Colonel S.H. Wattles, who was now in command of Fort Gibson since the departure of Colonel Phillips. Hopkins requested that all available troops be dispatched north along the Texas Road to meet his southbound supply train. The five-mile long train was an over-sized, slow, inviting target for Confederate raiders.[15]

When the train arrived at Hudson's Crossing on the Neosho River, Major Hopkins detached the 50 troops of the 2nd Indian Home Guard as a rear guard since he feared an attack from that direction. The train arrived at the Horse Creek Crossing on the evening of September 17 and camped for the night. At midnight Major Hopkins received the dispatch from Fort Gibson's commander that reported Gano and Watie's Confederate forces north of the Arkansas River and heading for Cabin Creek. Wattles ordered Hopkins to make haste and get the train to the relative safety of the small defensive works at Cabin Creek. Wattles also informed the train commander that he was sending six companies of troops and two howitzers from Fort Gibson to reinforce the train's defense. Hopkins immediately got the train in motion, placed them in a double column, and traversed the final 15 miles to Cabin Creek by 9 a.m. on September 18.[16]

The sheer size of the wagon train made its protection extremely difficult. Hopkins ordered all buildings to be fortified and hay ricks placed in strategic locations to provide a small level of protection to the troops and train. Although Hopkins wanted the train formed in a quarter-circle, some wagons were placed behind the stockade wall but many were still stretched out along the Military Road on the north side of Cabin Creek. The teamsters that could released their animals to graze on the rich prairie grass surrounding the military post. But soon the entire area was crowded with soldiers, teamsters, cattle, horses, and wagon-stock. After a short rest period, Hopkins took 25 men with him from the 2nd Kansas Cavalry

Major Henry Hopkins, USA
Commnding Cabin Creek Post

under Capt. Cosgrove and moved southward down the Texas/Military Road on a recon-naissance to determine the location of the Confederate forces. He found that Gano's leading elements were approximately three miles south of the crossing posted in a hollow in the prairie, probably current Rock Creek. Col. DeMorse later wrote:

> "...Col. Gano called for 300 volunteers from the Texas Brigade, for the purpose of attacking them. This call was instantly responded to, and with the volunteer force, and two pieces of Howell's Battery, we moved forward, and arrived within a mile, and a half of the place at 3 o'clock in the evening...[17]

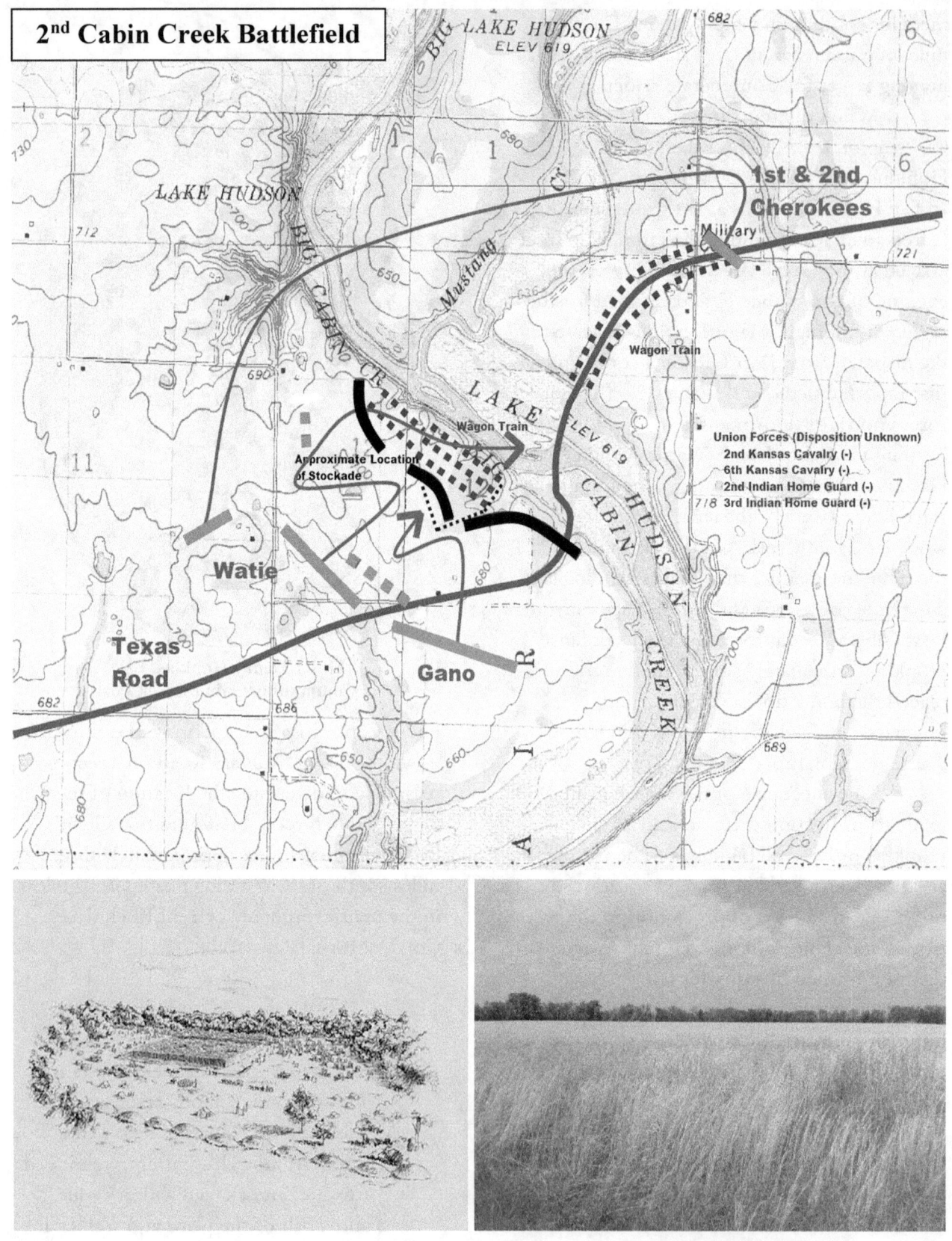

2nd Cabin Creek Battlefield

Artist Conception of Cabin Creek Stockade
(Artist Unknown)

Location of Union Stockade Walls
Photo by Author

The Federals quickly retreated to their defensive positions, and the commander posted heavy pickets along the southern approaches to the Cabin Creek crossing. Maj. Hopkins estimated the Confederate force to be approximately 1,200, which he believed was within his command's ability to defend the train. Wagon Master R.M. Peck wrote in *The National Tribune* in 1904 that Hopkins wasted the entire day of September 18

> "spreading out the wagon train for grazing as if it was a regular camp, instead of focusing on constructing defenses for the fight he knew was coming." Another teamster, Alfred Collins, noted that "…the train was allowed to camp in open order, scattered over all creation. It seemed like as though there wasn't an enemy within one hundred miles of us…"[18]

A dentist from Leavenworth, Kansas, Dr. George A. Moore, later stated:

> "…the loose disposition of the train wagons certainly indicated a lack of prudence by somebody…"[19]

Hopkins did not realize that Watie's brigade had been left at Wolf Creek in order to scout the east side of the Grand River, in case the train was on that route. When Watie arrived from his mission at midnight, the Confederate assault force would number over 2,000, giving the Southerners a two to one numerical advantage.[20]

Midnight brought a bright, moon-lit prairie. When Gano had observed Hopkins' reconnaissance in the afternoon, he sent word to Watie to bring up his brigade and artillery.

It must have been like "deja vu all over again" for Watie, who had been severely beaten by Union forces, including the 1st Kansas Infantry (Colored), at Cabin Creek just over a year prior. Col. DeMorse wrote to his hometown newspaper:

> "…This gallant officer, ever prompt, arrived at 12 o'clock, and the whole line was moved forward, and a partial investment of the place was made.— The enemy by this time infuriated with whiskey, would ride within a short distance of our lines, and defy us to move up, and give battle…"[21]

Together Watie and Gano moved up towards the Cabin Creek crossing and the wagon train prize. At approximately 1 a.m. under the moonlight, they deployed in a line of battle, cavalry in the front ranks and followed by infantry in the second rank. Gano and his Texans formed on the right and Watie and his Indian Brigade on the left. Captain Howell's six-gun artillery battery deployed at the center of the line in support of the Confederate advance.[22] At this point Federal Capt. Patrick Cosgrove came forward to determine who the troops were since they were waiting for reinforcements from Fort Gibson. When the Confederates identified themselves, Gano asked if the Federals would receive a flag of truce. Cosgrove said he would return with an answer in five minutes, and he returned to the Union lines. Cosgrove received Hopkins' decision to allow the flag of truce but apparently the Confederates decided against it, although both sides contend that the other side dismissed it. Gano waited for 15 minutes before beginning his advance although he could hear the wagons begin to move around behind the Union defenses.[23] At about 3 a.m. the entire Confederate force moved forward. Gano's Texans moved first and clashed with the Federal left. Howell's battery opened fire on the Union stockade, attempting to breach the walls. Watie's troops moved forward and skirmished with the Federal right in the area around the large hay ricks. Although both sides had the advantage of the bright moon, it was still too dark for either one to actually determine the size or disposition of the other. Gano pulled his forces back to wait for dawn.[24]

Major Hopkins had little verifiable information concerning the Confederate forces facing him. He assumed it was the large Confederate raiding unit under Watie and Gano that had attacked the Hay Station. One disheartening discovery was Howell's artillery battery. The Union force did not possess any cannons at all. The first volleys of standard infantry and cavalry weapons did not alarm the Federals, but when the artillery battery began lobbing shells into the Union lines the scene turned into pandemonium with mules, horses, soldiers, and teamsters running in all directions. Wagon Master Peck later recalled:

> "...Soldiers, teamsters, mules and wagons in a perfect cyclone of excitement, went flying back towards the timber.. some men mounted the unhitched mules, where they could catch them, only to be thrown off again by the frantic animals. Some ran on foot among the wildly rushing torrent of men, wagons and animals, teams and parts of teams dragging their wagons without guidance crashed into each other as they turned from the rebel fire in their front and

raced madly for timber in the rear..."[25]

Most teamsters immediately mounted mules from their teams and fled back up the road towards Fort Scott. Many tried to re-cross Cabin Creek, but the crossing became a logjam of wagons and mules trying to make it to the north side. A very small group of wagon-masters and teamsters attempted to quell the stampede to no avail. Wiley Britton stated:

> "...The bluff that rose almost abruptly from the creek in the rear of the camp, the stockade, and a narrow ravine on the Federal right afforded much protection to the Federal soldiers during the terrible artillery fire. Where their line was much exposed to this fire they were obliged to lie down prone upon the ground behind logs and felled trees, in depressions or behind elevations of the ground for protection. Having gained a position on the Federal right and rear, General Gano commenced driving in the Federal skirmishers from that quarter, when they came to a sudden stand, that halted the Confederates..."[26]

Due to the confusion Hopkins was only able to muster approximately 400 soldiers to hold the line, but they showed enough determination to make the Confederates back off until daylight. Lieutenant William B. Clark of the 14th Kansas Cavalry observed:

> "...Ordered to dismount and fight on

Gano & Watie's Retreat from Cabin Creek

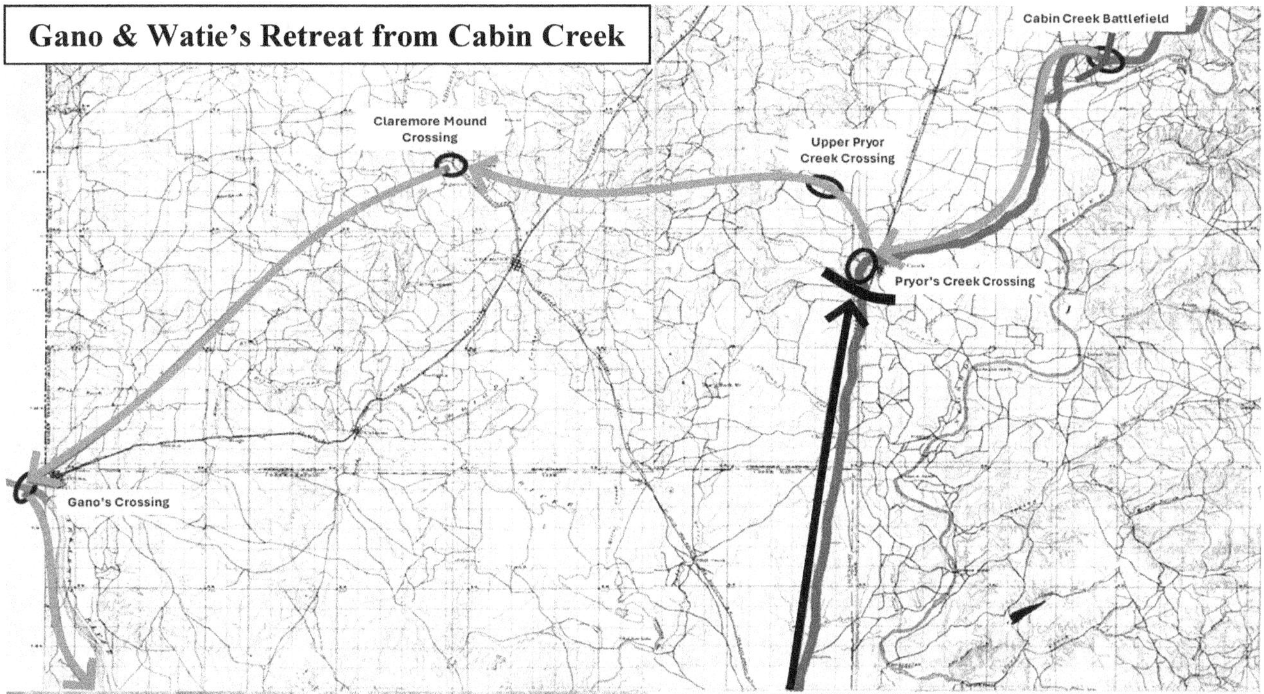

Texas Road Crossing of Cabin Creek
Photo by Author

Texas Road moving away from Cabin Creek Crossing. *Photo by Author*

Claremore Mound Crossing of Verdigris River

Gano's Crossing over the Arkansas River west of Tulsey Town (Tulsa)

foot and during the excitement of the stampeding of the train, the guard of the horses was driven back and the horses ridden off by soldiers, teamsters, etc. The Company holding their position until 9 a.m. fell back marching on foot to Fort Scott, Kansas a distance of 130 miles…"[27]

Unfortunately, the destroyed and damaged wagons at the crossing site made a retreat with the remaining wagons impossible and severely limited the Federal's options. Their only real hope was the arrival of Maj. Foreman and his infantry reinforcements and artillery from Fort Gibson, which could attack the Confederate lines from the rear. Hopkins did not know it at the time, but Foreman's force would not arrive at Cabin Creek until the next day. By then the Confederates would be gone.[28]

At daybreak the Confederates again moved upon the Union lines, heavily supported by their artillery. Col. DeMorse described the action:

"…At the dawn of day four pieces of Howell's Artillery opened upon the enemy, while the two remaining pieces were moved to our extreme left, and opened upon the right flank of the enemy's position. The Seminoles and Creeks, under Col. Jumper, conjointly with the 30th Texas Cavalry, engaged them upon this flank, whilst the remainder of the Brigade held its former position. Gano soon detected that this was the weakest point, and the entire command was immediately concentrated upon this position…"[29]

Watie ordered Howell to move one section of two guns to the far left flank, supported by the 1st and 2nd Creek Regiments. Prior to the final attack, the Indian troops began smearing potent "medicine" all over their bodies that would keep the bullets from penetrating their skin.[30] Fearing that the wagon train would escape, Watie sent the two Cherokee regiments across Cabin Creek upstream of the crossing to cut off any attempted escape. During the interlude, Hopkins had again attempted to rally some teamsters and wagon-masters to help clear the crossing. The attempt was abandoned when Hopkins discovered that Watie's Confederate force had swung around his right flank and had cut off his escape route towards Fort Scott, Kansas. The Union troops kept up a severe fire but began to fall back in small groups. A large group of Union Indians took refuge in the stockade, but Howell focused his artillery on the structure and drove the Federals from it. Pvt. D.O. Crane of the 5th Kansas Cavalry described the maelstrom within the stockade when the artillery began to fall among the wagons and teams:

"About 1 am, we were formed in a line of battle in front of the train. The teamsters had been ordered to hitch up. At about 2 am, the Rebels opened their fire with artillery and all the mules stampeded; some going over the high bank; leaving us with a handful of men out on the prairie with no protection. Soon a charge was made by the Rebels. We were lying flat and when the command was given a deadly fire was opened upon them.

They answered, but overshot and did no damage. They retreated in disorder. We fell back several yards and awaited results. I am sorry to say that several of our soldiers and officers put spurs to their horses toward Fort Scott at the first fire…"[31]

Troops from the Seminole Battalion and the 29th Texas Cavalry swung around the left side of the artillery section and swept through the Union lines and encampments. At this point the entire Federal line broke for the rear.[32] Maj. Hopkins rallied his remaining troops and led them downstream toward the Grand River, completely abandoning the train, supplies, and camp equipage to the Confederates. He hoped to finally meet up with Foreman's force and together retake the train. Unfortunately, by moving to the east towards the Grand River, Hopkins missed meeting with his reinforcements moving up the Texas/Military Road. Sgt. Charles Norhood Mumford, Company M, 3rd Wisconsin Cavalry, probably summed up the disaster the best by stating: "…they captured the entire train of wagons and scattered our men like chaff before the wind…"[33]

Gano and Watie could hardly believe their good fortune. They had captured approximately $1.5 million in Union supplies that were desperately needed by both sides but especially for the Southern Indians. Private Baze of the 13th Texas Partisan Rangers again remembered:

"…This was a God-send for us, for we were almost destitute of clothing and weapons. There was not enough clothing to go around so we drew for it, an overcoat fell to me…"[34]

Private W.T. Sheppard of the 5th Texas Partisan Rangers also recalled:

"…I drew a pair of boots (US) my size, but could wear only one of them for quite awhile, on account of James Yeary stepped on one of my home-made spurs. I being barefooted, it shaved off a good slice of the back part of my heel…"[35]

A lack of available transportation again hindered the Confederates and reduced the amount of supplies they could return south of the Arkansas River. A combination of battle-damaged wagons and missing draft animals meant that a significant amount of supplies would need to be destroyed lest they fall back into the hands of the Federals. They moved the wagons they could and burned what they could not along with the almost ten tons of hay and another set of haying equipment. Fearing the arrival of the new Union force, they quickly moved southwest from Cabin Creek on the Texas Road with approximately 130 wagons of captured booty, a small number compared with the nearly 300 wagons the Federals originally had in their wagon train. At approximately 11 am, at the Pryor's Creek crossing, the Confederates encountered Colonel James Williams and troops from the 1st and 2nd Kansas Infantry (Colored) with a battery of six Parrott guns enroute from Fort Gibson. Williams and his force had marched eighty-two miles in forty-six hours and were completely exhausted. Nevertheless, he deployed his Union troops in a line of battle and sent out skirmishers. The Confederate advance claimed to have made contact with the Federal skirmishers and drove them over three miles back towards the main Union line. Williams deployed his Parrott guns, which halted the Confederate advance. The Confederates deployed in a line of battle on some higher ground north of the Pryor's Creek crossing. This show of force helped make Williams reconsider a significant battle with Gano's and Watie's troops with his already exhausted brigade. Capt. John Graten, Company C, 1st Kansas Infantry (Colored) remarked on the artillery duel, claiming that the Confederate guns could not reach the Union lines:

> "…The second shot from our guns struck right in front of theirs… You had better believe that there was some scampering to get away from there… They could not reach us, while a few shots from our guns, completely broke up their lines, and they were glad to get behind some mounds nearby…"[36]

Both sides camped in lines of battle. But Gano had his troops light fires all along his line to fool the Union command that they were simply camping for the night. Instead, Gano and Watie moved the captured wagons under cover of darkness and crossed Pryor's Creek about six miles northwest of the Texas Road crossing. The Confederates recrossed the Verdigris River near Claremore's Mound and eventually crossed to the south side of the Arkansas River at Tulsey Town, current day Tulsa, at a hard rock shallow that is now known as Gano's Crossing. This retreat was made in a wide arc to avoid any more contact with the Union Army. They halted the train near the current day city of Okmulgee and distributed their captured goods and materials. Having been fooled by the Confederate

deception, there was no pursuit by the Union forces. Instead, the Federal command chose to heavily protect their supply trains, which began moving south from Fort Scott almost immediately after the Cabin Creek battle. The Confederates did not make any further attempts on the supply trains.[37]

For their success at Cabin Creek, Generals Gano and Watie received high honors from everyone in the Confederate command system including the Congress and President Jefferson Davis.[38] This one late-war action is remembered as the greatest victory for the Southern Confederacy in Indian Territory. But the capture of one supply train was not enough to salvage the Confederate war effort in the Indian Territory, especially since the remainder of the Confederacy was in the midst of collapsing. This is probably the main reason for the congratulatory message from Richmond. There was so little good news for the South at this point in the war. The victory was complete but not unexpected because it was a major-sized raid on a very soft target. The Union escort was out-numbered at least 3-to-1, had no artillery, and was encumbered with over 300 wagons manned by civilian teamsters, not to mention the hundreds of easily frightened horses, mules, and oxen. There really was no way Maj. Hopkins could have protected this train from Watie's and Gano's force once the artillery shells fell among the frightened animals and teamsters. The crossing of the Arkansas River by Gano and Watie's troops marked the end to any further significant military action by either side for the remaining course of the war.

At the same time as the Gano-Watie Raid, Price's Raid, also known as Price's Missouri Expedition, took place as planned by

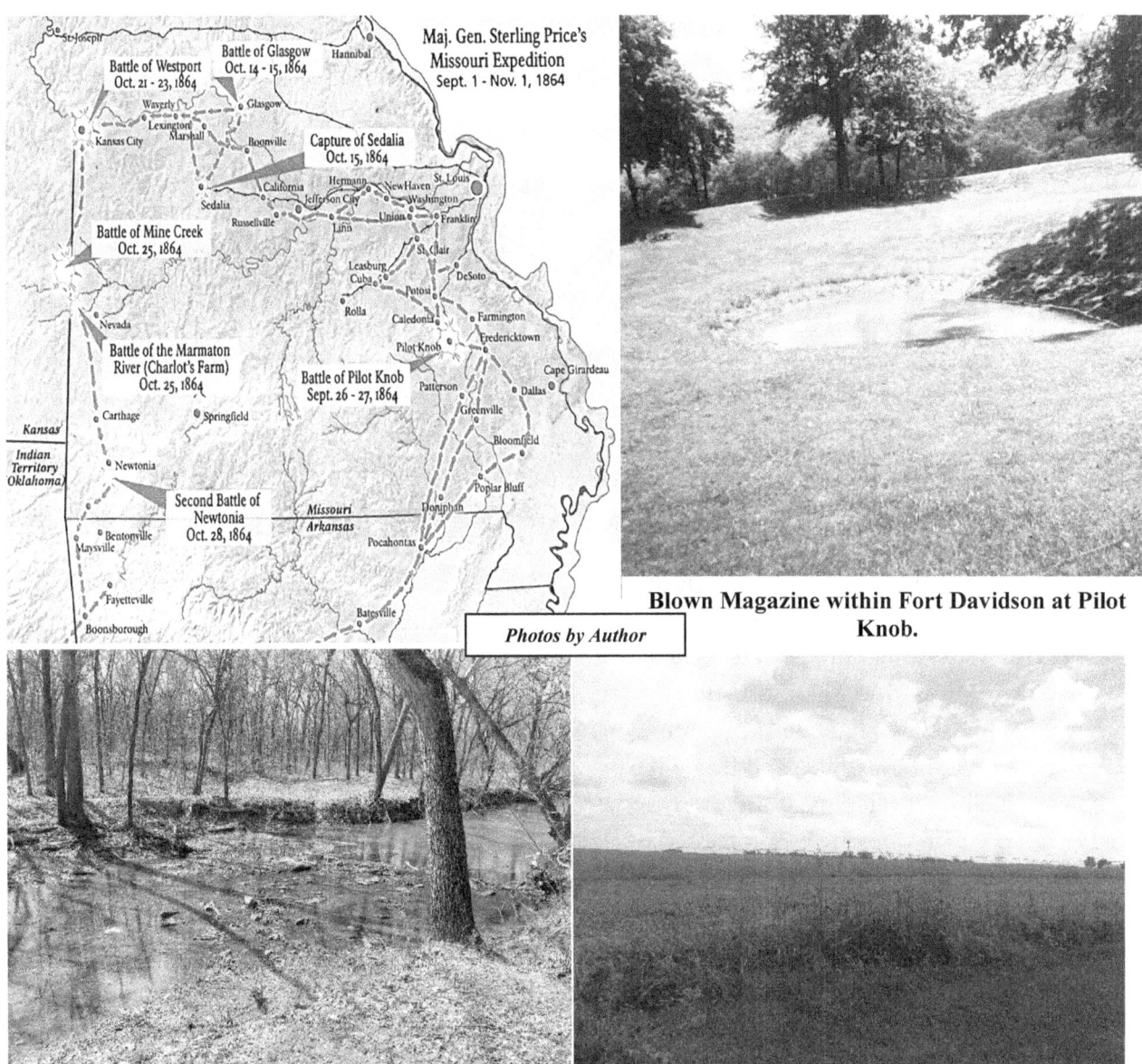

Maj. Gen. Sterling Price's Missouri Expedition Sept. 1 - Nov. 1, 1864

Photos by Author

Blown Magazine within Fort Davidson at Pilot Knob.

Historic Crossing of Mine Creek on Fort Scott-Fort Leavenworth Military Road

Fields over which the 2nd Battle of Newtonia was fought.

Kirby Smith and Major General Sterling Price. Samuel Maxey had agreed to create a diversion in the Indian Territory and did so. There were few major Confederate military operations in the Trans-Mississippi during the War. The raid was actually intended to be an invasion to permanently reoccupy Missouri for the Confederacy, obtain supplies and recruits, and disrupt Union efforts including President Abraham Lincoln's reelection campaign. Price's army consisted mostly of cavalry, focusing on swift movement and surprise attacks. The expedition included several key battles, such as the Battle of Pilot Knob, where Confederate forces initially succeeded, and the Battle of Westport, where they faced a decisive defeat. Their defeats continued at Mine Creek, Kansas, and Newtonia, Missouri. The failure of Price's Raid boosted Union morale and contributed to Lincoln's reelection. It also marked the end of significant Confederate military activity in the Trans-Mississippi Theater.

Starting in 1863, Price had argued to lead a raid on Missouri, believing its supporters would join him and provide access to weapons and supplies. His request was denied until 1864 when General E. Kirby Smith met with Thomas Reynolds, the exiled governor of Missouri. General Price was selected to lead this raid due to his popularity in the state, notwithstanding his political differences with Reynolds. The initial plan involved securing St. Louis to gather necessary supplies. Following the successful acquisition of St. Louis, they would travel by steamboat to Jefferson City and maintain control until elections could establish a pro-South government. Later, based on the strength of their assembled forces, they would endeavor to expel Union troops from the state. Price led three divisions into southern Missouri, commanded by Brig. Gen. James F. Fagan, Maj. Gen. John S. Marmaduke, and Brig. Gen. Joseph O. Shelby, all veterans of the Trans-Mississippi conflict. These forces consisted of 15,000 to 20,000 troops. They were inadequately supplied, with 4,000 of Price's men lacking weapons. The plan was to advance quickly to St. Louis, gathering supplies and recruits enroute.

Price and his troops assembled at Pocahontas, Arkansas, and departed on September 19, 1864, with each of the three columns following slightly different routes toward St. Louis. Generals John Marmaduke and Joseph Shelby urged the immediate capture of St. Louis, fearing that any delay would render the city impenetrable. Price soon learned that St. Louis was being reinforced by General Andrew Jackson Smith's veteran forces. By September 26, defenses were set up in Kirkwood. The swift defense of St. Louis led to disagreements among Price's army leaders. Price and Fagan

wanted to take Fort Davidson at Pilot Knob, while Marmaduke and Shelby pushed for capturing St. Louis immediately. Nevertheless, General Price decided to proceed with his plan to assault Fort Davidson. He probably feared leaving a substantial Union force in his rear that could cut his communications with Arkansas.

Union forces at Fort Davidson, led by General Thomas Ewing, numbered approx.-imately 2,000. After inflicting significant casualties on Price's army, Ewing successfully withdrew his troops after blowing up the fort's magazine and spiking its heavy guns and then fell back, unmolested, to Rolla. After deciding that St. Louis would be too difficult to attack, Price then redirected his efforts toward Jefferson City. As he advanced westward, he learned that Jefferson City had been heavily fortified, requiring him to reconsider his strategy again. As a result, Price opted to move further west toward Kansas City.

Meanwhile, Federal forces from Kansas and Missouri organized to intercept Price before he could reach Kansas City. General Samuel Curtis's Army of Kansas was positioned ahead of Price in Westport. A force of 2,000 under Major General James G. Blunt, who was brought out of exile to assist in the defense of Kansas City, moved toward Lexington, Missouri, about 30 miles east of Kansas City. On October 19, Price's army reached Lexington, clashing with Union scouts and pickets around 2:00 p.m. The Confederates quickly pushed them back and engaged the main Federal force. The Union troops resisted briefly, but Price's men eventually drove them through the town to its western edge. To complicate the issue for the Confederates, General Alfred Pleasonton and his 10,000 Union cavalrymen were closing in on Price from behind. At the Little Blue River, a determined Blunt was ordered to move his volunteers to the river. Upon approaching the stream, Blunt discovered that Col. Thomas Moonlight's brigade had engaged Price's advance guard at sunrise and burned the bridge as they had been instructed earlier. Price's main force had arrived and was fiercely engaging Moonlight's men, who were resolutely defending every ford in the vicinity. Blunt then launched an attack, attempting to push Price back beyond the defensive positions he intended to reclaim. A five-hour battle ensued during which Union troops managed to force the Confederates to retreat, entrenching themselves behind rock walls and preparing for an anticipated counterattack. Despite being outnumbered, the Federals made their adversaries fight for every inch of ground, but Confederate numerical superiority eventually compelled the Federals to withdraw. After two more rear guard actions at Independence and Blue River against Pleasanton, Price decided initially to confront Curtis' Army of Kansas at Westport, before addressing Pleasonton in his rear. However, Curtis maintained robust defensive positions, and despite multiple charges during the four-hour battle, Price was unable to breach the Union line. After Pleasonton crossed the Big Blue River at Byram's Ford, the outcome for Price was determined. His army retreated south through Kansas toward Arkansas, pursued by Pleasonton's cavalry. Price's army would not recover from these defeats.

In three further actions on October 25, Pleasanton and his mounted forces kept pressure on Price to attempt to destroy his army. The first two of these actions were related to getting Price's supply trains across the multiple plains rivers and creeks. The first was the Marais des Cygnes River, not far from the location of the 1856 Massacre that began the movement toward civil war. Pleasonton's forces initiated a strong attack. Price instructed his soldiers to cross the flooded river, assigning Fagan to delay the Federal troops until the wagon train could pass. After a sharp fight they were able to retreat.

About 6 miles south of the Marais des Cygnes River crossing, Col. Frederick W. Benteen (the one of Little Bighorn fame) and Col. John Philips' brigades overtook Price's Confederates crossing Mine Creek. Despite being outnumbered, the Federals launched a mounted attack led by the 4th Iowa Cavalry, which quickly broke Price's line. Superior Union firepower and aggressive tactics forced Price to retreat. Approximately 600 Confederate soldiers, including Generals Marmaduke and Brig. Gen. William L. Cabell, were captured along with six cannon. But once again, the remaining Confederates escaped.[39]

As Price's deteriorating Confederate Army fled south, the garrison of Fort Scott feared they would be attacked as the Southerners passed. But there was no danger since Price's army was in a demoralized rout, and he lost all ambition to try and strike Fort Scott and its massive Union Army supply depot. By this time, he simply wanted to get his army back to Arkansas as soon as possible. Later, on October 25, Price's supply train struggled again at the Marmiton River ford, about six miles east of Fort Scott in Missouri. Brig. Gen. John McNeil led Pleasonton's cavalry to engage Price's rallied troops, including many unarmed men. Misjudging their strength, McNeil avoided a full assault. After two hours of skirmishing, Price resumed his retreat after destroying his own supply wagons and the contraband collected by his troops. McNeil failed to pursue effectively.

On October 28, Price's weary army paused 2 miles south of Newtonia, Missouri, the location of the battle fought back in 1862 between Union General Salomon and Confederate General D.H. Cooper. Maj. Gen. Blunt's Union cavalry discovered the Confederates and soon attacked, catching the Confederates off guard. While many of Price's troops fled, Shelby's division, including the Iron Brigade, dismounted and fought the Federals to cover the retreat towards Indian Territory. Later, Brig. Gen. John Sanborn arrived with Union reinforcements, prompting Shelby to withdraw. Although the Union forced another Confederate retreat, they once again failed to capture or destroy them. This was the final battle of Price's Missouri campaign. Wiley Britton gives a very good description of their transit of the Indian Territory in their retreat:

> "…At Maysville [Arkansas] the Confederate army passed into the Cherokee Nation and marched south to the Illinois River near the State line, and thence southwest until it struck the Sallisaw River and marched down that stream some distance and then turned south and crossed the Arkansas River at Pleasant Bluff, after which Price with most of his Missouri troops marched southwest through the Choctaw and Chickasaw Nations, passing through Northfork, Perryville and Boggy Depot and other places on the Overland Mail Route and crossed Red River and passed into Texas north of Bonham. From that place his march was more leisurely down through the counties bordering on Red River to Laynesport, where he arrived December 2d, which ended

the expedition, and which, according to his Itinerary, embraced a march of 1,434 miles, after leaving Princeton…"[40]

General Sterling Price's raid into Missouri ended in disaster. They still took care to avoid any contact with the Union forces still operating in the region out of Fort Gibson and Fort Smith. Lt. Gen. Grant ordered Maj. Gen. Curtis, who had been on the Newtonia battlefield, to continue his pursuit of Price since Southwest Missouri and Northwest Arkansas had a successful growing season and was flush with corn and forage to replenish the Confederate Army. The pursuit by Curtis's cavalry prevented the Confederates from stopping to resupply. The final shots of the campaign occurred when Union guns fired on Price's rear guard as they crossed the Arkansas River at Pleasant Bluff. Although many of the units left to go back to their organic organizations, Price had still lost more than half of the number he had at the beginning of the campaign.

The lack of cooperation and coordination between Maj. Gen. Curtis of the Department of Kansas and Maj. Gen. Rosecrans of the Department of the Missouri, as well as their subordinate commanders, created an issue. There was a lack of a clear chain of command for this contingency because the battle spread into both departments. This failure to coordinate plans and information prevented the Union Army from promptly executing any actions to destroy Price's army. Moreover, as the retreat crossed into Northwestern Arkansas and the Indian Territory, the Union Army fell under the Department of Arkansas that was commanded by Maj. Gen. Thayer. Either he was not kept in the communications loop, or he opted to let Price go through unmolested. Whatever the reason, it did not matter in the end since the Confederate Army had been virtually destroyed.

It is not known how much of a diversion was created in the Indian Territory with the Gano-Watie Raid. Price's retreating forces did receive supplies and rations from the Confederate supply depot at Boggy Depot from Cooper and Watie as they passed through on their way to Texas. There were few Federal troops in the Fort Smith-Fort Gibson-Fort Scott corridor area of operations to counter Price or Gano outright. Most troops that were sent to counter the Confederate raiders moving on Cabin Creek were organic to those areas anyway. Price had failed to realize the substantial number of Union troops available in Missouri and Eastern Kansas to counter his actions. He had also believed that the masses would rise to support him, which did not happen. The people of Missouri were tired of the war and simply wanted to get on with their lives. In the Indian Territory, once the Confederate raiding party had crossed the Arkansas River with their prizes, it was the end of any more offensive operations by the Confederate Indians and Texans for the duration of the war.

Watie had planned another raid using his First Indian Brigade, but the disastrous results of Price's Expedition brought an end to that plan, and the Confederate Indians went into winter quarters. Although he wanted to continue his offensive raids, Stand Watie had a most difficult time organizing one since his Indian troops had taken what they could carry from the Cabin Creek wagon train, and headed for home after the battle. This left his force so depleted that they failed to venture out too far from their base

camps.[41] Stand Watie had established his headquarters at Boggy Depot, with the remaining parts of his brigade quartered at various camps around the Choctaw Nation and in Shawneetown, Texas, on the Red River. The remainder of the Confederate Indian Brigade went into winter quarters as well. Gano's Texans found themselves split up, some down in Louisiana and some in Texas, for the duration of the war. As an additional insult, all Texas Cavalry units were to be dismounted and perform as infantry.

In the late fall of 1864, Maj. Gen. Francis Herron conducted an inspection of Fort Gibson and immediately sent a scathing report to Maj. Gen. Canby about the appalling conditions there. Canby immediately restored Col. William Phillips as commander of the post and ordered him to clean up the mess. Most of the issues revolved around the McDonald & Fuller Trading Company and William Coffin, Southern Superintendent of the Bureau of Indian Affairs. Oklahoma historian Whit Edwards summarized the issues in clear language:

> "…While Fort Gibson may have been a haven against brutality and death, it was no refuge from corruption. Government agents and Union officers openly practiced fraud, deception, and robbery in Indian Territory. They stole livestock from civilians and the military alike, then resold the stock to the government to supply the people from whom the cattle were stolen. Salt, corn, haying, and milling operations were subject to fraud and theft. Some officers of the Indian Troops allegedly padded their rosters and collected pay for soldiers either killed, deserted, or never in

existence. The removal of both Gen, James G. Blunt, and Col. Williams did nothing to alleviate the situation…"[42]

Blunt and Curtis apparently did nothing more than turn a blind eye to some of the improprieties since neither ended the war as wealthy men. In the end, Coffin resigned from government service in March 1865 and returned to private business. [43]

On the Confederate side of Indian Territory, Stand Watie was having difficulty controlling some of his idle Indian troops. After having gone through the rations captured at Cabin Creek, the troops under Watie's command were again in destitute condition. The Union Army controlled most areas north of the Canadian River, and re-supply was very slow during the winter months due to travel as well as remnants of the previous harvest being depleted at a high rate. Now the Indian troops set their sights south and began to raid in Texas, which threatened a border war between the Indian Territory and the State of Texas. This was dangerous since so many Cherokee and Creek civilian refugees were living in and being supplied by Texas. Watie took actions with his troops by assigning punishments and moving them further away from civilian areas or back into the Indian Territory.[44] Due to the amount of equipment and supplies taken during the Cabin Creek raid, in late February 1865, Watie was equipping, mounting, and arming his Indian Division to resume offensive operations as soon as the prairie grass had grown enough to be able to sustain cavalry horses. The Confederates had correctly suspected that the Union Army units from Fort Gibson and Fort Scott had been stockpiling military and commissary supplies during the winter to also begin a spring

offensive within the Indian Territory. Lt. Gen. Grant had ordered all Indian troops to be mounted, so Phillips had acquired over 1,000 horses to mount a large and fast attack on Watie. The Federal plan was to drive the Confederate Indians permanently over the Red River by marching quickly to Boggy Depot and Fort Towson in the Choctaw Nation, then over to Fort Washita in the Chickasaw Nation. This would also dampen any hopes for Watie to conduct any offensive operations. Phillips planned to leave Fort Gibson on April 20, 1865, but he never got the chance to begin his offensive raid because he received news that the end of the war was near. All plans were put on hold.[45]

Almost all military activity over the winter of 1864-1865 took place east of the Mississippi River. Maj. Benjamin Butler moved his forces up the James River in Virginia towards Richmond. Maj. Gen Philip Sheridan and His Army of the Shenandoah had cleared the main Confederate Army under Maj. Gen. Jubal Early from the Shenandoah Valley during September and October of 1864. Maj. Gen. George Gordon Meade, commanding officer of the Army of the Potomac, accompanied by Lt. Gen. Ulysses Grant, the General-in-Chief of the United States Army, spent the winter laying siege to Petersburg, Virginia, until Gen. Robert E. Lee was able to break out on March 29, 1865, with Grant and Meade in close pursuit. The chase ended on April 9, at Appomattox Court House, Virginia, in the parlor of the Wilmer McLean house. Lee signed the generous surrender terms and sent his veterans home.

In the Western Theater, Maj. Gen. William T. Sherman had captured Atlanta, Georgia, in September 1864, after having bested Generals Braxton Bragg, Joseph E. Johnson, and John Bell Hood in the Atlanta

Campaign. Sherman began his March to the Sea on November 15 and captured Savannah, Georgia, during Christmas Week of that same year. After a short rest he began his pursuit of the remnants of Joseph Johnson's Army northward through South Carolina and North Carolina. In a final hurrah for the Army of Tennessee under John Bell Hood, in what was supposed to be simply a rear guard action for the Union Army, the Confederates almost destroyed themselves with head-on attacks on the Union earthworks at the battle of Franklin, Tennessee. In this fight the Union commander was our friend from the Trans-Mississippi Department, Maj. Gen. John Schofield. The effectiveness of the Army of Tennessee finally expired after they attempted to lay siege to Nashville, Tennessee. Unfortunately, Nashville was one of the most fortified cities in the country. The Union Army at Nashville, commanded by the "Rock of Chickamauga" Maj. Gen. George Thomas, struck Hood's Tennesseans on December 27, 1864, and drove them from their positions in a matter of hours. Those that remained with the skeleton of the once mighty Army of Tennessee made the long trek to the Carolinas where they joined up with Johnson's army for the last few months of the war. Sherman's pursuit of Johnson's rag-tag army ended on April 14, 1865, at Bennett Place, near Durham, North Carolina. There in a log cabin, Johnson surrendered his Army to Maj. Gen. Sherman.

In the war's final months, the Union forces at Fort Gibson settled into a routine of escorting supply trains between Fort Scott and Fort Smith and working the salt kettles at Mackey's Salt Works or other salt mills. The army still had to eat, so they also spent time hauling corn and milling flour and cornmeal at Hildebrand's Mill in the Cherokee Nation,

Rhea's Mill in Arkansas, or other mills in the Indian Territory. Phillips did maintain a healthy number of mounted patrols to watch the crossings and roads. It was said that after Price's army came through in October 1864, the Union forces in the area rarely saw a Confederate within 100 miles of Fort Gibson.

Back in Richmond the Confederate government and bureaucracy was still operating as usual. In March 1865 Stand Watie was notified by Elias Boudinot, the Cherokee representative to the Confederate Congress, that his son Saladin Watie had been nominated for a cadet slot at the Confederate States Army academy.[46] Regardless of how well the war was going, with plans for academies and officer selection boards, the Confederate States looked and acted like it was in for the long haul. Perhaps if the Confederacy had been able to hold out for another year or two, they would have been determined to be too big to fail.

The final shots of the Civil War in the Indian Territory were fired on April 24, 1865, when a Union cavalry patrol from Fort Gibson ran into a Confederate cavalry detachment carrying mail from the Boggy Depot Confederate Post through the Choctaw Nation. The small fight resulted in at least three Confederate deaths. The captured mail indicated that the Confederates were unaware that Generals Robert E. Lee and Joseph Johnston had surrendered the main Confederate armies and that the war was over. This was the final engagement/battle report from the Indian Territory:

Hdqrs. Third Division,
Seventh Aemy Corps,
Fort Smith, April 27,1865.
(Received 11.15 a. m. 28th.)
Twenty rebels going north from

Boggy Depot were attacked by my scouts fifty miles south three days since and three rebels killed and a small mail captured. The letters all speak of a combined movement of the rebel army to Missouri, to start about the 1st of May—33,000 infantry and 7,000 cavalry under Kirby Smith, Price, Parsons, Shelby, and company; at the same time a large force of wild Indians were to move into Kansas from Fort Arbuckle. An order was read to the rebel troops at Boggy announcing that General Lee had assumed command of all the Confederate forces, and that there was no hope of peace, except to fight for it. The letters were written by soldiers, and the reports are mere camp rumors. A letter from a rebel paymaster states that Stand Watie is coming up to cross the Arkansas River as soon as grass will subsist their horses, and that their horses would arrive at Boggy from forage camps in Texas by the 25th instant. They had no news of the capture of Richmond or Lee's army. I have scouts in that direction. I expect a considerable force will soon be up this way, but have no idea the rebel army will try to go to Missouri. In my opinion the line of the Arkansas should be strengthened by the addition of more troops, if they can be had.
Respectfully, &c.,
CYRUS BUSSEY,
Brigadier- General, Commanding.[47]

As the news spread, one by one the remaining Confederate commands lay down their arms and accepted surrender. On May 8,

1865, Lt. Gen. Richard Taylor surrendered the Department of Alabama and Mississippi at Citronelle, Alabama. On June 2, Lt. Gen. Edmund Kirby Smith surrendered the Trans-warship in Galveston harbor. He had given a final address to his troops in early May in which he had implied that they should continue to fight on:

> "...Soldiers of the Trans-Mississippi Army: the crisis of our revolution is at hand. Great disasters have overtaken us. The Army of Northern Virginia and our Commander-in-Chief are prisoners of war. With you rest the hopes of our nation, and upon your action depends the fate of our people. I appeal to you in the name of your firesides and families, so dear to you – in the name of your bleeding country, whose fate is in your hands... you possess the means of long resisting invasion. You have hopes of succor from abroad. Protract the struggle, and you will surely... secure the final success of our cause..."[48]

Unfortunately for General Smith, his soldiers decided to go home. By May 20 Camp Groce, the headquarters camp for the Trans-Mississippi Army near Hempstead, Texas, was completely empty of troops. Leaving each other after three years of war was difficult for them, but they realized the end had come and it was now the time for peace and not for war. General Smith made one final address to the Trans-Mississippi Army:

> "...Soldiers! I am left a commander without an army – a general without troops. You have made your choice. It

Stand Watie's Surrender Monument at Doaksville, Choctaw Nation
Photo by Author

Brig. Gen. Watie actually surrendered here at Rose Hill Plantation near Doaksville. This was the home of Robert M. Jones, Choctaw Delegate to CSA
Photo courtesy of the Oklahoma Historical Society

was unwise and unpatriotic, but it is final. I pray you may not live to regret it. The enemy will now possess your country, and dictate his own laws. You have voluntarily destroyed your organizations, and thrown away all means of resistance… Your present duty is plain, Return to your families. Resume your occupations of peace. Yield obedience to the laws. Labor to restore order. Strive both by counsel and example to give security to live and property. And may God, in his mercy, direct you alright, and heal the wounds of our distracted country…"[49]

Federal officials moved west and collected the surrenders of the Confederates still in the field. The Civil War in the Indian Territory finally ended on a whimper when Federal authorities finally got around to receiving the surrenders of the Five Civilized Tribes. But the surrender did not happen overnight. On May 15, 1865, Israel Vore, who was serving as the Confederate Indian Agent for the Creek Nation, attempted to gather representatives of the Nations at Council

Grove near Okmulgee. Federal officials were notified and threatened to break up any such meeting. Representatives of the Five Civilized Tribes and the Plains tribes met together at Confederate Camp Napoleon on the Washita River near present-day Verden, in Southwestern Oklahoma. They adopted a peace compact among the different tribes to present a unified front to any surrender. The Camp Napoleon Council laid the groundwork for the surrender agreements with the Union.[50]

Lt. Col. Asa C. Matthews and Adjutant William V. Vance were appointed by the United States Government to receive the surrender of the Confederate Indian forces. They met with Brig. Gen. Stand Watie at the Rose Hill Plantation of Col. Robert M. Jones, the Choctaw delegate to the Confederate Congress, located approximately 12 miles west of Doaksville, Choctaw Nation, on June 23, 1865. It was agreed that there would be a cessation of hostilities and that all Cherokees were to return to their respective homes and were not to commit any acts of violence against whites or Indians who had served with or remained loyal to the United States. The agreement between Stand Watie and the Federal delegation was applicable to all Five Civilized Tribes. At the signing of the agreement, Brigadier General Stand Watie officially surrendered the Confederate Indian Division to the United States. It is notable that Stand Watie holds the distinction of being the last Confederate general to lay down his sword, not necessarily due to defiance, but simply due to bureaucratic process. The members of the Indian Division were sent home. They were also allowed to take with them any property belonging to the Confederate States. With this last surrender, the Civil War was over.

1 Maxey, S.B. Orders, September 5, 1864, Vol. XLI, (P3) pp. 911.

2 Cooper to Scott, OR, Vol. XLI, (P1) pp. 781.

3 Warner, General in Gray, pp. 96.

4 Gano to Cooper, OR, Vol. XLI, (P1) pp. 789.

5 Britton, *Brigade*, pp. 437-438.

6 Barker to Adjutant General, *OR*, Vol. XLI, (P1) pp. 772.

7 Yeary, Mamie. *Reminiscences of the Boys in Gray, 1861-1865*. Dallas, Texas: n.p., 1912. 684.

8 Gano to Cooper, *OR*, Vol. XLI, (P1) pp. 789.

9 Baird, ed. *A Creek Warrior*, pp. 96.

10 Yeary, Mamie. *Reminiscences,* pp. 46.

11 Warren, Steven, Brilliant Victory: The Second Civil War Battle of Cabin Creek, Indian Territory, September 19, 1864. Wyandotte, Oklahoma: The Gregath Publishing Company, 2002. pp. 25

12 Baird, ed. *A Creek Warrior*, pp. 95-96.

13 Gano to Cooper, *OR*, Vol. XLI, (P1) pp. 789.

14 Watie to Cooper, *OR*, Vol. XLI, (P1) pp. 786.

15 Hopkins to Thomas, *OR*, Vol. XLI, (P1) pp. 769-770.

16 Ibid.

17 DeMorse, *Clarksville (Texas) Standard*, September 24, 1864. pp. 2.

18 Peck, R.M. *The National Tribune*, November 3, 1904, pp. 8.

19 Warren, *Brilliant Victory*, pp. 32-33.

20 Grady & Felmly, *Suffering to Silence*, pp. 159-160.

21 DeMorse, *Clarksville (Texas) Standard*, September 24, 1864. pp. 2.

22 Gano to Cooper, *OR*, Vol. XLI, (P1) pp. 789-790.

23 Jennison to Hampton, *OR*, Vol. XLI, (P1) pp. 772

24 Gano to Cooper, *OR*, Vol. XLI, (P1) pp. 790.

25 Peck, R.M. *The National Tribune*, November 3, 1904, pp. 8.

26 Britton, *Brigade*, pp. 443.

27 U.S. War Department, Record of Events, September and October 1864, *Supplement OR*, Vol. 21, Pt. 2: 445-446.

28 Hopkins, *OR*, Vol. XLI, (P1) pp. 770.

29 DeMorse, *Clarksville (Texas) Standard*, September 24, 1864. pp. 2.

30 Grady & Felmly, *Suffering to Silence*, pp.162.

31 Warren, *Brilliant Victory*, pp. 40.

32 Gano to Cooper, *OR*, Vol. XLI, (P1) pp. 790.

33 Charles N Mumford, Letter, September 27, 1864. Mumford Civil War Papers, Cat. #MisMSS1295. Wisconsin State Archives

34 Yeary, Mamie. *Reminiscences*, pp. 46.

35 Ibid, 684.

36 Spurgeon, *Army of Freedom*, pp 240-241.

37 Watie to Cooper, *OR*, Vol. XLI, (P1) pp. 783-788.

38 *Journal of the Congress of the Confederate States of America*, Senate Document 234, VII, pg. 495.

39 Interpretive Marker, Mine Creek Battlefield State Historic Site.

40 Britton, *Brigade*, pp. 448.

41 The story of Price's Raid is beyond the scope of this book and is only included since many of the same players were involved in the Indian Territory actions. What is presented here is a synopsis of the campaign and its battles gathered from numerous sources including: St. Clair, Edgar J. *The Battle of Pilot Knob*; Suderow, Bryce & Scott House. *The Battle of Pilot Knob: Thunder in Arcadia Valley*; Wood, Larry. *The Two Civil War Battles of Newtonia*; Busch, Walter E. *Fort Davidson and the Battle of Pilot Knob: Missouri's Alamo*; Kirkman, Paul. *The Battle of Westport: Missouri's Great Confederate Raid*.

42 Edwards, *Prairie Fire*, pp 130.

43 Tom Holman, "William G. Coffin, Lincoln's Superintendent of Indian Affairs for the Southern Superintendency," Kansas Historical Quarterly, Vol. XXXIX (Winter 1973), pp. 491-510.

44 Dale & Litton, ed's. *Cherokee Cavaliers*, pp. 210-213.

45 Franks, *Stand Watie*, pp. 177-179.

46 Ibid.

47 Bussey to Pope, *OR*, Vol. XLVIII, (P1) pp. 202.

48 Joseph Howard Parks, General Edmund Kirby Smith, C.S.A. Baton Rouge: Louisiana State University Press, pub date unk.

49 Parks, *General Edmund Kirby Smith*

50 Gibson, *Oklahoma*, pp. 127.

Chapter 19
The Reconstruction of the Five Civilized Tribes

Reconstruction would be hard for the Indian Nations. The war had stripped the entire northeastern portion of the Indian Territory of persons or tilled and pastured lands. The refugee Indians were concentrated in Kansas, the immediate vicinity of Fort Gibson, and along the Red River in either the southern Chickasaw and Choctaw Nations or northern Texas. In addition, since the United States Government regarded their alignment with the Confederate States as a clear-cut violation of all previously negotiated treaties, they were forced to submit to the negotiation and dictation of new treaties. These new treaties provided for tribal citizenship for slaves owned by the Indians as well as providing rights-of-way through the Territory for travelers and railroads. Eventually these treaties led to the dismemberment of the independent Indian Nations in Indian Territory, which in turn resulted in the creation of the State of Oklahoma in 1907.[1]

During the spring of 1865, the Confederate Indians were fairly certain that the war was going to end in the near future. The Federals also presumed the war was winding down. There was little military activity on either side even though communications with the eastern theaters was sparce and many times inaccurate. Both sides were going forward with plans for spring and summer offenses, but other than patrols and supply trains, most Union and Confederates just patiently waited for events to unfold. Col.

Delegates to the Post-War Five Civilized Tribe's Reconstruction Treaties

William Phillips at Fort Gibson was attempting to correct the food issues confronting the refugees:

"…I employed all the spare force of my command putting in crops or assisting the refugees to put in crops. I shall do so for two or three weeks. There is great suffering among the refugees. I shall endeavor to have beef driven from the south for them. Sales of subsistence, under the division commander's orders, had been made to those actively engaged raising crops, but General Bussey directs me to stop it, or issue to but few parties… I countermanded the order on the instructions from Fort Smith. I wish that at once. If possible, I desire to see this community self-sustaining another year…"[2]

The main issue came down to who was paying for and supplying the food since both the Army and the Interior Department were involved in the supply chain.

At the conclusion of the war, the Cherokee Nation had a population of 13,566. Approximately 2,200 Cherokees served in the Union Army, mostly within the 2nd and 3rd Indian Home Guard regiments, and about 1,400 in the Confederate forces, primarily in the 1st and 2nd Cherokee Mounted Rifles. Official records do not accurately reflect the number who lost their lives in battle; however, one-third of Cherokee women were widows, and one-fourth of Cherokee children were orphans.

The Creek Nation also experienced significant losses, with 1,675 serving in the Union Army, mainly in the 1st and 2nd Indian Home Guard regiments, and 1,575 in the Confederate Army, primarily with the 1st and 2nd Creek Mounted Rifles and various Confederate Creek battalions. The Creek Nation's population at the end of the war was listed as 13,500, with one-third identified as "entirely destitute."

The Seminole Nation in the Indian Territory was small at the onset of the Civil War and managed to field one battalion for the Confederate Army while also contributing some members to the Union Indian Home Guards. As Confederates, their combat experience was limited beyond the Battle of Middle Boggy in February 1864. After that, they served as makeshift military police within their own territory. Union members faced similar action as the rest of the Indian Home Guards.

The Choctaw Nation had an estimated population of 15,000 to 17,000 at the end of the war, while the Chickasaw Nation numbered approximately 6,000. Roughly 3,000 tribal members served during the war. Together, these two nations fielded two complete regiments for the Confederate Army, the 1st and 2nd Choctaw and Chickasaw Mounted Rifles, along with various smaller units throughout the war. In addition, there was a small representation of these two tribes in the Union Indian Home Guards.[3]

The Reconstruction Program for the United States had its birth early in the Civil War, with plans beginning as early as 1862. It began as a discussion as to how the various Confederate States of America would be brought back into the Union once they were defeated. With their larger manufacturing industry and soaring population due in part from European immigration, the United States did not harbor any doubts that they would eventually defeat the "rebels." The Federal Government firmly believed that the Confederacy had committed treason and would treat the separated Southern states as conquered provinces, subject to the jurisdiction of the United States. The Indian Territory was considered as one of these provinces after having broken their treaties with the United States and aligning with the Confederate States. Although there was a Reconstruction Plan particularly for the Indian Territory, there were differences from the plan used for the other separated states. The plan for the Indian Territory had been developed by Senators James Lane and Samuel Pomeroy of Kansas. Neither man cared about the Five Civilized Tribes or their homelands in the Indian Territory. Their interests lay in the vast lands owned and occupied by the tribes. This was a culmination of the plan discussed in an earlier chapter in which all Indians would be removed from Kansas and relocated to the Indian Territory. The poor conditions and supply issues within the Territory during the war had prevented the execution of this plan. This legislation passed by Congress in 1863 became the basis and blueprint for the reconstruction of the Indian Territory. The law reflected the hostility and maliciousness that Kansans felt about Indians in general, and especially those from the Five Civilized Tribes after four years of warfare. The intent of the Lane-Pomeroy legislation was to punish the Five Civilized Tribes for aligning with the Confederacy by removing all Indians in Kansas to the Indian Territory, confiscating all Indian lands within the State, and opening these lands to white settlement, a continuing process that had begun with the British colonists in the 17th century.

The Cherokee, Creek, and Seminole governments faced significant challenges due to internal factionalism. Confederate and Union partisans were vying for control within each of these nations. This division severely weakened tribal cohesion, almost leading to anarchy, and resulted in widespread disorder and lawlessness. For those committed to post-war recovery, this was a particularly arduous period. Of the Five Civilized Tribes, the Cherokee, Creek, and Seminole Nations were the most divided, with each faction consisting of approximately 50% of their populations. The Choctaw and Chickasaw Nations, on the other hand, were almost 100% supportive of the Confederacy, although as early as March 1864 there had been movements within the Choctaw Nation to renegotiate a new treaty with the United States. Moreover, these two

nations were not as directly affected by the destruction of the war as were the Cherokee and Creek Nations where the militaries of both sides, as well as roaming bushwhackers, ravaged the countryside for almost four years.

On June 28, 1865, Stand Watie, as the Principal Chief of the Cherokee Nation's Southern faction, sent six delegates to Fort Gibson to attempt to make contact with the Northern faction of the Nation. With an escort of about fifty armed men, the delegates approached Fort Gibson, sending an advance guard ahead with a flag of truce. Colonel John A. Garrett of the 40th Iowa Infantry was surprised by their arrival and questioned how they could be traveling in military formation given Kirby Smith's surrender conditions. Garrett doubted the wisdom of allowing them to meet other Indians without first disarming the escort and taking their parole. The loyal Indians were also wary, suspecting that the secessionists did not simply want to return home and reclaim property. Years of conflict had bred distrust and resentment, and the Union-supporting Cherokees were aware that their tribal rights might be affected as punishment for the actions of some half-breed members of the Nation. The result was that this action met with little success although the Southern Cherokees had been authorized to return to the Cherokee Nation and resettle according to the surrender agreements. Unfortunately, those returning were subject to the same passions, anger, and hostility that had existed before and during the war. The same result occurred in the Creek and Seminole Nations as well. The actions of the Southern Cherokees must have made some impact on the Nation as a whole because, on July 13, the Cherokee National Council, which was made up of mostly

Union-supporting Cherokees, passed an amnesty and pardon act for those who had fought against those tribal members who had remained loyal to the United States:

"*Whereas*, The National Council did adopt an act, dated and approved July 13th, 1865, which said act is in the following language, to wit:
"*Whereas*, Certain citizens of the Cherokee Nation became involved in the war in the United States on the side of the Rebellion and,
"*Whereas*, The success of the Union arms has closed the war, and now offers a suitable opportunity for the return to the Cherokee Nation of such citizens as became enemies to the Government of the Cherokee Nation, by joining and adhering to said rebellion, and,
"*Whereas*, The National Council are sincerely desirous of restoring peace and harmony, and of reuniting, as far as practicable, all the Cherokee people in the support of the Constitution and Laws of the Cherokee Nation, regardless of past differences; therefore,
"Be it enacted by the National Council, That the Principal Chief be and he is hereby authorized and directed to offer amnesty and pardon to all citizens of the Cherokee Nation, who have directly or indirectly participated in the rebellion in the United States, and against the existing Government of the Cherokee Nation, except as hereinafter excepted, on condition that every such person shall subscribe, on his return to the

Cherokee Nation, to the following oath, which shall be registered and preserved in the office of the clerk of the Supreme Court of the Cherokee Nation, to wit:

"I do solemnly swear or affirm, in the presence of Almighty God, that I will thereafter faithfully abide by, support and defend the Cherokee Nation."

"And all such persons so subscribing to said oath are allowed to be thereby readmitted to citizenship in the Cherokee Nation, and restored to all rights and privileges enjoyed by other citizens, except the right to possess or recover any improvements, or other property, that has or may be sold under provisions of any act confiscating the effects of persons declared to be disloyal to the Cherokee Nation..."

While it would have been expected that this act of the Cherokee National Council would have begun the process of bringing the Southern Confederates back into the folds of the Cherokee Nation, the second part of the legislation included some provisions that offended most of the leaders of the Southern Cherokees:

"*Be it further enacted*, That the following persons are excepted from the benefits of this act:
"1st All who have been military officers in the rebel service above the rank of captain since the first day of March, 1865.

"2nd All persons who held the pretended offices of Principal Chief and Assistant Principal Chief, Treasurer, and members of the National Council, in opposition to the existing Government of the Cherokee Nation.

"3rd All those citizens of the Cherokee Nation who may have violated their parole as prisoners of war, or deserted to the enemy from the 2nd and 3rd Regiments, I.H. Guards, or killed or otherwise maltreated loyal citizens or soldiers, while prisoners of war, and all white men, citizens of the Cherokee Nation by intermarriage, who may have joined and adhered to the rebellion.

"*Be it further enacted*, That persons who may wish to return to the Cherokee Nation, under the provisions of this act, shall do so on or before the 1st of January, 1866, or be thereafter debarred the benefits of the same.

"*Be it further enacted*, That special application may be made by any person excepted from the benefits of this act to the National Council for pardon and readmittance to citizenship in the Cherokee Nation.

"Now, therefore, be it known, that I, Lewis Downing, Assistant and Acting Principal Chief of the Cherokee Nation, do hereby offer amnesty and pardon to all citizens of the Cherokee Nation who participated in the Rebellion in the United States, and against the existing Government of the Cherokee Nation, upon the conditions set forth in the foregoing act, and earnestly invite all such citizens to return to the Cherokee Nation, comply with the requirements of said act, and henceforth lend their support to law and order in the Cherokee Nation.

> LEWIS DOWNING,
> Ass't and Acting Principal
> Chief, C.N.
> Executive Department,
> July 14, 1865.[4]

It is a common thread throughout history that to the victor go the spoils, which in this case would mean that the Union-aligned Cherokees would take a certain level of revenge against those they fought against. There was to be no discussion or acknowledgement of the Confederate Cherokee National Council led by Stand Watie. As far as the National Council was concerned, John Ross was and had remained the Principal Chief of the Cherokee Nation throughout the Civil War.

The Loyal Creeks in Kansas had already begun moving back into the Indian Territory during the winter of 1864-1865. During this period, almost 6,000 Loyal Creeks had been relocated to the areas around Fort Gibson. Sands, the recognized co-leader of the Loyal Creeks until the death of Opothleyahola in 1863, petitioned Col. Phillps at Fort Gibson for military protection that let them cross the Verdigris River and resettle in the Creek Nation. Most of the Loyal Creeks then proceeded to Tullahassee Mission where Phillips had established a military outpost for their protection. Another Creek leader known as Ispokogee Yahola controlled a separated group of Loyal Creeks who wished to remain independent of the main group. They refused to cross the Verdigris and instead camped on the east side of the river. Ispokogee Yahola of the Tuckabatchee clan claimed he had been designated by Opothleyahola as his successor. This separate group believed that the Creek Nation should insist to Federal officials that the previous treaties were still in force. Sands, Micco Hutke, and other Loyal Creek leaders were open to negotiating new treaties with the United States. This rift created another tear in the Creek Nation that would need to be repaired. The Creek Indian Agent, William Coffin, along with the Superintendent of Indian Affairs, William Dole from Washington, negotiated a new treaty with the Loyal Creeks. The new treaty was signed on September 3, 1863. The Treaty with the Creeks ended up as an unratified agreement between the United States and the Creek Nation, primarily focused on negotiating the terms of a potential peace and alliance during the Civil War. It aimed to establish a peaceful relationship, with the Creeks agreeing to remain neutral and the U.S. guaranteeing their protection. Even though the Creeks had accepted the Emancipation Proclamation, accepted the colonization of the freedmen, set aside lands for their settlements as well as for other relocated Indians, and provided a mechanism for accepting the Confederate Creeks back into the Creek Nation "upon such conditions as said nation may impose," the treaty was rejected by the U.S. Senate. While this treaty was never fully ratified, it exposed the selfish desires of the U.S. Congress in regard to annuities and land cessions that were expected to be surrendered by the Creek Nation. The war ended before a new treaty could be negotiated.[5]

Most Confederate Indian Refugees were reluctant to return to their homes north of the Arkansas and Canadian rivers due to concerns about retaliation from Loyal Indians. Upon returning, they found their towns, plantations, farms, schools, and churches severely damaged or destroyed. Houses and barns were reduced to rubble; stone chimneys remained as markers of former homesites, becoming "monuments to the old homeplace." All of the fields and pastures were overgrown with weeds and scrub; fences were dismantled; all cattle and other domestic farm animals, including horses, mules, and oxen, were gone; and movable agricultural and mechanical property, such as plows and tools, were missing.[6]

With the conclusion of the hostilities of the war, by mid-1865, approximately half of the Seminole refugees were relocated to Indian Territory. However, they were settled on Creek and Cherokee lands near Fort Gibson. The remaining loyalists, about five hundred individuals, that primarily comprised the families of volunteers in the Union Army, had stayed in Kansas near Neosho Falls. Plans were made to move them soon to Creek land south of Fort Gibson. Meanwhile, around one thousand Southern Seminoles remained on Chickasaw lands, as conditions in the Seminole country were too unsettled for their return at that time. In 1865, the Seminoles, like the Cherokee and Creeks, faced significant reconstruction challenges. They needed to rebuild homes and farms, address issues such as illegal cattle driving, crime, and general lawlessness. It was necessary to somehow reestablish tribal government, foster political harmony between Union and Confederate factions, and promote economic prosperity. While some reconstruction problems would be addressed later, immediate attention focused on reestablishing political relations with the United States.[7]

Chapter 19: The Reconstruction of the Five Civilized Tribes

It had been decided at the Camp Napoleon Conference in May 1865 that the Grand Council of the United Nations of the Indian Territory, made up of primarily Southern-supporting Indian tribes as well as many of the Confederate-supporting Plains tribes, would meet again before meeting with the United States officials in order to present a united front when the negotiations began. They intended to meet at Armstrong Academy, or Chahta Tamaha, in the Choctaw Nation during the first part of June. Since the Grand Council consisted only of Confederate-supporting Indians, they decided to meet again at Armstrong Academy on the first Monday of September, and feelers were sent out to the Union Indians. They also invited the Federal commissioners to meet to begin the process of reconciliation. The Federal representatives instead wanted to meet with all the tribes at Fort Smith on September 8, 1865. There is a question as to why the Indian Peace Commissioners refused to travel to Armstrong Academy and instead insisted on Fort Smith. It may have been that they felt they would be in a hazardous place without much support. Another belief was that they had other business to transact that could be better handled at Fort Smith. More than likely however, they wanted to meet at Fort Smith due to their ability to communicate with Washington via the telegraph as well as the town being a place of more comfort than a prairie in the Choctaw Nation.[8]

At Fort Smith, the U.S. Commissioner of Indian Affairs convened in the main barracks building with representatives from 13 Indian nations. This gathering, known as the Fort Smith Council, aimed to reestablish formal relations between the tribes and the United States Government following the war.

Although many tribes had signed treaties with the Confederacy at the beginning of the war, they had divided into factions supporting either the Union or the Confederacy. Indian delegates from the Caddo, Cherokee, Chickasaw, Comanche, Creek, Osage, Quapaw, Seminole, Seneca, Shawnee, Wichita, and Wyandotte nations attended the council meeting.[9]

President Andrew Johnson, who assumed the presidency after the assassination of Abraham Lincoln, appointed five representatives to meet with the Indian nations. The panel was led by Dennis. N. Cooley, U.S. Commissioner of Indian Affairs. Following the resignation of Commissioner William Dole on July 6, 1865, and the earlier departure of Interior Secretary John P. Usher in March, President Andrew Johnson appointed Senator James Harlan (Republican, Iowa) as Secretary of the Interior, with his nomination confirmed in May. Senator Harlan held specific views on Indian affairs and aimed for a comprehensive overhaul of the Indian Office. He sought a leader for Indian affairs whom he could trust and who would adhere to orders; prior experience in Indian matters was not essential. Senator Harlan identified his close personal friend, Dennis Cooley, also from Iowa, as a suitable candidate. Within four days of Dole's resignation, President Johnson—acting on Senator Harlan's recommendation—nominated Cooley as Commissioner of Indian Affairs. Cooley's policy agenda was significantly influenced by Harlan, focusing on the expedited transfer of land from the Indians. In Indian Territory, Cooley, under Harlan's direction, aimed to establish civil governance. With support from Major General John Pope, Cooley endorsed the use

Dennis N. Cooley
Commissioner of Indian Affairs
Photo Courtesy of State of Iowa

James Harlan
Secretary of the Interior
Photo Courtesy of Library of Congress

of military force to suppress any tribal resistance. Harlan and Stanton concurred that the War Department should oversee hostile tribes while the Bureau of Indian Affairs would manage relations with peaceful tribes. Harlan also prohibited employees of the Indian Office from publicly sharing any information regarding Indian affairs.

The one thing that Harlan and Cooley had in common was that they had no experience in dealing with Indians. Their concerns were rather centered about railroads and land acquisition.[10]

Other commission members included Elijah Sells, recently selected Southern Superintendency of Indian Affairs, Brig. Gen. William S. Harney, retired but called back to active duty for this commission, Thomas Wister, a Pennsylvania Quaker who had worked among the Comanche, Kiowa and Osage Nations, Col. Ely S. Parker, a Senaca

Indian serving on the staff of Lt. Gen. Ulysses S. Grant, and Charles Mix, long-time chief clerk at the Bureau of Indian Affairs who served as secretary of the council. Major General Francis Herron was also appointed but turned down the appointment.[11]

The council proceedings began on September 8 and lasted 13 days. The proceedings were dominated by the representatives of the Five Civilized Tribes. Both Union and Confederate Indian delegations represented the Cherokees, Creeks, and Seminoles. Many Confederate Indian delegates received late notice about the change in meeting location from Armstrong Academy to Fort Smith and did not arrive until near the end of the official council. In fact, the Confederate Indians held a council that adjourned on September 6 without the U.S. representatives and passed a series of resolutions signed by the tribal leaders.

Map Showing Lands Surrendered by Five Civilized Tribes Under the 1866 Reconstruction Treaties

Fort Smith at the Time of the 1866 Indian Council
Drawing Courtesy of National Park Service

FORT SMITH, INDIAN TERRITORY, ARKANSAS, THE PLACE WHERE THE GREAT INDIAN COUNCIL WAS HELD, AND TREATY OF PEACE SIGNED, SEPT. 16

Without any word from the Federal commission, they decided to meet at Fort Smith on September 15. They decided to take care of some tribal business and meet at "Mrs. Plaxe's" place on the Middle Boggy in the Choctaw Nation on September 9 and together head out to Fort Smith.[12]

At the council the Cherokees had strong representation for both sides. The Union perspective was presented by Principal Chief John Ross, who had recently arrived from Philadelphia, Smith Christie, Thomas Pegg, White Catcher, H.D. Reese, and Lewis Downing. And once they arrived, the Confederate Cherokee viewpoint was provided by Stand Watie, Elias C. Boudinot, Richard Fields, William Penn Adair, and James M. Bell. The Confederate Creeks were represented by D.N. McIntosh and James M.C. Smith, while the Loyal Creeks were represented by Principal Chief Sands, Coweta Micco, Micco Hutke, and Cotchoche, or Little Tiger. The small contingent of Union Choctaws and Chickasaws, approximately 220 tribal members, were represented by William S. and Robert B. Patton. The larger Confederate Choctaw and Chickasaw delegation arrived with the other Southern Indians on September 15 and was led by Robert M. Jones, the owner of the Rose Hill Plantation where Stand Watie had surrendered to the United States.[13]

It was evident from the first day of the Fort Smith Council that the United States had absolutely no interest in simply reinstituting the pre-war treaties in force with the Five Civilized Tribes. At the opening of the council meeting, the Loyal Indians were informed by the commissioners that all tribal members would be treated as Confederate supporters. They were told that their annuities and land rights in the Indian Territory were forfeited. However, the President expressed willingness to negotiate new treaties with specific stipulations.[14] Micco Hutke of the Creeks stated:

> "...Our people at home supposed that we came to meet and come to terms with our rebel brothers, and we thought that was all we had to do at this council... We expect to find out fully what the government wants us to do from your commission, and then be able to answer... [Now] we have learned what the government wants us to do, but are not ready at this time to reply..."[15]

The preamble to the proposed new treaties stated that the Indians had been "induced by the machinations of the emissaries" of the Confederacy. Most of the tribal delegates, especially those of the Choctaw and Chickasaw Nations, denied that they had been misled by the Confederates. Instead, they insisted that they chose the Confederacy because it was a way to maintain their national identity.[16]

The conditions for resuming relations with the United States were outlined as follows: each tribe must enter into an agreement for lasting peace and friendship with the United States; slavery must be abolished, and measures taken to include freedmen into the tribes as citizens with guaranteed rights; each tribe must consent to cede a portion of its lands to the United States for the relocation of tribes from Kansas and other areas, and the tribes must agree to the policy of creating a unified government for all tribes in the Indian Territory. In March 1865, the U.S. Government's stance on tribes in Indian Territory was made clear with the introduction of the Harlan Bill, introduced by Senator Harlan. The bill proposed re-organizing Indian Territory by dismantling tribal land ownership from pre-Civil War treaties, effectively opening the land to new white settlers.

When the Indian delegates were summoned, they had not been informed of the meeting's purpose and assumed it would focus on restoring harmony among various tribes. They were unprepared for Cooley's address and argued that they lacked the authority to negotiate treaties for their respective nations. The Loyal Cherokee and Loyal Creek delegates contended that they had not voluntarily relinquished their allegiance to the United States but did so under duress when the Confederacy asserted control by force along their border. As the Choctaws and Chickasaws conveyed to the commission: "…the United States, upon commencement of hostilities, had withdrawn all her troops from our territory and borders, thus failing to protect us as stipulated in her treaties with us…" The commissioners remained unmoved by these arguments and refused to compromise when several delegates expressed strong opposition to certain treaty stipulations. Furthermore, despite his previous collaboration with President Lincoln in Washington, the commission refused to acknowledge John Ross as the Principal Chief of the Cherokee Nation. The reason for this is unclear but probably harkened back to the letters Ross had written to the chiefs in the Creek, Seminole, and Osage tribes who were hesitant to sign with the Confederate States, copies of which were discovered by Union troops in Tahlequah. It is also not known if

the commissioners, none of whom had been involved in the Interior Department back in 1861, fully understood or cared about the pressure Ross was under at the time with Albert Pike and Ben McCulloch present at Park Hill and Stand Watie's newly recruited battalion of Cherokee Mounted Rifles hovering nearby.[17] This marked a complete departure from the agreement Ross had established with the Lincoln Administration during his time as a refugee in the East. At that time, Commissioner of Indian Affairs William P. Dole had acknowledged the forced nature of the Cherokees' secession and had recognized the legitimacy of Ross's leadership and his personal loyalty. Dole observed that Ross appeared to have resisted secession for as long as he could and consequently believed that Lincoln should be lenient towards the Cherokees.

The understanding developed between Ross and Dole however, did not extend into the new administration, especially since Vice President Andrew Johnson was not part of the Lincoln administration until the second term and most likely had no knowledge of Lincoln's intentions. Hoping to stir division within the Cherokees, Cooley claimed that Ross was "still at heart an enemy of the United States and…not the choice of any considerable portion of the Cherokee nation for the office which he claims." Confederate Congress Delegate Elias Boudinot, an eloquent speaker who saw John Ross in a weakened position, took this opportunity to drag Ross's name through the mud by bringing up issues going back to the Treaty of New Echota that sealed the Cherokees removal to the Indian Territory. Boudinot stated to the commissioners: "…I will show the deep duplicity & falsity that have followed him [Ross] from his childhood to the present day, when the winters of 65 or 70 years have silvered his head with sin, what can you expect of him now…" At this point Cooley, even with the blessings of President Andrew Johnson and Secretary of the Interior James Harlan, decided this was not the time to get tied up in that inter-tribal conflict and tried to let the issue rest. His attacks would show up again in the Washington meetings.[18] In addition, both Harlan and Cooley knew that John Ross had been in Washington as head of the Cherokee delegation and had been present in the chambers when the U.S. Senate debated then-Senator Harlan's Indian Territory bill, and had full knowledge of the plans the government had intended to execute on the Five Civilized Tribes and their lands. This gave Ross insider knowledge that neither Cooley or Harlan did not necessarily want spread among the different Indian nations, and the best possible way to accomplish this was by attempting to humiliate and dismiss Ross before the council.[19]

During the council's deliberations, the Confederate Cherokees were opposed primarily to the Union Cherokee law that confiscated their property. In a bold move, Stand Watie suggested dividing the Cherokee Nation into two jurisdictions, for Union and Confederate Cherokees. Ross and the Union Cherokees rejected this idea, and the commissioners promised to include a provision nullifying the confiscation law in the new treaty. Boudinot began to please the Federal commissioners by beginning to verbally accept the conditions of the new treaty. He advocated the consolidation of the Indian tribes into a territorial government and the opening of "excess" lands to white settlement. This stance shocked the other members of the Cherokee delegation, both Union and Confederate. None of the other delegates wanted to give up any of their lands without a fight. Yet, Cooley began to create a cozy relationship with Boudinot and the Confederate Cherokees because they were much more likely to agree to the government's desires.

In addition, the vast majority of Indians were opposed to the establishment of a single consolidated government, which they understood would dissolve tribal governance and effectively eliminate their distinct identities. Euro-Americans in many ways completely misunderstood Native American cultures, believing that the cultures were basically identical. However, these Indian nations for the most part were as distinct from one another in language, culture, religion, and temperament as a Hindu in Nepal would be from an Irish Catholic in Chicago. This also ignored the fact that large portions of the Cherokees and Creeks were mixed-blood, a divide that contributed to their split at the beginning of the war. Taking these matters into consideration, it seems there were at least eight factions within the Five Civilized Tribes: two Cherokee, two Creek, two Seminole, and the Choctaw and Chickasaw.[20]

Furthermore, although the Indian nations understood that slavery was to be abolished, they were not prepared for the requirement that freed slaves (freedmen) be granted full membership within the tribe that previously owned them, believing that the tribe's alone should be able to determine tribal citizenship. The Indian Nations believed that simply granting the freedmen their freedom and setting aside lands for their settlements was enough to satisfy the requirements of the government. As mentioned in an earlier chapter, the question of freedmen citizenship within the Cherokee Nation was still being debated into the 21st century.

Although Commissioner Cooley did not succeed in signing formal treaties at the Fort Smith Council, he achieved several important outcomes. An agreement of cooperation between the tribes and the United States was established and signed and the tribes accepted U.S. jurisdiction and annulled their treaties with the Confederate States. In return, the United States promised peace, friendship, and renewed protection for the tribes. Additionally, the Choctaw Principal Chief Allen Wright, introduced the name "Oklahoma," a Choctaw word meaning "Red People," as a potential name for a future state. The council concluded on September 21 with an agreement that authorized tribal delegates to convene in Washington, D.C. the following year.[21]

During the winter of 1865-1866, Federal officials continued to engage in negotiations for a definitive treaty with the Five Civilized Tribes. While they aimed to address the tribes' alliance with the Confederacy, they extended numerous concessions to the Southern agents who had persuaded the tribes to join their cause. In November, Interior Secretary Harlan presented his territorial bill to Albert Pike, the Confederate leader who had previously secured treaties favorable to the Indians. In a complete reversal of his earlier positions, Pike expressed his full support for Harlan's plans and suggested amendments to further diminish Indian sovereignty.[22]

Much of the period between the Fort Smith Council and the Washington Council was spent attempting to mend fences between

each of the wartime divisions of the Five Civilized Tribes. The United States Government wanted to negotiate with only one delegation from each of the tribes, although some Federal officials hoped to exploit these division, such as in the case of John Ross, to negotiate better deals for the government. Each of the tribes conducted meetings over the winter between their Union and Confederate factions in order to settle their differences enough to present a unified front at the negotiating table. As had been done to the Cherokees in the Treaty of New Echota, the United States was more than willing to sign a treaty with representatives of a small population of a tribe and then force its execution upon the entire nation.

The Washington Indian Council that began in January 1866 consisted of off-and-on meetings with the individual tribes. Commissioner Cooley, Southern Superintendent Sells, and Colonel Parker, along with Secretary of the Interior James Harlan, represented the Federal government. The Cherokees had two delegations: the Union delegation, led by John Ross, who passed away shortly after the treaty's completion, and the Confederate delegation, headed by former Confederate leaders Elias C. Boudinot and Stand Watie. The Creeks and Seminoles also had both Northern and Southern delegations, while the other two tribes each had only one delegation. Most discussions and conflicts revolved around cession of lands, railroad rights-of-way, equal distribution of annuities, funding for delegate travel and lodging, and military access and posts. This council did not include as much drama as the previous one. For the most part the Federal commissioners dealt with each tribe individually to consider the needs of all parties. Yet, all the treaties contained:

- amnesty for all crimes committed against the United States prior to the treaties;
- specific provisions of peace and friendship toward the United States;
- notice that all previous treaties were null and void;
- the Tribes acknowledgement of the supremacy of the United States Government, its Constitution, and its laws: past, present and future;
- clause stating that no federal legislation could interfere with their tribal organization;
- tribes would provide land grants in their various domains for rights-of-way for railroad (and sometimes telegraph) construction through Indian Territory.

All treaties represented a series of compromises between the Union factions, the Confederate factions, and the Federal government. Concessions were made by all tribes during the negotiations. The first was the abolition of slavery and the grant of tribal rights to the freedmen. The treaties with the Cherokee, Creek, and Seminole provided the Freedmen with unqualified rights, whereas the Choctaw and Chickasaw treaty offered them the choice of adoption into their nations or removal by the federal government for resettlement elsewhere. The second compromise established an intertribal council in which each tribe would have one representative, with an additional representative for every thousand tribal members. The Southern Superintendent of Indian affairs would serve as the council's

chief executive. The third concession involved granting rights-of-way for railroad and telegraph construction through Indian Territory by all tribes. The final concession required each tribe to cede a significant amount of land as a penalty for supporting the Confederacy. This was the prize that the U.S. Congress and the Johnson administration considered the most important.

In preparation for the Washington Council, the National Council of the Cherokee Nation, made up of mostly Union Cherokees, met on November 3 to discuss the Fort Smith Council and begin the process of selecting delegates for the council. The ailing John Ross would lead the Union delegation supported by Lewis Downing, Smith Christie, Daniel H. Ross, S.H. Benge, James McDaniel, and Thomas Pegg. The long-time Baptist missionary John B. Jones also accompanied the delegates as an unofficial advisor. Ross still traveled to Washington even though he was in very poor health. The Union delegation wanted to ensure that they were recognized as the legitimate government of the Cherokee Nation and able to prevent the splitting of that nation into two parts. They arrived in Washington in January 1866 and immediately began negotiations with the U.S. Commissioners.[23]

Stand Watie had called a meeting of the Confederate Cherokees on October 5 also to discuss the recently concluded Fort Smith Council. Joseph Vann was elected as the meeting's chairman. The purpose of the meeting was to take "appropriate measures to renew and perfect friendly relations with the Cherokee people from whom they had been divided." Unfortunately, the war was still too fresh in the memories of both sides to so quickly mend their differences and go to

Washington as a unified group. The Southern Cherokees chose Stand Watie, Elias Boudinot, Richard Fields, Saladin Watie (Stand Watie's son), Joseph A. Scales, J. Woodward Washbourne, and John Ridge as delegates to the Washington Council. Boudinot and Adair were sent in advance of the rest of the delegation. They arrived in early January 1866 and immediately began to try and discredit the Union Cherokees.

On April 7, Watie and the other Southern delegates joined the Northern Cherokees in a cordial meeting aimed at resolving their differences in a manner satisfactory and beneficial to both parties. Despite both groups expressing a desire for peace, their perspectives on achieving it were fundamentally opposed. Watie opposed the Union Cherokees' plans because he believed it would subject them to their government and laws, bringing together two communities with lingering memories of mutual grievances into direct, personal contact. The Southern Cherokee delegation feared that such a reunion would lead to renewed violence, despite efforts to prevent bloodshed. As a consequence, Watie and the Confederate Cherokees rejected the Union Cherokees' proposals. They believed that the Southern Cherokee plan for separation would foster peace and potentially lead to kind and friendly relations between the factions over time.[24] The two Cherokee factions submitted treaty drafts to the U.S. government, each receiving twelve stipulations from Cooley. The Union Cherokee accepted all but four of these terms. Despite some support for the Southern Cherokee treaty, ultimately, only the Union faction finalized treaty terms with the U.S. government. Although Cooley attempted to disregard the Union Cherokee treaty in favor

of the Confederate Cherokee version, which aligned more closely with his preferences regarding lands and railroads, the President refused to sign this version. The Union Cherokee treaty was chosen after John Ross met with President Johnson with whom he had a previous relationship. Issues such as the status of Cherokee freedmen and nullifying the Confederate treaty had been previously resolved, and both sides reached a compromise on amnesty for Cherokee individuals who fought for the Confederacy.

When the Cherokees signed their Reconstruction Treaty on July 19, 1866, they were forced to relinquish their "Neutral Lands" in southeastern Kansas and the Cherokee Strip. These lands were to be sold to the highest bidder for no less than $1.25 per acre. They also agreed to allow the Federal government to settle other tribes in the Cherokee Outlet in exchange for payment to the Cherokee Nation. In a clever move, Stand Watie requested an official and correct map of the Indian Territory:

> "…between the 37th parallel and Red River, and between the western boundary of Missouri and Arkansas and the 100 and 103 degree Meridians, giving the precise latitude and longitude of the northern Cherokee boundary… [this map] was of great importance to us in the cessions of land contemplated to be made, and the grants of land for Railroads, which we tribes desire to make…"[25]

This map would give the Cherokee Nation the ability to know and mark boundaries without being misled by dishonest government or railroad surveyors. The treaty was approved by the U.S. Senate on August 11, 1866.

The Loyal Creeks met together in December 1865 and chose their Principal Chief Sands, Coweta Micco, and Cotchoche as delegates with Harry Island serving as interpreter. They arrived in Washington on January 7, 1866, and immediately set to work on negotiations with the U.S. Commissioners. The Confederate Creeks did not believe that these full-blood men chosen were sufficiently able to carry on negotiations for the Creek Nation. So, they met in council on January 17 with their Principal Chief Samuel Checote along with Echo Harjo, and Yahkinhar Micco as second chiefs. The council chose Daniel N. McIntosh and James M.C. Smith as delegates representing the Southern faction. Checote issued instructions which directed them to cooperate with the Union Creeks which in part said, "[a spirit] of harmony and friendship for the best interest of the whole without reference to former difficulties." When the Confederate Creek delegates arrived in Washington on February 22, the U.S. Commissioners refused to recognize them as official delegates. Both sets of delegates sought attorneys for legal representation. The commissioners were taking advantage of the relative inexperience of the Loyal Indians, knowing that they were not as adept at negotiation as the mixed-blood Southern faction members, who tended to be better educated and articulate, especially in language. The Loyal Creeks also tended to believe in "The Great Father" in Washington and placed all their trust in him. When the Confederate Creeks were finally able to join the negotiations, they forced the commissioners to take a step back from the pressure they were exerting on the Loyal Creeks. McIntosh

and Smith, having seen their initial purpose thwarted, formally presented their objections to United States Commissioners Cooley, Sells, and Parker on March 18, 1866. They opposed the following specific issues: their exclusion from the treaty-making process; the confiscation of property; the political recognition of former enslaved individuals as equals to the Creeks; and the jurisdiction of Congress. Regarding the latter point, they expressed that they were "not opposed to a proper and safe organization of the Indian Country into a Territory, giving due consideration and weight to the rights and welfare of the Indians." These objections were reiterated and more clearly articulated in an address to the president on March 31, 1866. McIntosh and Smith wrote to Interior Secretary Harlan and stated:

> "... Our chief objection to that treaty grows out of the fact that in the title and in the body - not in the Preamble - of that instrument, the majority of the Creek people are practically ignored and their rights confiscated, and as that objection, in our opinion, was paramount to all others, we considered it -and still consider it- useless to attempt to discuss minor details, until the main difficulty is disposed of by the recognition of the great body of the Nation, and of their right to be heard in the settlement of questions affecting their very existence ..."[26]

Eventually, the government renegetiated some of the provisions that they had forced on the Loyal Creeks.[27] However, in the fundamental aspects, the old structures remained unchanged. Regarding the Creeks, the government found the Union faction to be more cooperative and consequently collaborated with them to the tribe's detriment. In later years, difficulties for the Creek population increased significantly, and investigations revealed that these issues stemmed from the Reconstruction Treaty of 1866. The Creek Nation representatives signed their renegotiated Treaty of Washington in June 1866, and it was ratified by the U.S. Senate on August 11, 1866. In doing so, they ceded the western half of their lands for $975,168, with some of the land designated for rebuilding and the remainder held in trust. The Creek delegates agreed to the money being distributed as follows: $200,000 in cash with up to $2,000 set aside to repair buildings damaged or destroyed during the war, $100,000 was to be paid to the Loyal Creeks and freedmen to help recover personal property destroyed by the Confederacy or their Creek allies; $275,168 would be retained in the U.S. treasury, in which the government would pay 5% interest per year; the remaining $400,000 was a debt that would also require 5% interest to be paid to the Creek Nation every year. Only a few days after the Creek signed their treaty, Congress granted franchises to two railroads to cross the Cherokee and Creek Nations. This opened up another issue as to the ownership of the land under the tracks.[28]

The Seminole Council sent John Chupco, whom the United States recognized as the principal chief of the Seminole Nation, Cho-cote Harjo, Foos Harjo, and John F. Brown, to Washington for the council to represent the Seminole Nation. John Chupco was a Union Seminole who had evacuated the Seminole Nation and went north with

Opothleyahola during his exodus. During the war he was regarded as the "Northern" Seminole chief, and he also served as the first sergeant of Company F, 1st Indian Home Guard. He had represented the Loyal Seminole at the Fort Smith Council of 1865. John Jumper led the Southern faction at the Fort Smith Council with little success and was not invited to the Washington Council. During the Civil War, Jumper served as major of the 1st Battalion, Seminole Mounted Rifles and as colonel of the 1st Regiment, Seminole Volunteers. The Seminole Nation, represented by Principal Chief John Chupco, signed the Seminole Reconstruction Treaty in March, and it was ratified by the U.S. Senate on August 16, 1866, the first of the Five Civilized Tribes to do so. John Chupco and his fellow delegates were not great negotiators and basically agreed to whatever the government wanted to do. Annie Abel had an interesting comment on the Seminole treaty:

> "…The Seminoles were the first to capitulate. Weak and impoverished as they were, they were no proof against intimidation. In them the hope of the extortionists was abundantly realized. The measure of the intimidation can be found in the preamble of their treaty, in which the commissioners, Cooley, Sells, and Parker, reminded the Seminoles that by throwing off their allegiance they had "incurred the liability of forfeiture of all lands and other property held by grant or gift of the United States… To call their western home a gift was a mockery. A grant it was only in the sense that it was held by title from the

United States given in exchange for a better and more ancient claim. How much the new title was worth as against the old title of ancient occupancy the Indians were soon to know to their cost."[29]

Along with the same conditions given to the other tribes, the Seminole were forced to cede all of their land to the federal government for fifteen cents per acre, subsequently repurchasing two hundred thousand acres at fifty cents per acre from the government, which had originally acquired it for thirty cents per acre from the Creeks. This netted the U.S. Government a net profit of $40,000 from the destitute Indian nation. John Chupco continued serving as principal chief from 1866 until his death in 1881. He was succeeded as principal chief by John Jumper.

The Choctaw and Chickasaw Nations took a different route in their negotiations by signing a joint treaty. Since the Union-factions of these two tribes were so small, numbering less than 250, the Confederate faction's overwhelming numbers muted any foes. The Chickasaw Nation was represented in Washington by Principal Chief or Governor Winchester Colbert, Edmund Pickens, Holmes Colbert, Colbert Carter, and Robert Love. The Choctaw delegation in Washington included Allen Wright, Robert M. Jones, Alfred Wade, James Riley, and John Page.[30] (For an unknown reason, Jones's signature does not appear on the treaty.) The delegates were instructed by their tribal councils not to surrender any of their occupied lands.

These two nations that bordered the Red River had been at the forefront of the secession movement and, despite some minor

hesitation, remained consistently loyal to the Confederacy until its conclusion. They constituted a powerful faction, with their leading mixed-blood members being astute politicians. The key to understanding the complexities of their reconstruction treaty lies in these two facts; particularly since their treaty was the least transformative among the entire series. Their treaty was simply titled, *"Articles of Agreement and Convention."* The document lacked an introductory statement, which was standard at the time for treaties, that would indicate any liability for forfeiture, any declaration of debt, or any presumption of guilt. The treaty recognized the abolition of slavery and also placed an obligation on the two tribes to use their influence and make every effort to persuade Plains Indians to maintain peaceful relations with each other, with other Indian nations, and with the United States.[31]

Through their reconstruction treaty, the Choctaw and Chickasaw ceded to the United States the Leased District in the western half of their domain for $300,000. The Leased District was, under the treaty, to be handled differently and in a manner that would integrate it with the future of the freedmen. Therefore, they agreed to its sale despite the significant reduction in the Choctaw-Chickasaw territory beyond the ninety-eighth parallel. The treaty also provided for north-south and east-west railroad rights-of-way across both nations.[32] The money for the Leased District would form a trust fund held by the United States under specific regulations: the money would be held in trust for the Choctaws and Chickasaws with an annual interest rate of 5%. For a period of two years, the two tribes were granted the privilege to make provisions for their freedmen by granting them a pre-emptive title to forty acres of land—suitable for residence and cultivation—and certain civil and political rights, including the right-to-vote. However, the Choctaw and Chickasaw Nations were specifically exempt from integrating the freedmen as bona fide members of the tribes and from admitting them as community participants in tribal lands, annuities, and other funds. Upon the expiration of the stipulated period, one of two arrangements was to be made concerning the $300,000 trust fund. If the Indians had adequately provided for their freedmen according to the outlined plan, then the funds were to be handed over to them as compensation for their lands, three-fourths to the Choctaws, and one-fourth to the Chickasaws.[33]

This treaty, ratified by the U.S. Senate on June 28, and proclaimed by President Johnson on July 10, included a clause referring to the entire area of the Five Tribes as the "territory of Oklahoma," marking the first official use of the name "Oklahoma" in a U.S. document.

As a result of these final concessions, the Five Civilized Tribes lost the western half of present-day Oklahoma. These treaties significantly impacted future relations between the federal government and each of the Five Tribes. The influx of whites into Indian Territory after the Civil War led to pressures for statehood during the 1890s and early 1900s. However, tribal members continually fought to preserve their sovereignty based on a Reconstruction Treaties clause stating that no federal legislation could interfere with or annul their tribal organization. That clause would be continually tested by the United States.

After the final closing and signing of the new Reconstruction Treaties, most of the delegates made their way home to the Indian Territory and their tribal nations. John Ross, who had been deathly sick and bedridden throughout most of the negotiations, finally passed away on August 1, 1866, in Washington, D.C. His body was brought back to the Cherokee Nation and buried on the hill that overlooks the remnants of his beloved Rose Cottage. In a small honor, the United States Government accepted John Ross's signature on the treaty as "Principal Chief of the Cherokees."

Stand Watie, after all his service for the Confederate States and as a Cherokee delegate, was tired of the meetings and negotiations, so he did not stay in Washington for the signing of the new Cherokee treaty. He instead returned to the Indian Territory, where he began to salvage the pieces of his for the signing of the new Cherokee treaty. He instead returned to the Indian Territory, where he began to salvage the pieces of his gristmill and sawmill businesses to find a way to make a living for himself and his recently returned family.

All of the delegates eventually returned to the Indian Territory and their respective Indian Nations. They carried treaties and some cash that had been provided to their tribes, only to return to a region filled with destroyed buildings, farms, plantations, homes, factories, mills, and sparse populations. The Cherokee Nation experienced the most significant hardship during the Civil War in the Indian Territory. Both the Union and Confederate armies, as well as bushwhackers who were indifferent to either side, devastated the area. The tribal families faced severe food shortages, as neither the Federal nor Confederate governments made substantial efforts to support the Indian nations. The United States withdrew military protection and financial support from the Five Civilized Tribes. The pre-war treaties did not anticipate the national crisis of the Civil War, which had required all U.S. Army forces for the defense of Washington, D.C. Additionally, there was considerable distrust within the military regarding individual allegiances, as demonstrated by Lieutenant Averill's escape from Fort Smith. Southern-supporting Indian agents for the Five Civilized Tribes exploited the absence of Federal oversight to negotiate new treaties by fostering fear and promising support and representation with the Confederate States. However, the Confederacy quickly neglected the promises made to the Indian Nations after signing these treaties.

The Union's war effort in the Indian Territory, particularly within the Cherokee and Creek Nations, focused primarily on relocating refugee Indians from Kansas to the Territory. After Fort Gibson was captured, most offensive operations for the Union ceased except those required to retake Fort Smith. From that point until the end of the war, the Union aimed to deliver supplies and Indian refugees to Fort Gibson, guard the supply trains, and protect Union Indian refugees from raids by Confederate forces and bushwhackers.

The Confederate States allocated limited resources to the Indian Territory. Following the events at Pea Ridge, the Confederate Army tried to distance itself from its Indian troops, who were continually inadequately supplied and often engaged the enemy without weapons. The strategic interest in Richmond was to ensure that the Confederate Indian Brigade kept the Union Army in Kansas away from Northern Texas. Militarily, Confederate Indian forces, even when supported by Texas regiments, struggled against the Union Army. Even the Union Indian Home Guard regiments performed better than the Confederate Indian Brigade in most engagements. Commanders on both sides experienced internal conflicts; for instance, Maj. Gen. James Blunt's disagreements with Maj. Gen. John Schofield hindered the Union war effort in the Indian Territory. On the Confederate side, Col. Douglas Cooper focused more on undermining Brig. Gen. Albert Pike rather than organizing operations against the Union. Despite Pike's limitations as a military leader, he exhibited integrity, unlike Cooper, who continually sought personal gain during the war.

As the months passed, the armies were slowly de-mobilized and the Regular Army officers returned to their regiments, most at their pre-war rank. The volunteer officers and men returned to their homes and attempted to restart their lives as civilians. There was no Department of Veterans Affairs for the thousands of veterans who were maimed in battle, lost legs or arms, or suffered from what we now call Post Traumatic Stress Disorder (PTSD). They were expected to return home and continue on as if nothing had happened. They found their only therapy in the reunions of the Grand Army of the Republic and the United Confederate Veterans on the various battlefields until all had passed away, the last passing away in 1959.

James Blunt had a difficult time adjusting back to civilian life. He had regained his command of the Indian Territory in the spring of 1865 and was in the process of organizing a 5,000 man strike force at Fort Gibson. This new brigade would be made up of the 2nd, 6th, 14th, and 15th Kansas Cavalry regiments along with reactivating some of the Indian Home Guard troops, against Kirby Smith's Confederates in Texas who were reluctant to surrender. The force was called off once Smith's surrender was secured on June 2. Blunt remained at Fort Gibson until June 18 when he trekked back to Fort Leavenworth to be mustered out. He submitted his resignation from the Army, which was accepted on July 29. He tried to go back to practicing medicine and did so for four years in Leavenworth, Kansas. Failing at this he eventually relocated to Washington, D.C. and became a claims clerk for the U.S. government. Eventually his mental state declined to the point in which he was placed into a government insane asylum where he died in 1881. They brought his body back for burial in Leavenworth.[34]

As for those who served in the Indian Territory, Fort Gibson remained an active post and distribution center for the re-building of the Indian nations. The Scotsman, Colonel William Phillips, stayed on for a short time to continue his oversight of the distribution of goods. He never received the general's star that he so richly deserved. He later served as prosecuting attorney of Cherokee County, Kansas, and served in the Kansas House of Representatives in 1865. Phillips was elected as a Republican to the Forty-third, Forty-fourth, and Forty-fifth Congresses (March 4, 1873 – March 3, 1879). He was an unsuccessful candidate for renomination in 1878. After leaving Congress, he was an attorney for the Cherokee Indians at Washington, D.C. and spent some of his final years as president of the Kansas

Historical Society. He died at Fort Gibson, Muskogee County, Indian Territory (now Oklahoma), November 30, 1893. He was interred in Gypsum Hill Cemetery, Salina, Kansas. The city of Phillipsburg, Kansas, is named in his honor.[35]

The saddest ending came from Senator James Lane of Kansas. As the center of all things in Kansas politics. he had been a strong supporter of Abraham Lincoln, and despite his rhetoric, he held the same "let 'em up easy" beliefs as the president regarding the reconstruction of the former Confederate states. After Lincoln was assassinated, Lane fully supported the new president's actions, which mirrored Lincoln's. Unfortunately, although he belonged to the Republican Party, he was not a strong supporter of the "Radical Republicans" who pushed for harsh treatment of the former Confederates. When President Andrew Johnson quarreled with the Radical Republicans, Lane deserted the latter and defended the Executive. Angered by his defection, certain senators accused him of being implicated in Indian contracts of a fraudulent character, In a fit of depression following this accusation he took his own life, dying near Fort Leavenworth, Kansas, on July 11, 1866, ten days after he had shot himself in the head.[36]

Douglas Cooper never attained the high honors he sought. The collapse of the Confederacy accelerated following General Robert E. Lee's surrender at Appomattox in April 1865. The Choctaw and Chickasaw regiments he had raised and commanded surrendered at that time as well, with their troops returning home immediately. In June 1865, Cooper ordered the surrender of all the remaining white Confederate troops in Indian Territory. He later pledged allegiance to the

United States government and received a formal pardon in April 1866. Post-war, Cooper aided the Choctaw and Chickasaw tribes in reconstruction negotiations, ensuring the viability of the Choctaw Net Proceeds claim (refers to a long-standing dispute between the Choctaw Nation and the United States regarding the net proceeds from the sale of Choctaw lands in Mississippi). He became entangled in the hearings regarding the claim and faced allegations of mis-appropriating funds during his tenure as an agent before the war, but he successfully proved his innocence of those charges. His actions as an agent under the Confederate States was never investigated. The fate of the gold he intended to conceal from the sale of horses early in the war remains unknown. Cooper passed away from pneumonia at Fort Washita on April 30, 1879, and was interred there in an unmarked grave.[37]

After the treaty was signed, Stand Watie had initially gone into exile in the Choctaw Nation. The Cherokees were still strongly divided over the treaty issues, and a new chief was elected, Lewis Downing, a full-blood and compromise candidate. He was a shrewd and politically savvy Principal Chief, bringing about reconciliation and reunification among the Cherokee. Shortly after Downing's election, Watie returned to the Cherokee Nation. He tried to stay out of politics and rebuild his fortunes. He returned to his Honey Creek home, where he died on September 9, 1871. Watie was buried in the old Ridge Cemetery, now called Polson's Cemetery, in what is now Delaware County, Oklahoma, as a citizen of the Cherokee Nation.[38]

After resigning from the Confederate Army, Albert Pike retired to Greasy Cove in Montgomery County, Arkansas. He was appointed as a judge of the Arkansas Supreme Court in 1864, but not much is recorded about his activities on the court. At the end of the Civil War, Pike moved to New York City, and later briefly to Canada. After receiving an amnesty from President Andrew Johnson on August 30, 1865, he returned to Arkansas to resume practicing law. In 1867, he relocated to Memphis, Tennessee, and entered a new law partnership with General Charles W. Adams. He also edited the Memphis Appeal. After he stopped practicing law, Pike's main focus was the Masonic Lodge. He had joined the Masons in 1850 and was involved in the establishment of the Masonic St. Johns' College in Little Rock that same year. In 1851, he helped form the Grand Chapter of Arkansas and served as its Grand High Priest from 1853 to 1854. Pike passed away at the Scottish Rite Temple in Washington DC on April 2, 1891, and was buried in Oak Hill Cemetery there. On December 29, 1944, the anniversary of his birth, his body was trans-ferred from Oak Hill Cemetery to a crypt in the temple.[39]

The quarter century between 1865 and 1889 brought significant social, economic, and political changes to the Indian Territory. During this period, the federal government relocated various tribes from across America to lands acquired from the Five Civilized Tribes through the Reconstruction Treaties, transforming the territory into a diverse mix of tribal cultures. The Five Civilized Tribes were joined by the Modocs, Osages, Pawnees, Poncas, Kickapoos, Shawnees, Delawares, Wichitas, and Caddos.

This era also marked rapid economic development in what would become Oklahoma. The construction of railroads revolutionized transportation and fostered new businesses, including mining, lumbering, ranching, expanded farming, and related services, bringing significant changes to the Indian nations. The cattle trails crossing the Indian Territory, such as the East Shawnee Trail (Texas Road), West Shawnee Trail, Chisholm Trail, and Dodge City Trail, facilitated the movement of thousands of beef cattle from Texas to Kansas railheads. These activities provided economic opportunities for the Indian nations and helped them financially endure until the Five Civilized Tribes and their lands were combined with the western Indian Territory to form the State of Oklahoma in 1907.[40]

In the larger scope of the Civil War, the Indian Territory was a backwater district, to a backwater department, to a backwater region. Neither side cared much about this area. The primary actions of the military on both sides during the entire war was the continual hunt for forage for the animals and grain and flour for the troops. One point that often gets overlooked when examining the actions of the Five Civilized Tribes during the war is that the Northern-supporting Indians held no loyalty to the United States, and the Southern-supporting Indians held no loyalty to the Confederate States. They were not citizens of either. Their loyalties lay in their own Indian nations, to protect them, and maintain their history, and promote their own national identities. They still had hard trail ahead of them.

[1] For a complete, in-depth review of the Fort Smith and Washington Councils, as well as an extensive review of the Reconstruction Treaties forced on the Five Civilized Tribes, see Annie Abel: The American Indian Under Reconstruction, Cleveland, Ohio: A.H. Clark Company, 1925.

[2] Phillips to Reynolds, OR, Vol. XLVIII, (P2) pp. 27.

[3] McReynolds, Oklahoma Sooner, pp.224-225.

[4] Able, Annie, Reconstruction, pp. 158-159.

[5] Debo, Road to Disappearance, pp.160-164.

[6] Gibson, Oklahoma, pp. 131

[7] Henslick, Harry E. "The Seminole Treaty of 1866," article, Chronicles of Oklahoma, Autumn 1970; Oklahoma City, Oklahoma. pp. 284.

[8] Able, Annie, Reconstruction, pp. 173-174.

[9] Frost, Lisa, "The Fort Smith Council of 1865," Blue and Gray, Vol. 31. #2. 2015, pp. 54.

[10] Dejong, David H. Paternalism to Partnership: The Administration of Indian Affairs, 1786–2021. University of Nebraska Press, 2022. pp. 121.

[11] Gibson, Oklahoma, pp. 131.

[12] Franks, Stand Watie, pp. 183.

[13] Arrell Gibson, The Chickasaws, Norman: University of Oklahoma Press, 1972. pp. 273-274.

[14] Gibson, Oklahoma, pp. 128; Franks, Stand Watie, pp. 183.

[15] Debo, Road to Disappearance, pp. 168.

[16] Angie Debo, The Rise and Fall of the Choctaw Republic, Norman: The University of Oklahoma Press, 1934. pp. 85-86.

[17] Frost, pp. 54.

[18] Gary E. Moulton "John Ross and the 1865 Fort Smith Council," Proceedings: War and Reconstruction in Indian Territory: History Conference in Observance of the 130th Anniversary of the Fort Smith Council September 14-17, 1995 Fort Smith, Arkansas. pp. 94.

[19] Moulton, Gary E. John Ross: Cherokee Chief. Athens and London: Brown Thrasher Books, University of Georgia Press, 1978. pp. 184-187.

[20] Debo, Choctaws, pp. 85-87.

[21] Frost, pp. 54.

[22] Debo, Road to Disappearance, pp. 171.

[23] Moulton, John Ross: Cherokee Chief, pp. 190-192.

[24] Franks, Stand Watie, pp. 186-187.

[25] Franks, Stand Watie, pp. 186.

[26] McIntosh and Smith to Harlan, May 14, 1866, Interior Department Files, Bundle, no. 56

[27] Debo, Road to Disappearance, pp. 173.

[28] Ibid, pp. 174-175.

[29] Able, Annie, Reconstruction, pp. 318-319.

[30] Gibson, The Chickasaws, pp. 274-275.

[31] Able, Annie, Reconstruction, pp. 329-330.

[32] Debo, *Choctaws*, pp. 89.

[33] Able, Annie, *Reconstruction*, pp. 330-331.

[34] Collins, *Tarnished Glory*, 211-222.

[35] William E. Connelley, <u>A Standard History of Kansas and Kansans</u>, Chicago : Lewis, 1918

[36] *"James Henry Lane". NNB. Retrieved May 15, 2025.*

[37] Delashaw, Corie. *"Cooper, Douglas Hancock (1815 - 1879)."* Encyclopedia of Oklahoma History and Culture. Oklahoma Historical Society

[38] Kenny A. Franks, *"Stand Watie,"* Encyclopedia of Oklahoma History and Culture. Oklahoma Historical Society

[39] Brown, Walter L. <u>A Life of Albert Pike</u>. Fayetteville: University of Arkansas Press, 1997

[40] Gibson, *Oklahoma*, pp. 130.

Bibliography

United States Government Records

Department of the Interior, Office of Indian Affairs
Letters Sent
Letters Received
Annual Reports of the Commissioner of Indian Affairs
General Files, Southern Superintendency, 1861-1865.
War Department
Official Records of the Union and Confederate Armies in the War of the Rebellion:
Series 1, Vol. 8; Vol. 13; Vol. 22, Parts 1 & 2; Vol. 34, Parts 1 - 4; Vol. 41, Parts 1 – 4;
Vol. 48, Parts 1 – 3; Vol. 53 Supplement. Washington, D.C.: U.S. Government Printing Office, 1888-1898.

Confederate States of America, *Journal of the Congress, 1861-1865*, Washington: Government   Printing Office, 1904.

United States Senate. *Report of the Joint Committee on the Conduct of the War.* 38th Congress,   1865.

Choctaw Nation, *Acts and Resolutions of the General Council of the Choctaw Nation*, 1858, published by authority of the General Council, by Josephus Dotson, printer for the Nation   (Fort Smith, Ark., 1859).

Hewitt, Janet, ed. *Supplement to the Official Records of the Union and Confederate Armies._* Wilmington: Broadfoot Pub. Co. 2001

National Park Service, *Civil War Soldiers and Sailors Database.* www.nps.gov

Manuscript Collections

University of Oklahoma
Western History Collection
Indian-Pioneer Collection, Works Progress Administration
Gilcrease Institute of American History and Art, Tulsa, Oklahoma
Grant Foreman Papers
Samuel Bell Maxey Papers
John Ross Papers
Northeastern State University/University Archives, Tahlequah, Oklahoma.
T.L. Ballenger Collection
Collections of the Kansas State Historical Society, Topeka, Kansas

Autobiographies, Diaries, Books as Primary Sources

Bates, James C. <u>A Texas Cavalry Officer's Civil War: The Diary and Letters of James C. Bates.</u> Richard Lowe, editor, Baton Rouge: Louisiana State University, 1999

Blunt, James G. "Civil War Experiences," <u>Kansas State Historical Quarterly</u>, May 1932: 211-265.

Sherman Bodwell Diary, Kansas State Historical Society, Topeka.

Baird, W. David, Editor. <u>A Creek Warrior for the Confederacy: The Autobiography of Chief G.W. Grayson</u>, Norman and London: University of Oklahoma Press. 1988.

Crawford, Samuel J. <u>Kansas in the Sixties</u>. Chicago: A.C. McClurg and Company, 1911.

Dale, Edward Everett, & Gaston Litton, eds. <u>Cherokee Cavaliers: Forty Years of Cherokee History as told in the Correspondence of the Ridge-Watie-Boudinot Family,</u> Norman: University of Oklahoma Press, 1939.

Duffield, George C. "The Long Drive: Driving Cattle from Texas to Iowa, 1866." Diary entry, <u>Kansas State Historical Society</u>

Folsom, Edward, "Reminisces of E.A. Folsom." E.E. Dale Collection, Box 218, F17, University of Oklahoma/Western History Collection. n.d.

Gardner, Theodore. "The First Kansas Battery: An Historical Sketch, With Personal Reminiscences of Army Life, 1861-1865," *Collections of the Kansas State Historical Society, 1915-1918*, 14 (1918)

Gause, Issac. <u>Four Years with Five Armies: Army of the Potomac, Army of the Missouri, Army of the Ohio, and the Army of the Shenandoah</u>. New York: Neale Publishing Company, 1908.

Grabill, Mary. Letter entitled "To My Daughters." Transcribed copy on file at Wilson's Creek National Battlefield, Republic, Missouri.

Grant, Ulysses S. <u>Personal Memoirs of U.S. Grant</u>. New York: Dover Publications, 1995.

Grisom, George L. <u>Fighting with Ross' Texas Cavalry Brigade, CSA: The Diary of George L. Grisom, Adjutant, Ninth Texas Cavalry Regiment</u>. Homer L. Kerr, editor, Hillsboro, Texas: Hill Junior College Press, 1976.

Hitchcock, Ethan Allen. <u>A Traveler in Indian Territory: The Journal of Ethan Allen Hitchcock</u>. Norman: University of Oklahoma Press, 1996.

Horn, Robert Cannon. <u>The Annals of Elder Horn: Early Life in the Southwest</u>. Ed. John Wilson Bower and Claude Harrison Thurman. New York: R.R. Smith, Inc. 1930.

Irving, Washington. <u>A Tour on the Prairies,</u> New York: Skyhorse Publishing, 2013.

Johansson, M. Jane, Editor <u>Albert C. Ellithorpe, The First Indian Home Guards, and the Civil War on the Trans-Mississippi Frontier</u>. Baton Rouge: Louisiana State University Press, 2016.

Killgore, G.S. "Spirited Actions Not in History." *The National Tribune*, Washington, DC. November 1, 1883, pp.7. Newspapers.com

Kitts, John H. "The Civil War Diary of John Howard Kitts," *Collections of the Kansas State Historical Society, 1915-1918*, 14 (1918)

Lindberg, Kip and Matt Matthews, eds., "The Eagle of the 11th Kansas: Wartime Reminiscences of Colonel Thomas Moonlight", *Arkansas Historical Quarterly*, 62 (2003)

Marple, Silas Hough. Burlingame, Kansas, Transcribed Letters, 1855-1862. ArchivesSpace at the University of Arkansas, Fayetteville, Arkansas. Box: 66, Folder: 15.

Robert T. McMahan Diary, University of Missouri-Columbia; Nov. 27, 1862

"Nassau" letter in St. Louis *Daily Missouri Democrat*, Dec. 11, 1862; Silas H. Marple to his wife, Nov. 25/26, 1862, Marple Family Collection, Shiloh Museum, Springdale, Arkansas.

Peck, Robert Morris. "Wagon Boss and Mule Mechanic," *National Tribune*, 1904.

Bibliography

Sparks, A.W. Recollections of the Great War: The War Between the States as I Saw It, Tyler: Lee & Burnett, Printers. 1901

Tenney, Lumen Harris. War Diary of Lumen Harris Tenney, 1861-1865. Cleveland: Evangelical Publishing House, 1914.

Books and Articles as Secondary Sources

Abel, Annie Heloise. The American Indian as Slaveholder and Secessionist: An Omitted Chapter in the Diplomatic History of the Southern Confederacy. Cleveland, Ohio: A.H. Clark Company, 1919.

--The American Indian as Participant in the Civil War, Cleveland, Ohio: A.H. Clark Company, 1919.

--The American Indian Under Reconstruction, Cleveland, Ohio: A.H. Clark Company, 1925.

Adair, James. The History of American Indians. Pantianos Classics, Online Publisher, 2020. Original Printing, 1775.

Agnew, Brad. Fort Gibson: Terminal on the Trail of Tears, Norman & London, University of Oklahoma Press, 1980

Bahos, Charles, "On Opothleyahola's Trail: Locating the Battle of Round Mountains," Chronicles of Oklahoma, Spring 1985.

Baker, William D. The Camden Expedition of 1864. Little Rock: Arkansas Historic Preservation Program

Ball, Durwood. Army Regulars on the Western Frontier, 1848-1861, Norman: University of Oklahoma Press, 2001.

Barry, Louise. "The Fort Leavenworth-Fort Gibson Road and the Founding of Fort Scott." Kansas State Historical Quarterly, May, 1942 (Vol. XI, No. 2), pp. 115 to 129

Bearss, Edwin C. and Gibson, Arrell M. Fort Smith: Little Gibraltar on the Arkansas. Norman and London: University of Oklahoma Press. 1969.

Bisel, Debra Goodrich. The Civil War in Kansas: Ten Years of Turmoil. Charleston: The History Press, 2012.

Bond, John W. "The Pea Ridge Campaign," The Battle of Pea Ridge, 1862, Eastern National Park Booklet, Date Unk

Breihan, Carl W. Quantrill and His Civil War Guerillas. New York: Promontory Press, 1959.

Britton, Wiley. The Union Indian Brigade in the Civil War. Kansas City: Franklin Hudson Publishing, 1922.

--The Civil War on the Border, Volumes I & II, New York and London: G.P. Putnam's Sons, 1899.

--Memoirs of the Border, 1863. Chicago: Cushing & Thomas, 1882.

Brown, Walter L. A Life of Albert Pike. Fayetteville: University of Arkansas Press, 1997

Busch, Walter E. Fort Davidson and the Battle of Pilot Knob: Missouri's Alamo. Charleston: The History Press, Civil War Sesquicentennial Series, 2010

Cantrell, M.L. and Harris, Mac. Editors. Kepis & Turkey Calls: An Anthology of the War Between the States in Indian Territory. Oklahoma City: Western Heritage Books. 1982. Selected articles and comments.

--Clifford, Ray A. "The Indian Regiments in the Battle of Pea Ridge." 62-73.

--Fischer, LeRoy H. and Franks, Kenny A. "Confederate Victory at Chusto-Talasah." 30-54.

--Heath, Gary N. "The First Federal Invasion of Indian Territory." 79-89.

--Rampp, Lary C. "Negro Troop Activity in Indian Territory." 186-214.
     "Civil War Battle of Barren Creek, Indian Territory, 1863." 122-130.

--Shirk, George H. "The Place of Indian Territory in the Command Structure of the Civil War." 2-9.

--Shoemaker, Arthur. "The Battle of Chustenahlah." 56-61.

--Willey, William J. "The Second Federal Invasion of Indian Territory." 94-104.

--Wright, Murial H. "Lieutenant Averell's Ride at the Outbreak of the Civil War." 12-26.

Castel, Albert. Civil War Kansas: Reaping the Whirlwind. Lawrence: The University of Kansas Press, 1958.
    --William Clarke Quantrill: His Life and Times, Norman: The University of Oklahoma Press, 1962.

Cheek, Charles D., Editor. Honey Springs: Search for a Confederate Powder House, An Ethnohistorical and Archeological Report. Series in Anthropology. Oklahoma City: Oklahoma Historical Society, 1976.

Christ, Mark A. Civil War Arkansas 1863: The Battle for a State. *Volume 23 of the Campaigns and Commanders Series.* Norman: University of Oklahoma Press, 2010.

Christ, Mark A. "War to the knife": Union and Confederate Soldiers' Accounts of the Camden Expedition, 1864." Arkansas Historical Quarterly,73, no. 4 (2014): Little Rock, pp. 381–413.

Clampitt, Bradley R. Editor. The Civil War and Reconstruction in Indian Territory. Lincoln and London: University of Nebraska Press, 2015

Collins, Charles D. Jr. *Battlefield Atlas of Price's Missouri Expedition of 1864.* Fort Leavenworth, Kan.: Combat Studies Institute Press, 2016

Collins, Robert. General James G. Blunt: Tarnished Glory, Gretna, Louisiana: Pelican Publishing Company, 2005.

Confer, Clarissa W. The Cherokee Nation in the Civil War, Norman: University of Oklahoma Press, 2007.

Connole, Joseph, "A Terrible Truth: The Tonkawa Massacre of 1862." Chronicles of Oklahoma, Volume 97, Number 4, Winter 2019, pp. 450-467

Corbett, William P. "Rifles and Ruts: Army Road Builders in Indian Territory," Chronicles of Oklahoma, Autumn 1982.

Cottrell, Steve. Civil War in the Indian Territory. Gretna, Louisiana: Pelican Publishing Company, 1998.

Cox, Dale. The Battle of Massard Prairie: The 1864 Confederate Attacks on Fort Smith, Arkansas. Bascom, Florida: William Cox, Publisher. 2008.

Crowe, Clint. Caught in the Maelstrom: The Indian Nations in the Civil War, 1861-1865. El Dorado Hills, CA: Savas Beatie, 2019.

Cunningham, Frank. General Stand Watie's Confederate Indians, Norman: The University of Oklahoma Press, 1998.

Debo, Angie. The Road to Disappearance: A History of the Creek Indians, Norman: The University of Oklahoma Press, 1941.

--The Rise and Fall of the Choctaw Republic, Norman: The University of Oklahoma Press, 1934.

Dejong, David H. Paternalism to Partnership: The Administration of Indian Affairs, 1786–2021. Lincoln: University of Nebraska Press, 2022.

DeMorse, Colonel Charles. "Indians for the Confederacy". The Chronicles of Oklahoma. Oklahoma City: Oklahoma Historical Society. Winter 1972-73.

DeRosier, Arthur H. Jr. The Removal of the Choctaw Indians, Knoxville: The University of Tennessee Press, 1970.

Duke, Larry D. "Nebraska Territory" The Western Territories in the Civil War. Manhattan, Kansas: Sunflower University Press, 1977.

Eakin, Joanne Chiles, *Battle of Independence, August 11, 1862*, Two Trails Publishing, 2000.

   --*Battle of Lone Jack, August 16, 1862*, Two Trails Publishing, 2001.

Edwards, Whit. The Prairie was on Fire: Eyewitness Accounts of the Civil War in Indian Territory. Oklahoma City: Oklahoma Historical Society, 2001.

Ehle, John. Trail of Tears: The Rise and Fall of the Cherokee Nation, New York: Anchor Books. 1988.

Evans, Clement A. Confederate Military History: A Library of Confederate States History. 12 vols. Atlanta, 1899.

Farlow, Joyce and Louise Barry, eds., "Vincent B. Osborne's Civil War Experiences," *Kansas Historical Quarterly*, 20 (1951)

Faulk, Odie B., Franks, Kenny A., Lambert, Paul F. Editors. Early Military Forts and Posts in Oklahoma. Oklahoma City: Oklahoma Historical Society. 1978.

   --Corbett, William P. "Confederate Strongholds in Indian Territory: Forts Davis and McCulloch." 65-77.

   --Rohrs, Richard C. "Fort Gibson: Forgotten Glory." 26-38.

   --Howard II, James A. "Fort Washita," 54-64

Felmly, Bradford K. and Grady, John C. Suffering to Silence: 29th Texas Cavalry, CSA. Regimental History. Quannah, Texas: Nortex Press, 1975.

Fischer, LeRoy H. "The Battle of Honey Springs." Oklahoma Today. Winter, 1970-71. 15-18.

Foreman, Grant. The Five Civilized Tribes, Norman: University of Oklahoma Press, 1934.

   --Indian Removal: The Emigration of the Five Civilized Tribes, Norman: The University of Oklahoma Press, 1972.

Foreman, Grant & Ross, Allen. "The Murder of Elias Boudinot" Chronicles of Oklahoma, Volume 12, No. 1, March, 1934

Forsyth, Michael J. The Camden Expedition of 1864: and the Opportunity Lost by the Confederacy to Change the Civil War. Jefferson, North Carolina: McFarland & Company, 2003.

Franks, Kenny A. Stand Watie and the Agony of the Cherokee Nation. Memphis: Memphis State University Press. 1979.

Frost, Lisa, "The Fort Smith Council of 1865," Blue and Gray Magazine, Vol. 31. #2. 2015

Furry, William, ed. The Preacher's Tale: The Civil War Journal of Rev. Francis Springer, Chaplain, U.S, Army of the Frontier. Fayetteville: University of Arkansas Press, 2001.

Gaines, W. Craig. The Confederate Cherokees: John Drew's Regiment of Mounted Rifles. Baton Rouge: Louisiana State University Press, 1989.

Gibson, Arrell Morgan. Oklahoma: A History of Five Centuries, 2nd ed. Norman and London: University of Oklahoma Press. 1981.

--The Chickasaws, Norman: University of Oklahoma Press, 1972.

Grummond, Elizabeth de and Hamlin, Christine. Horseshoe Bend National Military Park: Archeological Overview and Assessment, Tallahassee, Florida: Southeast Archeological Center, National Park Service, 2000.

Hale, Douglas, "Rehearsal for Civil War: The Texas Cavalry in the Indian Territory, 1861." Chronicles of Oklahoma, article, Autumn 1990; Oklahoma City, Oklahoma.

Hardee, W.J. Hardee's Rifle and Light Infantry Tactics, Title Fourth, School of the Battalion, Article Fourteenth, *Dispositions against Cavalry.*

Heidler, David S. and Heidler, Jeanne T. Indian Removal, New York & London: W.W. Norton Co. 2007.

Hess, Earl J., Richard W. Hatcher III, William Garrett Piston, and William L. Shea. Wilson's Creek, Pea Ridge, & Prairie Grove: A Battlefield Guide with a Section on the Wire Road. Lincoln: University of Nebraska Press, 2006.

Hicks, Brian. Toward the Setting Sun: John Ross, The Cherokees, and the Trail of Tears, New York: Atlantic Monthly Press, 2011.

Hill, Luther B. A History of the State of Oklahoma, with assistance of Local Authorities, Volumes 1 & 2, New York and Chicago: The Lewis Publishing Company, 1908.

Hood, Fred. "Twilight of the Confederacy in Indian Territory," Chronicles of Oklahoma, Winter 1963; Oklahoma City, Oklahoma.

Howard, Bess. "Frivolous History of Fort Gibson," T.L. Ballenger Collection, Northeastern State University/University Archives. Tahlequah, Oklahoma.

Hoxie, Frederick E. Editor. Encyclopedia of North American Indians: Native American History, Culture, and Life From Paleo-Indians to the Present, New York: Houghlin Mifflin Company, 1996.

Hudson. Charles, The Southeastern Indians, Knoxville: The University of Tennessee Press, 6th printing, 1992.

Jackson, W. Turrentine, Wagon Roads West: A Study of Federal Road Surveys and Construction in the Trans-Mississippi West, 1846-1869. New Haven & London: Yale University Press, 1964.

Jirikowic, Christine; Gwen J. Hurst; Tammy Bryant. "Archeological Investigation at 206 North Quaker Lane (44AX193)" (PDF). p. 2. Archived from the original (PDF) on October 25, 2021. Retrieved August 13, 2021 – via City of Alexandria, VA.

Johnson, Ludwell H. Red River Campaign: Politics & Cotton in the Civil War. Kent, Ohio: The Kent State University Press, 1993.

Josephy, Alvin M. War on the Frontier. (The Civil War) New York, New York: Time Life Books, 1986.

Bibliography

"Second Brigade" letter in Lawrence *Kansas State Journal*, Dec. 18, 1862; Newspapers.com

Kelman, Ari. "Deadly Currents: John Ross's Decision of 1861," <u>The Chronicles of Oklahoma</u>, Vol. 73, #1, Spring 1995, pp. 80-103.

Keun Sang Lee, "The Capture of the J.R. Williams." <u>Chronicles of Oklahoma</u>, 60(1), Oklahoma Historical Society, 1982

Kirkman, Paul. <u>The Battle of Westport: Missouri's Great Confederate Raid</u>. Charleston and London: The History Press: Civil War Sesquicentennial Series, 2011

Kremm, Thomas W. and Diane Neal, "Crisis of Command: The Hindman/Pike Controversy over the Defense of the Trans-Mississippi District," <u>The Chronicles of Oklahoma</u>, article, Spring 1992; Oklahoma City, Oklahoma.

Lale, Max S. "The Boy-Bugler of the Third Texas Cavalry: The A.B. Blocker Narrative." <u>Military History of Texas and the Southwest</u>, Vol. XIV, No. 2. 1978.

Lause, Mark A. <u>Price's Lost Campaign: The 1864 Invasion of Missouri</u>. Columbia and London: University of Missouri Press, 2011

Litlefeld,Jr. Daniel F, and Lonnie E. Underhill, "Fort Coffee and Frontier Affairs, 1834-1838." <u>Chronicles of Oklahoma</u>, article, Autumn 1976; Oklahoma City, Oklahoma.

Lothrop, Charles H, M.D. <u>A History of the First Regiment Iowa Cavalry Veteran Volunteers: From Its Organization in 1861 to Its Muster Out of United States Service in 1866</u>. Lyons, Iowa: Beers & Eaton, Printers, Mirror Office, 1890.

Manning, Michael J. "They Fought Like Veterans: The Civil War in Indian Territory, April 1861 – September 1863" <u>Blue & Gray Magazine</u>, David Roth, Editor. Volume 28, #3, 2011.

--"They Fought Like Veterans: The Civil War in Indian Territory, September 1863 – June 1865" <u>Blue & Gray Magazine</u>, David Roth, Editor. Volume 31, #2, 2015.

Matthews, Matt and Lindberg, Kip, "Shot All to Pieces, the Battle of Lone Jack, Missouri, August 16, 1862", *North and South*, Vol. 7, No. 1, January, 2004.

McCaffery, Isaias. "We-He-Sa-Ki (Hard Rope): Osage Band Chief and Diplomat, 1821–1883," <u>Kansas History: A Journal of the Central Plains</u>. Volume 41, Spring 2018.

McLean, William E. <u>Forty-Third Regiment of Indiana Volunteers: An Historic Sketch of its Career and Services</u>. Terre Haute: C.W. Brown Printer and Binder, 1903.

McMurry, Richard M. <u>Two Great Rebel Armies: An Essay on Confederate Military History</u>, Chapel Hill: University of North Carolina Press, 1989.

McPherson, James M. <u>The Battle Cry of Freedom: The Civil War Era</u>, New York: Oxford University Press, 1988.

McReynolds, Edwin C. <u>Oklahoma: A History of the Sooner State</u>, Norman: University of Oklahoma Press, 1964.

Mildfelt, Todd, and David D. Schafer. <u>Abolitionist of the Most Dangerous Kind: James Montgomery and His War on Slavery</u>. Norman: University of Oklahoma Press, 2023.

Minges, Patrick Neal. <u>The Keetoowah Society and the Avocation of Religious Nationalism in the Cherokee Nation, 1855-1867</u>, Unpublished Thesis, Union Theological Seminary, 1999.

Missall, John and Mary Lou Missall, The Seminole Wars: America's Longest Indian Conflict, Gainesville, University Press of Florida, 2004.

Monaghan, Jay. Civil War on the Western Border, 1854-1865. Lincoln and London: University of Nebraska Press. 1955.

Moore, Frank, ed. The Rebellion Record: A Diary of American Events, with Documents, Narratives, Illustrative Incidents, Poetry, etc. Vol. 7. New York: D. Van Nostrand, 1864.

Morton, Ohland, "The Confederate States Government and the Five Civilized Tribes, Part II" Chronicles of Oklahoma, Autumn 1953. pp. 299-323.

Moulton, Gary E. John Ross: Cherokee Chief. Athens and London: Brown Thrasher Books, University of Georgia Press, 1978.

       --"John Ross and the 1865 Fort Smith Council," Proceedings: War and Reconstruction in Indian Territory: History Conference in Observance of the 130th Anniversary of the Fort Smith Council September 14-17, 1995 Fort Smith, Arkansas.

Ness, George T. Jr. The Regular Army on the Eve of the Civil War, Baltimore: Toomy Press, 1990.

Newell, Clayton R. & Charles R. Shrader. Of Duty Well and Faithfully Done: A History of the Regular Army in the Civil War, Lincoln: University of Nebraska Press, 2011.

Nichols, Cheryl. "Construction of the Military Road Between Little Rock, Arkansas and Fort Gibson, Oklahoma." Arkansas Historic Preservation Program, May 2003.

Nichols, David A. Lincoln and the Indians: Civil War Policy and Politics. St. Paul: Minnesota Historical Society Press, 1978.

Oates, Steven B. Confederate Cavalry West of the River, Austin: University of Texas Press, 1961.

Oklahoma Historical Society, The Encyclopedia of Oklahoma History and Culture.
       --Steffen, Jerome O. "Stokes Commission," 2010.
       --Weaver, Bobby D. "Texas Road," 2010
       --Everett, Dianna, "Butterfield Overland Mail," 2010

Oliva, Leo E. Fort Scott, Topeka: Kansas State Historical Society, 1984.

Oswalt, Wendell H, and Sharlotte Neely, This Land Was Theirs: A Study of North American Indians, 5th Ed. Mountain View, California: Mayfield Publishing Company, 1996.

Owsley, Frank Lawrence. Struggle for the Gulf Borderlands: The Creek War and the Battle of New Orleans, Gainesville, Florida: University of Florida Presses, 1981.

Pelzer, Louis, ed. "A Journal of Marches by the First United States Dragoons, 1834-1835" Iowa Journal of History and Politics, Vol. VII, July 1909.

Perdue, Theda. Slavery and the Evolution of Cherokee Society, 1540-1866, Knoxville: The University of Tennessee Press, 1979.

Rein, Christopher M. The Second Colorado Cavalry: A Civil War Regiment on the Great Plains. Norman: University of Oklahoma Press, 2020.

Richards, Ira Don. "The Battle of Poison Spring." The Arkansas Historical Quarterly 18, no. 4 (1959): pp. 338–49.

Bibliography

Richards, Ira Don. "The Engagement at Marks' Mills." The Arkansas Historical Quarterly, vol. 19, no. 1, 1960, pp. 51–60.

Richards, Ira Don. "The Battle of Jenkins' Ferry." The Arkansas Historical Quarterly, vol. 20, no. 1, 1961, pp. 3–16.

St. Clair, Edgar J. The Battle of Pilot Knob. High Ridge, MO: American Gold Label & Printing Company, 1989.

Sallee, Scott E. "The Battle of Prairie Grove: War in the Ozarks, April '62 – January '63.' Blue & Gray Magazine. David Roth, Ed. Volume XXI, Issue 5. Fall 2004.

Satz, Ronald N. American Indian Policy in the Jacksonian Era, Lincoln: University of Nebraska Press, 1975.

Sedgwick, John. Blood Moon: An American Epic of War and Splendor in the Cherokee Nation, New York: Simon & Schuster, 2018.

Shea, William L. Fields of Blood: The Prairie Grove Campaign, Chapel Hill: The University of North Carolina Press, 2009.

Shea, William L. and Earl J. Hess. Pea Ridge: Civil War Campaign in the West, Chapel Hill and London: The University of North Carolina Press, 1992.

Spencer, John D. The American Civil War in the Indian Territory, New York and Oxford: Osprey Publishing, Ltd. 2006.

Sperry, A.F. History of the 33rd Iowa Infantry Volunteer Regiment. Des Moines: Mills & Company, 1866.

Spurgeon, Ian Michael. Soldiers in the Army of Freedom: The 1st Kansas Colored, the Civil War's First African American Combat Unit, Norman: University of Oklahoma Press, 2014.

Spurgeon, Ken. A Kansas Soldier at War: The Civil War Letters of Christian & Elise Dubach Isley. Charleston: The History Press, 2013.

Stevens, Walter Barlow. Centennial History of Missouri S.J. Clarke Publishing. 1921.

Suderow, Bryce & Scott House. The Battle of Pilot Knob: Thunder in Arcadia Valley. Cape Girardeau, MO: Southeast Missouri State University Press, 2014

Joseph B. Thoburn, ed., "The Cherokee Question," Chronicles of Oklahoma (Oklahoma City), Vol. II, 1924.

Trickett, Dean. "The Civil War in the Indian Territory, 1861," Chronicles of Oklahoma, Volume 17, No. 3. September 1939.

United States Military Academy. The West Point History of the Civil War. Rogers, Clifford J., Ty Seidule, and Samual J. Watson, Editors. New York: Simon & Schuster, 2014.

Unrau, William E. (October 1972). "The Civilian as Indian Agent: Villain or Victim?". Western Historical Quarterly. 3 (4): 405–420. doi:10.2307/966865. JSTOR 966865

Utley, Robert M. Frontiersmen in Blue: The United States Army and the Indian, 1848-1865, Lincoln: University of Nebraska Press, 1967.

Walker, Joe. Harvest of Death: The Battle of Jenkin's Ferry, Arkansas. Self-Published, 2011.

Warde, Mary Jane. When the Wolf Came: The Civil War and the Indian Territory, Fayetteville: The University of Arkansas Press, 2013.

Ware, James W. "Indian Territory", The Western Territories in the Civil War. Fischer, LeRoy H. ed. Manhattan, Kansas: Sunflower University Press. 1977.

Warner, Ezra J. Generals in Blue: Lives of Union Commanders, Baton Rouge and London: Louisiana State University Press. 1987.

--Generals in Gray: Lives of Confederate Commanders, Baton Rouge and London: Louisiana State University Press. 1987.

Warren, Steven L. Brilliant Victory: The Second Civil War Battle of Cabin Creek, Indian Territory, Wyandotte, Oklahoma, The Gregath Publishing Company. 2002.

Waugh, John C. Sam Bell Maxey and the Confederate Indians, Abilene, Texas: McWhiney Foundation Press. 1995.

Weeks, Philip. Farewell My Nation: The American Indian and the United States, 1820-1890, Wheeling, Illinois: Harlen Davidson Press, 1990.

White, Christine Schultz, and White, Benton R. Now the Wolf Has Come: The Creek Nation in the Civil War, College Station: Texas A&M University Press, 1996.

White House Historical Association, The Biography for President Buchanan, www.whitehouse.gov

Wilson, Terry Paul, "Delegates of the Five Civilized Tribes to the Confederate Congress;" Chronicles of Oklahoma, Volume 53, Number 3, Fall 1975.

Wood, Larry. The Two Civil War Battles of Newtonia, Charleston & London: The History Press, 2010.

Woodward, Grace Steele. The Cherokees, Norman: The University of Oklahoma Press, 1963.

Wright, Murial H. "General Douglas Hancock Cooper, CSA." Chronicles of Oklahoma, article, Summer 1954; Oklahoma City, Oklahoma

--"Colonel Cooper's Report on the Battle of Round Mountain," article, Winter 1961; Oklahoma City, Oklahoma.

--"Wapanucka Academy, Chickasaw Nation," Volume 12, December 1934

Yarbrough, Fay A. Choctaw Confederates: The American Civil War in Indian Country. Chapel Hill: The University of North Carolina Press, 2021.

Yates, Catherine H., Wyckoff, Don G., Baugh, Timothy G., Harrington, John A. Jr. A Survey of Prehistoric and Historic Sites in the Honey Springs Area, McIntosh and Muskogee Counties, Oklahoma. Oklahoma Archeological Survey, Archeological Resource Survey Report No. 11. Norman, Oklahoma. 1981.

Yeary, Mamie. Reminiscences of the Boys in Gray, 1861-1865. Dallas, Texas: n.p., 1912.

Index

**Note: Due to formatting and late edits of the text the page numbers for the Index may be off by one or two pages. The Author apologizes for any inconvenience.